THE PICTURE OF THE TAOIST GENII PRINTED ON THE COVER
of this book is part of a painted temple scroll, recent but traditional, given to
Mr Brian Harland in Szechuan province (1946). Concerning these four divinities,
of respectable rank in the Taoist bureaucracy, the following particulars have been
handed down. The title of the first of the four signifies 'Heavenly Prince', that
of the other three 'Mysterious Commander'.

At the top, on the left, is Liu *Thien Chün*, Comptroller-General of Crops and
Weather. Before his deification (so it was said) he was a rain-making magician
and weather forecaster named Liu Chün, born in the Chin dynasty about +340.
Among his attributes may be seen the sun and moon, and a measuring-rod or
carpenter's square. The two great luminaries imply the making of the calendar, so
important for a primarily agricultural society, the efforts, ever renewed, to reconcile
celestial periodicities. The carpenter's square is no ordinary tool, but the gnomon
for measuring the lengths of the sun's solstitial shadows. The Comptroller-General
also carries a bell because in ancient and medieval times there was thought to be
a close connection between calendrical calculations and the arithmetical acoustics
of bells and pitch-pipes.

At the top, on the right, is Wên *Yuan Shuai*, Intendant of the Spiritual Officials
of the Sacred Mountain, Thai Shan. He was taken to be an incarnation of one of
the Hour-Presidents (*Chia Shen*), i.e. tutelary deities of the twelve cyclical characters
(see p. 262). During his earthly pilgrimage his name was Huan Tzu-Yü and he was
a scholar and astronomer in the Later Han (b. +142). He is seen holding an
armillary ring.

Below, on the left, is Kou *Yuan Shuai*, Assistant Secretary of State in the Ministry
of Thunder. He is therefore a late emanation of a very ancient god, Lei Kung.
Before he became deified he was Hsin Hsing, a poor woodcutter, but no doubt an
incarnation of the spirit of the constellation Kou-Chhen (the Angular Arranger),
part of the group of stars which we know as Ursa Minor. He is equipped with
hammer and chisel.

Below, on the right, is Pi *Yuan Shuai*, Commander of the Lightning, with his
flashing sword, a deity with distinct alchemical and cosmological interests. According
to tradition, in his earthly life he was a countryman whose name was Thien Hua.
Together with the colleague on his right, he controlled the Spirits of the Five
Directions.

Such is the legendary folklore of common men canonised by popular acclamation.
An interesting scroll, of no great artistic merit, destined to decorate a temple wall,
to be looked upon by humble people, it symbolises something which this book has
to say. Chinese art and literature have been so profuse, Chinese mythological
imagery so fertile, that the West has often missed other aspects, perhaps more
important, of Chinese civilisation. Here the graduated scale of Liu Chün, at first
sight unexpected in this setting, reminds us of the ever-present theme of quantitative measurement in Chinese culture; there were rain-gauges already in the Sung
(+12th century) and sliding calipers in the Han (+1st). The armillary ring of
Huan Tzu-Yü bears witness that Naburiannu and Hipparchus, al-Naqqās and
Tycho, had worthy counterparts in China. The tools of Hsin Hsing symbolise that
great empirical tradition which informed the work of Chinese artisans and technicians
all through the ages.

SCIENCE AND CIVILISATION IN CHINA

"I THINK that if we are to feel at home in the world...we shall have to admit Asia to equality in our thoughts, not only politically but culturally. What changes this will bring about I do not know, but I am convinced that they will be profound and of the greatest importance."

BERTRAND RUSSELL
History of Western Philosophy (1946), p. 420.

中國科學技術史

李約瑟 著

萬朝鼎

SCIENCE AND CIVILISATION IN CHINA

BY

JOSEPH NEEDHAM F.R.S., F.B.A.

SOMETIME MASTER OF GONVILLE AND CAIUS COLLEGE, CAMBRIDGE,
DIRECTOR OF THE NEEDHAM RESEARCH INSTITUTE, CAMBRIDGE,
HONORARY PROFESSOR OF ACADEMIA SINICA

With the research assistance of

WANG LING PH.D.

EMERITUS PROFESSORIAL FELLOW, DEPARTMENT OF FAR EASTERN HISTORY,
INSTITUTE OF ADVANCED STUDIES
AUSTRALIAN NATIONAL UNIVERSITY, CANBERRA

VOLUME 2

HISTORY OF
SCIENTIFIC THOUGHT

The right of the
University of Cambridge
to print and sell
all manner of books
was granted by
Henry VIII in 1534.
The University has printed
and published continuously
since 1584.

CAMBRIDGE UNIVERSITY PRESS

CAMBRIDGE
NEW YORK   PORT CHESTER
MELBOURNE   SYDNEY

Published by the Press Syndicate of the University of Cambridge
The Pitt Building, Trumpington Street, Cambridge CB2 IRP
40 West 20th Street, New York, NY 10011-4211, USA
10 Stamford Road, Oakleigh, Melbourne 3166, Australia

Copyright Cambridge University Press 1956

Corrected reprint © Cambridge University Press 1991

First published 1956

Reprinted 1962 1969 1972 1975 1977 1980 1991

Printed in Great Britain at the University Press, Cambridge

ISBN 0 521 05800 7 hardback

*The Syndics of the Cambridge University Press
desire to acknowledge with gratitude certain financial aid
towards the production of this book, afforded
by the Bollingen Foundation*

CONTENTS

13 THE FUNDAMENTAL IDEAS OF CHINESE
   SCIENCE . . . . *page* 216

LIST OF ILLUSTRATIONS

LIST OF TABLES

LIST OF ABBREVIATIONS

The following abbreviations are used in the text. For abbreviations used for journals and similar publications in the bibliographies, see p. 586.

B   Bretschneider, E., *Botanicon Sinicum*.

B & M   Brunet, P. & Mieli, A., *Histoire des Sciences* (*Antiquité*).

CIB   *China Institute Bulletin* (New York).

CSHK   Yen Kho-Chün (ed.), *Chhüan Shang-Ku San-Tai Chhin Han San-Kuo Liu Chhao Wên* (complete collection of prose literature (including fragments) from remote antiquity through the Chhin and Han Dynasties, the Three Kingdoms, and the Six Dynasties, 1836).

CTCS   Li Kuang-Ti (ed.), *Chu Tzu Chhüan Shu* (collected works of the philosopher Chu Hsi).

CTYL   Li Ching-Tê (ed.), *Chu Tzu Yü Lei* (classified conversations of Chu Hsi).

ECCS   Hsü Pi-Ta (ed.), *Erh Chhêng Chhüan Shu* (collected writings and conversations of the brothers Chhêng I and Chhêng Hao), containing *Honan Chhêng shih I Shu* and *Wai Shu, I-Chhuan I Chuan, Sui Yen*, etc.

G   Giles, H. A., *Chinese Biographical Dictionary*.

HCCC   Yen Chieh (ed.), *Huang Chhing Ching Chieh* (monographs by Chhing scholars on classical subjects).

HWTS   Chhêng Jung (ed.), *Han Wei Tshung-Shu* (collection of works of the Han and Wei Dynasties); first completed in the Ming.

K   Karlgren, B., *Grammata Serica* (dictionary giving the ancient forms and phonetic values of Chinese characters).

KSP   Ku Chieh-Kang & Lo Ken-Tsê (ed.), *Ku Shih Pien* (discussions on ancient history and philosophy); a collective work.

M   Mathews, R. H., *Chinese-English Dictionary*.

N   Nanjio, B., *A Catalogue of the Chinese Translations of the Buddhist Tripiṭaka*, with index by Ross (3).

R	Read, Bernard E., Indexes, translations and précis of certain chapters of the *Pên Tshao Kang Mu* of Li Shih-Chen. If the reference is to a plant, see Read (1); if to a mammal see Read (2); if to a bird see Read (3); if to a reptile see Read (4); if to a mollusc see Read (5); if to a fish see Read (6); if to an insect see Read (7).
RP	Read & Pak, Index, translation and précis of the mineralogical chapters in the *Pên Tshao Kang Mu*.
SCTS	*Chhin-Ting Shu Ching Thu Shuo* (imperial illustrated edition of the Historical Classic, 1905).
SPTK	Ssu Pu Tshung Khan edition.
TH	Wieger, L., *Textes Historiques*.
TP	Wieger, L. (2), *Textes Philosophiques*.
TPYL	Li Fang (ed.), *Thai-Phing Yü Lan* (the Thai-Phing reign-period (Sung) Imperial Encyclopaedia, +983).
TSCC	*Thu Shu Chi Chhêng* (the Imperial Encyclopaedia of 1726). Index by Giles, L. (2).
TT	Wieger, L. (6), *Tao Tsang* (catalogue of the works contained in the Taoist Patrology).
TTC	*Tao Tê Ching* (Canon of the Virtue of the Tao).
TW	Takakusu, J. & Watanabe, K., *Tables du Taishō Issaikyō* (*nouvelle édition* (*Japonaise*) *du Canon bouddhique chinoise*), Index-catalogue of the Tripiṭaka.
YHSF	Ma Kuo-Han (ed.), *Yü Han Shan Fang Chi I Shu* (Jade-Box Mountain Studio Collection of (reconstituted and sometimes fragmentary) Lost Books, 1853).

ACKNOWLEDGMENTS

LIST OF THOSE WHO HAVE KINDLY READ THROUGH SECTIONS IN DRAFT

This list, which applies to this volume only, brings up to date the list printed in
Vol. I on pp. 15–16.

Mr S. Adler (Cambridge)	All sections.
Dr Shackleton Bailey (Cambridge)	Buddhism.
Dr Etienne Balazs (Paris)	Taoism, Confucianism, Natural Law.
Prof. Derk Bodde (Philadelphia)	Natural Law.
Mrs Margaret Braithwaite (Cambridge)	Mohists, Logicians, Neo-Confucians.
Mr Derek Bryan (Cambridge)	All sections.
Prof. K. Bünger (Tübingen)	Natural Law.
Dr Chêng Tê-Khun (Cambridge)	All sections.
Mr E. Conze (London)	Buddhism.
Prof. W. A. C. H. Dobson (Toronto)	Taoism.
Prof. E. R. Dodds, F.B.A. (Oxford)	Natural Law.
Prof. Dorothy Emmet (Manchester)	All sections.
Mrs Martha Kneale (Oxford)	Fundamentals, Neo-Confucianism.
Mr Arnold P. Koslow (New York)	All sections.
Mr D. Leslie (Cambridge)	All sections.
Miss Liao Hung-Ying (Mrs Bryan) (Cambridge)	All sections.
Dr P. van der Loon (Cambridge)	Confucianism.
Dr Lu Gwei-Djen (Paris)	All sections.
Dr Stephen Mason (Oxford)	Fundamentals.
Rev. H. W. Montefiore (Cambridge)	Confucianism.
Dr Walter Pagel (London)	Taoism.
Prof. Luciano Petech (Rome)	All sections.
Prof. E. Pulleyblank (Cambridge)	All sections.
Dr Dorothea Singer (Par)	Taoism.
Dr Otto van der Sprenkel (London)	All sections.
Prof. E. S. Wade (Cambridge)	Legalists, Natural Law.
Dr Arthur Waley (London)	Natural Law.
Prof. J. H. Woodger (London)	All sections.
Dr Wu Shih-Chhang (Oxford)	All sections.
Prof. W. P. Yetts (Amersham)	Fundamentals (Etymologies).

We are also greatly indebted to Mr D. M. Dunlop for help with the vocalisation of
Arabic and Persian names; to Mr Shackleton Bailey for assistance, equally kind, in the
accenting of Sanskrit words; to Mr R. L. Loewe for aid on Hebrew matters, and to
Dr J. R. McEwan for Japanese transliterations.

AUTHOR'S NOTE

W E are very conscious of the great range of territory surveyed in this volume, yet since Chinese cultural history is as complex as that of Europe, nothing less would have sufficed. The reader whose interests lie in the contrasting general development of thought at the two ends of the Old World will not consider superfluous one single note of this symphony. But we cannot but have in mind the reader who, perhaps himself a busy experimentalist, wishes to appreciate with minimum expenditure of time how far the scientific thought of ancient and medieval China differed from that of ancient Greece and medieval Europe. For such an enquirer the first necessity is to apprehend the deeply organic and non-mechanical quality of Chinese naturalism. This first appears in the -4th century with the Taoists (Section $10c$), the Mohists (11) and the nature-philosophers of Yin and Yang ($13c$). Later it achieves formulation and stability in the Chinese medieval world-picture (Section $13f$). The freshness of the original theme is reinforced by Buddhist contributions ($15e$), and reaches its definitive synthesis in $+12$th-century Neo-Confucianism (Section $16d$, with which $18f$ (§10) should be read at the same time). Two other aspects particularly concern the natural scientist, the strong tradition of scepticism ($14b-i$), and the Chinese attitude to the juristic analogy regarding 'Laws of Nature' (18). For philosophical readers, too, this last will be of equal importance with the tradition of organic naturalism, of which indeed it constitutes one particular aspect, for it reveals how the Chinese concept of order could and did (in Granet's felicitous phrase) positively exclude the concept of law. How far Chinese influences affected the thought of Leibniz and the development of organic naturalism in Europe is a question also raised ($13f$ (§1), $16f$). Lastly, one of the most important features of nearly all Chinese natural philosophy was its immunity from the perennial debate of Europe between the theistic world-view and that of mechanical materialism—an antithesis which the West has not yet fully resolved.

8. INTRODUCTION

ALL THE NECESSARY PRELIMINARIES having now been completed, we are free to consider the part which Chinese philosophy played in relation to the development of scientific thought. It is a commonplace that in China even the word 'philosophy' did not mean quite what it came to mean in Europe, being much more ethical and social than metaphysical. Nevertheless, the Taoists and Mohists worked out a naturalistic world-view of great importance, and the Logicians began the study of a logic which unfortunately did not develop. We shall examine first the various schools of thought in the classical period of Chinese philosophy, namely, the Warring States time ($-$4th and $-$3rd centuries).[a]

We shall begin with the *Ju Chia*[1] (Confucians[b]), giving them pride of place on account of their dominance over all later Chinese thought, although their contribution to science was almost wholly negative. From them the transition is easy to their mortal enemies the *Tao Chia*[2] (Taoists), whose speculations about, and insight into, Nature, fully equalled pre-Aristotelian Greek thought, and lie at the basis of all Chinese science. It will be necessary to emphasise an aspect of this antagonism usually overlooked, namely, the political, for while Confucianism accepted feudal society Taoism was strongly opposed to it. A third element was the *Fa Chia*[3] (Legalists), devoted to codification of law and largely responsible for the replacement of feudalism by the feudal-bureaucratic State. Proponents of an authoritarianism almost fascist, they came to grief, as we have seen (Sect. 6*b*), when the dynasty of Chhin overreached itself and was replaced by that of the Han. The ultimate bureaucratic ideology and social structure was a synthesis of Legalist and Confucian principles.[c] Then there were the *Mo Chia*[4] (Mohists), chivalrous military pacifists with an interest in scientific method and even experimentation arising out of war techniques; and the *Ming Chia*[5] (Logicians), who have often been compared to the Greek Sophists, with their paradoxes and definitions; together with a number of lesser schools. Last, but not least, came the

---

[a] Certain contemporary and almost contemporary accounts of the schools have come down to us, and these will be found well worth reading. The earliest are (*a*) the chapter 'against the twelve philosophers' (ch. 6) of *Hsün Tzu* (tr. Dubs (8), p. 77) which would be of about $-$250; and (*b*) the spurious ch. 33 of Chuang Tzu (tr. Legge (5), vol. 2, p. 214 and Chhen Tai-O, 1), not likely to be much later. But the best (*c*) is the essay of Ssuma Than[6] (d. $-$110), father of the historian Ssuma Chhien, preserved in ch. 130, pp. 3*a* ff. of the latter's *Shih Chi* of $-$90 (tr. Chavannes (1), vol. 1, pp. ix ff.; Porter (1), p. 51). We also have (*d*) the *catalogue raisonné* of books completed by Liu Hsin[7] around $-$6, which gives an account of the philosophical schools; this was incorporated in abridged form in the *Chhien Han Shu* of *c.*+100 as its bibliographical chapter (*I wên chih*,[8] tr. Porter (1), p. 57).

[b] The terms Confucians and Confucianism are Westernisms; *ju* simply means scholars. The followers of Confucius were considered the scholars *par excellence*.

[c] Cf. pp. 29, 212, 215 below, and Dubs (10).

¹ 儒家　　　² 道家　　　³ 法家　　　⁴ 墨家　　　⁵ 名家　　　⁶ 司馬談
⁷ 劉歆　　　⁸ 藝文志

School of the Naturalists (*Yin-Yang Chia*[1]), which developed a philosophy of organic naturalism and gave to Chinese proto-scientific thinking its characteristic fundamental theories.

Later sections take up the tradition of sceptical rationalism whose greatest exponent was Wang Chhung of the Han; the philosophy of Buddhism, favourable to science by its belief in causation, but inimical to it by its doctrine of illusion; and the Neo-Confucianism of the Sung. This was the school which brought the *philosophia perennis* of China to its highest expression, and in many ways anticipated the organic naturalism of our own time. The discussion, and the volume, ends with the anti-scientific idealism of Wang Yang-Ming, the historical materialism of Wang Chhuan-Shan in the +17th century, the\coming then of the new, or experimental, philosophy, and a general survey of the development of the concept of Laws of Nature in Europe and in China.[a]

[a] The most complete exposition of the history of Chinese philosophy available in English is that of Fêng Yu-Lan (1), but for a shorter survey the brilliant essay of Hu Shih (3) is to be recommended. Outlines and bibliographies by Porter (1) and Chhen Jung-Chieh (3).

[1] 陰陽家

9. THE *JU CHIA* (CONFUCIANS) AND CONFUCIANISM

(*a*) INTRODUCTION

AS ALREADY MENTIONED (Sect. 5*c*) the early iron age in China (−6th century) was a time at which bronze-age proto-feudalism was beginning to decay. Wars and diplomatic *tours-de-force* took place among the various feudal States, each of which had the ambition to conquer all the others, as Chhin ultimately succeeded in doing. The breaking-up and re-formation of feudal courts led to turmoil among that small middle class of specialists which had previously occupied for generations reasonably secure posts at State capitals.[a] These were the scribes and secretaries, experts in rites, sacrifices, music and military training, even metal- and wood-workers. Some of them may have been descendants of the old families of the Shang, debarred from feudal rank and social importance under the Chou. By the −3rd century, however, the term 'Ju', which was widely applied to them, and which had originally in all probability connoted in some sense 'weakling', had become an appellation proudly accepted.[b] For this floating population of wandering specialists succeeded, as time went on, in dominating the hereditary aristocracy by the superiority of its own ideas and interests. It provides all the names which have come down to us from the 'period of the philosophers'. Apart from the proto-Taoists, who lived a solitary life in mountain hermitages, all these men sought for employment at the courts of the feudal princes. Among them there may have been some before Confucius who taught doctrines similar to his, but none who, by force of character and originality of mind, succeeded as he did in impressing their conceptions and personality upon all following generations.

We are rather well provided with traditions concerning the life of Confucius; the only difficulty is to know which of them should be accepted. But many things are not in doubt. Khung[1] was his family name; his given name was Chhiu[2] and his style Chung-Ni,[3] but he is always referred to by his title of honour, Khung Fu Tzu,[4] i.e. Master Khung, whence the Latinised form Confucius. Born in −552[c] in the small State of Lu in modern Shantung, of a family which traced its descent from the imperial house of Shang through the Sung State, he spent his life in developing and propagating a philosophy of just and harmonious social relationships. He was constantly seeking for opportunities (without much success) of putting it into practice from the vantage point of an official position.[d] Certain it is that from about −495 he spent a number of

[a] A clear echo of this in *Lun Yü*, XVIII, ix, has been pointed out by Fêng Yu-Lan (*3*).
[b] Cf. Hu Shih (*8*, 8).
[c] Or the following year.
[d] An account of his life on the lines generally accepted until recently will be found in R. Wilhelm (*5*).

[1] 孔   [2] 丘   [3] 仲尼   [4] 孔夫子

years in enforced exile from his native place, wandering from State to State with a group of disciples, conversing with feudal princes and hoping for a chance to employ his great talents. The last three years of his life, however, were spent in Lu on literary work and the instruction of his students; he died in −479. Although his life might have seemed at the time somewhat of a failure, his subsequent influence was so far-reaching as to justify the title often attributed to him of the 'uncrowned emperor' of China.[a]

Opinions differ greatly on the question whether Confucius ever held any official position. Some accept the tradition that he was at first a minor administrator in charge of granaries and then of public lands, while later, after an absence in Chhi State, he became Minister of Justice and Chancellor in Lu for a short while about −501.[b] Others reject these statements, admitting only that he may have held a nominal advisory post at the turn of the century.[c] If some traditions are to be believed, there were two focal points in the life of Confucius. The first was when he successfully saved his prince by clever diplomacy from an ambush of ritual Pyrrhic dancers at an interview with the Duke of Chhi.[d] The second, which led to his exile, was when he tried to arrange for the dismantling of certain fortifications in Lu, so as to restore authority to the prince and reduce the power of the three great aristocratic families, who were maintaining a kind of 'shogunate'.[e] Their enmity pursued him long afterwards. It may be significant that on two occasions Confucius was tendered a position of authority by commanders who were in rebellion against their feudal superiors, and though he declined he did so with reluctance.[f] Apparently they also were fighting for the idea of bureaucratic monarchy, against feudalism.

Modern scholarship no longer insists that Confucius edited the *Shih Ching* (Book of Odes)[g] or the *Shu Ching* (Historical Classic). Nor did he write any part of the *I Ching* (Book of Changes),[h] the *Li Chi* (Record of Rites) or the *Chhun Chhiu* (Spring and Autumn Annals),[i] still less the *Yo Ching*[1] (Music Classic, long lost).[j] No doubt he used in his teaching such parts of these books or their prototypes as were already

[a] The idea that he received (in principle) an imperial mandate was developed by Tung Chung-Shu; see Fêng Yu-Lan (1), vol. 2, pp. 65, 71, 129.

[b] For a detailed exposition of this view, see Dubs (9).

[c] For a detailed exposition of this view, see Creel (4).

[d] *Tso Chuan*, Duke Ting, 10th year (Couvreur (1), vol. 3, p. 558). See Granet (1), pp. 171 ff., who brings out the ancient ritual magic background of these dancers and the sacrifice which was made of them.

[e] *Tso Chuan*, Duke Ting, 8th to 12th years; description in Dubs (9).

[f] The first of these was Kungshan Fu-Jao, who was holding a city for the prince against the 'shogun' Chisun family (*Lun Yü*, XVII, v) about −500. The second, some ten years later, was Pi Hsi (*Lun Yü*, XVII, vii). See Creel (4), pp. 41, 56.

[g] He may have rearranged the order of the pieces in it (cf. *Lun Yü*, IX, xiv).

[h] *Lun Yü*, VII, xvi, may be a late interpolation (Dubs, 17), and in any case the reading is uncertain.

[i] Mencius says (*Mêng Tzu*, III (2), ix, 11) that Confucius wrote a *Chhun Chhiu*, but no one knew of this at the time of the Analects. His role in connection with the text of the classics is still much debated.

[j] Cf. particularly Fêng Yu-Lan (1), p. 46, (4), (2).

[1] 樂經

extant in his time.[a] The *Lun Yü* (Conversations and Discourses, generally known as the Analects), however, was certainly put together in written form soon after his death, and preserves the most reliable information about him, hence the frequency of quotation from it in the following pages.[b] The long chapter devoted to Confucius by Ssuma Chhien and his father in the *Shih Chi* is, on the contrary, suspect, since some find ground for thinking that parts of it may have been intended satirically. Both great historians were Taoist in sympathy, and as they had to include a chapter of biography, they used it to damn with faint praise the hypocritical Confucianism of their own time.[c] Less reliable still is the *Khung Tzu Chia Yü*[1] (Table-Talk of Confucius), edited by Wang Su[2] about the beginning of the +3rd century. This contains much obviously Taoist material, and ideas characteristic of that Han Confucianism which had fused with the School of Naturalists (see on, Sect. 13c). What, then, was the essence of the Confucianism of Confucius himself and his immediate disciples?

(b) GENERAL CHARACTERISTICS OF THE SCHOOL

It was a doctrine of this-worldly social-mindedness. In so far as social justice could be conceived of within the framework of the feudal, or feudal-bureaucratic, social order, Confucius strove for it. He probably did not believe that the faults of his age could be cured by any system other than feudalism,[d] but rather that there should be a return to what he conceived it to have been in its purest form, the ancient 'way of the Sage Kings'.[e] Of course it was natural in his time to clothe ethical insights with legendary historical authority. Confucius called himself a transmitter, not an originator.

In order to understand Confucius it is indispensable to visualise what the world of his time was like.[g] His interest in the orderly administration of affairs may seem dry

[a] The Confucian school later systematically read into the old Book of Odes (essentially a collection of ancient Chou folk-songs) moralising symbolism, in a way quite analogous to the treatment of the Song of Songs by Christian theologians (see Ku Chieh-Kang, 4; Hu Shih, 2). Confucius started this himself (*Lun Yü*, I, xv; III, viii).

[b] It is generally agreed that chs. 16, 17, 18 and 20 are later than the rest of the book, and contain Taoist material. Many references to discussions of the authenticity of passages throughout the book, by both Chinese and Western scholars, will be found in Creel (4).

[c] So Creel (4), pp. 9, 266 ff. The biography is in ch. 47 of the *Shih Chi* (tr. Chavannes (1), vol. 5, pp. 283 ff.).

[d] It must, however, be admitted that some of the passages (e.g. *Lun Yü*, XVI, ii) on which this attachment to feudalism is based occur in parts of the book which are of disputed authenticity (see Creel (4), pp. 159, 239). This particular passage seems to be Legalist (see below, Sect. 12).

[e] This implied subordination of the feudal princes to the Chou emperor, branches of whose family or supporters their houses had originally been, according to the *tsung fa*[3] system whereby the eldest son inherited and the younger sons were given separate fiefs.

[f] *Lun Yü*, VII, i.

[g] Confucianism often seems unnecessarily bewildering to Europeans who approach it by way of some of the older translations of classical texts. Good guides are the relevant chapters in Fêng Yu-Lan (1) and the extracts and explanations given in Hughes (1). I must say that I myself received much help from the books of Liang Chhi-Chhao (1) and Hsü Shih-Lien (1), though the latter particularly contains many mistakes. The articles by Wu Tsê-Ling (1) are worth looking at. Recently, Creel (4) has given us an elaborate study, which, however, is devoted to proving a particular case.

[1] 孔子家語　　　[2] 王肅　　　[3] 宗法

and unromantic, but he lived in an environment where generally chaos reigned. Between the feudal States there was constant war, the smaller ones serving as battle-fields for the larger. There was little law and order save what each man could enforce by personal strength, armed followers, or intrigue. Aristocratic pastimes, hunting, war and extravagant living, laid crushing burdens on the common people, while at all levels human life was cheap. For the world of his time, Confucius' ideas were revolutionary. Read today, many of his speeches in the *Lun Yü* sound like 'pedantic little homilies' addressed to various nobles and rulers. Yet when the background is understood, it is clear that some of these remarks were 'pointed denunciations of weaknesses, not to say crimes, made directly to men who would have felt as much compunction about having Confucius tortured to death as about crushing a fly'.ᵃ

Confucius was certainly greatest as an educator. Before his time there is mention only of schools of archery. As has often been pointed out,ᵇ he was the first who stated clearly that in teaching there should be no class-distinctions.ᶜ No qualifications of birth were necessary in acceptance for the administrative and diplomatic training which Confucius gave. In this we see one of the germs of the bureaucratic system, according to which whoever was teachable and ambitious for letters could become a scholar and serve his prince (later, the imperial State) as an official, no matter what the social position of his family might have been. Confucius had much to say of the honour of such officials. The quality of his general teaching may be felt from the following remark of one of his chief students, Tsêng Shen:ᵈ

A *chün-tzu*,ᵉ in following the Tao, values three things above all others. From every attitude and every gesture he removes all trace of violence or arrogance; every expression of his face betokens sincerity; and from every word he utters, he eliminates all uncouthness or vulgarity.ᶠ

ᵃ We owe the thought of the above paragraph, and some of its wording, to the admirable pages of Creel (4), pp. 3, 14, 17 ff.

ᵇ E.g. by Fêng Yu-Lan (1), p. 49.

ᶜ *Lun Yü*, xv, xxxviii: '*Tzu yüeh: yu chiao wu lei.*'¹

ᵈ See below, pp. 11, 268.

ᵉ Like Tao, 'the Way', *chün-tzu*² is one of those words which I have come to the conclusion are better not translated. Originally the prince or ruler, and rendered most unsatisfactorily by Legge and others as 'the superior man' or by Waley as 'the gentleman', it has meant throughout Chinese history the man of sympathetic character, high attainments and moral greatness who may be (though no one of these attributes is essential to the meaning) well born, a scholar, an official, a soldier, a martyr. One can only point to certain European individuals, for example, Sir Thomas More, in order to show what is implied. The opposite of *chün-tzu* is *hsiao-jen*,³ but it connotes not only the man of low social station, but also meanness, boorishness, etc. (cf. 'villein', 'villain', etc.). The great difficulty of translating these phrases is brought out in the book of Chêng Thien-Hsi. Cf. Boodberg (3).

ᶠ *Lun Yü*, VIII, iv; tr. auct. adjuv. Legge (2), Creel (4). But the passage continues, 'As for the details of the sacrifices, that can be left to the clerks'. Here is our first indication of Confucian indifference to techniques.

子曰有敎無類   ² 君子   ³ 小人

And in another place,[a] the Master defined perfect love (of our neighbour) as:

When you go forth, to behave to every one as if you were receiving a great guest; to employ the people as if you were assisting at a great sacrifice, not to do to others what you would not wish done to yourself, and to give no cause for resentment either at home or abroad.[b]

The earlier groups of Confucius' students mostly became high officials of the feudal States; those of the later groups generally became teachers and social philosophers. The names of many of them have been preserved.[c]

The freeing of education from all barriers of privilege and social class was undoubtedly revolutionary doctrine, and if it paved the way for the mandarinate of feudal bureaucratism, it embodied also some of the essential elements of modern democratic thought. Opinions have differed greatly as to the extent to which consciously 'democratic' ideas can be attributed to Confucius, and the matter is not unimportant for us because there are close sociological connections between democracy and the natural sciences.[d] Ku Chieh-Kang (7) thought that the support of Master Khung for feudalism was fundamental; Mei Ssu-Phing (1) considered him a great counter-revolutionary. The question in China has of course been intimately bound up with current political questions of modern times, and the support of backward-looking groups for traditional Confucianism. But other scholars, such as Kuo Mo-Jo,[e] have emphasised the revolutionary ideas in Master Khung's life and teaching, pointing out, for example, his sympathy for those officials who had taken up arms against the feudal nobles. Certainly his followers were accused (in the *Mo Tzu* and the *Chuang Tzu*) of being fomenters of disorder, and no one knew more about rebellion under insufferable conditions than the Mohists and the Taoists. Certain it is that when Chhen Shêng (Chhen Shê[1]) led the first rebellion against the Chhin dynasty, taking the title of Chang Chhu Wang,[f] he had the direct descendant of Confucius in the eighth generation as his adviser, and this scholar (Khung Fu[2]) died with him when he was defeated in −208. Confucians and Mohists had flocked to his standard.[g]

Confucius seems to have believed that the true aim of government ought to be the welfare and happiness of the whole people, and that this would be brought about not by rigid adherence to enacted arbitrary law, but by subtle administration of

[a] *Lun Yü*, XII, ii.   [b] Tr. auct. adjuv. Legge (2).

[c] Such as Jan Chhiu, who rose to high office in Lu State; or, as an example of the second type, Yu Jo, who may have succeeded Confucius as the leader of the school, and passed on its traditions to the 'apostle' Mencius (Mêng Kho). See below, p. 16.

[d] This will be discussed more fully below (pp. 103, 130 ff.) in connection with the Taoists.

[e] (1), pp. 63 ff.

[f] Prince entrusted with the Expansion of Chhu; also called Chhen Wang. Cf. Vol. 1, p. 102 above, in Sect. 6a.

[g] The whole story of the unsuccessful predecessor of the successful Han is told in *Shih Chi*, ch. 48 (tr. Haenisch, 1). Khung Fu is mentioned in ch. 47, p. 30a (Chavannes (1), vol. 5, p. 432).

[1] 陳涉   [2] 孔鮒

customs generally accepted as good and having the sanction of natural law.[a] Since men of real intelligence, sympathy and learning were necessary for such administration, they would have to be sought for far afield. Capacity to govern had no necessary connection with birth, wealth or position; it depended solely on character and knowledge, i.e. upon qualities generated only by right education. Education should therefore be universally available.[b]

From this there followed a conclusion important for science. If every man was potentially educable, then every normal man was potentially as good a judge of truth as every other, the qualifications which added value to his judgement being only education, experience and demonstrated competence. He could be a member of the 'community of observers'. The group of Confucius understood this intellectual democracy. The Master himself, moreover, often counselled suspended judgement, saying that one should leave on one side what is doubtful,[c] and that scribes should follow the good old practice of leaving a blank space in texts when copying, instead of faking a character of which they were not sure.[d]

> The Master said: '(Chung) Yu, shall I tell you what knowledge is? When you know a thing, to say that you know it; and when you do not know a thing, to admit that you do not know it—this is true knowledge.'[e]

As good a device as could be found, surely, for any modern scientific academy. Yet traces of interest among the early Confucians in natural science as opposed to human affairs, are few. Confucius recommended the study of the Book of Odes, apart from other reasons, because it would widen one's acquaintance with the names of birds, beasts, plants and trees.[f] He said he agreed with the southern proverb that a man without constancy would not make even a good wizard or a good physician.[g] There are indications[h] that one of his chief students, Tsêng Shen, had scientific interests approximating to those of the later School of Naturalists. But that is all.

While believing in a moral order of the universe (*Thien*;[1] Heaven),[i] the Confucians used the word *Tao*[2] (the 'Way') primarily if not exclusively as meaning the ideal way or order of human society. This comes out clearly in their attitude to the world of spirits, and to knowledge. While not separating man from social man, nor social man

---

[a] On this, see Sect. 18 below.

[b] Part of the thought of this paragraph is derived from the excellent discussion of Creel (4), pp. 177 ff. Cf. Creel (6).

[c] *Lun Yü*, II, xviii. Cf. the famous remark of Mencius (VII (2), iii, 1) that it would be better to be without the Book of History altogether than to believe all there is in it.

[d] *Lun Yü*, xv, xxv.   [e] *Lun Yü*, II, xvii, tr. Legge (2), mod.

[f] *Lun Yü*, XVII, ix.

[g] *Lun Yü*, XIII, xxii.

[h] See below, p. 268.

[i] Confucius thought of Heaven 'as an impersonal ethical force, a cosmic counterpart of the ethical sense in man, a guarantee that somehow there is sympathy with man's sense of right in the very nature of the universe'; Creel (4), p. 126. See also Creel (5).

[1] 天   [2] 道

from the whole of Nature, they always considered that the only proper study of mankind was man. They were thus, throughout Chinese history, in opposition to those elements which groped for a scientific approach to Nature, and for a scientific interpretation and extension of technology.

Fan Hsü requested to be taught agriculture, but the Master said 'I am not so good for that as an old farmer'. He also requested to be taught horticulture, but the Master said 'I am not so good for that as an old gardener'.[a]

This might have been modesty in regard to traditional technicians but unfortunately:

When Fan Hsü had gone out, the Master said 'What a small-minded man is Fan Hsü!'... 'If a ruler or an official loves good customs, righteousness and sincerity, people will flock to him from all quarters, bearing their children on their backs. So what does he need to know about agriculture?'

But two thousand years have shown that (like patriotism) good customs, righteousness and sincerity are not enough for the solution of all humanity's problems.

Still, it was magnificent as far as it went. A few further quotations will illustrate Confucian social-mindedness.

The Duke of Shê[b] asked about government. The Master said 'Good government obtains when those who are near are made happy, and those who are far off are attracted.'[c]

When the Master went to Wei, Jan Chhiu acted as driver of his carriage. The Master observed, 'How numerous the people are!' Jan Chhiu said, 'Since they are thus numerous, what more shall be done for them?' The Master replied, 'Enrich them.' Jan Chhiu said, 'And when they have been enriched, what more shall be done?' The Master said, 'Educate them.'[d]

Fan (Tzu-) Chhih (Fan Hsü) asked about benevolence. The Master said, 'It is to love men.' He asked about knowledge. The Master said, 'It is to know men.'[e]

Thus in early Confucianism there was no distinction between ethics and politics. Government was to be paternalistic. If the prince was virtuous the people would also be virtuous. And there was to be no equivocation about what virtue, peace and justice really were. Basing themselves upon certain passages in the Analects,[f] later (but still pre-Han) Confucians developed a doctrine of the 'rectification of names' (chêng

---

[a] *Lun Yü*, XIII, iv, tr. Legge (2), mod.
[b] A feudal lord of Chhu, whom Confucius probably met in Tshai; cf. pp. 92, 94 in Vol. 1 and p. 545 below.
[c] *Lun Yü*, XIII, xvi, tr. Legge (2).
[d] *Lun Yü*, XIII, ix, tr. Legge (2) and Ku Hung-Ming (1).
[e] *Lun Yü*, XII, xxii, tr. Legge (2).
[f] E.g. *Lun Yü*, XII, xi and xvii. XIII, iii has long been suspected of being a late interpolation.

ming[1]), i.e. the precise definition of actions and relations.[a] This was particularly associated with the school of Hsün Tzu (Hsün Chhing[b]) in the – 3rd century.[c] The nicety of the distinctions made in the rectification of names may be seen from the fact that in the traditional text of the *Chhun Chhiu*, of the thirty-six acts of regicide there recorded some are qualified as *shih*[2] (murder, implying the guilt of the assassin), while others are termed *sha*[3] (killing, implying that the act was legally justified).[d] Legally justified, because Confucian teaching also contained the democratic idea that the prince (and later, the emperor) derived his power primarily from the will of the people, expressing Heaven's will or mandate. This was much developed by the great Confucian apostle Mencius (Mêng Tzu), some hundred years later.[e] Thus that 'right of rebellion against unchristian princes' which so exercised the minds of the + 16th and + 17th century theologians of Europe had already been laid down two thousand years before by the Confucian school. Its revolutionary tendency, combined with the desire to uphold the established order (which predominated in most Confucian circles later), was perhaps one of the factors which permitted the Confucian bureaucracy of subsequent ages to rise superior to every change of dynasty by espousing popular causes and then presenting itself to each new ruler as the only possible instrument whereby government could be carried on. But the democratic element was real.

These points are illustrated by the following passages:

Chi Khang Tzu asked Master Khung about the art of ruling. The Master said, 'Ruling (*chêng*,[4] governing) is straightening (*chêng*,[5] rectifying). If you lead along a straight way, who will dare go by a crooked one?'[f]

Chi Khang Tzu was troubled by robbers. He asked Master Khung what he should do. Master Khung replied, 'If only you were free from desires, they would not steal even if you paid them to.'[g]

Chi Khang Tzu asked the Master about government, saying, 'Supposing we liquidated all those people who have not the Tao in order to help those who do have the Tao, what would you think of it?' Master Khung replied, 'You are there to rule, not to kill. If you desire what is good, the people will be good. The *chün-tzu* has the virtue of wind, the people have the virtue of grass. The grass must needs bend when the wind blows over it.'[h]

---

[a] This might be described as the determination to call a spade a spade, no matter what powerful influences might be desirous of having it called something else. This was particularly important in Chinese culture, where social courtesy and face-saving brought euphemism from early times to the level of a fine art. But there are parallels among the early sophists especially Prodicus of Ceos (– 5th century); Freeman (1), p. 372. And Jeremy Bentham's Theory of Fictions is not far removed from it.

[b] See below, pp. 19, 26 ff. Cf. Boodberg (3).

[c] The famous ch. 22 of *Hsün Tzu* is entitled 'The Correct Use of Terminology'. Cf. especially Duyvendak (4). The doctrine was also acceptable to the Legalists (cf. pp. 204 ff. below), as appears from *Shang Chün Shu*, ch. 26, and *Han Fei Tzu*, ch. 2.

[d] These interpretations of the historical classics were systematised chiefly by Tung Chung-Shu (cf. Fêng Yu-Lan (1), vol. 2, p. 71). They are very marked in the *Kuliang Chuan* and the *Kungyang Chuan*.

[e] The *locus classicus* is *Mêng Tzu*, I (2), viii; cf. IV (1), ii, 4. See p. 16 below.

[f] *Lun Yü*, XII, xvii, tr. Waley (5).   [g] *Lun Yü*, XII, xviii, tr. Waley (5).

[h] *Lun Yü*, XII, xix, tr. auct. adjuv. Legge (2), Waley (5).

[1] 正名   [2] 弑   [3] 殺   [4] 政   [5] 正

Tzu-Yu (Yen Yen) said, 'Formerly, Master, I heard you say, "When the *chün-tzu* learns well the Tao, he loves men. When the *hsiao-jen* learns well the Tao, he is easily led".'[a]

Duke Ting asked if there were any one phrase that sufficed to save a country. Master Khung replied, 'No phrase could ever be like that. But here is one that comes near to it. There is a saying among men; "It is hard to be a prince and not easy to be a minister". A ruler who really understood that it was hard to be a prince would have come fairly near to saving his country by a single phrase.' Duke Ting said, 'Is there any one phrase that could ruin a country?' Master Khung replied, 'No phrase could ever be like that. But here is one that comes near it. There is a saying among men: "What pleasure is there in being a prince unless one can say whatever one chooses and no one dares to disagree?" So long as what he says is good, it is of course also good that he should not be opposed. But if what he says is bad, will it not come near to ruining his country by a single phrase?'[b]

As to the honour of the feudal-bureaucratic officials whom Confucius knew how to train, one may quote from many interesting passages:

Tzu-Lu asked the Master how a prince should be served. He answered, 'When it is necessary to oppose him, withstand him to his face, and do not take refuge in disingenuous expedients.'[c]

The Master said, 'The determined scholar and he who is filled with the love of man[d] will not seek to live at the expense of injury to love. They will even suffer the death of the body in order that love may be accomplished.'[e]

We noted just now that the Confucians used the word Tao, the Way, to mean the proper way of life for social humanity. Thus:

The Master said, 'Shen, my Tao is woven on a single principle.' (Tsêng) Shen assented, and the Master went out, whereupon others who were present asked what he had meant. Master Tsêng said, 'Our Master's doctrine is just Loyalty (*chung*[1]) and Forgiveness (*shu*[2])— nothing more.'[f]

In other words, the doctrine of a cooperative society in which men's interests complemented each other without conflict. But this was thought of as embedded in the whole range of Nature so that goodness and social virtue among men were congruent

---

[a] *Lun Yü*, XVII, iv, tr. auct. adjuv. Legge (2).

[b] *Lun Yü*, XIII, xv, tr. Waley (5).

[c] *Lun Yü*, XIV, xxiii, tr. auct.

[d] The Chinese word here is *jen*,[3] which again is almost impossible to translate. Legge chose 'benevolence' or 'virtue', but both are too colourless; Waley uses 'goodness', but this does not convey (to my mind) the warmth of the conception, which comes very near the ἀγάπη τοῦ πλησίον, the love of our neighbour, of the Gospels. I find that in this I have the support of Graf (2), vol. 1, pp. 83, 266ff., who for the Neo-Confucian use of the term (cf. p. 488 below) suggests *humanitas*, and the *amor* of the scholastics. Chhen Jung-Chieh (5) and Chou I-Chhing (1) use an acceptable word, 'human-heartedness'. Cf. Boodberg (3).

[e] *Lun Yü*, XV, viii, tr. auct.

[f] *Lun Yü*, IV, xv, tr. auct. adjuv. Waley (5), Creel (4).

[1] 忠   [2] 恕   [3] 仁

with the will of the highest powers in the universe, powers which had, by the time of Confucius, lost whatever personalisation they had anciently possessed, and were referred to by the awe-inspiring name of *Thien* (Heaven). Confucius himself believed that Heaven 'knew' and 'approved' of his activities. In the *Chung Yung*,[a] the unity of the universe is clearly stated:

The Tao of Heaven and Earth may be completely declared in one sentence—it is without any doubleness; and so they produce things in a manner that is unfathomable. The Tao of Heaven and Earth is large and substantial, high and brilliant, far-reaching and long-enduring. The sky now before us is only a bright shining area, but when viewed in its inexhaustible extent, the sun, moon, stars and equatorial constellations are suspended in it, and all things are overspread by it. The earth before us seems only a handful of soil, but when regarded in its breadth and thickness it sustains mountains like Hua and Yo, without feeling their weight, and contains the seas and rivers without their leaking away. The mountain now before us appears only a stone, but when contemplated in the vastness of its size, we see how grass and trees are produced on it, birds and beasts dwell on it, and precious things which men treasure are found in it. Water may be but a ladleful, but when we think of what unfathomable depths it has, what enormous tortoises, dragons, fishes, etc., are produced in it, what articles of wealth and value....[b]

And from this picture of Nature, we should read straight on: 'Man is born for uprightness. If he lose it and yet live, it is the effect of mere good fortune.'[c] Here is the beginning of that great debate about the essential goodness or badness of human nature, to which so much of the thought of later Chinese philosophers was devoted. We shall shortly return to it, for it had considerable connection with scientific thought.

(c) THE AMBIVALENT ATTITUDE TOWARDS SCIENCE

Now appear the two fundamental tendencies which paradoxically helped the germs of science on the one hand and injured them on the other. On one side Confucianism was basically rationalistic and opposed to any superstitious or even supernatural forms of religion. Examples will be given later (pp. 365, 386 ff.) from widely different epochs of Chinese history. But on the other side its intense concentration of interest upon human social life to the exclusion of non-human phenomena negatived all investigation of Things, as opposed to Affairs. Hence, not for the last time in history, nor only in China, rationalism proved itself less favourable than mysticism to the progress of science. It will not be long before we shall have abundant demonstration of this; here I need only illustrate what has just been said.

---

[a] Some parts of this book are still thought to be due to the grandson of Confucius, but the part here quoted is probably the work of an unknown Confucian of the Chhin dynasty (mid −3rd century). Much of the book, which contains Legalist material, is of this date.

[b] *Chung Yung*, XXVI, 7, tr. Legge (2).

[c] *Lun Yü*, VI, xvii, tr. Legge (2).

There are two chief passages about the spirits:

Fan (Tzu-) Chhih asked what constituted wisdom. The Master said, 'To give one's self earnestly to securing righteousness and justice among the people, and while respecting the gods and demons, to keep aloof from them, that may be called wisdom.'[a]

Chi-Lu asked about serving the ghosts and spirits. The Master said, 'While you are not yet able to serve men, how can you serve ghosts?' Chi-Lu then ventured upon a question about the dead. The Master said, 'You do not yet know about the living, how can you know about the dead?'[b]

And we shall add two revealing passages about rites. The first shows that Confucius was far from having an *ex opere operato* theory of the efficacy of ritual. His school, indeed, always believed in the value of the rites (*li*[1]) for the living human beings who participated in them rather than in their magical power to affect the spirits of ancestors or the local and minor deities.[c]

The Master said, 'If a man be without the virtues proper to humanity, what can he have to do with ritual? A man without the virtues proper to humanity, what can he have to do with music?'[d]

The second passage, though long, is worth including *in extenso*, for it shows so well the relations of Confucius with his disciples, and his deep appreciation of the numinous quality of traditional rites and ceremonial.[e]

Once when Tzu-Lu, Tsêng Hsi, Jan Chhiu and Kunghsi Hua were seated in attendance on the Master, he said, 'You consider me as a somewhat older man than yourselves. Forget for a moment that I am so. At present you are out of office, and feel that your merits are not recognised. Now supposing someone were to recognise your merits, what employment would you choose?'

Tzu-Lu promptly and confidently replied, 'Give me a country of a thousand war-chariots, hemmed in by powerful enemies, or even invaded by hostile armies, with drought and famine to boot—in the space of three years I could endow the people with courage, and teach them in what direction right conduct lies.'

Our Master smiled at him and said, 'What about you, Chhiu?'

---

[a] *Lun Yü*, VI, xx, tr. Legge (2), modified. Waley (5) and Creel (4) reverse the sense of this passage by translating 'he who by respect for the Spirits keeps them at a distance'. This principle of *do ut abias* was certainly familiar to ancient Chinese religious thought, as elsewhere in the world, and Waley's interpretation therefore has some plausibility, but I prefer to retain Legge's, since the idea of 'keeping aloof' from the gods and demons, while not positively denying their existence, was so characteristic of Confucianism in all subsequent times. Cf. Creel (5).

[b] *Lun Yü*, XI, xi, tr. Legge (2), mod., Waley (5).

[c] So also *Hsün Tzu* (see Dubs (7), pp. 144, 152).

[d] *Lun Yü*, III, iii, tr. Legge (2) and Waley (5).

[e] That is to say, when it passed the test of his advanced humanitarian morality. In *Lun Yü*, III, xxi, there may be a reference to the human sacrifices of the earlier chthonic religion; Confucius does not wish even to discuss it.

[1] 禮

Chhiu replied saying, 'Give me a domain of fifty to seventy square leagues, and in the space of three years I could bring it about that the common people should lack for nothing. But as to rites and music, I should have to leave those to a real *chün-tzu.*'

'What about you, Chhih?'

Kunghsi Hua answered, 'I do not say that I could do this; but I should like at any rate to be trained for it. In ceremonies at the ancestral temple, and at the audiences of the Princes with the High King, I would like, dressed in the dark square-made robe and the black linen cap, to act as a junior assistant.'

'Tien, what about you?'

Tsêng Hsi laid aside the lute on which he had been softly playing, rose and said, 'I fear my words will not be so well chosen as those of the other three.'

The Master said, 'What harm is there in that? All that matters is that each should name his desire.'

Tsêng Hsi said, 'At the end of spring, when the making of the Spring Clothes has been completed, to go with five or six newly capped young men and six or seven boys, perform the lustration and bathe in the River Yi, enjoy the breeze among the Rain Dance altars, and return home singing.'

The Master sighed and said, 'I agree with Tien.'[a]

This gives us a taste of the way in which Confucianism was typical of all Chinese civilisation in its combination of the romantic and the rational. Favourable to, even emphasising, traditional rites and ceremonies, it remained unshakably sceptical and averse to any kind of supernaturalism. This was the element referred to above which could have helped the growth of the scientific view of the world. But it was more than counterbalanced by the Confucian attitude to knowledge, which never wavered from the standpoint that man and human society were alone worthy of investigation. Here are type-passages:

The Master's frequent themes of discourse were: the Odes, the History and the maintenance of the Rites. On these he frequently discoursed. (*Tzu so ya yen, Shih, Shu, chih Li, chieh ya yen yeh.*[1])[b]

The Master took four subjects for his teaching: culture (letters), the conduct of affairs, loyalty to superiors and the keeping of promises. (*Tzu i ssu chiao, wên, hsing, chung, hsin.*[2]) [c]

The subjects on which the Master never talked were: extraordinary things (natural prodigies), unnatural strength, disorders (in Nature) and spiritual beings. (*Tzu pu yü, kuai, li, luan, Shen.*[3])[d]

---

[a] *Lun Yü,* XI, xxv, tr. Legge (2) and Waley (5), mod.

[b] *Lun Yü,* VII, xvii, tr. Legge (2). There is another interpretation, in which *ya*[4] is taken to mean *chêng*[5] instead of *chhang*;[6] this, adopted by Waley (5), would say that Confucius used the correct pronunciation, instead of the Lu dialect, when reciting these classics. We adhere to the traditional interpretation.

[c] *Lun Yü,* VII, xxiv, tr. Waley (5). Cf. XI, ii, a parallel list.

[d] *Lun Yü,* VII, xx, tr. auct. We diverge here from the generally accepted interpretations of the commentators.

[1] 子所雅言詩書執禮皆雅言也   [2] 子以四教文行忠信
[3] 子不語怪力亂神   [4] 雅   [5] 正   [6] 常

The third is particularly important. Interest in natural phenomena is first awakened by surprising or startling departures from the normal course of things—comets, seismic or volcanic phenomena, teratological productions (monstrous births), unusual oecological distributions, snowfalls in summer, owls hooting by day, thunder from an apparently clear sky, and so on. Many examples of such observations will occur later in the book (Sects. 20, 21, 43), and can be paralleled from all ancient civilisations. The characters *kuai*[1] and *luan*,[2] and indeed *shen*,[3] since spiritual beings of low rank would naturally be brought in as explanations, are therefore easily understandable. *Li*,[4] however, has always been translated as 'feats of strength', which seems quite meaningless as it has no connection with any other Confucian standpoint; surely it refers to the superhuman force of Nature as shown in natural convulsions such as earthquakes, tidal waves, avalanches, hot springs or geysers, and the like. Confucius had no intention of being drawn into a discussion of such phenomena, which seemed to have no bearing on the problems of human society. And for two thousand years his followers adopted his example, to the despair of Taoists and technologists.

The eighteenth book of the *Lun Yü* is widely recognised as a later Taoist interpolation of legendary character, but there are two stories in it which simplify the transition which we shall shortly make from Confucianism to Taoism. First the Madman of Chhu:

> Chieh Yü, the madman of Chhu, came past Master Khung, singing as he went
>
> > 'O Phoenix, Phoenix
> > How dwindled is your power!
> > As to the past, reproof is idle,
> > But the future may yet be remedied.
> > Hopeless! Alas!
> > Dangerous in these days are those who fill office!'
>
> Master Khung got down from his chariot, desiring to speak with him, but the madman hastened his step and got away, so that the Master did not succeed.[a]

Then the Irresponsible Hermits:

> Long Rester and Firm Recluse[b] were working at the plough together. Master Khung, happening to pass that way, sent Tzu-Lu to ask them where the river could be forded. Long Rester said, 'Who is that person you are driving for?' Tzu-Lu replied, 'Khung Chhiu.' 'What, Khung Chhiu of Lu?' and Tzu-Lu assented. Long Rester said, 'In that case he already knows where the ford is.' So Tzu-Lu turned to Firm Recluse...(but he would only say), 'Under heaven there is none that is not swept along in the same flood. Such is the world and who can change it? As for you, instead of following one who flees from this man

---

[a] *Lun Yü*, XVIII, v, tr. Waley (5).

[b] The Taoist provenance of this story is obvious from the facetious naming of the imaginary characters, an amusing custom which we shall meet again.

[1] 怪   [2] 亂   [3] 神   [4] 力

and that, you would do better to follow one who shuns this whole generation of men.' And with that he went on covering the seed.

Tzu-Lu went and reported their remarks to the Master, who sighed and said, 'It is impossible to associate with birds and beasts. If I do not participate in the social life of man, what else is there with which to associate? If the world were as it ought to be (lit. if the true Tao of social life prevailed everywhere under Heaven) I should not be wanting to change it.'[a]

One cannot help sympathising with the real mind of the social reformer or revolutionary beautifully represented by Confucius in this passage. But the hermits, whom we may at once recognise as Taoists or proto-Taoists, though socially irresponsible, are not yet showing their true colours. They were out of the society of Man partly at any rate in the interests of the study of Nature, as will appear below.

At this point it would be desirable to embark without delay upon the description of Taoism, but before doing so the picture of Confucianism must be completed by some account of its later development. Though this has little connection with science, except in an inimical way, there is one department of Confucian thought, namely, that concerned with the status and characteristics of human nature, which deserves elucidation. This has two aspects, first the debate about the intrinsic goodness or badness of human nature, and secondly the development of thought as to the relation of human nature to animal and plant natures.

(d) DOCTRINES OF HUMAN NATURE

Mêng Kho[1] (Mencius, c. − 374 to − 289) was Master Khung's greatest disciple, though a hundred years separated the death of the latter from the birth of the former. Born in the tiny State of Tsou[2] on the southern borders of Lu, he spent most of his life teaching in Liang and Chhi and advising their rulers. There was nothing essentially new in his doctrines, so far as we are concerned.[b] He developed the democratic conception that the goodwill of the people was essential in government, emphasising the famous saying in the *Shu Ching*[c] that Heaven sees according as the People see, Heaven hears according as the People hear.[d] The voice of the people was to have predominant weight over other advice,[e] and they were to be considered the most important element in a State, the spirits of the land and grain coming next, and the prince last.[f] Rites and usages were made for man, not vice versa,[g] and were bad if they became empty conventions, as in the hands of 'your good careful people of the villages'.[h] The right of rebellion against tyrants was laid down in the most uncompromising way.[i]

a *Lun Yü*, XVIII, vi, tr. Legge (2) and Waley (5), modified.
b See the book of Yuan Cho-Ying (1).   c Ch. 21 (Thai Shih), Legge (1), p. 128.
d *Mêng Tzu*, V (1), v, 8.   e *Mêng Tzu*, I (2), vii, 4, 5; x, 3.
f *Mêng Tzu*, VII (2), xiv, 1.   g *Mêng Tzu*, IV (1), xvii.
h *Mêng Tzu*, VII (2), xxxvii, 8 ff.
i *Mêng Tzu*, V (2), ix. See on this the fine disquisition of Legge (3), p. [48].

¹ 孟軻   ² 鄒

But for the history of scientific thought the most interesting aspect of Mencius is the doctrine of human nature.

Mêng Tzu said, 'All men have a mind which cannot bear to see the sufferings of others. The ancient kings had this commiserating mind, so as a matter of course they had likewise a commiserating government. When this was practised the government of the empire was as easy a matter as twirling something in one's hand. When I say that all men have a mind which cannot bear to see the sufferings of others, my meaning may be thus illustrated—even nowadays, if men suddenly see a child about to fall into a well, they will without exception experience a feeling of alarm and distress. They will feel it, not because they think to gain the favour of the child's parents, nor in order to seek the praise of neighbours and friends, nor from a dislike from the reputation they might get as having been unmoved by it. From this we may perceive that the feeling of commiseration is essential to man, the feeling of shame and dislike is essential to man, the feeling of modesty and complaisance is essential to man, and the feeling of approving and disapproving is essential to man.'[a]

Mencius thus maintained that man's nature had a natural tendency to good. But he had to dispute the matter with his contemporaries, as we know from what follows:

Kao Tzu[1] said, 'Man's nature is like the willow-tree, and righteousness is like a cup or bowl. Fashioning benevolence and righteousness out of man's nature is like carving cups and bowls from willow wood.'

Mêng Tzu replied, 'Can you make cups and bowls while leaving untouched the nature of the wood? You must do violence and injury to the willow, before you can make cups and bowls with it.[b]...On your principles you must in the same way do violence and injury to humanity to fashion from it benevolence and righteousness.... If this were so, men would think them calamities.'

Kao Tzu said, 'Man's nature is like swirling water. Open a passage for it to the east and it will flow to the east; open a passage to the west and it will flow west. Man's nature makes no distinction between what is good and what is evil, just as water makes no distinction between east and west.'

Mêng Tzu replied, 'Water will indeed flow indifferently to east and west, but will it flow indifferently up or down? The tendency of man's nature to good is like the tendency of water to flow downhill.... By striking water and causing it to leap up you make it go over your head, and by damming it and leading it you may force it uphill, but are such movements according to water's nature? No, it is the force applied which causes them. When men are made to do what is not good, their nature is subjected to force.'

Kao Tzu said, 'That which at birth is so, is called nature.'[c]

Mêng Tzu replied, 'Do you mean that what at birth is so, is to be called nature, as whiteness is called whiteness?'

'Yes', he said.

---

[a] *Mêng Tzu*, II (1), vi, tr. Legge (3).

[b] The similarity of Mencius' thought here to Taoist ideas should be noticed; see below, pp. 106 ff.

[c] This, *sêng chih wei hsing*,[2] has been held to be one of the most ambiguous sentences in all ancient Chinese texts. It is considered in detail by I. A. Richards (1), p. 23.

[1] 告子   [2] 生之謂性

Mêng Tzu continued, 'Is the whiteness of a white feather like the whiteness of white snow, or the whiteness of snow like that of a white gem?'

'Yes', he said.

'Very well', said Mêng Tzu, 'Is the nature of a dog like the nature of an ox, and the nature of an ox like the nature of a man?'

Kao Tzu answered, 'Hunger and sexual desire are human nature'.[a]

And the different opinions were summarised thus:

The disciple Kungtu said, 'Kao the philosopher says that man's nature is neither good nor bad. Some say that the nature may be made to be good or may be made to be evil. Thus under Kings Wên and Wu the people loved what was good, while under Kings Yu and Li they loved what was cruel. Others say that the nature of some individuals is good while the nature of others is evil. Therefore it was that under (so good) a sovereign as Yao there yet appeared the (evil) Hsiang; that with (such a bad father as) Ku Sou there yet appeared the (excellent) sovereign Shun; and with the (wicked) Chou (Hsin) as emperor there were (such virtuous men as) Chi the viscount of Wei, and the prince Pi-Kan.'[b]

Whereupon Mencius expounds his views once again, saying that if men do what is not good, the blame cannot be imputed to their natural powers.

Summing up the situation so far, then, Confucius himself had simply said that by nature men, formed for uprightness, were closely alike, but in the course of their practice and experience they grow to be far apart.[c] Kao Tzu had affirmed the moral neutrality of human individuals, considering that the training and moulding of the social inheritance was necessary for the development of virtuous or social behaviour. Mêng Tzu (Mencius) had boldly maintained that there was in human nature a bias towards the good, which not only rendered education easier, but permitted an optimistic view of the possibilities of human society. Lastly, other thinkers, whose names are not given, suggested that in fact human individuals were very differently endowed with tendencies to good or bad.

We thus have in germinal form all the elements of those vast controversies which have lasted right through European history as well as Chinese.[d] Augustinianism against Pelagianism, eighteenth-century optimism in revolt against theological pessimism, Neo-Lamarckianism against Mendelian genetics, the varying estimates of the roles of Nature and Nurture in human biology—vital questions still in hot dispute in our own time—have one at least of their origins in this ancient Chinese discussion. We have to remember that all these controversies, except those of our own day, took place in the absence of any clearly defined understanding of organic evolution. In the light of biological evolution theory and of modern psychology we should presumably say that anti-

[a] *Mêng Tzu*, VI (1), i, ii and iii, tr. Fêng Yu-Lan (Bodde) (1), vol. 1, pp. 125, 145, and Legge (3), mod.

[b] *Mêng Tzu*, VI (1), vi, tr. Fêng Yu-Lan (Bodde) (1), vol. 1, p. 147, and Legge (3).

[c] *Lun Yü*, XVII, ii; cf. Dubs (15).

[d] Among books and articles on the history of the Chinese controversy there are monographs by Fu Ssu-Nien (2) and Chiang Hêng-Yuan (1), interesting papers by Inouye (1) and Lau (2), and a good discussion by A. C. Graham (1), pp. 134 ff.

social (evil) impulses and actions arise from the animal inheritance of human individuals,[a] while social (good) impulses arise from that tendency to cohere in social organisms which is peculiar to man's nature. Kao Tzu recognised neither, placing all the responsibility on nurture, Mencius recognised simply the latter; it only remained for someone to emphasise the former, and so to take up a position similar to that of the European theologians who developed the Christian doctrine of the Fall and of Original Sin. This was the contribution of Hsün Tzu. Before stating it, I wish only to compare the attitude of the fourth group of Chou philosophers mentioned above, who thought that individuals had mixed inheritances, with the conception which we find in medieval Europe illustrated in one of the visions of St Hildegard of Bingen (+1098 to +1180), in which devils and angels are seen preparing the respective constitutions of unborn human beings, adding strength or weakness, virtue or corruption.[b] I shall go on to show that during the course of Chinese history, the Chinese themselves, without the assistance of modern evolution theory, finally came to a position in which the specifically human and 'animal' elements in human psychology were recognised, and the views of the Chou philosophers therefore reconciled. Only the time factor was missing. But first Hsün Tzu.

Hsün Chhing[1] (c. −305 to −235) was a native of Chao[2] (in modern Shansi) and wandered about the courts of the feudal princes as other Chou thinkers had done. His official life was spent as magistrate of the city of Lanling in the State of Chhu to the south, and he was widely known as a writer and teacher.[c] The characteristic word in his philosophy was Nurture,[d] precisely because he was convinced that human nature had intrinsic tendencies to evil, and that everything good depended upon education, in the widest sense. This was not a kind of Calvinism, since he believed that all human beings had an infinite capacity for development in the direction of good. Yet chapter 23 of his book *Hsün Tzu* bears the title 'That the Nature of Man is Evil'. It opens thus:

The nature of man is evil—his goodness is only acquired by training. The original nature of man today is to seek for gain. If this desire is followed, strife and rapacity results and courtesy (*tzhu jang*[3])[e] dies. Man originally is envious and naturally hates others. If these tendencies are followed, injury and destruction result, loyalty and faithfulness are destroyed. Man originally possesses the desires of the ear and the eye; he likes praise and is lustful. If these are followed, impurity and disorder result, and the rules of proper conduct (rites, *li*[4]), justice (*i*[5]) and refined culture (*wên li*[6]) are done away with. Therefore to give rein

[a] Cf. Tennyson's 'Let the ape and tiger die'.

[b] Her *Liber Scivias*; see Singer (3, 4, fig. 106) and Needham (2), p. 66, fig. 7. This was the intellectual precursor of Mendelian genetics, save that the action of the genes in the developing organism, as we understand it today, is always subject to an environmental factor. The genotype is never fully expressed in the phenotype.

[c] Fêng Yu-Lan (1), vol. 1, pp. 279 ff.; Dubs (7).　　　　[d] Dubs (7), p. xiii.

[e] The literal meaning of these characters is 'declining and yielding'. *Jang* is a very ancient Chinese conception which is found in all the Chou philosophers of whatever school, but particularly in the Taoists, as we shall see (p. 61).

[1] 荀卿　　[2] 趙　　[3] 辭讓　　[4] 禮　　[5] 義　　[6] 文理

to man's original nature, to follow man's feelings, inevitably results in strife and rapacity, together with violations of good customs and confusion in the proper way of doing things; there is reversion to a state of violence. Hence the civilising influence of teachers and laws, the guidance of the rites and justice, is absolutely necessary. Thereupon courtesy appears, cultured behaviour is observed, and good government is the consequence. By this line of argument it is evident that the nature of man is evil and his goodness is acquired.[a]

During the Han dynasty these great problems continued to be discussed, as by the sceptical thinker Wang Chhung[1] (+27 to +97).[b] Here we meet with a statement of the opinion that all human beings are born with mixed endowments of good and bad tendencies, i.e. that human nature does not always tend to good (Mêng Tzu), nor to bad (Hsün Tzu), nor is it neutral (Kao Tzu), nor are all human beings either completely good or completely bad (the unnamed school). The scientific approach was developing. The new view is attributed by Wang Chhung to a Chou philosopher Shih Shih,[2] of whom we otherwise know almost nothing, though traditionally numbered among the disciples of Confucius.

Shih Shih held that human nature is partly good and partly bad, and that if the good nature in man be cultivated his goodness increases, whereas if his bad nature be cultivated his badness increases.[c]

This view can be traced into the generation before Wang Chhung, since Yang Hsiung[3] (−53 to +18) discussed the problem in his *Fa Yen*[4] (Model Sayings). He wrote:

Man's nature is a mixture (*hun*[5]) of good and bad. If he cultivates the good part he becomes good, if he cultivates the evil he becomes evil. (Riding) one's *chhi*[6] (*pneuma*) one can direct it towards good or evil, as if one were riding a horse.[d]

And a century earlier, Tung Chung-Shu had spoken[e] of the 'rudiments of goodness' (*shan chih*[7]) in each individual, which would not fully manifest themselves without training.

Continual attempts were made to penetrate further into the matter, but we need only mention the famous Thang Confucian Han Yü[8] (+762 to +824)[f] whose *Yuan Hsing phien*[9] (Essay on the Original Nature) Legge[g] translated.[h] He thought,

---

a Tr. Dubs (8) and Chêng Chih-I (1).   b *Lun Hêng*, ch. 13.
  c Tr. Fêng Yu-Lan (Bodde) (1), vol. 1, p. 147; Forke (4), vol. 1, p. 384. A few philosophers associated with Shih Shih are known to us by name: Fu Tzu-Chien,[10] Chhitiao Khai,[11] and Kungsun Ni-Tzu.[12] Cf. Kuo Mo-Jo (1), p. 127.   d Ch. 2, p. 12b (cf. Forke (12), pp. 60, 90), tr. auct.
  e *Chhun Chhiu Fan Lu*, ch. 35, end (cf. Forke (12), p. 59).
  f G632.   g (3), p.[92].
  h From the *Han Chhang-Li hsiensêng chhüan chi*, ch. 11, p. 7a. Han Yü's teaching was known as the 'three-grade theory', *hsing san phin shuo*.[13]

¹ 王充   ² 世碩   ³ 揚雄   ⁴ 法言   ⁵ 混   ⁶ 氣
⁷ 善質   ⁸ 韓愈   ⁹ 原性篇   ¹⁰ 宓子賤   ¹¹ 漆雕開
¹² 公孫尼子   ¹³ 性三品說

adapting the Analects, that men's natures could be divided into three classes, the superior, which will do good whatever the circumstances, the middle, which may do either good or evil according to training and environment, and the inferior, irretrievably bad. He thought that Mêng Tzu had had the first class in mind and Hsün Tzu the third (a fatalism which misinterpreted Hsün Tzu), while Yang Hsiung had referred to the middle class. This was really no advance, and such classifications had been attempted before, as by Hsün Yüeh[1] in his *Shen Chien*[2] [a] about +190.[b]

Much more important is the fact that precisely contrary to the orthodoxy of Augustine and the heresy of Pelagius in the West, in China Mencius came to be orthodox and Hsün Tzu 'heretical'. Hsün Tzu was formally condemned by the Neo-Confucian school in the Sung, perhaps, as Dubs says,[c] because the word *hsing*[3] (nature) had by then taken on a cosmological significance, and because the word *wei*,[4] which Hsün Tzu used for 'artificial training', could also mean falsity, perversion, or a lie.[d] In any case, this difference was a cardinal one for the whole of Chinese culture, and we shall refer to it again (Sect. 49).[e]

(e) THEORIES OF THE 'LADDER OF SOULS'

We must now return to the time of Hsün Tzu in order to follow the other line of thought which meets the human nature problem at the end of the Sung (+13th century), namely, the question of the 'Ladder of souls'.

As has often been explained,[f] Aristotle adopted the word ψυχή (*psyche*) for the principle which differentiated living from non-living substance, but was forced to the conclusion that there were different kinds or orders of *psyche* or 'soul'. According to

---

[a] Ch. 5 (cf. Forke (12), p. 133).

[b] In fact, it was a mere development of thoughts found in Han works, e.g. *Huai Nan Tzu*, ch. 19; *Chhun Chhiu Fan Lu*, ch. 36; and *Lun Hêng*, ch. 13. It is interesting, however, that in many of these formulations the number of individuals in the middle majority was recognised as being much larger than those at the very good and very bad ends of the range. Was this not an intuitive appreciation of phenomena which would now be represented by a Gaussian distribution curve?

[c] (7), p. 82.

[d] Cf. Fêng Yu-Lan (1) in Bodde (3), p. 32; and below, pp. 109, 393, 450.

[e] The Mencian doctrine was crystallised in universally used school books such as the famous *San Tzu Ching*,[5] a kind of catechism in verses of three characters each, intended for memorisation, produced by Wang Ying-Lin[6] (+1223 to +1296) in the Sung (G 2253). The opening lines assert, 'Men at their birth are naturally good; their natures are much the same, their habits become widely different' (tr. H. A. Giles, 4). A characteristic attitude of later thinkers is that of Tsou Shou-I[7] (+1491 to +1562) who considered evil actions to be in the category of diseases, i.e. aberrations analogous to affections of the eye which impede the sight (cf. Forke (9), p. 407). So modern is this view that European civilisation is only now arriving at it. When Matteo Ricci came to talk with Chinese scholars at the beginning of the 17th century, he had no small difficulty in making the doctrine of original sin intelligible (Trigault (1), tr. Gallagher, p. 341).

[f] W. D. Ross (1); Singer (1), pp. 37 ff.

[1] 荀悅　　　[2] 申鑒　　　[3] 性　　　[4] 僞　　　[5] 三字經　　　[6] 王應麟
[7] 鄒守益

the doctrine which grew up out of Aristotle's work and which dominated all the biology of subsequent ages, plants possessed only a vegetative or nutritive soul, animals possessed in addition an animal or sensitive soul, while man was further

Table 10. *The doctrines of the 'ladder of souls'*

ARISTOTLE (−4th century)

    plants   ψυχὴ θρεπτική
             vegetative soul

    animals ψυχὴ θρεπτική   ψυχὴ αἰσθητική
             vegetative soul + sensitive soul

    man   ψυχὴ θρεπτική   ψυχὴ αἰσθητική ψυχὴ διανοητική
             vegetative soul + sensitive soul + rational soul

HSÜN CHHING (−3rd century)

    water and fire   *chhi* 氣
    plants   *chhi* 氣 + *sêng* 生
    animals   *chhi* 氣 + *sêng* 生 + *chih* 知
    man   *chhi* 氣 + *sêng* 生 + *chih* 知 + *i* 義

LIU CHOU (+6th century)

    plants   *sêng* 生
    animals   *sêng* 生 + *shih* 識

WANG KHUEI (+14th century)

    heaven, sky, rain, *chhi* 氣
      dew, frost and
      snow
    earth   (*chhi* 氣) + *hsing* 形
    plants (and some *chhi* 氣 + *hsing* 形 + *hsing* 性
      minerals)
    animals   *chhi* 氣 + *hsing* 形 + *hsing* 性 + *chhing* 情
    (man   *chhi* 氣 + *hsing* 形 + *hsing* 性 + *chhing* 情 (+ *i* 義)

endowed with a rational soul. If these terms have dropped out of modern science, it is only because increasing precision of experimentation and terminology has rendered them unnecessary, not because they were so far off the mark in describing the various manifestations of the activity of living things.

We do not think that it has hitherto been pointed out[a] that the Chinese developed a remarkably similar scheme for the same purpose. It is compared with that of Aristotle in Table 10. Here is the passage from Hsün Tzu which justifies this.

Water and fire have subtle spirits (*chhi*[1]; somewhat analogous to the *pneuma* of the Greeks)[b] but not life (*sêng*[2]). Plants and trees have life (*sêng*[2]) but not perception (*chih*[3]); birds and animals have perception[c] (*chih*) but not a sense of justice (*i*[4]). Man has spirits, life, and perception, and in addition the sense of justice; therefore he is the noblest of earthly beings. In strength he does not equal the ox, nor in power of running the horse, and yet he uses them; how can this be? Man is able to form social organisations (*chhün*[5]) and they are not. How is it that men can do this? Because they can cooperatively play their parts and receive their portions (*fên*[6]). How is it that they can carry this out? Because of justice and righteousness (*i*), which unite the parts into a harmony, and therefore a unity, and lead to strength, and in the end to triumph.[d]

And this raises one more of those awkward time confrontations which have already been referred to, for Aristotle's life (-384 to -322) was only very little anterior to that of Hsün Tzu (*c.* -305 to -235). Since this was a century and a half before the opening of the Silk Road, I confess to much difficulty in believing that the one system could have been derived from the other, and would prefer to suppose that both were independent, though very similar, results of reflection on the same phenomena. It is typical of Chinese thought that what particularly characterised man should have been expressed as the sense of justice rather than the power of reasoning.

In later Chinese literature there are many other statements of similar views[e] but one of the best we have found is that of the Ming biologist Wang Khuei[7] whose *Li Hai Chi*[8] may be of the late $+14$th century. He says:

The heavens (*thien*[9]) have subtle spirits (*chhi*[10]), but these spirits have no natural endowments[f] (*hsing*[11]) or sensitivity (*chhing*[12]), nor have rain, dew, frost and snow any endowments or sensitivity. The earth (*ti*[13]) has form (*hsing*[14]). Substances possessing form may have endowments without sensitivity; thus herbs, wood and some minerals have endowments but no sensitivity. The intercourse of heaven and earth combines subtle spirits (*chhi*[10]) with form (*hsing*[14]), thereby giving rise both to endowments (*hsing*[11]) and sensitivity

---

[a] Except by Lu & Needham (1) (see Sarton (1), vol. 3, p. 905).

[b] See on, pp. 228, 242, 250, 275.

[c] 'Instinct' might be a better translation here, or perhaps Hsün Tzu meant unconscious reflex action.

[d] *Hsün Tzu*, ch. 9, p. 13 *a*, tr. auct., adjuv. Dubs (8), p. 136; Hughes (1), p. 246.

[e] Just as there are in European writings, e.g. Bartolomaeus Anglicus, fl. $+1230$; Sarton (1), vol. 2, p. 586.

[f] Of rhubarb, for example, the laxative property is its 'natural endowment'. See especially p. 569 below. *Hsing*[11] is usually (and enigmatically) translated 'nature', but here we must take it as something like 'active principle' or 'outstanding property'.

[1] 氣   [2] 生   [3] 知   [4] 義   [5] 羣   [6] 分   [7] 王逵
[8] 蠡海集   [9] 天   [10] 氣   [11] 性   [12] 情   [13] 地
[14] 形

(*chhing*[1]). Birds, beasts, insects and fishes (thus) possess both endowments (*hsing*[2]) and sensitivity (*chhing*[1]). Their watery secretions and excretions have spirits (*chhi*[3]) like those of the heavens; their feathers, fur, scales and carapaces have forms (*hsing*[2]) like those of earth. How can we deny that spirits (*chhi*[3]) and forms (*hsing*[2]) have to combine, in order that endowments (*hsing*[2]) and sensitivity (*chhing*[1]) may (together) be present?[a]

This seems to be much more than an elaboration of the ideas of Hsün Tzu.

It will be noticed that we have put in the table, as a connecting link, the formulation of Liu Chou[4] (+519 to +570), whose book, the *Liu Tzu*,[5] is included in the *Tao Tsang* (no. 1018). His use of the word *shih*[6] betrays Buddhist influence, for it is one of the twelve *nidānas* (see p. 400).[b] Between his time and that of Wang Khuei the Neo-Confucians did a good deal of thinking about this question (see on, p. 568). Chhêng I[7] followed the same system as Hsün Chhing and Liu Chou, adding 'good instinct' (*liang nêng*[8]) to the *chih*[9] (perception) of animals and man. Chu Hsi[10] had a somewhat more complicated view[c] (cf. on, pp. 488, 569). Apart from the *chhi*[11] or matter-energy[d] of which everything is composed, and the universal organising principle, *Li*,[12] inorganic things possessed only substances and qualities (lit. form, substance, smell and taste; *hsing chih chhou wei*[13]). Plants, in addition, possessed *sêng chhi*[14] or vital force. But to this was added, in animals and man, the *chhi* of blood, and perception and sensation associated with it (*hsüeh chhi chih chio*[15]).

After Wang Khuei's time, the question continued to be discussed, as by Hsüeh Hsüan[16] (+1393 to +1464), who wondered if the vital force in plants was not a 'governing impulse' (*chu-tsai hsin*[17]).[e] Ku Hsien-Chhêng[18] (+1550 to +1612) engaged in arguments with the Buddhists about such questions.[f]

We are now in a position to examine the synthesis proposed during the latter half of the +13th century by Tai Chih.[19] It had become evident that the composition of man's nature was much more complex than classical philosophy had thought. Tai Chih in his *Shu Pho*[20] (Rats and Jade),[g] which has to be placed somewhere about +1260, saw that the more highly social tendencies of man were peculiar to him, while his anti-social tendencies had to do with those elements of his nature which he shared with the lower animals. He wrote:

People talk about human nature—some say it is good, others that it is bad. Generally they prefer Mêng Tzu's view and reject Hsün Tzu's. After studying both books I realised that

---

[a] P. 50*b*, tr. auct.   [b] For more information about him, see Forke (12), p. 250.

[c] In the *Chu Tzu Chhüan Shu*,[21] ch. 42, p. 34*a*; tr. Bruce (1), p. 69.

[d] We shall discuss later the place of *chhi* in the philosophy of the Neo-Confucians; this translation indicates the development which took place between the time of Hsün Chhing and Chu Hsi.

[e] Cf. Forke (9), p. 327.   [f] Forke (9), p. 427.

[g] This curious title comes from an ancient story that people of different dialects confused these two things; hence the need which was felt for standardisation of terms.

[1] 情	[2] 性	[3] 氣	[4] 劉晝	[5] 劉子	[6] 識	[7] 程頤
[8] 瓦能	[9] 知	[10] 朱熹	[11] 氣	[12] 理	[13] 形質臭味	
[14] 生氣	[15] 血氣知覺		[16] 薛瑄	[17] 主宰心	[18] 顧憲成	
[19] 戴埴	[20] 鼠璞		[21] 朱子全書			

Mêng Tzu is talking about the heaven-nature (*thien-hsing*[1]) and what he calls the goodness of human nature referred to its (innate) uprightness and greatness. He wished to encourage it. That is what the *Ta Hsüeh* calls '(developing) sincerity' (*chhêng i*[2]).

But Hsün Tzu is talking about the matter-nature (*chhi-hsing*[3]), and what he called the badness of human nature referred to its (innate) wrongness and roughness. He wished to repair and control it. This is what the *Chung Yung* calls 'forceful checking' (*chhiang chiao*[4])....

Thus Mêng Tzu's teaching is to strengthen what is already pure, so that defilement tends to disappear of itself. While Hsün Tzu's teaching is to remove defilement actively. Both are equally helpful to later students.[a]

Nothing was now lacking but the time factor, which only a knowledge of biological evolution could provide.[b] But in spite of certain other sprouts of ideas, soon to be alluded to (p. 78), Chinese culture never of itself reached this knowledge. Tai Chih was not the only Sung scholar who thought along these lines. Huang Hsi[5] (d. *c.* + 1060), in his *Ao Yü Tzu Hsü Hsi So Wei Lun*[6] (Whispered Trifles by the Tree-stump Master), had compared[c] Mencian innate goodness with the inoffensiveness of plants. Hsün Tzu, he said, had seen only that element in human nature which corresponded with the aggressive savagery of tigers and wolves. Yang Hsiung, with his encouragement of good tendencies and suppression of bad, had been too interfering. Huang Hsi seems to have taken the Taoist attitude that all things would work together for good if Nature was allowed to take its course.

There were, perhaps, some premonitions of this synthesis already in the earliest writers. Fêng Yu-Lan[d] draws attention to an obscure passage in Mêng Tzu,[e] where he distinguishes between that part of man which is great and that part which is small;[f] the former is specifically human, the latter is shared with the animals. The discussion in Tung Chung-Shu's *Chhun Chhiu Fan Lu* (cf. p. 20 above) of about − 135 comes nearer the point.

In this age [he says], people have taken up different positions about man's nature, and are not at all clear about it.... The natural inborn endowment (*tzu-jan chih tzu*[7]) of a man we may call his (congenital) nature (*hsing*[8]) or raw material (*chih*[9]). How can it fit the facts to call that good?...

If the raw material of man is considered in relation to the (congenital) nature of birds and beasts, then the nature of the people is good; but if it is considered in relation to the goodness of the Tao of human society, then it is not good.... The raw material which I regard as the (congenital) nature is different from that of Mencius. He thought of the raw material in comparison with what birds and beasts do below, and therefore he called it good. But I

---

a *Shu Pho*, p. 44*a*, tr. auct.   b Cf. S. F. Mason (1).
c Ch. 1, p. 11*a*.
d (1), vol. 1, p. 122.
e *Mêng Tzu*, VI (1), xiv, 2 and xv; cf. also IV (2), xix, 1.   f *Ta thi*[10] and *hsiao thi*[11] respectively.

---

[1] 天性   [2] 誠意   [3] 氣性   [4] 強矯   [5] 黃晞
[6] 鰲隅子歔欷瑣微論   [7] 自然之資   [8] 性   [9] 質
[10] 大體   [11] 小體

think of the raw material in comparison with what the sages achieve above, and therefore I call it not yet good....

To name the (congenital) nature rightly, we should not use too high or too low (a standard), but (one) exactly in the middle. The (congenital) nature is like a cocoon or an egg, the egg awaiting the change which brings the chick, the cocoon awaiting the winding which makes the thread. The (congenital) nature awaits (authoritative) teaching; so (only) can it become good. This (state of potentiality) is what may be called 'true naturalness' (chen thien [1]).[a]

The relativity of this passage is characteristic of Tung Chung-Shu as a Confucian greatly influenced by Taoism,[b] but it stands a little off the main line of development of thought on human nature and the ladder of souls.

Discussion of this has taken us rather far away from the main stream of Confucianism. There are still a few words to be said concerning Hsün Tzu's contribution.

(f) THE HUMANISM OF HSÜN CHHING

Hsün Tzu exemplifies perfectly that ambivalent relation of Confucianism to science already emphasised. While, on the one hand, he preached an agnostic rationalism and even a denial of the existence of spirits,[c] on the other he strongly objected to the efforts of the Logicians and the Mohists to work out a scientific logic, and insisted on the practical application of technological processes while denying the importance of theoretical investigation. In this way he struck a blow at science by emphasising its social context too much and too soon.

His scepticism is illustrated in the following passage:

When a person walks in the dark, he sees a stone lying down and takes it to be a crouching tiger; he sees a clump of trees standing upright and takes them to be standing men. Darkness has perverted his clearsightedness. A drunken man crosses a canal a hundred paces broad and takes it to be a ditch half a step wide; he bends down his head when going out of a city-gate, taking it to be a small private door—the wine has confused his spirit. When a person sticks his finger in his eye and looks, one thing appears as two; when he covers his ears and listens, he hears a noise though all around is silent—the circumstances have confused his senses. So when viewed from a mountain, a cow looks like a sheep (but whoever wants a sheep knows better than to go down and lead it away)—distance obscures size.[d] Seen from the foot of a mountain, a sixty-foot tree looks like a chopstick (but whoever wants a chopstick knows better than to go up and break it off)—heights obscure lengths. When the water moves, the shadows dance; men cannot decide whether they are good-looking or ugly—the state of the water is confused....

[a] Ch. 35, p. 14a; tr. Hughes (1), p. 304, mod.; cf. Bodde in Fêng Yu-Lan (1), vol. 2, p. 36.
[b] We shall return to him in the section on the basic ideas of Chinese scientific thought; cf. Wieger (2), p. 181. He said that animals think only of their own self-preservation (sêng[2]) and their own well-being (li[3]); and compared egoism (than[4]) and altruism (jen[5]) in man with the Yin and Yang influences respectively.
[c] Cf. Dubs (7), pp. 65 ff.   [d] Cf. Lucretius, De Rerum Natura, II, 317–22.

[1] 真天   [2] 生   [3] 利   [4] 貪   [5] 仁

South of the mouth of the Hsia river there was a man called Chüan Shu-Liang. In disposition he was stupid and timorous. When the moon was bright and he was out walking he bent down his head and saw his shadow, and thought it was a devil following him. He looked up and saw his hair and thought it was a standing ogre. He turned around and ran. When he got home he lost his breath and died. Wasn't that too bad?

Whoever says that there are demons and spirits, must have made that judgment when they were suddenly startled, or at a time when they were not sure, or confused. This is thinking that something exists when it does not, or that it does not when it does, and so making a judgment.

Thus when a person, having got rheumatism from dampness, beats a drum and boils a sucking-pig (as an offering to the spirits to obtain a cure), then there will necessarily be the waste resulting from a worn-out drum and a lost pig, but he will not have the happiness of recovering from his sickness. So although he may not live south of the Hsia, he is no different from Chuan Shu-Liang.[a]

Here, then, is Confucian agnostic rationalism, which should have been favourable to early science. A whole chapter of Hsün Tzu,[b] moreover, is devoted to an attack on one of the superstitions of that time, physiognomy (*hsiang shu*[1]), or fortune-telling from a person's appearance (cf. Sect. 14a).

But Hsün Tzu's humanism was too humanistic. He was sufficiently influenced by the Taoists to use the word Tao sometimes to mean the Order of Nature, including the right Way of human society,[c] but he exalted *li*,[2] the essence of rites, good customs, traditional observances, into a cosmic principle, as if men in human society were but imitating at their own level the numinous dance of the stars and the seasons.[d] Thus:

Li is that whereby Heaven and Earth unite, whereby the sun and moon are brilliant, whereby the four seasons are ordered, whereby the stars move in their courses, whereby rivers flow, whereby all things prosper, whereby love and hatred are tempered, whereby joy and anger keep their proper place. It causes the lower orders to obey, and the upper classes to be illustrious; through a myriad changes it prevents going astray. If one departs from it, one will be destroyed. Is not *Li* the greatest of all principles?[e]

This pantheistic statement is reminiscent of the mystical rhapsodies to love as the motive power of the universe, which, stemming perhaps from the pre-Socratic philosophers, such as Empedocles,[f] and recurring in such passages as the Orphic hymn in Hellenistic *Daphnis and Chloe*,[g] still remain among the profoundest insights of earlier thinkers into the cosmic processes of attraction and repulsion. In an organic view of the world, such as developed later in China, and such as we have in mind

---

[a] *Hsün Tzu*, ch. 21, p. 15a, tr. Dubs (8), p. 275, mod.
[b] Ch. 5.   [c] As, for example, in ch. 17.
[d] Cf. Dubs (7), p. 52; Boodberg (3), and particularly pp. 151, 283, 287 ff., 488, 548 below.
[e] *Hsün Tzu*, ch. 19, p. 7b, tr. Dubs (8), p. 223.
[f] B & M, p. 137; Diels-Freeman (1), pp. 51 ff.
[g] Cf. Needham (3), p. 39.

[1] 相術   [2] 禮

today, Hsün Chhing's conception of human society as part of the cosmic order[a] would be acceptable enough, indeed not without sublimity.[b]

But any value it might have had for science in his own time was destroyed by his refusal to admit the necessity for the detailed pedestrian processes of scientific logic and investigation, while praising the social value of techniques. In a famous metrical passage directed against the Taoists, he says:

> You glorify Nature and meditate on her;
> Why not domesticate her and regulate her?
>
> You obey Nature and sing her praises;
> Why not control her course and use it?
>
> You look on the seasons with reverence and await them;
> Why not respond to them by seasonal activities?
>
> You depend on things and marvel at them;
> Why not unfold your own abilities and transform them?
>
> You meditate on what makes a thing a thing;
> Why not so order things that you do not waste them?
>
> You vainly seek into the causes of things;
> Why not appropriate and enjoy what they produce?
>
> Therefore I say—To neglect man and speculate about Nature
> Is to misunderstand the facts of the universe.[c]

The crux of the matter is in the penultimate verse. Chuang Tzu, said Hsün Chhing, saw only Nature and failed to see Man.[d] Hsün Tzu considered that the arguments of the Logicians and Taoists were full of fallacies. He said:

All perverse theories and heretical notions which have been invented in open contradiction to truth can be dealt with under (one or other of) these three fallacies (which he had just described). Wise rulers realise this, so that they do not care to argue about them. They know that the people can be united by the royal Tao but cannot be expected to reason about things

[a] His selection of the term *li* to describe his cosmic principle, though at first sight strange, is less so when we remember that Liu Hsi's *Shih Ming* dictionary of about +100 etymologised *li*[1] as being related to *thi*,[2] 'living body', and gave its meaning as 'the way in which the affairs of the body of human society are (or should be) handled'. Here the thought came close to the idea of society as a supra-human organism. We know now that what really unites the two characters is the right-hand element, which is an ancient drawing of a ritual vessel containing some unidentified object (p. 230). But this does not affect the organic quality manifested in the thought of Hsün Chhing and Liu Hsi (cf. pp. 294 ff. below).

[b] There is some doubt as to whether these passages on *li* as a cosmic principle should really be attributed to Hsün Chhing. They occur again almost verbatim in texts such as the *Li Chi* and the *Ta Tai Li Chi*. Some are therefore inclined to think (e.g. Bodde (14), p. 78), following Yang Yün-Ju, that they were produced by Ritualists of the Han dynasty and subsequently became incorporated into the *Hsün Tzu* book. For convenience, however, we continue to accept the usual attribution.

[c] Ch. 17, p. 23b, tr. Hu Shih (2), p. 152. These lines always remind me of Dr Lo Chung-Shu, whose appreciation of them much impressed me long ago.

[d] Ch. 21, p. 5b (Dubs (8), p. 264).

[1] 禮   [2] 體

in the same manner. Therefore a wise ruler establishes authority over them, guides them by truths, reminds them from time to time by ordinances, makes truth clear to them by expository treatises, and forbids their deviation by penalties. Thus the people can be converted to truth as readily as if by divine magic. What use could they have for argument and dialectic?[a]

Authority, then, was to be the last resort for the rectification of names. Again:

All those things which have nothing to do with the distinction of right and wrong, truth and falsehood, good government and misrule, or with the ways of mankind, are things the knowledge of which does not benefit men, and ignorance concerning which does no harm to men....They belong to the speculations of unruly persons of a degenerate age....

As to the displacement of body and empty space, or the separation of whiteness and hardness, or the distinction of agreement and difference, they are things beyond the power of the eye and the ear, and are inexplicable even by the most eloquent dialecticians. Even the wisdom of the sages does not always comprehend them. Not knowing them does not make one less of a *chün-tzu* (gentleman), knowing them does not raise one from being a small-minded man. Without them, artisans can be just as good artisans. And the sages can very well govern a state without them.[b]

There was no room for science, therefore, only traditional technology. And in these passages Hsün Tzu, though exhibiting his Legalist leanings, crystallised the position of all subsequent Confucians. Their fundamental mistake was not the belief that the State should be organised according to natural laws (as Chêng Chih-I well puts it),[c] but rather the conviction that these laws could be ascertained by the study of human tradition and history alone. We shall see shortly how Hsün Tzu's argument about the social significance of techniques looked from the Taoist side (p. 98). Hu Shih (2) is surely right in saying that Hsün Tzu's codification of the Confucian position was a sign of the downfall of the most glorious era of Chinese thought. In handing over the Confucian logic of definitions (rectification of names) to political authority,[d] Hsün Tzu approximated closely to the Legalists, and it is not therefore surprising that one of his pupils was no other than Li Ssu, the minister of the arch-authoritarian first emperor Chhin Shih Huang Ti.

[a] Ch. 22, p. 9*b*, tr. Hu Shih (2), p. 168.
[b] Ch. 8, tr. Hu Shih (2), p. 169.
[c] (1), p. 55; cf. the comments of Pott (1).
[d] Cf. Duyvendak's (4) translation of ch. 22 of the *Hsün Tzu*.

(g) CONFUCIANISM AS THE ORTHODOXY OF FEUDAL BUREAUCRATISM

As we have seen in the historical introduction, Confucianism became, during the Han dynasty, the official doctrine of the bureaucratic society.[a] Although less study has been given to the specifically Confucian thinkers of that time and for several centuries afterwards, it does not appear that anything substantial was altered in the teaching of its founders, at any rate in so far as concerns its bearing on science and scientific thought. The tendency was rather to synthesise the opposed ideas of former times and to accept new influences such as Taoism and Buddhism. Scholars of these ages were eclectic commentators rather than original philosophers,[b] e.g. Ma Jung[1] (+79 to +166),[c] Chêng Hsüan[2] (+127 to +200)[d] and Chia Khuei[3] (+30 to +101).[e] The assurance of public office led often to an empty formalism which forgot the original educational and equalitarian elements in Confucianism, and concentrated on the craft of bureaucratic government. Gradually the Confucians diverged either into sceptical rationalism (here the greatest representative was Wang Chhung[4] of the +1st century, to whom we shall devote a special section), or, influenced by the Taoists and the Yin-Yang experts, into somewhat superstitious semi-political number-mysticism and manipulations of the Five Element theories and the *I Ching* hexagrams (see below, pp. 380 ff., for the rise of the 'Apocryphal Classics', the *Chhan-Wei*[5]).

What true sciences could a Confucian scholar legitimately study during the early middle ages? Mathematics was essential, up to a certain point, for the planning and control of the hydraulic engineering works, but those professing it were likely to remain inferior officials. With astronomy a man might hope to rise higher. The practice of medicine was possible and agricultural studies were always respectable. But alchemy was severely frowned upon, and familiarity with the crafts of smiths, millwrights or other artisans, was considered unbecoming to a Confucian.

The carving of the orthodox texts of the Confucian classics on stone first took place in +171 at the order of Han Ling Ti;[6] and when these had been broken up and lost another carving was made during the Wei dynasty (San Kuo) about +245.[f] A third fixation of the text on stone was made in +837 in the Thang, just before its first printing between the Thang and the Sung. Thang Confucianism represented a return to older forms, but was more moralistic than philosophical,[g] and something will be said of its greatest representative, Han Yü, under the head of the Sceptical Tradition. When we reach the Sung we come, of course, to that second flowering known as Neo-Confucianism, and this will demand a section to itself, on account of its great scientific and cosmological interest.

---

[a] Vol. 1, pp. 103 ff. Cf. Nagasawa (1), pp. 109 ff., 125 ff.
[b] Cf. Nagasawa (1), pp. 131 ff., 135 ff.   [c] G1475. A great teacher.
[d] G274. An eminent commentator.   [e] G323. Also an astronomer.
[f] Cf. Nagasawa (1), p. 136.   [g] Cf. Nagasawa (1), p. 175.

[1] 馬融   [2] 鄭玄   [3] 賈逵   [4] 王充   [5] 讖緯   [6] 漢靈帝

(h) CONFUCIANISM AS A 'RELIGION'

More than a short reference to the development of Confucianism as a 'religion' would take us outside the scope of this book. The origin and development of the State cult of Confucius has been rather thoroughly investigated by famous Chinese scholars such as Ku Chieh-Kang (3) and by Shryock (1), to whose excellent book the reader is referred. The best work on the position of Confucianism in this sense under the Chhing dynasty is that of Legge (6) and for modern China that of Johnston (1). The belief that there was a cult of Confucius in the State of Lu rests upon the biography in the *Shih Chi*[a] (written more than four centuries after his lifetime). It appears certain, however, that the first Han emperor performed important sacrifices at the Khung family temple in honour of the sage in −195, and that Ssuma Chhien visited it,[b] but it was not until +37 that the descendants of Confucius were ennobled. In +59 Han Ming Ti ordered official sacrifices to him in all the schools of the country. 'It was this act', says Shryock, 'which took the worship of Confucius outside the Khung family, and changed him from the model of scholars into their patron saint.' The cult of Confucius thus became, what it remained through the centuries, a hero worship, celebrated everywhere but with especial ceremony at the sage's tomb-temple in Shantung,[c] and a symbol of the power and prestige of a non-hereditary social group, the literati, in the framework of society. It borrowed from the cults of nature-deities on the one hand and from ancestor-worship on the other.

During the course of the centuries every city and town came to have its Confucian or 'literary' temple (*wên miao*[1]). The conception of priesthood being entirely foreign to Confucian thought, it was natural that the guardians and celebrants should be none other than the local scholars and officials.[d] The character of the temple swung slowly between the two poles of nature-deity cult and ancestor-cult, so that while from the +8th to the +16th centuries Confucius and his seventy-two disciples were represented as images, subsequently these were replaced by carved and gilded tablets bearing their names. The designation of the Confucian cult as a religion, depends, of course, on the definition of religion adopted; but if the sense of the holy (the 'numinous' of Rudolf Otto) be the criterion, there are no more numinous and beautiful places in the world than the Confucian temples (though in recent times often sadly neglected). A Confucian temple consists of a series of courtyards, surrounded with buildings in the Chinese style containing inscribed stone tablets of former ages, or empty rooms in which visitors formerly stayed; each court is on a higher level than the last, and on the topmost level, approached by ceremonial steps up to the terrace, stands the Great Hall

---

[a] Ch. 47 (Chavannes (1), vol. 5, pp. 428).
[b] *Shih Chi*, ch. 47 (Chavannes (1), vol. 5, p. 435).
[c] For the beliefs of the mass of the people concerning Confucianism, see Doré (1), pt. III, vols. 13 and 14; Maspero (11) and Watters (2).
[d] See Biallas (1).

[1] 文廟

containing the name-tablets of the Sage and his followers.[a] There are gardens, with a ritual bridge over a semicircular pool,[b] and usually many fine old trees; in former times Confucian temples often included a library where the local scholars assembled, and where school was held.[c] Still, to this day,[d] once a year, on the Sage's traditional birthday, the officials and scholars of the district assemble between midnight and dawn, there to make the *thai-lao*[1] sacrifice (an ox, a sheep and a pig), to read liturgical essays and listen to speeches. Music and a solemn ritual dance were part of the ceremony until contemporary times.

Shryock, whose book has been quoted above, had, like G. E. Moule (2), the opportunity of being present several times at the annual sacrificial ceremony in Confucian temples, and once acted as assistant to the celebrant. In a memorable passage, which those who (like the present writer)[e] have spent much time in these beautiful buildings will particularly appreciate, he sketches the scene as it must have appeared when the cult was in its prime.

The account of the rubrics of the service, as it existed in the +14th century, is largely devoid of colour, whereas it is in reality one of the most impressive rituals that has ever been devised. The silence of the dark hour, the magnificent sweep of the temple lines, with eaves curving up toward the stars, the aged trees standing in the courtyard, and the deep note of the bell, make the scene unforgettable to one who has seen it even in its decay. In the days of Khubilai the magnificence and solemnity of the sacrifice would have required the pen of a Coleridge to do it justice. The great drum boomed upon the night, the twisted torches of the attendants threw uncertain shadows across the lattice scrolls, and the silk embroideries on the robes of the officials gleamed from the darkness.... Within the hall, the ox lay with its head towards the image of Confucius. The altar was ablaze with dancing lights, which were reflected from the gilded carving of the enormous canopy above. Figures moved slowly through the hall, the celebrant entered, and the vessels were presented towards the silent statue of the Sage, the 'Teacher of Ten Thousand Generations'. The music was grave and dignified.... Outside in the court the dancers struck their attitudes, moving their wands tipped with pheasant feathers in unison as the chant rose and fell. It would be hard to imagine a more solemn or beautiful ceremonial.

Shryock adds, however, his conviction that no one would have been more surprised at it, and perhaps even shocked, than Master Khung himself.

But all this has nothing to do with the history of science. Confucianism as a 'religion' had no theologians who could resent the intrusion of the scientific view of the world into their preserves. It simply turned away its face, in accordance with the attitude of its founding fathers, from Nature and the investigation of Nature, to concentrate a millennial interest on human society and human society alone.

[a] Some photographs will be found in Needham (4), Figs. 6, 17, 82–5. Hett (1) was able to photograph some of the ceremonies in the elaborate form continued in Korea until recent times.
[b] This copies the custom of the ancient *phan-kung*[2] or schools of the feudal aristocracy.
[c] Cf. Fig. 37.   [d] This refers to the time immediately following the second World War.
[e] I particularly regret that I was not able to take advantage of an invitation which I received to attend the ceremony in the Confucian temple of Meithan in Kweichow (1944).

[1] 太牢   [2] 泮宮

PLATE XIV

Fig. 37. One of the courts of the Confucian temple at Chhêng-kung, Yunnan (orig. photo., 1945).

10. THE *TAO CHIA* (TAOISTS) AND TAOISM

(a) INTRODUCTION

WE ARE NOW TO LOOK AT THE WORLD through the eyes of the opponents of
Confucius, the 'madman of Chhu' and the 'irresponsible hermits' who have already
been mentioned. The Taoist system of thought, which still today occupies at least as
important a place in the background of the Chinese mind as Confucianism, was a
unique and extremely interesting combination of philosophy and religion, incorporating
also 'proto'-science and magic. It is vitally important for the understanding of all
Chinese science and technology. According to a well-known comment (which
I remember hearing from Dr Fêng Yu-Lan himself at Chhêngtu), Taoism was 'the
only system of mysticism which the world has ever seen which was not profoundly
anti-scientific'.

Taoism had two origins. First there were the philosophers of the Warring States
period who followed a Tao[a] of Nature rather than a Tao of Human Society and therefore, instead of seeking for employment at the courts of the feudal princes, withdrew
into the wildernesses, the forests and mountains, there to meditate upon the Order of
Nature, and to observe its innumerable manifestations. Two of them we have already
met—the hermits—irresponsible from the Confucian point of view. But the philosophers of the Tao of Nature[b] may be said to have felt 'in their bones', for they
could never fully express it, that human society could not be brought into order, as
the Confucians strove to bring it, without a far greater knowledge and understanding
of Nature outside and beyond human society. They attacked 'knowledge', but what
they attacked was Confucian scholastic knowledge of the ranks and observances of
feudal society, not the true knowledge of the Tao of Nature. Confucian knowledge
was masculine and managing: the Taoists condemned it and sought after a feminine
and receptive knowledge which could arise only as the fruit of a passive and yielding
attitude in the observation of Nature. These differences we shall shortly analyse.

The other root of Taoism was the body of ancient shamans and magicians which
had entered Chinese culture at a very early stage from its northern and southern
elements respectively (cf. Sect. 5b) and which later concentrated on the north-eastern
coastal regions especially in the States of Chhi and Yen. Under the names of *wu* and
fang shih[c] they played an important part in ancient Chinese life as the representatives
of a kind of chthonic religion and magic (basically shamanistic), closely connected with

[a] I agree fully with Forke (13), p. 271, and others that this word can only be left untranslated. The
ancient graph has a 'head' and the sign for 'going', hence the meaning 'Way', but it early became
a technical term charged with philosophical and numinous significance.

[b] As to the manifold meanings of the word Nature, more than one of which is applicable to Taoism,
I would refer to the surveys made by Lovejoy (2), and Lovejoy & Boas (1), p. 447.

[c] See on, pp. 132 ff.

the masses of the people[a] and opposed to the ouranic State religion encouraged by the Confucians.

It may at first sight be puzzling to understand how these two different elements in ancient Chinese society could have combined so completely to form the Taoist 'religion' of later times. But really there is little difficulty. Science and magic are in their earliest stages indistinguishable. The Taoist philosophers, with their emphasis on Nature, were bound in due course to pass from the purely observational to the experimental. Later we shall study the first beginnings of this in the history of alchemy, a purely Taoist proto-science; and the beginnings of pharmaceutics and medicine, too, were very closely associated with Taoism. But as soon as observation passed over to experimentation (which after all means no more than changing the conditions and observing again) the crucial step had been taken out of the charmed circles of feudal aristocratic philosophy and of later bureaucratic literary culture, because *manual operations* were involved. Nothing was left therefore by which the people could distinguish the Taoist philosopher—based on the high abstractions of Lao Tzu and Chuang Tzu but stoking his alchemical furnace in order to prepare the pill of immortality and acquiring peace of mind by meditation on the workings of the Five Elements and the Yin and Yang—from the Taoist magician, writing out mysterious charms or engaging in liturgical rites for the control of dragon-spirits. That the mastery of Nature by manual operations is possible was the firm belief of magicians and early scientists alike, and the world was divided into the mystical operators who believed in this view and the rationalists who did not. The possibility of distinguishing between magic and science does not arise until a relatively late period in the history of human society, for it depends on sufficient continuity of test conditions and sufficient experimental scepticism to note unflinchingly the real effects of operations. Even the early Royal Society found it difficult to distinguish between science and what we should now call magic. In the +16th century science was commonly called 'Natural Magic'. Kepler was active as an astrologer and even Newton has with justice been called 'the last of the magicians'.[b] Indeed, the differentiation of science and magic did not occur before the birth of modern science and technology in the early +17th century—at a point, in fact, which Chinese civilisation never independently reached. Such considerations may help us to understand how Taoist philosophy combined with *wu* magic to form Taoist 'religion'.

It is necessary to say that, for one reason or another, Taoist thought has been almost completely misunderstood by most European translators and writers. Taoist religion has been neglected and Taoist magic has been written off wholesale as superstition, Taoist philosophy has been interpreted as pure religious mysticism and poetry. The scientific or 'proto'-scientific side of Taoist thought has been very largely overlooked, and the political position of the Taoists still more so.[c] One would not

---

[a] The shamanist element in Taoism was recognised more than seventy years ago by Eitel (3).
[b] Keynes (1).
[c] Almost the only sinologist who has done justice to the political aspect of the Taoists is Balazs (1).

wish to deny that ancient Taoist thought had strong elements of religious mysticism [a] and that the most important thinkers of Taoism were among the most brilliant writers and poets in history. But the Taoists not merely withdrew from the courts of the feudal lords, where Confucian humanitarian, if sententious, moralism battled with Legalist justifications for tyranny; on the contrary, they launched bitter and violent attacks on the whole feudal system. In the interests of exactly what they thus inveighed, I shall try to explain below. But this highly characteristic anti-feudal element has been ignored by Western, as also by most Chinese, expositors of Taoism. Here was another reason for the conjunction of Taoist philosophy and *wu* magic, for the representatives of Shamanism were, as has been said, closely associated with the most ancient folk-practices of the people, and at some enmity with the more rational ouranic worships of Heaven and Shang Ti (the Ruler Above). Taoism was religious and poetical, yes; but it was also at least as strongly magical, scientific, democratic and politically revolutionary.

In what follows I shall quote from a number of the sources of Taoist philosophy, and it will be convenient to discuss their dates shortly here. The *Tao Tê Ching*[1] (Canon of the Virtue (in the sense of power or even *mana*) of the Tao), which may be regarded as without exception the most profound and beautiful work in the Chinese language,[b] has as its author Lao Tzu,[2] one of the most shadowy figures in Chinese history. There has been extensive discussion concerning his probable date.[c] The authoritative view, expressed by Fêng Yu-Lan (1), is that the old accounts (such as that in the *Shih Chi*, ch. 63) which made Lao Tzu a −6th century contemporary of Confucius, must be given up, and that the *Tao Tê Ching* must be considered a Warring States document. It cannot be later, since it was commented on by Han Fei Tzu[3] (d. −233), criticised by Hsün Tzu (−305 to −235), and paralleled by Chuang Tzu[4] (−369 to −286). Fêng Yu-Lan (1) thinks that Ssuma Chhien confused a historical person, Li Erh,[5] with a legendary person, Lao Tan.[6] In the most recent discussion Dubs (11) has tentatively identified the son of Lao Tzu as a certain general Tuankan Tsung[7] whose *floruit* was −273. Lao Tzu would thus have been of a noble Honan family, the hereditary position of which he refused to accept. The subsequent discussion[d] is worth reading, but the general conclusion is that the life of Lao Tzu is to be

[a] Cf. the parallels brought out by K. J. Spalding (1).

[b] The text is often very obscure and, like all other ancient Chinese texts, somewhat corrupt; the most recent work on the establishment of the best readings is that of Kao Hêng (1), which has been helpful. The *Tao Tê Ching* has been translated into nearly all living languages; in the library of my friend the late Mr J. van Manen at Calcutta I counted more than thirty. The words of Duyvendak (5) about the 'host of dilettantes who have preyed on its text in order to make it say what best suited themselves' are to be borne in mind; whether the present interpretation is another such subjective approach, or something more, must be left to further investigation and research. The versions which I have felt able to adopt will be found below; often they are Waley's, but the critique of Erkes (6) on Waley's approach should be consulted. We regret that the translation of John C. H. Wu (Wu Ching-Hsiung) has not been available to us. On *Tê* cf. Boodberg (3).

[c] The chief symposium on the subject, to which Hu Shih, Liang Chhi-Chhao and many of the most eminent Chinese scholars contributed, is in *KSP*, vol. 4, pp. 303 ff.   [d] Dubs (12); Bodde (2).

[1] 道德經   [2] 老子   [3] 韓非子   [4] 莊子   [5] 李耳   [6] 老聃
[7] 段干宗

placed within the −4th century, and that the *Tao Tê Ching* may be dated not long before −300, i.e. about the time when Aristotle was old and Epicurus and Zeno were young.

The next greatest Taoist book[a] is the *Chuang Tzu*[1] of Chuang Chou,[2] whose dates, which have just been given, make its appearance contemporary with or very shortly after the *Tao Tê Ching*. Two other important texts are much more difficult to date. The *Lieh Tzu*,[3] named[b] after a semi-legendary writer, Lieh Yü-Khou,[4] is certainly late, and partly post-Han, but it is thought to contain much Warring States (−5th to −3rd century) material.[c] The most miscellaneous source is the *Kuan Tzu*[5] book, named after the historical figure Kuan Chung,[6] a statesman of pre-Confucian time (d. −645), but actually put together probably in the State of Chhi just before −300 by the scholars of the Chi-Hsia Academy, with later Han interpolations.[d] Quotations from it can therefore best be made when they are sufficiently near the known opinions of distinct schools to warrant their use as marginal illustrations. With the *Lü Shih Chhun Chhiu*[7] (Master Lü's Spring and Autumn Annals) and the *Huai Nan Tzu*[8] we are back in the realm of more precisely datable material, and both of them are extremely important for the scientific aspects of Taoism. Both were the compilations of groups of more or less Taoist scientists gathered together under the patronage of a powerful person; the former under Lü Pu-Wei[9] (d. −235) who was associated with the first emperor Chhin Shih Huang Ti; the latter under Liu An,[10] prince of Huai Nan (d. −122) in the Former Han dynasty.[e]

(b) THE TAOIST CONCEPTION OF THE *TAO*

It has already been made clear that for the Taoists the Tao or Way was not the right way of life within human society, but the way in which the universe worked; in other words, the *Order of Nature*.[f] This is what Lao Tzu said of the creature and the Tao:

[a] Called, since +742, the *Nan Hua Chen Ching*[11] (True Classic of Nan Hua); latest edition with collected commentaries by Liu Wên-Tien (1). Much discussion has ranged round the authenticity of the various chapters. The so-called 'Inner' chapters are generally accepted as genuine, but some of the rest are thought to be by later writers, none of whom, however, could be later than the Early Han; cf. Fu Ssu-Nien (1), Lo Kên-Tsê (2), Hu Chih-Hsin (1).

[b] Called, since +742, the *Chhung Hsü Chen Ching*[12] (True Classic of Upwelling Emptiness).

[c] Cf. Forke (13), p. 287.

[d] Cf. Vol. 1, p. 95. Forke (13), p. 74, gave an evaluation of the contents and probable date of the various chapters. Grube (4) studied its style. Incomplete translation by Than Po-Fu *et al.*

[e] Latest edition, with collected commentaries, by Liu Wên-Tien (2).

[f] Almost the only Western writer who has appreciated this is T. Watters, whose essay (3), written in 1870, is still worth reading today. The early Russian sinologist N. Y. Bichurin would have agreed with him. But long after this Section was written, we were glad to find that Graf (2) shares our interpretation; and Yang Chin-Shun (1) has recently given a strong statement of it. The work of Huang Fang-Kang (1) is one of the best analyses of the conception from the more metaphysical point of view;

[1] 莊子   [2] 莊周   [3] 列子   [4] 列禦寇   [5] 管子   [6] 管仲
[7] 呂氏春秋   [8] 淮南子   [9] 呂不韋   [10] 劉安
[11] 南華眞經   [12] 冲虛眞經

The Tao gave birth to it
The Virtue (of the Tao) reared it
Things (within) endowed it with form,
Influences (without) brought it to its perfection.
Therefore of the ten thousand things there is not one that does not worship the Tao and do
    homage to its Virtue. Yet the worshipping of the Tao, and the doing of homage to its
    Virtue, no mandate ever decreed.
Always this (adoration) was free and spontaneous.
Therefore (as) the Tao bore them, and the Virtue of the Tao reared them, made them grow,
    fostered them, harboured them, fermented them, nourished them and incubated them—
    (so one must)
'Rear them, but not lay claim to them,
Control them but never lean upon them,
Be chief among them, but not lord it over them;
This is called the invisible Virtue.'[a]

Immediately notes are struck which we shall hear again and again. The Tao as the
Order of Nature, which brought all things into existence and governs their every
action, not so much by force as by a kind of natural curvature in space and time,
reminds us of the *logos* of Heracleitus of Ephesus, controlling the orderly processes of
change.[b] Heracleitus was a contemporary of Confucius, but while, as we have seen,
Lao Tzu's date is later, there is no doubt that the germinal ideas of Taoist thought
existed at the beginning of the − 5th century, or even before.[c] The sage is to imitate
the Tao, which works unseen and does not dominate. By yielding, by not imposing
his preconceptions on Nature, he will be able to observe and understand, and so to
govern and control.

    Lao Tzu says again:

> The supreme Tao, how it floods in every direction!
> This way and that, there is no place where it does not go.
> All things look to it for life, and it refuses none of them;
> Yet when its work is accomplished it possesses nothing.
> Clothing and nourishing all things, it does not lord it over them.

he compares the Tao with the One of Parmenides (Freeman (1), p. 140) beneath Heracleitus' Flux of
Things (Freeman (1), p. 104). Fêng Yu-Lan's expositions were always of that kind. Misch (1) has gone
further along this path, describing the Tao as a metaphysical Absolute; even Pure Act, equivalent to the
Being of European philosophers (pp. 180, 209). Our tendency is precisely the opposite; we believe that
the Chinese mind throughout the ages did not, on the whole, feel the need for metaphysics; physical
Nature (with all that that implied at the highest levels) sufficed. The Chinese were extremely loth to
separate the One from the Many or the 'spiritual' from the 'material'. Organic naturalism was their
philosophia perennis. It is hardly necessary to point out, with Maspero (13), p. 213, that although the word
Tao meant 'Way', it had nothing whatever in common with the 'Way' of Christian and Muslim mystics.

  [a] *TTC*, ch. 51, tr. Waley (4), Duyvendak (18), Chhu Ta-Kao (2), mod. On the denial of any
'mandate', see p. 561 below.

  [b] Diels-Freeman (1), pp. 24 ff.; Freeman (1), pp. 115, 116. Cf. Rémusat (8); Amiot (3), pp. 208 ff.,
and see p. 476 below. One may also read what Whitehead (2), p. 192, has to say on the 'receptacle'
of Plato.

  [c] Cf., for example, Fêng Yu-Lan (1), vol. 1, p. 135.

Since it asks for nothing from them
It may be classed among things of low estate;
But since all things obey it without coercion
It may be named Supreme.
It does not arrogate greatness to itself
And so it fulfils its Greatness.[a]

Or, in the words of Chuang Tzu:

The Tao has reality and evidence, but no action and no form. It may be transmitted but cannot be received. It may be attained but cannot be seen. It exists by and through itself. It existed before Heaven and Earth, and indeed for all eternity. It causes the gods to be divine and the world to be produced. It is above the zenith, but it is not high. It is beneath the nadir but it is not low. Though prior to heaven and earth it is not ancient. Though older than the most ancient, it is not old.[b]

We have, therefore, a naturalistic pantheism, which emphasises the unity and spontaneity of the operations of Nature. The Taoist texts are full of questions about Nature. Thus *Chuang Tzu*:

How (ceaselessly) heaven revolves! How (constantly) earth abides at rest! Do the sun and the moon contend about their respective places? Is there someone presiding over and directing these things? Who binds and connects them together? Who causes and maintains them, without trouble or exertion? Or is there perhaps some secret mechanism, in consequence of which they cannot but be as they are? Is it that they move and turn without being able to stop of themselves? Then how does a cloud become rain, and the rain again form clouds? What diffuses them so abundantly? Is there someone with nothing to do who urges them on to all these things for his enjoyment? Winds rise in the north, one blows to the west, another to the east, while some rise upwards, uncertain of their direction. What is it sucking and blowing like this? Is there someone with nothing to do who thus shakes the world? I venture to ask about the causes.[c]

And in one of those famous imaginary interviews between Lao Tzu and Confucius:

'We have a little time today', said Confucius to Lao Tzu. 'May I ask about the Great Tao?' Lao Tzu replied, 'Give a ceremonial bath to your mind! Cleanse your spirit! Throw away your sage wisdom! The Tao is dark and elusive, difficult to describe. However, I will outline it for you. Light (*chao*[1]) came from darkness (*ming*[2]), order (*yu lun*[3]) from the formless (*wu hsing*[4]). The Tao produces vital energy (seminal essence, *ching-shen*[5]),[d] and this gives birth to (organic) forms; all the myriad things (reproduce their kind) shape giving rise

---

[a] *TTC*, ch. 34, tr. Hughes (1), Chhu Ta-Kao (2), Waley (4), Duyvendak (18), mod.
[b] Ch. 6, tr. Fêng Yu-Lan (5), p. 117.
[c] Ch. 14, tr. Legge (5), vol. 1, p. 345; Lin Yü-Thang (1), p. 146.
[d] There is room for a valuable monograph comparing these Chinese concepts with the *semina* and spermatic *logos* of the Stoics.

[1] 昭   [2] 冥   [3] 有倫   [4] 無形   [5] 精神

to shape[a] (*wan wu i hsing hsiang sêng*[1]). Thus it is that animals with nine orifices are born from the womb, and those with eight from eggs. Life springs into existence without a visible source and disappears into infinity. It stands in the middle of a vast expanse, without visible exit, entrance or shelter. Those who seek for and follow (the Tao) are strong of body, clear of mind, and sharp of sight and hearing. They do not load their mind with anxieties, and are flexible in their adjustment to external conditions. Heaven cannot help being high, the earth cannot help being wide, the sun and moon cannot help going round, and all things of the creation cannot help but live and multiply (*Thien pu tê pu kao, ti pu tê pu kuang, jih yüeh pu tê pu hsing, wan wu pu tê pu chhang*[2]). Such is the operation of the Tao. The most extensive "knowledge" does not necessarily know it, "reasoning" will not make men wise in it; the sages eschewed these things (*Chieh fu po chih pu pi chih, pien chih pu pi hui, shêng jen i tuan chih i*[3]). However you try to add to it, it will not increase, whatever you try to take from it, it admits of no diminution so the sages have spoken of it. Fathomless, it is like the sea. Awe-inspiring, beginning again in cycles ever new. Sustaining all things, it is never exhausted. In comparison with it, do not the teachings of the "gentlemen" deal merely with (superficial) externals (*Tsê chün-tzu chih tao pi chhi wai yü*[4])? What gives life to all creation and is itself inexhaustible—that is the Tao.'[b]

Thus the biological no less than the inorganic comes under the operation of the Tao of all things. In this passage a new element enters (to be examined soon more closely), namely, the contrast between this true knowledge, and the superficial scholastic social learning of the feudal scholars. And Necessity, like the *anangke* (ἀνάγκη) of Anaximander[c] (mid −6th century), Parmenides[d] and Empedocles[e] (mid −5th century), governs all. Another element, frequently to be met with again, is the reference to the physical as well as mental benefits to be obtained by those who follow the Tao. Later, this formed a large part of Taoism, crystallising as the search for a kind of material immortality, in which the body would be so preserved and rarefied as to take its place among the *hsien* (or *genii*, but the word is untranslatable, see on, p. 141). For this purpose the adept would have recourse to drugs and alchemical preparations, to yogistic breathing exercises, to sexual techniques and to gymnastics.

Reminiscent of the 'love and hate', the attraction and repulsion, which Empedocles of Akragas[f] placed so prophetically as the most important force in the operations of Nature, is the following passage from *Chuang Tzu*:

Little Knowledge[g] said, 'Within the four cardinal points and the six boundaries of space, how did the myriad things take their rise?' Thaikung Thiao replied, 'The Yin and the Yang

---

[a] In these few words, Chuang Tzu condenses an appreciation of the fact that individual life-cycles of animals and plants may comprise stages (eggs, larval forms, seeds, bulbs, etc.) which are in shape and appearance almost unrecognisably different from the familiar adult.

[b] *Chuang Tzu*, ch. 22, tr. Legge (5), vol. 2, pp. 63, 64; Lin Yü-Thang (1), p. 65, mod.

[c] Diels-Freeman (1), p. 19; Freeman (1), p. 63.

[d] Diels-Freeman (1), p. 44; Freeman (1), p. 152.   [e] Freeman (1), p. 187.

[f] Diels-Freeman (1), p. 51; Freeman (1), pp. 182 ff.

[g] Cf. the 'Idiota' as a dialogue-character in Nicholas of Cusa.

[1] 萬物以形相生   [2] 天不得不高地不得不廣日月不得不行萬物不得不昌
[3] 且夫博之不必知辯之不必慧聖人以斷之矣   [4] 則君子之道彼其外輿

reflected on each other, covered each other and reacted with each other.[a] The four seasons
gave place to one another, produced one another and brought one another to an end.
Likings (*yü* [1]) and dislikings (*o* [2]), avoidings of this (*chhü* [3]) and movements towards that
(*chiu* [4]) then arose in all their distinctness, hence came the separation and union of male and
female. Then were seen now safety, now danger, in mutual change; misery and happiness
produced each other; slow processes and quick jostled each other; and the motions of
collection (or condensation, *chü* [5]) and dispersion (or rarefaction, scattering, *san* [6]) were
established. These names and processes *can* be examined, and however minute, *can* be
recorded. The principles determining the order in which they follow one another (*sui hsü
chih hsiang li* [7]), their mutual influences, now acting directly, now revolving; how, when they
are exhausted, they revive; and how they come to an end only to begin all over again—these
are the properties belonging to things (*tzhu wu chih so yu* [8]). Words *can* describe them and
knowledge *can* reach them—but not beyond the extreme limit of the natural world (*Yen chih
so chin, chih chih so chih, chi wu erh i* [9]). Those who study the Tao (know that) they cannot
follow these changes to the ultimate end, nor search out their first beginnings—this is the
place at which discussion has to stop.[b]

Here we find, not only *yü* resembling the *philia* (φιλία) of Empedocles, and *o* his
neikos (νεῖκος), but also the conception of condensation and rarefaction so common
among the pre-Socratics beginning with Anaximenes[c] (fl. −546). Condensation
(*pyknosis*, πύκνωσις) appears here as *chü*; rarefaction (*manosis*, μάνωσις) as *san*. It
would seem, therefore, that one of the oldest and most important of all physical
discoveries, that of differences of density, was independently made in ancient China
and ancient Greece, for one would be most reluctant, in view of all that has been
said above (Section 7), to accept transmission of such ideas before the −1st century.
And just as the conception lived on into later European thought, so we shall find it
subsequently in China (e.g. in Chang Chan's +4th-century commentary on *Lieh
Tzu*)[d] and appealed to in cosmological speculation by the Neo-Confucians in the +11th
century (pp. 373, 414, 483).[e] Moreover, one may note in the above passage the
characteristic distaste for metaphysics; the ultimate beginning and the ultimate end
are the Tao's secret, all that man can do is to study and describe phenomena; it is indeed
a profession of faith in natural science.

As a commentary on this we may take the story in the *Lieh Tzu* book about the man
of Chhi:

There was a man in the State of Chhi [10] who was so afraid that the universe would collapse
and fall to pieces, leaving his body without a lodgment, that he could neither sleep nor eat.

[a] Contrast the Platonic primacy of contemplation over action.
[b] Ch. 25, tr. Legge (5), vol. 2, p. 128, mod.
[c] Diels-Freeman (1), p. 19; Freeman (1), pp. 65 ff.
[d] Ch. 1, p. 9*b*.
[e] Not only so, but the same ideas persist in the systems of living Chinese philosophers (see Chhen
Jung-Chieh (4), pp. 37, 247, 248, 258).

[1] 欲   [2] 惡   [3] 去   [4] 就   [5] 聚   [6] 散   [7] 隨序之相理
[8] 此物之所有   [9] 言之所盡知之所至極物而已   [10] 杞

Another man, pitying his distress, proceeded to enlighten him. 'Heaven', he said, 'is nothing more than an accumulation (*chi*[1]) of air (*chhi*[2]), and there is no place where this air is not. It is as if there were bendings (*chhü*[3]) and stretchings (*shen*[4]), exhalations (*hu*[5]) and inhalations (*hsi*[6]), continually taking place up in the heavens. Why then should you be afraid of a collapse?' The man replied, 'If it be true that the heavens are only an accumulation of air, then why do the sun, the moon, and the constellations not fall down upon us?' His informant replied, 'Those bright lights are only shining masses of condensed air themselves. Even if they did fall down they would not hurt anybody.' 'But what if the earth itself should fall to pieces?' 'The earth too is only an accumulation (*chi*[1]) of matter (*khuai*[7]) which fills up the four corners of space, and there is no part where this matter is not. It is what you walk on. All day long there is a continual treading and trampling on the surface of the earth. Why then should you be afraid of its falling to pieces?' So the man was relieved of his fears and rejoiced exceedingly. And his instructor was also very pleased.

But Chhang Lu Tzu,[8] hearing of it, laughed at them both, saying, 'Rainbows, clouds and mist, wind and rain, the four seasons—these are all forms of agglomerated (*chi*[1]) air, and go to make up the heavens. Mountains and cliffs, rivers and seas, metals and rocks, fire and timber, these are all forms of agglomerated matter, and constitute the earth. Knowing that they have been thus formed, who can say that they will never be destroyed? Heaven and Earth form only a small speck in the midst of the Void (*khung chung chih i hsi wu*[9]), but they are the greatest of all existing things. Certain it is that even as their nature is hard to fathom and to understand, slow they will be to come to an end. He who fears lest they should fall to pieces is indeed far off the mark, but on the other hand he who says that they will never be destroyed has not got the truth either. Heaven and earth must of necessity pass away in the end. Whoever has to face that day may well be alarmed.'

Master Lieh heard of these discussions, smiled, and said, 'He who maintains that heaven and earth will pass away, and he who maintains the contrary, are both at fault. Whether they will or not is something we can never know. If they go, we shall go with them; if they stay, we shall stay (and not know the end). The living and the dead, the going and the coming, know nothing of each other's state. Why should we worry about whether destruction awaits the world or no?'[a]

Again we meet with 'agglomeration' and 'dispersion'. Lieh Tzu is typically Taoist in his aversion from cosmogony and eschatology, from the 'Creation' and the 'Last Things'; he emphasises the operation of the Tao here and now. Chhang Lu Tzu represents the calm reckoning of the scientific mind, aware of compositeness and prepared to face decomposition. But the man of Chhi and his comforter are the most interesting of all, since they demonstrate the peace of mind brought about by at least formulating hypotheses and theories concerning Nature. To this we shall have to return before long.

It must not be thought that all the ancient Taoist writers were as sublime and brilliant as Chuang Chou or as entertaining as the author of the *Lieh Tzu* book. In order to give an example which may convey more of the matrix of the thought of

[a] Ch. 1, p. 16*a*, tr. auct. adjuv. L. Giles (4), p. 29; Wieger (7), p. 79.

[1] 積　　[2] 氣　　[3] 屈　　[4] 伸　　[5] 呼　　[6] 吸　　[7] 塊
[8] 𪕲盧子　　　　[9] 空中之一細物

the age when they lived, we present the thirty-ninth chapter of the *Kuan Tzu*.[a]
At the same time this gives us yet another parallel with the pre-Socratic Greek
philosophers, since the chapter is devoted to the doctrine that water is the original
element of all things and the ground of change—in other words a doctrine analogous
to that of Thales of Miletus (fl. − 585), first of the pre-Socratic nature-philosophers.[b]
Although it is almost sure that the *Kuan Tzu* book did not reach final form until the
Han, and that a passage such as that which follows[c] cannot possibly be earlier than the
− 5th century, I am not disposed (in view of the discussion in Section 7 on Contacts)
to believe that there can be any question of transmission of the ideas. Similar minds
working on similar problems would be expected to come to similar results.

(1) The earth (*ti*[1]) is the origin of all things, the root and garden of all life; and the place
where all things, the beautiful, the ugly, the good, the bad, the foolish and the clever, come
into being. Now water is the blood and breath (*hsüeh chhi*[2]) of the earth, flowing and
communicating (within its body) as if in sinews and veins.[d] Therefore we say that water is
the preparatory raw material of all things (*chü tshai*[3]).

How do we know that this is so?

The answer is that water is yielding, weak and clean, and likes to wash away the evils of
man—this (may be called) its 'benevolence'. It looks sometimes black, sometimes white—
this (may be called) its 'essence'. When you measure it you cannot force it to level off (at
the top), for when the vessel is full it does that by itself—this (may be called) its 'rectitude'.
There is no space into which it will not flow, and when it is level it stops—this (may be
called) its 'fairness'.[e]

People all like to go up higher, but water runs to the lowest possible place. This principle
of going down to the bottom is the Palace of the Tao, and the instrument of (true) rulers.
The bottom is where water goes and lives.[f]

The water-level instrument (*chun*[4]) is the ancestor of the five measurements. The white
(or colourless) is the base of the five colours. The insipid is the centre of the five tastes. Thus
water is the standard level (*chun*[4]) of all things, and the common factor of all life. It is the
medium (*chih*[5]) in which all gains and losses take place.[g] Therefore there is nothing which
water cannot fill and dwell in. It is collected (*chi*[6]) in the heavens and on earth, and stored
up (*tshang*[7]) in all things. It is produced amidst metal and stone,[h] and collected (*chi*[6]) in all
living beings. It is thus mysterious and magical (*shen*[8]). Being collected in herbs and trees,
their roots grow in measured increase (*tu*[9]), their flowers in due profusion blossom,

---

[a] Its significance was first brought to my attention by my late friend Professor G. Haloun.

[b] Diels-Freeman (1), p. 18; Freeman (1), p. 49.

[c] As it is rather long, we shall confine ourselves mainly to footnote comments. The numbering
of the sections is ours. Tr. auct. adjuv. Than Po-Fu *et al.* (*1*), pp. 86 ff.

[d] If this is not a later interpolation, it must be one of the earliest statements of the theory underlying
geomancy (*fêng-shui*[10]); (see Sect. 14*a* below).

[e] Ethical undertone. Water occupies spaces of all shapes impartially, and knows where to stop.

[f] Cf. the Taoist use of water as a symbol for feminine receptivity in scientific observation, and
equality in political theory (see on, pp. 57 ff., 444).

[g] Cf. the drying up of a broken egg or the swelling of a ripening plum.

[h] In mountain springs.

---

[1] 地   [2] 血氣   [3] 具材   [4] 準   [5] 質   [6] 集   [7] 藏
[8] 神   [9] 度   [10] 風水

and their fruits get measured ripeness. (Being collected in) birds and animals, they get their form and flesh, their feathers and furs, their clearly marked fibres and veins.[a]

Thus there is nothing which cannot achieve its germination (*chi*[1]).[b]

(2) What are the nine virtues which make jade precious? Jade is warm, agreeable, and enriched with favours, this (may be called) its 'benevolence'.[c] Its lines run back and forth near each other, communicating systematically, this (may be called) its 'wisdom'. It is hard, but not over-compacted (*tsu*[2]), this (may be called) its 'righteousness'. It is sharp but its angles are not hurtful, this (may be called) its 'conduct'. It is fresh and bright, but cannot get dirty, this (may be called) its 'purity'. It can be broken but not bent, this (may be called) its 'courage'. Its cracks and spots all appear on the exterior, this (may be called) its 'refined quality' (i.e. it does not try to cover up its weak points). Its flourishing, shining, agreeable lights reflect each other but do not trespass upon one another, this (may be called) its 'tolerance'. Upon being struck it gives a clear, far-away and pure sound, not screaming, this (may be called) its 'gentleness'. These are the reasons why the rulers of men appreciate and value it for making auspicious seals.

(3) Human beings are made of water. The seminal essence of the man, and the *chhi* of the woman unite, and water flows, forming a new shape.[d]

(The mouth of a foetus) three months old can already function.[e] How does it do so? It receives the five tastes. What are they? They come from the five viscera. The sour governs the spleen, the salt governs the lungs, the acrid governs the kidney, the bitter governs the liver, the sweet governs the heart. After the five viscera have been formed, then the flesh develops. The spleen produces the diaphragm, the lungs produce the bones, the kidney produces the brain, the liver produces the skin, and the heart produces the muscles.[f] After the five fleshes have been completed, the nine orifices of the body appear (*fa*[3]). The spleen gives the nose, the liver the eyes, the kidney the ears, the lung the other orifices. In five months the foetus is complete and in ten months it is born. After birth, the child sees with its eyes, hears with its ears, and thinks with its heart. Its eyes can see not only great mountains but also small and indistinct things. Its ears can hear not only thunder and drums but also quietness. Its mind (heart) can think not only gross (*tshu*[4]) but also subtle (*wei miao*[5]) things. Carefully studying these facts, we obtain important and mysterious secrets.

(4) Thus if water collects (*chi*[6]) in the form of jade, the nine virtues of jade appear. If water congeals (*ning chien*[7]) to form human beings, the nine orifices and five organs appear.

---

[a] The correct appreciation of the importance of water in living creatures is striking. Cf. the dictum of Sir Arthur Shipley: 'Even a bishop is only eighty per cent water.'

[b] This word will be met with again in Chuang Tzu's famous passage on 'evolution', see on, pp. 78, 469, 470, 500.

[c] The following list of nine properties shows the ancient Taoists struggling with the problem of scientific nomenclature. At this time they were still unable to think of anything better than a technical use of names of virtues already familiar in human society. This is surely connected with the fact that the crucial distinction between animate and inanimate is one which arises relatively late not only in the mental development of humankind, but in the ontogeny of each human individual. Personality is not projected into things; they are simply apprehended as persons. Cf. Frankfort (1); Gordon Childe (14).

[d] We shall refer again to this embryological passage in the relevant Section (43).

[e] Lit. there can already be a chewing or sucking. The writer must have had dimly in mind the umbilical cord, since the nine orifices are said to be formed much later.

[f] Strange anticipation of our modern conception of induction phenomena in embryonic development.

[1] 幾   [2] 蹙   [3] 發   [4] 麤   [5] 微眇   [6] 集   [7] 凝蹇

These are (part of its) essence. Such essence, being thick and viscous (*tshu chien*[1]), can continue living and not die.[a]

> A digression on fabulous animals follows, which provide a new category, logically interesting.

(5) Now there are two things which are able to continue living while looking as if they were dead—the (oracle-) turtle and the dragon. Though the turtle lives in the water, when (its shell is) put on the fire, it can predict correctly the bad and good fortune in all things. The dragon also lives in the water, but it acquires the five colours of water, so it becomes a spirit. If it wishes it can make itself as small as a silkworm or caterpillar. Alternatively, it can make itself so large that it can cover the whole world. If it wishes to go up it can fly among the clouds, if it wishes to go down it can visit the deepest springs. Constantly changing, it can go up or down whenever it likes....

> Next further examples are adduced, from religious-demonological folklore, in the style of the *Shan Hai Ching*, showing further what water can do.

(6) There are two other Things, which men occasionally see. One is the *Chhing-Chi*[2] and the other is the *Wei*.[3] The *Chhing-Chi* comes into being in watery marshy places where the water never disappears. It is shaped like a man four inches long, dressed in yellow clothes with a yellow hat, riding rapidly on a small horse. If you can call it by its name, it will come to you in one day from a distance of a thousand miles. This is the Spirit of the watery fens.

On the other hand the *Wei*[b] comes into being in dry river-beds. It has one head and two bodies. It is shaped like a serpent eight feet long. If you can call it by its name, you can make it fetch fish and turtles.[c] This is the Spirit of the dry river-bed.

> The writer implies that though it has never seen water, it still has a magic power because it is of the essence and spirit of water.

(7) The essence of water is thick, viscous and congealed (*tshu cho chien*[4]).[d] It confers continuity of living, and not death. It gives rise to jade, turtles, dragons, the *Chhing-Chi* and the *Wei*. All are connected with water. People all drink water, but I alone take it as my model. People all have water, but I alone know how to make use of it. Why do we call water the preparative element? Because the myriad things get their life from it. So those who know on what water depends can know the true way in which water is preparatory to all things. People ask what water is. It is the *origin of all things*, and the ancestral temple of all Life. Water produces the beautiful and the ugly, the virtuous and the wicked, the foolish and the clever.

> Now the writer seeks to establish a correlation between habitat and the character of the populations.[e]

(8) How can this be shown to be so?

The water of Chhi flows rapidly and the streams are always turning backwards (in rocky

---

[a] Note here the dependence of the manifestation of specific qualities upon underlying processes taking place in one universal medium.

[b] Cf. Granet (1), p. 317.

[c] Very convenient in that sort of country.

[d] As in paragraph (4) above, the writer had in mind essentially what we should call the consistency of protoplasm.

[e] Cf. Hippocrates, *Airs, Waters and Places* (tr. F. Adams, 1). Cf. p. 84. Parallel passages in *Huai Nan Tzu*, ch. 4, p. 7 b (tr. Erkes (1), p. 64, and *Ku Wei Shu*, ch. 32, p. 7 a). Cf. the memorial of Chhao Tsho to the emperor about − 160 (*Chhien Han Shu*, ch. 49, p. 12 a).

[1] 麤蹇   [2] 慶忌   [3] 螖   [4] 麤濁蹇

gorges or meanders); thus its people are covetous, rough and brave. The water of Chhu is soft, weak and pure; thus its people are light and sure of themselves. The water of Yüeh is turbid and heavy, soaking through the land; thus its people are foolish, unhealthy and dirty. The water of Chhin is laden with sediment, muddy and clogged with dust; thus its people are greedy, deceptive and given to machinations. The water west of Chhi and east of Chin is often low, stagnant and dull; thus the people there are flatterers, cunning and eager for profit. The water of Yen collects in low places but is weak, slow-moving and turbid; thus its people are simple, chaste, quick and willing to lay down their lives. The water of Sung is light, strong and pure; thus its people are quiet, easy-going and like things to be done in the proper way.[a]

(9) Therefore the sage's transformation of the world arises from solving the problem of water.[b] If water is united, the human heart will be corrected.

If the water is pure and clean the heart of the people will readily be unified and desirous of cleanliness. If the people's heart is changed their conduct will not be depraved. So the sage's government does not consist of talking to people and persuading them family by family. The pivot (of his work) is Water.

This interesting chapter puts the profound sayings of the best of the Taoists in a better perspective regarding the current thought of their environment.[c]

If anyone should have any remaining doubts that by the term Tao the Taoists meant the Order of Nature, I would refer him to the magnificent rhapsodical passage with which the *Huai Nan Tzu* opens (ch. 1), but which is unfortunately rather too long to quote here.[d] I prefer to cite a passage which indicates that the Tao was thought of not only as vaguely informing all things, but as being the naturalness, the very structure, of particular and individual types of things. It is the famous story of the butcher of King Hui of Liang:

Ting, the butcher of King Hui, was cutting up a bullock. Every blow of his hand, every heave of his shoulder, every tread of his foot, every thrust of his knee, every sound of the rending flesh, and every note of the movement of the chopper, were in perfect harmony— rhythmical like the *Mulberry Grove* dance, harmonious like the chords of the *Ching Shou* music.

'Admirable', said the prince. 'Yours is skill indeed!'

'Sir', said the cook, laying down his chopper, 'what your servant loves is the Tao, which is higher than mere skill. When I first began to cut up oxen, I saw before me the entire carcasses. After three years' practice I saw no more whole animals. Now I work with my mind and not my eyes, my spirit having no more need of control by the senses. Following the natural structure,[e] my chopper slips through the deep crevices, slides through the great

---

a Can one conclude that the writer came himself from Yen or Sung?

b Water-conservation and hydraulic engineering undertone?

c The 'water' theme runs on through Chinese history probably partly because of the choice of it as a symbol by the Taoists (see pp. 57 ff.); e.g. in Su Tung-Pho[1] (+1036 to +1101), Forke (9), p. 142; and in the *Tshao Mu Tzu*[2] of Yeh Tzu-Chhi[3] (fl. +1378), Forke (9), p. 331.

d Tr. E. Morgan (1), pp. 2 ff., de Harlez (3), pp. 174 ff.

e *Thien li.*[4]

---

[1] 蘇東坡   [2] 草木子   [3] 葉子奇   [4] 天理

cavities, taking advantage of what is already there. My art avoids the tendinous ligatures, and much more so the great bones. A good cook changes his chopper once a year, because he cuts. An ordinary cook needs a new chopper once a month, because he hacks. But I have had this chopper for nineteen years, and although I have cut up many thousands of bullocks, its edge is as if fresh from the whetstone. For where the parts join there are interstices, and since the edge of the chopper has no thickness, one can easily insert it into them. There is more than enough room for it....Nevertheless, when I come to a complicated joint, and see that there will be some difficulty, I proceed with caution. I fix my eyes on it. I move slowly. Till by a very gentle movement of my chopper, the part is quickly separated, and yields like earth crumbling to the ground. Then standing up with the knife in my hand I look around and pause with an air of triumph. I wipe my chopper and put it in its sheath.'

'Excellent', cried the prince. 'From the words of Ting the Cook we may learn how to nourish (our) life.'[a]

Thus the anatomy of an ox and the skill of an anatomist are no less part of the Order of Nature than the movements of the stars. All things have their part in the Tao.

(c) THE UNITY AND SPONTANEITY OF NATURE

If there was one idea which the Taoist philosophers stressed more than any other it was the unity of Nature, and the eternity and uncreatedness of the Tao. In chapter 22 of the *Tao Tê Ching* we read:

Therefore the sage embraces the Oneness (of the universe) (*shih i shêng jen pao i*[1]), making it his testing-instrument[b] for everything under Heaven (*wei thien hsia shih*[2]).[c]

This conception has echoes everywhere in Taoist writings. One may quote, for example, chapter 49 of the *Kuan Tzu* book:

Only the *chün-tzu* (gentleman)[d] holding on to the idea of the One (*chün-tzu tê i chih li*[3]) can bring about changes in things and affairs. If this holding on is not lost, he will be able to reign over the ten thousand things. The *chün-tzu* commands things and is not commanded by things, for he has gained the principle of the One.[e]

And many other examples could be cited.

[a] *Chuang Tzu*, ch. 3, tr. Legge (5), vol. 1, p. 198; Fêng Yu-Lan (5), p. 67; Lin Yü-Thang (1), p. 216; Waley (6), p. 73, mod. Concerning the chopper's 'no-thickness', see on, Sect. 19*h*; it was a concept of the Mohist geometers. For the significance of the final words, see on, p. 143.
[b] We modify the translation in this way, for, as will later be shown, this word *shih*[2] could mean, from high antiquity, a diviner's board, and stands at the very origin of the discovery of the magnetic compass. See Sect. 26*i* below. Duyvendak (18) has also appreciated this.
[c] Tr. Waley (4), mod.
[d] The use of the word *chün-tzu* in this passage shows that it must be a fragment of early date.
[e] Tr. Haloun (2).

[1] 是以聖人抱一   [2] 爲天下式   [3] 君子得一之理

These passages have generally been regarded as affirmations of religious mysticism, analogous to superficially similar expressions used by Islamic and Christian mystics. But the point is that at this early stage of development of thought in China we are standing at a point before the differentiation of religion and science. While there was without doubt a numinous element in these early Chinese statements, they may more readily be interpreted, in view of all else that we know about the Taoists, as affirmations of that unity in Nature which is the basic assumption of natural science. But we must also not forget that yet a third element may be present in them, namely, the political. The Taoists were, as we shall see, in favour of a primitive undifferentiated form of society, and against the differentiated feudal form. I mention this by way of preparing the reader's mind, at this point, for what will be better understood at a somewhat later stage of the analysis.

The unity of the Tao runs through everything. *Chuang Tzu* says:

The Master (perhaps Lao Tzu) said, 'The Tao does not exhaust itself in what is greatest, nor is it ever absent from what is least; therefore it is to be found complete and diffused in all things. How wide is its universal comprehensiveness! How deep is its unfathomableness!'[a]

and, more imaginatively,

Tungkuo Shun-Tzu said to Chuang Tzu, 'Where is this so-called Tao?'
Chuang Tzu answered, 'Everywhere.'
The other said, 'You must specify an instance of it.'
Chuang Tzu said, 'It is here in these ants.'
Tungkuo replied, 'That must be its lowest manifestation, surely.'
Chuang Tzu said, 'No, it is in these weeds.'
The other said, 'What about a lower example?'
Chuang Tzu said, 'It is in this earthenware tile.'
'Surely brick and tile must be its lowest place?'
'No, it is here in this dung also.'
To this Tungkuo gave no reply.[b]

From this again we derive the point of view strictly characteristic of science, that nothing is outside the domain of scientific inquiry, no matter how repulsive, disagreeable or apparently trivial it may be. This is a really important principle, for the Taoists, who were orienting themselves in a direction which would ultimately lead to modern science, were to have to take an interest in all kinds of things utterly disdained by the Confucians and their descendants—in seemingly worthless minerals, wild plants and animal and human parts and products.

Something of this same idea may lie behind another phrase which one finds constantly recurring in Taoist writings, namely, that the sage should 'cover all things

[a] Ch. 13, tr. Legge (5), vol. 1, p. 342.
[b] Ch. 22, tr. Legge (5), vol. 2, p. 66.

impartially', without private preferences and prejudices.[a] In *Chuang Tzu*, chapter 17, for example:

> Be severe and strict, like the ruler of a State who does not bestow his rewards with favouritism (*wu ssu*[1]). Be scrupulous, yet gentle, like the spirits of the land and grain, who accept sacrifices and give blessings without favouritism (*wu ssu*[1]). Be large-minded like space, the four directions of which are limitless, forming no particular enclosures[b] (*wu so chen yü*[2]). Hold all things in your love, favouring and supporting none specially. This is called being without any local or partial regard; all things are equally esteemed; there is no long or short among them.[c]

Sometimes the word used is *hsü*[3] (emptiness): 'The mind should be an emptiness, ready to receive all things.'[d] This approximates to Hsün Tzu's use of the same word to mean clearing the mind of prejudice before entering into an argument.[e] But the usual word is that which we have just met with, *ssu*, which means private as opposed to public interest, and personal prejudices or preconceived opinions as opposed to what may be observed in the whole range, omitting nothing, of natural phenomena. Thus *Kuan Tzu*, chapter 37: 'The sage is like heaven, he covers everything impartially (*wu ssu fu yeh*[4]); he is like earth, bearing up everything impartially (*wu ssu tsai yeh*[5]).'[f] Other examples could readily be adduced.

From this it is but a short step to the definitive rejection of ethics from the scientific world-view now in the forging. For Confucianism and indeed all the other schools this was a frontal attack. It was part of the relativistic attitude of Taoist thought, to which we shall return, and Taoist texts never tire of insisting that the human (and the individual human) is not the only criterion. The *Tao Tê Ching* does not hesitate to say:

> Heaven and Earth are not benevolent;
> They treat the ten thousand things like straw dogs.[g]
> Nor is the Sage benevolent;
> To him also the hundred clans are but as straw dogs.
> Yet Heaven and Earth and all that lies between
> Is like a bellows, empty but not collapsed,
> The more you work it, the more comes forth.
> (Inexhaustible! Silent!)
> Whereas the force of words is soon spent,
> It is better to hold fast to the (certainty) within.[h]

[a] Of course the Confucians also used the phrase, in more humanistic senses.
[b] Again perhaps a political innuendo (see below, pp. 109 ff.).
[c] Tr. Legge (5), vol. 1, p. 382.
[d] *Chuang Tzu*, ch. 4; Fêng Yu-Lan (5), p. 80.
[e] Dubs (7), p. 92.   [f] Tr. Haloun (2).
[g] Straw dogs were part of the ancient sacrificial ritual, perhaps substitutes for earlier living sacrificial animals. They were brought in with pomp but thrown away afterwards.
[h] *TTC*, ch. 5, tr. Waley (4); Carus (1), p. 99; Strauss (1), p. 28; Hughes (1); Duyvendak (18). Cf. *Chuang Tzu*, ch. 14. On the silence of the Tao, see below, pp. 70, 448, 546, 563, 564.

[1] 無私   [2] 無所畛域   [3] 虛   [4] 無私覆也   [5] 無私載也

No one can understand this unless it is realised that the expulsion of ethical judgments from natural science was an essential step in its development.[a] Though the search for truth is itself an ethical value, nature cannot be divided into the edifying, which can be written about, and the unedifying, which should be hushed up. Natural phenomena cannot be divided into the noble and the ignoble; ethical criteria have no application outside social relations; science must be ethically neutral. It is to the credit of the Taoists that they should have seen this in spite of the ethical character of the culture from which they came. In recalling the experimental methods which modern medical and epidemiological science, in spite of all its ultimate 'benevolence', is logically forced to adopt in order that the knowledge, and hence the power, of mankind may be increased,[b] one is irresistibly reminded of the third and fourth lines of the fifth chapter of Lao Tzu. Ultimate benevolence may require temporary non-benevolence.

The expulsion of partiality and human weakness in the investigation of the more disgusting or terrible aspects of Nature, and the expulsion of human ethical criteria and preconceptions from the human approach to Nature, lead naturally to a realisation that human standards are irrelevant outside humanity. *Chuang Tzu* has several parables illustrating this. For example, in chapter 2 there is a discussion about standards of good and bad.

If a man sleep in a damp place, he gets lumbago and may die. But what about an eel? And living up a tree is frightening and tiring to the nerves. But what about monkeys? What habitat can be said to be 'absolutely' 'right'? Then men eat flesh, deer eat grass, centipedes enjoy small worms, owls and crows delight in mice. Whose is the 'right' taste, 'absolutely'? Monkeys mate with apes, bucks with does, eels consort with fishes, while men admire great beauties such as Mao Chhiang and Li Chi. Yet at the sight of these women the fish plunged deep into the water, birds flew from them aloft, and deer sped away. Who shall say what is the 'right' standard of beauty? In my opinion, the doctrines of benevolence and righteousness and the paths of right and wrong are inextricably confused. How could I discriminate among them?[c]

And there are parallel passages in *Chuang Tzu*[d] and elsewhere. Man could not be considered the measure of all things, as the Confucians wished to believe.

So far we have spoken of the unity of Nature and its independence of human standards. But Nature was also self-sufficient and uncreated. Here the key phrase

[a] As I have tried to explain elsewhere (Needham (5), pp. 104, 170). Dubs (19) appositely quotes Spinoza's *Ethics* here: 'The perfection of things is to be judged by their nature and power alone; nor are they more or less perfect because...beneficial or prejudicial to human nature.' Pt. 1, App. Lao Tzu was certainly conscious of the smallness of human affairs in comparison with the immensity of the universe.

[b] One might refer to the theme of Sinclair Lewis' novel *Martin Arrowsmith*, in which is recounted the testing of a new vaccine in the field by its application to half only of the sufferers in an epidemic; for scientific proof of effectivity demands a controlled experiment. Cf. Beveridge (1), p. 18, and the popular expositions of de Kruif.

[c] Tr. Legge (5), vol. 1, p. 192; Fêng Yu-Lan (5), p. 59; Lin Yü-Thang (1), p. 259.

[d] Ch. 18 (Legge (5), vol. 2, p. 8).

was *tzu-jan*,[1] spontaneous, self-originating, natural. The *locus classicus* is in Lao Tzu:

(In the beginning) there was something undifferentiated (*hun*[2])[a] and yet complete (*chhêng*[3])
Before Heaven and Earth were produced,
Silent! Empty!
Sufficient unto itself! Unchanging!
Revolving incessantly, never exhausted.
Well might it be the mother of all things under heaven.
I do not know its name.
'Tao' is the courtesy-name we give it.
If I were forced to classify it, I should call it 'Great'.
But being great means being penetrating (in space and time),[b]
And penetrating implies far-reaching
And far-reaching means coming back to the original point....
The ways of men are conditioned by those of earth, the ways of earth by those of heaven,
    the ways of heaven by those of the Tao, and the Tao came into being by itself (*jen fa ti,*
    ti fa thien, thien fa Tao, Tao fa tzu-jan[4]).[c]

This affirmation is the basic affirmation of scientific naturalism. One remembers Lucretius:[d]

> Quae bene cognita si teneas, natura videtur
> libera continuo dominis privata superbis
> ipsa sua per se sponte omnia dis agere expers.[e]

Here is Chuang Tzu's naturalistic account of wind noises, phenomena most inviting to the ancients to postulate the activity of spirits, naiads and dryads, etc.

'The breath of the universe', said Tzu-Chhi, 'is called wind. At times it is inactive. But when it rises, then from a myriad apertures there issues its excited noise. Have you never listened to its deafening roar? On a bluff in a mountain forest, in the huge trees, a hundred spans round, the apertures and orifices are like nostrils, mouths or ears, like beam-sockets, cups, mortars, or pools and puddles. And the wind goes rushing through them, like swirling torrents or singing arrows, bellowing, sousing, trilling, wailing, roaring, purling, whistling in front and echoing behind, now soft with the cool breeze, now shrill with the whirlwind, till the tempest and the apertures are all empty (and still). Have you never observed how the trees and branches shake and quiver, twist and twirl?'

---

[a] Note especially this word, on account of its political significance, to be later explained.
[b] There is a pun in the text here.
[c] *TTC*, ch. 25, tr. Waley (4), Hughes (1), Lin Yü-Thang (1), p. 145.
[d] −99 to −50, therefore a younger contemporary of Ssuma Chhien.
[e] *De. Rer. Nat.* II, 1090–2:
    Nature, delivered from every haughty lord
    And forthwith free, is seen to have done all things
    Herself, and through herself, of her own accord
    Rid of all gods...(tr. Leonard).

[1] 自然   [2] 混   [3] 成   [4] 人法地地法天天法道道法自然

Tzu-Yu said, 'The notes of earth then are simply those which come from its myriad apertures, and the notes of man may be compared to those (which issue from tubes of) bamboo—allow me to ask about the notes of heaven?'

Tzu-Chhi replied, 'When (the wind) blows, the sounds from the myriad apertures are each different, and its cessation makes them stop *of themselves* (*tzu i*[1]). Both these things arise from themselves—what other agency could there be exciting them?'[a]

Later, the expression *tzu-jan* became universally adopted for speaking of natural phenomena, as in *Huai Nan Tzu*, where one finds such passages as this:

He who conforms to the course of the Tao (*hsiu Tao li chih shu*[2]), following the natural processes of Heaven and Earth (*yin thien ti chih tzu-jan*[3]), finds it easy to manage the whole world. Thus it was that Yü the Great was able to engineer the canals by following the nature of water and using it as his guide (*yin shui i wei shih*[4]). Likewise Shen Nung, in the sowing of seed, followed the nature of germination and thus obtained instruction (*yin miao i wei chiao*[5]). Water-plants root in water, trees in earth; birds fly in the air and beasts prowl on the ground; crocodiles and dragons live in the water, tigers and leopards dwell in mountains—such is their inherent nature (*hsing*[6]). Pieces of wood when rubbed together generate heat, metal subjected to fire melts, wheels revolve, scooped-out things float. All things have their natural tendencies (*tzu-jan chih shih yeh*[7])....Thus all things are by themselves so (*wan wu ku i tzu-jan*[8]).[b]

It is clear that the Taoists were close to an appreciation of the problems of causality, though they never embodied it in formal propositions as Aristotelians did. The best passage to illustrate this is probably *Chuang Tzu*, chapter 2:

Penumbra said to Shadow, 'At one moment you move, at another you are at rest. At one moment you sit down, at another you get up. Why this instability of purpose?'

'Do I have to depend', replied the Shadow, 'upon something which causes me to do as I do? (lit. do I have to wait (for something else), so that (my movement) may come about; *wu yu tai erh jan chê yeh*[9]). And does that something have to depend in turn upon something else, which causes it to do as it does? (lit. does what I wait for have to wait in its turn (for something else), so that (its movement) may come about; *wu so tai yu yu tai erh jan chê yeh*[10]). Is not my dependence (more like the unconscious movements of) the scales of a snake or the wings of a cicada? How can one tell whether movement is dependent or independent?'[c]

Here Chuang Tzu adumbrates a principle of non-mechanical causation. Later he reveals a veritable organic philosophy. From the operation of the natural processes in

---

[a] Ch. 2, tr. Legge (5), vol. 1, p. 177; Lin Yü-Thang (1), p. 141.
[b] Ch. 1, p. 5*b*, tr. Morgan (1), p. 9, mod.
[c] Tr. auct. adjuv. Legge (5), vol. 1, p. 196; Lin Yü-Thang (1), p. 255. See also on this below, in Section 26*g* on optics.

[1] 自已   [2] 脩道理之數   [3] 因天地之自然   [4] 因水以爲師
[5] 因苗以爲敎   [6] 性   [7] 自然之勢也   [8] 萬物固以自然
[9] 吾有待而然者邪   [10] 吾所待又有待而然者邪

an animal or human body, uncontrolled by consciousness, he suggests that in the whole universe the Tao needs no consciousness to bring about all its effects.

It might seem as if there were a real Governoi (*tsai*[1]), but we find no trace of his being. One might believe that he could act, but we do not see his form. He would have (to have) sensitivity (*chhing*[2]) without form (*hsing*[3]).[a] But now the hundred parts of the human body, with its nine orifices and six viscera, all are complete in their places. Which should one prefer? Do you like them all equally? Or do you like some more than others? Are they all servants? Are these servants unable to control each other, but need another as ruler? Or do they become rulers and servants in turn (*hsiang wei chün chhen*[4])? Is there any true ruler (*chün*[5]) other than themselves?[b]

These words are indeed striking when we think of what is now known about the complex interrelations of stimulators and reactors in living organisms and their development, or the mutual influences of the glands of the endocrine system.[c] The same note is struck often elsewhere, for example in *Kuan Tzu*, chapter 55, where we have: 'Though the heart (consciousness) does not regulate the nine orifices, the nine orifices are well ordered.'[d] The Taoists concluded that neither in the microcosm nor in the macrocosm was there need to postulate a conscious controller. There will be more to say later[e] concerning the organic nature of their thought, which was part of a prevailing Chinese trend.

It would be desirable for someone to scrutinise more closely than is here possible, the parallels between the organic, spontaneous and unconscious Tao, and the *physis* ($\phi \acute{v} \sigma \iota \varsigma$) of the Greeks.[f] In Galen at any rate, the *physis* of a living organism is an indwelling agent propelling and throwing, giving and receiving, but working entirely unconsciously and untaught (*adidaktos*, $\dot{\alpha} \delta \acute{\iota} \delta \alpha \kappa \tau o \varsigma$). But the idea of a demiurge or *logos* was never far from such Greek thinking, and any concept of that kind vitiates the pure organicism which the Taoists sought.[g]

---

[a] Note the paradoxical nature of this requirement from the point of view of the 'ladder of souls', discussed on pp. 22 ff. above.

[b] *Chuang Tzu*, ch. 2, tr. Fêng Yu-Lan (5), p. 46; Legge (5), vol. 1, p. 179; Wieger (7), p. 217. Cf. Hughes (7), p. 225, who gives a rendering not here adopted. It is well to be aware of the context of this remarkable passage. It comes in the chapter which opens with the description of the storm in the forest, quoted just now (p. 50). From the myriad notes of heaven, which Tzu-Yu has asked about, the discussion turns to the myriad notes of man, his fleeting moods and emotions, also his opinions and convictions. Tzu-Chhi shows how they are organically connected: 'If there were no others, there would be no "me"; if there were no "me", they would not be perceived. This seems to be approximately the truth, but we do not know what makes it so.' Then, after examining and rejecting the idea of a personal Maker, he gives the analogy of the living organism, all the parts of which (normally) 'work together for good' without any conscious oversight. The rest of the chapter, one of the finest in the book, is devoted to relativity and dialectical logic.

[c] Chuang Chou's views on organism are developed in ch. 23. Hsün Chhing combated them in his insistence that the mind (*hsin*[6]) was the absolute governor of the body (ch. 17, p. 17*b*; ch. 21, p. 9*b*; Dubs (8), pp. 175, 269).

[d] Tr. Haloun (5).   [e] Pp. 77, 153.

[f] See particularly Heidel (1) and Pagel (7).   [g] Cf. pp. 302 ff. below.

[1] 宰   [2] 情   [3] 形   [4] 相爲君臣   [5] 君   [6] 心

(1) AUTOMATA AND THE PHILOSOPHY OF ORGANISM IN CHUANG CHOU

It is in connection with this that we may, if in a sense paradoxically, seek the significance of certain Taoist parables about automata. Inventors were supposed to have constructed automata in human guise, which, when opened, revealed nothing but mechanisms. Was this intended to suggest the organismic concept of Chuang Tzu? The most striking of these stories, which demands citation, occurs in the *Lieh Tzu*.

King Mu of Chou made a tour of inspection in the west... and on his return journey, before reaching China, a certain artificer, Yen Shih by name, was presented to him. The king received him and asked him what he could do. He replied that he would do anything which the king commanded, but that he had a piece of work already finished which he would like to show him. 'Bring it with you tomorrow', said the king, 'and we will look at it together.' So next day Yen Shih appeared again and was admitted into the presence. 'Who is that man accompanying you?' asked the king. 'That, Sir', replied Yen Shih, 'is my own handiwork. He can sing and he can act.' The king stared at the figure in astonishment. It walked with rapid strides, moving its head up and down, so that anyone would have taken it for a live human being. The artificer touched its chin, and it began singing, perfectly in tune. He touched its hand, and it began posturing, keeping perfect time. It went through any number of movements that fancy might happen to dictate. The king, looking on with his favourite concubine and other beauties, could hardly persuade himself that it was not real. As the performance was drawing to an end, the robot winked its eye and made advances to the ladies in attendance, whereupon the king became incensed and would have had Yen Shih executed on the spot had not the latter, in mortal fear, instantly taken the robot to pieces to let him see what it really was. And, indeed, it turned out to be only a construction of leather, wood, glue and lacquer, variously coloured white, black, red and blue. Examining it closely, the king found all the internal organs complete—liver, gall, heart, lungs, spleen, kidneys, stomach and intestines; and over these again, muscles, bones and limbs with their joints, skin, teeth and hair, all of them artificial. Not a part but was fashioned with the utmost nicety and skill; and when it was put together again, the figure presented the same appearance as when first brought in. The king tried the effect of taking away the heart, and found that the mouth could no longer speak; he took away the liver and the eyes could no longer see; he took away the kidneys and the legs lost their power of locomotion. The king was delighted. Drawing a deep breath, he exclaimed, 'Can it be that human skill is on a par with that of the great Author of Nature (*tsao hua chê*[1])?' And forthwith he gave an order for two extra chariots, in which he took home with him the artificer and his handiwork. Now Pan Shu, with his cloud-scaling ladder, and Mo Ti, with his flying kite, thought that they had reached the limits of human achievement. But when Yen Shih's work was brought to their knowledge, the two philosophers no longer dared to talk of their mechanical skill, and hesitated often when they had the square and compasses in hand.[a]

If this passage is not a declaration of faith in naturalistic explanations of life phenomena, it looks extraordinarily like it. If, as may be probable, it is of the −3rd century,

[a] Ch. 5, pp. 20*a*ff., tr. L. Giles (4), p. 90. Often afterwards repeated, as in the +6th-century *Chin Lou Tzu*, ch. 5, p. 14*b*.

[1] 造化者

one could not insist upon a clear distinction between organismic and mechanistic conceptions, since no definitive sciences of the inorganic world had then developed, and the problem of the relation between the organic and the inorganic could not have been posed. The essence of it is the denial of a conscious guidance of the affairs of the microcosm. The Tao, whether organising the structure of the joints of a bullock, or guiding the movements of the heavenly bodies, did not need to be conscious.

The thought occurs again very strikingly in the (perhaps +8th century) *Kuan Yin Tzu*.

A dried tortoiseshell has no will (*wu wo*[1]), yet it can predict the far future. A lodestone has no will, yet it can exercise great attractive power. Bells and drums have no will, yet they can make a great noise. Boats and carriages have no will, yet they can travel far. So with regard to our bodies, the fact that we can perceive, act, walk about, and talk, does not prove that we have some intrinsic will (as distinct from our responses to the external world).[a]

And elsewhere in the *Lieh Tzu*,[b] there is a curious story about the semi-legendary physician Pien Chhio[2] performing an operation in which the hearts of two men were transplanted, a process which exchanged their minds while leaving their appearance and will-power unaltered. It must also belong to this very old mechanistic-naturalistic tradition.[c]

The mention of Mo Ti (Mo Tzu) and his artificial flying wooden kite or bird (*mu yuan*[3]) is particularly interesting,[d] since exactly the same story is told of Archytas of Tarentum.[e] Their dates are very similar; the *floruit* of Archytas being about the time of the death of Mo Ti (−380). One version is in the book *Mo Tzu*,[f] where the invention of the flying automaton which stayed aloft three days is attributed to Kungshu Phan,[4] who came and showed it to Mo Ti along with other war machines. Another is in the *Han Fei Tzu*,[g] where it is said that Mo Tzu himself made it. These stories have been discussed by Laufer (4) who, in an interesting paper on the pre-history of aviation, was perhaps inclined to take them too seriously. The point here is that it is quite natural that they should have arisen among thinkers in ancient times who were developing a scientific view of the world, both the pre-Socratics in Greece and the Taoists and others in China.[h]

[a] Ch. 2, p. 19*b*, tr. auct. adjuv. Wieger (2), p. 342. The thought here anticipates remarkably that of the 'reflexological' school in modern physiology, of which the Czech George Prochaska (1749 to 1820) was one of the first founders.   [b] Ch. 5, pp. 16*b* ff. (tr. Wieger (7), p. 141).

[c] Moreover, this theme lived on. Cf. the story of a magical exchange of organs between bodies in Wieger (8), no. 64.

[d] We shall mention it again later in Section 27*j* on mechanical engineering. Whether it was a kite or rocket can only be speculation.

[e] Freeman (1), p. 234, Sarton (1), vol. 1, p. 116; Feldhaus (1), col. 46. The main reference is Aulus Gellius, *Noctes Atticae*, x, 12, ix ff.

[f] Ch. 49; tr. Mei Yi-Pao (1), p. 256.   [g] Ch. 11, p. 2*a*.

[h] There are, moreover, Indian (Levi (3); Tawney & Penzer (1), vol. 3, pp. 56 ff., 281), Tibetan (Schiefner, 1) and even Tocharian (Sieg, 1) parallels, but it would take us too far to examine the folklore of the world for examples of this theme. A thorough treatment in a special paper would be well worth while. Meanwhile, cf. Section 27*b* on mechanical toys below.

[1] 無我   [2] 扁鵲   [3] 木鳶   [4] 公輸般

(2) TAOISM, CAUSALITY AND TELEOLOGY

'*Cognoscere causas*' thus became the motto of the Taoists. Through all the convulsions caused by the substitution of feudal bureaucratism for feudalism at the time of the unification of the empire by Chhin Shih Huang Ti, they continued to pursue it. Here the *Lü Shih Chhun Chhiu* is particularly interesting because it shows the way in which the more powerful forces of ethical Confucian culture constrained Taoist naturalism either to amalgamate with it, or to go underground altogether. In this book (the first part of which was completed in −239), while there is much scientific argument, as we shall see, it usually ends with some application to human society. For example:

All phenomena have their causes. If one does not know these causes, although one may happen to be right (about the facts), it is as if one knew nothing, and in the end one will be bewildered. It was through this knowledge that the ancient kings, the famous men, and clever scholars, distinguished themselves from the mass of the people. The fact that water leaves the mountains and runs to the sea is not due to any dislike of the mountains and love for the sea, but is the effect of height as such. The wheat that grows on the plains and is gathered into granaries has no desire for this; it happens because men want to use it....

All this is so, too, with regard to the endurance or fall of States, and to the goodness or badness of individuals. For everything there must be a reason. Therefore the sage does not inquire about endurance or decay, nor about goodness or badness, but about the reasons for them.[a]

The denial of general teleology in this passage is echoed in a lively story in *Lieh Tzu*[b] about a banquet at which the complacent speech of one of the elders is interrupted by a twelve-year-old *enfant terrible* who ventures to suggest that man is only one species among other animal species in the world, and that fish and game had certainly not been created for his benefit any more than he had been created for the benefit of mosquitoes and tigers. Such anti-anthropocentrism was a typical Taoist *coup de patte* (as Wieger would say) at the Confucians:

Mr Thien, of the State of Chhi, was holding an ancestral banquet in his hall, to which a thousand guests had been invited. As he sat in their midst, many came up to him with presents of fish and game. Eyeing them approvingly, he exclaimed with unction; 'How generous is Heaven to man! Heaven makes the five kinds of grain to grow, and brings forth the finny and the feathered tribes, especially for our benefit.' All Mr Thien's guests applauded this sentiment to the echo, except the twelve-year-old son of a Mr Pao, who, regardless of seniority, came forward and said; 'It is not as my Lord says. The ten thousand creatures (in the universe) and we ourselves belong to the same category, that of living things, and in this category there is nothing noble and nothing mean. It is only by reason of size,

a Ch. 44, tr. R. Wilhelm (3), p. 111; eng. auct.
b Ch. 8, pp. 20a ff., tr. R. Wilhelm (4), p. 108; L. Giles (4), p. 119, mod.

strength, or cunning, that one particular species gains the mastery over another, or that one feeds upon another. None of them are produced in order to subserve the uses of others. Man catches and eats those that are fit for (his) food, but how (could it be maintained that) Heaven produced them just for him? Mosquitoes and gnats suck (blood through) his skin; tigers and wolves devour his flesh—but we do not therefore assert that Heaven produced man for the benefit of mosquitoes and gnats, or to provide food for tigers and wolves.'

The following excerpt shows further clear traces of scientific mentality in the *Lü Shih Chhun Chhiu*:

Enlightened men who understand the Tao consider it valuable to infer the far from the near, the old from the new, so that they may know those things which they have not seen from those things which they have seen. Thus if you study the shadow in the court you may know the course of the sun and moon and the changes of light and darkness. If you see ice in a vessel you know that it has been cold all over the earth, and that the fishes and the turtles are hiding. If you taste a sample of meat you know the taste of the whole stew throughout the cauldron.[a]

This invites us to have a closer look at the rather special precepts in which the Taoists enshrined their doctrine of the observation of Nature.

(d) THE APPROACH TO NATURE; THE PSYCHOLOGY OF SCIENTIFIC OBSERVATION

As Maspero (11) has reminded us, there is much significance in the fact that throughout the centuries Taoist temples have been known as *kuan*;[1] and other terms, such as *ssu*[2] or *miao*,[3] which have been applied to temples of other religions, have not been used for those of the Taoists. Now the original meaning of *kuan*[b] was 'to look'. It combines Rad. 147, which indicates 'seeing', with a graph which in its most ancient form was a drawing of a bird, probably a heron. The meaning contained in the word, therefore, was essentially to observe the flight of birds, no doubt with the object of making predictions from the omens so obtained. Already, in the *Tso Chuan*, *kuan*[1] has acquired the meaning of a watch-tower, and is the regular word used for the observation of natural phenomena for divination. Interpretations which later derived *kuan*[1] from the meditative duty of the recluse to look within himself are quite

---

[a] Ch. 84, tr. R. Wilhelm (3), p. 231; eng. auct. It should be noted here that 'Master Lü's Spring and Autumn Annals', Taoist though it is, belongs to a rather special milieu, that of the proto-urban proto-industrial mercantile group which flourished during the Chhin and Han, and about which more will be said later (Sect. 48). The men who acquired great wealth by preparing salt or as iron-masters were closely connected with Taoism on account of its technological aspects. It is one of the paradoxes of history that those who are often not unjustly called the 'capitalists' of the −3rd and −2nd centuries stood much closer to the collectivist Taoists than to the feudal bureaucrats. By contrast all the other Taoist books show a stronger peasant-agricultural character.

[b] K 158 i.

[1] 觀   [2] 寺   [3] 廟

secondary and fanciful.[a] Embodied therefore in the common present-day name for a Taoist temple is the ancient significance of the observation of Nature, and since in their beginnings magic, divination and science were inseparable, we cannot be surprised that it is among the Taoists that we have to look for most of the roots of Chinese scientific thought.[b] What benefits the ancient Taoist philosophers hoped might be derived from the observation of natural phenomena, wider than the satisfaction of the desire of rulers and feudal lords to know what the future held for them, will shortly appear.

(1) THE WATER SYMBOL AND THE FEMININE SYMBOL

The observation of Nature, as opposed to the management of Society, requires a receptive passivity in contrast to a commanding activity, and a freedom from all preconceived theories in contrast to an attachment to a set of social convictions. This is the sense (though doubtless not the only sense) in which we may interpret the symbols of 'water' and 'the feminine' so dear to the early Taoist schools and so perplexing to later commentators. Perhaps it was precisely because of the failure of experimental science to develop in China that the later Chinese commentators failed to understand these passages, and their failure in turn set Western exponents almost without exception on the wrong track. Thus the *Tao Tê Ching*:

> The highest good is like that of water. The goodness of water is that it benefits the ten thousand creatures, yet itself does not wrangle, but is content with the places that all men disdain. It is this that makes water so near to the Tao.[c]

Water is yielding and assumes the shape of whatever vessel it is placed in, it seeps and soaks through invisible crevices, its mirror-like surface reflects all Nature.[d] Compare also chapter 43:

> What is of all things most yielding
> Can overwhelm that which is most hard,
> Being substanceless it can enter in even where there is no crevice.
> That is how I know the value of action which is actionless.[e]
> But that there can be teaching without words,
> Value in action which is actionless
> Few indeed can understand.[f]

[a] The Sung author of the *Shih Wu Chi Yuan* (ch. 8, p. 33b) was well aware of the explanation here given, and endorsed it.

[b] There is no room here to say anything of the immense influence which Taoist observation of Nature had on Chinese art, but reference has to be made to it, and the reader is referred to the monograph of Petrucci (2) on the subject.   [c] *TTC*, ch. 8, tr. Strauss (1); Waley (4).

[d] Cf. *Chuang Tzu*, ch. 7 (Legge (5), vol. 1, p. 266), and the chapter of *Kuan Tzu* translated above.

[e] Evidence will later be adduced which suggests that what the writer here had in mind was the slow and insensible chemical changes going on during dyeing or retting, i.e. when a substance is steeped in a solution, or, as in the case of drugs, extracted from plants or minerals by water or wine (see on, p. 383 and Sects. 31, 32, 45).

[f] Tr. Waley (4), mod. The theme is repeated in ch. 78.

Water also, running down to the valleys, receives all kinds of defilements, but cleanses itself and is never defiled.[a]

The symbols of water and the female have not only philosophical, but also great social significance. Instead of the Confucian or Legalist conception of leadership from above, we reach the Taoist principle of leadership from within. Thus:

> How did the great rivers and seas get their kingship over the hundred lesser streams?
> Through the merit of being lower than they; that was how they got their kingship.
> Therefore the sage, in order to be above the people,
> Must speak as though he were lower than they,
> In order to guide them
> He must put himself behind them.
> Thus when he is above, the people have no burden,
> When he is ahead, they feel no hurt.
> Thus everything under heaven is glad to be directed by him
> And does not find (his guidance) irksome.
> The sage does not enter into competition[b]
> And therefore no one competes with him.[c]

Before analysing this complex of thought further, some examples of the 'feminine' symbol must be given. The *locus classicus* is the *Lao Tzu* book, chapter 6:

> The Valley Spirit never dies (*ku shen pu ssu*[1]).
> It is named the Mysterious Feminine (*shih wei hsüan phin*[2]).
> And the Doorway of the Mysterious Feminine
> Is the root (from which) Heaven and Earth (sprang).
> It is the thread for ever woven;
> And those who use it can accomplish all things.[d]

Again, chapter 28:

> He who knows the male, yet cleaves to what is female
> Becomes like a ravine, receiving all things under heaven[e]
> (Thence) the eternal virtue never leaks away.
> This is returning to the state of Infancy.

[a] On comparative water symbolism in ancient religions see Eliade (2), pp. 168 ff.
[b] This theme reappears in ch. 22.
[c] *TTC*, ch. 66, tr. Waley (4); Chhu Ta-Kao (2); Duyvendak (18).
[d] Tr. Waley (4), mod. The conception of the 'Valley Spirit' has afforded endless matter for the uninspired discussions of subsequent Chinese commentators (whose opinions have been assembled by Neef, 1), and of Western sinologists. Erkes (7) has drawn attention to an +8th-century poem by Lü Yen which indicates that by that time the phrase had acquired an alchemical significance.
[e] This occurs also in *Chuang Tzu*, ch. 33 (Legge (5), vol. 2, p. 226). The sexual aspect of the symbolism, male convexity contrasted with female concavity, should not be overlooked.

[1] 谷神不死   [2] 是謂玄牝

He who knows the white, yet cleaves to the black,
Becomes the instrument (*shih*[1]) by which all things are tested
(And so has) a constant virtue which never errs.
This is returning to the Limitless.
He who knows glory, yet cleaves to ignominy
Becomes like a valley receiving into it all things under heaven,
(For him) the immutable virtue all-sufficient.
This is returning to the Undifferentiated (*phu*[2]).[a]
Now when the Undifferentiated is broken up (dispersed, differentiated, *san*[3])
      it separates into discrete objects (*chhi*[4])
But if the sage uses it, it becomes the Chief of all Ministers.
Truly 'The greatest carver does the least cutting.'[b]

Many other passages using the symbol of the feminine could be brought together.[c]

There has been a great failure in subsequent ages to understand this psychological symbolism. The level of poetry with which the Taoists expressed it was clearly very high, presenting analogies with Goethe's 'ewig weibliche',[d] and though it may well be true that they built on ancient Chinese 'Urmutter' creation-goddess myths,[e] such demonstrations do not reach the heart of the matter. The point is that they intuitively went to the roots of science and democracy alike. The Confucian and Legalist social-ethical thought-complex was masculine, managing, hard, dominating, aggressive, rational and donative—the Taoists broke with it radically and completely by emphasising all that was feminine, tolerant, yielding, permissive, withdrawing, mystical and receptive.[f] Their very exaltation of the 'Valley Spirit' was an affront to the Confucians; for is it not said in the *Lun Yü*: 'The superior man hates to dwell in a low-lying situation, where all the evil in the world will flow down upon him.'[g] And the female receptiveness which the Taoists desired to display in their observation of Nature was inextricably connected with the feminine yieldingness which they believed should be prominent in human social relations. Inevitably they were in opposition to feudal society because the yieldingness in which they believed was incompatible with that society; it was suited for, and in a real sense the poetical expression of, a cooperative collectivist society.[h] Such a society had once existed, in the primitive

---

[a] The use of this word is to be particularly noted, as also the concluding lines in general, with reference to the Taoist political tendency described below (p. 114).

[b] Tr. Waley (4); Chhu Ta-Kao (2); Duyvendak (18), mod.

[c] E.g. *Chuang Tzu*, ch. 33 (Legge (5), vol. 2, p. 226).

[d] Which really goes back to Paracelsus.

[e] See on (p. 78), in connection with the cosmogonic *TTC*, ch. 42; and Conrady (3); Forke (13), p. 265.

[f] Compare the remarkable words of W. E. Hocking (1), who, in a study of the Neo-Confucian empirical-rationalist philosophy of the +12th century (see below, p. 474), said: 'It is quite possible to regard the whole modern scientific effort, from the sixteenth century onward, as an effort inspired by an ethical consideration. Empiricism is itself a form of self-denial, a moral will to let the object speak for itself. Empiricism holds that if we allow it to do so, the object will speak, i.e. that truth is accessible. Pragmatism does not.'   [g] XIX, xx, tr. Legge (2).

[h] There has been much debate as to whether ancient Chinese society was matriarchal. We shall recur to the matter presently (pp. 108, 134, 151 below).

[1] 式   [2] 樸   [3] 散   [4] 器

collectivism of the villages before the full differentiation of lords, priests and warriors
in bronze-age proto-feudalism; it may still have existed on the fringes of Chinese
culture in the centuries immediately preceding the rise of Taoist thought;[a] and it was
to exist again (though the Taoists could not know that millennia must elapse before
humanity would return to their ideals). How profound Taoist insight was may be
appreciated by reading the brilliant essays of William Morton Wheeler, the great
American entomologist,[b] and Ernst Bergmann, which urge that the liquidation of
masculine aggressiveness is one of the most important limiting factors for the success
of that cooperative and collectivist society towards which mankind is inevitably
moving as the scope and potentialities of the highest social organisations continue to
increase.[c] For the moment these words[d] must serve to indicate the social truths
embodied in the *Lao Tzu* quotations just given, for we shall have to return to the
political position of the Taoists on account of its enormous importance for the whole
history of the development of scientific thought in China. Conducting a socialist
holding action for two thousand years and condemned to perpetual heterodoxy,
Taoism had to retain, unborn within itself, science in the fullest sense.

The *Kuan Tzu* book contains certain sentences which elucidate further the role of
the sage with respect to Nature.

Ch. 37: Things in general carrying with them their names come forward. The sage follows
after them and judges them (semi-magical control by knowing their true names). As thus
reality is not harmed, there will be no disorder in Nature, and Heaven and Earth will be
brought into order (controlled).

Ch. 55: The sage follows after things, therefore he can control them (*shêng jen yin chih,
ku nêng chang chih*[1]).[e]

Ch. 37: Collecting (*tsuan*[2]) and selecting (*hsüan*[3]) is the way to grade matters. Changing
to the utmost is the means by which to respond to things (*chi pien chê so i ying wu yeh*[4]).

Ch. 37: The sage commands things and is not commanded by things (this occurs again in
chapter 49).

a Cf. the well-known work of Margaret Mead in which the attitudes of several very different
primitive peoples were compared. With all due reservations it might be said that the Taoists would
have been strongly in favour of the Arapesh, and perhaps they actually knew some primitive societies
of that character; cooperative, unaggressive and responsive.

b I like to recall that it was Wheeler of whom another great American biologist, Lawrence J. Hender-
son, said that he was the only man he had ever met who would have been both capable, and worthy, of
conducting a conversation with Aristotle.

c I have elaborated this point elsewhere (Needham (6), p. 194).

d Compare also the words of another American biologist, one of the best minds of our age, G. Evelyn
Hutchinson (1): '...until some of the excessively masculine, sadistic, manipulative attitude of our
culture is corrected by a view more nearly like that of the Arapesh, most discoveries will turn out to be
admirably devised for destruction, not because the discoverers were wicked or because the discovery is
inherently destructive, but because in so much of the modern world, that is the "obvious" way in
which discoveries are used.'

e Almost identical statements occur in the *Kuei Ku Tzu*[5] and the *Têng Hsi Tzu*,[6] which are mainly
Legalist works. Haloun (5) gives 'The saint follows (events), therefore he is equal (to them),' preferring
tang[7] to *chang*,[8] but I retain his earlier view.

1 聖人因之故能掌之	2 纂	3 選	4 極變者所以應物也
5 鬼谷子	6 鄧析子	7 當	8 掌

Ch. 49: The highest degree of the spiritual is to know clearly the myriad things (*shen ming chih chi, chao chih wan wu*[1]).[a]

In reading the second of these we are reminded of the saying of Francis Bacon that 'we cannot command Nature except by obeying her',[b] and of the words of a modern philosopher about mankind stepping from the realm of necessity into the realms of freedom by the very study of necessity (the laws of Nature) itself.[c] And in connection with Taoist thought one may ponder the words of Thomas Henry Huxley:[d]

Science seems to me to teach in the highest and strongest manner the great truth which is embodied in the Christian conception of entire surrender to the will of God. Sit down before fact as a little child, be prepared to give up every preconceived notion, follow humbly wherever and to whatever abysses Nature leads, or you shall learn nothing.

Any ancient Taoist philosopher could have written that, and no Confucian would ever have understood it.[e]

(2) THE CONCEPT OF *JANG* (YIELDINGNESS)

If it were not unthinkable (from the Chinese point of view) that the Yin and the Yang could ever be separated, one might say that Taoism was a Yin thought-system and Confucianism a Yang one.[f] But this inseparability is well demonstrated by one of the basic conceptions common to all schools; the feminine yieldingness of masculine ownership, expressed in the word *jang*.[2] The dictionary meaning of this word is to yield up, to cede, to give up the better place, hence to invite. Karlgren (1) gives no oracle-bone form, but Granet (1) made an exhaustive examination of the concept from the mythological and folkloristic point of view, concluding that the custom of 'potlatch' must have been of great importance in the most ancient Chinese society.

[a] Tr. Haloun (2).

[b] 'Hominis autem imperium in res, in solis artibus et scientiis ponitur; Natura enim non imperatur, nisi parendo' (*Novum Organum*, aphorism 129).

[c] Friedrich Engels. His exact words (2, p. 82) were: 'the ascent of man from the kingdom of necessity to the kingdom of freedom.' And thus Plekhanov glossed it: 'Man would be "freer" if he could satisfy his needs without exertion (in food production). He submits himself to Nature even when he compels Nature to serve his purposes. But this submission is the condition of his enfranchisement. By submitting to Nature he increases his power over Nature, and thus enlarges his freedoms' (1), p. 87. Has it not a Taoist ring? Is not every experiment in pure science a submission to Nature from which is won that power seen in every technological enterprise? Cf. Marx (1), vol. 3, pt. 2, p. 355.

[d] L. Huxley (1), vol. 1, p. 316. The wonderful letter which includes this passage was written by T. H. Huxley to Charles Kingsley on 23 September 1860. It was in reply to a letter of sympathy occasioned by the death of Huxley's little son aged four.

[e] The only sinologist who has given evidence of any understanding of the Taoist attitude of humility towards Nature is Chhen Jung-Chieh (1), p. 255.

[f] After I had written this I was interested to find that Lin Thung-Chi (1) had made the same comparison.

[1] 神明之極照知萬物   [2] 讓

According to this custom, which still exists in some tribal communities,[a] the prestige of a leading man depends on the amount of food or other commodities which he can distribute to the community as a whole at periodical or seasonal feasts. In China the magical virtue, social prestige, and ultimately 'face', derived from ceding and yielding, became a dominant element in the culture, as everyone knows who has himself lived in China and experienced the difficulty of passing through any doorway with a group of people, or seen scholars positively struggling for the least honourable places at a dinner-party. The rest of the world seems the poorer for lack of a similarly deep-rooted tradition, and if it came originally from 'potlatch' it attained its greatest heights of expression in the Taoist texts, though it is by no means peculiar to them.[b]

This conception of *jang* reaches its highest point in *Lao Tzu*:

> ...Therefore the sage
> Puts himself in the background, yet is always to the fore.
> Remains outside, but is always here.
> Is it not just because he does not strive for any personal end
> That all his personal ends are fulfilled?[c]

Or where, in chapter 68, it is said: 'The greatest conqueror wins without joining issue; the best user of men acts as though he were their inferior.' The same lesson is repeated with all kinds of variations.[d] 'To remain whole, be twisted—to become straight, let yourself be bent.' And the idea is extended to the mutual relations of States; larger ones should win the adherence of smaller ones and not annex them or dominate them by force.[e]

From this yieldingness, this giving up in order to get, this renunciation in order not to lose, this profound non-possessiveness; it was but a short step for the Taoist scholars to a refusal to accept State offices even when called upon to do so. '*Nolo episcopari*' became, throughout history, the watchword of Taoists. Already in *Chuang Tzu* there are famous stories to this effect,[f] and the +4th-century *Lieh Hsien Chuan*[1] (Lives of Famous Hsien) offers many examples of the characteristic situation of Taoist sages declining office.[g]

---

[a] Cf. Benedict (1), ch. 6.

[b] The idea may be found, for example, in *Mêng Tzu*, II (1), vi, 4; *Hsün Tzu*, ch. 23.

[c] *TTC*, ch. 7, tr. Waley (4). One cannot help being reminded of some of the great Christian poetical paradoxes, e.g. 'having nothing and yet possessing all things'. And St Francis:
> C'est en donnant qu'on reçoit;
> C'est en s'oubliant qu'on trouve;
> C'est en pardonnant qu'on est pardonné;
> C'est en mourant qu'on ressuscite à l'éternelle vie!

[d] *TTC*, chs. 2, 22, 36, etc. Right Taoist was 'Crazy Jane' in Yeats, *Collected Poems*, p. 294.

[e] *TTC*, chs. 61, 69, 73.

[f] Ch. 17 (Legge (5), vol. 1, p. 390) and the whole of ch. 28.

[g] Cf. p. 141 below. There are parallels for some of these things in other cultures; see, for example the paper of Wensinck (1) on the 'Refused Dignity' in Hebrew-Arabic tradition.

[1] 列 仙 傳

One or two writers have come near to the realisation that the Taoists were exponents of a pre-feudal form of society, but they have not quite stated it. Dubs[a] has recognised that Lao Tzu looked *backwards*, but was content with the supposition that it was to a legendary 'golden age' such as one finds in Greek mythology. H. Wilhelm[b] has put the matter a stage more concretely by suggesting that the Taoists represented a kind of small middle-class, between the feudal lords of the Chou and the mass of peasant-farmers, which had originated from the remnants of the aristocracy of the conquered Shang dynasty. It is certainly true that the word *shang* came to mean merchant. Wilhelm suggests that it was natural that this group of literate scribes developed into specialists of all kinds, philosophers and artificers, magicians and priests; and that their world outlook should have given a high place to the virtues of humility and yieldingness. He is less easy to follow when he suggests that the form of society to which they looked back was in fact the dynasty of the Shang, and that they cherished a superstition that a great Shang emperor would return 'to make all things well'. Their doctrines were too radical for this, and while we may accept as plausible some degree of Shang social origin, the Taoists went far beyond it in attacking feudal society root and branch.

(3) ATARAXY

We are now in a position to ask a central question, what was the main motive of the Taoist philosophers in wishing to engage in the observation of Nature? There can be little doubt that it was in order to gain that peace of mind which comes from having formulated a theory or hypothesis, however provisional, about the terrifying manifestations of the natural world surrounding and penetrating the frail structure of human society. Whether the phenomena be those of natural convulsions, earthquakes, eruptions, storms or floods, or of the varied forms of disease, man at the beginning of the path of science feels stronger and more confident when once he has differentiated and classified them, and especially named them and formulated naturalistic theories about their origins, nature and likely future incidence. This distinctively proto-scientific peace of mind the Chinese knew as *ching hsin*.[1] The atomistic followers of Democritus and the Epicureans knew it as ἀταραξία, ataraxy.[c] We have already noticed (p. 40) the story in *Lieh Tzu* of the man of Chhi who was afraid that the sky would collapse. This parable was by no means the joke that so many have taken it to be; reflective spirits who were not content with Confucian concentration on the affairs of human society stood in great need of assurance, and the Taoists, like the Epicureans, were determined to 'pass beyond the flaming ramparts of the world'[d] to seek it.

---

[a] (7), p. 209.   [b] (1), p. 50, following Hu Shih (8) but contrary to Fêng Yu-Lan (3).
[c] Freeman (1), pp. 293, 316, 325 ff., 330 ff., 336, 351; Needham (3), p. 9; (5), p. 33. Among Western writers on Taoism the only one who has appreciated the scientific nature of Taoist ataraxy is the Czech Emanuel Rádl. In this connection one should not overlook that very ancient component of human thought, the belief that magical power could be obtained over objects and processes by a knowledge of their 'true names' (cf. Edgerton (1) on the Upanishads). But assurance of such power would lead to ataraxy.   [d] Lucretius speaking of Epicurus in *De Rerum Natura*, I, 73.

[1] 靜心

The most important relevant passage in the *Tao Tê Ching* is:

> Push on to the ultimate Emptiness (*chih hsü chi* [1]),
> Guard the unshakable Calmness (*shou ching tu* [2]),
> All the ten thousand things are moving and working (*wan wu ping tso* [3])
> (Yet) we can see (the void, whither they must) return (*wu i kuan fu* [4]).
> All things howsoever they flourish
> Turn and go home to the root from which they sprang.
> This reversion to the root (*kuei kên* [5]) is called Calmness
> It is the recognition of Necessity (*fu ming* [6])
> That which is called Unchanging (*chhang* [7])
> (Now) knowing the Unchanging means Enlightenment (*ming* [8]),
> Not knowing it means going blindly to disaster.... [a]

And *Chuang Tzu* says:

> The ancients who regulated the Tao nourished their knowledge by their calmness (*thien* [9]), and all through life refrained from employing that knowledge in action (contrary to Nature); moreover they may also be said to have nourished their calmness by their knowledge. [b]

So also in many other places, e.g. chapter 13 [c] or chapter 6, [d] where the 'True Men of Old' had no anxiety when they awoke, forgot all fear of death, and 'composedly went and came'; or chapter 18 where the stories are told of Chuang Tzu's lack of mourning for his wife, [e] and of the calmness of the two imaginary characters Deformed and One-Foot. [f] Innumerable are the places in the Taoist writings where ataraxy is described. A particularly good example is the imagined scene at the death of Lao Tzu. [g] Chhin Shih having entered into the chamber and wailed in a perfunctory way, he was taken to task by one of the disciples, but replied:

> All this (excessive) grief is 'violating the principle of Nature' (*tun thien* [10]) and doubling the emotion of man, forgetting what we have received from Nature. This was called by the ancients the penalty of violating the principle of Nature. When the master came it was because he had occasion to be born. When he went he simply followed the normal course. Those who are quiet (*an* [11]) at the proper times, and follow the course of Nature cannot be affected by grief or joy. These were considered by the ancients as the men who were *released from bondage* by the Ruler (Above) (*Ku chê wei shih ti chih hsüan chieh* [12]). [h]

---

[a] Ch. 16, tr. auct. adjuv. Waley (4); Hughes (1); Chhu Ta-Kao (2); Duyvendak (18).
[b] Ch. 16, tr. Legge (5), vol. 1, p. 368.
[c] Legge (5), vol. 1, pp. 330 ff.
[d] Legge (5), vol. 1, p. 238.   [e] Legge (5), vol. 2, p. 4.
[f] Legge (5), vol. 2, p. 5.
[g] *Chuang Tzu*, ch. 3.   [h] Tr. Fêng Yu-Lan (5), p. 70.

[1] 致虛極   [2] 守靜篤   [3] 萬物並作   [4] 吾以觀復   [5] 歸根
[6] 復命   [7] 常   [8] 明   [9] 恬   [10] 遁天   [11] 安
[12] 古者謂是帝之懸解

The parallel with the Epicureans and Lucretius is indeed quite close and unmistakable. The *De Rerum Natura* speaks of science as the only remedy for the multitudinous fears of men:

> nobis est ratio, solis lunaeque meatus
> qua fiant ratione, et qua vi quaeque gerantur
> in terris, tum cum primis ratione sagaci
> unde anima atque animi constet natura videndum,
> et quae res nobis vigilantibus obvia mentes
> terrificet morbo adfectis somnoque sepultis,
> cernere uti videamur eos audireque coram,
> morte obita quorum tellus amplectitur ossa.[a]

And Lucretius repeats (at least three times) in the course of his long poem, a passage which might be taken as the very battle-song of the earliest hosts of science:

> hunc igitur terrorem animi tenebrasque necessest
> non radii solis neque lucida tela diei
> discutiant, sed naturae species ratioque.[b]

Chuang Tzu rises to his greatest heights in a series of passages,[c] paralleled in other Taoist writers,[d] where he speaks of 'Riding on the Normality of the Universe' or on the 'Infinity of Nature', and thus describes the sense of liberation which could be attained by those who could abstract themselves from the trivial quarrels of human society and unify themselves with the great world of Nature. This unification embodied no doubt a strong religious element, for, as has already been said, the religious experience had not yet differentiated itself from the scientific conviction of the unity of Nature. But there was also a magical element, a half-belief that the perfected sage might really ride on the winds and the clouds, a wish-fulfilment psychology embodying too a distinct flavour of Baconian affirmation concerning the future powers over Nature which might await the investigators of Nature. This

[a] I, 128 ff.
> Then be it ours with steady mind to grasp
> The purport of the skies, the law behind
> The wandering courses of the sun and moon;
> To scan the powers that speed all life below;
> But most to see with reasonable eyes
> Of what the mind, of what the soul, is made,
> And what it is so terrible that breaks
> On us asleep, or waking in disease,
> Until we seem to mark, and hear at hand
> Dead men whose bones earth bosomed long ago (tr. Leonard).

[b] I, 146; III, 91; VI, 39.
> These terrors, then, this darkness of the mind
> Not sunrise with its flaring spokes of light
> Nor glittering arrows of morning can disperse
> But only Nature's aspect, and her Law (tr. Leonard).

[c] E.g. ch. 1, the title of which is 'The Happy Excursion'; ch. 2 (Legge (5), vol. 1, p. 192); ch. 24 (Legge (5), vol. 2, p. 96); ch. 32 (Legge (5), vol. 2, p. 212).

[d] E.g. *Huai Nan Tzu*, ch. 1, p. 3a (Morgan (1), p. 5), an elaborate passage; *Tao Tê Ching*, ch. 10.

element was of course congenial to the shamanist magicians who mingled with the Taoists, and grew in importance as the 'adepts' succeeded the philosophers.[a] The passages in question may be termed the *Chhêng*[1] passages, since 'riding' is the key word of all of them; though an allied technical term is *yu*,[2] 'making excursion in'.[b]

The idea is thus introduced:

A boat may be hidden in a creek, a trap-basket in a lake. These may be said to be safe enough. But at midnight a strong man may come and carry them away on his back. The ignorant do not see that no matter how well you hide things, smaller ones in larger ones, there is always a chance for them to be lost. But if you hide the universe in the universe, there will be no room for it to be lost. This is a great truth. To have attained the human form is a source of joy. But in the infinity of changes (*wan hua*[3]) there are thousands of other forms equally good. What an incomparable bliss to undergo these countless transitions! Therefore the sages make excursion in what things cannot escape, and thus they always endure (*Ku shêng jen chiang yu yü wu chih so pu tê tun, erh chieh tshun*[4]).[c]

Chuang Tzu is here suggesting that it is unwise to set the affections on anything short of the totality of Nature. Only the contemplation of Nature can free man from fear and from disappointment. He who can plunge into Nature determined to flinch from nothing as too trivial, too painful, too disgusting, too horrible to be named and investigated, will conquer fear, become invulnerable, and 'ride upon the clouds'. Thus:

Lieh Tzu could ride upon the wind. Cool and skilfully sailing, he would go on for fifteen days before returning. He could attain this happiness because he was not always seeking for it. Yet although he was able to dispense with walking, he still had to depend upon something (the wind). But supposing there is one who rides upon the normality of the universe, and drives before him the changes of the six energies (of the seasons) as his team, roaming thus through the realm of the inexhaustible (*jo fu chhêng thien ti chih chêng, erh yü liu chhi chih pien, i yu wu chhiung chê*[5])? What would he need to depend upon?[d]

Lest it should be thought that this interpretation is putting into the minds of the ancient Taoists ideas which they did not have, here is a decisive passage from *Huai Nan Tzu*, which states the situation quite plainly and without poetical imagery:

He who is of an intelligent nature is not terrified by any of Nature's operations; he who is wise by experience is not disturbed by any strange phenomena. The sage infers the far from the near, and concludes that the myriad things are based upon a single principle.[e]

---

[a] Indeed, the whole idea may well have been suggested by shamanism, for as Eliade (3) has shown in his comprehensive treatment of the subject, journeys to, and in, the heavens, ascensions, magical flight, and so on, were always a prominent element in the thought and beliefs of Asian shamans.

[b] Waley (6), p. 60, makes the good point that for the Confucians this word signified the peripatetic journeys of the philosophers from one feudal court to another.

[c] Ch. 6, tr. Fêng Yu-Lan (5), p. 116; Legge (5), vol. 1, p. 242.

[d] *Chuang Tzu*, ch. 1, tr. Legge (5), vol. 1, p. 168; Fêng Yu-Lan (5), p. 33; Lin Yü-Thang (1), p. 96, mod.

[e] Ch. 8, p. 3*a*, tr. Morgan (1), p. 84.

[1] 乘   [2] 遊   [3] 萬化   [4] 故聖人將遊於物之所不得遯而皆存
[5] 若夫乘天地之正而御六氣之辨以遊無窮者

And this theme never died out in Taoism, however much its philosophy became overlaid by magic and superstition. Third-century thinkers such as Ho Yen[1] and Chung Hui[2] maintained the doctrine of the impassibility of the sage. The *Kuan Yin Tzu* (probably the work of a +8th-century Taoist) says, many years later:

Minds occupied with fortune and misfortune may be invaded and controlled by devils. Minds occupied with love affairs may be attacked by lustful ghosts. Minds worried about deep waters may be subjected to the ghosts of the drowned. Minds prone to unrestrained activity may be attacked by mad ghosts. Minds occupied with oaths may be attacked by magical ghosts. Minds concentrated on drugs and tempting food (lit. baits) may be attacked by the ghosts of material things. Such ghosts take the shapes of shadows, wind, *chhi*, clay images, paintings, old animals or old vessels....Minds possessed by spirits sometimes see things strange and abnormal, or even fortunate, and successfully prognosticate by them; they are often proud, saying that they are not possessed by spirits but that they have a special Tao, yet later they die by wood, metal, rope, or falling into wells. Only the sage can control the spirits and not be controlled by the spirits. Only he can make use of all things, grasp their mechanisms, connect all things, disperse all things, defend all things. For every day the sage faces the facts of Nature, and his mind is untroubled.[a]

The analogy between the Confucians and the Stoics on the one hand, and the Taoists and Epicureans on the other, is not new.[b] In making it, however, a distinction has to be drawn between Epicureanism in the strict, atomistic, Lucretian sense, and Epicurea ism in the vulgar sense of the hedonistic pursuit of pleasure or avoidance of pain. This latter school does indeed seem to have had representatives in ancient China, and is associated with the name of Yang Chu[3] or Yang Sêng,[4] who lived shortly before Mencius and whose doctrines were vigorously combated by him. The traditional exposition of Yang Chu's views is contained in the seventh chapter of the *Lieh Tzu* book, translated partially by Legge (3)[c] and fully by Forke (2). Fêng Yu-Lan urges,[d] however, that this is a late +3rd-century interpolation, and basing his view on the fragments of descriptions of Yang Chu's views which remain in other books,[e] regards him as one of the earliest of the Taoists, who withdrew from the world for purely selfish reasons, considering the preservation of one's own tranquillity of mind and health of body as the most important thing in life. This doctrine was known as *chhüan sêng*,[5] 'completeness of living', or 'preservation of the intactness of life'. It was not an asceticism, for it aimed at the harmonious function of all the senses, avoiding both deprivation and excess. It considered that all inclinations, however gross or indefensible, were better than the perverse inclination for interference with others, for rule, power and authority. But since it had nothing of that interest in

[a] Ch. 4, p. 8*b*, tr. auct. Note the implicit reference to the terrors of mental diseases, and the influence of Buddhism.
[b] Cf. Dubs (7), p. 176.   [c] Pp. [95] ff.
[d] (1), vol. 1, pp. 133 ff.; vol. 2, pp. 195 ff.
[e] *Mêng Tzu*, VII (1), xxvi; *Lü Shih Chhun Chhiu*, ch. 99; *Han Fei Tzu*, ch. 50; *Huai Nan Tzu*, ch. 13.

[1] 何晏   [2] 鍾會   [3] 楊朱   [4] 陽生   [5] 全生

Nature which makes the Taoists so important for our present theme, it need not be enlarged upon here. Its only possible connection with the development of science lies in the possibility that it stimulated medical and hygienic practices, and for this we lack evidence.

(4) ACTION CONTRARY TO NATURE (*WEI*) AND ITS OPPOSITE (*WU WEI*)

So far we have spoken of Taoist contemplation, but what of Taoist action? In the passage from *Chuang Tzu* cited a few pages back, it was said that those who 'nourished their knowledge by their calmness' refrained all their lives from 'employing their knowledge in action contrary to Nature'. The words 'contrary to Nature' were bracketed, since they do not appear in the translation by Legge of the words *sêng erh wu i chih* wei *yeh*;[1] and the point at issue is the translation of the word *wei*.[2] Practically every translator and commentator has adopted the unmodified word 'action', so that the expression *wu wei*,[3] which became one of the greatest Taoist slogans, appears as 'non-action' or 'inactivity'. I believe that the majority of sinologists have been wrong here,[a] and that the meaning of *wu wei* was, so far as the early proto-scientific Taoist philosophers were concerned, 'refraining from activity contrary to Nature', i.e. from insisting on going against the grain of things, from trying to make materials perform functions for which they are unsuitable, from exerting force in human affairs when the man of insight could see that it would be doomed to failure, and that subtler methods of persuasion, or simply letting things alone to take their own course, would bring about the desired result. In support of this view I would quote the following passage from the *Huai Nan Tzu*:

> Some may maintain that the person who acts in the spirit of *wu wei* is one who is serene and does not speak, or one who meditates and does not move; he will not come when called nor be driven by force. And this demeanour, it is assumed, is the appearance of one who has obtained the Tao. Such an interpretation of *wu wei* I cannot admit. I never heard such an explanation from any sage....
> The configuration of the earth causes water to flow eastward, nevertheless man must open channels for it to run in canals. Cereal plants sprout in spring, nevertheless it is necessary to add human labour in order to induce them to grow and mature. If everything were left to Nature, and birth and growth were awaited without human effort, Kun and Yü[b] would have acquired no merit, and the knowledge of Hou Chi[c] would not have been put to use. What is meant, therefore, in my view, by *wu wei*, is that no personal prejudice (or private will) interferes with the universal Tao (*ssu chih pu tê ju kung Tao*[4]), and that no desires and obsessions lead the true courses of techniques astray (*shih yü pu tê wang chêng shu*[5]). Reason

---

[a] I can, however, claim Forke (12), p. 39, as on my side, and to some extent Duyvendak (7).
[b] Significantly Kun as well as Yü is said here to have acquired merit; see on, p. 117, and in Sect. 28*f*.
[c] The Lord of the Millet, an ancient agricultural folk-hero.

[1] 生而無以知爲也   [2] 爲   [3] 無爲   [4] 私志不得入公道
[5] 嗜欲不得枉正術

(li[1])[a] must guide action ($shih$[2]), in order that power may be exercised according to the intrinsic properties and natural trends of things (tzu-jan $chih$ $shih$[3])....

Now were there such a thing as using fire to dry up a well, or leading the waters of the Huai River uphill to irrigate a mountain, such things would be personal effort, and actions contrary to Nature (lit. turning one's back on Nature, pei tzu-jan[4]). This could be called yu wei[5] (action with useless effort). But using boats on water, sledges on sand, sleighs on mud, or litters on mountain-paths; digging channels for summer floods, arranging protections against winter cold, making fields on high ground and reserving low ground for marshes—such activities are not what may be called wei (or yu wei). The sages, in all their methods of action, follow the Nature of Things.[b]

If all the passages concerning wu wei in Taoist writings be re-examined from this point of view, they will be found to fit in well with the general proto-scientific character of the school. Thus the *Tao Tê Ching*, with its usual gemlike brevity, says: 'Let there be no action (contrary to Nature), and there is nothing that will not be well regulated (wu wei, $tsê$ wu pu $chih$[6]).'[c] *Chuang Tzu*[d] calls the refraining from activity contrary to Nature, wu wei, the lord of all fame, the treasury of all plans, able to bear all offices, and to make him who practises it the lord of all wisdom. And of the ancient kings he says that by not acting (contrary to Nature) they could use the whole world in their service, and might have done yet more, but by acting (contrary to Nature) they would not have been sufficient for the service required of them by the world.[e]

The *Kuan Tzu* book says: 'Heaven helps him who works according to (the sense of) Heaven, and opposes him who works in opposition to (the sense of) Heaven',[f] which reminds us of the definition of the Devil given by one of the Christian fathers (Hippolytus) as 'he who resists the cosmic process'. And about +300 Kuo Hsiang,[7] in his commentary on *Chuang Tzu*, wrote: 'Non-action does not mean doing nothing and keeping silent. Let everything be allowed to do what it naturally does, so that its nature will be satisfied.'[g] *Huai Nan Tzu* couples this with a statement of the macrocosm-microcosm doctrine, saying, 'To promote plans which do not comply with the will of Heaven, is to fight against man's own nature.'[h]

Kuo Hsiang's point of view is exaggerated in an amusing story related in the *Chin Shu*[i] about a +4th-century magician, Hsing Ling.[8] When a boy, he was told by his father to guard some growing rice against the depredations of oxen, but he failed to do

[a] Much will be said later about the full meaning of this word. Here we leave it as Morgan translated it, though some such expression as 'natural pattern' would be much better; see below, pp. 472 ff.

[b] Ch. 19, pp. 1*a*, 3*b* ff., tr. Morgan (1), pp. 220, 224, 225, mod. The *Wên Tzu* book speaks similarly, cf. Forke (13), p. 345.

[c] Ch. 3, tr. auct.

[d] Ch. 7; Legge (5), vol. 1, p. 266.

[e] Ch. 13; Legge (5), vol. 1, p. 333.

[f] Ch. 2, p. 1*b*, tr. auct.

[g] Ch. 11 (*Pu Chêng* ed. ch. 4C, p. 4*a*), tr. Dubs (19). See also ch. 13 (*Pu Chêng* ed. ch. 5B, p. 5*b*). Dubs, however, thinks that this was a later interpretation, and not, as I believe, in the minds of the Taoist thinkers from the first.

[h] Ch. 3, p. 16*b*, tr. Chatley (1).

[i] Ch. 95, p. 10*b*.

[1] 理   [2] 事   [3] 自然之勢   [4] 背自然   [5] 有爲
[6] 無爲則無不治   [7] 郭象   [8] 幸靈

so, and then replanted the damaged rice. On being reproached, he said that it was in the nature of the ox to eat, but equally it was in the nature of the rice to grow up; he had therefore allowed nature to take its course and repaired the damage as well as he could.

This refraining from going against the grain of things, which only wears people out fruitlessly, was consciously modelled on the operation of the Tao of Nature itself, which does nothing and yet accomplishes everything. Thus the *Chuang Tzu* says:

(The operations of) heaven and earth (proceed in the) most beautiful (way), but they do not talk about them. The four seasons observe the clearest laws, but they do not discuss them.[a] All things have their intrinsic principles, but they say nothing about them. The Sages trace out the beautiful (operations of) heaven and earth, and penetrate into the intrinsic principles of all things. Therefore the Perfected Man does nothing (contrary to Nature) and the Sage makes no invention (contrary to Nature). They look to Heaven and Earth as their model.[b]

At a later stage[c] I shall suggest that one of the deepest roots of the concept of *wu wei* may lie in the anarchic nature of primitive peasant life; plants grow best without interference by man; men thrive best without State interference.

As usual, other schools adopted the same phraseology but gave it a different meaning. The Legalists, for example, interpreted *wu wei* as the non-activity of a ruler who devolves his power on his high officials, or who arranges for automatic settling of disputes by the imposition of a code of written law.[d] The Han Confucians emphasised the quality of effortlessness in *wu wei*.[e] By later vulgarisation, and under the influence of Buddhist meditation techniques, it did come to mean the avoidance of every kind of activity, and the historical records give many descriptions of statesmen such as Tshao Tshan[1] in −193 who sat in his bureau minimising business to the utmost (*fu chung wu shih*[2]),[f] or Chi Yen[3] in −134.[g] No doubt in many cases the methods of 'letting alone' were much more successful than activist policies would have been, but the ultimate misunderstanding of the words *wu wei* led without doubt to abuses, and helped to bring Taoism into discredit. Nevertheless, the +3rd-century commentary of Hsiang Hsiu and Kuo Hsiang on the *Chuang Tzu* upheld the true view, saying that everything should be allowed to follow its natural tendency, and that the artisan's 'non-activity', for instance, consisted in the pursuit of his art.[h]

a So we write, following Legge, but there is much more to be said about this seemingly innocent sentence, as will appear below in Section 18 on the Laws of Nature (p. 546).
b Ch. 22, tr. Legge (5), vol. 2, p. 60; Lin Yü-Thang (1), p. 68, mod.
c In the section on Laws of Nature (p. 576).
d Fêng Yu-Lan (1), vol. 1, pp. 292, 330 ff.   e Fêng Yu-Lan (1), vol. 1, p. 375.
f TH, p. 312.   g TH, p. 431.
h *Pu Chêng* ed. ch. 5B, p. 5b, tr. Fêng Yu-Lan (1), vol. 2, p. 216.

1 曹參   2 府中無事   3 汲黯

(5) TAOIST EMPIRICISM

Wei, then, was 'forcing' things, in the interests of private gain, without regard to their intrinsic principles, and relying on the authority of others. *Wu wei* was letting things work out their destinies in accordance with their intrinsic principles. To be able to practise *wu wei* implied learning from Nature by observations essentially scientific. Hence we find ourselves, by an insensible transition, at the beginning of that thread of *empiricism* which was of capital importance for the whole development of science and technology in China. Here there are two passages, which, though rather long, cannot be omitted. The first, from the *Huai Nan Tzu*, opens with an attack on the Legalists, which will be better appreciated by reference to the section on them below (p. 204):

Now as regards the methods of government of Shen (Pu-Hai), Han (Fei Tzu) and Shang Yang, they pulled things up by the roots, neglected their origins (causes, *pên*[1]), and did not thoroughly investigate their coming into being. Why did they act thus? Increasing the five punishments to a severity which was contrary to the foundations of the virtue of the Tao (*pei Tao tê chih pên*[2]) they sharpened the points of weapons and cut down the greater part of the people like straw. Filled with satisfaction, they considered that they had put the world in order. But this was like adding fuel to fire or trying to empty an ever-flowing spring; planting *tzu*-trees[a] round wells so that the buckets cannot go up and down; or willow-trees along canals, so that boats cannot go past—in three months they will be cut down.

Why make such mistakes? These wild and masterful men cared nothing for origins. The (Yellow) River, though it has nine bends, is always flowing to the sea, because it has an inexhaustible source in the Khun-lun mountains. Floodwaters spread all over the land, but if there is no rain for ten days or a month, they dry up because they have no source. For example, Yi the Archer went to ask Hsi Wang Mu[b] for the medicine of immortality (*pu ssu chih yao*[3]), but Chang Wo (his wife) stole it, ate it and flew to the moon; thus he was very sad at his irreparable loss. Why? Because he did not know where the medicine of immortality grew. So therefore, rather than begging or borrowing fire, you had better take a burning mirror, and rather than drawing water from other people's wells, you had better dig one yourself.[c]

In other words, the methods of the draconic Legalists were basically contrary to the mainsprings of human behaviour and so were bound to fail. Those who pay insufficient regard to what may be ascertained about causes and intrinsic principles in Nature will exhaust themselves in trying to do the impossible. Yi the Archer, relying on authority, travelled far out of his way, yet the medicine of immortality was all the time growing just outside the door of his house. He should have investigated what was near at hand and not have embarked upon a long and fruitless journey to the

---

[a] *Catalpa* spp., B II, 508; III, 319. Their branches or roots would interfere with the wells.
[b] Legendary goddess in the West.   [c] Ch. 6, p. 10*a*, tr. auct.

[1] 本   [2] 背 道 德 之 本   [3] 不 死 之 藥

goddess in the West.[a] He practised *wei*, his wife practised *wu wei*. Finally, go to Nature and not to Authority, make your own fire and dig your own well.

The second passage is a remarkable statement of empiricism; it comes from the *Lü Shih Chhun Chhiu* and may be considered one of the finest affirmations of the ancient Taoist technologists against the politicians and sophists of their time.

To know that one does not know—that is high wisdom. The fault of those who make mistakes is that they think they know when they do not know. In many cases phenomena seem to be of one sort (alike) when they are really of quite different sorts. This has caused the fall of many States and the loss of many lives.

Among the vegetables there are *hsing*[b] and *lei*.[c] If you eat of either alone, you may die. If you mix them with other vegetables, they may prolong life. If you mix together many species of *chin*[d] (hemlock) the mixture is not poisonous.

Lacquer is liquid, water is also liquid, but when you mix the two things together, you get a solid. Thus if you moisten lacquer it will become dry. Copper is soft, tin is soft, but if you mix both metals together they become hard. If you heat them they will again become liquid. Thus if you wet one thing it becomes dry and solid; if you heat a (hard) thing it becomes liquid. Thus one may see that you cannot deduce the properties of a thing merely by knowing the properties of the classes (of its components) (*lei ku pu pi kho thui chih yeh*[1]).

A small square is of the same class as a big square. A little horse is of the same class as a big one. But little knowledge is not of the same class as great knowledge. In the State of Lu there was a man called Kungsun Cho who said he could raise the dead. When they asked him how, he replied, 'I can heal hemiplegia (apoplexy). If I gave a double dose of the same drug, I could therefore raise the dead.' But among things there are some which can have small-scale effects, but not large-scale ones, and other things which can perform the half but not the whole.

A swordsmith said, 'White metal (tin) makes the sword hard, yellow metal (copper) makes it elastic. When yellow and white are mixed together, the sword is both hard and elastic, and these are the best ones.' Somebody argued with him, saying, 'The white is the reason why the sword is not elastic, the yellow is the reason why the sword is not hard. If you mix yellow and white together the sword cannot be both hard and elastic. Besides, if it were soft it would easily bend, and if it were hard it would easily break. A sword which easily bends and breaks, how could it be called a sharp one?' Now a sword does not change its nature, yet some may call it good and some bad; that is only a matter of opinion. If you know how to distinguish between good and bad arguments, nonsense will cease. If you do not, then there is no difference between Yao and Chieh.[e] This is what true officials are always worrying about, and the reason why many good men have been dismissed....

Kaoyang Ying was having a house built. His mason said, 'It won't do to use wood that is too green; when it is plastered it will warp. If you use fresh wood, the house may look all right for a short time, but it will be certain to fall down before long.' Kaoyang Ying replied,

---

[a] 'We carry with us the wonders that we seek without us', Sir Thomas Browne was to say centuries later; 'there is all Africa and her prodigies in us.'

[b] *Asarum* spp., B III, 40.　　　　　　　　　[c] *Rubus* spp., B II, 131.

[d] *Aconitum* spp., B II, 134.

[e] Types of good and bad emperors.

[1] 類固不必可推知也

'(On the contrary) according to your own statement, the house cannot fall down. The drier the wood the harder it will be, the drier the lime the lighter it will be. If you put something which is always getting harder with something which is always getting softer, they could not possibly hurt each other.' The mason did not see how to answer this, so he accepted the order and built the house. When it was done it looked well, but very soon it fell to pieces. Kaoyang Ying liked such small sophistries, but had no understanding of the great principles (of Nature) (*pu thung hu ta li yeh*[1]).[a]

If Chi Ao and Lu Erh (two of the fastest horses) started running due west, so that they had the sun behind them, they would find, when evening came, that it had got ahead of them. So there are things which the eye cannot see, and which the understanding cannot apprehend, and which cannot be reckoned up in numbers. We do not know their how and their why. (Therefore) the Sage follows (Nature) in establishing social order, and does not invent principles out of his own head (*shêng jen yin erh hsing chih, pu shih hsin yen*[2]).[b]

We shall think again of the swordsmith and the mason who knew better than the sophists who argued with them when we study the history of Chinese metallurgy and engineering. However fast the horses of logical rationalism might run, Nature would get ahead of them in the end, confounding them and justifying Taoist empiricism.

This theme goes sounding down through the centuries of Chinese thought. In the book *Shen Tzu*,[3] attributed to Shen Tao,[4] one of the lesser-known philosophers just anterior to Chuang Tzu in the Warring States period, but most of which was probably written much later[c] (perhaps sometime between Han and Thang), we have:

As regards the people who protect and manage the dykes and channels of the nine rivers and the four lakes, they are the same in all ages; they did not learn their business from Yü the Great, they learnt it from the waters.[d]

And in the *Kuan Yin Tzu*, a Taoist book of the Thang (perhaps +8th century) it is said:

Those who are good at archery learnt from the bow and not from Yi the Archer. Those who know how to manage boats learnt from boats and not from Wo (the legendary mighty boatman). Those who can think learnt for themselves, and not from the Sages.[e]

From the same time, too, we have a famous story regarding the artist Han Kan.[5] The greatest painter of horses in the Thang period, he had been called to court in his youth by the emperor, who offered him the tuition of the most celebrated painters

[a] The mason would have appreciated the discussion in Vitruvius VII, i, 2.
[b] Ch. 150 (vol. 2, p. 158), tr. R. Wilhelm (3), p. 434; eng. auct. mod.
[c] Fêng Yu-Lan (1), vol. 1, p. 155.   [d] P. 11*b*, tr. auct.
[e] Ch. 5, p. 11*a*, tr. auct.

[1] 不通乎大理也   [2] 聖人因而興制不事心焉   [3] 慎子   [4] 慎到
[5] 韓幹

of the age, but he turned his back upon them, and asked only to be allowed to frequent the imperial stables.[a] We shall shortly recur to this theme in considering the curious series of 'knack' passages which occur in Taoist writings (p. 121). It must in any case be regarded as very significant of the whole emphasis of Chinese culture on practical technology rather than abstract science, and it cannot be fanciful to connect it with the great gifts of that culture to the West during the first thirteen centuries of our era.

(e) CHANGE, TRANSFORMATION AND RELATIVITY

Concentrating their interest upon Nature as they did, it was inevitable that the Taoists should be obsessed by the problem of Change. This extended to certain other schools also, especially the Naturalists and the Logicians, of whom we shall speak below (pp. 232 and 185). A number of technical terms were developed, such as *pien*,[1] *hua*,[2] *fan*[3] and *huan*,[4] the exact meanings of which are sometimes difficult to differentiate.[b] *Fan* and *huan* both have the significance of 'reaction' or 'return', as when some kind of reverse change takes place as the result of a former action, or when a cyclical process brings back the phenomena to a state similar to that at the beginning, or identical with it. The exact difference between *pien* and *hua* is perhaps more uncertain. In modern Chinese usage, *pien* tends to signify gradual change, transformation or metamorphosis; while *hua* tends to mean sudden and profound transmutation or alteration (as in a rapid chemical reaction)—but there is no very strict frontier between the words. *Pien* could be used of weather changes, insect metamorphosis,[c] or slow personality transformations; *hua* may refer to the transition points in dissolving, liquefying, melting, etc., and to profound decay. *Pien* tends to be associated with form (*hsing*[5]) and *hua* with matter (*chih*[6]). When a snowman melts, the form changes (*pien*[1]) as the snow melts (*hua*[2]) to water. In the Sung dynasty, Chhêng I explained *pien* as implying inward change with full or partial conservation of the external Gestalt or form, and *hua* as fundamental change in which the outward appearance is also altered.[d] The difficulty is, of course, to know how far these distinctions can be applied to the terms as used by the Taoist philosophers of the −4th century. The problem seems to be connected with the two trends which Demiéville (3c) has descried throughout Chinese intellectual history, and which he characterises as 'subitisme' and 'gradualisme'; these may later appear in the form of social revolution against slow development, or more often of sudden conversion (as in Buddhism and Taoism) against Confucian sobriety and this-worldliness.[e]

[a] I am grateful to my old friend Dr Chi Chhao-Ting for remembering this. For some of Han Kan's paintings see Sirén (6), vol. 1, plates 60, 61, 62.

[b] In this paragraph I owe thanks for clarifications to Dr E. Balazs and Dr Lu Gwei-Djen. Another word, *kai*,[7] will be discussed later on.

[c] But in ancient times *hua* was also used of all biological transformations (cf. *Pao Phu Tzu (Nei Phien)*, chs. 2 and 3; *TPYL*, chs. 887, 888).

[d] Forke (9), p. 93.   [e] Cf. Fêng Yu-Lan (1), vol. 2, pp. 283, 387.

[1] 變   [2] 化   [3] 反   [4] 還   [5] 形   [6] 質   [7] 改

Take a typical phrase from the *Kuan Tzu* book (chapter 49): '*Shêng jen pien erh pu hua*';[1] should we say 'The sage changes inwardly but is not transformed outwardly'? Or 'slowly and deliberately, without jumping to conclusions'? The passage goes on: 'He follows the (ramifications of) things, but does not abandon all his former principles (*tshung wu erh pu i*[2]).' The meaning may thus be that the sage slowly adapts himself to the world of Nature, as experience dictates, without modification of his fundamental outlook. But it was on the changes in Nature, even more than those in the sage, upon which the Taoists concentrated their attention. They were especially impressed by cyclical change, not only of the seasons and of birth and death, but as visible in all kinds of observable cosmic and biological phenomena. This is what Hou Wai-Lu calls the doctrine of cyclically recurring differences, *hsün huan i pien lun*.[3]

We can immediately find a parallel once more between Lao Tzu and Lucretius. The *Tao Tê Ching* says:

> ...Among the creatures of the world some go in front, some follow;
> Some blow hot when others would be blowing cold,
> Some are feeling vigorous just when others are worn out
> Some are loading just when others are delivering,
> Therefore the sage discards the 'absolute', the 'all-inclusive', the 'extreme'.[a]

And the *De Rerum Natura*:

> omnia migrant,
> omnia commutat natura et vertere cogit.
> namque aliut putrescit et aevo debile languet,
> porro aliut clarescit et e contemptibus exit.[b]

This is part of the dialectics of Nature, the ever-recurring opposition between the old decaying factors and the new arising factors at any given stage. Cornford[c] gives further striking parallels from Greek writers.

The *Tao Tê Ching* describes the cyclical changes in no uncertain terms. For example, chapter 58:

'Prosperity tilts over to misfortune, and good fortune comes out of bad.' Who can understand this extreme turning-point? For it recognises no such thing as normality (*wu chêng*[4]). Normality changes into abnormality (*chêng fu wei chhi*[5]). The good changes into the diabolical. Too long has mankind been bewildered by these changes....[d]

[a] Ch. 29, tr. Waley (4).

[b] v, 830 ff.

> All things depart;
> For Nature changes all, and forces all
> To transmutation; lo, this moulders down,
> Aslack with weary eld, and that, again,
> Prospers in glory, issuing from contempt (tr. Leonard).

Cf. Needham (3), p. 191.   [c] (1), p. 165.

[d] Tr. auct. following Hou Wai-Lu. Duyvendak (18) now gives a similar rendering.

[1] 聖人變而不化   [2] 從物而不移   [3] 循環異變論   [4] 無正

[5] 正復爲奇

and chapter 40:

'Returning' is the (characteristic) movement of the Tao (*fan chê Tao chih tung*[1]).

Chuang Tzu also emphasises this:

Life is the follower of death, and death is the predecessor of life; but who knows their cycles (and the connections between them, i.e. the Tao)? Man's life is due to the conglomeration (*chü*[2]) of the *chhi*;[3] and when they are dispersed (*san*[4]) death occurs. Since death and life thus attend upon each other, why should I account (either of) them an evil?...(Life) is accounted beautiful because it is spirit-like and wonderful. (Death) is accounted hateful because it is foetid and putrid. But the foetid and putrid, returning, is transformed again into the spirit-like and wonderful; and then the reverse change occurs once more. Therefore it is said that all through the universe there is one *chhi*,[3] and therefore the sages prized that unity.[a]

One can see the tendency to incorporate an appreciation of, and resignation to, these changes, as part of that understanding of Nature which was at the basis of Taoist ataraxy, or calmness of mind.[b] The Tao is that which accompanies all other things and meets them, which is present when they are overthrown and when they come to their perfection; it is the Tranquillity at the centre of all Disturbances.[c] It produces fullness and emptiness (*ying hsü*[5]), but it is neither fullness nor emptiness; it produces withering and killing (*shuai sha*[6]), but it is neither withering nor killing; it produces roots and branches (*pên mo*[7]), but it is neither root nor branch; it produces accumulation and dispersion (*chi san*[8]), but it is itself neither accumulated nor dispersed.[d] Numerous parallel passages may be found, e.g. in *Lieh Tzu*,[e] where every 'end' is to be seen as the 'beginning' of something else; *Kuan Tzu*,[f] *Huai Nan Tzu*,[g] etc.

The great difficulty about change is that it is so hard to know when the limits of one category have been overpassed and the next category entered. Such insensible transitions have therefore always been the thorn in the flesh of formal logic, as we can see by comparing the Procrustean beds of Victorian science with the paradoxical yet powerful and mathematically expressible conceptions of science today, or, going further back, the flexibility of 17th-century science with the rigid Aristotelian formalism of the Middle Ages from which it had successfully struggled to free itself.[h] The dialectical reconciliation of contradictions in a higher synthesis, which is so often

---

a Ch. 22, tr. Legge (5), vol. 2, p. 59.
b More explicit statements are in *Chuang Tzu*, ch. 6 (Legge (5), vol. 1, p. 249) and ch. 27 (Legge (5), vol. 2, p. 144). In some of them the word *chhan*[9] is used as a technical term for linked successions of changes.
c *Chuang Tzu*, ch. 6 (Legge (5), vol. 1, p. 246).
d *Chuang Tzu*, ch. 22 (Legge (5), vol. 2, p. 67).
e Ch. 5, p. 1a (L. Giles (4), p. 82).   f Ch. 49.
g Esp. ch. 1.   h Cf. Needham (3).

[1] 反者道之動   [2] 聚   [3] 氣   [4] 散   [5] 盈虛   [6] 衰殺
[7] 本末   [8] 積散   [9] 禪

seen in science, appears with much clarity in the Taoist writings, especially in the second chapter of *Chuang Tzu*.[a] Among the numerous affirmations and denials, the sage does not take sides, but recognising that truth may be distributed among many opinions, forms his judgment in the light of Heaven[b] (*shih i shêng jen pu yu, erh chao chih yü Thien*[1]). Those who put forward views fail to realise that all are partly right and partly wrong; they can only be judged from the 'axis' of the Tao, around which all Nature moves.[c] The harmonising of conflicting opinions is only to be found in the invisible operations of Heaven, and by following these back into the illimitable past.[d] The *locus classicus* on the subject is the parable of the monkeys.

To wear out one's spirit and intelligence in order to unify things without knowing that they are already in agreement—this is called 'Three in the Morning'. What is meant by this? A keeper of monkeys said with regard to their rations of nuts that each monkey was to have three in the morning and four at night. But at this the monkeys were very angry. Then the keeper said they might have four in the morning but three at night, and with this arrangement they were all well pleased. His two proposals were substantially the same, but one made the creatures angry and the other pleased. Thus the sages harmonise the affirmations 'it is' and 'it is not', and rest in the natural equalisations of Heaven (*Shih i shêng jen ho chih i shih fei, erh hsiu hu thien chün*[2]). This is called 'following two courses at once' (*Shih chih wei liang hsing*[3]).[e]

This dialectical quality in Chuang Tzu, and in the Taoists generally, has been the subject of an interesting discussion by Thang Chün-I (*1*), who attempts a detailed comparison with Hegel. Both Chuang Chou and Hegel would have subscribed to the view of change as eternal, and reality as process, and both would have denounced that perennial philosophy which has sought to deny the reality of change or to interpret it solely in terms of a changeless eternal. We shall later find traces at least of a similar recognition of process and dialectic in the logic of the Mohist school, and these tendencies should be borne in mind with a view to the ultimate casting of accounts with which the present book must end. There we shall have to discuss, for example, the extent to which the structure of the Chinese language itself encouraged these ancient thinkers to develop an approach, not only to the type of thinking usually called Hegelian, or approximating to that of Whitehead, but even more fundamentally and exactly, to what is now being investigated under the head of combinatory logic.

On the whole the Taoists avoided the elaboration of a cosmogony, wisely considering that the original creative operations of the Tao must remain for ever

---

a Cf. Waley (6), p. 26.   b Legge (5), vol. 1, p. 183.
c Legge (5), vol. 1, p. 184.   d Legge (5), vol. 1, pp. 195, 196.
e Ch. 2, tr. Legge (5), vol. 1, p. 185; Fêng Yu-Lan (5), p. 52; Lin Yü-Thang (1), p. 244, mod.

[1] 是以聖人不由而照之於天   [2] 是以聖人和之以是非而休乎天均
[3] 是之謂兩行

unknowable. The *Tao Tê Ching*, however, has one passage embodying a cosmogonic myth (chapter 42):

The Tao produced one, the one produced two, the two produced three, and the three produced the ten thousand things (everything). The ten thousand things are all backed by the Yin and embrace the Yang[a] (i.e. stand between these two forces), and are harmonised by the *Chhi* (*pneuma*) of the void.... Reducing a thing (*sun*[1]) often increases it (*i*[2]); and making it flourish often leads to its decay....[b]

Apart from the reference to cyclical change, coming-into-being and passing-away, which again reminds us of Aristotle's περὶ γενέσεως καὶ φθορᾶς, the meaning of the statement is not at all clear. Erkes (3), who has devoted special attention to the passage, sees in it a reference to the idea of the Cosmic Egg, which appears to have existed in ancient Chinese, as in ancient European, thought.[c] Another and longer, but rather confused, cosmogonic passage occurs at the beginning of chapter 2 of *Huai Nan Tzu*. These ancient ideas have not yet, it seems, received detailed comparative examination and elucidation.

A story known among the people in modern China tells of a Buddhist monk who was offered, when nearly starving, only an egg. For a long time he refused to eat it, but finally did so, and wrote the following poem on the wall:

> Chaos (*hun-tun*) and Alpha and Omega (*Chhien Khun*) I take into my mouth
> Still without skin, without flesh, without down;
> Let an old monk, then, bring you to the Western Heaven,
> And spare you the edge of the knife which you would find among men.[d]

Much more interesting from the scientific point of view is the fact that the Taoists elaborated what comes very near to a statement of a theory of evolution. At the least, they firmly denied the fixity of biological species. The principal passage occurs in the eighteenth chapter of *Chuang Tzu*. It has been the despair of translators, but fortunately we have a version from the master-hand of Hu Shih:

All species (*chung*[3]) contain (certain) germs (*chi*[4]).[e] These germs, when in water, become *chüeh*.[5][f] In a place bordering upon water and land they become (lichens or algae, like what we

---

[a] Waley (4), considering that the book of Lao Tzu was too early in date to have been influenced by the *Yin-Yang* theory, translates the words still as 'sunny side' and 'shady side'. We shall meet with an echo of the idea in Wang Khuei (cf. Section 39 on zoology below).

[b] Tr. auct. following Hou Wai-Lu. Cf. Duyvendak (18) and Waley (4).

[c] On the Chinese side see A. Kühn (1), p. 29, and W. Eberhard (6). There is apparently a connection with the conception of *hun-tun*, which is of great importance, see below, p. 115. On the Orphic cosmic egg, cf. Needham (2), pp. 9, 10, with references to A. B. Cook (1); together with Freeman (1) and Diels-Freeman (1).

[d] This was remembered from his grandfather by one of us (W.L.). See below, pp. 107, 313.

[e] Needless to say, the word 'germ' is not to be taken here in any precise modern sense, such as that of bacteria.

[f] A minute organism, not further specified, as tiny as a cross-section of a silk fibre.

[1] 損   [2] 益   [3] 種   [4] 幾   [5] 蠨

call the) 'clothes of frogs and oysters'. On the bank they become *ling-hsi*.[1][a] Reaching fertile soil the *ling-hsi* become *wu-tsu*.[2][b] The roots of this give rise to the *chhi-tshao*;[3][c] the leaves become *hu-tieh*,[4][d] or *hsü*.[5][e] The *hu-tieh* later changes into an insect, born in the chimney-corner, which has the appearance of newly formed skin. Its name is *chhü-to*.[6][f] After a thousand days the *chhü-to* becomes a bird called *kan-yü-ku*;[7][g] the saliva of which becomes the *ssu-mi*.[8][g] The *ssu-mi* becomes a wine-fly (*shih-hsi*[9]),[g] and from this in turn comes the *i-lu*.[10][g] The *huang-kuang*[11][g] are produced from the *chiu-yu*.[12][g] Mosquitoes (*mou-nei*[13]) are produced from rotting *huan*.[14][g] *Yang-hsi*,[15][g] paired with the *pu-hsün-ju-chu*,[16][h] produces the *chhing-ning*,[17][i] which produces the *chhêng*,[18][j] which (ultimately) produces the horse, which (ultimately) produces man. Man again goes back into the germs.[k] All things come from the germs and return to the germs.[l]

The Taoist observers were certainly acquainted with such phenomena as insect metamorphosis, and no doubt drew the same inaccurate conclusions as the early Europeans from the appearance of insects in decaying animal bodies and vegetable matter ('spontaneous generation').[m] They then extended their conceptions of the astonishing transformations which may take place in Nature to other, more imaginary and less well-based, examples,[n] which we shall examine later in Section 39. Once this conviction of radical transformation became established, it was not a far cry to a belief in slow evolutionary changes, whereby one species of animal or plant arose from another. This idea comes out quite clearly from the remarkable passage just quoted, and was, moreover, applied also (as the *Huai Nan Tzu* shows)[o] to the slow growth and generation by successive changes of ores and metals in the earth. Such an application of the concept of transformations to what we should now call the inorganic world was also found in European thought, but appeared very early in

[a] A now unidentifiable plant, if it was a plant.

[b] Lit. 'crows'-feet', but the plant is not now identifiable.

[c] This name-combination is now applied to cerambycid wood-boring beetle larvae, but there is no telling what it meant in Chuang Tzu's time.

[d] Now any papilionid or pierid butterfly. Chuang Tzu may have noted (and misunderstood) the mimicry of leaves by certain Lepidoptera.

[e] Now means a preparation of salted crabs; perhaps here the particular species of crab that was so used.

[f] Inexplicable. *Chhü* since the +16th century has meant a bird, one of the mynahs, cf. R 296.

[g] Unidentifiable. All the animals it turns into are supposedly insects.

[h] An insect parasitic on bamboos?

[i] If written with the 'insect' radical (Rad. 142), as they are not in the text, *chhing* means dragonfly and *ning* cicada. What insect Chuang Tzu had in mind cannot be stated.

[j] Later meant the leopard, cf. R 352.   [k] Reading *chi*[19] for *chi*.[20]

[l] Tr. Hu Shih (2), p. 135; cf. Legge (5), vol. 2, p. 9. There is a closely parallel passage in *Huai Nan Tzu*, ch. 4, p. 11*b* (tr. Erkes (1), p. 77), but the names of the plants and animals are all quite different. Also *Lieh Tzu*, ch. 1, p. 6*b* (tr. Wieger (7), p. 73; but omitted by R. Wilhelm (4), p. 4); here the names agree but the text is expanded.

[m] See on, pp. 481, 487, for statements of belief in it (Chu Hsi). For current views see Pirie (1).

[n] Extensive lists in *TPYL*, chs. 887, 888.   [o] Ch. 4, tr. Erkes (1), p. 79.

[1] 陵舄	[2] 烏足	[3] 蠐螬	[4] 蝴蝶	[5] 胥	[6] 鴝掇
[7] 乾餘骨	[8] 斯彌	[9] 食醯	[10] 頤輅	[11] 黃軦	[12] 九猷
[13] 瞀芮	[14] 蠸	[15] 羊奚	[16] 不筍入竹		[17] 青寧
[18] 程	[19] 幾	[20] 機			

China, and furnishes a link between Chuang Tzu's biological conceptions and the attempt to hasten these changes by active interference, i.e. Alchemy (see on, Sect. 33).[a] Another point of interest in this passage is the use of the word 'germs', meaning the smallest imaginable particles of living matter; the term employed, *chi*, is not one of common occurrence, but appears in the *I Ching* (Book of Changes) with the sense of the minute embryonic beginnings of things, out of which good and evil come.[b] Etymologically it derives from a pictorial representation of two embryos. The fact of its use by Chuang Tzu is important in view of the general absence of atomistic ideas in Chinese thought, a question which will have to be referred to again (Section 26 b).

This is not all that may be found in *Chuang Tzu* concerning biological change.[c] Several passages show a recognition of different aptitudes as having arisen by adaptation to different environments, e.g. chapter 17 (horses, wild cats and owls),[d] and chapter 2 (the 'right' habitat) which we have already quoted (p. 49). But there is also even an approach to the idea of natural selection. This arises from a number of passages[e] in which the advantages of being useless are pointed out. Trees attain great size and longevity only by being of no use to anyone; thus do they avoid being cut down. An old sacrificial ritual is referred to, forbidding the sacrifices of animals and men having certain defects. A spirit-like tortoise would have preferred to enjoy its quiet life, but its shell was judged useful for divination, so it was killed to be hung up in the royal ancestral temple. Pigs, rather than being fattened for ceremonies in which their bodies would play a leading role, would prefer to live, though on poor food. While these passages were no doubt part of the argument for the withdrawal of Taoists from active social life, they nevertheless represent a certain appreciation of the 'survival of the fittest'.[f]

We doubt whether these aspects of ancient Taoist thought have been taken into account by those who have written on the history of evolution theories.[g]

Among the aspects of the world of living creatures which particularly impressed the Taoists was the great differences between their forms and functions. What was good for one was bad for another (cf. the oecological passage, already quoted, in which

[a] Note also how Ko Hung, the great Taoist alchemist (*c.* +300), seeks to refute any fixity of species and emphasises the numerous semi-legendary zoological transformations, in order to support his belief in the possibility of changing (i.e. prolonging) man's natural span of life (*Pao Phu Tzu*, ch. 2, tr. Feifel (1), pp. 142 ff.).

[b] Great Appendix, part 2, sect. 5 (tr. R. Wilhelm (2), vol. 2, p. 261); see Hu Shih (2), p. 34.

[c] From its title, the short monograph of Stadelmann (1) would appear to be an appreciation of the strong biological interests of the Taoists, but unfortunately it is fanciful and not very helpful.

[d] Legge (5), vol. 1, p. 381.

[e] *Chuang Tzu*, ch. 1 (Legge (5), vol. 1, p. 174); ch. 4 (Legge (5), vol. 1, pp. 217, 220); ch. 17 (Legge (5), vol. 1, p. 390); ch. 19 (Legge (5), vol. 2, p. 18); ch. 20 (Legge (5), vol. 2, p. 27); ch. 26 (Legge (5), vol. 2, p. 137). Parallel passage in *Mo Tzu*, ch. 1 (Mei (1), p. 3).

[f] Cf. *Lieh Tzu*, ch. 8, and *Lun Hêng* (Forke (4), vol. 1, pp. 92, 105; vol. 2, p. 367), where we find a clear appreciation of the struggle for existence in Nature, sharpness of claw and quickness of pounce being valuable for survival. Echoes in later Tibetan folklore, R. Cunningham (1), p. 53. Elaborations by Emperor Yuan of the Liang (+550) in *Chin Lou Tzu*, ch. 4, p. 19a.

[g] E.g. Osborn (1). They have been pointed out, however, by Chinese philosophers such as Hu Shih (3), and geologists such as Chang Hung-Chao (2) in his article on Darwin and Chuang Tzu.

anthropocentric judgments were shown to be absurd when applied to the non-human world), and it was realised that the universes and time-scales of the various animal species were extremely different. This denial of anthropocentrism expressed itself most strongly in that famous passage with which the *Chuang Tzu* book opens (chapter 1):

In the Northern Ocean there is a fish, by the name of *khun*,[1] which is many thousand *li*[a] in size. This fish metamorphoses into a bird by the name of *phêng*,[2] whose back is many thousand *li* in breadth. When the bird rouses itself and flies, its wings obscure the sky like clouds. When the bird moves itself in the sea, it is preparing to start for the Southern Ocean, the Pool of Heaven. A man named Chhi Hsieh,[3] who recorded marvellous things, said 'When the *phêng* is moving to the Southern Ocean, it flaps along the water for three thousand *li*. Then it ascends on a whirlwind up to a height of ninety thousand *li*, for a flight of six months' duration.' (Its movement is just as natural as that of) the dust-devils in the fields (*yeh ma*,[4] lit. wild horses), or that of the motes of dust (*chhen ai*[5]) (in sunbeams), or that of the living things which are blown against one another in the air. We do not know whether the blueness of the sky (for example) is its original colour, or is simply caused by its infinite height. What the *phêng* sees (as the earth) from above (probably) looks just the same (as the sky does to us from below).

Without sufficient water, a large boat cannot be floated. But when a cupful of water is upset into a small hole, a mustard-seed will float on it. Try to float the cup on it, and it will stick, because the water will be too shallow to support so large a vessel. Without sufficient density (*chi hou*,[6] lit. thickness of condensation), the wind would not be able to support the large wings. Therefore (when the *phêng* has ascended to) the height of ninety thousand *li*, the wind is all beneath it. Then, with the blue sky above, and no obstacle ahead of it, it mounts upon the wind, and starts for the south.

A cicada and a young dove laughed at the *phêng*, saying, 'When we make an effort, we fly up to the trees. Sometimes, not able to reach them, we fall to the ground midway. What is the use of going up ninety thousand *li* in order to start for the south?'...But what should these two small creatures know about the matter? Small knowledge is not to be compared with great, nor a short life to a long one. The morning mushroom does not know what happens between the beginning and the end of a month, the *hui-ku*[7b] knows nothing of the alternation of spring and autumn. These are instances of short spans of life. But south of Chhu State there is the *ming-ling*[8c] whose spring is five hundred years and whose autumn is equally long. Anciently there was the *ta-chhun*[9c] whose spring and autumn were each eight thousand years. And among men Phêng Tsu[10d] was especially renowned for his length of life—if all men were to wish to match him, would they not be miserable?[e]

[a] A *li* is now ½km. Whatever it was in Chuang Tzu's time, he meant to indicate that the fish was unimaginably enormous.

[b] Since the +16th century this word-combination has been stabilised to mean the mole-cricket; whether it meant exactly that to Chuang Tzu cannot be said.

[c] Supposed to be names of trees, not now identifiable.

[d] The Chinese Methuselah.

[e] Tr. Fêng Yu-Lan (5), pp. 27ff., mod.

[1] 鯤   [2] 鵬   [3] 齊諧   [4] 野馬   [5] 塵埃   [6] 積厚   [7] 蟪蛄

[8] 冥靈   [9] 大椿   [10] 彭祖

A parallel passage, longer and not quite so entertaining, is in *Lieh Tzu*,[a] and others with the same aim could be cited.[b] A point which should not be lost sight of is a possible political, as well as scientific, significance of this relativism—the refusal to make distinctions between great and small, each being free to function in their naturalness (see below, p. 103).

Relativity was thus understood to be partly a question of the observer's standpoint.[c] The *Lü Shih Chhun Chhiu* says this in so many words:

If a man climbs a mountain, the oxen below look like sheep and the sheep like hedgehogs. Yet their real shape is very different. It is a question of the observer's standpoint.[d]

The Taoist observers were thus aware of the danger of optical illusions. What they said was taken up again by Wang Chhung in his *Lun Hêng* of about +80, where [e] one can see the close relevance of such arguments to early astronomical speculations on the distance of the heavenly bodies from the earth. It is interesting to recall that a very similar warning was given by Lucretius in the beautiful lines beginning:

> nam saepe in colli tondentes pabula laeta
> lanigerae reptant pecudes quo quamque vocantes
> invitant herbae gemmantes rore recenti,
> et satiati agni ludunt blandeque coruscant;
> omnia quae nobis longe confusa videntur
> et velut in viridi candor consistere colli....[f]

---

[a] Ch. 5 (R. Wilhelm (4), p. 49).

[b] There is a brilliant one in the +6th-century *Chin Lou Tzu*, ch. 4, p. 19*a*.

[c] The proto-scientific significance of this and other aspects of early Taoist thought has long been appreciated in China, as may be seen, for example, by the speech which Lin Yü-Thang puts into the mouth of one of the characters in his excellent novel *Moment in Peking*, p. 714.

[d] Ch. 141 (vol. 2, p. 144), tr. R. Wilhelm (3), p. 413.

[e] Ch. 32, tr. Forke (4), pp. 262, 274.

[f] *De Rerum Natura*, II, 317.

> ...Often thus
> Upon a hillside will the woolly flocks
> Be cropping their goodly food and creeping about
> Whither the summons of the grass, begemmed
> With the fresh dew, is calling, and the lambs,
> Well filled, are frisking, locking horns in sport;
> Yet all, for us, seem blurred and far confused
> A glint of white at rest on a green hill.
> Again, when mighty legions, marching round,
> Fill all the quarters of the plains below,
> Rousing a mimic warfare, there the sheen
> Darts to the sky, and all the fields about
> Glitter with brass, and from beneath, a sound
> Goes forth from feet of stalwart soldiery,
> And mountain walls, smote by the shouting, send
> The voices onwards to the stars of heaven,
> And hither and thither darts the cavalry,
> And of a sudden down the midmost fields
> Charges with onset stout enough to rock
> The solid earth—and yet some post there is
> On the high mountain, seen from which they seem
> To stand, a gleam at rest along the plain (tr. Leonard).

Noteworthy, too, is the contradiction between Taoist appreciation of change, transformation, 'evolution', perhaps social evolution,[a] on the one hand, and Confucian-Legalist belief in stability and permanence. Several commentators have pointed this out.[b] Thus, for Hsün Tzu, things only appeared to change and did not really do so, terms meant now the same things as in ages past, human nature does not alter, and between ancient and modern times there was no real difference. Similarly, the Taoist denial of anthropocentrism was diametrically at variance with those whose interests were focused on human society, and for whom man was the measure of all things.[c]

(1) TAOISM AND MAGIC

In the light of the foregoing we are much better able to understand the close connection which developed between the Taoists and magic. Reference has just been made to the beginnings of alchemy which we shall examine in the proper place. Without anticipating what will there be said, one cannot emphasise too much that in their initial stages there is nothing to distinguish magic from science. The complex processes of control and statistical analysis which alone can unravel the differences between the effectiveness of diverse manual operations do not become available until much later.

Here we need only note that the philosophy of natural changes was connected already in the Warring States time with experimentation on natural changes. There is the testimony of the historian Ssuma Chhien,[d] which will be quoted later (Sect. 13c) in connection with Tsou Yen and the School of Naturalists, closely allied to the Taoists. Alchemy originated in both. It was connected with a belief in the existence of islands in the Eastern Sea where lived genii who possessed, and might be persuaded to transmit, the secrets of the drugs which would confer immortal life. The search for the islands, where, it was thought, great treasures of magical knowledge existed, came to its climax with the expedition sent out by Chhin Shih Huang Ti in −219 (see on, Sects. 10h, 13c, 33). Yetts (4) has drawn a comparison with the searches of the Carthaginians for the 'Fortunate Isles' which had a foundation in fact (Madeira, Canaries), as did the Chinese missions (Japan). But the point here is the early association of Taoism with practical magic and techniques. The *Hsi Ching Tsa Chi*[1] which, though probably (as we have it now) of the +6th century, is considered rather well informed concerning later Han events, says:

Liu An, Prince of Huai Nan, liked (to surround himself with) magicians (*fang shih*[2]) who all distinguished themselves with various techniques (*shu*[3]). Some could make a river flow simply by drawing a line on the ground, some could gather up earth to form mountains

---

[a] See below, p. 167.
[b] Hu Shih (2), p. 153; Dubs (7), p. 75; Chhêng Chih-I (1), p. 56.
[c] Especially aristocratic man. The Taoist denial of anthropocentrism was related to their social attitude.
[d] *Shih Chi*, ch. 28, pp. 10b ff.

[1] 西京雜記   [2] 方士   [3] 術

and precipices, others used their breathing to influence the temperature inducing winter or summer at will, others again by sneezing and coughing formed rain or fog. In the end the Prince disappeared with those magicians.[a]

Chhen Mêng-Chia has sought some significance, plausibly enough, in the fact that so much of these early magical-scientific traditions came from the eastern sea-coast States of Chhi and Yen. Tsou Yen's school of Naturalists spread out from there, and the magical operators who surrounded the Emperor Wu of the Han dynasty were mostly of eastern maritime origin. For Chhin Shih Huang Ti himself the sea had an irresistible attraction, and when in −210 he died it was at the mouth of the Yellow River where he had been hunting imaginary sea-monsters, alleged to be preventing access to the mysterious islands.[b] The coastal peoples, said Chhen Mêng-Chia, whose ideas had sprung from years of observation of the changing moods of the ocean, tended naturally to concentrate on the importance of change in Nature. Inland groups, never subjected to the sudden and terrifying movements of a great body of water, thought more naturally in terms of stability. Thus the changelessness of the learned and immortal islanders would have been antithetical to the unstable vagaries of the sea and the shore. Here we cannot but be reminded of the part played by the oceans in Greek and all European development, the undoubted impulsion to the recognition and study of change given by the changeful sea to those who lived beside it and sailed upon it.

But the Taoists never developed a systematic theoretical account of Nature, analogous to that of Aristotle. The Yin and the Yang, the various forms of *chhi*, the Five Elements, were insufficient for the task assigned to them. But that did not prevent great progress in all practical technology, interpenetrated though it continued to be by distinctively magical beliefs. Technologists lacking scientific background to their thought have a habit of doing the right thing for the wrong reasons, and this was very true in China.

The following passage from *Chuang Tzu* seems to show the Taoists in the very act of resigning the possibility of a detailed theoretical interpretation of Nature, and falling back on the observation of the properties of things as found *in* use, and *for* further use:

It was separation (*fên*[1]) (that led to) completion (*chhêng*[2]); and from completion (ensued) dissolution (*hui*[3]). But all things, without regard to their completion and dissolution, return again into the Unity (of Nature). Only the far-reaching in thought can know how to comprehend them in this Unity. This being so, let us give up devotion to our own (preconceived)

---

[a] Ch. 3, p. 1*a*, tr. auct. There is a parallel, but longer, account in the *Shen Hsien Chuan* of Ko Hung (+4th century), which has been translated by L. Giles (6), p. 42. For the body of legends which grew up in later ages about the Prince of Huai Nan and his magicians, see Doré (1), part II, vol. 9, pp. 582, 604; and Maspero (11). The puffing and blowing magic obviously derives from the breathing exercises complex (p. 143 below).

[b] *TH*, vol. 1, pp. 222, 223. Cf. Sect. 30*h*.

[1] 分   [2] 成   [3] 毀

views, and follow the 'common' and 'ordinary' views, which are grounded on the *use* of things. The study of that *use* leads to comprehension, and that secures success. That success gained, we are near (to the object of our search) and there we (have to) stop. When we stop, and yet do not know how it is so, we have what is called the Tao.[a]

The spirit of technology without theoretical science seems thus to be found within Taoist philosophy itself. Compare the following deliciously cooked anecdote from chapter 14:

Confucius went to see Lao Tan, who said to him, 'I hear you are a wise man from the North; have you also found the Tao?' 'Not yet', replied Confucius. 'How did you go about to search for it?' said Lao Tan. 'I sought it in measures and numbers', Confucius answered, 'but after five years I still hadn't got it.' 'And how then did you seek it?' 'I sought it in the Yin and the Yang, but after twelve more years I did not find it.' 'Just so,' said Lao Tan; 'if the Tao could be offered from person to person, all men would present it to their rulers; if it could be served up (in bowls) men would all have given it to their parents; if it could be talked about, everybody would have told their brothers; if it could be inherited, men would have bequeathed it to their sons and grandsons. But no one could do (any of these things). Because if you have not already got it in you, you cannot receive it....'[b]

Even allowing for the general view that this chapter is a later interpolation, and that the account of the imaginary interview was written by someone influenced by Buddhist meditation practice and the more obscurantist aspects of Taoist mysticism, the passage serves as an epitaph on the great scientific movement which Taoism might have become.

It is always interesting to look at what happened to these thought-complexes in much later times. The following passage gives a remote echo of Taoist proto-scientific observation of natural change, in the Sung dynasty. It will be seen that the soaring thought of Chuang Tzu has become watered down into physical exercises and weather-lore. Yet a germ of the good old doctrine still persists. Yeh Mêng-Tê,[1] in the *Pi Shu Lu Hua*[2] of +1156, says:

The truths (of Nature) in the world appear before us every day, and are distinctly apprehensible by men. But men are enslaved by outside (social) affairs, so they do not see them; they rush to and fro and notice nothing. It is only quietly observing men who can obtain truth. When I was young I used to talk about the nourishing of life (*yang sêng*[3])[c] with Taoists, and discussed the rising and falling of the *chhi* at noon and midnight with the magicians (*fang shih*[4]). Even after long discussion there are secrets which they do not tell ordinary people. I knew a Taoist who laughed and said, 'It's very easy; often when sitting in meditation I feel the *chhi* rising and falling within me at those hours, like the succession

---

[a] Ch. 2, tr. Legge (5), vol. 1, p. 184. The italics are ours.
[b] Tr. auct. adjuv. Legge (5), vol. 1, p. 355; Lin Yü-Thang (1), p. 316.
[c] Technical term for cultivation of the body, see above, pp. 46, 67, below, p. 143.

[1] 葉夢得   [2] 避暑錄話   [3] 養生   [4] 方士

of hunger and satiety. I don't understand it either, but if the mind is empty (one can apprehend these things). The same applies to cold and heat, dryness and wetness, which attack people with diseases.' I myself tested this, and believe that what he said was right. When I lived in the mountains I often saw that old farmers could predict rain and sunshine, proving right seven or eight times out of ten. I asked them their methods, but they said that there was nothing but experience. If you ask those who live in cities, they know nothing. Since at that time I had plenty of leisure, I often rose very early in the morning, and with an empty mind concentrated on the clouds, mountains, river, fields and trees in all their beauty, and found I could predict the weather aright seven or eight times out of ten.... Thus I realised that it is only in quietness that the cosmos can be observed, the body's moods felt, and real truth obtained.[a]

(f) THE ATTITUDE OF THE TAOISTS TO KNOWLEDGE AND TO SOCIETY

We now approach the question of the political position of the Taoists. It cannot be separated from those proto-scientific tendencies which have already been described. If these have been overlooked by nearly all European expositors of Taoism, its political significance has been understood by none.

The Taoists 'walked outside society'. 'They travel outside the human world', the *Chuang Tzu*[b] makes Confucius say, 'I travel within it. There is no common ground between these two ways.' They 'roam beyond the limits of the mundane world'.[c] But if they did so, it was not only because they wished to observe Nature free from the encumbrances and trivialities of social life, it was also because they were in complete opposition to the very structure of feudal society, and their withdrawal was part of their protest.

Let us first take up certain points which lie on the borderline between the scientific and the political. Of these the most important is the question of the Taoist attitude to 'knowledge'. Those who have given attention to Taoist texts have often been puzzled by the numerous and strongly worded diatribes against 'knowledge' which occur in them, and have drawn the facile conclusion that these could only be interpreted in the traditional sense of religious mysticism, inveighing against rational thought and empirical learning alike. At least seven chapters[d] of the *Tao Tê Ching* exemplify this. For example, chapter 3 (I write 'knowledge' with quotation-marks for a reason which will be explained immediately):

> ...Therefore the sage rules (the people)
> By emptying their minds and filling their stomachs
> By weakening their ambitions and strengthening their bones
> Ever striving to make them without 'knowledge' and without desire (for
>   private gain)

[a] Ch. 2, p. 17a, tr. auct.
[b] Ch. 6, tr. Fêng Yu-Lan (5), p. 124. Cf. ch. 13 (Legge (5), vol. 1, p. 340).
[c] Ch. 2 (Lin Yü-Thang (1), p. 154)   [d] Chs. 3, 19, 20, 48, 65, 71, 81.

And if there be any who have 'knowledge'
He sees to it that they do not interfere.[a]

And chapter 19:

Banish 'wisdom'; discard 'knowledge',
And the people will be benefited a hundredfold.
Banish 'benevolence'; discard 'morality'
And the people will be dutiful and compassionate.
Banish 'skill'; discard 'profit',
And thieves and robbers will disappear....[b]
Banish 'learning' and there will be no more grieving.[c]

Or chapter 65:

In olden times the best practisers of the Tao
Did not use it to awaken the people to 'knowledge',
But to restore them to 'simplicity'.
People with much 'knowledge' are difficult to govern,
So to increase the people's 'knowledge' is to destroy the country....[d]

Such statements are obviously in apparent contradiction with the interest of the Taoists in natural knowledge already demonstrated. But the clue to the puzzle is immediately evident if we turn to *Chuang Tzu*, chapter 2. False social 'knowledge' is to be contrasted with true natural knowledge.

Chuang Tzu says:

Men in general bustle about and toil; the sage seems unlettered (*yü*[1]) and without 'knowledge' (*chhun*[2])....When people dream they do not know that they are dreaming. In their dream they may even interpret dreams. Only when they wake they begin to know that they dreamed. By and by comes the great awakening, and then we shall find out that life itself is a great dream. All the while the fools think that they are awake, and that they have knowledge. Making nice discriminations, they differentiate between princes and grooms (*chün hu! mu hu!*[3]). How stupid![e]

Scornfully Chuang Tzu describes Confucian scholastic social knowledge as the 'distinctions between princes and grooms'; this is 'knowledge' as distinguished from that true knowledge of the Tao and of Nature for which the Taoists sought. Once we have the thread in our hand we can explain a large number of passages which would

---

[a] Tr. Chhu Ta-Kao (2); Waley (4), mod.
[b] Tr. Waley (4). As in the previous passage, the words placed in quotation-marks are so written by us, and not by the translators whose version is used.
[c] A line from the following chapter.
[d] Tr. Chhu Ta-Kao (2), mod.   [e] Tr. Fêng Yu-Lan (5), p. 62.

[1] 愚   [2] 芚   [3] 君乎牧乎

otherwise be puzzling. This interpretation must be correct, for the same idea occurs again in the middle of one of the most violently anti-feudal chapters (10), where it is said that if (Confucian-Legalist) 'sageness' and 'wisdom' were put away, great robbers would cease to arise.[a] Moreover, Chuang Tzu quotes the *Tao Tê Ching* in this connection.[b] He dismisses Confucian learning as the 'vestiges left by former kings',[c] and speaks of 'vulgar learning' (*su hsüeh*[1]) and 'vulgar thinking' (*su ssu*[2]).[d] 'From the standpoint of the Tao,' he makes one of his characters say, 'what is noble and what is mean?'[e] The attacks on 'knowledge' are therefore not anti-rational mysticism, but proto-scientific anti-scholasticism. Almost the only European writer who has appreciated this cardinal point is Wulff (1), who speaks of the 'falscher Schmuck und nutzloser Plunder' of feudal philosophy, and recognises these passages as an 'Angriff auf den Konfucianismus und dessen Ethik'. It has deceived even the elect.[f]

I am not saying that there was not an extremely strong mystical element in Taoism, but only that a certain amount of knowledge of the operations of the Tao was believed to be attainable, and that Confucian-Legalist social scholasticism was definitely no help. 'It would be useless', says the *Huai Nan Tzu*, 'to discuss the great Tao with a narrow-minded scholar; he is bound to the conventional and tied to his own (orthodox) doctrine'[g] (*su, chiao*[3]). And also, 'Those who follow the natural order flow in the current of the Tao. Those who follow men become involved with conventional society' (*yü su chiao*[4]),[h] and with 'commonplace worldly knowledge' (*su shih chih hsüeh*[5]).[i] Occasionally the Confucians are mentioned by name (*ju*[6]).[j] And of the Tao Chuang Tzu says:[k] 'The most extensive "knowledge" will not necessarily know it; reasoning will not make men wise in it (*po chih pu pi chih, pien chih pu pi hui*[7]).'[l]

'Emptying their minds and filling their stomachs' (*hsü chhi hsin, shih chhi fu*[8]), says chapter 3 of the *Tao Tê Ching*, which we have just quoted. Many who might be willing to discard the usual interpretation of these words as praise of ignorance, might think we were reading too much into them if we interpreted them as meaning that people should be trained to abandon preconceived ideas and prejudices, and that

---

[a] Legge (5), vol. 1, p. 286.   [b] Ch. 11; Legge (5), vol. 1, p. 297.
[c] Ch. 14; Legge (5), vol. 1, p. 361.   [d] Ch. 16; Legge (5), vol. 1, p. 368.
[e] Ch. 17; Legge (5), vol. 1, p. 382.

[f] As, for instance, Duyvendak (7); Creel (4); H. Maspero (2), p. 493. Elsewhere Maspero (26), p. 73 suggested that the 'knowledge' against which the Taoists fulminated was the 'foreknowledge' claimed as possible by the *I Ching* diviners and other groups associated with the School of Naturalists (see below, p. 234). In so far as their predictions were concerned with social affairs, the Taoists would certainly not have been interested in them; in so far as their studies of the Book of Changes seemed to throw light on the nature of the universe, the Taoists would have adopted them, as in fact later happened.

[g] Ch. 1, p. 7*a* (tr. Morgan (1), p. 11).
[h] Ch. 1, p. 6*b* (tr. Morgan (1), p. 11); see also ch. 2, p. 4*a* (Morgan (1), p. 36).
[i] Ch. 2, p. 10*b* (Morgan (1), p. 48).   [j] Ch. 7, p. 11*b* (Morgan (1), p. 75).
[k] *Chuang Tzu*, ch. 22; Legge (5), vol. 2, p. 64.
[l] Parallel words in *TTC*, ch. 81, and in *Lü Shih Chhun Chhiu*, ch. 94.

[1] 俗學   [2] 俗思   [3] 俗敎   [4] 與俗交   [5] 俗世之學   [6] 儒
[7] 博之不必知辯之不必慧   [8] 虛其心實其腹

if this could be accomplished, the resulting increase in knowledge of nature would have the effect of multiplying many times the available amount of true knowledge, and hence indeed of food. Yet such was the interpretation of a scholar who lived towards the end of the Sung, Lin Ching-Hsi.[1] In his *Chi Shan Chi*[2] (Poetical Remains of the Old Gentleman of Chi Mountain) he wrote:

Scholars of old time said that the mind is originally empty, and only because of this can it respond to natural things (*ying wu*[3])[a] without prejudices (lit. traces, *chi*,[4] left behind to influence later vision). Only the empty mind (*hsü hsin*[5]) can respond to the things of Nature. Though everything resonates with the mind, the mind should be as if it had never resonated, and things should not remain in it. But once the mind has received (impressions of) natural things, they tend to remain and not to disappear, thus leaving traces in the mind. (These affect later seeing and thinking, so that the mind is not truly 'empty' and unbiased.) It should be like a river gorge with swans flying overhead; the river has no desire to retain the swan, yet the swan's passage is traced out by its shadow without any omission. Take another example. All things, whether beautiful or ugly, are reflected perfectly in a mirror; it never refuses to show anything, nor retains anything afterwards. It always remains 'empty'. The mind should be like this....Lao Tzu wrote about 'empty minds and full stomachs', and people often criticise this, asking how he could have called for emptiness and fullness at the same time. The answer is that because the empty mind seems to have no natural things (*wu wu*[6]), the full stomach possesses everything in the world (*wan wu*[7]). The meaning is that through emptiness, fullness is achieved. The words of Lao Tzu indeed embodied the true principle (*li*[8]) of Nature, though he did not develop his thought fully....[b]

Thus 'emptying the mind' did not mean emptying it of that true natural knowledge which Chuang Tzu contrasted with the false knowledge of feudal social distinctions, but rather emptying it of distorting memories, prejudices and preconceived ideas, so that true practical knowledge might flourish and all abundance come in its train.[c] The absolute justification of this complex of thought is seen in the great inventions of ancient China, as, for example, the use of water-power.[d]

(1) THE PATTERN OF MYSTICISM AND EMPIRICISM

It is not possible to understand the full import of this situation without comparing it with a somewhat analogous one which arose at the time of the Renaissance in Europe. In modern science the relation between the rational and the empirical seems obvious, but this was not always so. W. Pagel, in a monograph now classical, *Religious Motives in the Medical Biology of the Seventeenth Century*, has traced the alliance

[a] The use of this word, the technical term for 'resonance', is significant. Cf. p. 304 and Sect. 26*h* below. [b] Ch. 4, p. 5*b*, tr. auct.

[c] Of course in this passage Lin Ching-Hsi did not do justice to the preservation of necessary memories compatible with sound judgments on Nature, yet he seems to have brilliantly understood the 'fresh and seeing eye' of the inventor and the naturalist.

[d] Cf. below, in Mechanical Engineering (Section 27*f*).

[1] 林景熙   [2] 霽山集   [3] 應物   [4] 迹   [5] 虛心   [6] 無物
[7] 萬物   [8] 理

which existed in the +16th and +17th centuries between mysticism and experimental science, and has shown how modern science in its early stages had to struggle against scholastic rationalism. The proper marriage of rational processes to empirical observations had not then been consummated. It was considered (in the ironic words of Robert Boyle)[a] 'much more high and Philosophical to discover things *a priore* than *a postiore*'.

In his monograph Pagel demonstrated that the origins of modern science are to be found in the strife between four factors or tendencies in that couple of centuries; two on one side and two on the other. In the first place theological philosophy allied itself with Aristotelian rationalism (as following the orthodox scholastic synthesis it could not but do), and opposed the first gropings of modern science. In the second place, experimental empiricism, reacting against this intellectual pride, found an ally in religious mysticism. Once this cleavage is fully appreciated, much that would otherwise be inexplicable falls into its place in the early history of modern science. Christian theology, as one would expect from its then universal domination over men's minds, was on both sides in the struggle; but while rational theology was antiscientific, mystical theology proved to be pro-scientific. The explanation of this apparent paradox is that rational theology was anti-magical while mystical theology tended to be pro-magical. And the fundamental cleavage here lay not between those who were prepared to make use of reason and those who felt that it was totally insufficient, but between those who were prepared to make use of their hands and those who refused to do so. The theologians who declined, when invited, to look through Galileo's telescopes, were certainly scholastics, and therefore already, as they thought, in possession of sufficient knowledge about the material universe.[b] If Galileo's findings agreed with Aristotle and St Thomas there was no point in looking through a telescope; if they did not they must be wrong. But others were then to be found who (although full of what now seem to us the wildest and darkest ideas about witchcraft, sympathetic attraction, and the like) would have been perfectly willing to look through the telescope, and to judge what they saw on the merits of the case. This elucidation explains many curious facts, such as the title of books like J. B. da Porta's *Natural Magic*, which included many scientific matters; the interest taken by the early Royal Society in what we can now see were magical claims; the standpoints of men such as Sir Thomas Browne and Joseph Glanville, that staunch anti-Aristotelian, on witchcraft; and the addiction which seventeenth-century biologists had for dipping into the *Kabbalah*, believing that its ancient mysticism might contain ideas of value for them.[c]

[a] *Sceptical Chymist* (1661), p. 20. No one who had been a laboratory colleague of my ingenious friend Mr N. W. Pirie, F.R.S. could ever forget this devastatingly Taoist remark, so much appreciated by him.

[b] Actually the story is told of Cremonius who was an Averroist and already doubted Aristotle so much that he was afraid he would find confirmation of his fears; for this point I am indebted to Dr W. Pagel. But the orthodox attitude was as stated above.

[c] For such exemplificative cases as those of Robert Fludd, Agrippa of Nettesheim, Quercetanus, etc., the reader is referred to Pagel (1).

Most typical of this period was the Flemish chemist John Baptist van Helmont (+1577 to +1644) to whom Pagel (2, 3, 8) has devoted profound studies. Van Helmont, one of the founders of biochemistry, was among the first to use the balance in quantitative experiments, devised one of the earliest thermometers, demonstrated the acid of the stomach and the alkali of the duodenum, and by introducing the concept of 'gas' and making experiments on fermentation, initiated that pneumatic chemistry the implications of which were to be so far-reaching. Yet here was a figure deeply anti-'rational', embodying what Pagel has called a 'religious empiricism'. Two whole chapters of his works[a] (translated into English by J. Chandler in +1662) are devoted to an attack on hair-splitting formal logic, which had, he felt, nothing to do with reality, and led the mind round in a circle, teaching nothing new. He was thus strongly anti-scholastic ('Logica est inutilis ad inventionem scientiarum').[b] He was also opposed to the theoretical formulations of traditional thought, to the four Aristotelian elements no less than to the three alchemical principles (thus anticipating Boyle). He had no use for Galenic humours and qualities, or for atoms, and combated the view of disease as an intrinsic lack of *krasis*, preferring the idea of a specific external entity, an alien ferment, a *contagium vivum* (cf. Singer, 7). On the positive side, he laid great emphasis on the specificity of living organisms (anticipating modern immunology and protein chemistry), but thought that the various forms of gas (which he, for the first time, distinguished from ordinary air, and recognised as different substances in the gaseous state) were the material carriers of this specificity.[c] Such pneumatised 'Form' permitted matter to lose its coarse corporeality and to meet the odour-like ferments half-way. It contained 'concrete semen', a conception doubtless derived from the Stoic 'seeds', and destined to mould, through van Helmont's son, F. M. van Helmont, the monads of Leibniz.[d] In sum, J. B. van Helmont was a remarkably Taoist character, and if one had to select that quality in which he (and other 17th-century scientists whom he typifies) differed most from the Taoists, it would be his strong belief in a personal God.[e]

It may be said, then, that at the initial phases of modern science in Europe, the mystical approach was often more helpful than the rationalist. In China at the time of the ancient philosophical schools we meet with exactly the same phenomenon.[f] It is clearly not a purely intellectual one, but rests on the value placed on manual operations. A man such as van Helmont was an active laboratory worker as well as

[a] Chs. 8 and 9.

[b] The full phrase is Francis Bacon's, but the first three words were used by van Helmont as a chapter heading.

[c] One is strongly reminded of the 'principles of organisation' (*Li*) and the 'rarefied matter' (*Chhi*), usually concreted (*ning*, cf. p. 43), but capable of existence in a highly subtle state; with which we shall meet in considering the Taoist-influenced Neo-Confucian thinkers of the Sung (pp. 472 ff. below.)

[d] It is impossible not to be reminded of the 'seminal essence' (*ching*) spoken of by the Taoists (cf. pp. 38, 146).

[e] The significance of this will be better appreciated after perusal of the section on natural law and Laws of Nature (Sect. 18).

[f] The only sinologist who appears to have seen even dimly this fact is R. Wilhelm (1), p. 248.

a thinker and writer. In Florence the apparatus used by Galileo and Torricelli may still be seen. The Confucian social scholastics, like the rationalist Aristotelians and Thomists nearly two millennia later, had neither sympathy for, nor any interest in, manual operations. Hence science and magic were driven into mystical heterodoxy together.

In order to complete the parallel with the Taoists, it is necessary to show that some at least of the chief figures of mystical naturalism in its + 16th-century Western phase manifested revolutionary political tendencies. It would take us too far to investigate here the somewhat complex relations of the new, or experimental, science, with the conflict between old and new in the Reformation, though there is a wealth of evidence that most scientific men in northern Europe were on the Protestant-Puritan side, with all its progressive political implications.[a] But for one outstanding figure of mystical naturalism, Paracelsus, an investigation of social tendencies is possible, since his social, ethical and political writings have recently been collected, published and annotated by Goldammer. We can now see that Paracelsus, the standard-bearer of alchemy applied to medicine, the introducer of mineral drugs despite all opposition of Galenists, the first observer of the occupational diseases of miners—was an equalitarian, almost an Anabaptist, in fact a Christian socialist. He knew nothing, of course, of socialism based on economic theories, but he found himself opposed to the accepted institutions of his time because he had a vision of the charismatic community of goods in the commonwealth of Christian brotherhood. Himself an intense individualist, he saw the salvation of society only in a thorough reformation along collectivist lines. 'It was not God's will', he said (unknowingly echoing Chuang Chou), 'that there should be lords and commoners, but all brothers.' Like most of the left-wing democratic leaders of the + 16th and + 17th centuries, he was opposed to the rising merchant class as well as to the feudal lords, and therefore (though part of a Protestant movement) in sympathy with certain medieval ideas including the suspicion of wealth as such and the condemnation of usury. Not afraid of tyrannicide and the prosecution of just wars against unjust princes, Paracelsus was yet in some moods strongly pacifist, and, as we now know, one of the earliest opponents of capital punishment. Sometimes he wrote favourably of the Emperor, to whom he looked for fundamental land reforms, but he did not consider any kingly or princely office divinely ordained. In the third decade of the + 16th century Paracelsus was active as a political leader in Salzburg, whence he was lucky to escape with his life, but it seems doubtful whether he had later any close connections with the Anabaptists, Hutterian Brethren, and other communist organisations of the Peasant Wars. Clearly Paracelsus would have had a great deal in common with the Taoist alchemists.

It is not generally realised that some of these trends can be seen clearly in the thought of the great instaurator of modern science, Francis Bacon. The 18th-century flavour of the admirable phrase often applied to him, 'The Bell that call'd the Wits together', has perhaps obscured the profoundly religious character of his thought. It is the merit

[a] Cf. the well-known book of L. C. Miall (1).

of Farrington (6) to have revealed this in a recent book, in which he discusses, *inter alia*, the extreme hostility of Bacon to Aristotle in particular, and to some extent to all the Greeks except Democritus and some of the other pre-Socratics.[a] Farrington writes:[b]

It has been observed that his quarrel seems to be not with their intellectual (i.e. philosophical) but with their *moral* position, but the ground of it has not been made clear. In the strange writing called *The Masculine Birth of Time*[c]—which it must be admitted is violent and intemperate—he speaks of his concern with Greek philosophy as a pollution. What did he mean by this? What did he mean when he said about Plato and Aristotle that no denunciation could be adequate for their monstrous guilt (*pro ipsorum sontissimo reatu*[d])? The answer is simple. He believed that the type of philosophy for which they stood was the great obstacle to a divinely promised revolution in human affairs. They held up that blessing which was the subject of Bacon's fervent orisons. The following was found among his papers:

'To God the Father, God the Word, and God the Spirit, we pour out our humble and burning prayers, that they would be mindful of the miseries of the human race, and of this pilgrimage of our life, in which we wear out evil days and few, and would unseal again the refreshing fountain of their mercy for the relief of our sufferings.'

To be held fit to receive this blessing, Bacon believed, it was necessary to reject the false philosophy of the Greeks.

For, however unjustly, Bacon did consider Aristotle and the rest to be in some sense guilty. In the *Refutation of Philosophies* and again in the *De Augmentis* he even compares Aristotle to Anti-Christ. The philosophy of Aristotle is involved in guilt, and its punishment is to be fruitless in works. Various passages define the nature of the guilt. In the *De Augmentis* Solomon and St Paul are cited to testify that all knowledge which is not mixed with love is corrupt, and the proof of love in philosophy is that it should be designed not for mental satisfaction but for the production of works.

The same theme is the subject of the second of his *Sacred Meditations*. Here his argument is that while the doctrine of Jesus was for the benefit of the soul, all his miracles were for the body. 'He restoreth motion to the lame, light to the blind, speech to the dumb, health to the sick, cleanness to the lepers, sound mind to them that were possessed of devils, life to the dead. There was no miracle of judgment, but all of mercy, and all upon the human body.'[e]

What, then, precisely, was the nature of the sin which had rendered Aristotelianism and so much else of Greek philosophy fruitless for good? It was the sin of *intellectual pride*, manifested in the presumptuous endeavour *to conjure the knowledge of the nature of things out of one's own head, instead of seeking it patiently in the Book of Nature*. In almost the last thing that Bacon published—the preface to the *History of the Winds* (+1623)—he sets forth at length his understanding of the matter.

'Without doubt we are paying for the sin of our first parents and imitating it. They wanted to be like gods; we, their posterity, still more so. We create worlds. We prescribe laws to

---

[a] Here Bacon was closely paralleled by Spinoza; see Letter no. 60, to Hugo Boxel (+1674), in the correspondence edited by van Vloten & Land.

[b] Pp. 146 ff.

[c] Cf. the special study by Farrington (7).

[d] Apparently a piece of law Latin.

[e] Cf. the Taoist insistence on material and bodily immortality.

Nature and lord it over her. We want to have all things as suits our fatuity, not as fits the
Divine Wisdom, not as they are found in Nature. We impose the seal of our image on the
creatures and works of God, we do not diligently seek to discover the seal of God on things.
Therefore not undeservedly have we fallen from our dominion over the Creation; and,
though after the Fall of Man some dominion over rebellious Nature still remained—to the
extent at least that it could be subdued and controlled by true and solid arts—even that we
have for the most part forfeited by our pride, because we wanted to be like gods and follow
the dictates of our own reason.'[a]

The attitude is perfectly clear. False philosophy is due to man's intellectual pride,
inherited from his first parents, and punished by loss of dominion over Nature. How
strongly this echoes the attacks of Chuang Tzu on the Confucians—who, indeed,
were one stage worse than the Aristotelians, since their rationalism was limited to
human society and did not even admit that the world of Nature was worth theorising
about at all. In the same preface just quoted Bacon rises to heights of eloquence:

Wherefore, if there be any humility towards the Creator, if there be any reverence and
praise of his works; if there be any charity towards men, and zeal to lessen human wants and
human sufferings; if there be any love of truth in natural things, any hatred of darkness,
any desire to purify the understanding; men are to be entreated again and again that they
should dismiss for a while, or at least put aside, those inconstant and preposterous philosophies, which prefer theses to hypotheses, have led experience captive, and triumphed over
the works of God; that they should humbly and with a certain reverence draw near to the
book of Creation; that they should there make a stay, that on it they should meditate, and
that then washed and clean they should in chastity and integrity turn them from Opinion.
This is that speech and language which hath gone out to all the ends of the earth, and has not
suffered the confusion of Babel[b]—this must men learn, and resuming their youth, become
again as little children, and deign to take its alphabet into their hands.[c]

The scientific reforms urged by Bacon were thus in all sincerity put forward as
part of a mystical interpretation of the Christian religion. His opposition to the
rationalism of the medieval Christian scholastics comes out in a famous passage:

This kind of degenerate learning did chiefly reign among the schoolmen; who, having
sharp and strong wits, and abundance of leisure, and small variety of reading (their wits
being shut up in the cells of a few authors, chiefly Aristotle their Dictator, as their persons
were shut up in the cells of monasteries and colleges), and knowing little history, either of
Nature or Time, did, out of no great quantity of matter, and infinite agitation of wit, spin
out unto us those laborious webs of learning, which are extant in their books. For the wit and
mind of man, if it work upon matter, which is the contemplation of the creatures of God,
worketh according to the stuff, and is limited thereby; but if it work upon itself, as the spider

---

[a] I.e. ratiocination not modified by the humble observation of nature. Cf. Farrington (14).

[b] A most striking anticipation of the universal understanding which exists among scientists of all
nations.

[c] Cf. the words of T. H. Huxley quoted on p. 61.

worketh his web, then it is endless, and brings forth indeed cobwebs of learning, admirable for the fineness of thread and work, but of neither substance nor profit.[a]

The association between nature-mysticism and science is therefore to be found embedded in the very foundations of modern (post-Renaissance) scientific thought.

It would take us much too far to pursue this theme to its origins in European history; Plotinus and Dionysius the Areopagite would have to be mentioned. Before the time of Bacon there had been not a few precursors who had revolted against the orthodoxy of the rational intellect—Nicholas of Cusa, for example (+1401 to +1464), with his 'coincidence of contraries' in the *De Docta Ignorantia*; and especially Giordano Bruno (+1548 to +1600) to whom an excellent book has recently been consecrated by D. W. Singer, and whose writings abound in Taoist echoes.[b]

One may ask whether similar situations have occurred in civilisations other than those of Europe and China. The answer is clearly in the affirmative. In Islamic culture mystical theology was closely associated with some of the developments at the beginning of science. About +950 there started a movement, soon developing into an organisation, at Basra in modern Iraq, called the Ikhwān al-Ṣafā', the Brethren of Sincerity.[c] Like the Taoists, this semi-secret society had at one and the same time mystical, scientific and political tendencies. The men who thus gathered together acknowledged the existence of mysteries transcending reason; and believed in the efficacy of manual operations. All scientists in this early mystical phase of science recognise that effects may be brought about by specific manipulations without our being able to say exactly how or why, and they think that information ought to be accumulated about these things; while their opponents, the rationalists, whether Christian, Muslim or Confucian, consider that the nature of the universe can be apprehended by ratiocination alone, that quite sufficient information concerning it has already been made available by the sages, and that in any case the use of the hands is unworthy of any persons claiming to be scholars. The early scientists are in a dilemma, for they must either set up a rationalism of their own consisting of obviously inadequate theories, or rest in the simple thesis that 'there are more things in heaven and earth, Horatio, than are dreamt of in your philosophy'. Only prolonged experimentation and hypothesis can escape from this situation. It is perhaps in such a light that we may interpret the imaginary interview between Confucius and Lao Tan quoted a few pages back.

The Brethren of Sincerity embodied their thoughts and experimental results in a series of epistles, the *Rasā'il Ikhwān al-Ṣafā'*,[d] which had a great influence on Islamic thought, and which has come down to us. Besides ethics and metaphysics,

[a] *Advancement of Learning* (1605).

[b] E.g. his doctrine of 'inherent necessity', the continuity of contraries, the universality of change and motion; and his 'praise of asininity' (i.e. simplicity and humility of mind in the confrontation of Nature). D. Singer (1), pp. 84, 122.

[c] Al-Jalil (1), p. 180; Hitti (1), p. 372; Sarton (1), vol. 1, p. 660.

[d] Sarton (1), vol. 1, p. 661. There is no English translation, and the German of Dieterici (1) is only a partial one.

it included writings on nearly all the sciences, including mathematics and music,[a] astrological astronomy,[b] geology and mineralogy[c] (in which they were greatly in advance of their time), physics and chemistry. Their general philosophy has been called an eclectic gnosticism, and they undoubtedly drew from very widespread sources, Greek, Hebrew, Syriac, Iranian and Indian.

It is clear, however, that the Brethren of Sincerity had a strongly political aspect.[d] They flourished at a time when, though the caliphate was still Abbassid, the Buwayhid amirs (or shoguns) were in power; and they were probably ultra-Shi'ite and Ismailite, and certainly Qarmatian in sympathy.[e] Qarmatianism was an extreme socialist, even communist, movement which began about +890, kept perpetual war with the caliphate throughout the +10th century, and even after being crushed at the beginning of the +11th, bequeathed much of its equalitarian doctrine to the Egyptian Fatimids, the Druses of Lebanon and the Neo-Ismailites. The Qarmatians stressed tolerance and fraternity, organised workers and artisans into guilds, and themselves had the ritual of a guild. The writings of the Brethren of Sincerity give the oldest extant account of Muslim guilds. That an alliance of this kind should have existed between the mystical scientists and the organised workers is not in the least surprising, since, as cannot be too often repeated, the great cleavage lay between those who were prepared to engage in manual operations and those who considered them unworthy of a gentleman. Between techniques and magical recipes there was no wide gap. We shall shortly see to what extent went the parallel alliance between the Taoists and the people.

Lastly, the Brethren of Sincerity were closely connected with the whole mystical movement in Islam known as Sufism. It had been precisely in Basra that the oldest beginnings of this movement had had their origin,[f] with al-Ḥasan al-Baṣrī about +728, though the centre had shifted to Baghdad after +864. The Ikhwān al-Ṣafā' flourished both at Baghdad and Basra, and the movement of the sufis[g] (tasawwuf) was still going on in the +11th and +12th centuries,[h] though by then it had become somewhat dissociated from the scientific current.

Such a cleavage never became complete, however. Sayyad Nurul Ḥasan has written of the mystics and sufis of India as propagators of science. There was, for instance, in the late +12th century Hazrat Niẓām ud-Dīn Aulia, a mystical scholar of Delhi under Muḥammad Ibn Ṭughlaq, who spoke of laws of movement in a manner somewhat adumbrating Newton.

When in the Lebanon in the autumn of 1948 at an international conference I happened to be discussing this association of science and mysticism with an Indian friend, Abdul Rahman. In the evening we dined with the then Indian Ambassador

---

[a] Hitti (1), p. 427.   [b] Hitti (1), p. 373.
[c] Hitti (1), p. 386.
[d] On their general background, see Gibb (2).
[e] Hitti (1), p. 445.   [f] Al-Jalil (1), p. 147.
[g] See Massignon (1, 2); Arberry (1); R. A. Nicholson (1).
[h] Al-Jalil (1), p. 185.

in Cairo, Syed Hossain, who quite spontaneously enlarged upon the nature of the 'true hakim' in Muslim conception. 'The true hakim', he said, 'was a physician, yes, but he was also a professional philosopher, a student of Nature, indeed a sufi, a mystic....' Abdul Rahman and I looked at each other and applauded the ambassador, who had, from the natural background of his thought, confirmed the conclusions of the afternoon.

There is of course a great distinction between the mystical naturalism with which this section is concerned, and other forms of mysticism which are focused in purely religious concentration upon a God or gods. All that the former characteristically asserts is that there is much in the universe which transcends human reason here and now, but since it prefers the empirical to the rational, it adds that the sum total of incomprehensibility will diminish if men humbly explore the occult properties and relations of things. Religious mysticism (in the usual sense) is very different; it dotes upon the arbitrary residuum and seeks to minimise or deny the value of investigations of natural phenomena.[a]

The question may then be asked, under what social conditions do mysticism and rationalism[b] have respectively the role of progressive social forces? We usually think of rationalism as the characteristically progressive element, fighting against superstition and irrationalism where the latter has become the habitual bulwark of entrenched irrational privilege. This was presumably the state of affairs in western Europe before the French Revolution, and most significantly it was just at that time that Confucianism impinged upon the Encyclopaedists, contributing no small help to revived Pelagian optimism, and to the conception of morality without supernaturalism. But there have perhaps existed quite other situations, in which mysticism[c] has played the part of a progressive social force. When a certain body of rationalist thought has become irrevocably tied to a rigid and outdated system of society, and has become associated with the social controls and sanctions which it imposes, then mysticism may become revolutionary. Law as a whole might be considered a special case of this association of rationalism with reaction; for often esoteric, authoritarian and inaccessible, its function has generally been to act as a brake upon inevitable change.[d] The converse association of mysticism with revolutionary social movements has been constantly seen in European history, as, for example, in the apocalyptic, millenniarist or chiliastic tendencies in early Christianity, the Donatists and other schisms, the Hussites and Taborites of Bohemia, the Anabaptists of the German peasant wars, the Levellers and Diggers of seventeenth-century England, and so on.[e] In Islamic history, as we have just seen, there was the example of the Qarmatians,[f] outstanding

---

[a] Presumably it is hardly necessary to mention the classical books of Inge and James on religious mysticism.

[b] My friend Mr S. Adler questions whether 'formalism' would not be a better word here. Rationalist systems sometimes leave room for a measure of non-rationalisable arbitrariness ('the illogical core of the universe'). The real contrast would then be between formalised orthodoxy and liberal open-mindedness.

[c] Non-obscurantist mysticism, be it understood.

[d] The phrase is due to Eggleston (1). Of course, the legal process may be an agency of change.

[e] Cf. Lewis & Polanyi (1); Needham (6), p. 14, etc.   [f] Massignon (3).

among others. And as has also been pointed out, it is authority-denying mysticism, not rationalism, which at certain times in world history aids the growth of experimental science.

After the inevitable climax, when the wave of progressive social action is defeated by the governing power, usually not unmodified in its turn by the rebellion which it has suppressed, the mystical systems tend to go over into purely religious and unworldly forms. Thus it was not a far cry from the socialist doctrines of the Levellers in the English Revolution (the Civil War and the Cromwellian Republic, 1649 to 1660) to the equalitarian religious mysticism of the Quakers; we know, for instance, that John Lilburne, one of the greatest of the Leveller leaders, ended his life as one of the first of the Society of Friends.

This transition from revolutionary social activity to religious mysticism, which has not given up its conceptions of the world and society, but which has abandoned all hope in the possibility of actually establishing them within the lifetime of those then living, has a very immediate bearing on what happened to the Taoists in ancient China. Essentially an anti-feudal force, they glided imperceptibly, when it more and more appeared that there could be no going back to their ideals, and that feudal bureaucratism was destined to be the characteristic form of Chinese society, into a heterodox religious mysticism. In the light of this analysis it is not in the least surprising that Taoism was, as we shall see, associated with all revolts endeavouring to overthrow the established order, for more than a thousand years.

In the preceding paragraphs I have hoped to show that there can be, in the opening phases of the history of science, an intimate connection between science and mystical faith. These considerations arose from the contention that the Taoists made a sharp distinction between the social 'knowledge' of the Confucians and Legalists, rational but false; and that knowledge of, or insight into, Nature, which they wished to acquire, empirical, perhaps even liable to transcend human logic, but impersonal, universal and true.

(2) SCIENCE AND SOCIAL WELFARE

This contrast is detectable in certain passages of Chuang Tzu which almost take a form reminiscent of modern discussions about science and social welfare. These parables and imaginary conversations seem surely intended to imply that the application of science to human benefit was premature, and that what the Confucians should do if they really wanted to apply human knowledge for the improvement of the conditions of the life of Man was to become Taoists and devote themselves first to the observation of Nature. To help Man without understanding Nature was impossible. Thus, chapter 11:

Huang Ti[a] had been on the throne for nineteen years, and his writ was running everywhere in the empire, when he heard that Kuang Chhêng Tzu[b] was living on the top of Mount Empty-togetherness, so he went there to see him.

---

[a] Legendary emperor.   [b] Imaginary hermit.

'I have heard', he said, 'that you, Sir, are profoundly learned in the perfect Tao. May I ask what is its essence? I wish to take the subtlest essences of heaven and earth and assist with them the (growth of the) five cereal grains, for the (better) nourishment of the people. I also wish to direct the (operations of the) Yin and the Yang, so as to secure the comfort of all living beings. How should I proceed?'

Kuang Chhêng Tzu replied, 'What you are asking about is the material basis of things (*wu chih chih yeh*[1]); what you desire to control can only be the scattered fragments of these things (*wu chih tshan yeh*[2]) (which have been destroyed by your previous interference). According to your government of the world, the vapours of the clouds, before they were collected, would descend in rain; the herbs and trees would shed their leaves before they became yellow; and the light of the sun and moon would hasten to extinction. You have the shallow mind of a glib talker; it is not fit that I should tell you about the perfect Tao.'[a]

Kuang Chhêng Tzu reproaches Huang Ti for the superficial approach to Nature whereby immediate advantages are sought from the broken fragments of the material manifestations of things. He hints that the only way really to benefit human society is to go back and elucidate the fundamental principles of Nature. Huang Ti's attitude is compared to that of a greedy plunderer of Nature, who would allow neither clouds nor crops to ripen, instead of waiting to find out and apply the basic principles of Nature. Bearing in mind what mankind knows today about soil conservation and nature protection, and all the experience we have gained as to the proper relations between pure and applied science, this passage of Chuang Tzu seems as profound and prophetic as any he ever wrote. An analogous story occurs later in the same chapter, where General Clouds comes to Great Nebulous with a request similar to that of Huang Ti to Kuang Chhêng Tzu, and is even more rudely put off.[b] And in chapter 26, Chuang Tzu discusses with Hui Tzu 'the usefulness of what is (apparently) of no use'.[c]

(3) THE RETURN TO COOPERATIVE PRIMITIVITY

What then was the attitude of the Taoists to society? Had they some ideal, other than that of the Confucians, to which they wished human society to conform? They had, and it was at first sight a rather odd one. Its classical expression is found in the 80th chapter of the *Tao Tê Ching*:

Take a small country with a small population. The sage could bring it about that though there were contrivances which saved labour ten or a hundred times over, the people would not use them.[d] He could make the people ready to die twice over for their country rather

[a] Tr. Legge (5), vol. 1, p. 297, mod. Everyone has missed the significance of this; thus Lin Yü-Thang (1) places the passage among those which treat of life and death, and of how Huang Ti became an immortal.

[b] Legge (5), vol. 1, p. 301.

[c] Legge (5), vol. 2, p. 137. 'One must understand the use of uselessness before one can understand the use of usefulness' (Lin Yü-Thang (1), p. 88). These ideas seem to have a relation to the 'natural selection' theme referred to above (p. 80); cf. Fêng Yu-Lan (5), p. 93.

[d] See below, p. 124.

[1] 物之質也   [2] 物之殘也

than emigrate. There might still be boats and chariots but no one would ride in them. There might still be weapons of war but no one would drill with them. He could bring it about that 'the people should go back (from writing) to knotted cords,[a] be contented with their food, pleased with their clothes, satisfied with their homes, and happy in their work and customs. The country over the border might be so near that one could hear the cocks crowing and the dogs barking in it, but the people would grow old and die without ever once troubling to go there.'[b]

Pondering once over this passage, the words of our English seventeenth-century Leveller (or rather Digger) thinker, Gerrard Winstanley, came into my mind, that 'all the world's evils had come about from the dreadful device of buying and selling'.[c] Indeed, the only occasions on which the people from one country would have needed to go to another in ancient Taoist times would have been to buy and sell; or to make war under the leadership of one of the feudal lords. This passage gives the clue, therefore, that the Taoists were the spokesmen of some kind of primitive agrarian collectivism, and were opposed to the feudal nobility and to the merchants alike. Significantly, Ssuma Chhien quoted these very words of Lao Tzu at the beginning of the 129th chapter of the *Shih Chi*, that on the Rich Merchants and Industrialists of the Chhin and Han.[d]

(g) THE ATTACK ON FEUDALISM

It is surprising that the enmity of the Taoists not only for Confucianism but for the whole feudal system has not been more widely understood. The extreme emphasis and even violence of their language ill assorts with the common conception of them as milk-and-water mystics delivering the 'Wisdom of the East'.

At least fifteen chapters of the *Tao Tê Ching* have a clear political significance.[e] The very opening words of chapter 1 state that the Tao (of human society) which can be discussed is not the unvarying Tao (of Nature), hinting that the immutable law of the Legalists (see below, p. 205) is an impossibility.[f] Chapter 9 warns the feudal lords:

> When bronze and jade fill the hall
> It can no longer be guarded.
> Wealth and place breed insolence
> Which brings ruin in its train....

[a] Cf. the Amerindian *quipu*. Cf. pp. 327, 556 below.
[b] Tr. Waley (4); Hughes (1). The passage in quotation-marks occurs also in *Chuang Tzu*, ch. 10 (Legge (5), vol. 1, p. 288).
[c] See Sabine ed. p. 511.
[d] Swann (1), p. 419.
[e] Chs. 1, 9, 13, 14, 16, 17, 24, 38, 39, 49, 53, 57, 58, 74, 79. The translations in this paragraph are those of Waley (4), except the last.
[f] So runs, at least, one widely accepted interpretation. Duyvendak (18) suggests another which is attractive though perhaps not entirely convincing.

and chapter 53 suggests that property is robbery:

> So long as the court is in order,
> (Rulers are content to) let the fields run to weeds
> And the granaries stand empty.
> They wear patterns and embroideries,
> Carry sharp swords, glut themselves with drink and food, have more possessions than
>   they can use—
> These are the riotous ways of brigandage (*tao*[1]); they are not the Tao.

'When the ruler looks depressed', we read in chapter 58, 'the people will be happy and satisfied; when the ruler looks lively and self-assured, the people will be carping and discontented.' Chapter 79 says:

When great wrongs are thought to be righted, there will surely be some remaining bitterness left behind. How can this bring about anything good? (i.e. within feudalism things cannot be set right). Therefore the sage holds the left tally (the less honourable or inferior side)[a] (i.e. takes the part of the people), and does not demand from the people the impossible. (The ruler who) has the Virtue (of the Tao) is (benevolent like the) Grand Almoner; (the ruler who) has not got it is (an oppressor like the) Comptroller of Taxes. The Tao of Heaven has no likes and dislikes; wherever the good are, it is there in their midst.[b]

The attitude of the Taoists is stated in *Chuang Tzu* without circumlocution. An entire chapter, chapter 29, is devoted to an imagined interview between Confucius and a famous brigand, the Robber Chih (*tao* Chih[2]). With his nine thousand followers he marched here and there, devastating houses and farms, stealing movable property, and carrying off people's wives and daughters. Confucius decided to go and see him, and to offer him his services as counsellor. With bitter sarcasm on the origin of kingship, Confucius is made to say:

If you, General, are inclined to listen to me, I should like to go as your commissioner to Wu and Yüeh in the south, to Chhi and Lu in the north...and to Chin and Chhu in the west. I will get them to build for you a great city several hundred *li* in size, to establish under it towns containing several hundred thousands of inhabitants, and honour you there as a feudal lord. Thus you will begin your career afresh, you will disband your soldiers and cease from war, you will collect and nourish your brothers, sacrificing with them to your ancestors—this will be a course befitting a sage and an officer of ability, and will fulfil the wishes of the whole empire.[c]

---

[a] Cf. Granet (5), pp. 364 (6), pp. 261 ff., for a long discussion of left and right in ancient China. My friend Dr O. v. d. Sprenkel has pointed out to me evidence from the *Chhien Han Shu* that in Chhin and former times the poorer people lived on the 'left' side of the villages—the 'wrong side of the railroad tracks' as some might say. This was called *lü tso*.[3] See *Shih Chi*, ch. 48, p. 1*a*, and Dubs (2), vol. 1, p. 123.

[b] Tr. auct. following Hou Wai-Lu (1).   [c] Tr. Legge (5), vol. 2, p. 169.

[1] 盜   [2] 盜跖   [3] 閭左

The Robber, however, will have none of Confucius' advice, and after haranguing him in a long speech of highly Taoist flavour, sends him away completely discomfited. By such stories, scarcely disguised as parables, did the Taoists satirise the tendency of the Confucians to flock around the worst robber barons, vying with one another to become their counsellors. Later in the same chapter, Man Kou-Tê[a] says, 'The shameless become rich, and good talkers become high officials....Small robbers are put in prison, but great robbers become feudal lords, and there in the gates of the feudal lords will your "righteous scholars" be found.'[b] This is reminiscent of the English 18th-century rhyme:

> The law condemns the man or woman
> Who steals the goose from off the common,
> But leaves the greater felon loose
> Who steals the common from the goose!

Admittedly chapter 29 is regarded by some as a later interpolation,[c] but as Legge points out, it is specifically referred to by Ssuma Chhien, so that if it is not from Chuang Tzu, it must be by an early hand. Not even slight suspicion attaches, however, to chapter 10, where exactly the same sentiments are expressed. Its title is Chhü Chhieh,[1] The Cutting Open of Sacks, and Chuang Tzu comes quickly to the point:

Do not those who are vulgarly called wise prove to be but collectors for the great thieves? (*Shih su chih so wei chih chê yu pu wei ta tao chi chê hu*[2]). And do not those who are considered sages then prove to be but guardians in the interest of the great thieves? (*So wei shêng chê yu pu wei ta tao shou chê hu*[3]).[d]...Here is one who steals a buckle (for his girdle)—he is put to death for it. Here is another who steals a State—he becomes its prince. And it is at the gates of the princes that we find benevolence and righteousness (most strongly) professed—is not this stealing benevolence and righteousness, sageness and wisdom? Thus they hasten to become great robbers, carry off princedoms, and steal benevolence and righteousness, with all the gains springing from the use of pecks and bushels, weights and steelyards, tallies and seals....[e] Therefore, if an end were put to 'sageness' and 'wisdom', great robbers would cease to arise.....[f]

In the face of words so strong, there is hardly need to press the point.

---

[a] Legge (5), vol. 2, p. 177. Man Kou-Tê goes on to give detailed examples. Legge regards Man Kou-Tê as a fictitious name, meaning 'Full-of-Gain-recklessly-Got'. Cf. the conversation later in the same chapter between Mr Dissatisfied and Mr Know-the-Mean.

[b] Tr. Lin Yü-Thang (1), p. 80. Compare St Augustine, *De Civ. Dei*, IV, 4, 'Elegant and excellent was that pirate's answer to the great Macedonian, Alexander, who had taken him: the King asking how he dared molest the seas so, he replied, with a free spirit, "How darest thou molest the whole world? Because I do it with a little ship only, I am called a thief; thou doing it with a great navy art called an emperor!".'

[c] Cf. Forke (13), p. 312.

[d] Legge (5), vol. 1, p. 281. The theme is repeated word for word in the course of the chapter.

[e] Note this expression in view of what will follow below, p. 124.

[f] Tr. Legge (5), vol. 1, pp. 281, 283, 285, 286; Vacca (10), mod.

[1] 胠篋   [2] 世俗之所謂知者有不爲大盜積者乎   [3] 所謂聖者有不爲大盜守者乎

Chuang Tzu also takes up the theme of the *Tao Tê Ching* (ch. 79) that the true sages act *with* the people; 'such men live in the world in closest union with the people, going along abreast with them (*tho sêng yü min ping hsing*[1])'.[a] They 'bury themselves' among the people.[b] 'That which is low but must be let alone, is matter. That which is humble but still must be followed, is the people.'[c] Men should not be dealt with as if they were 'things'.[d] The ambition to 'attain one's aim' by getting 'chariots and crowns' is a distorted one.[e] And the following passage brings together many threads in the argument—the ejection of scholastic feudal ethics from developing science allied with primitive democracy, and the assertion of a relativistic view of the universe.

River Spirit said, 'When we are considering either the externals of things, or that which is internal to them, how do we come to make distinctions between them as to noble and mean (*kuei chien*[2]) or as to great and small?' The God of the Northern Sea answered, 'When we look at them in (the light of) the Tao, they are all neither noble nor mean. Among themselves, each thinks itself noble, and despises the others. According to common opinion, their being noble or mean does not depend on themselves. But examining their differences, if we call those great which are greater than others, there is nothing which is not great; and in the same way there is nothing that is not small. To know that heaven and earth are no bigger than a grain of the smallest rice, and that the tip of a hair is as big as a mountain mass—that is to understand the relativity of standards (*tsê chha shu tu i*[3]). Again, examining the services they render, if we call those useful which are more useful than others (for some particular purpose), there is not one which is not useful; and in the same way there is nothing that is not useless. So also we know that East and West are opposed to each other, and yet the (idea of) one cannot exist without the (idea of) the other—thus is their *mutual service* determined (*tsê kung fên ting i*[4])'....[f]

What the Taoists were attacking may remind us, for instance, of the biological statement of Albertus Magnus, made at the height of European feudalism, that male chicks hatch from those eggs which are most spherical, since the sphere is the 'noblest' of all figures in solid geometry.[g] The Taoists were against the concepts of noble and mean as applied to Nature, but they were also against them as applied to Man, and thus they affirmed their science and their democracy at the same time. Just as there was no real greatness and smallness in Nature, so there should be none in human society. The accent should be on mutual service.

[a] Ch. 12 (Legge (5), vol. 1, p. 321); cf. *Huai Nan Tzu*, ch. 1 (Morgan (1), p. 19).
[b] Ch. 25 (tr. Legge (5), vol. 2, p. 121; Waley (6), p. 83). This is characteristic also of European religious pragmatists such as Paracelsus.
[c] Ch. 11 (Lin Yü-Thang (1), p. 77).
[d] Ch. 11 (Legge (5), vol. 1, p. 304; see also p. 378).
[e] Ch. 16 (Legge (5), vol. 1, p. 372; see also p. 379).
[f] Ch. 17, tr. Legge (5), vol. 1, p. 379; Lin Yü-Thang (1), p. 50; Wieger (7), p. 341, mod.
[g] Cf. Needham (5), p. 170; Balss (2), p. 67.

[1] 託生與民並行   [2] 貴賤   [3] 則差數睹矣   [4] 則功分定矣

(1) TAOIST CONDEMNATION OF CLASS-DIFFERENTIATION

What, then, did the Taoists propose as an alternative to feudal society? They proposed nothing new, they did not look *forward*, and strictly speaking, therefore, they were not revolutionary; they looked *back*, and the type of society to which they wished to return can have been nothing other than primitive tribal collectivism. Their ideal was the undifferentiated 'natural' condition of life, before the institution of private property, before the appearance of proto-feudalism with its lords and 'high kings', its priests, artisans and augurs, at the beginning of the bronze age. If it is hard to believe that the memory of this ancient feeling of social solidarity, prior to the development of classes,[a] could have persisted sufficiently long to have inspired the Taoists, one may remember that groups following this way of life are likely to have persisted at the fringes of Chinese society far down into the feudal period.[b] No doubt the 'barbarians', against whom the feudal lords so frequently fought, followed it.

The ideal society of the Taoists was cooperative, not acquisitive. Instead of being subjected to corvée labour and ordered about by the feudal lords, the people in ancient society carried out their activities communally and according to custom.[c] The crafts had not so far differentiated as to preclude communal collaboration at tasks such as house-building. The people spontaneously came together in those annual mating festivals which Granet (2) so carefully reconstructed from the most ancient Chinese folklore, instead of being regimented to assist at periodical sacrifices to the altars of the spirits of the land and grain associated with particular feudal houses and states. In the ancient society there was little need for division of labour, and we shall probably not be far wrong in seeing the great turning-point here as the introduction of bronze-metallurgy, in which a complex technique was associated with the making of superior weapons. The ancients had no use for weapons, for there were no organised wars; they had no use for transportation contrivances such as chariots or boats, for there was no commerce and no need for journeys. Their chiefs half-apologetically exercised leadership from within, and vied with one another in the distribution of the products of the chase or of agriculture in potlatch ceremonies; unlike the feudal lords, whose pleasure it was to tyrannise from above. There was spontaneous cooperation instead of directive force—have we not here the oldest secret of the distinction between *wu wei*[1]

---

[a] Eberhard (9) gives evidence that the Lung-Shan culture (Vol. 1, p. 83 above) was the first to show class-differentiation, judging from extant remains.

[b] Just as Maspero (12), p. 156, was able to make personal observations on mating festivals among the Thai peoples of Indo-China, closely similar to those which are believed to have existed during the Chou period. Indeed, it is possible to quote from Chinese writers of many ages who read back their observations of the customs of environing peoples into their own antiquity. When in +1221 the Taoist Chhiu Chhang-Chhun was on his way across Central Asia to the court of Chingiz Khan, he was much impressed by the Mongol tribesfolk whom he met. 'They have indeed preserved', he said, 'the simplicity of primeval times' (*Chhang-Chhun Chen Jen Hsi Yu Chi*, tr. Waley (10), p. 68).

[c] For an analysis of ancient Chinese society in terms of social anthropology, see Quistorp (1).

[1] 無為

and *wei*[1]?[a] Finally, that ancient society was in all probability matriarchal—is this perhaps not the oldest meaning surviving in that symbol of the Feminine so dear to the Taoists, of which we have already spoken?

To obtain a view of that primitive society one has only to open any anthropological work, e.g. that of Forde, in which a wide variety of primitive societies of food-gatherers, hunters, agriculturists and pastoral nomads is described. Here we may see the transition from communal ownership conditions to feudal tenures and private property, with its concomitants of leases and rents, land-owning lords and landless serfs. The kind of society which the Taoists had in mind might tentatively be likened to the New Guinea society of the Arapesh, described by Mead, which prizes an extreme of non-aggressiveness. It is not necessary to rake over the old controversies associated with the names of L. H. Morgan and F. Engels on 'primitive communism';[b] all that we need to admit is that there was a stage of early society before the development of bronze-age proto-feudalism and the institution of private property, and that the ideals of this society were those which inspired the Taoists. There is, moreover, an analogy for the Chinese case in the fact, demonstrated in a remarkable work by George Thomson (1), that Greek democracy was partly the reassertion by the common people of their lost tribal equality, and that Greek tragedy, when fully understood, shows abundant evidence of the existence of these memories and these processes. In what follows we shall see whether this general interpretation is justified by what the Taoists themselves said.

In the earliest material which we possess, the songs in the *Shih Ching*, much of which must be well before −600, complaints against the feudal lords are already to be found. Two examples may be given:

> Men had their land and farms
> But you (the feudal lord) now have them,
> Men had their people and their folk
> But you have seized them from them.
> Here is one who ought to be held guiltless
> But you keep him (in prison);
> There is one who ought to be held guilty
> But you let him escape and go free.[c]

> Khan, Khan, sings my axe on the *tan*-trees,
> Here on the river's bank I'll lay what I hew,
> Ah, how clear the waters flow, and rippling!
> But you (the feudal lord) sow not nor reap,
> Where do you get the produce of those three hundred farms?

[a] Cf. *Tao Tê Ching*, ch. 29, 'Those that would gain what is under Heaven by force (*wei*[1]); we have seen that they do not (in the long run) succeed. For that which is under Heaven is a holy instrument; it cannot be taken by force. Force ruins it. To grab at it is to lose it.'

[b] For a recent re-examination of these, see Stern (1).

[c] III, iii, 10 (Chan Ang), tr. Legge (8) mod. Mei Yi-Pao (2); cf. Karlgren (14), p. 236.

[1] 爲

You do not follow the chase;
How is it we see those deer hanging up in your hall?
You are a gentleman
And (you say) you do not eat the bread of idleness![a]

Great rats, great rats,
Keep away from our wheat!
These three years we have worked for you
But you despised us;
Now we are going to leave you
And go to a happier country,
Happy land, happy land,
Where we shall find all that we need.[b]

This was the ever-living tradition of protest.

But the Taoists maintained the tradition of something better. Here is the *Chuang Tzu*'s description of primitive collectivism:

(Anciently) the people had a constant nature, they wove themselves clothes and tilled the ground for food. This was what we call the Virtue of the Common Life (*shih wei thung tê*[1]). They were united, forming one single group (*i erh thung tang*[2]) (not separated into different classes); this was what we call Natural Liberty (*ming yüeh thien fang*[3])....In the age of perfect virtue, men lived in common with birds and beasts, and formed one family with all creatures—how could they know of such distinctions as 'princes' and 'villeins' (*chün-tzu, hsiao-jen*[4])? All living without 'knowledge' they kept (to the path of) their natural virtue. This was what we call the state of Pure Simplicity. In that state, the people retained their 'constant nature'. But when the 'sagely men' appeared, cringing and fawning in the imposition of 'benevolence', jostling and tiptoeing in the enforcing of 'righteousness', then men began everywhere to be suspicious. With extravagant orchestras and gesticulating ceremonies, men began to be separated from one another (*thien hsia shih fên i*[5]). The pure solidarity of wood (*shun phu*[6]) was cut about and hacked to make sacrificial vessels. The white jade was broken and injured to make libation-cup handles. The virtues (of the Tao) were disallowed in favour of 'benevolence' and 'righteousness'. The natural instincts were departed from in favour of ceremonies and music. The five colours were confounded to make ornamental patterns. The five notes were confused to make the six pitch-pipe sounds. Now the cutting and hacking of the pure solidarity of raw materials (*phu*[7]) to make vessels was the crime of the 'skilful workmen'; and the injury done to the virtue of the Tao in order to enforce 'benevolence' and 'righteousness' was the transgression of the 'sages'.[c]

[a] I, ix, 6 (Fa Than); tr. Legge (8), mod. Mei Yi-Pao (2); cf. Karlgren (14), p. 71. The songs are of course difficult to translate, but it is astonishing that Waley (1), in his version, makes this come out in praise of the feudal lord. We shall shortly give some remarkable examples of how different translations can be when they are based on different general attitudes to the (often rather ambiguous) Chinese text.
[b] I, ix, 7 (Shih Shu); tr. Legge (8), mod. Alley (2); cf. Karlgren (14), p. 73.
[c] Ch. 9, tr. Legge (5), vol. I, pp. 277 ff.; Vacca (10); mod. Balazs et auct.

[1] 是謂同德   [2] 一而同黨   [3] 命曰天放   [4] 君子小人
[5] 天下始分矣   [6] 純樸   [7] 樸

(2) The Words *Phu* and *Hun-tun* (Social Homogeneity)

Thus we have both the description of primitive collectivism and an account of how it was destroyed.[a] Perhaps it was hard on the 'sages' that the inevitable changes in society arising from changes in productive relationships and the progress of inventions should all be blamed on them, but the Taoists probably had good ground for their denunciations of complacent Confucian social 'wisdom'. The reader is asked to take particular note of the word *phu*[1] in the above passage. It is one of the most important technical terms in Taoist political thought, and we shall immediately meet with it again. The word *fên*[2] may be compared with the Greek *moira* (μοῖρα: Cornford (1), pp. 12 ff.; Thomson (1), p. 38), meaning the part allotted for each person to play in the community, or the goods distributed to each person, but as used here (and in other Chou books such as the *Têng Hsi Tzu*[3] and the −4th-century fragment known as the *Shih Tzu*[4]) it relates to the allotment of duties to persons by the ruler or lord, though its primary meaning is simply separation.[b]

The next passage, from the *Huai Nan Tzu*, continues Chuang Tzu's identical doctrine:

The (true) sages inhaled the *chhi* of the Yin and the Yang, and among all the hosts of the living there was nothing that did not depend upon their virtue, in general likemindedness. At that time there was no special governing authority to give decisions, the people lived their lives in quiet retirement, and things came to fruition of themselves. The world was an undifferentiated unity (*hun-hun tshang-tshang*[5]), the pure collectivity (*shun phu*[6]) had not been broken up and dispersed (*wei san*[7]), the different sorts of people formed a oneness, and all creation flourished exceedingly. Hence if a man with the 'knowledge' of Yi himself had appeared, the world could not have made use of him....(later, as the complexity of society increased) the collectivity was dispersed (*phu san*[8]).[c]

This introduces us to another word, *hun*, which more usually appears in conjunction with *tun* (written in a variety of ways, *hun-tun*[9]), also a cardinal technical term.

Chuang Tzu returns to the decay of primitive society in chapter 11, and now he brings in for the first time not only the legendary benefactors of humanity typical of Confucian sermon-texts, but also the legendary rebels, the execrated monsters,

---

[a] Cf. a parallel passage in *Chuang Tzu*, ch. 12 (Legge (5), vol. 1, p. 315).

[b] *Moira* came in Greek thought to mean Fate, and was considered as superior to the Gods. *Moira*, says Cornford (1), as Fate, came to be supreme in Nature over all the subordinate wills of men and gods, because she had first, as Customary Communal Distribution, been supreme in human society, which was thought of as continuous with Nature. Thus we find Liu An saying that day is the *fên* of Yang and night is the *fên* of Yin (*Huai Nan Tzu*, ch. 3, p. 10b). *Fên*, however, did not have such a successful career, and never usurped or paralleled the position of *Tao*[10] or *Ming*.[11] See below, pp. 109, 112, 461, 479, 528, 550. Cf. Demiéville (3b).

[c] Ch. 2, p. 9b, tr. auct., adjuv. Morgan (1), pp. 46, 47. Note the use of the same technical term for 'dispersion' as we have met with in the physical naturalistic speculations (p. 40).

---

¹ 樸   ² 分   ³ 鄧析子   ⁴ 尸子   ⁵ 渾渾蒼蒼   ⁶ 純樸
⁷ 未散   ⁸ 樸散   ⁹ 混敦 混沌 渾敦 渾沌   ¹⁰ 道   ¹¹ 命

prowling in the folklore as half-beasts, half-men, against whom the 'sage-kings' battled and whom they put to death. I mention them now, to return to them later.

Anciently, Huang Ti was the first to meddle with 'benevolence' and 'righteousness' and to disturb men's minds with them. Afterwards Yao and Shun wore the hair off their legs and the flesh off their arms endeavouring to feed the world. They tormented its economy to enforce 'benevolence' and 'righteousness', and exhausted its circulation to make laws and statutes. Even so they did not succeed. Then Yao shut up Huan-Tou[1] on Mount Chhung, exiled the Three Miao people (San Miao[2]) to the San Wei Mountains, and banished Kung-Kung[3] to Yu-Tu—this was no true conquest of the world....When the Great Virtue lost its togetherness, men's lives were frustrated. When there was a general rush for 'knowledge' men's covetousness outran their possessions. Then the next thing was to invent axes and saws, to kill by laws and statutes set up like carpenters' measuring-lines, to disfigure by hammers and gouges. The world seethed with discontent, and the blame rests upon those who interfered with the (natural goodness of the) heart of man. Hence virtuous men sought refuge in mountain caves, while rulers of great States sat trembling in their ancestral halls. And now, when dead men lie around pillowed on each other's corpses, when cangued prisoners jostle each other in crowds, and condemned 'criminals' are to be seen everywhere, the Confucians and Mohists bustle about and flick back their sleeves, wringing their hands in the midst of the manacled crowd. Alas, they know not shame, nor what it is to blush![a]

A parallel passage in chapter 29 gives a picture of primitive life before even the invention of clothes. The people were innocent, peaceful and cooperative. They 'knew their mothers, but did not know their fathers' (note that this is Chuang Chou speaking about matriarchy, not a modern theoretical archaeologist).[b]

This was the time of Perfected Virtue. Huang Ti, however, was not able to attain to it. He fought with Chhih-Yu[4] (another of the legendary rebels) in the wilderness till the blood flowed for a hundred *li*. When Yao and Shun arose, they instituted the host of officials.... Since that time the strong have tyrannised over the weak....The rulers have all been promoters of disorder and confusion....[c]

Lastly there is a passage in the *Huai Nan Tzu*[d] which is too long to quote, but which must be referred to. After giving in enlarged form a picture of primitive collectivism, it describes all the evils which came from the growing complexity of later ages. Mentioning the introduction of metallurgy and the greatly increased exploitation of all Nature, it points the contrast, saying that in spite of all this 'the instruments of the

---

[a] Tr. auct. adjuv. Legge (5), vol. 1, p. 295; Lin Yü-Thang (1), p. 126; Waley (6), p. 104.

[b] Kungsun Yang says exactly the same thing in a brief sketch of social evolution (seen from the Legalist point of view) at the beginning of ch. 7 of the *Shang Chün Shu* (cf. Duyvendak (3), p. 225). It seems very probable that matriarchal systems existed in ancient Chinese society; cf. Erkes (15) and Rousselle (3).

[c] Tr. Legge (5), vol. 2, p. 171.

[d] At the opening of ch. 8 (Morgan (1), p. 81). Cf. similar passages in ch. 2 (Morgan (1), p. 35) and ch. 13 (Morgan (1), p. 144).

[1] 驩兜   [2] 三苗   [3] 共工   [4] 蚩尤

people were not sufficient for their use, while the storehouses (of the rulers) were overfull (*jen hsieh pu tsu, hsü tsang yu yü*[1])'. The wealthy (*chi*[2]) thought only of enrichment (*wei li*[3]). It seemed as if nothing could satisfy the desires of the rulers (*wei nêng tan jen chu chih yü yeh*[4]). Then:

the mountains and streams were divided (*fên*[5]) with boundaries and enclosures, censuses of the populations were made, cities were built and dykes dug, barriers were erected and weapons forged for defence. Officials with special badges were ordained, who differentiated the people into the classes of 'noble' and 'mean' (*i kuei chien*[6]), and organised rewards and punishments. Then there arose soldiers and weapons, giving rise to wars and strifes. There was the arbitrary murder of the guiltless and the punishment and death of the innocent....[a]

It would be needless to enlarge further on this theme or to adduce more evidence. All is summed up by Lao Tzu (*Tao Tê Ching*, chapter 18):[b]

> It was when the Great Tao declined
> That 'benevolence' and 'righteousness' arose;
> It was when 'knowledge' and 'wisdom' appeared
> That the Great Lie began.
> Not till the six near ones had lost their harmony
> Was there talk of 'filial piety',
> Not till countries and families were dark with strife
> Did we hear of 'loyal ministers'.[c]

There are other parts of the *Tao Tê Ching* which are of great interest in this connection, and I cannot forbear from mentioning them because they show what gulfs can separate translations made from different points of view.[d] For example, Waley (4) has given a well-known version of chapter 17:

> Of the highest the people merely know that such a one exists,
> The next they draw near to and praise,
> The next they shrink from, intimidated, but revile.
> Truly, 'It is by not believing people that you turn them into liars.'
> But from the sage it is so hard at any price to get a single word
> That when his task is accomplished, his work done,
> Throughout the country everyone says,
> 'It happened of its own accord.'

[a] Tr. Morgan (1), pp. 82, 83.
[b] Echoed in *Wên Tzu* and elsewhere (cf. Forke (13), p. 347).
[c] Tr. auct. adjuv. Waley (4); Duyvendak (18).
[d] The versions of Hou Wai-Lu which will follow may easily be considered rather 'forced', and even too greatly to 'read back' into the text distinctively modern ideas. But I consider that it is worth while to run the risk of overstating the case, in order to redress a balance till now too heavily the other way.

[1] 人械不足蓄藏有餘   [2] 積   [3] 爲利   [4] 未能澹人主之欲也
[5] 分   [6] 異貴賤

The versions of Hughes, Chhu Ta-Kao and others are not substantially different. But a modern Chinese scholar, appreciating the position just outlined, manifests his realisation of it in the following translation of the passage:

> In highest antiquity (the people) did not[a] know private property.
> Later on families acquired it and held it in high repute;
> Still later this led to fear and reviling.
> Truly it is by not trusting people that mistrust is generated.
> (How remote from this were the) sages, brief of speech!
> For when their tasks were accomplished, their work done
> Throughout the country the people said, 'It all came to us quite naturally.'[b]

Whatever may be thought of his choice of terms, it is justified to this extent, that the words *thai shang*[1] may mean either the highest or the oldest,[c] and *yu*[2] may mean to exist or to have. Here is another case. Waley accepts chapter 11 in the usual mystical sense:

> We put thirty spokes together and call it a wheel;
> But it is on the space where there is nothing that the utility of the wheel depends.
> We turn clay to make a vessel;
> But it is on the space where there is nothing that the utility of the vessel depends.
> We pierce doors and windows to make a house;
> And it is on these spaces where there is nothing that the utility of the house depends.
> Therefore just as we take advantage of what is,
> We should recognise the utility of what is not.

Far different is Hou Wai-Lu's interpretation:

> Thirty spokes combine to make a wheel;
> When there was no private property carts were made for use.
> Clay is formed to make vessels;
> When there was no private property vessels were made for use.
> Windows and doors go to make a house;
> When there was no private property houses were made for use.
> Thus having private property leads to profit (*li*[3]) (for the feudal lords),
> But not having it leads to use (*yung*[4]) (for the people).[d]

Unconsciously perhaps, in what might be considered a strange translation, the author recognises the opposition between *li* and *yung*, which is obscured in the conventional version, and interprets *wu* and *yu* as not-having and having (private property)

---

[a] Emending *hsia*[5] to *pu*.[6] Chhu Ta-Kao (2) and Duyvendak (18) also thus.
[b] Tr. auct. following Hou Wai-Lu (*1*), p. 164.
[c] Hou Wai-Lu's interpretation here was subsequently adopted also by Duyvendak.
[d] Tr. auct. following Hou Wai-Lu (personal conversation, March 1946).

[1] 太上   [2] 有   [3] 利   [4] 用   [5] 下   [6] 不

respectively, instead of not-existing and existing. While admitting that the conventional interpretation is sanctioned by Chinese commentators in subsequent ages, there can be little doubt as to which is more in line with the general political position of the ancient Taoists.[a]

Before finally fixing on the meanings of the technical terms to which allusion has been made, we may glance at the *Lieh Tzu* book and its contribution. Without much overlap in parallel passages, it takes just the same line. In chapters 2 and 5 there are delightful descriptions of the Taoist paradise, where class distinctions and rulers are unknown and where the people with ageless bodies wander beside the Waters of Life;[b] and chapter 1 has a significant passage on property, in which it is denied that even a man's own body is his own property, still less his 'possessions'—all belongs to Heaven.[c] This leads on to a story which demands quotation, on account of its linked scientific and political significance; it might be entitled 'Rob Nature and not Man':

A man called Kuo of the State of Chhi was very rich, while a man called Hsiang of the State of Sung was very poor. The latter travelled from Sung to Chhi to ask the former the secret of his prosperity. Kuo said, 'It is because I am a good robber. The first year I started I got already something, the second year I got enough, and the third year I had great lands. In the end I found myself the owner of whole villages and districts.' Hsiang was delighted, having understood the words but not the sense. So he started to climb over walls and break into houses, grabbing everything he could set eye or hand upon. But before long his robberies brought him into trouble, and he was stripped even of what he had previously had. Thinking that Kuo had basely deceived him, he went to him and bitterly complained. Kuo asked him how he had gone about being a robber, and when he had explained, Kuo said, 'Alas, what a misunderstanding. Now I will tell you how to do it. I heard that heaven has its seasons and earth gives its increase. These are what I rob, the moisture of the clouds and rain, the fruitfulness of mountain and valley, to grow my grain and ripen my crops, to build my walls and make my houses. Fowls and game I rob from the land, fish and turtles from the waters. There is nothing that I do not steal. For all these things are Nature's products; how could I claim them as my private property (yu^1)? But this kind of robbery is not ill-omened. On the other hand, gold, jade, precious stones, stores of grain (got by feudal tenures), silk stuffs, and other kinds of property are things accumulated by men and not the free gifts of Heaven. So who can complain if those who rob them get into trouble?'

Hsiang was much perplexed, and fearing to be led astray a second time by Kuo, went and consulted Master Tungkuo. Tungkuo said to him, 'Are you not already a robber in respect of your own body? You steal the harmony of the Yin and the Yang in order to keep alive and maintain your bodily form. How much more, then, are you a thief with regard to external possessions! Assuredly Heaven and Earth cannot be dissociated from the myriad objects of Nature. To claim any of these as private property (yu) betokens confusion of thought.

[a] Wulff (1) also rejects the common interpretation, and proposes another along the lines of 'before spoked wheels existed, carts were in use', suggesting that the passage is a polemic against undue luxury or complexity—still with a political undertone. Wulff agrees with Hou Wai-Lu in emphasising the contrast between *li* and *yung*.

[b] Tr. R. Wilhelm (4), p. 53; see p. 142 below.

[c] Tr. R. Wilhelm (4), p. 9.

[1] 有

Kuo's robberies are carried out in the spirit of the Tao of the common life (*kung Tao*[1]) and therefore bring no retribution. But your robberies were carried out in a spirit of self-seeking (*ssu hsin*[2]) and therefore landed you in trouble. He who aligns his private interests with those of the common weal (*yu kung ssu*[3]) is (in one sense) a robber; he who does not (*wang kung ssu*[4]) is also (in quite another sense) a robber. Community breeds community and selfishness breeds selfishness, such is the principle of Heaven and Earth. If we know this principle, can we not say who is a robber (truly so called) and who is a robber (falsely so called)?'[a]

The moral of this striking passage obviously is that it is legitimate to take and enjoy the spoils of Nature for the good of the whole community, but that accumulating wealth for private ends is an anti-social characteristic of the feudal lords, and only leads to the appearance of true robbers who destroy their predecessors and become feudal lords in their turn. For its modern flavour it deserves to be set beside the passage of Chuang Tzu on 'pure and applied science', pointing as it does to the socialist-capitalist arguments of more than two thousand years later, on production for profit as opposed to exploitation of Nature for use.

The Taoists, then, condemned the differentiation of society into classes. Rightly they associated the process with increasing artificiality and complexity of life, and urged a return to the pure Primitive Solidarity (*shun phu*[5]). Surely this is the sense in which we should take the famous parable at the end of chapter 7 of *Chuang Tzu*, which is always conventionally interpreted in a mystical sense:

The Ruler of the Southern Ocean was called Reckless-Change (*Shu*[6]); the Ruler of the Northern Ocean was called Uncertainty (*Hu*[7]), and the Ruler of the Centre was called Primitivity (*Hun-tun*[8]). Reckless-Change and Uncertainty often used to meet on the territory of Primitivity, and being always well treated by him, determined to repay his kindness. They said, 'All men have seven orifices, for seeing, hearing, eating, breathing, etc. Primitivity alone has none of these. Let us try to bore some for him.' So every day they bored one hole; but on the seventh day Primitivity died.[b]

The boring of the holes symbolises the differentiation of classes, the institution of private property, and the setting up of feudalism. With this in mind we can understand chapter 56 of the *Tao Tê Ching*:

> Block up the 'apertures',
> Close the 'doors' (which Change and Uncertainty made in Primitivity),
> Blunt the edges (of weapons),
> Dissolve the feudal class-distinctions (*fên*[9])
> Harmonise the brilliances (the talented, for the community),

[a] Tr. R. Wilhelm (4), p. 10, eng. auct.; adjuv. L. Giles (4), p. 32.
[b] Tr. Fêng Yu-Lan (5), p. 141. 'Fuss' and 'Fret' are the two Rulers in Waley (6), p. 97.

[1] 公道   [2] 私心   [3] 有公私   [4] 亡公私   [5] 純樸   [6] 儵
[7] 忽   [8] 渾沌   [9] 分

Unite the dusts (the rank and file, for the community)—
This is called the mysterious Togetherness (*hsüan thung*[1]),
    (For in this community) there can be no likings nor dislikings,
    No private profit (*li*[2]) and no loss (*hai*[3]),
    There can be no 'honourable persons' and no 'mean ones',
    And therefore it is the most honourable thing under Heaven.[a]

A glance at the conventional translation, e.g. Waley (4), will show how it is possible to interpret a passage of this kind purely in terms of mystical meditation. This is too restricted a view, though once the original meaning was lost, it was natural and inevitable that it should arise (cf. above, p. 98, below, p. 140).

All this suggests that the Taoists believed in the practicability of action in their own time to restore the state of Primitivity. Like the other schools they sought for rulers who would be prepared to put their principles into practice—and it goes without saying that they were remarkably unsuccessful. But, as chapter 14 of the *Tao Tê Ching* says: 'By grasping the Tao that was of old (*chih ku chih Tao*[4]), you can master the present era of private property (*i yü chin chih yu*[5]).' And a whole programme is mapped out in chapter 57:

Governing a country (the Confucians say) needs rectification,
Command of soldiers in the field (they say) requires strategy,
But the adherence of all under Heaven can only be won by those who have no private ends
    in mind (*wu shih*[6]).
How do we know that this is so? By looking at the facts.
The more tabus there are, the poorer the people will be,
The more contrivances for private profit (*li chhi*[7]) there are, the more benighted the whole
    land will grow,
The more cunning craftsmen there are, the more monstrous inventions there will be,
The more laws are promulgated, the more bandits will abound;
Therefore the sages have said,
If we do nothing (for private ends, and contrary to Nature, *wu wei*[8]), the people will be
    spontaneously transformed,
If we love calmness of mind (ataraxy), the people will set themselves in order,
If we take no action (for private ends), the people will spontaneously grow prosperous
If we have no personal ambitions (*wu yü*[9]) the people will spontaneously achieve cooperative
    simplicity (*min tzu phu*[10]).[b]

We have now reached the point when we can understand the technical terms which most Chinese and all European expositors of Taoism have failed to appreciate. The crucial passage, which connects *phu* and *hun-tun* together, is in *Chuang Tzu*,

[a] Tr. auct. following Hou Wai-Lu's interpretation.
[b] Tr. auct. following Hou Wai-Lu. The subsequent version of Duyvendak (18) is not very dissimilar.

[1] 玄同   [2] 利   [3] 害   [4] 執 古 之 道   [5] 以 御 今 之 有
[6] 無事   [7] 利 器   [8] 無 爲   [9] 無 欲   [10] 民 自 樸

chapter 12.[a] Tzu-Kung, one of the disciples of Confucius, had been on a journey, and had met a Taoist farmer to whom we shall shortly have to refer again (p. 124) in a rather different connection.

When he returned to (the State of) Lu, he told Confucius about the interview and conversation. Confucius said, 'That man pretends to cultivate the arts of the Primitive Homogeneity School (*pi chia hsiu Hun-Tun shih chih shu chê yeh*[1]). He is acquainted with the first (stage), but does not know the second (*shih chhi i, pu chih chhi erh*[2]) (i.e. he understands primitive collectivist society, but not feudal society). He can regulate what is internal to himself (i.e. his own cooperativeness), but not what is external to himself (i.e. the government of men). He understands how to be unsophisticated (*ju su*[3]), how to avoid acting contrary to Nature (*wu wei*[4]), how to return to the Primitive Undifferentiatedness (*fu phu*[5]); embodying his (true) nature and cherishing his spirit, he wanders among the people as one of themselves (*i yu shih su chih chien chê*[6])—you may well be alarmed at his depravity. But as for the arts of the Primitive Homogeneity School, what should you or I find worth knowing in them?'[b]

These ironical words are of course put into the mouth of Confucius by Chuang Chou or whoever wrote the passage. It should by now be quite clear that we are dealing with a definite political system, which used certain technical terms as a half-disguise already perhaps necessary as a protection against the enmity of the feudal lords. What should Confucius, indeed, the counsellor-in-chief of feudal kingship, find worth knowing in the Primitive Homogeneity School?

The dictionary meaning of the word *phu* is 'sincere, simple, the raw substance of things, things in the rough'. It occurs at least half a dozen times in the *Tao Tê Ching*,[c] and is generally englished by Waley (4, 6) as 'the uncarved block' with a purely mystical sense, a rendering which has had much approbation. I suggest, on the contrary, that though later doubtless it was understood in this way, its original meaning contained a very strong political element, referring to the solidarity, homogeneity and simplicity of primitive collectivism. This, indeed, appears from chapter 15, where the 'good leaders *of old*' (*ku chih shan wei shih chê*[7])[d] are described as 'honest, as in the time of primitive solidarity' (*tun hsi chhi jo phu*[8]),[e] 'receptive, like a valley' (cf. 'the Valley Spirit never dies'); and 'all-embracing, like a turbid stream' (suspending all particles within its homogeneity) (*hun hsi chhi jo cho*[9]).[f] Chapter 19 urges

---

[a] We have already noted a similar juxtaposition in the quotation from *Huai Nan Tzu*, ch. 2, on p. 107.
[b] Tr. auct. adjuv. Legge (5), vol. 1, p. 322.
[c] Chs. 15, 19, 28, 32, 37, 57.
[d] Of course Waley translates, 'Of old those that were the best officers of Court.' We write 'leaders' with hesitation; perhaps just 'wise men' were meant.
[e] 'Blank, as a piece of uncarved wood' (Waley, 4). Note that *tun* is a *tun* of *hun-tun*.
[f] 'Murky, as a troubled stream' (Waley, 4). Note that *hun* is the *hun* of *hun-tun*.

[1] 彼假修渾沌氏之術者也   [2] 識其一不知其二   [3] 入素   [4] 無爲
[5] 復樸   [6] 以遊世俗之間者   [7] 古之善爲士者   [8] 敦兮其若樸
[9] 混兮其若濁

the people to cherish solidarity (*pao phu*[1]); chapter 28 urges the sage to return to the solidarity principle (*kuei yü phu*[2]); chapter 32 says:

> The Tao is eternal, but has no fame (*ming*[3]) (feudal glory).
> As for the commonwealth of equals (*phu*[4]), though seemingly of small account,
> There is no being under Heaven that could look down upon it.
> If lords and princes were willing to guard it,
> The ten thousand creatures would spontaneously do them homage.
> Heaven and Earth would be at one, and sweet dew would descend,
> Without law or compulsion, men would dwell in harmony....[a]

And the other places where the word is used are quite similar. It also occurs twice in the *Lü Shih Chhun Chhiu* and nine times in *Huai Nan Tzu*, always with the same significance.[b]

Its appearances are frequently associated with the term *hun-tun*,[c] normally translated 'chaos' but which has also meant 'turbid, confused, disorderly, etc.' Like *phu*, as an ancient Taoist political technical term, I am convinced that it signified 'undifferentiated, homogeneous' and hence implied the state of primitive pre-feudal collectivism. *Hun* and *tun*, alone or together, are found several times in the *Tao Tê Ching*, and at least five times in *Huai Nan Tzu*, always with approximately the same significance. Granet[d] brings evidence that one of the meanings of *hun-tun* was the bag or bellows of the earliest metallurgists (like the term *huan-tou* discussed immediately below).[e]

(3) THE LEGENDARY REBELS

Now it is extremely interesting that there is a group of other terms, the names in fact of mythological beings, which, as we have seen, are mentioned in Taoist texts, and which denote the legendary rebels against whom the earliest legendary kings had to fight, and whom they destroyed. A wealth of evidence concerning them has been collected by Granet (1, 2, 4), Karlgren (2) and Maspero (8). While there remains much disagreement concerning their origin, exact mythological position, and the various cults connected with them, there is no doubt that they played an important part in ancient Chinese religious thought and practice (Fig. 38). Let us list them:

(1) CHHIH-YU[5] (already mentioned), minister and rival of Huang Ti, an ox-headed monster or sea-dragon, legendary inventor of metallurgy and metal weapons, connected in

---

[a] Tr. auct. adjuv. Waley (4). We have all unconsciously fallen here into the metre of old William Langland, who would not, I think, have entirely disowned the sentiments of Lao Tzu.

[b] Cf. Wieger (2), p. 333.

[c] This is written in a number of alternative ways,[6, 7, 8, 9] as we saw above.

[d] (1), pp. 543 ff.

[e] A bellows bag is also called *nang*.[10] Perhaps significantly, this term occurs again much later in the Taoist tradition, in the titles of a number of books in the *Tao Tsang*.

[1] 抱樸   [2] 歸於樸   [3] 名   [4] 樸   [5] 蚩尤   [6] 混敦

[7] 混沌   [8] 渾敦   [9] 渾沌   [10] 囊

四山服罪圖

伯鯀

三苗

共工

驩兜

Fig. 38. A ate Chhing representation of the expulsion of the Legendary Rebels (left to right: San Miao, Kung-Kung, Po Kun (Lord Kun) and Huan-Tou). From *SCTS*, ch. 2, Shun Tien, (Karlgren (12), p. 5). The military uniforms are of course anachronisms.

some way with the Nine Li and the Three Miao (see below).[a] Granet (1), p. 351, etc.;
Karlgren (2), p. 283; H. Maspero (8), pp. 55, 79. Killed by Huang Ti.

(2) HUAN-TOU[1] (already mentioned), identified with Hun-Tun, a monster banished by
Huang Ti. Granet (1), pp. 240, 248, 258, 267, etc.; Karlgren (2), pp. 249, 254.

(3) KUN,[2] father of Yü the Great who succeeded in constructing the necessary water-
conservation works where Kun had failed.[b] Minister of Huang Ti and rebel against him,
banished, executed and cut into pieces by him. Changes into various animals (bear, yellow
dragon), and is eaten by various animals (owl, tortoise). Legendary inventor of embankments and walls. Granet (1), pp. 240-73, etc.; Karlgren (2), pp. 249, 254.

(4) THAO-WU,[3] monster banished by Huang Ti, sometimes identified with Kun. Granet (1),
pp. 240 ff.; Karlgren (2), p. 248.

(5) KUNG-KUNG[4] (already mentioned), chief of the artisans ('Minister of Works'),
banished and killed by Shun or by Yü. Granet (1), pp. 240, 318, 368, 523, etc.; Karlgren (2),
pp. 218, 309, 349; H. Maspero (8), pp. 54, 75.

(6) THAO-THIEH,[5] banished and killed by Huang Ti. Connected with the Three Miao and
with copper metallurgy. Represented as an ox or an owl. Granet (1), pp. 240, 244, 248, 258,
491, etc.; Karlgren (2), p. 248.

(7) CHIU LI,[6] the Nine Li, tribes or confraternities of people connected with Chhih-Yu,
overcome by Huang Ti, disturbers of time and calendar. Granet (1), pp. 242, 350, etc.

(8) SAN MIAO[7] (already mentioned), the Three Miao, tribes or confraternities of people,
overthrown and banished by various legendary kings, disturbers of time and calendar,
connected with Chhih-Yu and Thao-Thieh, associated with metallurgy, symbolised by an
owl with three bodies. Granet (1), pp. 239-69, 494, 515, etc.; Karlgren (2), pp. 249, 254;
H. Maspero (8), pp. 97 ff. (Fig. 39).

It is of much interest that some of these names have a distinctly similar ring to the
technical terms just mentioned. Huan-Tou means literally 'peaceable bellows', *tou*
having the significance of an empty bag, i.e. with a homogeneous content of air;
he or it was identified with *hun-tun*, homogeneous 'chaos'. Thao-Wu means an
untrimmed stake, post, beam or log. Other names have an obvious connection with
the working people. Kung-Kung means literally 'communal labour', and Thao-
Thieh is always translated 'glutton', which may well have been an expression used
by the feudal lords for the mass of the people, whom they considered were consuming
too much of the available agricultural product. In later times this name became
a technical term for a certain kind of ornamental design found on bronzes, jades,
buckles, etc. (Ferguson (2), p. 9; Bushell (2); Lemaitre (1)) and persisting in
Tibetan religious art till today (Cammann, 1). Hentze (3) has shown that it must be
the head and pelt of a bear skinned in such a way that the lower jaw is split in two and
retracted; this was undoubtedly worn by shamans in their rites—again a proto-Taoist
connection (Hopkins, 33).

---

[a] Chhih-Yu, later deified, became a favourite god of the Han military (Liu Ming-Shu, 1).
[b] We shall meet with Kun again in connection with hydraulic engineering (Sect. 28*f*).

[1] 驩兜   [2] 鯀   [3] 檮杌   [4] 共工   [5] 饕餮   [6] 九黎
[7] 三苗

苗民詛盟圖

Fig. 39. A late Chhing representation of the swearing of an oath (of mutual alliance) by the Confraternity of the Three Miao. From *SCTS*, ch. 47, Lü Hsing, (Karlgren (12), p. 74).

I suggest, therefore, as a hypothesis for further research, that we should see behind these legendary symbols the leaders of that pre-feudal collectivist society which resisted transformation into feudal or proto-feudal class-differentiated society. The Three Miao and the Nine Li would represent metal-working confraternities. It is striking that in every case the legends attribute to the rebels the character of great metal-workers. It is striking that the bag or bellows (tho[1]) comes prominently into the picture, for a great deal of ancient Chinese folklore gathered round that primitive contrivance, much of it relating to owls, which would seem to have been the tabu-animal of the earliest Chinese metallurgists. The leaders of pre-feudal collectivist society would then have attempted to resist the earliest feudal lords, and to prevent them from acquiring metal-working as the basis of their power. The failure of Kun and the success of Yü may indicate that the relatively unorganised collectivist tribal society was unable to master the task of constructing the minimal requirement of water-conservancy and flood-protection works, and that the institution of forced corvée labour was necessary for this. A vast mass of folklore is available from Han and pre-Han texts, and a full working out of the views here suggested would embrace such diverse subjects as the origins of towns in Chinese proto-feudalism,[a] the position of totemism and ritual dances, the secret societies of the first bronze-founders, human and other sacrifices, drums, potlatch, ordeals, rain- and foam-magic, etc.

Granet, as the result of his researches, convinced himself that the beginning of bronze-working was connected with the rise of Chinese proto-feudalism, but he did not notice the connection between these legendary rebels and their subsequent favourable mention in Taoist texts. Karlgren, venturing on no interpretative hypo-thesis, has strongly criticised Granet's methods, which did not differentiate between the various bodies of pre-Han and Han legends, but there is perhaps something to be said for taking the whole of ancient Chinese folklore as a unity. Recent books by Chinese scholars, such as Hsü Ping-Chhang (1), throw little further light on the problem here raised, but those of Hou Wai-Lu (1) and Kuo Mo-Jo (3) contain hints of the present interpretation, and I know that these two Chinese scholars, among others, are in general agreement with the description of the political position of the Taoists here outlined. The interesting recent book of Yang Hsing-Shun (1) also supports it.

Already before the Han, and for centuries afterwards, the legendary rebels had become spirits of various kinds which received worship and sacrifices. By the +4th century Hun-Tun and Thao-Wu had become miraculous animals in the *Shen I Ching*.[2] The great Taoist, Ko Hung, of the +4th century, adopted 'pao Phu' as his name[b] (Pao Phu Tzu[3]), but it is unlikely that by then it had retained much of its original political significance.[c] Hun-Tun, used as a term for primeval chaos, is found, for

---

[a] We shall return to this; Sect. 48.   [b] 'Embracing' or 'Preserving, Solidarity'.
[c] He may have copied it from the sobriquet of Master An Chhi, a pharmaceutical magician of Chhin Shih Huang Ti's time. See p. 134 below.

[1] 橐   [2] 神異經   [3] 抱樸子

instance, in the +7th-century *Pên Chhi Ching*[1] (Book of Origins);[a] and by the +13th century it has become a technical term for the tenuous matter out of which the adept can form, by uniting seminal essence and *chhi*, through breathing and other exercises, an embryo of immortality within himself (*Shu Chü Tzu*[2]).[b]

(4) THE 'DIGGERS', HSÜ HSING AND CHHEN HSIANG

If the Taoists really held the political views which I am suggesting that they did hold, one would expect to find traces of some close connection between them and the working people. Such traces exist. In the practice of Hsü Hsing[3] and Chhen Hsiang,[4] two 'philosophers' who appear in *Mêng Tzu*, and whose date must therefore be somewhat before −300, we can dimly see traces of cooperative agricultural units reminiscent of the Digger Movement in the English Revolution of the 17th century.[c] The school with which these names are associated was given an independent existence by the compiler of the bibliography of the *Chhien Han Shu*, who termed it the Nung Chia[5] (School of Agriculturists), but we can see that they must have been extremely close to the Taoists.

There came from Chhu (in the south) to Thêng one Hsü Hsing, who gave out that he acted according to the words of Shen Nung.[d] Coming right up to his gate, he addressed Duke Wên, saying, 'A man of a distant region, I have heard that you, Prince, are practising a virtuous government, and I wish to receive a site for a house, and to become one of your people.' Duke Wên gave him a dwelling-place. His disciples, amounting to several tens, all wore coarse hempen cloth, made sandals of hemp and wove mats for a living.

At the same time Chhen Hsiang, a disciple of Chhen Liang, together with his younger brother Hsin, came from Sung to Thêng, with their plough handles and shares on their backs...(and settled in the same way)....Chhen Hsiang became the follower of Hsü Hsing.

At an interview with Mencius, Chhen Hsiang thus reported the words of Hsü Hsing: 'The ruler of Thêng is indeed a worthy prince, but nevertheless he has not heard of the Tao. Real leaders cultivate the ground in common with the people, and so eat. They prepare their own morning and evening meals, carrying on government at the same time. But now

---

[a] Wieger (2), p. 342.

[b] Wieger (2), p. 349. The oddest trace which Hun-Tun has left is the dish still commonly eaten in China today, called *hun-tun*,[6] the characters being written with the 'eat' radical, no. 184. It is a soup containing meat wrapped in paste ravioli with very thin walls. Tai Chih, in *Shu Pho* (Rats and Jade), p. 8*b*, of *c.* +1260, examined the question of its origin, and could not trace it back before the Thang, though from that time onwards it had been popular. But he knew of a pharmaceutical book (not specified) which stated that if these ravioli were fried with *ai* (*Artemisia vulgaris*, B III, 72), which drives away all demon *chhi*'s, the effect would be enhanced, and this shows, he said, that they must be connected with the ancient ideas of *hun-tun*. In other words, some very old sacrificial or exorcistic custom must be involved. How few who enjoy this dish today realise its roots in the ancient past! (The passage is included in *Shuo Fu*, ch. 99, p. 3*a*.) Cf. *Thang Yü Lin*, ch. 8, p. 28*a*.

[c] Another, more contemporary, parallel would be the 'self-sufficiency' doctrine of Hippias of Elis (fl. −5th century); see Freeman (1), p. 381; Lovejoy & Boas (1), p. 115. Li Mai-Mai (1) has emphasised the social significance of the followers of Hsü Hsing.

[d] The Heavenly Husbandman, a legendary culture-hero.

[1] 本起經   [2] 叔苴子   [3] 許行   [4] 陳相   [5] 農家   [6] 餛飩

the ruler of Thêng has his granaries, treasuries and arsenals, which is oppressing the people to nourish himself. How can he be deemed a real leader?'[a]

Mencius then engages Chhen Hsiang in an argument on the division of labour, and maintains that just as a man cannot be an artisan and a farmer at the same time, so some must labour with their minds (at governing) while others labour with their bodies, both having a right to their daily bread. He glosses over the inequality of the rewards, and does not hesitate to abuse Hsü Hsing as 'that shrike-tongued barbarian of the south'. Chhen Hsiang, however, returns to the attack and claims that if Hsü Hsing's doctrines were followed, there would be a standardisation of prices and no deceit in markets. Mencius gives himself the last word, declaring that 'it is of the nature of things to be of unequal quality'. The compiler of the bibliography in the *Chhien Han Shu* says of the followers of the Agriculture School that 'they could see no use for sage-kings. Desiring both ruler and subject to plough together in the fields, they overthrew the order of upper and lower classes.' But he listed nine books of this school. All have long been lost, and doubtless some of them were technical.

In any case, throughout the subsequent centuries material production and manual labour continued to be a trait of Taoist communities.[b]

(5) THE 'KNACK-PASSAGES' AND TECHNOLOGY

Another connection of the Taoists with manual work and technology is seen in a type of story which is so frequent that one may call them 'knack-passages'. Their general burden is that wonderful skills cannot be taught or transferred, but are attainable by minute concentration on the Tao running through natural objects of all kinds. We have already seen in the *Chuang Tzu* a typical 'knack-passage' in the story of Ting, the butcher of Prince Hui (cf. p. 45). But there are many more, concerning musicians,[c] cicada-catchers,[d] boatmen,[e] swimmers,[f] sword-makers,[g] bellstand-carvers,[h] arrow-makers,[i] and wheelwrights.[j] The *Lieh Tzu* is full of them too, and speaks of animal-tamers,[k] boatmen,[l] cicada-catchers,[m] swimmers,[n] and mathematicians.[o] The *Huai Nan Tzu* adds a story about buckle-makers.[p] It is hard at first to see exactly what was the purport of this recurring theme, but one cannot overlook a connection with that

[a] *Mêng Tzu*, III (1), iv, 1, tr. Legge (3), mod.
[b] Cf. Chhen Jung-Chieh (4), pp. 148, 150. The writer has clear recollections of the iron-foundry which was an important part of the great Taoist abbey of Miao-thai-tzu, in Shensi, visited by him several times during the second World War.
[c] Ch. 2 (Legge (5), vol. 1, p. 186). This passage includes a criticism of the Logicians, who, failing to master techniques themselves, chop logic about them.
[d] Ch. 19 (Legge (5), vol. 2, p. 14).　　　　　[e] Ch. 19 (Legge (5), vol. 2, p. 15).
[f] Ch. 19 (Legge (5), vol. 2, p. 21).　　　　　[g] Ch. 22 (Legge (5), vol. 2, p. 70).
[h] Ch. 19 (Legge (5), vol. 2, p. 22).　　　　　[i] Ch. 19 (Legge (5), vol. 2, p. 23).
[j] Ch. 13 (Legge (5), vol. 1, p. 343).
[k] Ch. 2 (L. Giles (4), p. 47). In this case, the expert explains his methods.
[l] Ch. 2 (R. Wilhelm (4), p. 18).　　　　　[m] Ch. 2 (R. Wilhelm (4), p. 19).
[n] Ch. 2 (R. Wilhelm (4), p. 19).　　　　　[o] Ch. 8 (R. Wilhelm (4), p. 107).
[p] Ch. 12 (Morgan (1), p. 125).

empiricism which we have already remarked (p. 73) and which has its echoes as far down as the Thang in, for example, *Kuan Yin Tzu*. The Taoists probably saw in those who exhibited these skills a certain admirable self-forgetfulness arising out of an extremely close contact with the processes of Nature. It was perhaps their substitute for the theoretical and analytic-synthetic approach of the Greeks, and one cannot fail to view it against the background of the great contributions of early Chinese technology. The Taoists felt, moreover, that these workers by hand and brain had much to teach the rulers of society.

In view of its joint technological and political importance, the story of Duke Huan and the Wheelwright demands quotation. It is found both in *Chuang Tzu*[a] and in *Huai Nan Tzu*;[b] I use the former:

Duke Huan (of Chhi), seated above in his hall, was (once) reading a book, and the wheelwright Pien was making a wheel (in the courtyard) below. Laying aside his mallet and chisel, Pien went up the steps, and said, 'I venture to ask, Sir, what you are reading?' The duke said, 'The words of the sages.' 'Are those sages, then, alive?' Pien continued. 'They are dead', was the reply. 'Then', was the reply, 'what you, my ruler, are reading are only the dregs and refuse of bygone men.' The duke, angered, said, 'How should you, a wheelwright, have anything to say about the book which I am reading? If you can explain yourself, very well; if you cannot, you shall die!' The wheelwright said, 'Your servant looks at the matter from the point of view of his own art. If my stroke is too slow, then the tool bites deep but is not steady; if my stroke is too fast, then it is steady but does not go deep. The right pace, neither (too) slow nor (too) fast, is the hand responding to (some influence which) the heart (sends forth). But I cannot tell (how to do this) by word of mouth—there is a knack in it. I cannot teach the knack to my son, nor can my son learn it from me. Thus it is that though in my seventieth year, I am (still) making wheels in my old age. But these ancients, and what it was not possible for them to convey, are dead and gone—so then what you, my ruler, are reading, is but their dregs and refuse!'[c]

In this remarkable passage the Taoist artisan counsels the feudal lord. An unexplainable knack is obtained by following the Tao of things, so instead of looking in the books of dead Confucians, study the Tao of the people and acquire the knack of governing, of leading from within. See as they see and hear as they hear. Do not interfere with the fulfilment of the people's natural human needs and desires. Do not set yourself above them, but return to the ideal of the Common Life. Everyone who has at one time or another borne the burden of command will recognise the truth of the words of Pien the wheelwright.[d]

Something has been contributed to our understanding of the meaning of the 'knack'-passages by Huard (2). He has pointed out that in modern machine tech-

---

[a] Ch. 13.
[b] Ch. 12 (Morgan, pp. 114, 116).   [c] Tr. Legge (5), vol. 1, p. 344; Waley (6), p. 32, mod.
[d] Cf. a passage which seems to continue this, in ch. 25 (Legge (5), vol. 2, p. 123). Elsewhere (ch. 14, tr. Waley (6), p. 37) the 'former kings' are compared with the straw dogs (cf. p. 48), which ought to be thrown away when the sacrifices are over. For another, much later, passage of a similar kind, see below, p. 577.

nology the scientific comprehension of productive processes has so much deepened that there is relatively little chance of their deviating from the normal and failing to give the results sought. Mysterious causes which upset them have mostly been eliminated.[a] Mastery of the technical control is transferred from instructor to apprentice in an impersonal and objective way. But in the days of eotechnic craftsmanship, when the Taoists were philosophising, the situation was very different; personal skill and flair had to strain to the utmost to bring into existence products which today may pour automatically from machines practically untended. Lacking the fruits of scientific analysis of productive processes, the Taoist artisans had to hold fast to empirical tricks and *tours-de-main* which were often hardly explainable in logical language to their apprentices; they were helped by a background of legend and myth, and they had to cultivate, by techniques of meditation and imagination, a state of tense emotion and an iron will to successful accomplishment. Religious rites often preceded the work, as was the case with the swordsmiths of Japan down to very recent times;[b] and in common with metallurgists among many early peoples, Chinese smiths in Chuang Chou's time certainly undertook prior procedures of purification and ascesis. In view of all this, the transmission of the arts and crafts from one generation to another naturally involved a total education of the body and spirit of the learner. It will be evident that this complex of concerns and attitudes had much in common with the world-view of Taoism.

Besides, at bottom, the artisan and the Taoist stood together in the conviction that the Tao was *in* natural things and not something other-worldly and transcendent. The inscriptions existing in the Taoist Tung-Hsiao temple at Hangchow in the +13th century have been preserved for us by Têng Mu[1] in his book *Tung-Hsiao Thu Chih*,[2] and there we find an appropriate statement inscribed in +1289 in one of the halls by a Taoist, Shen To-Fu,[3] of whom otherwise little is known. It reads:

All the labour (of building) throughout the months and years is devoted to the handling of natural things. But people say that to be enslaved to natural things (*i yü wu*[4]) is not the right Tao. However, I believe that the right Tao lies exactly in being the servant of the things of Nature. If this were not so, the people would not know the use of all simple everyday things. Even the Confucian scholars cannot depart from practical things for a single moment....[c]

---

[a] A whole history of technology could almost be written around this theme alone. The 'gremlins' of aircraft in World War II were the most recent representatives of that host of incomprehensible unanalysed 'things gone wrong' which bedevilled production throughout the ages. Naturally the fermentation industries would provide a wealth of illustration of the seemingly odd procedures which grew up empirically (cf. Sects. 34, 40, below). The final phase came with the Royal Society and the encyclopaedia of Diderot, when natural science established permanent sovereignty over the techniques of production out of which it had itself largely arisen. Craft-lore and magic, ritual and technique, were all present at its origins (cf. Childe, 14).

[b] Cf. Inami, pp. 78, 91.   [c] Ch. 6, p. 45b, tr. auct.

[1] 鄧牧   [2] 洞霄圖志   [3] 沈多福   [4] 役於物

Contrasting with their emphasis on the skill of artisans, the Taoist texts show a distinct prejudice against technology and inventions which seems at first sight very curious. Like the question of their attitude to 'knowledge', this has put many upon a false scent, since it has seemed hard to reconcile it with Taoist naturalistic philosophy and the known connections of Taoism with science and technology. We have already noticed traces of it in several quotations; it usually takes some such form as, 'The more cunning inventions there are, the more evils will arise.'[a] While many examples could be quoted from the *Huai Nan Tzu*[b] and other books,[c] the *locus classicus* is in *Chuang Tzu*, chapter 12,[d] and concerns the swape or counterbalanced bailing bucket for raising water (frequently known under its Arabic name of shadūf, see Sect. 27*e*):

Tzu-Kung had been wandering in the south in Chhu, and was returning to Chin. As he passed a place south of the Han (river), he saw an old man working in a garden. Having dug his channels, he kept on going down into a well, and returning with water in a large jar. This caused him much expenditure of strength for very small results. Tzu-Kung said to him, 'There is a contrivance (*chieh*[1]) by means of which a hundred plots of ground may be irrigated in one day. Little effort will thus accomplish much. Would you, Sir, not like to try it?' The farmer looked up at him and said, 'How does it work?' Tzu-Kung said, 'It is a lever made of wood, heavy behind and light in front. It raises water quickly so that it comes flowing into the ditch gurgling in a steady foaming stream. Its name is the swape (*kao*[2]).' The farmer's face suddenly changed and he laughed, 'I have heard from my master', he said, 'that those who have cunning devices use cunning in their affairs, and that those who use cunning in their affairs have cunning hearts. Such cunning means the loss of pure simplicity. Such a loss leads to restlessness of the spirit, and with such men the Tao will not dwell. I knew all about (the swape), but I would be ashamed to use it.'[e]

In reality the reasons behind this attitude are not far to seek. If the power of feudalism rested, as it must certainly have done, on certain specific crafts, such as bronze-working and irrigation engineering; if, as we have seen, the Taoists generalised their complaint against the society of their time so that it became a hatred of all 'artificiality'; if the differentiation of classes had gone hand in hand with technical inventions—was it not natural that these should be included in the condemnation?

Here our most valuable clue has already been given in the passages quoted above (p. 102) about the feudal lords 'gaining the advantages springing from the use of pecks and bushels, weights and steelyards, tallies and seals', and (p. 108) about the 'invention of axes and saws, to kill by laws and statutes set up like carpenters'

---

[a] Cf. *Tao Tê Ching*, ch. 57.

[b] Esp. the opening part of ch. 8 (Morgan (1), pp. 81 ff.).

[c] E.g. *Wên Tzu* (Forke (13), p. 349).

[d] See also ch. 17 (Legge (5), vol. 1, p. 384); ch. 9 (Legge (5), vol. 1, p. 279); ch. 10 (Legge (5), vol. 1, pp. 286 ff.).

[e] Tr. Legge (5), vol. 1, p. 320; Lin Yü-Thang (1), p. 267, mod. An almost identical version of this story is told in the Han by Liu Hsiang, with Têng Hsi Tzu as the principal character, in the *Shuo Yuan*, ch. 20.

[1] 橰   [2] 槹

measuring-lines, to disfigure with hammers and gouges...'.[a] One can see, in fact, that mechanical inventions have always been double-edged, their effects depending on what people have used them for. Espinas and Schuhl (1) have drawn attention to the remark of the author of the Hippocratic treatise on Joints that the apparatus for reducing dislocations was so powerful that if anyone wanted to use it for doing evil instead of good he would have an almost irresistible force at his disposal. This was, in fact, the origin of the rack. No wonder the Taoists were suspicious. Their *méfiance* sprang from the (not unjustified) impression that all machines were infernal machines, or very liable to be so.

A striking example of this, enshrined in the structure of the Chinese language, was noted by the Chhing scholar Chang Chin-Wu[1] in his *Kuang Shih Ming*[2] (Enlargement of the 'Explanation of Names' Dictionary).[b] The word *chieh*[3] (variously pronounced *chiai, hsieh, hsiai*) means an implement, and participates in combinations which mean 'mechanism' (*chi-chieh*[4]) and 'apparatus' (*chhi-chieh*[5]), but its original significance was that of 'fetters' or 'shackles'. It derives from the phonetic *chieh*[6] (or *chiai*) (K 990), meaning to warn, and composed (in Shang times) of two hands and a dagger-axe; the addition of the wood radical would therefore imply a material 'warner'. The theme of warning is intensified in *chieh*,[7] and the addition of the horse radical gives 'to frighten, to overawe' (*hsieh*[8] or *hsiai*). Nothing could better exemplify the mental association between machines as such and the interests of the dominant social group. Though there is no telling how the association grew up in this particular case, one might conjecture that the art of the locksmith was involved, the earliest 'mechanism' being the padlock with which the contumacious peasant was confined.

In their anti-technology complex the Taoists surely represented the popular feeling that whatever machines or inventions might be introduced it would be only for the benefit of the feudal lords; they would either be weighing-machines to cheat the peasant out of his rightful proportion, or instruments of torture with which to chastise those of the oppressed who dared to rebel.[c] Although in eotechnic times there could be no question of technological unemployment, the social pattern repeated itself in a certain sense at the time of the machine-wreckers' riots in the early 19th-century in the West,[d] and doubtless other parallels could be found. Notwithstanding this aspect

---

[a] Cf. Lucretius, *De Rer. Nat.* III, 1017, 'verbera carnifices robur pix lammina taedae', a terrifying consort of nouns, and with the same social implications.

[b] Ch. 2, p. 16*b*.

[c] This explanation is further sustained by the fact that it was Tzu-Kung's report of the swape episode to Confucius (in the *Chuang Tzu* book) which led to the crucial passage connecting *phu* with *hun-tun* (p. 114 above). The anti-technological farmer was said to belong to the Primitive Homogeneity School. Later Chinese writers understood these matters no better than modern Europeans; thus the +12th-century author of the *Mêng Chai Pi Than* (ch. 1, p. 14*b*) takes the old farmer seriously to task, and urges that there can be no possible harm in swapes or other labour-saving devices.

[d] A recent interpretation (Hobsbawm, 1) shows that these movements were by no means so disadvantageous to the working class as has usually been supposed.

[1] 張金吾   [2] 廣釋名   [3] 械   [4] 機械   [5] 器械   [6] 戒
[7] 誡   [8] 駴

of Taoism, however, the technologists of later ages continued to venerate Taoist genii whose names became associated with the crafts, and the inevitable alliance between various kinds of manual operations, whether for magical or practical ends, continued on its course, as the rise of alchemy and other proto-sciences shows.

These correlations become still more convincing when one considers the large 'technological' element in those later apotheoses of neurotic obsessional states, the Buddhist hells. More than fifty years ago, F. W. K. Müller (2) was able to show that one kind of torment, that of burning iron wires, was directly derived from the innocent inked thread of Asian carpenters and joiners, with its box and spool.[a] Recently Duyvendak (20) has studied several Chinese texts which present remarkable parallels to those visits to the underworld of which the *Divina Commedia* of Dante is the most famous. Here again it is possible to identify, in the processes to which the bodies of sinners are subjected, a great variety of human techniques, not only gripping, compressing, cutting, piercing and pounding, but also many aspects of rotatory motion. In all this there is nothing specifically Chinese, for parallels are easily found in many other cultures (Iran, India, Islam, Europe), and the genre is so old that an Egyptian or Mesopotamian origin seems inescapable. But the point is that what the demons did in the underworld was only an image of what the minions of 'the powers that be' were capable of doing in the world above. It is easy to excuse the Taoists, representatives of a cooperative society, for having an ambivalent attitude to those techniques which the society of force and dominance could use for its own ends. They saw that the tools of mastery over the inanimate world could be turned against the flesh and blood of their creators. Their insight was part of the whole history of man's relations with machinery, sometimes health-giving, sometimes oppressive, sometimes lethal; one of the greatest social themes to which justice has never yet been done.[b]

Nor did Lao Tzu himself wish to chastise the technicians. Chapter 74 contains masterly and prophetic words:

The people are not frightened of death. What then is the use of trying to intimidate them with death-penalties? And even supposing that they were, and that the makers of ingenious contrivances could be seized and slain, who would dare to do it? There is the Lord of Slaughter always ready for the task, and to act in his stead is like thrusting oneself into the Master-Carpenter's place and doing his chopping for him. Whoever 'tries to do that will be lucky if he does not cut his hand'.[c]

It was indeed natural, in view of all that has been said, that the Taoists should have been a strongly pacifist school. The relevant chapters of the *Tao Tê Ching*[d] are too well known to quote. Reference may be made to the interesting monograph of

---

[a] Further reference to this will be found in Sect. 27a.
[b] We have only the brilliant introduction by Stuart Chase.
[c] Tr. Waley (4), and Duyvendak (18) whose discussion of this chapter is excellent.
[d] Chs. 30, 31, 46 especially.

Tomkinson on pacifist doctrines in ancient China. A striking example of them in action during the Han is found in the memorial addressed to the emperor by Liu An, Prince of Huai Nan, in −135,[a] and indeed they formed an important element in the Confucian-Taoist synthesis of that period, impressing themselves on all subsequent scholarly thinking.

(6) EUROPEAN PARALLELS; THE 'GOLDEN AGE'

At an earlier stage we got some light on the attitude of the Taoists towards 'knowledge' by examining the position in Europe at the beginning of the modern scientific movement, when science was helped by mystical faith and hindered by scholastic rationalism. Now, looking over the several aspects of the attitude of the Taoists towards society, their retrospective faith in primitive collectivism, and their hatred of the feudal institutions of their own time, we seem again to find a flavour not altogether unfamiliar to students of the history of thought in Europe. What they have called 'primitivism' may be said to have taken three outstanding forms: (a) the repudiation of civilised life by the Cynics and Stoics, (b) the Christian doctrine of the Fall of man, and (c) the 18th-century admiration of the Noble Savage.

An immense collection of texts has been brought together by Lovejoy & Boas (1) to illustrate these tendencies.[b] From the writings of European classical antiquity it is possible to find parallels for most of the qualities ascribed by the Taoists to the primitive collectivism which they so much admired. The European authors referred them to a 'Golden Age', or an age of Saturn or Cronos. Though Lovejoy & Boas do not consider the possibility of any concrete basis for these ideas, it seems not unreasonable to suppose that to some extent, at any rate, they originated from memories of primitive collectivist society (cf. G. Thomson). In Greece and Rome such memories seem to have been kept alive by certain festivals, the Cronia and the Saturnalia,[c] at which there was a temporary social equality, including even the slaves, and during which accounting was forbidden. According to the classical texts, land in ancient times had been held in common,[d] there had been no enclosures[e] or surveyors to measure them,[f] and the doors of houses were always left open[g] (cf. the Mohist passage quoted below, p. 167). The operations of mining, and the erection of walls, fortifications and boundary-stones, were seen as signs of increasing degeneration of society[h]—'omne nefas' (cf. the passage from *Huai Nan Tzu* quoted above, p. 109). Like the Taoists, too, some of the writers of classical antiquity looked for a return of the age of primitive collectivism; the most famous example is Virgil's *Fourth Eclogue*.

---

[a] *TH*, p. 419.   [b] Cf. also Eliade (2), pp. 338 ff.
[c] Description and abundant references in Lovejoy & Boas (1), pp. 66, 67.
[d] Virgil, *Georgics*, I, 125–55; Seneca, *Epist. Mor.* xc, 34.
[e] Tibullus, *Elegies*, II, iii, 35–46.
[f] 'Communemque prius, ceu lumina solis et auras,
   cautus humum longo signavit limite mensor'
                    (Ovid, *Metamorphoses*, I, 76–215, esp. lines 135, 136).
[g] Tibullus, *Elegies*, I, iii, 43, 44.
[h] Pseudo-Seneca, *Octavia*, 388–448; Maximus Tyrius, *Diss.* xxxvi.

Although both the exact significance of this, and its sources, remain uncertain (Mayor), similar texts exist.[a] The ethics of primitivism were put fully into practice by the Cynics,[b] and in a more organised way by the Epicureans.[c] And its scheme of values was perpetuated by the Stoics, combining the Socratic ideal of self-sufficiency with the maxim of conformity to Nature.[d]

The older the civilisation studied, the further back we should expect to find traces of the lament for the lost cooperative and collectivist form of society. Albright (1) has shown that these are present in the Sumerian story of Engidu, and the poem of Uttu, dating from the end of the -3rd millennium. They describe a state of life when there were no canals, no arrogant overseers, no liars, no sickness nor old age. In one direction, the tradition passed to India, where Dumont recognises it in the Yuga and Kaliyuga successions of social degeneration; in the other it gave rise, perhaps, to the conception of the Fall of man (Begrich, 1) and some of the themes of the prophets, such as Amos (Roll, 1), in the thought of Israel. Thence this passed into the patrimony of Christian doctrine, where it fused with Graeco-Roman primitivism.[e] We have from Boas (1) an interesting collection of texts dealing with the medieval aspects of this. As was well understood at the time, a Cynic philosopher could turn into a Christian monk with very little alteration either of ideas or of external appearance. From this point of view, the communism of the early Church,[f] or of Lactantius[g] or of Ambrose,[h] could be considered an attempt to reverse the degenerative tendency in human history, quite analogous to the hopes of the Taoists for a similar improvement.

As has already been suggested, the Taoists may have been acquainted with contemporary social organisations in the more primitive communities of 'barbarians', equivalent to the Miao, the Lolo, Chia-jung, and so on, of today, who lived around the outskirts of developed Chinese civilisation, and in enclaves within it. The parallel for this in European culture would doubtless be the third of the three types of primitivism

[a] *Oracula Sibyllina*, III, 743–59, 787–95; probably mid –2nd century.

[b] Lovejoy & Boas (1), pp. 117, 145.

[c] The whole of the latter part of Bk. v of the *De Rerum Natura* is a description, first of the ancient tribal life, the *vita prior* (lines 925–1104); then of the growth of class-differentiated society (lines 1105–1457). On this Farrington (9–13) has written. Lucretius did not idealise primitivity, but he seems to have thought that what followed was even worse. Organised war, for instance. Lines 999–1001 rang in my mind for years:

> But not in those far times
> Would one lone day give over unto doom
> A soldiery in thousands marching on
> Beneath the battle-banners, nor would then
> The ramping breakers of the ocean dash
> Whole argosies and crews upon the rocks   (tr. Leonard).

The Epicureans were at one with the Taoists in thinking that mutual agreement had been replaced by imposed order, *concordia* by *justitia*; and that the change was for the worse.

[d] Lovejoy & Boas (1), pp. 260ff. Here we cannot survey the theme of the Golden Age in later European literature, but it may be remarked that a striking parallel to ch. 80 of the *Tao Tê Ching* may be found in Shakespeare's *Tempest*, Act 2, Scene 1.

[e] Lovejoy & Boas (1), p. 381.　　　　　　　　　[f] A. Robertson (sen.) (1); Needham (10).

[g] Boas (1), pp. 33, 91.　　　　　　　　　　　　[h] Lovejoy (4).

mentioned above, namely, the admiration for the 'Noble Savage'. This did not begin, as is often supposed, in the 18th century, but may be found represented in ancient classical authors.[a] The 'Fortunate Islands' or 'Islands of the Blest', a theme which suggested that the life of primitive collectivism continued to be lived in some remote part of the world, occurs in Homer,[b] Pindar,[c] Horace,[d] Pliny,[e] Lucian[f] and others, i.e. from the −6th century downwards. Numerous authors ascribe great virtues to the Hyperboreans[g] (cf. Sect. 7e), the Scythians,[h] the Arcadians[i] and so on. In the Middle Ages the tradition continued,[j] with a tendency to locate the inhabitants of the Earthly Paradise somewhere in the east. The *Gesta Alexandri* of pseudo-Callisthenes, of uncertain date, the first link in the long chain of romances which constituted later the *Alexander Romance*, made the Brahmins of India occupy, from the +4th century onwards, the same position as the Scythians had had for the Greeks.[k] Later there were the complexes of Brendan's Paradise (+6th to +10th centuries),[l] Tnugdal's Paradise (+12th century),[m] and the impressive legend cycle of the Country of Prester John (first mention +1145).[n]

The primitivism of the 18th century, the admiration for the Noble Savage, one of the chief components of the romantic reaction against the classical tradition, is more familiar, and has been described by many historians of thought (Whitney (1); Gonnard (1); Lovejoy (1), etc.).[o] One of its earliest manifestations was the *Voyage au Brésil* of de Léry (+1556 to +1558). Its climax, perhaps, was the beautiful and witty 'Supplement to the Voyage of Bougainville' by Denis Diderot (+1772) (tr. Stewart & Kemp, 1). It was still vigorous in the work of de Chateaubriand on the Natchez Indians at the beginning of the 19th century (cf. Honigsheim, 1).

What has been said, therefore, in this Section, concerning the social and political attitudes of the Taoists, might serve, in a way, as an extension of the work of the historians of primitivism to the field of Chinese culture. Nevertheless, the Taoists show certain characteristic differences from any analogous groups in occidental history. They formed a much more organised element than the Cynics or the Stoics, and their combination of political anti-feudalism with the beginnings of a scientific movement has no parallel in the West. This is understandable to the extent that the Graeco-Roman primitivists were living in a context of city-state civilisation which had no parallel in China. The city-state was, as we may see later on, basically favourable to the development of science, and the Graeco-Roman primitivists were reacting against it. Their anti-intellectualism was therefore of a quite different type, for the

---

[a] Lovejoy & Boas (1), pp. 287 ff.
[b] *Odyssey*, IV, 561–8.
[c] *Olymp.* II, 68–76.
[d] *Epod.* XVI, 40–end.
[e] *Nat. Hist.* VI (202–5), 32 (37).
[f] *Verae Narr.* II, 4–16.
[g] Texts collected in Lovejoy & Boas (1), pp. 304 ff.
[h] Texts collected in Lovejoy & Boas (1), pp. 315 ff.
[i] Texts collected in Lovejoy & Boas (1), pp. 344 ff.
[j] Boas (1), pp. 129 ff.
[k] Boas (1), p. 139.
[l] Boas (1), p. 158.
[m] Boas (1), p. 166.
[n] Boas (1), p. 161; cf. Sect. 6h (Vol. 1, p. 133) above.
[o] Its parallels with Taoist thought account in part for the interest now being taken by Japanese scholars in Rousseau (e.g. in the collective work edited by T. Kuwabara, 1).

Taoists attacked 'knowledge' only so far as it was social and conventional, leaving place for the study of natural phenomena, while the Cynics and Stoics admitted only an ethical and personal philosophy. Cynic anti-intellectualism turned into the medieval Christian doctrine (or tendency, for it was never orthodox) of the *cultus ignorantiae*, the 'vanity of all arts and sciences',[a] which may be traced from Tertullian to its climax in Bernard of Clairvaux. With this the Taoists had little or nothing in common.

(7) SCIENCE AND DEMOCRACY

This long section can now be brought to a close. Taoist thought is basic to Chinese science and technology, but there has often been a failure to appreciate this on account of the ambivalent Taoist attitude to 'knowledge', which lent itself later on to a domination of that mystical element which had always been present. In order to explain what kind of knowledge they were in favour of, therefore, it was necessary to explain what kind of 'knowledge' they were against. And this could not be done without elucidating their political position.

But there is, moreover, a great intrinsic interest in the anti-feudal attitude of the Taoists, since it raises the general question of the relations between science and democracy (whether that be in its most ancient tribal collectivist form or in its modern representative or socialist forms). As has already been mentioned, elements of primitive tribal collectivism have been traced in the foundations of Greek democracy.[b] Several scholars have drawn attention to the correspondence between the rise of Ionian and Milesian pre-Socratic science and the democratic (even mercantile) character of the city-states of Greece.[c] The rise of generalised thinking, as Crowther puts it, was perhaps due, among other things, to the necessity for persuasion in an equalitarian community. Acceptance of assertions on authority may pass in a proto-feudal or feudal milieu, but is not acceptable to a cooperative social entity, whether composed of Greek citizen-merchants, or of Chinese peasant-farmers.

Much must have been thought and written on the theoretical connections between science and democracy, but I had the occasion of putting my own thoughts in order on the subject during an enforced stay (far from all books) at Wa-yao in Yunnan on the China-Burma border during the second world war.[d] Historically, it is evident that modern science and modern democracy grew up together, as parts of that great movement in European development which included the Renaissance, the Reformation and the rise of capitalism. Some relation between Greek democracy and Greek science has long been recognised. We can now add to it a new parallel drawn from the roots of Chinese science and technology. But more interesting are the theoretical and even psychological connections, among which I may mention two. First, Nature is no respecter of persons. The status of an observer, if competent, as to age, sex, colour, creed or race, is, as we know today, irrelevant. This was appreciated among the

a Boas (1), pp. 121 ff.
b G. Thomson (1).   c Farrington (1–5); Crowther (1) and others.
d Cf. Needham (7).

ancient Chinese. Authority, even that of the lord of a State in feudal China, is not enough. Force will not accomplish its end. Neither kings nor sages can withstand or reverse the Tao of Nature. The *Lü Shih Chhun Chhiu* says:

If you force someone to laugh, he will not thereby be amused; if you force someone to weep, he will not thereby be sad....If you try to attract mice with a cat or flies with ice, you may give yourself much trouble but you will certainly not succeed....Bait cannot be used to drive things away. When tyrants like Chieh and Chou tried to govern the people by terror, they could make the punishments as draconic as they liked; it was no good. In seasons of cold, the people try to warm themselves; in seasons of heat they seek for coolness....Whoever wishes to be a ruler of this world will fail if he does not consider the principles on which the people move.[a]

Throughout the passage the words *pu kho, pu kho*,[1] impossible, impossible,[b] repeat like the strokes from a drum-tower, delivering the characteristic Taoist message, not only that the 'humblest' human being can observe Nature as well as the 'highest', but also that even the 'highest' courts ruin if he acts contrary to Nature (*wei*[2] as against *wu wei*[3]). It might be said that in their personal veneration for age, the Chinese fell into the pitfall from the social aspect of which the Taoists had wished to guard them, but it was always recognised that no one, however aged and venerable, could escape the consequences of *wei* and *wu wei*. As the *Lü Shih Chhun Chhiu* says again:[c]

Although the feudal lord is honoured, if he calls black white, his servants will not hear him. Although a father is honoured by his son, if he calls black white, his son will not hear him.

And if age and sageness could not change the facts of Nature, neither were they dependent upon the differences between different peoples. The *Huai Nan Tzu* says excellently:[d]

At the present time the balance and the scales, the square and the compass, are fixed in a uniform and unvarying manner (*i ting erh pu i*[4]). Neither (the people of) Chhin nor Chhu (can) change their specific properties—neither the northern Hu barbarians nor the men of Yüeh in the south can modify their appearances. These things are for ever the same and swerve not, they follow a straight path and do not meander (*chhang i erh pu hsieh, fang hsing erh pu liu*[5]). A single day formed them, ten thousand generations propagate them. And the action of their forming was non-action (*i jih hsing chih, wan shih chhuan chih; erh i wu wei wei chih*[6]).

Secondly, the birth of science requires the bridging of the gap between the scholar and the artisan. It is a point to which we shall return, but it must be mentioned here,

---

[a] Ch. 10 (vol. 1, p. 21), tr. R. Wilhelm (3), p. 25; eng. auct.
[b] Cf. p. 175 below.
[c] Ch. 63 (vol. 1, p. 123), tr. R. Wilhelm (3), p. 162; eng. auct.
[d] Ch. 9, p. 5*a*. (tr. Escarra & Germain, p. 23; eng. auct.; mod.).

[1] 不可不可　　　[2] 爲　　　[3] 無爲　　　[4] 一定而不易
[5] 常一而不邪方行而不流　　　[6] 一日形之萬世傳之而以無爲爲之

for the Confucians were entirely on the side of the literate administrators,[a] and lacked all sympathy with artisans and manual workers. The Taoists, on the other hand, were, as we have seen, in close contact with them (here is another parallel with the pre-Socratic Greek nature-philosophers). These attitudes run through all later Chinese history. Ko Hung seeks for an official post in Annam, of rank much lower than that to which he is entitled, in order to collect cinnabar there for his alchemical experiments. Thao Hung-Ching gathers and identifies plant drugs, an early example of a long line of scholars who excluded themselves from the ranks of the Confucian bureaucratic hierarchy and earned their living by selling medicinal herbs.

Other points of connection might be raised, but enough has been said to indicate that it was probably no coincidence that Taoism in its ancient form was connected both with the earliest Chinese science and technology, and with the ideals of ancient equalitarian pre-feudal Chinese society.

(h) SHAMANS, *WU*, AND *FANG-SHIH*

We turn now to a widely different, but almost equally important, element in the development of Taoism, namely, its connections with the most primitive sorcery of the North Asian peoples. Shamanism has been termed the native religion of the Ural-Altaic peoples from the Behring Straits to the borders of Scandinavia, including Lapps and Eskimos. Amerindian medicine-men have often been called shamans by anthropologists, and not without reason, as their practices are analogous. The cult, which may still be observed in many tribes today, is one of polytheistic or poly-daemonistic nature-worship, sometimes involving a supreme god, but often not. The 'priest', whose equipment consists characteristically of drums, spears and arrows,[b] is, and always was, primarily occupied with magical healing (the expulsion of evil spirits which have possessed the patient) and divination (still employing scapulimancy). Aided by abnormal neurotic or epileptic-like states, the shaman, who is a mediator between the spirits and men, goes into autohypnotic trances, during which he is supposed to journey to the abodes of gods and demons, afterwards announcing the results of his conversations with them. Dancing has always been a particularly important element in shamanic rites, but ventriloquy appears to have been used also, as well as juggling and tricks whereby the shaman releases himself from bonds.[c]

[a] My friend Professor Li Fang-Hsün has been heard to draw attention to the fact that in Chinese paintings Confucian sages and scholars have always been depicted holding their hands hidden in their long sleeves, while Taoist sages and scholars frequently brandish magic swords, fan alchemical furnaces, or carry out other manual operations. The tradition was well justified. Maspero (13, p. 64) has described how Hsi Khang, one of the +3rd-century Seven Sages of the Bamboo Grove, exercised himself with iron-working. This scandalised Confucian visitors.

[b] Biologists will recall the famous picture of Linnaeus dressed in shaman's clothes and holding the ritual instruments which he collected on his visits to Lapland.

[c] The shortest summary of shamanism is that of McCullogh (1), but much interesting information is contained in the papers and books of Shirokogorov (1), Mikhailovsky (1), Nioradze (1), Ruben (1), Ohlmarks (1) and König, Gusinde, Schebesta & Dietschy. Perhaps the most convenient book is the recent one of Eliade (3).

There has been some controversy about the origin of the word shaman, and its transliterations into Chinese. There is no doubt that *sha-mên*[1] is the transliteration of the Sanskrit *śramaṇa*, which meant in pre-Buddhistic times an ascetic, and later a Buddhist monk.[a] Mironov & Shirokogorov (1) believe that this word got up into the Tarim basin from India at a very early date, and then spread through all the North Asian tribes as the term for their own medicine-men. I have not found their arguments convincing, however, and prefer the view of Laufer (5) that shaman is a very ancient Tungusic word, and that its identification with *sha-mên* was an 18th-century error. I believe I am right in saying that the Chinese never confused the two,[b] and that no instance exists of the use of either *sha-mên* (or *shih-mên*,[2] a variant form derived from the Chinese transliteration of Śākyamuni) to indicate Taoist magicians, celebrants or exorcists. The most widely used term for Taoists of all kinds in medieval and modern use was of course *Tao-shih*.[3] Laufer believed that Tungusic *shaman* entered Persian as *saman* (e.g. in Firdausī), and Nemeth has traced it as *kam-* in Turkish and Uighur.

If Laufer's view is right, it might well be that we should find some early transliteration of shaman in Chinese. Although I have not seen the suggestion made, we might perhaps recognise it in the term *hsien-mên*,[4] which does occur, in very significant context, in the Chhin and Han periods.[c] Both the *Shih Chi*[d] and the *Chhien Han Shu*[e] list a certain Hsienmên Kao as among the followers of Tsou Yen's magical-scientific Yin-Yang school. He came from the State of Yen in the far north, and appears in the former text as Hsienmên Tzu Kao,[5] which might mean that Hsienmên was a family name and his given names were Tzu-Kao, or that he was Kao the Hsienmên Master. His date would be the latter half of the −4th century. But later the term seems to have a more generalised meaning, which suggests that it had always had it. In the same chapter of the *Shih Chi* Ssuma Chhien tells us that 'Chhin Shih Huang Ti wandered about on the shore of the eastern sea, and offered sacrifices to the famous mountains and the great rivers and the eight Spirits; and searched for *hsien*[6] and *hsien-mên*[7] and the like'.[f] Here then the word has a generic sense side by side with the *hsien*[g] (K 193) which we shall speak of shortly, and could easily mean magicians possessed of supernatural powers. In −215 the same emperor sent a Master Lu[8] to search in the mountains for Hsienmên Tzu Kao (who was thought to be still living),

---

[a] Clement of Alexandria's *samanaei* were certainly Buddhists. Cf. Section 7*f* (Vol. 1, p. 177).
[b] See the interesting paper of Schott (1).
[c] My friend Dr Li An-Chê has signified in conversation his agreement with this. Dr Balazs later pointed out to me the early suggestion of Terrien de Lacouperie (1), who thought that *hsien-mên* was a transliteration of *śramaṇa*, and was criticised by Chavannes (1), vol. 2, p. 165, accordingly. It is noteworthy that on the first occasion when a transliteration for the Sanskrit word was needed (*Hou Han Shu*, ch. 72, p. 6*a*; cf. p. 398 below), the form *sang-mên*[9] was used. This persisted till the +7th century (Ware (4), p. 114).
[d] Ch. 28, p. 8*b*.   [e] Ch. 25, p. 10*b*.   [f] p. 10*a*, tr. auct.
[g] A magician or adept who has attained material immortality.

[1] 沙門   [2] 釋門   [3] 道士   [4] 羨門   [5] 羨門子高   [6] 仙
[7] 羨門   [8] 盧   [9] 桑門

and this emissary duly brought back a letter concerning the fall of the dynasty.[a] By the time of Han Wu Ti the term may have become a title, since Luan Ta (see below, Sect. 26*i*) spoke of having met Master An Chhi and the Hsienmên,[b] in −113.

If this word did have the significance of shaman it seems to have been quite forgotten, for Pelliot (11) has shown that when in +1139 it was necessary to find a transliteration for what an important Jurchen, Wanyen Hsi-Yin, was called by his own people, the characters *shan-man*[1] were used. Under the Chhing dynasty, the shamans of the imperial Manchu house were termed *ssu-chu*.[2]

The Chinese had a word of their own for shaman, however, namely, *wu*,[3] and it is interesting that the idea of dancing is what binds all these words together. Hopkins (2) showed that both *wu*[3] (K 105) and *wu*,[4] to dance or to posture (K 103), go back to the same oracle-bone forms, which all depict a dancing thaumaturgic shaman, holding plumes, feathers, or other ritual objects in his, or her, hands.[c] Sometimes, as exorcist, he has a bearskin mask (Hopkins, 33).[d] But the same idea is also present in the character *hsien*[5] (an alternative of *hsien*[6]), which means to caper or hop about (K 206), and since the phonetic here means to rise high into the air, one is irresistibly reminded of the belief of English country-folk within living memory that the higher the morris-dancers sprang into the air, the better the harvest would be. It seems that there were two kinds of *wu*, the *wu*[7] proper, who were women, and the *hsi*,[8] who were men. The prominence of women here seems very significant, in view of (*a*) the connection of the Taoist ideal society with matriarchal memories, (*b*) their Feminine Symbol, (*c*) their emphasis on sex techniques (see on, p. 146), etc. The author of the *Shuo Wên* says that all kinds of *wu* were similar to *chu*,[9] professional Invokers or Imprecators.[e] The only remaining important term is *fang-shih*,[10] which some like to translate as 'gentlemen possessing magical recipes'—we think they were just straight magicians. It should be mentioned here that in view of the close connection between shamanist exorcism and early medicine it is of much interest that the earliest way of writing the character *i*[11] (medicine) was *i*,[12] in which we see the *wu* component appear instead of the wine radical (no. 164). In later use, combined with radical no. 149, *wu*[13] acquired the meaning of imposture.

So large a selection of translated texts concerning the nature and activities of the *wu* has been made available in the great work of de Groot[f] that it will not be necessary to

---

[a] *TH*, p. 214. *Shih Chi*, ch. 6, pp. 20*b*, 21*b*; Chavannes (1), vol. 2, pp. 164, 167.

[b] *Chhien Han Shu*, ch. 25, p. 23*b*. An Chhi was one of the magicians who were supposed to live on the islands in the eastern sea.

[c] The character *wu*,[14] meaning 'not, negative', is also derived from them, but it is unclear whether as a phonetic loan or because after the shaman had carried out his rites the evil would *not* come to pass.

[d] Head-dresses, etc., of shamans of Han date have been found in Korean tombs: Hamada & Umehara (1); Hentze (2).

[e] On the whole subject see the monograph of Schindler (1).

[f] (2), vol. 6, pp. 1187 ff. A more condensed series (without references, of course) is in *TP*, pp. 93, 118 ff. In Chinese there is a mass of material collected in the *Thu Shu Chi Chhêng* (*I shu tien*, chs. 809, 810; *Shen i tien*, chs. 283–91). See also the résumé by Chhü Tui-Chih (1).

[1] 珊蠻	[2] 司祝	[3] 巫	[4] 舞	[5] 僊	[6] 仙	[7] 巫
[8] 覡	[9] 祝	[10] 方士	[11] 醫	[12] 毉	[13] 誣	[14] 無

do more than sketch out the main references to them. An excellent summary will be found in the book of H. Maspero,[a] and there are articles by L. Giles (7) and others. We have just met the *wu* on the oracle-bones, and further evidence that they go back to the highest Chinese antiquity lies in the fact that they are mentioned at least twice in the *Shu Ching* (Historical Classic).[b] The second of these mentions suggests the existence of State magicians, and this is borne out by the much later *Chou Li* (Record of the Rites of Chou),[c] which clearly has *wu* employed in the State religion. It may be significant, in view of the probable northern steppe component in this element of Chinese culture, that a special category of *wu*, expert in the care and cure of horses, is described.[d] Confucius in the *Lun Yü*[e] quotes with approval a southern proverb that a man without constancy will not make either a good *wu* or a good physician.

The *wu* were certainly concerned with rain-making magic (the oracle-bones call it *chhih*[1]), and the *Tso Chuan*[f] has a story of −638 about a feudal prince who wanted to expose to the sun or scorching fire one or more *wu* in order to bring an end to a drought, but was dissuaded from doing so. Similar practices were later referred to in the *Li Chi* (Record of Rites).[g] Schafer (1) has recently given us an exhaustive account of this magic, which involved ritual nakedness (a social phenomenon which lasted remarkably late in Chinese history; well into the Thang),[h] and probably the copious sweating of the dancing shaman within a ring of fire under the blazing sun as sympathetic magic. Drops of sweat, it was hoped, would induce drops of rain. The ceremony was an exposure (*pu*[2] or *lou*[3]) or a scorching (*fên*[4]) of the naked (*lo*[5]) shaman (*wu*[6]), and some kind of king-substitute scapegoat (*wang*[7]) seems also to have been involved.[i] In later centuries, so strong was the tradition, the rites were followed even by Confucian officials themselves, in times of need.

The mentions of *wu* in the *Shan Hai Ching* (Classic of Mountains and Rivers)[j] are particularly interesting because the *wu* are there associated with the elixir of immortality and with drugs in general. For example, we read[k] that east of Khaiming there is the place where the Six Wu live—they 'carry the corpse of Cha-Yü[8],[l] and have

---

[a] (2), pp. 187 ff.

[b] E.g. ch. 13, I Hsün (Medhurst (1), p. 142) and ch. 36, Chün Shih (Medhurst (1), p. 268; Karlgren (12), p. 61).

[c] Ch. 17 (Biot (1), vol. 1, pp. 412, 413), ch. 25 (Biot (1), vol. 2, pp. 102 ff.), chs. 28, 32 (Biot (1), vol. 2, pp. 157, 259).

[d] Wuma became a family name. One of the disciples of Confucius bore it. Wu also of course, and perhaps significantly, the legendary founder of medicine, physician of the emperor Yao, was named Wu Phêng,[9] while one of the three important astronomers of the −4th century (see Sect. 20*f*) took the name of a legendary minister Wu Hsien.[10]

[e] XIII, xxii. Cf. the mention in *Mêng Tzu*, II (1), vii, 1.

[f] Duke Hsi, 21st year (Couvreur (1), vol. 1, p. 327); see also another mention for −543 (Couvreur (1), vol. 2, p. 520).   [g] Ch. 4 (Legge (7), vol. 1, p. 201).

[h] Schafer gives also occidental and other parallels.   [i] Cf. *Po Wu Chih*, ch. 5, p. 2*b*.

[j] On this, see below in Section 22*b* on geography.

[k] Ch. 11, p. 5*b*.

[l] A man-eating dragon; cf. Granet (1), p. 378, etc.

---

[1] 赤   [2] 暴   [3] 露   [4] 焚   [5] 裸   [6] 巫   [7] 尪

[8] 窫窳   [9] 巫彭   [10] 巫咸

in their hands death-banishing medicinal herbs with which to drive him away (*chieh tshao pu ssu chih yao i chü chih*[1])'. Elsewhere[a] there is a list of ten *wu* living on a certain mountain where the 'hundred medicinal plants' grow. In a third place[b] there is mention of a country in the north full of *wu*, and taking its name from one of the *wu* 'officials' mentioned in the *Shu Ching*. Here then is a close link with pharmaceutics and alchemy. It is borne out by the reputation which the *wu* had in connection with poisonous drugs. The ancient Chinese seem to have had particular fear of a virulent poison known as *ku*.[2] One of the earliest mentions of it, if not the first, occurs in a discussion between a feudal prince and a physician, dated −540.[c] It appears in the *I Ching* as hexagram no. 18 (see on, p. 316). In −91 there occurred a dreadful witch-hunt in the palace of Han Wu Ti, large numbers of people being put to death on suspicion of being involved in the preparation of this poison.[d] According to the tradition recorded by Li Shih-Chen in the *Pên Tshao Kang Mu* (the great pharmacopoeia of the +16th century), the poison was prepared by placing many toxic insects in a closed vessel and allowing them to remain there until one had eaten all the rest—the toxin was then extracted from the survivor.[e] Chhen Tshang-Chhi, the author of the *Pên Tshao Shih I* of *c.* +725, who says the same, adds the particularly interesting information that *ku* could also be used as a cure or preventive, thus suggesting that someone had stumbled on an immunisation process.[f] In any case, exact information has long been lost, in so far as it was ever known outside Taoist circles in which it passed down from adept to adept. All one can say is that grave fear was always inspired by it. In +598, for example, there was an imperial decree forbidding its use.[g] The chief point to note is that the connection of *wu* and Taoists with pharmaceutics as well as alchemy was a close one.[h]

When we come to the *Shih Chi*[i] and *Chhien Han Shu*[j] there is a great deal of information about the *wu* and *fang-shih*, some of which has already been quoted (p. 83) and more of which we shall quote later (p. 240 and Sects. 26*i*, 33), especially in connection with the origins of alchemy and the beginning of the knowledge of magnetism in the circle of magicians surrounding Han Wu Ti, in the −2nd century. It was not every emperor, however, who believed in the claims, and took an interest in the

---

[a] Ch. 16, p. 3*b*.   [b] Ch. 7, p. 5*a*.

[c] *Tso Chuan*, Duke Chao, 1st year (Couvreur (1), vol. 3, p. 39).

[d] *TH*, p. 467; Dubs (2), vol. 2, p. 114; cf. the collection of stories about *ku* in Pfizmaier (40).

[e] It is strange to think that this same method has been successfully employed in our own times for the isolation of strains of soil bacteria capable of attacking the tuberculosis bacillus (the grammicidin of Dubos & Avery).

[f] Read (7), no. 99. There are many possible toxins which it might have been, such as scorpion-venom and centipede-venom. Chhen Tshang-Chhi says that for antidotes *ku* from animals other than those suspected of having caused the poisoning was used. The earliest medical book which describes the preparation seems to be the *Chu Ping Yuan Hou Lun* of Chhao Yuan-Fang (*c.* +607), ch. 25, p. 1*a*.

[g] De Groot (2), vol. 5, p. 825. For further background, largely folkloristic, see Fêng Han-Chi & Shryock (2).

[h] Cf. Harvey (1), pp. 143, 155.

[i] Chs. 28 and 12.   [j] Chs. 25 and 63. Cf. Chhen Phan (7).

[1] 皆操不死之藥以拒之   [2] 蠱

techniques, of the *wu* and the *fang-shih*; and the more influential Confucianism became during the Han as the cult of the imperial bureaucracy, the worse it was for the shamanistic and experimental side of Taoism. De Groot has occasion to note repeatedly that the *wu* were by no means always on good terms with the ruling authorities,[a] so that the magical as well as the political-philosophical aspects of the Taoist system drove it inevitably into general opposition to the government. In some respects, as on the famous occasion when Hsimên Pao[1] of the State of Wei stopped the *wu* custom of sacrificing girls as brides of the Yellow River,[b] about −415,[c] one cannot but sympathise with the Confucian rationalists who must often have had opportunity to demonstrate their humanitarianism in such ways.

That the female *wu* were still numerous in the +2nd century we know from the *Chhien Fu Lun*[2] (Essays of a Hermit)[d] of Wang Fu,[3] who bitterly complained of the large number of women in his time who took up the profession. The *Chin Shu* (History of the Chin Dynasty)[e] records several remarkable stories of them in the +4th century, from which it appears that *wu* were at that time commonly engaged to perform the ancestral sacrifices of families, much as Taoists were hired (at least until recent years) to carry out funeral and other domestic ceremonies. From this time onwards the accounts of the *wu* blend more and more with stories of marvellous occurrences in general, especially such as were recorded in connection with the immortals (*hsien*) (see on, p. 141). About +460, one of the Liu Sung emperors engaged *wu* to evoke the spirit of his dead consort for him, and they partially succeeded, just as Shao Ong had done for the emperor Han Wu Ti in the −2nd century, in a famous incident which we shall note later (Sect. 26*g*) and which was doubtless the inspiration for this new attempt. How completely the *wu* were by now incorporated in the Taoist system is not clear, but they were still being mentioned as *wu* in the Thang.[f] For this period we have a story showing clearly the use of ventriloquism by a *wu* sorceress,[g] dated +825; and numerous accounts of exorcistic medical practice.[h] But the *wu* had been excluded from the State sacrifices in +472, and though the process of gradual severance which ended their employment by the emperors and the orthodox Confucian bureaucrats was rather slow, it was nearly complete by the end of the Thang. In the Sung they were definitely persecuted by governors and prefects,[i] and down to the end of the Chhing provisions against sorcerers and wizards remained in the Penal Code.[j]

---

[a] Especially (2), vol. 6, pp. 1188, 1199.

[b] Significantly, the general relation of ancient Chinese shamans to their Spirits was that of lovers; cf. Waley (23), pp. 13 ff., 19, 40, 49.

[c] *Shih Chi*, ch. 126, p. 10*b*; *TH*, p. 155; de Groot (2), vol. 6, p. 1196.

[d] Ch. 12 (tr. de Groot (2), vol. 6, p. 1210).

[e] Ch. 94 (tr. de Groot (2), vol. 6, p. 1213).

[f] *Hsin Thang Shu*, ch. 210 (tr. de Groot, vol. 6, p. 1217).

[g] In the *Hsü Yu Kuai Lu*[4] (Supplementary Record of Things Dark and Strange) of Li Fu-Yen.[5]

[h] De Groot (2), vol. 6, p. 1228. And techniques of mass suggestion (van Gulik, 4).

[i] De Groot (2), vol. 6, p. 1238.   [j] Staunton (1), pp. 175, 179, 273, 548.

[1] 西門豹   [2] 潛夫論   [3] 王符   [4] 續幽怪錄   [5] 李復言

More and more, then, the *wu* aspect of Taoism was driven underground, and tended to take the form of those secret societies among the people which in later centuries played such an important part in Chinese life. Already in the −3rd century a remarkable cult flourished for a short time, a Dionysian, orgiastic and soteriological devotion to Hsi Wang Mu,[1][a] which has been studied by Dubs (13).[b] Throughout the subsequent centuries, the Taoists were always associated with that succession of subversive secret societies which played prominent parts at the changes of dynasties.[c] The *Hou Han Shu*[d] describes the 'Red Eyebrows' who first formed part of the army which restored the Han dynasty after the interregnum of Wang Mang, and then continued in rebellion.[e] Later there were the 'Yellow Turbans' in +184, who so greatly weakened the Later Han dynasty.[f] In subsequent centuries some of the secret societies seem to have cloaked political activity under forms which were closely related to religions other than Taoism, such as Manichaeism and Buddhism (e.g. the 'White Cloud' and 'White Lotus' societies[g] respectively). They were also important in movements of a distinctively nationalist character, as in the case of the 'Red Turbans' who prepared the way for the expulsion of the Mongols and the rise of the Ming dynasty,[h] and the Thai-Phing rebellion of the last century. The +12th century saw the rise of a number of Taoist societies which formed an underground movement against the Chin (Jurchen) State.[i] In our own time there has been important activity on the part of the 'Elder Brothers' society, especially in Szechuan, and everyone who lived for some time in China before the revolution was sure to come upon the traces, in one way or another, of the Hung Pang[2] and the Chhing Pang[3] ('Red' and 'Green' Associations).[j]

Unfortunately, doubtless owing to the difficulty of the subject, these secret societies have not yet been subjected to the thorough investigation which would elucidate their roots in ancient Taoism.[k] One must remember that shamanism in China was continually reinforced by fresh waves of primitive religion from the north, as, for example, in the case of the Chhi-tan people who founded the Liao dynasty (cf. Wittfogel,

[a] Perhaps an ancient mother goddess, and certainly a prominent figure in Chinese mythology, cf. H. A. Giles (5), vol. 1, pp. 1, 298.

[b] See *Chhien Han Shu*, chs. 11, 26, 27. All three passages now tr. Dubs (2), iii, 33 ff.

[c] This point was made to me in conversation with much emphasis by Professor Fu Ssu-Nien at Lichuang in 1943. Recognition of this political role of later Taoism can be found in Neo-Confucian writings of the +12th century; e.g. Chu Hsi agreeing with Chang Wên-Chhien (cited in *TSCC*, Ching chi tien, ch. 433, p. 10*a*) that Taoism led to rebellions and stratagems.

[d] Ch. 41.

[e] *TH*, p. 623. See also Bielenstein (2).

[f] *TH*, p. 773.

[g] Cf. Chhen Jung-Chieh (4), pp. 158 ff. Nestorian Christianity may have contributed to the 'Golden Pill' society.

[h] *TH*, p. 1734.

[i] Cf. Chhen Jung-Chieh (4), pp. 148 ff.

[j] Cf. Chhen Jung-Chieh (4), pp. 170 ff.

[k] We can do no more here than cite the monographs of Favre (1), Stanton (1), Ward & Stirling (1) Glick & Hung Shêng-Hua (1), de Korne (1) and the paper of Brace (1).

[1] 西王母   [2] 紅帮   [3] 青帮

Fêng Chia-Shêng *et al.*). And there was always the influence of neighbouring countries with strong shamanistic traditions, such as Tibet (Li An-Chê, 1).

As regards the practices of *wu* shamanism in later times, it is clear that they fused imperceptibly with the numerous pseudo-sciences (divination, astrology, fate-calculation, geomancy, oneiromancy, etc.) which we shall separately examine.[a] A large part was played by the preparation of written charms and talismans (cf. the investigations of Chhen Hsiang-Chhun, 1) to which a whole volume of Doré's compilation is devoted.[b] For the study of the remnants of *wu* shamanism in modern China (on which far more research is needed), recourse must be had to the works of Doré (1),[c] v. d. Goltz (1), Hodous (1), Dennys (1), etc. That of E. D. Harvey is interesting as the study of a trained sociologist who lived for some years in China. The terms stabilised in late medieval and modern use for shamanistic practices have been 'diabolistic' systems (*yao tao*[1]), techniques and methods (*fa shu*[2]) of sorcery and exorcism; or 'depraved' techniques or methods (*hsieh shu*,[3] *hsieh fa*[4]) of divination (cf. Chatley, 5, 6). These terms reflect Confucian orthodoxy and rationalism.

(*i*) THE AIMS OF THE INDIVIDUAL IN TAOISM; THE ACHIEVEMENT OF MATERIAL IMMORTALITY AS A *HSIEN*

From the beginning Taoist thought was captivated by the idea that it was possible to achieve a material immortality. We know of no close parallel to this in any other part of the world.[d] It was of incalculable importance to science, since, as will be seen later on (Sect. 33), this ideal stimulated the development of the techniques of alchemy almost certainly earlier in China than anywhere else. But one cannot help being struck by a seeming inconsistency between this individual discipline[e] and the emphasis laid by the Taoist philosophers on social collectivism, which we have elucidated earlier in this Section. It is doubtful whether, during the slow development of Taoism through the centuries, this paradox or inconsistency was ever felt. There can be little hesitation in saying that it arose because of the dual origin of Taoism, that strange association

---

[a] See on, pp. 346 ff.   [b] Pt. 1, vol. 2.
[c] Pt. 1, vol. 4, pp. 332 ff.

[d] Of course the doctrine of the resurrection of the body, which Christianity and Islam inherited from Tanaitic Judaism, comes immediately to mind. But though in some forms, such as the Muslim paradise, felicity might be material enough, it would not be experienced until after a prolonged period of absolute separation of soul from body, nor would any part of this world be the scene of it. Living belief in their ultimate reunion survives in the hostility of the Latin church to cremation, and of orthodox Judaism to anatomical dissection. On the other hand, the interest in longevity was not quite absent from Europe, as may be seen by the treatises of Roger Bacon and Arnold of Villanova on the subject (Förster, 1), but the question did not have the same significance. It is interesting that certain classical authors attributed exceptional longevity to the Seres (the Chinese), which might be an echo of Taoist ideas (Strabo xv, i, 34, 37; Lucian, *Makrobioi*, 5; Coedès (1), pp. xii, xxvi, 7, 75).

[e] This section is concerned with arts concerning the cultivation of the body. It will be obvious from what has gone before that an equally great part was to be played by the cultivation of the mind, e.g. ataraxy, *wu wei*, etc.

[1] 妖道   [2] 法術   [3] 邪術   [4] 邪法

between the hermit-philosophers of mystical naturalism on the one hand and the tribal shaman-magicians on the other. Both were in perpetual opposition to feudal lords and later bureaucratic officials, whose 'gentry' mentality had no room for the primitive collectivism admired by the philosophers, and whose ouranic State Confucianism disliked the techniques of the shamans. The more impotent Taoist philosophy became to liberate Chinese society as a whole, the more success accrued to Taoist adepts and their methods of liberation of the individual.

The Taoists were fascinated by youth with its firmness of flesh and exquisite skin-complexion, and they believed that techniques could be found out whereby it would be possible to arrest the processes of ageing, or to return to the physical condition of the young organism. In chapter 55 of the *Tao Tê Ching* we have a meditation on the human organism at the beginning of its development:

> He who possesses abundant virtue may be likened to a babe,
> Poisonous insects will not sting it,
> Fierce beasts will not seize it,
> Clawing birds will not attack it.[a]
> Its bones are weak,
> Its sinews tender,
> Yet its grasp is strong;
> It has known nothing of the union of male and female
> Yet its penis is sometimes erect
> Showing that its vitality is perfected;
> It may cry all day long without growing hoarse,
> Showing that its harmony is accomplished;
> To understand this harmony is (to understand) the unfailing (vital force)
> To understand the unfailing, is to be enlightened.
> Now by intensifying one's (worldly) living (one invites) the ominous,
> By allowing (the emotions of) heart and mind to dominate over the life-
>   breath (*chhi*), (one succumbs to the) rigidity (of death).
> Whatever has force and violence will dwindle to decay,
> For (excessive vigour) is against the Tao
> And whatever goes against the Tao is destroyed.[b]

In later presentations, such as are contained in the medieval books of the *Tao Tsang*, the inherent factors of senescence were 'personified' as the Three Worms (*san chhung*[1]), or the Three Cadavers (*san shih*[2]). To eject these from the body was one of

---

[a] One of the oldest ideas about those who had made some progress on the way to becoming *hsien* was that they would be invulnerable against wild animals or attacks of man. This is found more than once in the *Tao Tê Ching* (cf. ch. 50), and often in *Chuang Tzu* (Legge (5), vol. 1, pp. 192, 237, 383; vol. 2, p. 13), *Huai Nan Tzu* (Morgan (1), p. 66), and other similar books. The superstition, if such it was, rather than an attempt to describe a certain psychological state of mystical union with Nature, lasted on a very long time, and reappeared in the beliefs of the secret societies (e.g. 'Harmonious Fists' (Boxers), etc.). Cf. Waley (6), pp. 74 ff. who mentions Indian parallels; and Berthold (1).

[b] Tr. Huang Fang-Kang (1), mod.

[1] 三蟲   [2] 三尸

the great objects of all the techniques.[a] To become a *hsien*,[1] or a 'True Man' (*chen jen*[2]) meant that one would go on living for ever (*chhang sêng*[3]) with a youthful body in a kind of earthly paradise. Representations of the *hsien*, often in the form of feathered men, are not uncommon in Han art. One's body might appear to be left behind in the coffin, but it would be only a simulacrum feigned by an object such as a sword or a piece of bamboo, which had previously been prepared with special rites. This was called the 'deliverance of the corpse' (*shih chieh*[4]), or the 'transmutation of the *hun*-soul' (*lien hun*[5]). The process was thought of as similar to insect metamorphosis.[b] Thus the *Yün Chi Chhi Chhien*[6] (Seven Bamboo Tablets of the Cloudy Satchel)[c] collected by Chang Chün-Fang[7] about + 1000 and edited by Chang Hsüan[8] in the + 17th century, says:[d] 'When men use a precious sword for the deliverance of the body, this is the highest example of metamorphic transformations (*Shih jen yung pao chien i shih chieh chê, shan hua chih shang phin*[9])'.[e]

Feathered *hsien*. From an inlaid bronze basin in the
Hosogawa Collection (Rostovtzev (3), Pl. XII).

But apart from these final rites, the perfected body had to be prepared, like an embryo in the womb, by a lifetime of actual *practices*. Some of these might answer to Indian or European conceptions of asceticism, and others might not; in any case, the basic ideas of sacrificial masochism (as among the Amerindian Aztecs and others), or of gaining magic power over the gods (as among the Indian *rishis*), or of pleasing the supreme deity by elaborate abstinences (as in the fasts of Jewish or Christian theology) were all absent. It was a question of preparing oneself for a further life, after 'death', equally material but subtler and purer, holy and beautiful yet comprising all the pleasurable forms of experience which man can have in his present life and freed from the anxieties of disease, old age and dissolution. The *hsien* would be able, it was thought, to revisit the ordinary world more or less at will, but their own

---

[a] Maspero (13), pp. 20, 98. He points out that Taoist vegetarianism was quite different from that of the Buddhists. It did not spring, like the latter, from a ban on the taking of life, but from a belief that the *chhi* of blood and meat was inimical to the spirits inhabiting the body, and favourable to the senescence factors. Cf. Kubo (1).

[b] Cf. Fig. 47 below.   [c] TT 1020.

[d] Ch. 84, p. 4*b*.

[e] *Shan* is a common name for cicada. Pfizmaier (88) has translated four chapters from the *Thai-Phing Yü Lan* encyclopaedia on these magical procedures.

[1] 仙   [2] 眞人   [3] 長生   [4] 尸解   [5] 鍊魂   [6] 雲笈七籤
[7] 張君房   [8] 張萱   [9] 世人用寶劍以尸解者蟬化之上品

would be much more desirable. Whatever means would effect this transformation were to be followed. Here it is impossible not to quote one of the descriptions of Taoist paradises, from which the condition of the perfected *hsien* may be visualised. There are two in *Lieh Tzu*;[a] the first is of the more ascetic kind, where the *hsien* have been purged from all desire, but the second is more poetic.[b]

After Yü the Great had set the waters and the land in order, he lost his way and came to a country which lay on the north shore of the northern ocean.[c] I can't say how many hundreds of thousands of miles it was from the State of Chhi. It was called 'Northend-land' and we don't know what lay on its boundaries. There was neither wind nor rain there, neither frost nor dew. It did not produce the birds, animals, insects, fishes, plants and trees of the (same) species (as ours). All round it seemed to rise into the sky. In the midst of it there was a mountain called 'Amphora', shaped like a vase, at the top of which there was an opening, in the form of a round ring, called 'Hydraulica', because streams of water came out of it continually. This was called the 'Divine Spring'. The perfume of the water was more delicious than that of orchids or pepper, and its taste was better than that of wine or ale. The spring divided into four rivers which flowed down from the mountain and watered the whole land. The *chhi* of the earth was mild, there were no poisonous emanations causing sickness. The people were gentle, following Nature without wrangling and strife; their hearts were soft and their bodies delicate; arrogance and envy were far from them. Old and young lived pleasantly together, and there were no princes nor lords. Men and women wandered freely about in company; marriage-plans and betrothals were unknown. Living on the banks of the rivers, they neither ploughed nor harvested, and since the *chhi* of the earth was warm, they had no need of woven stuffs with which to clothe themselves. Not till the age of a hundred did they die, and disease and premature death were unknown. Thus they lived in joy and bliss, having no private property; in goodness and happiness, having no decay and old age, no sadness or bitterness. Particularly they loved music. Taking each other by the hand, they danced and sang in chorus, and even at night the singing ended not. When they felt hungry or tired, they drank of the water in the rivers, and found their strength and vitality restored. If they drank too much, they were overcome as if drunk, and might sleep for ten days before awaking. They bathed and swam in the waters, and on coming out their skins were smooth and well-complexioned, with a perfume which remained perceptible for ten days afterwards.

King Mu of Chou, when he was on his journey to the north, also found this country, and forgot his kingdom entirely for three years. After he had returned home, he yearned for that country with such a yearning that he lost all consciousness of his surroundings. He took no

---

[a] Ch. 2, p. 2*a*, and ch. 5, p. 12*a*.

[b] For this vision one may find a European parallel in the triptych painted by Hieronymus Bosch about +1490 and known as the 'Garden of Earthly Delights'. Fränger (1), who has recently given an analysis of its symbolism, believes that it was made for the Homines Intelligentiae (or Brothers and Sisters of the Free Spirit), one of the late medieval sects which sought salvation through a return to primal innocence and a diffuse half-sublimated sexuality. If these movements were fully studied, one would not be surprised to find Tantric and hence ultimately Taoist elements in them (cf. pp. 427 and 151 below).

[c] I.e. on one of those other continents which Tsou Yen described (see p. 236 below) as being separated from the continent in which China lay. It seems rather significant that most of the versions of this 'paradise legend' refer to the distance of the place from the State of Chhi, the homeland of Tsou Yen and the general headquarters of magicians, Taoists and Naturalists.

interest in wine or meat, and would have nothing to do with his concubines and servitors. It took him months to recover himself.[a]

No wonder that the Taoist aspirant for *hsien*-ship was prepared to undergo a considerable amount of training.

The practices just mentioned fall into several categories:

(1) respiratory techniques;
(2) heliotherapeutic techniques;
(3) gymnastic techniques;
(4) sexual techniques;
(5) alchemical and pharmaceutical techniques;
(6) dietary techniques.

The last two of these will be reserved for the Sections on alchemy and nutritional science, but I shall briefly discuss the others here. Much gratitude is owing from all scholars to H. Maspero (7), who in a classical series of papers brought a beginning of order and interpretation to the mass of obscure and difficult material contained in that vast patrology[b] of Taoism, the *Tao Tsang*.[1] All these techniques went under the collective name of 'nourishing the *chhi*, or the nature' (*yang chhi*,[2] *yang hsing*[3]).[c] Some of them must certainly have been very ancient, for in the *Chuang Tzu* book there is a passage distinctly antagonistic to the respiratory techniques,[d] and so also in *Huai Nan Tzu*,[e] while the *Tao Tê Ching*, on the other hand, seems to recommend them (chapter 10).

(1) RESPIRATORY TECHNIQUES

First, the breathing exercises undoubtedly go back to a high antiquity in China. H. Wilhelm (6) has drawn attention to an inscription on twelve pieces of jade, which may have formed part of the knob of a staff, and which in date are certainly Chou and may be as early as the middle of the −6th century. This is what the inscription says:

In breathing one must proceed (as follows). One holds (the breath) and it is collected together. If it is collected it expands. When it expands it goes down. When it goes down it becomes quiet. When it becomes quiet it will solidify. When it becomes solidified it will begin to sprout. After it has sprouted it will grow. As it grows it will be pulled back again

---

[a] Tr. R. Wilhelm (4), p. 53; eng. auct.

[b] The catalogue of Wieger (6) is indispensable. Books mentioned from the *Tao Tsang* will be identified here according to his numbering, e.g. *TT* 233. The index of Ong Tu-Chien (Yin Tê series, no. 25) is also important. The patrology took its present form first in the Sung dynasty, but was in after times somewhat expurgated. It was first printed in the Sung and under the Chin about +1190, later also in Yuan and Ming, about +1445. The best recent analyses of it in a Western language are those of Gauchet (2, 3) and Maspero (13). Much the oldest catalogue of Taoist books is in *Pao Phu Tzu* (*Nei Phien*), ch. 19.

[c] Pfizmaier (89) has translated four relevant chapters of the *Thai-Phing Yü Lan* encyclopaedia.

[d] Ch. 15 (Legge (5), vol. 1, p. 364).   [e] Ch. 7 (Morgan (1), p. 67).

[1] 道藏   [2] 養氣   [3] 養性

(to the upper regions). When it has been pulled back it will reach the crown of the head. Above, it will press against the crown of the head. Below, it will press downwards.

Whoever follows this will live; whoever acts contrary to it will die.[a]

We see here the characteristic sorites reasoning (cf. Sect. 49), closely parallel to the most ancient epigraphic evidence for the theory of the five elements, the sword-inscription reproduced on p. 242. As in that case also, the inscription closes with a promise to followers and a commination of opponents.

The great aim of the breathing exercises was to try to return to the manner of respiration of the embryo in the womb. Knowing nothing of the gases in the maternal and foetal circulations, this could have been for the Taoists but a fantasy; they tried to keep the inspiration and expiration as quiet as possible, and, above all, to hold the breath closed up (*pi chhi*[1]) for as long a time as possible. There can be little doubt that the subjective effects which they experienced, and which they believed were so good for them, were due largely to anoxaemia, since they experienced asphyxic symptoms, buzzing in the ears, vertigo and sweating. There are many books in the *Tao Tsang* which discuss the techniques; the *Thai Hsi Ching*[2] (Manual of Embryonic Respiration),[b] of uncertain date, is particularly to be mentioned, and an important source is Ko Hung's early +4th-century *Pao Phu Tzu*.[c] In the Thang dynasty there was a considerable remodelling of ideas on the subject, for the details of which Maspero must be consulted; the earlier theories had envisaged the inspired air as nutritive as well as respiratory, while the new ones developed the idea of a special inner breath, or *nei chhi*,[3] the circulation and transformation of which had to be effected and accelerated by imaginative meditation.[d] Needless to say, there were many precepts about the proper times and places for the breathing exercises.[e]

[a] Tr. H. Wilhelm (6), eng. auct.
[b] *TT* 127, tr. Balfour (1), cf. Forke (9), p. 456.
[c] *TT* 1171–1173. *Nei Phien*, ch. 8, p. 2*b*; *Wai Phien*, ch. 2, p. 7 (cf. Forke (12), p. 219).
[d] A Chhing manual, the *Thai I Chin Hua Tsung Chih*[4] (apparently 17th century) was translated by R. Wilhelm and provided with a curious commentary by the eminent psychologist C. G. Jung. There is an English version of this collaborative work. The text was apparently connected with one of the Taoist secret societies, the Chin Tan Chiao[5] ('Golden Pill doctrine').
[e] The question of the relation, if any, between these and other Taoist practices and those of the Indian yogis is an extremely difficult one (see above, Sect. 7*b*). Maspero (13), p. 194, has pointed out that Buddhist technique sought for regularity and slowness of inspiration and expiration; unlike the Taoist, which sought to retain air in the lungs as long as possible. The facts concerning Indian Yogism are very hard to get at, since the practices have always been transmitted personally from guru to disciple, and outsiders have to approach them through a haze of uncritical mystification. A recent résumé is tha tof Abegg, Jenny & Bing (1) (with bibliography), and reference may be made to the works of Garbe (1), Woodroffe (1, 2), Behanan (1), Rele (1), and J. H. Woods (1). On the interpretation of yogistic exercises and feats in terms of modern physiology there is a large literature; from it I mention only the interesting paper of Laubry & Brosse (1). Yogistic training for meditation and rapt contemplation seems to have included hypnotism, autohypnotism, and especially an extension of conscious control over the functions of the autonomic nervous system which are normally exempt from it. To what extent Taoist techniques paralleled the rather well authenticated phenomena of yogism (e.g. suspension of respiratory movements and heart-beat, etc.) remains to be determined. Such phenomena are after all no more extraordinary physiologically than the catatonia seen, for instance, in certain cases

[1] 閉氣   [2] 胎息經   [3] 內氣   [4] 太一金華宗旨   [5] 金丹教

(2) HELIOTHERAPEUTIC TECHNIQUES

Secondly, the Taoists seem to have discovered some of the virtues of heliotherapy, not recognised by European medicine until our own time. The 'method of wearing the sun rays' (*fu jih mang chih fa*[1]) consisted in the exposure of the body to the sunlight, while holding in the hand a special character (the sun within an enclosure) written in red on green paper. Quite logically, according to their lights, women adepts were to expose their bodies likewise to the moon, holding a special character (the moon within an enclosure) written in black on yellow paper—unfortunately their vitamin-D content cannot have been much enriched thereby. The principal treatises in the *Tao Tsang* concerning these methods are the *Shang Chhing Wo Chung Chüeh*[2] (Explanation of the High and Pure Method of Grasping the Central Ones),[a] attributed to Fan Yu-Chhung[3] of the Later Han; and the *Têng Chen Yin Chüeh*[4] (Instructions for Ascending to the True Concealed Ones)[b] by Thao Hung-Ching[5] of the late +5th century. In the light of what we saw above (p. 135) concerning the ritual nudity of the rain-bringing shaman, this therapeutic technique may well have been a development of ancient magic. It has persisted until the present time.[c]

(3) GYMNASTIC TECHNIQUES

Thirdly, various kinds of comparatively mild gymnastics were practised; this was called *tao yin*,[6] i.e. extending and contracting the body. Perhaps it derived from the dances of the rain-bringing shaman. In later times the names *kung fu*[7] and *nei kung*,[8] implying work, or inwardly-directed work, came into use for it. It undoubtedly originated from the idea, very old in Chinese, as in Greek, medicine, that the pores of the body were liable to become obstructed, thus causing stasis and disease (see below, Sect. 44). Massage (*mo*[9]) was also performed. These techniques generated a very large literature, of which the principal books may be cited, first the *Thai-Chhing Tao Yin Yang Sêng Ching*[10] (Manual of Nourishing the Life by Gymnastics)[d] of uncertain date, and secondly the *Tsun Sêng Pa Chien*[11] (Eight Chapters on Putting Oneself in Accord with the Life Force) by Kao Lien,[12] of +1591. The latter has been analysed at considerable length by Dudgeon (1) who also describes some of the minor works on the subject.[e] Chinese boxing (*chhüan po*[13]), an art with rules different from that of the

---

of dementia; the only remarkable thing would be the power to produce them at will in the normal person. Maspero's considered opinion (14), p. 46, was that the Taoist techniques were an indigenous development of ancient pneumatic physiology, and not derived from contacts with Indian yogism.

[a] *TT* 137.   [b] *TT* 418.

[c] Information from an old Taoist of the Chungshan temple in the Chhilien Mountains west of Shantan, Kansu (through Mr R. Alley).   [d] *TT* 811.

[e] Such as Hu Wên-Huan's *Pao Sêng Hsin Chien* (Mirror of Medical Gymnastics) of +1506, and Wang Tsu-Yuan's late nineteenth-century opuscule, both of which I happened to pick up in Peking in 1952.

[1] 服日芒之法   [2] 上清握中訣   [3] 范幼沖   [4] 登眞隱訣
[5] 陶弘景   [6] 導引   [7] 功夫   [8] 內功   [9] 摩
[10] 太清導引養生經   [11] 遵生八牋   [12] 高濂   [13] 拳搏

West, and embodying a certain element of ritual dance (cf. H. A. Giles),[a] probably
originated as a department of Taoist physical exercises.

This whole subject will again be referred to in the Section on Medicine, but it is
well to note here that a knowledge of Chinese therapeutic gymnastics came to Europe
in the 18th century, and seems to have played a part of capital importance in the
development of modern hygienic and remedial methods (Dudgeon (1), McGowan (2),
Peillon[b]). The main link was an elaborate article by P. M. Cibot (3)[c] in 1779 which
stimulated the work of the Swedish pioneer of medical gymnastics, P. H. Ling.
Some of the postures shown in the Chinese works strongly recall positions
which have been widely used in modern medicine.[d] And one is also tempted to
wonder whether the heliotherapeutic ideas of the Taoists, transmitted in similar Jesuit
articles and books, did not exert an effect on the growth of modern physiotherapy.[e]

(4) SEXUAL TECHNIQUES

Fourthly, there were the sexual techniques. Owing to Confucian and Buddhist
antagonism these have remained much the most recondite, yet they have considerable
physiological interest.[f] It was quite natural, in view of the general acceptance of the
Yin-Yang theories, to think of human sexual relations against a cosmic background,
and indeed as having intimate connections with the mechanism of the whole universe.[g]
The Taoists considered that sex, far from being an obstacle to the attainment of
hsien-ship, could be made to aid it in important ways. Techniques practised in private
were called 'the method of nourishing the life by means of the Yin and the Yang'
(*Yin Yang yang sêng chih tao*[1]), and their basic aim was to conserve as much as possible
of the seminal essence (*ching*[2]) and the divine element (*shen*[3]), especially by 'causing
the *ching* to return' (*huan ching*[4]). At the same time, the two great forces, as

---

[a] (5), vol. 1, p. 132.

[b] (1), see especially p. 639.

[c] 'Notice du Cong-Fou des Bonzes Tao-Sse' in the fourth volume of *Mémoires concernant l'Histoire,
les Sciences...des Chinois*.

[d] For example, the posture shown on p. 492 of Dudgeon (1) which resembles that for the draining of
pus from the lungs in bronchiectasis.

[e] Cf. Delherm & Laquerrière (1). Modern treatment by forms of radiant energy did not start until
the eighteenth century.

[f] This section was substantially in its present form long before the appearance of the excellent book
of van Gulik (3) on Chinese ideas concerning sex physiology and practice. He was moved to embark
on this study by discovering a set of blocks for one of those books of erotic colour-prints which were
produced in the Ming dynasty between +1560 and +1640. This was the *Hua Ying Chin Chhen*
(Varied Positions of the Flowery Battle) of +1610, which he has now reproduced and translated. The
only difference in our conclusions is that I think van Gulik's estimate of the Taoist theories and practices
in his book (e.g. pp. 11, 69) was in general too unfavourable; aberrations were few and exceptional.
Dr van Gulik and I are now in agreement on the subject (personal communication).

[g] Cf. the vision of Lao Tzu described in *Chuang Tzu*, ch. 21 (tr. Waley (6), p. 34). It will be remembered that earlier on (p. 23), in connection with the 'ladder of souls', we noted Wang Khuei's theory
of how heaven and earth had to combine in order that the higher levels should be produced.

¹ 陰陽養生之道   ² 精   ³ 神   ⁴ 還精

incarnated in separate human individuals, were to act as indispensable nourishment the one for the other;[a] *i yin i yang hsiang hsü*,[1] as the *Hsüan Nü Ching* says.

All the books concerning these arts disappeared from the *Tao Tsang* during the Ming dynasty, if not earlier, but long fragments were preserved in Japanese medical works from the +10th century onwards. Of these the most important was the *Ishinhō*[2] of Tamba no Yasuyori,[3] composed in +982 but not printed till +1854. The chief Chinese source is the *Shuang Mei Ching An Tshung Shu*[4] (Double Plum-Tree Collection), a group of books and fragments assembled by Yeh Tê-Hui[5] in 1903. A single chapter (perhaps on account of its being only a chapter and not a whole book) survived into the modern *Tao Tsang*; this is chapter 6 of the *Yang Shêng Yen Ming Lu*[6] (Delaying Destiny by Nourishing the Life),[b] attributed both to Thao Hung-Ching[7] of the +5th and Sun Ssu-Mo[8] of the +7th centuries. Among the fragments brought together by Yeh Tê-Hui are the *Su Nü Ching*[9] (Immaculate Girl Canon) and *Hsüan Nü Ching*[10] (Mysterious Girl Canon), the *Yü Fang Pi Chüeh*[11] (Secret Instructions concerning the Jade Chamber), the *Tung Hsüan Tzu*[12] (Book of the Mystery-Penetrating Master), and the *Thien Ti Yin Yang Ta Lo Fu*[13] (Poetical Essay on the Supreme Joy). Other ancient fragments are found mainly in the Japanese collection, e.g. the *Yü Fang Chih Yao*[14] (Important Matters of the Jade Chamber).[c]

No sharp line of distinction can be drawn between arts specific to the Taoists and the general techniques of the lay bedchamber (*fang shu*[15]), which they, as well as others, taught and transmitted. Van Gulik (3) has rightly emphasised that the texts in question are entirely devoid of pathological aberrations, such as sadism and masochism, while only in the later books do practices appear which may be considered unusual or ancillary, though not abnormal. The numerous references to mythical and other emperors in the early texts suggest that some of the techniques may have originated in the situation of ancient kings and princes who found themselves possessed, according to custom, of a large number of concubines. To a lesser degree this persisted throughout the centuries in all families of importance, where the problem of organising a healthy sexual life in the polygamous household must have been a very real one.

---

[a] It is instructive to compare Chinese ideas of sex with Indian and Japanese ideas. For the former there are the works of R. Schmidt (1, 2, 3); and translations by Basu (1), Tatojaya (1), Ray (1), etc. For the latter there is the book of Krauss, Sato & Ihm. At a later stage certain Indian conceptions will have to be discussed (pp. 426 ff. below) in connection with Tantrism, which may have been partly Chinese in origin.

[b] *TT* 831.

[c] Popular versions of some of these still circulate (or did so until recently) in the lending-libraries of pedlars in China, and others are also passed privately from hand to hand. I always remember the reply given to me by one of the deepest students of Taoism at Chhêngtu when I asked him how many people followed these precepts: 'Probably more than half the ladies and gentlemen of Szechuan.'

[1] 一陰一陽相須　　　[2] 醫心方　　　[3] 丹波康賴　　　[4] 雙梅景闇叢書
[5] 葉德輝　　　[6] 養生延命錄　　　[7] 陶弘景　　　[8] 孫思邈
[9] 素女經　　　[10] 玄女經　　　[11] 玉房祕訣　　　[12] 洞玄子
[13] 天地陰陽大樂賦　　　[14] 玉房指要　　　[15] 房術

That some of the texts are ancient can hardly be doubted. The bibliography of the *Chhien Han Shu* lists eight relevant books, all now lost, which must have been current in the − 1st century. Two of them bore the title *Yin Tao*[1] (The Tao of the Feminine), but we know nothing of their authors, Jung Chhêng[2] and Wu Chhêng.[3] Others were called after various emperors of antiquity. The names of certain men who were regarded as great experts in these matters have come down to us, notably Lêng Shou-Kuang,[4] the contemporary[a] and associate of the famous + 3rd-century physician Hua Tho, and Kan Shih,[5] who lived at about the same time.[b] Significantly, emphasis is placed on the value of their techniques for longevity. Most typical perhaps of all the documents is the *Su Nü Ching*, the style of which is certainly similar to that of the Han medical classic *Huang Ti Nei Ching* (cf. Sect. 44 below). Although it is not given in the Han bibliography, it must have existed in some form in the + 1st century because it is referred to both by Wang Chhung[c] and Chang Hêng.[d] By the time of Ko Hung (early + 4th century), three other wise women are mentioned,[e] including Tshai Nü[6] (the Chosen Girl),[f] which has led to the suggestion[g] that they may have been originally an order of female magicians (*wu*[7]).

The official Sui bibliography (+ 7th century) lists seven books, among them the *Yü Fang Pi Chüeh*, which we still have. Though the *Tung Hsüan Tzu*[h] does not appear until the bibliography of the Thang, its text is rather archaic, and like all the rest, its explanations, which are elaborate, are medically and physiologically sound.[i] Among the most remarkable of the documents is the *Thien Ti Yin Yang Ta Lo Fu* (Poetical Essay on the Supreme Joy), written by Pai Hsing-Chien[8] (d. + 826), the younger brother of Pai Chü-I, and preserved only in manuscript form in the monastic library of Tunhuang, until it was recovered in our own time.[j]

The *Yün Chi Chhi Chhien*,[k] with a remarkable echo of Aristotle,[l] says that the seminal essence is held in the seminal vesicles (*ching shih*[9]) in the lower part of the

---

[a] *Hou Han Shu*, ch. 112B, p. 10*b*.

[b] *Hou Han Shu*, ch. 112B, p. 18*a*. Other late Han and San Kuo experts were Tungkuo Yen-Nien,[10] Fêng Chün-Ta[11] and Wang Chen,[12] as also the famous magician Tso Tzhu.[13]

[c] *Lun Hêng*, ch. 6 (Forke (4), vol. 1, p. 141), where the reference is unfavourable to the Su Nü. 'The Immaculate Girl, in describing to the Yellow Emperor the methods (of love-making) of the Five Girls, (set forth that which) does harm not only to the bodies of the parents, but to the natures of the male and female children also' (tr. Leslie, 1). Wang Chhung did not explain what he meant.

[d] In his beautiful epithalamion, *Thung Shêng Ko*[14] (The Song of Harmony), just before + 100. From this poem it is clear that brides were given a scroll in Han times, with pictures of the positions of intercourse, and an accompanying text. One position is referred to metaphorically in ch. 61 of the *Tao Tê Ching*.   [e] *Pao Phu Tzu* (*Nei Phien*), ch. 4/146.

[f] This was also the lowest rank of imperial concubines.

[g] Van Gulik (3), p. 15.

[h] Translated by van Gulik (3).

[i] Thirty positions are described.

[j] Summarised in paraphrase by van Gulik (3).

[k] Ch. 58, p. 6*a*.   [l] Cf. Needham (2), pp. 24 ff.

---

[1] 陰道	[2] 容成	[3] 務成	[4] 冷壽光	[5] 甘始	[6] 采女
[7] 巫	[8] 白行簡	[9] 糟室	[10] 東郭廷年	[11] 封君達	
[12] 王眞	[13] 左慈	[14] 同聲歌			

abdomen (*hsia tan thien*[1]), and that while the sperm is stored there in men, menstrual blood accumulates in the corresponding part of the female body (*nan jen i tshang ching, nü tzu i yüeh shui*[2]). The purpose of the Taoist techniques was to increase the amount of life-giving *ching* as much as possible by sexual stimulus, but at the same time to avoid as far as possible the loss of it. Moreover, if the Yang force in man were continually fed by the Yin force, it would not only conduce to his health and longevity, but its intense maleness would ensure that when emission did take place, the sex of the resulting child would be male. Continence was considered not only impossible, but improper, as contrary to the great rhythm of Nature, since everything in Nature had male or female properties. Celibacy (advocated later by the Buddhist heretics) would produce only neuroses. The technique consisted first, therefore, of frequent *coitus reservatus*,[a] numerous intromissions with a succession of partners occurring for every one ejaculation.[b] Female orgasms (*khuai*[3]) strengthened man's vital powers, hence the male act was to be prolonged as much as possible so that the Yang might be nourished by as much Yin as possible.[c] It is at first sight puzzling that *coitus reservatus* should have been considered so valuable for mental health, since as a method of contraception *coitus interruptus* is widely condemned in modern medicine. But the psychological conditions were quite different; the object was not to prevent conception but to ensure the nourishment of the two forces, especially the Yang.[d] Much emphasis was laid on the succession of partners, and many (and conflicting) directions for their choice appear, but owing to an elaborate system of prohibitions depending on the seasons, phases of the moon, the weather, the astrological situation, and so on, suitably propitious occasions for the Taoist adepts could not have occurred very often. Within families, where the attainment of *hsien*-ship was not the primary aim, less attention was paid to times and seasons.

Another method, that of 'making the *ching* return', consisted of an interesting technique which has been found among other peoples in use as a contraceptive device,[e] and still appears sporadically among European populations.[f] At the moment of ejaculation, pressure was exerted on the urethra between the scrotum and the anus, thus diverting the seminal secretion into the bladder, whence it would later be voided with the excreted urine. This, however, the Taoists did not know; they thought that the seminal essence could thus be made to ascend and rejuvenate or revivify the

---

[a] Termed by van Gulik, inadvertently, *coitus interruptus*.

[b] *Su Nü Ching*, p. 1b; *Yü Fang Chih Yao*, p. 1b.

[c] *Su Nü Ching*, pp. 2a, 4a. The physiological soundness of the procedure needs no comment, whatever may be thought of the old Chinese theories.

[d] Since the same techniques were recommended to women aspirants to *hsien*-ship, we can understand the case of the *wu*, unearthed by de Groot (2), vol. 6, p. 1235, from the *Chiu Thang Shu*, ch. 130—'a beauty of mature age', who travelled about by imperial order offering sacrifices to various local deities, attended by a troop of 'depraved young men'.

[e] Notably among the Turks, Armenians and Marquesan Islanders (private communication from Dr Gene Weltfish). Seventeenth-century physicians, such as Sanctorius, recommended their patients to refrain sometimes from ejaculation in coitus.

[f] Cf. Griffith (1), p. 95.

[1] 下丹田   [2] 男人以藏精女子以月水   [3] 快

upper parts of the body—hence the principle was termed *huan ching pu nao*,[1] making the *ching* return to restore the brain.[a] One should note the close parallel between *huan ching* and *pi chhi*. Since the spinal cord, in Taoist physiology, was likened to the Yellow River in its downward-radiating trophic influence, the process is recognisable under the phrase 'making the "Yellow River" flow backwards' (*Huang Ho ni liu*[2]), found in late books.[b] All this is allusively described in the *Thai Shang Huang Thing Wai Ching Yü Ching*[3] (Excellent Jade Classic of the Yellow Court),[c] which is mentioned in the *Lieh Hsien Chuan* and in *Pao Phu Tzu*, and must therefore be not later than the +2nd or +3rd centuries. Perhaps the oldest reference, however, is that in the *Hou Han Shu*,[d] where the text says that Lêng Shou-Kuang practised the arts of Jung Chhêng and lived to a great age. The commentary quotes the *Lieh Hsien Chuan* as saying: 'The art of commerce with women consists in refraining from ejaculation and causing the sperm to return and nourish the brain (*Yü fu-jen chih shu wei wo ku pu hsieh, huan ching pu nao*[4]).'[e]

The most astonishing aspect of this whole department of Taoist philosophical-religious practice (astonishing too for most modern Chinese) is that it comprised public ceremonies as well as ordinary conjugal life and private exercises for *hsien*-ship candidates. These liturgies were called 'The True Art of Equalising the *Chhi*'s' (*Chung Chhi Chen Shu*[5]), or 'Uniting the *Chhi*'s' (*ho chhi*,[6] *hun*[f] *chhi*,[7] *ho chhi*[8]) of male and female. Their origin was attributed to the great Chang family of Taoists of the +2nd century (the 'san Chang'[9]), and they were certainly in common practice about +400 under the leadership of Sun Ên.[10] Much of what we know about them comes from Chen Luan[11] the mathematician (fl. +566), who was a convert from Taoism to Buddhism, and wrote the *Hsiao Tao Lun*[12] (Taoism Ridiculed). The ceremony was intended for 'deliverance from guilt' (*shih tsui*[13])[g] and occurred on nights of new moon and full moon, after fasting. It consisted of a ritual dance, the 'coiling of the dragon and playing of the tiger',[h] which ended either in a public hierogamy or in successive unions of the members of the assembly in the chambers along the sides

---

[a] *Su Nü Ching*, p. 2a; *Yü Fang Chih Yao*, p. 1b; *Pao Phu Tzu* (*Nei Phien*), ch. 6, p. 57b. This again is an interesting idea from the point of view of the history of embryology (cf. Needham (2), p. 60). The idea that 'the father sows the white, and the mother sows the red', i.e. that the white parts of the body, such as the brain and nerves, come from the seminal secretion, and the red parts from the menstrual blood, is one of the oldest speculations which biological thinkers have entertained.

[b] For example, the *Su Nü Miao Lun*[14] (Mysterious Discourses of the Immaculate Girl), about +1500; cf. van Gulik (3), p. 109.

[c] *TT* 329. Cf. Wilhelm & Jung, pp. 35, 69, 70.

[d] Ch. 112B, p. 11a.

[e] Tr. van Gulik (3).

[f] Note the survival of this ancient watchword of community life (cf. pp. 107, 115 above).

[g] This is commended as noteworthy to the diverse schools of modern psychology.

[h] It is important to notice the use of the male and female alchemical symbols (cf. pp. 330, 333 below).

---

¹ 還精補腦   ² 黃河逆流   ³ 太上黃庭外景玉經
⁴ 御婦人之術謂握固不洩還精補腦   ⁵ 中氣眞術   ⁶ 合氣   ⁷ 混氣
⁸ 和氣   ⁹ 三張   ¹⁰ 孫恩   ¹¹ 甄鸞   ¹² 笑道論   ¹³ 釋罪
¹⁴ 素女妙論

of the temple courtyard.[a] The couples were instructed in the techniques already mentioned. The liturgy seems to have been contained in a book called *Huang Shu*,[1] from which a fragment of high poetic quality has survived.[b] Naturally Buddhist asceticism and Confucian prudery were both scandalised, and a counter-movement was already under way by +415. By the middle of the +6th century it had made great inroads into Taoism, and there were probably no Ho-Chhi festivals after the +7th.[c] But private practices continued until well into the Sung so far as Taoists attached to temples were concerned, and until the last century for lay people in general, all the more as they were approved and counselled by the medical profession.[d]

The recognition of the importance of woman in the scheme of things, the acceptance of equality of women with men, the conviction that the attainment of health and longevity needed the cooperation of the sexes,[e] the considered admiration for certain feminine psychological characteristics, the incorporation of the physical phenomena of sex in numinous group catharsis, free alike from asceticism and class distinctions, reveal to us once more aspects of Taoism which had no counterpart in Confucianism or ordinary Buddhism. There must surely be some connection between these things and the matriarchal elements in primitive tribal collectivism, some reflection in the prominence of the Female Symbol in ancient Taoist philosophy.[f] It can be no coincidence that the Taoists were the supreme representatives in ancient China of social solidarity, of aggregation and unity, of all that was opposed to division and separation. Indeed, their thought and practice went so deep as to be universal, having Ionian and Orphic parallels; love, the power of affinity and union in the universe, commands elements, stars and gods; so it was a Greek commonplace to say,

[a] Maspero (13), p. 167, conjectures plausibly a connection with the primitive tribal mating-festivals described by Granet (2), but it might be hard to prove (cf. pp. 104 ff. above). One cannot fail to see a strong current of primitive community solidarity, characteristic of Taoism, running through these festivals in which sex itself was made numinous. It is highly significant that one of the Buddhist antagonists said that in these observances 'men and women unite in an improper way since they make no distinction between nobility and commoners' (Maspero (7), p. 406). The Taoist emphasis was indeed upon humanity as such (cf. pp. 112 ff., 130 above; 435, 448 below).

[b] Maspero (7), p. 408. Exactly what deities were worshipped during the Ho-Chhi liturgies is not easy to say, but they seem to have been star-gods, the gods of the five elements, and the spirits supposed to reside in, and to control, the various parts of the human body. Cf. Maspero (27).

[c] That is to say, within Chinese Taoism. But it may be that this numinous sexuality continued until much later in Tantric Buddhism and Lamaism. Later on it will be suggested that the origin of many Tantric ideas and practices was Taoist (cf. pp. 427 ff. below). As late as 1950, when a certain secret society was being dissolved in China, there were allegations of group sexual intercourse pursued as a way to health and immortality (van Gulik (3), p. 103). Ideas die hard.

[d] Cf. Dudgeon (1), where clear traces will be found (pp. 376, 440, 454, 494, 516). The great physician Sun Ssu-Mo (d. +682) has material in *Chhien Chin Fang* (The Thousand Golden Remedies) essential for the study of the medical traditions (cf. van Gulik (3), pp. 76 ff.). References in later times are quite numerous, for example the *Ming Tao Tsa Chih*[2] (Miscellany of the Bright Tao) by Chang Lei,[3] who was one of the circle of Su Tung-Pho.

[e] Perhaps this is symbolised in bronze by those Chinese (and Scythian?) boxes which have been described by Salmony (2). The lids are ornamented by two kneeling figures, naked, a man and a woman opposite each other.

[f] Cf. above, pp. 57 ff., 61.

[1] 黃書   [2] 明道雜志   [3] 張耒

as in the *Daphnis and Chloe* of Longus.[a] Thus it was that Lucretius dedicated his great poem to Venus,[b] for only by aggregation and union of particles, as of persons, can organisms at all levels be constructed and maintained in being. The physiology of the Taoists might be primitive and fanciful, but they had a much more adequate attitude to the male, the female and the cosmic background than the paternal-repressive austerity of Confucianism, so typical of a feudal property-owning mental state,[c] or the chilling other-worldliness of Buddhism, for which sex was no natural or beautiful thing, but a mere device of Māra the Tempter.

In several medieval dynasties there were still noted Taoist women adepts and preachers, and Han Yü, the great Thang Confucian, wrote a poem about one of them.[d] In certain places surviving local cults bear witness to old recognitions of the importance of women, for example, a flood-legend at Thaiyuan in Shansi which still generates an annual procession with girls playing the parts of triumphal and deified cult-heroines.[e] Here the feminine symbol and the water symbol are both present. All in all, the Taoists had much to teach the world, and even though Taoism as an organised religion is dying or dead, perhaps the future belongs to their philosophy.

(5) HAGIOGRAPHY OF THE IMMORTALS

It only remains to add that a large literature exists about the lives, achievements and 'miracles' of famous *hsien*. The earliest surviving book which contains material of this kind is the *Fêng Su Thung I*[1] (Popular Traditions and Customs) of Ying Shao,[2] written about +175. The series then begins with the *Lieh Hsien Chuan*[3] (Lives of Famous Hsien),[f] attributed to Liu Hsiang[4] (*c.* −50) but certainly the work of a Taoist who lived between the +2nd and early +4th centuries. To about the same period belong, therefore, Ko Hung's[5] *Shen Hsien Chuan*[6] (Lives of the Divine Hsien); and Kan Pao's[7] *Sou Shen Chi*[8] (Reports on Spiritual Manifestations) continued by Thao Chhien[9] (*Sou Shen Hou Chi*[10]).[g] Ko Hung's book was enlarged and amplified by Shen Fên[11] in the Thang dynasty as *Hsü Shen Hsien Chuan*[12] (Supplementary Lives

---

[a] Elsewhere (3, p. 39), in studying the thought of a great Victorian, Henry Drummond, I found how living these ideas still are; love, he thought, might be considered the social analogue of the physical bonds which unite particles at the molecular level. And indeed in the history of chemistry the first understanding of chemical reaction involved the sexual analogy.

[b] As Friedländer has pointed out, the Romans etymologised the name of the goddess as the force tying together fire and water, man and woman, 'horum vinctionis vis Venus' (Varro, *De Lingua Latina*, v, 61).

[c] Of course, Confucianism in its traditional forms was not essentially ascetic. When King Hsüan of Chhi confessed his interest in women, Mencius assured him that it was no sin as long as all his subjects could also satisfy their natural desires (*Mêng Tzu*, I (2), v, 5). And long afterwards Matteo Ricci, coming from a Europe still largely feudal, marvelled that in the choosing of concubines, beauty was the only rank (Trigault, tr. Gallagher, p. 75).

[d] Translated by Erkes (10).   [e] Described by Körner.
[f] Tr. Kaltenmark (2).   [g] Cf. Bodde (9).

[1] 風俗通義   [2] 應劭   [3] 列仙傳   [4] 劉向   [5] 葛洪
[6] 神仙傳   [7] 干寶   [8] 搜神記   [9] 陶潛   [10] 搜神後記
[11] 沈汾   [12] 續神仙傳

of the Hsien). In the Sung the tradition was continued with Li Fang's[1] *Thai-Phing Kuang Chi*[2] of +981. And so rapacious was the popular appetite for miracles and magical techniques that in the Ming and Chhing the genre was still added to; there was the *Shen Hsien Thung Chien*[3] (a title challenging comparison with the great Confucian historical work) by Hsüeh Ta-Hsün[4] in +1640, and finally the *Li Tai Shen Hsien Thung Chien*[5] (Survey of the Lives of the Hsien in All Ages) by Chang Chi-Tsung[6] in +1700. Selections of texts from these and other romantic stories of sages and immortals are available in translated form.[a]

(6) *HSIEN*-SHIP AND ORGANIC PHILOSOPHY

Leaving this mass of bizarre but fascinating detail, let us consider for a moment the philosophical significance of the Taoist ambition for a specifically *material* immortality. It was not that the Chinese lacked any conception of 'souls' or subtle spiritual essences; on the contrary, there were more of them than the European mind imagined— but it was not thought, as Maspero[b] has pointed out, that an individual personality could continue to exist without some bodily component. In other words, their conception of the living organism was an organic one, neither spiritualistic nor materialistic. Later on, in the Sections on the fundamental ideas of Chinese science, on the developed theories of the Neo-Confucians, and on the problem of laws of Nature, we shall see the vast significance of the organic outlook for all Chinese thought about natural phenomena. We need only note at this stage that the material immortality of the Taoists was no peculiar whim, but a belief with far-ranging implications.

If the Taoists [wrote Maspero][c] in their search for longevity, conceived it not as a spiritual but as a material immortality, it was not as a deliberate choice between different possible solutions but because for them it was the only possible solution. The Graeco-Roman world early adopted the habit of setting Spirit and Matter in opposition to one another, and the religious form of this was the conception of a spiritual soul attached to a material body. But the Chinese never separated Spirit and Matter, and for them the world was a continuum passing from the void at one end to the grossest matter at the other; hence 'soul' never took up this antithetical character in relation to matter. Moreover, there were too many souls in a man for any one of them to counter-balance, as it were, the body; there were two groups of souls, three upper ones (*hun*[7]) and seven lower ones (*pho*[8]), and if there were differences of opinion about what became of them in the other world, it was agreed that they separated at death. In life as in death, these multiple souls were rather ill-defined and vague; after death, when the dim little troop of spirits had dispersed, how could they possibly be

---

[a] E.g. L. Giles (6) and de Harlez (4). Pfizmaier (87) has translated four chapters of the *Thai-Phing Yü Lan* encyclopaedia on the subject.
[b] (12), p. 53.   [c] (13), p. 17.

[1] 李昉   [2] 太平廣記   [3] 神仙通鑑   [4] 薛大訓
[5] 歷代神仙通鑑   [6] 張繼宗   [7] 魂   [8] 魄

re-assembled into a unity? The body, on the contrary, was a unity, and served as a home for these as well as other spirits. Thus it was only by the perpetuation of the body, in some form or other, that one could conceive of a continuation of the living personality as a whole.

Nothing could better show that the material immortality of the Taoists was one facet of the whole organic character of Chinese thought, which did not suffer (to use a phrase which will find employment later) from the typical schizophrenia of Europe, the inability to get away from mechanistic materialism on the one hand, and from theological spiritualism on the other.

(j) TAOISM AS A RELIGION

One day in 1943 I made an excursion, with a number of eminent Chinese scientists, from Kunming, the capital of Yunnan, to the western hills to visit the three beautiful temples there and enjoy their magnificent views of the Kunming Lake. The first two you come to are Buddhist, but we were all more interested in the third, the Taoist one, being conscious of the interest for science of ancient Taoist thought. Known as the San Chhing Ko,[1] the Chamber of the Three Pure Ones, this exquisite rock-hewn shrine is built half-way up an almost perpendicular cliff (Fig. 40). But when I asked who exactly the Three Pure Ones were, no one in the party had any idea.[a]

This exemplifies the general lack of study which has been accorded to one of the most interesting phenomena in the whole of comparative religion.[b] How could it have come about that the high philosophy (at one and the same time scientific and mystical) of the Taoist fathers, which we have been examining—even with its strange marriage[c] to the primarily practical magic of the shamanistic *wu*—was transformed into a theist and supernaturalist religion,[d] heavily laden with superstition, and not without an element of conscious mystification? It is true that this question does not belong, *sensu stricto*, to the history of science, but its interest for the history of Chinese civilisation is so great, and the whole process has been so little elucidated, that we cannot entirely overlook it here. Besides, we require some explanation of the disappearance of those germs of scientific thought so prominent in ancient and early medieval Taoism. There can be no doubt that Maspero[e] is right in saying that

---

[a] Later I was much helped in the understanding of Taoism by my friends Dr Kuo Pên-Tao in Chhêngtu, and the late Dr Huang Fang-Kang in Chiating. I remember that one of the party at the Western hills was Dr Li Shu-Hua the physicist, to whom I remain indebted for many kindnesses when first in China.

[b] Unfortunately some of the best work on the history of Taoism is available only to those who read Japanese, e.g. Tsumaki (1); Tokiwa (1, 2), etc. (see Aurousseau, 1). But Maspero (13) (in French) is indispensable and brilliant.

[c] Cf. Erkes (5); Hsü Ti-Shan (2).

[d] The Taoist 'pantheon' can be studied in Doré and other accounts; I would add mention of the short general paper of Hayes (1), which can serve as an introduction. The old papers of Edkins (16) and Mueller (1) are still interesting. Cults of deified heroes and city-gods have always been served by Taoists: see Ayscough (1); Volpert (1); Pfizmaier (82).

[e] (12), pp. 35, 47; (13), p. 15.

[1] 三清閣

PLATE XV

Fig. 40. One of the galleries and shrines of the San Chhing Ko temple, Western Hills, Kunming, Yunnan. The shores of the Kunming lake can be seen far below.

religious Taoism was a reaction against the purely collective religion of ancient Chinese feudal society with its altars of the gods of soil and grain. The larger the State became, the more impossible was it for all the people to participate in their rites. Taoism, therefore, became China's indigenous individualistic religion of salvation.

The factual story begins at the beginning of the Han, and has a direct connection with the antagonisms of the philosophical schools. It is clear that although they had some aspects of thought in common, the Taoists hated the Legalists. The Legalists, in a sense, out-Confucianed the Confucians. Upholding the feudal system, but abandoning all pretence of humanising it, they essayed to establish the power of the ruler on a basis of draconic authoritarianism and terror. Although in the end this led them to go beyond the feudal system altogether and take the first essential steps towards feudal bureaucratism, they must nevertheless have been anathema to the primitive collectivist, 'democratic', political theorists of Taoism. Hence was it not natural that when the unified Legalist empire of Chhin rocked to its fall, there should have been important Taoist figures on the side of the adventurer who became the first emperor of the Han? Such a person was Chang Liang,[1] statesman and aspirant to *hsien*-ship.[a] He is said to have owed much to a semi-legendary figure, Huang Shih Kung[2] (the Old Gentleman of the Yellow Stone),[b] who is supposed to have written the (rather insipid) *Su Shu*.[3][c] Chang Liang died in −187.

The exact connection between Chang Liang and the famous Chang family of the +1st century who did so much towards making Taoism an organised religion is not clear, though tradition has long asserted them to have been his direct descendants. In any case Chang Ling (afterwards called Chang Tao-Ling[4]),[d] who was a Taoist and alchemist (fl. +156), acquired so many followers that he was able to set up a kind of semi-independent State in a strategic position on the borders of Szechuan and Shensi, which lasted till +215. A description of this State, which had its centre at Hanchung just south of the Chhinling Mountains, and which after some time became a province with Chang Tao-Ling confirmed as governor, is to be found in Ko Hung's *Shen Hsien Chuan*.[e] Chang Tao-Ling's government seems to have been much assisted by the belief of the people in his magical powers, but Confucians in other parts of the country termed his teachings the 'five-bushel rice Tao' (*wu tou mi tao*[5]) on account of the contribution which was fixed from each family; and the name stuck.[f] Just after Chang Tao-Ling's death, the prestige of Taoism had so much increased that in +165 official imperial sacrifices were for the first time offered to Lao Tzu.[g] The Taoist

---

[a] G 88. [b] G 866.

[c] Generally thought to have been fabricated in the Sung (see Wylie (1), p. 73); but recently Ku Chieh-Kang has expressed the opinion that it may be a genuine −2nd-century work. It discusses civil government and military tactics, from a more or less Taoist angle. The present writer has spent many happy hours in the Taoist temple of Huang Shih Kung at Miao-thai-tzu in Shensi.

[d] G 112.

[e] Tr. L. Giles (6), p. 60.

[f] *TH*, p. 784. Cf. Duyvendak's review of Maspero (13). [g] *TH*, p. 754.

[1] 張亮 [2] 黃石公 [3] 素書 [4] 張道陵 [5] 五斗米道

leadership seems to have descended to Chang Tao-Ling's son, Chang Hêng,[1][a] and in turn to his grandson Chang Lu.[2] It is thought extremely probable that this family of Changs was closely connected with Chang Chio[3] who[b] with his brothers organised the terrible revolution of the 'Yellow Turbans' in +184, which was a mass movement and not at all the uprising of a few alchemists.[c]

Recently it has been suggested that the new start made by Chang Tao-Ling was under strong Zoroastrian (Mazdaean) influence from Persia (Dubs, 19). The evidence for this, which so far does not seem at all convincing, may become more so when given in fuller form. Eberhard (7, 8), who has made some preliminary criticisms of it, believes that Indo-Iranian influence was active not only at this time, but also much earlier in the school of Naturalists of Tsou Yen (see pp. 232 ff.), which had its centre along the eastern coast and not in Szechuan. In agreement with many others, he views the theoretical geography of Tsou Yen as derivative from the ancient Indian system of nine continents (*dvīpa*) (cf. p. 236), though in our judgment considerable scepticism is still desirable here. Then we find, both in the Chhan-Wei divination books (cf. pp. 380, 382),[d] which, according to Chhen Phan, go back to the school of Naturalists, and in *Huai Nan Tzu*[e] and the *Lun Hêng*,[f] mention of nine stars or palaces important for astrology—these Eberhard identifies as the seven planets of the Iranian planetary week plus the Indian hypothetical planets Rahu and Ketu. But the subject is still very obscure and requires much further research. We need only to remember here that there may possibly have been some stimulus from abroad for the activities and doctrines of Chang Tao-Ling.

In any case there was, from the +2nd century onwards, a definite Taoist 'Church'. Maspero[g] has given a quite detailed account of it. We know the names of the various officiants and exorcists, and much material about the liturgy and ceremonial has come down to us. Thus its 'parochial' organisation is preserved in the *Hsüan Tu Lü Wên*[4] (Code of the Mysterious Capital)[h] and the rituals of several kinds of 'masses' are contained in a group of books in the *Tao Tsang*.[i] At least two among the books in the patrology seem to date from the +2nd century;[j] they contain a wealth of names of gods and spirits.[k] Taoist monasticism was no doubt partly modelled on Buddhist

---

[a] Not to be confounded with his contemporary of exactly the same name, the mathematician and astronomer.

[b] G 36.

[c] Maspero (13), p. 156, found difficulty in understanding the mainsprings of this mass movement. I believe this was because, though full of understanding for the religious side of Taoism, he failed to appreciate its strongly political aspect (cf. above, pp. 104, 138). I would align the Yellow Turban and other similar uprisings with the rebellions of the Donatists and Anabaptists in Europe—socialist upsurges with a religious mode of expression (cf. Needham (6), p. 14). Balazs (1) and H. Franke (3) have well emphasised the strongly equalitarian character of Taoism.

[d] Bibliography in Kaltenmark (1).   [e] Ch. 21, p. 8a.

[f] Ch. 15, p. 2a.   [g] (13), pp. 45, 48, 150 ff., 163.

[h] TT 185.   [i] TT 479–502.

[j] TT 329 and 7.

[k] Pfizmaier (99, 102) translated some of the chapters in the *Thai-Phing Yü Lan* encyclopaedia, which deal with spirits.

[1] 張衡   [2] 張魯   [3] 張角   [4] 玄都律文

practices, but much of it may be traced back, as Erkes (14) shows, to the hermit-philosophers of the Warring States and the early Han.

During the following century philosophical Taoism had as its principal representatives the men who formed in +262 the 'Seven Sages of the Bamboo Grove' (*chu lin chhi hsien*[1]).[a] Politically the most important was probably Hsi Khang[2] (+223 to +262);[b] but Hsiang Hsiu,[c] as we shall see in a later section, was associated with alchemy; and we shall meet again another member of the group, Wang Jung,[3] for he was an important patron[d] of the early water-mill technologists (see on, Sect. 27*f*). This was in the Wei State, the northern one of the three into which China was divided during the San Kuo period. But south in the State of Wu, more important things were happening. Ko Hsüan,[4] of whose personal details we know little except that he was the great-uncle of the outstanding alchemist Ko Hung (Pao Phu Tzu), was a friend of the reigning king. He had been a disciple, according to tradition,[e] of the famous Han magician Tso Tzhu[f] (+155 to +220), a man with whom we have already met (Sect. 7*i*), since, as will be remembered, he was cited as one of the greatest of Chinese thaumaturgists by one of the monk-ambassadors to India. During the period from +238 to +250 Ko Hsüan was the recipient of visions and revelations from the ruler of Heaven (who was now beginning to take on a distinctly more personalised character), Thai Shang,[5] or Thien Chen Wang,[6] who sent him four celestial visitants[g] with more than thirty divinely inspired texts, among which was the *Ling Pao Ching*[7] (Divine Precious Classic).[h] Ko Hsüan is supposed to have been the author himself, among other books, of the *Chhing Ching Ching*[8] (Classic of Pure Calm).[i] Through Chêng Ssu-Yuan[9] and other intermediaries, these new doctrines reached Ko Hung, who, though not himself mentioning the Ling Pao in his *Pao Phu Tzu*, talks a good deal of the heavenly Ssu Ming[10] (Controller of Destinies, or Rewards and Punishments), and thus contributed his share to the personalisation of the original completely naturalistic conception of the impersonal Tao, paving the way for the Jade Emperor.[j]

In the +4th century the series of revelations continued. During the Chin dynasty, between +326 and +342, the woman Taoist Wei Hua-Tshun[11] received all kinds of further information about the organisation of heaven and earth from a mysterious

---

[a] *TH*, p. 857.
[b] G 293.
[c] G 693.
[d] G 2188.
[e] Almost all that we know of Ko Hsüan is derived from the *Yün Chi Chhi Chhien*, the basic compendium on the early history of religious Taoism, which we had occasion to quote above (p. 141).
[f] G 2028.
[g] Wieger (4), p. 511, traces foreign influence in the names of certain of these but not very convincingly. Three of them seem to go together, and may be the first appearance of the Taoist 'Trinity' described in the following paragraphs.
[h] Presumably the original version of *TT* 1.
[i] *TT* 615, tr. Legge (5).
[j] Wieger (4) devotes the whole of his ch. 52 to an analysis of Ko Hung's 'theology' and magic, as seen in the *Pao Phu Tzu*; this is worth reading, though with caution.

[1] 竹林七賢　　[2] 稽康　　[3] 王戎　　[4] 葛玄　　[5] 太上　　[6] 天真王
[7] 靈寶經　　[8] 清靜經　　[9] 鄭思遠　　[10] 司命　　[11] 魏華存

figure Wang Pao;[1] and about the same time a similar corpus was worked up by Hsü Ying[2] and others of his name. A century later, about +489, all this material came into the hands of Thao Hung-Ching,[3] the famous Taoist physician[a] of the Liang dynasty, who published much of it in his book *Chen Kao*[4] (True Reports); the earliest dated elements in this being of +365 onwards. They consist of conversations with genii and heavenly visitants. The True Heavenly King (*Thien Chen Wang*) of Ko Hsüan has now become the First Original Heavenly Venerable One (*Yuan Shih Thien Tsun*[5]) (a first cause). The liturgies are now stabilised; they speak of the Great Mysterious Three in One (*Thai Hsüan San I*[6]), the Sagely Father (*Shêng Fu*[7]), the Lord and Master of the Human Spirit (*Jen Shen chih Chu-Tsai*[8]),[b] and the Pivot of all Transformations (*Tsao-Hua chih Shu Chi*[9]).[c] Wieger (4) was at first inclined to see the influence of Christianity in this trinitarianism, but when he afterwards found references to the Seven Citadels (*chhi yü*[10]) and the Eight Pure Ones (*pa Su*[11]) he proposed that in some way or another Gnostic doctrines such as those of Basilides had found their way to China. The problem is, so far as we know, completely unsolved. It seems to us that this +3rd- and +5th-century trinitarianism could equally well have been derived from the cosmogonic (42nd) chapter of the *Tao Tê Ching*.[d] Thao Hung-Ching's disciple Wang Yuan-Chih[12] is credited with the intro- duction of spoken charms (*shêng chüeh*[13]) and written talismans (*fu lu*[14]), but they must have been earlier because the latter at any rate occur in Ko Hung.[e]

Meanwhile developments had been proceeding in the north at the court of the Northern Wei dynasty, where in +423 the Taoist Khou Chhien-Chih[15] got himself[f] the title of *Thien Shih*[16] (Heavenly Teacher).[g] This, the so-called Taoist 'papacy', descended in an unbroken line until the present century. In +1016 its seat was moved to Chiangsi,[h] where it remained until the Red Army passed that way about 1930, dispersed the retinue, and broke all the jars in which, according to local belief, the Taoists had imprisoned the winds. We are particularly well informed about the period of origin, since Ware (1) has studied and translated the relevant chapters of the *Wei Shu* (chapter 114) and the *Sui Shu* (chapter 35) of +554 and +656 respectively. They show a steady development along the general lines here described.

Organised Taoism prospered under the Thang, making a good start[i] because the

---

[a] G 1896 (+451 to +536).

[b] Note that this character *tsai*, Lord, is precisely that found in the passage of Chuang Tzu where he denies that such a personalised power exists in the universe (see above, p. 52).

[c] This alone preserves the flavour of ancient Taoism.

[d] See above, p. 78. After I had written this, the posthumous works of Maspero (13) were published, and I was gratified to find that he had taken just the same view (p. 138). Already in −130 Han Wu Ti was sacrificing to the 'Three Unities' (*Shih Chi*, ch. 28; Chavannes (1), vol. 3, p. 467; and *Chhien Han Shu*, ch. 25A, p. 8b).

[e] *Pao Phu Tzu* (*Nei Phien*), ch. 17. Cf. Doré (1), pt. 1, vol. 5; de Groot (2), vol. 6, pp. 1024 ff.

[f] G 984.   [g] *TH*, pp. 1073, 1113.

[h] *TH*, p. 1582.   [i] *TH*, p. 1301.

---

[1] 王褒	[2] 許映	[3] 陶弘景	[4] 眞誥	[5] 元始天尊
[6] 太玄三一		[7] 聖父	[8] 人神之主宰	[9] 造化之樞機	[10] 七域
[11] 八素	[12] 王遠智	[13] 勝訣	[14] 符籙	[15] 寇謙之	[16] 天師

name of the imperial family was the same as that of Lao Tzu, i.e. Li.[1] The great
abbey of Loukuantai, in the district of Chouchih, near Sian, dates from that time.[a]
The foundations of the patrology, the *Tao Tsang*, were laid in +745. Many Taoist
books were written, such as the *Yin Fu Ching*[2] (Harmony of the Seen and the
Unseen)[b] of Li Chhüan.[3] Many distinguished men, such as Li Pai,[c] were practising
Taoist initiates. Under strong pressure to compete with Confucianism and Buddhism,
the Taoists now appeared in the role of preachers of conventional morality,[d] hence the
Thai Shang Kan Ying Phien[4] (Tractate of Actions and Retributions)[e] of the early
+11th century;[f] following the *Kung Kuo Ko*[5] (Examination of Merits and Demerits)
attributed to the famous alchemist and *hsien*[g] Lü Tung-Pin.[6] This was the time when
the great controversy over the *Hua Hu Ching*[7] (Book of Lao Tzu's Conversions of
Foreigners) reached new heights; the Taoists pretended that Lao Tzu riding away into
the West, had been the spiritual father of Buddhism. The Buddhists ultimately
(+1258) secured the suppression of this thorn in their flesh, and it is not to be found
in the present *Tao Tsang*.[h] Here there is no space to tell of the prolonged Taoist-
Buddhist controversies. The two sides exhausted each other, permitting so the social
and organisational triumph of Neo-Confucianism.

At the beginning of the Sung dynasty Taoism continued to hold a strong position,
and was further strengthened by the comedy (as we now see it) of the third Sung
emperor's[i] mystifications. A whole series of 'revelations' was arranged—the finding
of letters from Heaven congratulating him, the announcing of auspicious omens, the
conferring of titles on spirits and genii by the emperor, the sending of the magic
mushrooms (*chih*[8]) to the court, etc.[j] These events took place from +1008 to +1022.
But after the north had fallen to the Chin, Taoists were involved in sterner work.
A number of schools and secret societies grew up, some at least of which were
essentially resistance movements against the Jurchen domination.[k] The *Tao Tsang* was
first printed ca. +1190 under both the Sung, and the Chin.

[a] The present writer's visit to Loukuantai in 1945 was particularly enjoyable and fruitful.
[b] Tr. Legge (5).   [c] Waley (13), p. 30.
[d] Interesting here is the case of the 'kitchen god' (*tsao chün*[9]), ubiquitous in medieval and modern
folk practice, who is supposed to report once a year to his superiors in the heavenly bureaucracy the
good or evil deeds of the family from which he comes. I fully agree with Doré (1), pt. II, vol. 11,
pp. 901 ff. (where full details are given), in believing that this god is the lineal descendant of that 'spirit
of the stove' to whom Han Wu Ti sacrificed at the instigation of the alchemist Li Shao-Chün. He was
therefore originally the spirit both of cookery and chemistry—a very significant connection for the
history of science. The character *tsao* is the same in both cases (cf. the important passage from the
Chhien Han Shu quoted below, Sect. 33). And since the kitchen was ruled by feminine manual practica-
lity, we find yet another aspect of the Taoist exaltation of the female principle. See also Nagel (1).
[e] *TT* 1153.   [f] Tr. Legge (5).
[g] G 1461 (+755 to +805).   [h] *TH*, p. 1420; Pelliot (12); Maspero (12), p. 75.
[i] Chen Tsung.
[j] The whole story, as translated by Wieger from the *Thung Chien Kang Mu*, makes very amusing
reading (*TH*, pp. 1572 ff.).
[k] Details, with references, in Chhen Jung-Chieh (4), pp. 148 ff.

[1] 李   [2] 陰符經   [3] 李筌   [4] 太上感應篇   [5] 功過格
[6] 呂洞賓   [7] 化胡經   [8] 芝   [9] 竈君

The Taoist trinity—the Three Pure Ones with whom we started—was now stabilised.[a] The Jade Emperor (*Yü Huang*[1]) perhaps represented the Unity.[b] The Persons were as follows:

(*a*) The Precious Heavenly Lord (*Thien Pao Chün*),[2] the First Original Heavenly Venerable One (*Yuan Shih Thien Tsun*[3]), controlling time past, likened by some to God the Father.

(*b*) The Precious Spiritual Lord (*Ling Pao Chün*[4]), the Great Jade-Imperial Heavenly Venerable One (*Thai Shang Yü Huang Thien Tsun*[5]), controlling time present, likened by some to God the Son.

(*c*) The Precious Divine Lord (*Shen Pao Chün*[6]), the Pure Dawn Heavenly Venerable One appearing from the Golden Palace (*Chin Kuan Yü Chhen Thien Tsun*[7]), controlling time to come, likened by some to God the Holy Ghost.

There can be little doubt that the Taoists had intimate contact with Nestorian Christians at the capital during the Thang dynasty.[c] The really interesting question is where their trinity came from eight centuries previously.[d]

After the Sung there was a decline. Foreign dynasties such as the Mongols and the Manchus were suspicious of Taoism[e] on account of its continuing subversive political nature, which so easily took the form of anti-foreign agitation; all governmental circles were afraid of it because of its methods of divination, which could so easily be used to launch predictions of changes of dynasty. In spite of mild persecutions in the Yuan,[f] a copy of the canon was carved on stone about +1346, and the whole Tao Tsang was again printed in the Ming (+1445; +1596 ff.).[g] Books continued to be written till a late date, e.g. the *Yü Shu Ching*[8] (Jade Pivot Classic)[h] in the +13th century. One such late book, the *Chhuan Tao Chi*,[9] contained in the collection *Tao Yen Nei Wai Pi Shê Chhüan Shu*,[10] and giving a general account of Taoist philosophy and religion, is available in a translation by Pfizmaier (81).[i]

a Cf. Doré (1), pt. II, vol. 6, p. 7; pt. II, vol. 9, p. 468; Wieger (4), p. 544; Maspero (11).
b As Fêng Han-Chi & Shryock (1) point out, the origin of the Jade Emperor is very obscure, and occurred some time during the Thang (+8th century); see also Fêng Han-Chi (1).
c Thus it is generally accepted that the 'Blue Goat' Temple at Chhêngtu represents what became of the Paschal Lamb.
d Maspero (13), p. 140, has given a striking example of the use of Buddhist logic by a Taoist expounder of the Taoist Trinity about the +4th century It is quite reminiscent of Athanasius.
e They had good reason to be. The 'White Lotus' secret society, founded in +1133, played a considerable part in the expulsion of the Mongols in +1351. The Yuan period was one of growing Buddhist ascendancy, though Chingiz Khan himself had chosen a Taoist as his spiritual adviser. This adept, Chhiu Chhang-Chhun, made a celebrated journey from Peking to the Khan's court south of Samarqand, and back, between +1219 and +1224 (see Waley, 10).
f Cf. *TH*, p. 1703. Edicts of +1258 and +1281 ordered the burning of all Taoist books except the *Tao Tê Ching*, but they were probably not carried out. Rinaker ten Broeck & Yü Tung (1) have translated a Taoist inscription of the early +14th century.
g See Erkes (13) and Pelliot's criticism of Wieger (6).   h Tr. Legge (5).
i In our own time the remnants of liturgical Taoism have been conducting a hopeless rearguard action against modern medicine and hygiene. To Hsü Lang-Kuang (1) we owe a valuable sociological

¹ 玉皇   ² 天寶君   ³ 元始天尊   ⁴ 靈寶君   ⁵ 太上玉皇天尊
⁶ 神寶君   ⁷ 金闕玉晨天尊   ⁸ 玉樞經   ⁹ 傳道集
¹⁰ 道言內外祕設全書

When one looks at the whole picture, one comes inescapably to the conclusion that the entire development was fundamentally the working up of an indigenous opposition system to Buddhism. First, political Taoism was sent underground. Then Confucian feudal bureaucratism allowed no outlet for the scientific energies potentially present in the Taoist philosophers and the shamanist magicians. Thought thus being sterilised and experimental techniques despised, the shamans, from the +1st century onwards, found their living being taken away from them by the new foreign religion of salvation from India. Gauchet (4) has shown how this idea of 'salvation' was quickly incorporated into important Taoist texts such as the *Tu Jen Ching*,[1] which dates from the early part of the +4th century. Individualistic religion was not wholly of Indian origin; Maspero (12) could see it growing during the Han as a reaction against the purely collective nature of ancient Chinese folk- and State-religion. But now, with half-conscious resource, the Taoists copied theology, sūtras and discipline to such good effect that for many centuries they were able to hold their own in the form of an organised religious institution which satisfied equally well, if not better, the needs of the peasant farmers and a minority of heterodox scholars. We shall see later (Sect. 49) how the influences of Chinese social life induced in the developing structure of Taoist theology a vast system of celestial bureaucracy, reflecting as in a mirror the bureaucratism of the earthly world. An approximate analogy would be to imagine that after the first successes of the Christian evangelisation of England, the Celtic and Saxon pagans had elaborated an entirely parallel cult based on some such figures as Arthur or Merlin. But the Taoists had, of course, far more to build on, and were much more firmly rooted in numerous thought and behaviour patterns deep-seated in the Chinese people. As for the development of an organised religion from a primitive social-revolutionary movement, the whole of Christianity itself might be called upon to provide a parallel.

(k) CONCLUSIONS

The philosophy of Taoism, as we have seen it in this analysis, though containing the elements of political collectivism, religious mysticism and the training of the individual for a material immortality, developed many of the most important features of the scientific attitude, and is therefore of cardinal importance for the history of science in China. Moreover, the Taoists acted on their principles, and that is why we owe to them the beginnings of chemistry, mineralogy, botany, zoology and pharmaceutics in East Asia. They show many parallels with the scientific pre-Socratic and Epicurean philosophers of Greece. Unfortunately, they failed to reach any precise definition of the experimental method, or any systematisation of their observations of Nature.

study of the varying attitudes of the people during a cholera epidemic in a small Yunnanese town (Hsichow) in 1942. Since the poetry and symbolism of the rites had long been wholly subordinate to their apotropæic function, and in any case not 'understanded of the people', hypodermic injections were steadily replacing them.

[1] 度人經

So wedded to empiricism were they, so impressed by the boundless multiplicity of Nature, so lacking in Aristotelian classificatory boldness, that they wholly dissociated themselves from the efforts of their contemporaries of the Mo and Ming schools to elaborate a logic suitable for science. Nor did they realise the need for the formation of an adequate corpus of technical terms.

The Taoists were profoundly conscious of the universality of change and transformation—this was one of their deepest scientific insights. But they themselves proved to be not immune from it. The 'deification' of Confucius may be considered strange, but the strangest transformation of all was that which converted Taoist agnostic naturalism into full-blown mystical religion and ultimately theist trinitarian theology, Taoist proto-scientific experimentalism into fortune-telling and rustic magic, Taoist primitive communalism into a way of personal salvation, Taoist anti-feudalism into equalitarian secret societies of anti-foreign or anti-dynastic tendency. The result came very near to exemplifying the words of Antoine de Rivarol,[a] 'Que l'histoire vous rappelle que partout où il y a mélange de religion et de barbarie, c'est toujours la religion qui triomphe; mais que partout où il y a mélange de barbarie et de philosophie, c'est la barbarie qui l'emporte....'

A later section of this book will hope to show that the responsibility for these extraordinary transmutations is to be laid at the door, not so much of the Taoists' complacent and conventional rival, social-minded Confucianism, as of the socio-economic system of feudal bureaucratism itself. In so far as it was this which sterilised the sprouts of natural science, no opening was left for the growth and flowering of the scientific elements in Taoism; on the contrary, its empirical element tended to be emphasised, in natural conjunction with the primarily technological achievements of Chinese society from the −2nd to the +13th centuries. Taoist philosophy being thus inhibited, Taoist shamanism inherited its ideas, and in view of the competition soon offered by Buddhism, could perhaps have continued to exist in no other way than by following the course which we have just described.

It is an intriguing question to ask why the history of European thought shows no real parallel to the Taoist complex. I often feel that if we had a complete answer to this question much of the respective mechanisms of the civilisations of Europe and Asia would be laid bare. There are, of course, groups and figures in European history which have a Taoist flavour, for example, the Pythagoreans[b] and Gnostics[c] as schools, and Roger Bacon,[d] Nicholas of Cusa[e] and Giordano Bruno[f] as individuals. The circle of Lady Conway at Ragley in the mid 17th-century, which included Francis Mercurius van Helmont and Dr Henry More of Christ's College, was 'Taoist' in many ways. Among later thinkers William Blake stands out as exceedingly 'Taoist' in his religious naturalism, and many expressions of his spring naturally to the mind when reading

[a] Cited in Sainte-Beuve, *Causeries du Lundi*, vol. 5, p. 82.
[b] Freeman (1), pp. 73 ff., 244 ff.
[c] I recommend the excellent book of Burkitt (2).
[d] +1214 to +1292 (Sarton (1), vol. 2, p. 952).
[e] +1401 to +1464.   [f] +1548 to +1600.

Taoist writings.[a] So much have I found this to be the case that the question arises whether by any chance Blake could have been made aware of Taoist modes of thought— there seems a bare possibility of it.[b] The great influence of the Confucian classics and their Neo-Confucian commentaries on 18th-century Europe, initiated by the famous *Confucius Sinarum Philosophus* (+1687) of Intorcetta, Couplet and their colleagues, is well appreciated—how different might have been the effect if the classics of Taoism had also been translated.

We have spoken already of the cleavage between rational logic and experimental empiricism; this went far deeper and lasted much longer in China than in the West. The rationalist Confucians and Logicians had practically no interest in Nature, the Taoists were deeply interested in Nature but mistrusted reason and logic. As Wang Chhung said, in the *Lun Hêng*, about +80: 'The Taoist school argues about spontaneity (*tzu-jan*) but does not know how to substantiate its cause by evidence. Therefore their theory of *tzu-jan* has not found general acceptance.'[c] This state of affairs was quite foreign to Greek culture, where we find a continuous transition from the pre-Socratics through Aristotle to the Alexandrians. The Renaissance phase which gave us the clue to the Taoist position on 'knowledge' was one of rather short duration.

One wonders whether the dominance of Hebrew monotheism in Europe might not provide an important clue. If the conception of a single 'personal' creator deity is firmly held ('liberating the mind', as one of the Fathers said, 'from the tyranny of ten thousand tyrants'), the nature of Nature is as much an indication of God's rationality as is the nature of Man.[d] One thinks of the two books from which Sir Thomas Browne said he collected his divinity, one the scriptures, the other 'that open and publick manuscript which lies expans'd unto the eyes of all'. Europe had perhaps no parallel to the Confucian phenomenon, the refusal to look at Nature, and hence no parallel to the Taoist phenomenon, the disinclination to trust reason and logic. Confucianism could perhaps be considered a parallel with the Hebrew 'priestly'

[a] See, for example, pp. 47, 142. Others have also felt this (e.g. Waley (19), p. 21).

[b] At least two Jesuit translations of the *Tao Tê Ching* and its commentaries are known. One was made by Francis Noel some time between +1685 and +1711 (Pfister (1), p. 418), and another, by J. F. Foucquet, into Latin and French, some time between +1700 and +1720 (Pfister (1), p. 553); both were sent back to France, and the latter still exists in MS at Paris. A third (if it was different from either of these) was given by Fr. de Grammont to Matthew Raper, F.R.S., who presented it to the library of the Royal Society in January 1788. It was there in Blake's time, but afterwards found its way to the library of the India Office. It is in Latin, and some think (Dr A. Waley, personal communication) that it was made by a Portuguese about +1760. Quite apart from the question of whether Blake could have read any of these translations or heard echoes of their contents, it may be asked how far they could have transmitted, in view of the primitive state of sinology at the time, any kind of idea of the meaning of the text. Only one fragment has been printed, namely the rendering by the Grammont-Raper MS of *TTC*, ch. 72 (Legge (5), vol. 1, p. 115); from this it would seem that it did rather better than Legge himself, certainly no worse—but that would not have been enough to give all the characteristic atmosphere and world-outlook required. The first printed translation did not come till St. Julien (1842). *Chuang Tzu* was not translated until the eighties, by Legge, Balfour and Giles more or less simultaneously. Its most recent appearance is in a Polish translation, beautifully produced, by Jabłoński *et al*.

[c] Ch. 54, tr. Forke (4), vol. 1, p. 97.

[d] After I had written this I found that A. N. Whitehead (1), p. 18, had made very similar observations.

tradition in so far as it regularised and supported the State sacrifices, and with the Hebrew 'prophetic' tradition in so far as it attempted to humanise and ameliorate first feudalism and then feudal bureaucratism. But there remains the 'wisdom literature' tradition[a] which has been shown to have its sources both in ancient Egypt and Babylonia, and which embodied nature philosophy, observation and inquiry into phenomena, and the facing of a sceptical issue, as in Job. This again is not without a Taoist flavour, and it had great importance in Europe, being transmitted through the Arabs to the early Humanists. But all such comparisons are unsatisfying and must remain mere suggestions for further thought.

In any case, Confucianism and Taoism still form the background of the Chinese mind, and for a long time to come will continue to do so. 'Confucianism' as Dubs (19) has well said, 'has been the philosophy of those who have "succeeded" or hoped to succeed. Taoism is the philosophy of those who have "failed"—or who have tasted the bitterness of "success".' Taoist patterns of thought and behaviour include all kinds of rebellion against conventions, the withdrawal of the individual from society, the love and study of Nature, the refusal to take office, and the living embodiment of the paradoxical non-possessiveness of the *Tao Tê Ching*; production without possession, action without self-assertion, development without domination.[b] Many of the most attractive elements of the Chinese character derive from Taoism.[c] China without Taoism would be a tree of which some of its deepest roots had perished.

These roots are still vigorous today.[d] I willingly acknowledge personal experience of them. Taoist scholars can still turn out a paradox for you in the good old tradition. The Abbot of Loukuantai, a venerable and delightful old man, said to me: 'The world thinks that it is going forward and that we Taoists are going backwards, but really it is just the opposite; we are going forward and they are going backward.' And still there remains that ancient connection between Taoism and proto-scientific naturalism. At the oilfield in the Nan Shan, in the far north-west along the Old Silk Road, a temple was erected in years gone by at the site of the seepages, which were considered a natural wonder, and of course it was a temple to Lao Tzu, who best of all men understood Nature. And, moreover, it was kept in repair during World War II by the Kansu Petroleum Administration. And, finally, among the beautiful gardens of the Taoist temple at Heilungthan near Kunming, where the National Academy of Peiping had its wartime laboratories, if one ascended through all the lower halls with their various images one came at last to an empty hall where there were no images, nothing but a large inscribed tablet with the characters *Wan Wu chih Mu*[1]—Nature, the Mother of All Things.

---

[a] Cf. Peet (1); Kent & Burrows (1).

[b] These phrases are quoted by Bertrand Russell (1), p. 194, but we have not been able to ascertain from where he derived them.

[c] Interesting psychological analyses which bring this out are to be found in the earlier books of Lin Yü-Thang (3, 4), and the paper of Lin Thung-Chi (1), who distinguishes as four main Taoist types the rebel, the recluse, the rogue and the returnist.   [d] Cf. Rousselle (1); Hackmann (1).

[1] 萬物之母

11. THE *MO CHIA* (MOHISTS) AND THE *MING CHIA* (LOGICIANS)

DISCUSSIONS of ancient Chinese philosophy usually treat of these two schools separately, but here they may well go together since their greatest interest for us is the effort which they made to work out a scientific logic. The Mohists provide a convenient transition on account of their strong political interests, which the Logicians seem to have shared in less degree.

Unlike Confucianism and Taoism, Mohism was completely overwhelmed by the social upheavals at the end of the Warring States period, and Ssuma Chhien did not even know the approximate dates of the birth and death of its founder Mo Ti.[1] It is now certain, however, that his life fell wholly within the period −479 to −381, so that he died not long before the birth of Mencius who in due course wrote against him.[a] He was thus a contemporary of Democritus, Hippocrates and Herodotus. Master Mo was a native of the State of Lu, and is believed to have been for a short time minister in Sung. He seems to have kept, like Confucius, a kind of school for those who wished to become officials of the feudal princes. His great doctrines, which have made him one of the noblest of China's historical figures,[b] were those of universal love, and the condemnation of offensive war; we shall briefly examine them in a moment.

Paradoxically, in view of its later complete disappearance, Mohism seems to have been from the beginning better organised than either the Confucian or Taoist movements. Fêng Yu-Lan (3) says that the Mohists represented what might almost be called the 'chivalrous' element in Chinese feudalism; they preached pacifism only up to a certain point, and trained themselves in military arts, in order to rush to the help of a weak State attacked by a strong one. Indeed, their practice of the techniques of fortification and defence was probably what led them to take interest in the basic methods of science, and to those studies in mechanics and optics which are among the earliest records of Chinese science which we possess (see below, Section 26c, g). If the interest of the Taoists had been directed rather to biological changes, that of the Mohists was attracted to physics and mechanics. In the later −4th century the

---

[a] *Mêng Tzu*, III (1), v; III (2), ix, 9 ff.; VII (1), xxvi. Legge (3) translated the chapter on universal love in the *Mo Tzu* book in his Mencius, pp. [103] ff.

[b] In an excellent popular exhibition of Chinese archaeology in one of the galleries of the Imperial Palace at Peking in the autumn of 1952, the three historical characters on which emphasis was laid were: Mo Ti, Kungshu Phan the mechanic, and Hsimên Pao the humanitarian official and hydraulic engineer. But the bookshops had available many popular expositions of the ancient philosophies, such as that of Yang Jung-Kuo (1), which compares Confucius with Mo Tzu.

[1] 墨翟

Mohist schools seem to have split into a number of different groups (*pieh Mo*[1]), but all acknowledged as their head a Grand Master (*Chü Tzu*[2]).[a]

The *Mo Tzu* book as we have it today is undoubtedly a compilation of very different dates. Some scholars think[b] that Mo Ti probably wrote nothing himself, but chapters 8–39 (systematic expositions of doctrine) and chapters 46–50 (conversations or 'Analects') must come from very shortly after −400. On the other hand, the 'Canons', the 'Expositions of the Canons' and the 'Illustrations' (chapters 40–45) cannot be much earlier than −300. No one knows when the important chapters on fortification technology (52–71) were written,[c] but they may well date from between −300 and −250. So limited in outlook have classical sinological studies been that attention has been concentrated almost solely on the ethical chapters; they alone, for example, are included in Mei Yi-Pao's translation (1), and the scientific work of the Mohists is generally fortunate if it receives a passing reference.[d] Even those who have devoted careful study to the Canons and their Expositions (*Ching*[3] and *Ching Shuo*[4]) such as Fêng Yu-Lan (1) and Maspero (9) from the point of view of logic, omit the scientific propositions.[e] It is not surprising, therefore, that only a very one-sided view of the ideas of the Mohists has so far been available.

One may thus say roughly that the earliest Mohists were interested in ethics, social life and religion; while the later Mohists dealt rather with scientific logic, science and military technology. Nevertheless, this change in orientation was a gradual one, the steps of which can dimly be followed. Before discussing the later phase, which is of greater interest to the historian of science, a few words must be said of the doctrines of the earlier.

(a) MO TI'S RELIGIOUS EMPIRICISM

Let us take up the thread of the attitude of the scholar to feudal society. We have seen in the preceding section that according to the Taoists social evolution had gone off on the wrong track; what they advocated was a return to primitive collectivist society before the differentiation of feudal class distinctions. On this the Mohists had a somewhat ambiguous attitude. In certain places they condemn primitive society, saying that it had been a war of each against all (e.g. chapter 11):

Mo Tzu said: In the beginning of human life, when there was as yet no law and government, the custom was 'everybody according to his own idea'. Accordingly each man had

---

[a] Mei Yi-Pao (2), pp. 166 ff. We have the bare names of some of the 'Later Mohists' who were probably concerned with the logical and scientific propositions which have come down to us—Hsiang Li-Chhin,[5] Hsiang Fu,[6] Têng Ling,[7] Chi Chhih[8] and Khu Huo.[9]

[b] Fêng Yu-Lan (1), pp. 76 ff.; Forke (3); Mei Yi-Pao (2); Hu Shih (4).

[c] They are associated with the name of Chhin Ku-Li.[10]

[d] E.g. Maspero (2), p. 620; Dubs (7), p. 216; Rowley (1).

[e] Forke (3), in his complete translation of the *Mo Tzu*, is the only exception to this rule. Of course some scholars in modern China have well appreciated the scientific significance of the Mohists.

[1] 別墨   [2] 鉅子   [3] 經   [4] 經說   [5] 相里勤   [6] 相夫
[7] 鄧陵   [8] 己齒   [9] 苦獲   [10] 禽滑釐

his own idea, two men had two different ideas, and ten men had ten different ideas—the more people the more different notions. And everyone approved of his own view and disapproved of the views of others; so arose mutual disapproval among men. As a result, fathers and sons, and elder and younger brothers, became enemies and were estranged from each other, since they were unable to reach any agreement. Everyone worked for the disadvantage of others with water, fire and poison. Surplus energy was not spent for mutual aid, surplus goods were allowed to rot without sharing, the excellent Tao was hidden away and not taught among men. The disorder in the human world could be compared to that among birds and beasts. All this disorder was due to the want of a ruler. Therefore the virtuous in the world were chosen and made emperors.[a]

On the other hand, elsewhere the Mohists adopted an attitude to society similar to that of the Taoists. The most famous passage illustrating this does not occur in the *Mo Tzu* book at all, but in the *Li Chi* (Record of Rites),[b] a Confucian compendium. The grounds for considering it Mohist, however, are strong, for passages with the same words and phrases occur in *Mo Tzu*.[c] No one knows how it got inserted in the *Li Chi*, and many, from Legge to Nagasawa, have considered it Taoist rather than Mohist; in any case it is surely not Confucian.[d]

When the Great Tao prevailed, the whole world was one Community (*thien hsia wei kung*[1]) Men of talents and virtue were chosen (to lead the people); their words were sincere and they cultivated harmony. Men treated the parents of others as their own, and cherished the children of others as their own. Competent provision was made for the aged until their death, work for the able-bodied, and education for the young. Kindness and compassion was shown to widows, orphans, childless men and those disabled by disease, so that all were looked after. Each man had his allotted work, and every woman a home to go to. They disliked to throw valuable things away, but that did not mean that they treasured them up in private storehouses. They liked to exert their strength in labour, but that did not mean that they worked for private advantage. In this way selfish schemings were repressed and found no way to arise. Thieves, robbers and traitors did not show themselves, so the outer doors of the houses remained open and were never shut. This was the period of the Great Togetherness (*Ta Thung*[2]).

But now the Great Tao is disused and eclipsed. The world (the empire) has become a family inheritance. Men love only their own parents and their own children. Valuable things and labour are used only for private advantage. Powerful men, imagining that inheritance of estates has always been the rule, fortify the walls of towns and villages and strengthen them by ditches and moats. 'Rites' and 'righteousness' are the thread upon which they hang the relations between ruler and minister, father and son, elder and younger brother, and husband and wife. In accordance with them they regulate consumption,

---

[a] Here the words in the Chinese text do not say *who* chose these virtuous ones. Mei Yi-Pao inserts 'by Heaven' because he does not agree with Liang Chhi-Chhao (2) that the people chose them. Naturally the interpretation of Mo Tzu's thought as approximating to modern democratic socialism depends on points like this, but the ancient Chinese text gives no clue. See Mei Yi-Pao (2), p. 111.

[b] Ch. 9 (Li Yün).   [c] Chs. 11, 12, 14, 15. Mei Yi-Pao (1), pp. 55, 59, 80, 82.

[d] As writers such as Hsü Shih-Lien and Hsiao Ching-Fang have claimed it to be.

[1] 天下爲公   [2] 大同

distribute land and dwellings, raise up men of war and 'knowledge'; achieving all for their own advantage. Thus selfish schemings are constantly taking their rise, and recourse is had to arms; thus it was that the Six Lords (Yü, Thang, Wên, Wu, Chhêng and the Duke of Chou) obtained their distinction.... This is the period which is called the Lesser Tranquillity (*Hsiao Khang*[1]).[a]

It must have been by a very peculiar historical turn of events that this highly subversive account became embedded in one of the Confucian classics. The phrase *Ta Thung*, the Great Togetherness, was used as the title of a famous book on socialism (*Ta Thung Shu*) by Khang Yu-Wei (1858–1927), one of the greatest of modern Chinese scholars,[b] and since then has been adopted as the watchword of the Chinese communists.[c]

Elsewhere the Mohists spoke of the sage-kings as leading from within, and echoed Lao Tzu in saying that they had things made for use and not for display.[d] But on the whole, in spite of the magnificent passage just quoted, one might say that while they agreed to some extent with the ideals and practices of pre-feudal society, they did not emphasise it so much as the Taoists,[e] considering that the situation could be saved by their doctrine of universal love (*chien ai*[2]). They were therefore not fundamentally against feudalism as such. On the contrary, they aimed at making it work better, as in their doctrines of the exaltation of the virtuous (*shang hsien*[3]) and the praise of social solidarity (*shang thung*[4]). With their chivalrous aspect already referred to they almost remind us, therefore, of the Christian military orders in their later occidental feudal setting. Thus it is natural to find that the Taoist arguments about the different fates of the great robbers and the little robbers, while used by the Mohists, are transformed into the condemnation of offensive war only (*fei kung*[5]) and not the whole feudal system.[f] Similarly, though Mei Yi-Pao suggests[g] that Hsü Hsing's 'Digger' type of practical communism (cf. p. 120) should have been sympathetic to Mohism, we have an account in the *Mo Tzu* book (chapter 49) of a discussion between Mo Tzu and one Wu Lü[6] who was clearly a representative of these doctrines, and the former talks about the division of labour, etc., just like Mencius.

[a] Tr. Legge (7), vol. 1, pp. 364 ff. mod., adjuv. Hsü Shih-Lien (1), pp. 235 ff.

[b] Khang Yu-Wei, however, was under the misapprehension that this famous passage referred to the future and not to the past. This is often insufficiently explained, as, for example, in the article of Chhen Jung-Chieh (1). But Khang was right in thinking that the parallel passages in the *Kungyang Chuan* and its commentaries referred to the future. These speak of 'Three Ages', that of Disorder (*shuai luan*[7]) followed successively by that of Approaching Peace (*shêng phing*[8]) and Universal Peace (*thai phing*[9]). The idea of social evolution was certainly not absent from ancient Chinese thinking (cf. p. 83). See Fêng Yu-Lan (1), vol. 2, pp. 83, 680; and Wu Khang (1), pp. 94 ff., 162 ff., on Ho Hsiu[10].

[c] It will be found, for example, in Mao Tsê-Tung's essay *On the Democratic Dictatorship of the People* (1949).

[d] Ch. 6 (Mei (1), pp. 22–4).

[e] Significantly, the legendary rebels such as Kun (cf. p. 117 above) were not heroes to the Mohists (ch. 9; Mei Yi-Pao (1), p. 45).

[f] Chs. 17, 28, 46, 49, 50 (Mei (1), pp. 98, 157, 220, 246, 257).   [g] (2), p. 177.

[1] 小康   [2] 兼愛   [3] 尙賢   [4] 尙同   [5] 非攻   [6] 吳慮
[7] 衰亂   [8] 升平   [9] 太平   [10] 何休

It follows that in many ways the Mohists were more akin to the Confucians. But they differed from them in having much greater interest in all that would benefit the people (*li min*[1]).[a] Hence their minor doctrines of economy of expenditure (*chieh yung*[2]), economy in burial rites (*chieh tsang*[3]) and the condemnation of music (*fei yo*[4]). There was, moreover, a stronger religious element in Mohism than in any other of the ancient Chinese thought-systems. The Will of Heaven (*Thien chih*[5]), says Mo Tzu in chapter 27, 'abominates the large State which attacks small States, the large house which molests small houses, the strong who plunder the weak, the clever who deceive the stupid, and the honoured who disdain the humble'.[b] 'Obedience to the will of heaven is the standard of righteousness.'[c]

In agreement with their supernaturalism, the Mohists strongly asserted the existence of ghosts of the dead and other spirits, which they seem to have looked upon as watchers of the morality of the living; three chapters (*ming kuei*[6]) are devoted in the *Mo Tzu* to this subject. It is very interesting, however, that they were led to this standpoint by an insistence on the empiricism of which we shall soon see some distinctively scientific examples. In this case it depended on consensus of opinion. Thus chapter 31:

Mo Tzu said: The way to find out whether anything exists or not is to depend upon the testimony of the eyes and ears of the multitude. If some have heard it or some have seen it then we have to say it exists. If no one has heard it and no one has seen it then we have to say it does not exist. So why not go to some villages or districts and inquire? If from antiquity to the present, and since the beginning of man, there are men who have seen the bodies of ghosts and spirits, and have heard their voices, how can we say that they do not exist?...[d]

One is reminded of the complex situation in 17th-century Europe, when Joseph Glanville and Sir Thomas Browne both believed in witches, Harvey was doubtful, and Johannes Weyer did not; while many rationalist scholastics were entirely sceptical, both of witchcraft and of the new natural or experimental philosophy.[e]

This attitude of Mo Ti was by no means unscientific, though it led to the wrong conclusions.[f] The appeal to the community of observers is part of the structure of natural science. But he underestimated the role of the critical intellect. This was

---

[a] The Mohist symbolism for this was perhaps the claim that the Confucians did not go back far enough for their authority; Mo Tzu claimed to follow the practices of the Hsia, rather than the Chou, dynasty (ch. 48; Mei (1), p. 233).

[b] Mei (1), p. 142. [c] Mei (1), p. 150.

[d] Tr. Mei Yi-Pao (1), p. 161.

[e] Cf. Withington (1).

[f] Arguments of his disciples, especially Sui Chhao Tzu,[7] with rationalist opponents, have been preserved in the *I Lin*,[8] ch. 1 (cf. Forke (13), p. 397), a Thang philosophical encyclopaedia.

[1] 利民   [2] 節用   [3] 節葬   [4] 非樂   [5] 天志   [6] 明鬼

[7] 隨巢子   [8] 意林

pointed out in a discussion on the Mohists five centuries later by Wang Chhung,[a] who said:[b]

If in argument one does not exercise the purest and most undivided thought, but indiscriminately uses examples from the outside to establish the correctness or wrongness of things, trusting what one hears and sees from without (*hsin wên chien yü wai*[1]), and not interpreting it by one's internal (intellect) (*pu chhüan ting yü nei*[2]); this is to argue only with the ears and eyes, exercising no judgment of the intellect (*hsin i*[3]). Now such ears-and-eyes argument leads to the formulation of statements on the basis of empty semblances (*hsü hsiang wei yen*[4]). And when such empty semblances serve as examples, this results in fictions passing for actual things.[c]

The fact is that truth and falsehood do not depend (only) upon the ear and eye, but require the exercise of intellect (*pi khai hsin i*[5]). The Mohists, in making judgments, did not use their minds to get back to the origins of things, but indiscriminately believed what they heard and saw. Consequently, although their proofs were clear, they failed to reach the truth (*tsê sui hsiao yen chang ming, yu wei shih shih*[6]). Judgments which thus fail to reach the truth are difficult to impart to others, for though they may accord with the inclinations of silly people, they will not harmonise with the minds of learned men. Failing to reach the (truth of) things and yet insisting on using one's conclusions, is of no benefit to the world. This is perhaps one reason why the arts of the Mohists have not been handed down.[d]

Here we have in embryo many of the problems which have been in the foreground of the philosophy of science since the Renaissance. What status have sense-impressions compared with the synthesising mind? How far does the intellect impose upon Nature its own *a priori* patterns? How empty of hypotheses should the intellect be in asking questions of Nature? It is much to the credit of Wang Chhung that he appreciated the possibility of such questions.

The last important doctrine of the Mohist school to which we must refer is their denunciation of the belief in fate (*fei ming*[7]), which they considered led to irresponsibility and was detrimental to industry and frugality. This was a little inconsistent with the belief in causation which appears in the scientific writings of the school (cf. the contrary position of Wang Chhung in the +1st century, p. 378), but the relevant chapters present no arguments other than the pragmatic one of the evil effects of fatalism on human behaviour in society.

---

[a] See below, pp. 382 ff.   [b] Ch. 67 of the *Lun Hêng* (+83).
[c] The text reverses this contrast, but since the discussion is concerned with the Mohist argument in favour of ghosts, the phrase was probably originally worded as given here.
[d] Tr. auct., adjuv. Fêng Yu-Lan (1) (Bodde), vol. 2, p. 160.

[1] 信聞見於外   [2] 不詮訂於內   [3] 心意   [4] 虛象爲言
[5] 必開心意   [6] 則雖效驗章明猶爲失實   [7] 非命

(b) SCIENTIFIC THOUGHT IN THE MOHIST CANON

While everyone will agree with the high tributes which are paid to Mo Tzu for his doctrine of universal love preached as early as the −4th century, there is nothing in such beliefs of special interest for the history of science. It is when we come to examine the Canons and their Expositions that we realise how far the later Mohists went in their effort to establish a thought-system on which experimental science could be based. One can only suppose that this grew out of their practical interests in fortification technology, which must have waxed as the original ethical and social aspects waned or became the accepted tenets of the school. But it was doubtless also connected with their desire to place their social doctrines on a basis of logical reasoning which should bring them success in dialectical disputations with adherents of other schools.

The Canons (*Ching*[1]) and their Expositions (*Ching Shuo*[2]) form chapters 40–43 inclusive of the *Mo Tzu* text, and they are followed by two further chapters of logical content. Apart from a commentary by Lu Shêng[3] of the Chin dynasty (+3rd and +4th centuries) long ago lost, and a few references by Han Yü of the Thang, no attention was given to this important material until Pi Yuan[4] in +1783 produced[a] the first modern edition. But the text, which contains many words of uncertain ancient meaning, and which had become more garbled by copyists than perhaps any other in Chinese literature, needed much further work. The results of this were embodied in the *Mo Tzu Chien Ku* published by Sun I-Jang[5] in 1894, and it was largely on the basis of this[b] that the complete translation of Forke (3) was made (1922). Later, in 1937, a selection of the logical propositions was published by Fêng Yu-Lan,[c] whose interpretation of the text and choice of emendations often differed from that of Forke.[d] None of these scholars, however, had either training or orientation in the natural sciences, so that the work of Than Chieh-Fu in 1935 was of much importance. He renumbered the propositions, suggested further emendations, sorted out stray fragments of text into new positions, and proposed many illuminating interpretations on the basis of his scientific knowledge. His work was continued by the physicist Chhien Lin-Chao in 1940 with special reference to the propositions in optics and mechanics. These we shall treat of in the Section on physics, dealing here only with those which concern general scientific theory. In the earlier editions there is a curious arrangement, the scientific propositions alternating with logical and sociological ones. Than rearranged them so that the various groups are collected together. It is generally agreed that the mixed order arose from the fact that ancient or medieval copyists arranged the text in two parts on each page, all the upper ones being intended to be read before any of the lower ones, and that later scribes, who understood little or nothing of what they were writing, then confused the two into one continuous text.

Let us now permit the Mohists to speak for themselves.

---

[a] Hummel (2), p. 622. [b] Hummel (2), p. 677. [c] (1), vol. 1; tr. Bodde.
[d] For references to other earlier studies, see Forke (13), p. 409.

[1] 經 [2] 經說 [3] 魯勝 [4] 畢沅 [5] 孫詒讓

NOTE

There are two Ching (Canons) and two Ching Shuo (Expositions). We here adopt the numbering of Than, who has in the first or upper Canon (Ching Shang) ninety-six propositions or discussions, and in the second or lower Canon (Ching Hsia) eighty-two. To each one of these there corresponds an entry in the Expositions. Although the Expositions must have been intended as explanatory commentaries, they are now sometimes more obscure than the propositions themselves. There are various theories about the schools which elaborated them, for which see Forke (3). In the second Canon, each proposition ends with the phrase, 'The reason is given under so-and-so', as if referring to a glossary of definitions. But this is now lost, though frequently the word in question appears in the corresponding Exposition.

We identify the propositions in the following way:

Cs 84/250/72.76 means that the proposition is numbered 84 in Than's text of the Ching Shang, that it is translated or discussed on p. 250 of Fêng Yu-Lan (1), and that it is Forke's (3) Ching Shang no. 72 and Ching Shuo Shang no. 76. Similarly,

Ch 22/—/42.31 means that it is Ching Hsia no. 22 in Than's text, that it is not referred to by Fêng Yu-Lan (1), and that the Forke (3) numbers are as shown.

Two Forke numbers have to be given because he adhered to the medieval 'mixed' order, so that his Ching Shuo entries generally do not correspond in numbering with his Ching ones.

Our own comments are inset. We have added headings. It would be a very difficult task to arrange the entries in a manner pleasing to everyone; we have tried to do it in such a way as to let the thought run on as easily as possible.

Key to the translators of the following passages, and to the other passages of the *Mo Tzu* (*Ching*) in Sections 13*d*, 19*h*, 26*c*, *g* on mathematics, physics, etc., below:

auct.	Present author and his collaborators.
auct./CLC	Present author and his collaborators, following Chhien Lin-Chao.
auct./TCF	Present author and his collaborators, following Than Chieh-Fu.
F	Forke (3).
FYL/B	Fêng Yu-Lan (1), tr. Bodde.
H	Hughes (1).
M	Maspero (9).
mod.	Indicates modifications in the versions of any of the foregoing introduced here by the author and his collaborators.

N.B. C=Ching, CS=Ching Shuo.

Cs 32/257/65.31. *Speech*

C   Speech is the uttering of appellations (*chü*[1]).

CS   Thus speech is what all mouths are capable of, and that which utters names. Names are like painted tigers (i.e. hard to make look like real tigers). When we say of a thing 'it may be called' (so-and-so) the name should reach (the thing) (i.e. be appropriate to the thing). (F, mod.)

Ch 7/—/12.7. *Attributes*

C   An attribute (lit. a side, *phien*[2]) may be (added on to or) taken away from (something) without involving increase or reduction. The reason is given under 'origin' (or cause) (*ku*[3]).

[1] 舉   [2] 偏   [3] 故

CS Both are the same one thing and no change has occurred. (auct./TCF)

This refers to subjective judgments as of a 'beautiful' flower, which remains the same flower whether it is called beautiful or not.

Cs 66/265/34.59. *Hardness and whiteness*

C Hardness and whiteness are not mutually exclusive.

CS Within a stone, the (qualities of) hardness and whiteness are diffused throughout its substance; thus we can say that the stone has these two qualities. But when they are in different places they do not pervade one another; not thus pervading they are then mutually exclusive. (FYL/B, mod.)

The background of this will be appreciated later when we describe the discussions of the School of Logicians (p. 187).

Ch 47/270/13.39. *Sensations*

C Fire is hot. The reason is given under 'assimilation' (*shih*[1]).

CS Fire: when one says that fire is hot, this is not (only) on account of the heat of the fire; it is (because) I make the assimilation (or correlation) (of the visual sensation of) light (and the tactile sensation of heat). (M)

Cf. Ch/46 below. The work of the mind in sorting and ordering sensations and perceptions was much discussed by the Mohists. Perception (*chih*[2]) has for its object the sensible world (*tshai*[3]) which the sense organs, the 'five roads' (*wu lu*[4]), apprehend; their data are then subject to reflection (*lü*[5]), and through this conceptual or interpretative knowledge (*chih*[6]) is attained. It is interesting that this last character was apparently invented by the Mohists as a technical term; it has long disappeared from dictionaries. The general parallel is with the *intellectus agens* and the *nous poietikos* in Europe. Cf. Chang Tung-Sun (1, 4).

We now come to a set of propositions about a concept very important to the Mohists, namely, the *fa*.[7]

Cs 70/260/42.63. *The models or methods of Nature*

C A *fa*[7] (model or method) is that according to which something becomes (or gets the sum of its characteristic qualities, its 'so-ness').

CS Either the concept (of a circle), or the compasses, or an (actual) circle, may be used as the *fa* (for making a circle). (FYL/B; H)

Hu Shih (2), p. 95, has suggested that the meaning of *fa* (K 642) is almost identical with the Aristotelian 'form' as opposed to 'matter'; and thus he translates it. He points out that one of its earliest meanings was 'mould'. As we shall shortly see, chapter 45 in *Mo Tzu* says that 'imitation consists in taking a model (*fa*[7])', but this raises other questions concerning what exactly is meant by 'imitation'; Hu Shih regards it as deduction, others are not so sure.

I feel very doubtful whether the Mohist *fa* ought to be compared with the Aristotelian 'form'. The latter has quite precise and very important biological implications (see the exposé of W. D. Ross, 1). I would rather make the suggestion that the three forms of *fa* mentioned above correspond to three of the

[1] 視   [2] 知   [3] 材   [4] 五路   [5] 慮   [6] 恕   [7] 法

four Aristotelian causes (see the clear explanations of Peck (1), p. xxxviii; (2), p. 24). The concept of a circle here seems closely similar to the final cause, the compasses would be the efficient cause, and the actual circle would be the formal cause. Since the example chosen was a geometrical one, the absence of the material cause is not surprising, and I doubt whether the Mohists ever felt the necessity of stating it explicitly.

The whole flavour of these passages is Aristotelian—a remarkable parallelism in time since the Mohist logicians were working just about the time of his death (−322).

Ch 65/260/49.58. *The models or methods of Nature*

C   The mutual sameness of things of one *fa*[1] extends to all things in that class. Thus squares are the same, one to another. The reason is given under 'square' (*fang*[2]).

CS All square things have the same *fa*,[1] though (themselves) different, some being of wood, some of stone. This does not prevent their squarenesses mutually corresponding. They are all of the same kind, being all squares. Things are all like this. (FYL/B)

Cs 94/—/97.83. *The models or methods of Nature*

C   (Since different) *fa*[1] have similarities, these should be observed. If we investigate and turn them over, we can search out the (basic) causes.

CS Similarities of *fa* should be selected, observed, investigated and turned over (i.e. compared). (auct.)

I believe that this comes very near the principle of induction. It certainly describes a phase of classification.

Cs 95/—/99.83. *The models or methods of Nature*

C   (Since similar) *fa*[1] have differences, we should observe the point where they exclude each other (*i chih*[3]), and speak of that as the 'parting of the ways' (*pieh tao*[4]).

CS This is selected, or that is selected. The cause is inquired into, the exclusiveness is observed. For instance, some men are darker than others. There must be some exclusive point (at which lightness begins) and darkness stops. Or again, there is 'love for others' and 'no love for others'. There must be an exclusive point where one stops and the other begins. (auct.)

At first sight, this looks like an attempt to state the principle of excluded middle in the Aristotelian syllogism. But it would be better to regard it as another exercise in general processes of classification.

This ends the propositions about the 'methods' or *fa*. Those which follow discuss other aspects of classification and causation.

Ch 2/—/3.2. *Classification*

C   Applying the principles of classification (*thui lei*[5]) is difficult. The reason is given under 'broad and narrow' (*ta hsiao*[6]).

CS For instance, 'animals of four feet' form a broader group than that of 'oxen and horses', while the group of 'things' is broader still. Everything may be classified in broader or narrower groups. (auct.)

[1] 法   [2] 方   [3] 宜止   [4] 別道   [5] 推類   [6] 大小

Ch 12/—/22.12. *Classification*

C   Things can be separated into different groups (*chhü wu*[1]). The reason is given under 'responding respectively' (*wei shih*[2]). And different things can be combined into a single group (*i thi*[3]). The reason is given under 'common point' (*chü i*[4]).

CS   The 'common point' is, for example, like both oxen and horses having four feet. 'Responding respectively' is, for example, that the ox should be called ox and the horse horse. If the ox and horse are considered separately they make two things, but if the ox-horse group is considered, they make one thing. It is like counting fingers, each hand has five, but one can take one hand as one (thing). (auct.)

Cs 15/269/29.15. *Loose appellations*

C   Loose appellations (*khuang chü*[5]) are chosen and used by individuals (without regard to the criticism of others).

CS   One can agree with one's associates (with whom one discusses), but not with the vulgar crowd (each member of which has an uncritical opinion). (auct.)
     Khuang means wild, mad, private, uncritical.

Ch 66/269/51.59. *Loose appellations*

C   Loose appellations (false reasonings) are those which are not correct as to the knowledge of differences. The reason is given under 'not correct' (*pu kho*[6]).

CS   The horse and the ox are different, but if someone says that a horse is not an ox because the ox has teeth and the horse has a tail, that will not do. In fact, both have (teeth and tails), these not being (attributes) belonging to one and not the other. One has to say that horses are not the same as oxen because the ox has horns and the horse does not; it is in this that the species are not identical. If the reasoning 'an ox is not a horse because oxen have horns and horses do not' was a loose appellation, like saying that '(an ox is not a horse because) the ox has teeth and the horse has a tail', it would result that (the propositions) 'it is a non-ox' and 'it is not a non-ox' would simultaneously be correct; and that (the propositions) 'it is a non-ox' and 'it is an ox' would simultaneously be correct. (M)

Ch 67/268/52.60. *Universal and particular*

C   (To say that) an ox and a horse are not oxen, and to grant that they are, are both the same. The reason is given under 'the general' (*chien*[7]).

CS   It is not permissible to say that an ox and a horse are not oxen, nor to say that they are. In some ways it is permissible and in some ways it is not permissible. Moreover the ox is not two and the horse is not two, while the ox and the horse are two. Then there is no difficulty (in that) an ox is nothing but an ox, and a horse nothing but a horse, but an-ox-and-a-horse are not an ox and not a horse. (FYL/B)

     Fêng Yu-Lan (1) points out that here the emphasis is on the particular, contrary to that in the arguments of the School of Logicians (Kungsun Lung, see p. 189 below) where it is on the universal.

Cs 78/254/59.69. *Types of Names*

C   Names are general (*ta*[8]), classifying (*lei*[9]) and private (*ssu*[10]).

---

[1] 區物   [2] 唯是   [3] 一體   [4] 俱一   [5] 狂舉   [6] 不可

[7] 兼   [8] 達   [9] 類   [10] 私

CS   Names: 'Thing' is a general name. All actualities (*shih*[1]) must bear this term. 'Horse' is a classifying name. All actualities of that sort must have that name. '*Tsang*' (a man's name) is private. This name is restricted to this actuality. (FYL/B)

Cs 79/255/61.70. *Designation*

C   In the process of designation (*wei*[2]) there is that of transference (*i*[3]), of general appellation (*chü*[4]), and of direct designation (*chia*[5]).

CS   In designation, to name a puppy a dog is called transference. Puppies-and-dogs is a general appellation. To call out 'Puppy!' is a direct designation. (FYL/B)

Ch 6/264/10.6. *Comparisons between classes*

C   Different classes are not comparable. The reason is given under 'measurement' (*liang*[6]).

CS   Difference: what is longer, a tree or a night? Of what is there more, knowledge or rice? Of the four things, rank, parents, conduct or price, which is more valuable?... (FYL/B)

      Cf. the categories of Aristotelian logic (Ross (1), p. 21).

Ch 8/—/14.8. *Wrong use of terms*

C   Falsity must be due to confusion (*po*[7]). The reason is given under 'not-so' (*pu jan*[8]).

CS   Falsity must involve a negative, and after (we have found it out) we call it 'false'. (Thus) a dog is falsely called a crane. (For instance), some people (wrongly) called the *yu*[9] (a kind of timid monkey) a crane. (auct.)

Cs 1/258/1.1. *Causation*

C   A cause (*ku*[10]) is that with the obtaining of which something becomes (comes into existence, *chhêng*[11]).

CS   Causes: A minor cause is one with which something may not necessarily be so, but without which it will never be so. For example, a point in a line. A major cause is one with which something will of necessity be so (*pi jan*[12]) (and without which it will never be so). As in the case of the act of seeing which results in sight. (FYL/B)

      Clearly with discussions such as this we are in the very engine-room of scientific thinking. The minor cause here is a necessary condition, we should say, rather than a cause. The distinction reminds me, not so much of anything in Aristotle, as of the distinction made in modern biology between competence or reactivity to stimulus on the one hand, and the conjunction of competence and stimulus on the other.[a] Cf. the passage from the *Lü Shih Chhun Chhiu* on 'cognoscere causas' quoted above, p. 55. Wang Chhung in the +1st century often distinguishes implicitly between necessary and sufficient causes.[b]

[a] I would also call attention to the fact that Chrysippus (−280 to −208) and other Stoics distinguished between principal causes (αἴτια συνεκτικά) and subsidiary assistant causes (συνεργά and συναίτια) in a somewhat similar way (B & M, p. 571).

[b] E.g. ch. 15 (Forke (4), vol. 1, p. 322) on sages; ch. 28 (Forke (4), vol. 1, p. 405) on the phoenix and the Ho Thu. Mr Donald Leslie noticed these examples.

[1] 實	[2] 謂	[3] 移	[4] 舉	[5] 加	[6] 量	[7] 諍
[8] 不然	[9] 猶	[10] 故	[11] 成	[12] 必然		

Cs 2/—/3.2. *Part and whole*

C   The part (*thi*[1]) is to be distinguished from the whole (*chien*[2]).

CS   A part has the oneness of a piece of something which has been halved into two. It is also like a point (cut off from) a line. (auct.)

Cs 86/263/76.78. *Agreement*

C   In agreement (*thung*[3]), there is that of identity (*chhung*[4]), of part-and-whole relationship (*thi*[1]), of co-existence (*ho*[5]), and of generic relation (*lei*[6]).

CS   When there are two names for one actuality this is identity. Inclusion in one whole is part-and-whole relationship. Both being in the same region (lit. room) is co-existence. Having some points of similarity is generic relation. (FYL/B)

Cs 87/263/78.79. *Difference*

C   In difference (*i*[7]), there is that of duality (*erh*[8]), of not having part-and-whole relationship (*pu thi*[1]), of separation (*pu ho*[5]), and of generic otherness (*pu lei*[6]).

CS   Two (separate things) are bound to be unlike in some respect, that is duality. When things are not linked together and conjoined (*pu lien shu*[9]), that is the absence of part-and-whole relationship. When things are not in the same region, that is separation. When they have no similarities they are not classifiable together. (FYL/B; H)

Cs 89/263/80.80. *Agreement and Difference*

C   Agreement and difference being taken together, what exists in a thing and what does not can be set forth.

CS   For example, the practice in wealthy families of achieving reciprocity in the exchange of the good things they possess for those which they do not, by measurement, allowing so many oysters in return for so many silkworms. Or the difference between age and youth in the case of an unmarried girl or the mother of a child. Or white and black, centre and sides, long and short, light and heavy, etc....(H)

Cs 80/253/63.71. *What knowledge comprises*

C   Knowing (*chih*[10]) comprises hearing about something (*wên*[11]), making an inference from it or an exposition of it (*shuo*[12]), experiencing it personally (*chhin*[13]), a harmonising of names with the actualities, and then action (*wei*[14]).

CS   Receiving something transmitted is hearsay knowledge. (Classifying) unhindered by position in space (because the things concerned may be far apart) is inference or exposition. What is observed by one's own body is personal experience. What designate are names, what are designated are actualities; when names and actualities are yoked together like a ploughteam, that is (the required) harmony. So also will (*chih*[15]) mated to movement (*hsing*[16]) is action (*wei*[14]). (H)

     Note the absence of the prejudice against *wei* so characteristic of the Taoists. On Chinese epistemology in general cf. Chang Tai-Nien (*1*); Wu Khang (*1*).

Cs 81/—/66.73. *Hearing*

C   'Things heard' has two different senses, 'heard from other people' and 'heard by yourself'.

[1] 體	[2] 兼	[3] 同	[4] 重	[5] 合	[6] 類	[7] 異
[8] 二	[9] 不連屬		[10] 知	[11] 聞	[12] 說	[13] 親
[14] 爲	[15] 志		[16] 行			

CS 'Someone said' means 'heard from other people'. 'I saw it with my own eyes' means personally witnessed by yourself. (auct.)

Cs 82/—/68.74. *Seeing*

C 'Seeing' has two meanings, partially seeing and fully seeing.

CS Seeing one side means partially seeing; seeing both sides (all sides) means fully seeing. (auct.)

Cs 85/256/74.77. *Action (of Nature and of man)*

C Action comprises preservation (*tshun*[1]), destruction (*wang*[2]), exchange (*i*[3]), diminution or decay (*tang*[4]), accretion or growth (*chih*[5]), and transformation (*hua*[6]).

CS Fortifying a pavilion would be an example of preservation, disease of destruction, buying and selling of exchange. Smelting ore would be an example of diminution, and the growth (of the body) of accretion. The (metamorphoses of) frogs and rats would be an example of transformation. (FYL/B)

> Note the reference here to biological changes such as the Taoists discussed (see above, p. 79; below, Section 39).

Ch 70/253/58.63. *Inference*

C When one hears that what is not known is like what is known, then both are known. The reason is given under 'reporting' (*kao*[7]).

CS What is outside is known. Then someone says, 'The colour inside the room is like this colour (outside)'. Thus what is not known is like what is known.... Names serve, by what is understood, to make certain what was not previously known. They do not use the unknown to conjecture at what is understood. It is like using a foot-rule to measure an unknown length. (FYL/B)

Ch 46/252/11.38. *Knowledge of duration*

C There is knowledge which does not come through the 'five roads' (the five senses). The reason is given under 'duration' (*chiu*[8]).

CS We see by means of the eyes, and their vision is aroused by fire (light), but the fire is not perceived except through the five roads of sense. But in durational knowledge there is no necessity for seeing with the eyes or for a fire to be present. (FYL/B; H; mod.)

> As Hughes (1) points out, the Mohist writer did not clearly distinguish between knowledge *of* duration and the activity of memory *in* duration. Cf. the other passages on time and duration in Section 26*c* on physics.

Ch 58/—/35.50. *Knowledge and practice*

C If one has a general idea which one does not as yet understand, (what to do about it?). The reason is given under 'use' (*yung*[9]) and 'precedence' (*kuo*[10]).

CS A last, a hammer and an awl are all things used for making shoes. The ornamentation may be put on before the shoe is hammered, or afterwards. The process takes its course; the exact order of the operations may be a matter of chance (the precedence may be equivalent). (F)

> Only by practical experience (experiment?) can the essentials of a process be distinguished from the non-essentials.

---

[1] 存   [2] 亡   [3] 易   [4] 蕩   [5] 治   [6] 化   [7] 告   [8] 久
[9] 用   [10] 過

Ch 48/256/15.40. *Knowing what one does not know*

C A man may (seem to) know what he does not know. The reason is given under 'selection by means of names' (*i ming chhü*[1]).

CS Knowing: Mix what a man knows and what he does not know together, and ask him about them. Then he must say, 'This I know' and, 'This I do not know'. If he can select and reject, he knows them both. (FYL/B)

Cf. the quotation from the *Lun Yü*, p. 8 above.

Ch 9/274/16.9. *Investigation*

C Why a thing becomes so; how to find it out; and how to let others know it; these need not be the same. The reason is given under 'disease' (*ping*[2]).

CS There is something which injures; that is the way the thing is. Seeing this (injury) gives knowledge of it. Speaking about it is letting others know. (FYL/B)

Cs 14/—/27.14. *Belief and non-ethical evidence*

C Belief occurs when words are in accord with likely presuppositions (*i*[3]).

CS It does not matter whether words are in accord with (so-called) morality or not. (For instance, if someone says that he supposes there is gold in) a certain town, (the only way to find out is to) send someone to go and see. (If) the gold is obtained, (the report will have proved true, no matter what moral questions may have been involved). (auct./TCF)

Forke (3) suggests that there may be a reference here to the Legalist Wei Yang,[4] who about −350 proclaimed high rewards to any of the people who would carry certain logs of wood from one city-gate to another, and duly paid them, to accustom the people to the idea that the government would always be as good as its word. But I would prefer to take the entry more generally.

Ch 10/—/18.10. *Doubt*

C Doubt....The reason is given under 'unexpected turns of events' (*fêng*[5]), 'following hearsay' (*hsün*[6]), 'unexpected encounters with unforeseen facts' (*yü*[7]), and 'past experience' (*kuo*[8]).

CS If we see someone busy about certain affairs we (naturally) suppose that he is the manager. If we see someone making a mat-shed like a cattle-stall in summer, we (naturally) suppose that it is for a cool retreat. These may be unexpected turns of events, but there is no reason for doubting our conclusions. But sometimes one can lift a (supposedly heavy) thing as easily as if it were a feather; at other times one has to put down a (supposedly light) thing as if it were a heavy stone. It was not one's own strength which made the difference (there was room for doubt as to the reports). The character *hsia*,[9] cutting bamboo slips, came to be written *hsia*,[10]a which is wrong; this was not due to ingenuity, but to hearsay tradition. Or one may accidentally meet people fighting, and one may suspect that they may be drunk, or else may have quarrelled at the noon market—you cannot be certain what the reason is. This would be an unexpected encounter (in which there would be room for doubt). As for what we now know, is it not mostly derived from past experience? Really it is. (auct.)

Note that later, in Wang Chhung's views on causation, we shall meet with a word closely similar to the *yü* here (p. 385 below.).

a Now pronounced *hsiao* (K 1149 c).

1 以名取 2 病 3 億 4 衛鞅 5 逢 6 循
7 遇 8 過 9 梜 10 削

Cs 72/257/46.—. *Argument*

Cs 73/257/48.—.

Cs 74/257/50.65.

C   Statement (*shuo*[1]) is that whereby to bring understanding. If there is one person who denies, both will deny. Argument (*pien*[2]) is conflict over something. In argument (dialectic) the one who wins is right.

CS As to this 'something', if both persons deny that ox-trees (the name of a tree) are oxen, they will have nothing to dispute about. But one may say it is and the other may say it is not. This is conflict over something. They cannot both be right, and thus one must be wrong....(FYL/B)

Ch 71/277/60.64. *Argument*

C   To hold that all speech is perverse, is perverseness. The reason is given under 'speech' (*yen*[3]).

CS To hold that all speech is perverse is not permissible. If the speech of the man (who urges this doctrine) is permissible, then speech is not perverse. But if his speech is permissible, it is not necessarily correct. (FYL/B, mod.)

> This is, of course, a direct attack upon the Taoist distrust of reasoned argumentation. The proposition immediately preceding reveals the conviction of the Mohists that valuable results *could* be obtained by reasoning about Nature.

Cs 44/—/90.42. *Change*

C   Change (*hua*[4]) is the manifestation of (the principle of) transformation (*i*[5]), (which can be demonstrated but not explained).

CS For example, the *wa*[6a] turns into the *chhun*.[7b] (auct.)

> It is interesting that the Mohist writer took as his cardinal example of change in nature an entirely imaginary biological metamorphosis. This particular transformation was widely believed to occur however, for there are parallel passages in the *Lieh Tzu*, the *Lü Shih Chhun Chhiu*, the *Huai Nan Tzu*, the *Li Chi* and the *Lun Hêng*. Again see p. 79 above and Section 39 below.

Cs 51/—/4.48. *Action and reaction*

C   That which 'must be so' is not a terminus (*i*[8]).

CS Every affirmative is accompanied by a negative, every natural phenomenon meets another one behaving oppositely to it. Wherever there is a must-be-so there will also be a must-not-be-so. Wherever there is an 'is' there will also be an 'isn't'. And this is what really 'must-be-so'. (auct.)

> This approximates closely to the principles of Hegelian dialectical logic, and Than does not hesitate so to expound it. The Mohists could not of course have supposed that everything in the world was balanced by equal and opposite forces, for that would have frozen change and spontaneity, but they seem to have understood that the victory of one process over another only brings the victor face to face with a new antagonist on a higher level. What is lacking here is a statement that such victories are syntheses.

---

[a] Edible water-frog (R 80).   [b] Quail (R 278).

[1] 說   [2] 辯   [3] 言   [4] 化   [5] 易   [6] 蠹   [7] 鶉   [8] 已

Cs 96/—/—.—. *Contradictions*

Ch 1/—/1.1.

C   The fixed species (*chih lei*[1]) consists of changing individuals (*hsing jen*[2]). The reason is given under 'similarity' (*thung*[3]). But ultimate truth (*chêng*[4]) has no (more) contradictions (*wu fei*[5]).

CS   Some maintain that a certain thing is so, and say that it is. I may think that it is not so. Thus the matter is doubtful. But when the 'may-be-so' becomes the 'must-be-so', then (it will be found that the two propositions) have united (*chü*[6]) into one (i.e. both were partly wrong and partly right). Some affirm that certain things are so, and are convinced that their affirmation is right. Others deny it and raise questions about it. But (ultimate truth) is like the sage; it contains all the negations but has no (more) contradictions (*yu fei erh pu fei*[7]). (auct.)

> These remarkable words confirm the suspicion raised by the previous proposition that the Mohists came very near to a dialectical logic. Unfortunately it is difficult to be sure of this, owing to the corruption and misplacement of the text, particularly bad here, just at the end of the Ching Shang and the beginning of the Ching Hsia. Nevertheless, the suspicion cannot but be a strong one, especially as the Mohists, like the Taoists, were clearly very conscious of the movement of natural change and were unfettered by Aristotelian formal logic.

We conclude with a few further miscellaneous propositions.

Ch 61/271/41.53. *Indestructibility of events and things past*

C   There may be nothingness. But what has once existed cannot be done away with. The reason is given under 'what has happened' (*chhang jan*[8]).

CS   There may be nothingness, but what is already so is something which has happened (lit. which is given, *kei*[9]), and so cannot be non-existent. (FYL/B)

> Forke (3) suggests that this was an anticipation of the law of the conservation of matter. It would surely be better simply to regard it as an affirmation of strict (though not necessarily catenarian) causality.

Ch 49/276/17.41. *Non-existence*

C   Non-existence is not necessarily dependent upon existence. The reason is given under 'the existence of non-existence' (*yu wu*[10]).

CS   Non-existence: suppose there were no horses. There could (only be said to be none) after they had first existed. But the collapse of the sky is something (really) non-existent. It can be called non-existent without ever having first existed. (FYL/B, mod.)

> In this there is an anti-Taoist undertone. The *Tao Tê Ching* (chapter 2) says that existence and non-existence grow out of one another. But, as Fêng Yu-Lan rightly says, the Mohists did not consider them mutually dependent. They therefore drew a distinction between things which, they considered, can exist and have existed; and things which cannot and will not exist. Note that the example

---

[1] 止類   [2] 行人   [3] 同   [4] 正   [5] 無非   [6] 俱

[7] 有非而不非   [8] 嘗然   [9] 給   [10] 有無

of the collapse of the sky is the same as that which appeared in the long anecdote from *Lieh Tzu* already quoted (pp. 40, 63). This gives point to the following proposition about calmness of mind (cf. immediately below, p. 190).

Cs 25/—/49.24. *Ataraxy*

C   Calmness of mind (*phing*[1]) is (the acquirement of) knowledge without preferences or attractions (*yü*[2]) and without prejudices or repulsions (*o*[3]).

CS Calmness of mind: Tranquillity (in the acceptance of) the thus-ness of things. (auct.)

It is interesting to find that the scientific view of the world meant, for the Mohists as well as the Taoists, liberation from fear.

When we consider as a whole the preceding work of the Mohists, taking into account also the propositions in physics and biology which will be found in the relevant Sections (26 and 39) of this book later on, we feel in a totally different world from that of the Taoists. There is nothing of the Taoist poetry and vision, and there is less interest in life phenomena as such; but the Mohists, not mistrusting human reason at all, clearly laid down what could have become the fundamental basic conceptions of natural science in Asia. We see their work, of course, through the dark glasses of corrupted texts and ingenious emendations. But the minute details of it are not really so important as the broad fact that they sketched out what amounts to a complete theory of scientific method. They treated of sensation and perception, of causality and classification, of agreement and difference, and of the relations of parts and wholes. They recognised the social element in the fixing of terminology and nomenclature, and they distinguished first-hand from second-hand evidence, appreciating its independence of prevailing ethical beliefs. They spoke of Change and of Doubt. The one thing they did not do was to propose some general theory of natural phenomena alternative.to, and more satisfactory than, the five-element doctrine of Tsou Yen, though, as we shall see,[a] they criticised it, and on quantitative grounds. One feels how badly Mohist scientific logic needed some equivalent of the Epicurean atomic theory. And one is tempted to think that perhaps the greatest tragedy in the history of Chinese science was that Taoist naturalist insight could not be combined with Mohist logic.

The question will naturally arise as to how clearly the Mohists stated the principles of deduction and induction. Unfortunately there is here considerable disagreement. Readers can judge for themselves, from the extracts of the *Mo Tzu* (*Ching*) which have just been given, how near they came to it. But besides these there is an important passage in one of the immediately succeeding chapters which must detain us for a moment.

It occurs in chapter 45, i.e. the 'Minor Illustrations' which follow the Ching.[b]   It

[a] P. 259 below.

[b] The chapter as a whole has been translated and examined by D. C. Lau (1), who believes that the nature of the Chinese language acted in a very inhibitory way on these beginnings of logic. This question is highly disputable, and will be taken up in Sect. 49 below.

[1] 平   [2] 欲   [3] 惡

has been translated by Fêng Yu-Lan,[a] Hughes,[b] Hu Shih[c] and Maspero (9); the two latter scholars have given it very close study. Seven statements may be distinguished in it.

(1) What is limited (*yü*[1]) is that which is not universal (*pu chin*[2]).

(2) What is false is that which in fact is not so.

    Definition of reasoning by examples:

(3) Imitation (*hsiao*[3]) consists in taking a model (*fa*[4]).

    What is imitated is that which is taken for a model.

    Therefore (*ku*[5]) if it is adequate to the imitation (the reasoning) is correct.

    If it is not adequate to the imitation (the reasoning) is false. Such is *hsiao*.[3]

    The four kinds of reasoning by examples:

(4) Comparison (*pi*[6])[d] is taking one thing to explain another (i.e. Analogy).

(5) Paralleling (*mou*[7]) is comparing terms (or propositions) and finding that they are in complete agreement.

(6) Conclusion (*yuan*[8]) is saying, 'You are of such and such a nature, why should I alone refuse to admit that you are of such and such a nature?'

(7) Extension (*thui*[9]) is considering that that which one does not accept is identical with that which one does accept, and admitting it.[e]

It is quite evident that the Mohists were here trying to define various forms of scientific reasoning. Unfortunately they were, so to say, groping, and we have always to reckon with the uncertainty of the text. Maspero believes that the Mohists were interested mainly in public disputations (which the Taoists abhorred), and therefore did not seek to establish a general theory of all intellectual operations; but this view does not agree very well with the existence of the Mohist scientific propositions, e.g. on optics and mechanics, which, incidentally, Maspero made no attempt to translate. According to Hu Shih (2), *hsiao* here means definitely deduction, but he did not convince Maspero. So also he takes *thui* to mean definitely induction, which Maspero could not allow. The problem remains open.

The character *hsiao*[3] certainly has the meanings of form, mould, model and to imitate. Here it is associated with the mysterious *fa*[4] which set us a difficult translation problem a few pages back (Cs/70 and the following propositions). If Hu Shih were right in suggesting that its meaning is analogous to Aristotelian 'form' as such, the Mohists would hardly have said that a pair of compasses could be one of the *fa* of a circle. Perhaps it would be preferable to interpret *fa* as meaning the 'methods' of Nature, and so including all the Aristotelian causes. Paragraph 3 of the present passage would then mean that Nature's methods should be imitated in thought, and that if this imitation is adequate one's reasoning about causes will be correct. Part of the argu-

---

[a] (1), p. 259.
[b] (1), p. 137.
[c] (2), p. 99.
[d] *Pi* is here used for *phi*.[10]
[e] Tr. Maspero (9), eng. auct.

[1] 域   [2] 不盡   [3] 效   [4] 法   [5] 故   [6] 辟   [7] 侔

[8] 援   [9] 推   [10] 譬

ment has centred round the *ku* in the third line, which Hu Shih takes as meaning 'cause' in the philosophical sense, while Maspero allows it no more weight than a mere 'therefore'. But I believe that granting Maspero this, the passage may still prove Hu Shih right on the main issue, and I would suggest:

(3) 'Model-thinking' consists in following the methods (of Nature).
What is followed in 'model-thinking' are the methods.
Therefore if the methods are truly followed by the 'model-thinking' (lit. hit it in the middle), the reasoning will be correct.
But if the methods are not truly followed by the 'model-thinking', the reasoning will be wrong. Such is 'model-thinking'.

Hence, since recognised causes will be much fewer in number than the multiplicity of phenomena, the 'model-thinking' will be in fact deduction.

It will not escape the reader that these arguments of the Mohists about 'model-thinking' may bear a strong resemblance to considerations which are being advanced in contemporary discussions on the logic of scientific 'models', especially (though not exclusively) in the less exact sciences. These go back to the speculations of 19th- and early 20th-century scientists on the role of concrete models in physical thinking as opposed to the exclusive use of mathematical symbolism (Hertz and Clerk-Maxwell, Rutherford and Eddington). Taken in this way, the seven definitions of Mo Tzu or his disciples have a strangely modern ring. It might indeed be argued that the general attitude of Chinese thinkers towards conceptual model-making was induced in them by the structure of their language.[a] This perhaps enabled them to attain a sophistication in differentiating those intellectual operations which can be carried on with models from those which cannot, only now being rediscovered and developed by modern philosophers of science (e.g. Wittgenstein; Schrödinger; Braithwaite).

I have been less able to follow Maspero's objections to Hu Shih's interpretation of *thui* (extension) as induction.[b] It seems to me that the last sentence would read better:

(7) Extension is considering that that which one has not yet received (i.e. a new phenomenon) is identical (from the point of view of classification) with those which one has already received, and admitting it.

This would clearly be the formulation of a new generalisation based on many instances, and hence induction, as Hu Shih points out in a long discussion which is well worth reading.[c] It is generally agreed that paragraphs 4, 5 and 6 in the passage represent various forms of analogical reasoning, perhaps unnecessarily distinguished by the Mohists.

---

[a] On this subject, see the discussion below (Sect. 49) on Chinese language and logic as conditioning the growth and nature of Chinese scientific thought.
[b] Here Forke (13), p. 406, seems to side with Hu Shih rather than Maspero.
[c] (2), pp. 100 ff.

(c) THE PHILOSOPHY OF KUNGSUN LUNG

Something must now be said of another school, never very clearly differentiated from the Taoists and the Mohists, but sufficiently so to have been listed as an independent group, the *Ming Chia*[1] (School of Names, or the Logicians),[a] by Ssuma Than and Pan Ku (cf. p. 1). Its two greatest names were Hui Shih,[2] who lived during the −4th century, and Kungsun Lung,[3] whose life fell mostly in the first half of the −3rd. Both were therefore contemporaries of Chuang Chou, who in a vivid passage bemoans the fact that after Hui Shih died there was no longer anyone with whom he could talk.[b] Of the details of their lives little is known,[c] except that they gave counsel to various feudal princes in the manner of all Warring States scholars,[d] and doubtless they tried to interest students in their logical exercises, not meeting with much success. All their works are lost, with the exception of the partially preserved *Kungsun Lung Tzu*[4] book,[e] and the paradoxes recorded in chapter 33 of *Chuang Tzu* and elsewhere.

The *Kungsun Lung Tzu* book has been said to reach the highest point of ancient Chinese philosophical writing; its dialogue form, which resembles the Platonic style, is not a mere literary device, since the arguments of the interlocutor are always serious. The disappearance of the greater part of it since the Han must be considered one of the worst losses in the transmission of ancient Chinese books, and any judgment on the achievements of ancient Chinese thought must always take this into account. The fundamental idea in that part which is available to us is the recognition of what Western philosophy has called 'universals' (e.g. 'white', 'horse', 'hard', etc.) as distinct from concrete things. What Kungsun Lung called *chih*[5] were therefore distinguished from particular things, *wu*[6]; *chih*[f] meaning a finger, or designation, hence here a 'designated' universal common factor.[g]

---

[a] Also occasionally called Hsing Ming Chia,[7] the School of Forms and Names (*Chan Kuo Tshê* (*Chao Tshê*, ch. 2), ch. 19).

[b] *Chuang Tzu*, ch. 24 (Legge (5), vol. 2, p. 100).

[c] See Fêng Yu-Lan (1), vol. 1, pp. 192 ff.; and Forke (5).

[d] Kungsun Lung was strongly pacifist, as we see from his efforts to get the State of Chao to disarm (*Lü Shih Chhun Chhiu*, chs. 101 and 107; vol. 2, pp. 60, 73). His patron was the Prince of Phing-Yuan (see p. 233). The names of some associates and pupils are known—Chhiwu Tzu,[8] Huan Thuan,[9] Mao Kung[10] and Thien Pa.[11] One was a prince of Chungshan, Wei Mou.[12] An opponent was Yochêng Tzu-Yü.[13] A famous argument took place between Kungsun Lung and one of the direct descendants of Confucius, Khung Chhuan[14] at the court of the Prince of Phing-Yuan at Chao about −298.

[e] The only complete translation until recently was that of Forke (5), but Fêng Yu-Lan (1) expounded Kungsun Lung's meaning much better, and his interpretations are to be preferred. We now have complete versions by Ku Pao-Ku (1), Mei Yi-Pao (3), and (much less satisfactory) Perleberg (1). Extracts are in Hughes (1).

[f] It is not certain whether this word should rather be written *chih*,[15] which means an idea or a concept.

[g] It should be noted however that this interpretation is not universally accepted. Chang Tung-Sun (2), followed by Ku Pao-Ku (1), pp. 37 ff., 115, believes that the discourse has to do with the

---

[1] 名家	[2] 惠施	[3] 公孫龍	[4] 公孫龍子	[5] 指	[6] 物
[7] 形名家	[8] 綦毋子	[9] 桓團	[10] 毛公	[11] 田巴	
[12] 魏牟	[13] 樂正子輿	[14] 孔穿	[15] 恉		

The discourse on universals (*chih*) is contained in the *Kungsun Lung Tzu*, chapter 3:

There are no things (in the world) that are without *chih*, but these *chih* are without *chih* (i.e. they cannot be analysed further or split up into other *chih*).

If the world had no *chih*, things could not be called things (because they would have no manifested attributes).

If, there being no *chih*, the world had no things, could one speak of *chih*? *Chih* do not exist in the world. Things do exist in the world. It is impossible to consider what does exist in the world to be (the same as) what does not exist in the world. In the world there exist (materially) no *chih*, and things cannot be called *chih*. If they cannot be called *chih*, they are not (themselves) *chih*.

There are no (materially existing) *chih*, (and yet it has been stated above that) there are no things that are without *chih*. That there are no *chih* (materially existing) in the world, and that things cannot be called *chih*, does not mean that there are no *chih*. It is not that there are no *chih*, because there are no things that have not *chih*....

That there are no *chih* existing in the world (in time and space), arises from the fact that all things have their own names, but these are not themselves *chih* (because they are individual names, not universals)....

Chih, moreover, are what are held in common in the world (*chien*[1])[a] (because they are manifested in all members of the relevant class).

No *chih* exist in the world (in time and space), but no things can be said to be without *chih* (because every individual thing manifests an assortment of various universal qualities)....[b]

The applications of this attempt to think out the relation between individual phenomena and universal qualities are found in chapter 2 (The White Horse) and chapter 5 (Hardness and Whiteness). These titles will explain a number of references to hardness, whiteness and horseness, which have been cropping up in the preceding pages. The white horse discourse, abridged, runs thus:[c]

A white horse is not a horse....The word 'horse' denotes (*ming*[2]) a shape, 'white' denotes a colour. What denotes colour does not denote shape. Therefore I say that a white horse is not a horse (as such)....When a horse (as such) is required, yellow and black ones may all be brought forward, but when one requires a white horse, they cannot....Therefore

process of designating, i.e. the act of distinguishing an object or class from all other objects or classes of objects, and characterising it by a specific name. There would thus be initially (*a*) the indicative gesture sign (*chih*), (*b*) the object (*wu*), and (*c*) the relation between them (*wu-chih*). It will be seen that the scientific interest of Kungsun Lung's thought remains great, for on this view he was working at the foundations of all classification. Mei Yi-Pao (3) adopts 'attributes' for *chih*, believing that Kungsun Lung was discussing a distinction similar to that in Western philosophy between substance and qualities. The difficulty here is similar to that which sometimes arises in old mathematical texts, cuneiform as well as Chinese, namely to determine exactly what the writer was talking about—once that is sure, everything falls into place and even emendations assume plausibility.

[a] Note that this word is the same as that used in the Mohist technical term for 'universal love', p. 168.

[b] Tr. Bodde in Fêng Yu-Lan (1), vol. 1, p. 209, mod.

[c] It is in dialogue form, but that has been omitted in this condensation.

[1] 兼   [2] 命

yellow and black horses are things of the same kind, and can respond to the call for a horse, but not to the call for a white horse. Hence it results that a white horse is not a horse (as such; or horseness)....

Horses certainly have colour. Therefore there exist white horses. Suppose there could be horses without colour, then one would have only horses as such (*yu ma ju erh i*[1]). How could we then get a white horse? A white anything is not a horse. A white horse is 'horse' associated with 'white'. But 'horse' with 'white' is (no longer merely) 'horse'. Therefore I say that a white horse is not a horse (as such)....

The word 'white' does not specify (*ting*[2]) what is white....But the words 'white horse' specify of whiteness what it is that is white....[a]

No doubt Kungsun Lung's aim in stating an apparent absurdity, that a white horse was not a horse, was to attract the interest of prospective thinkers. The school of logicians had a particular interest in paradoxes, as we shall see.

In his epistemological discourse on hardness and whiteness Kungsun Lung wished to prove that these were two universals, separately apprehended.

Q. Is it possible that hard, white and stone are three?
A. No.
Q. Can they be two?
A. Yes.
Q. How?
A. When without hardness one finds whiteness, this gives two. When without whiteness one finds hardness, this gives two....Seeing does not perceive hardness, but finds whiteness without hardness. Touching does not perceive whiteness, but finds hardness without whiteness....Seeing and non-seeing are separate from one another. Neither can pervade the other, and therefore they are separate. Such separateness is called 'concealment' (*tsang*[3]).

> Concealment is Kungsun Lung's term for the subsistence of universals, out of which they emerge to manifest themselves in the existence of material things.

Hardness does not associate (*yü*[4]) itself only with stone and thus be hard; it is common to other things. It is not hard because it is associated with things; its hardness is necessarily hardness (in itself). Not being hardness (because of) stones and other things, but being hardness (as such), if nothing hard existed at all in the world (of time and space), it would (just) lie concealed....Hardness is perceived by the contact of the touching hand. Yet it is the mind, not the hand, which perceives. If it does not, we have 'separateness'. By such 'separateness' all the world is ordered.[b]

The point of interest here is not so much the detailed comparison between this kind of thinking and parallels in the history of European thought as the fact that it was

---

[a] Tr. Bodde in Fêng Yu-Lan (1), vol. 1, p. 204; Ku Pao-Ku (1), pp. 30 ff., mod.
[b] Tr. Bodde in Fêng Yu-Lan (1), vol. 1, pp. 207 ff.; Ku Pao-Ku (1), pp. 54 ff., mod. This version may be compared with the different ones given in Hughes (1), p. 126, and Perleberg (1), p. 110. Ku Pao-Ku (1), pp. 65, 101, 111, 113, 117, brings out the radical opposition between Kungsun Lung and Hui Shih in the last sentence here.

[1] 有馬如而巳   [2] 定   [3] 藏   [4] 與

being done so early in ancient China.[a] The problems which Kungsun Lung was attacking were of course similar to those which led 17th-century European thinkers to distinguish between primary and secondary qualities.

Perhaps the chapter of greatest interest from the point of view of natural science is the 4th, on the Explanation of Change (*thung pien*[1]). It is most significant that Kungsun Lung addressed himself to this central problem in the investigation of Nature. His aim apparently was to show that the universal is changeless, while the particular is ever-changing.

Q. Does two contain one?
A. Two(-ness) does not contain one.
Q. Does two contain right?
A. Two(-ness) has no right.
Q. Does two contain left?
A. Two(-ness) has no left.
Q. Can right be called two?
A. No.
Q. Can left be called two?
A. No.
Q. Can left and right together be called two?
A. They can.
> The universal of two is simply twoness and nothing else. But 'right' added to 'left' is two in number, and they can therefore be called two.

Q. Is it permissible to say that a change is not a change?
A. It is.
Q. Can 'right' associating itself (with something) be called change?
A. It can.
Q. What is it that changes?
A. It is 'right'.
> I.e. the universal of righthandedness manifesting itself in double things, and then disappearing again.

Q. If 'right' has changed, how can you still call it 'right'? And if it has not changed, how can you speak of a change?
A. 'Two' would have no right if there were no left. Two contains 'left-and-right'. A ram added to an ox is not a horse. An ox added to a ram is not a fowl.
> The discussion now approaches the problem of biological classification—the universals of species.

Q. What do you mean?

---

[a] It appears moreover that Kungsun Lung was by no means the first to argue about the 'hard and the white'. According to the *Han Fei Tzu* (ch. 32, p. 3*a*), there was a logician called Ni Shuo[2] who discoursed to the Chi-Hsia Academicians on such subjects. That would be about −315. Kuo Mo-Jo (*1*), p. 225, thinks that Ni Shuo was identical with Mao Pien,[3] a dialectician mentioned in the *Chan Kuo Tshê* (*Chhi Tshê*, ch. 1), ch. 8. Perhaps it was significant that Mao Pien came from the State of Chhi (cf. p. 241).

[1] 通變   [2] 倪說   [3] 貌辨

A. A ram and an ox are different. Since a ram has (upper front) teeth and an ox none, we cannot say that an ox is a ram, nor a ram an ox. They have different characteristics and belong to different species (*lei*[1]). But because a ram has horns and so does an ox, we cannot say that an ox is a ram or *vice versa*. They may both have horns and yet belong to quite different species. A ram and an ox have horns, a horse none, but a horse has a long tail, which the two others do not. This is why I say that a ram together with an ox does not make a horse. That means that there is no horse (in the present discussion). Consequently a ram is not two, and an ox is not two, but 'ram-and-ox' are two....

> In other words, though of different species, two individuals may combine to manifest the universal 'twoness'.

An ox and a ram have hair, while a fowl has feathers. Speaking about the legs of fowls makes one (i.e. the universal of 'fowl legs'). Each (individual) fowl has two legs. Two and one make three (i.e. the two real ones plus the universal idea). Speaking about the legs of ox or ram makes one (i.e. the universals of 'ox legs' and 'ram legs'). Each (individual) ram or ox has four legs. Four and one make five (i.e. the four real ones plus the universal idea). Thus when I say that an ox and a ram do not make a fowl, I have no other reason than this. If choosing for comparison a horse or a fowl, the horse is better (because also a quadruped).

What has certain qualities and what has not, cannot be put in the same species. To make such appellation is called a confusion of terms (*luan ming*[2]) and a loose appellation (*khuang chü*[3]).[a]

Q. Let us talk about something else.

(*d*) THE PARADOXES OF HUI SHIH

The writings of the Logicians always have as an undercurrent the wish to 'épater le bourgeois' (cf. the paradoxes below), as here in the statement that quadrupeds have five legs each, which was doubtless made to draw attention to the existence of the unchanging universal 'quadruped-leg-as-such'. In this they resembled the Taoists, who, however, did not care about such logical abstractions. The remainder of this chapter on Change is more obscure, but in general purpose it continues the affirmation that what changes in nature are the individual things, the universals remaining unchanged. Significantly, the five elements and the five colours are touched on, with some statements about their mutual conquests. This links the Logicians with the Yin-Yang school of Tsou Yen (see pp. 232, 243 below), and its system of unchanging medium and elements underlying all visible changes, like the play of atoms in Greek thought.

There are a few indications that Hui Shih (as might be expected from his friendship with Chuang Chou) was more akin to the Taoists.[b] He is said to have taught the desirability of 'abolishing positions of honour',[c] and like Kungsun Lung and the

---

[a] Note the same technical term already seen in the Mohist writings. Tr. Bodde in Fêng Yu-Lan (1), vol. 1, p. 213; Ku Pao-Ku (1), pp. 47ff.; Mei Yi-Pao (3), pp. 426ff.; Perleberg (1), p. 101, mod. There are wide divergences among the translators in the latter part of this passage, depending on which textual emendations they prefer.

[b] Kuo Mo-Jo (4), pp. 52ff., concurs.

[c] *Lü Shih Chhun Chhiu*, ch. 129 (vol. 2, p. 124). Cf. Ku Pao-Ku (1), pp. 4ff.

[1] 類   [2] 亂名   [3] 狂舉

Mohists, he advocated pacifist policies. He was clearly interested in science as well as logic, for we read:

In the south there was a queer man named Huang Liao,[1] who asked why the sky did not fall and the earth did not sink; also about the causes of wind, rain and the rolling thunder. Hui Shih answered without hesitation, and without taking time for reflection. He discussed all things continuously and at great length, imagining that his words were but few, and still adding to them strange statements.[a]

Thus he was no mere sophist, as might otherwise be supposed, for unfortunately none of his abundant writings have come down to us. All we possess attributed to him are ten paradoxes contained in the 33rd chapter (Thien Hsia) of *Chuang Tzu*. These are of great interest. Besides these ten, the same chapter lists twenty-one further paradoxes, as examples of the kind of startling things which the 'dialecticians' (*pien chê*[2]) propounded. In addition to these, the *Lieh Tzu* book preserves a further six (three of which are identical with three in the dialecticians' list), and the *Hsün Tzu* a further five (one of which is from Hui Shih and another from the dialecticians).[b] One is at once struck by the coincidence between the Logicians' paradoxes and those famous paradoxes in Greek history associated with the name of Zeno of Elea. Zeno's floruit was −450; the Chinese paradoxes must have been under discussion about −320. I find it very hard to believe in any transmission or influence at such a time (see Section 7). It would not be unnatural that the paradoxical form should arise spontaneously at that particular stage of thought, but the temporal coincidence remains remarkable. Let us now list the series.[c]

Chuang Tzu, ch. 33

HS/1   The greatest has nothing beyond itself, and is called the Great Unit (*ta i*[3]); the smallest has nothing within itself, and is called the Small Unit (*hsiao i*[4]).

HS/2   That which has no thickness cannot be piled up, but it can cover a thousand *li* (square miles) in area.

HS/3   The heavens are as low as the earth; mountains are on the same level as marshes.

HS/4   The sun at noon is the sun declining, the creature born is the creature dying.

HS/5   A great similarity (*ta thung*[5]) differs from a little similarity (*hsiao thung*[6]). This is called the little-similarity-and-difference (*hsiao thung i*[7]). All things are in one way all similar, in another way all different. This is called the great-similarity-and-difference (*ta thung i*[8]).

[a] *Chuang Tzu*, ch. 33, tr. Bodde in Fêng Yu-Lan (1), vol. 1, p. 196. This reminds us of the frightened man in *Lieh Tzu* (pp. 40, 63 above), and as regards thunder, see Wang Chhung below, p. 379.

[b] In the following list HS will stand for Hui Shih, PC for the dialecticians, LT for *Lieh Tzu*, HT for *Hsün Tzu*, and KT for *Khung Tshung Tzu*.

[c] We have compared the versions of Legge (5), Forke (5), Fêng Yu-Lan (1), Hughes (1), Wieger (7), Ku Pao-Ku (1), and others, but as it is too complicated here to give credit to each previous translator it must suffice to say that we have had them all at our side in studying the Chinese text.

[1] 黃繚   [2] 辯者   [3] 大一   [4] 小一   [5] 大同   [6] 小同
[7] 小同異   [8] 大同異

HS/6   The South has at the same time a limit and no limit.

HS/7   Going to the State of Yüeh today, one arrives there yesterday.

HS/8   Linked rings can be sundered.

HS/9   I know the centre of the world, it is north of the State of Yen and south of the State of Yüeh.

HS/10   Love all things equally; the universe is one body (*Fan ai wan wu, thien ti i thi*[1]).

PC/11   An egg has feathers.

PC/12   A fowl has three legs.

PC/13   Ying (the capital of the State of Chhu) contains the whole world.

PC/14   A dog can (be?, become?, be considered as?) a sheep.

PC/15   Horses have eggs.

PC/16   Frogs have tails.

PC/17   Fire is not hot.

PC/18   Mountains issue from mouths.

PC/19   Wheels do not touch the ground.

PC/20   Eyes do not see.

PC/21   The *chih* (universals) do not reach, but what reaches is endless.

PC/22   Tortoises are longer than snakes.

PC/23   Carpenters' squares are not square; compasses cannot make circles.

PC/24   Gimlets do not fit into their handles.

PC/25   The shadow of a flying bird has never yet moved.

PC/26   There are times when a flying arrow is neither in motion nor at rest.

PC/27   A puppy is not a dog.

PC/28   A brown horse and a dark ox make three.

PC/29   A white dog is black.

PC/30   An orphan colt has never had a mother.

PC/31   If a stick one foot long is cut in half every day, it will still have something left after ten thousand generations.

Lieh Tzu, ch. 4[a]

LT/1   There can be ideas (*i*[2]) without cogitation (*hsin*[3]).

LT/2=PC/21   (with slightly different wording).

LT/3=PC/25   (with slightly different wording).

LT/4   A hair can lift a thousand *chün* (30,000 catties weight).

LT/5   A white horse is not a horse (=Kungsun Lung's ch. 2).

LT/6=PC/30   (with one word different).

Hsün Tzu, ch. 3[b]

HT/1=HS/3   (the two sentences in reversed order, with slightly different wording).

HT/2   The States of Chhi and Chhin are coterminous.

HT/3   That which enters by the ear issues from the mouth.

HT/4   A woman can have a beard.[c]

HT/5=PC/11.

---

[a] P. 19*a*.   [b] P. 1*b*.

[c] Accepting an ingenious emendation by Yü Yüeh from *kou*[4] to *chhü*.[5]

[1] 氾愛萬物天地一體   [2] 意   [3] 心   [4] 鉤   [5] 姁

Khung Tshung Tzu[a]

KT/1   Chang has three ears (=PC/12).

Naturally, these statements need a commentary. Let us group them.

(i) *Relativity and the all-pervadingness of change*

This is the first of the divisions which we may make. As we have already seen, the idea of relativity,[b] of the difference between aspects of the universe as seen from different frames of reference in space-time, was clearly appreciated by the Taoists (pp. 49, 81), and here we find it among the Mohists and Logicians. Spatial relativity is seen in the paradox about the heavens and the earth, the mountains and the marshes (HS/3, HT/1), for there must be a frontier at which the heavens touch the earth, and from the point of view of the universe, the irregularities of the earth's surface are minimal.[c] The paradox of the limit, and yet no limit, of the south (HS/6) has been variously interpreted; Hu Shih suggests that it may betray a conviction of the Mo-Ming thinkers that the earth is spherical (for which other grounds will shortly appear), but Huang Fang-Kang and Fêng Yu-Lan think that it may have been meant to imply only that there were vast regions beyond the bounds of contemporary geographical knowledge. In any case it belongs to the spatial relativity group. The paradox about the centre of the world being north of Yen (the most northerly of the States) and south of Yüeh (the most southerly of the States) indicates again, according to Hu Shih, an appreciation of the sphericity of the earth.[d] Fêng Yu-Lan quotes in this connection the commentary of Ssuma Piao[1] (+3rd century) on *Chuang Tzu*: 'The world has no compass points; therefore (from one point of view) wherever we may happen to be is the centre; cycles have no starting-point, therefore whatever period we may happen to be in is the beginning.'

Besides this appreciation of relativity as applied to space, there is also appreciation of relativity as applied to time. Paradox HS/4 applies it both to astronomy and biology. The brief moment of noon seems illusory, and if a sufficiently broad period of time is taken, the sun is always declining, since noon occurs at different times at different places on the earth's surface. In the biological analogy, Hui Tzu hit the mark better perhaps than he dreamed of, if we may judge from the discussions of modern biologists about senescence, which goes on at its greatest rate the younger the organism is.[e] Going to Yüeh today, and arriving yesterday (HS/7) is a phrase

---

[a] A book of late Han or even later.

[b] The word 'relativity' is not used here, needless to say, in its strict scientific sense.

[c] Cf. the God of the Northern Sea talking to River Spirit in *Chuang Tzu*, ch. 17: '(Thus) we know that heaven and earth are as small as a grain of the smallest rice, and that the tip of a hair is as vast as a mountain mass' (Legge (5), vol. 1, p. 379). Cf. p. 103 above.

[d] Huang Fang-Kang, however, points out that the word *hsia* in *thien-hsia* may be an interpolation, in which case the meaning would simply be that the zenith of the celestial sphere is directly above us wherever we happen to be on the earth's surface. Cf. Sect. 20*b* below.

[e] Cf. Needham (1), pp. 400 ff., on senescence.

[1] 司馬彪

which sounds as if it came out of a modern text-book of physical relativity; it recognises the existence of different time-scales in different places. The linked rings have also been generally assumed to come in this category; the paradox (HS/8) may not have had a topological significance, but meant perhaps that whatever substance they were made of, it would in course of time decay, and the linking would be sundered. It may also have meant that each ring could be considered separately, by a geometer for example, as to its degree of approximation to a perfect circle, and may therefore have been an affirmation of the dissociability of phenomena in thought. Hu Shih suspects, however, a play upon words, in that the rings could be simply broken.[a]

These relativistic interpretations are not at all the same as that adopted in one of the most famous expositions of Hui Tzu, that of Chang Ping-Lin (1). He considered that the paradoxes aimed at the establishment of the view that all quantitative measurements and all spatial distinctions are unreal or illusory, and that time also is unreal.[b] It is easier to follow Hu Shih (2) when he suggests that the paradoxes were intended to prove a 'monistic theory of the universe', and points out that some of them at least had therefore the same purposes as those of Zeno. The Mohists (as we shall see in the Section on physics) distinguished duration (chiu[1]) from particular times (shih[2]), and space (yü[3]) from particular locations (so[4]); in Mo Ching Cs/39 and Cs/40 respectively. That time is constantly passing from one moment to another is obvious to common sense, but the Mohists also held that particular locations in space were also constantly changing. It is here that Hu Shih puts forward his view that they had recognised the sphericity and some kind of movement of the earth.[c] The two passages, which are certainly hard to explain on any other basis, are in the Mo Ching, Ch/13 and Ch/33 (I give Hu Shih's translations):

Ch 13/—/24.13. *Space and time*

C   The boundaries of space (the spatial universe) are constantly shifting. The reason is given under 'extension' (chhang[5]).

CS   There is the South and the North in the morning, and again in the evening. Space, however, has long changed its place.

Ch 33/—/63.24. *Space and time*

C   Spatial positions are names for that which is already past. The reason is given under 'reality' (shih[6]).

CS   Knowing that 'this' is no longer 'this', and that 'this' is no longer 'here', we still call it South and North. That is, what is already past is regarded as if it were still present. We called it South then and therefore we continue to call it South now.

[a] Cf. the Gordian knot of Greek legend; in the *Chan Kuo Tshê* (Records of the Warring States) there is a similar story. In Sect. 19h we shall find a topological puzzle of linked rings.

[b] He was perhaps influenced by Buddhist philosophy.

[c] Cf. Sects. 20 and 22 on astronomy and geography below. In the *Hun Thien* cosmological theory, dominant during the Han and later, the earth was frequently described as spherical in the centre of the celestial sphere.

[1] 久   [2] 時   [3] 宇   [4] 所   [5] 長   [6] 實

The assumption underlying the paradoxes would therefore be that within the universal space-time continuum there are an infinitely large number of particular locations and particular times constantly changing their positions with regard to one another. From the standpoint of an observer at any one of them, the universe will look very different from that which another observer sees. All the paradoxes so far considered fit without difficulty into this scheme. Its striking modernity, paralleling the dialectical traces which we have noted in the *Mo Ching*, invites one to wonder what Chinese science would have been capable of, without having to pass through the discipline of Aristotelian logic, had environmental conditions favoured its growth.

(ii) *Infinity and problems related to atomism*

The universal space-time continuum, just mentioned, must surely correspond to the 'Great Unit' of paradox HS/1. The particular times and locations might correspond to the 'Small Unit' found there also.[a] This seems to be one of those not infrequent places where the ancient Chinese thinkers paused at the door of atomism, without ever going in. The Small Unit, which has nothing within itself,[b] might well be thought of as an atom. Moreover, the idea of indivisibility is not far off, for we have it in PC/31, though stated in an anti-atomic form, in that the dividing-in-half of a stick would essentially never end. As will be seen later,[c] the *Mo Ching* has at least two propositions concerning geometrical 'atoms' in its definition of the geometrical point (Cs/61 and Ch/60). Their interpretation is difficult. Hu Shih (2) considered them anti-atomic, but I adhere to the view of Fêng Yu-Lan and others that they were really intended to define the geometrical point as the line which could not be cut into half any more. In this case there must have been a difference of opinion between Hui Shih and the Mohists. Certain it is at least that lively discussions verging very closely upon atomism were going on.[d]

If PC/31 is directed against the existence of points or particles then we have ready explanation of three other paradoxes, PC/19 on the wheel,[e] PC/25 (LT/3) on the bird's shadow, and PC/26 on the arrow in flight. All are concerned with motion, and the last is startlingly similar to Zeno.[f] Let us recall what Zeno's four paradoxes were:

(1) You cannot get to the end of a racecourse. You cannot traverse an infinite number of points in a finite time....

[a] Cf. River Spirit talking to the God of the Northern Sea in *Chuang Tzu*, ch. 17: 'The disputers of the world all say, "That which is most minute has no form; that which is most vast cannot be encompassed"; what is your opinion on this?' (Legge (5), vol. 1, p. 378).

[b] How extraordinary an anticipation this is of present-day conceptions of the atom, which has *practically* nothing within it.   [c] Section 19*h*.

[d] Cf. Forke (13), p. 429, who, however (p. 320), also finds atomism in the *Chuang Tzu* book, to my mind less convincingly.

[e] Which, as Chhien Pao-Tsung says (1), p. 12, only touches the ground at one point; and since the point has no size, cannot really be said to touch the ground at all.

[f] And since a tortoise is mentioned in PC/22, Hu Shih has proposed that it may be a corruption of a paradox of Eleatic type.

(2) Achilles will never overtake the tortoise. He must first reach the place where the tortoise started, and by that time the tortoise will have got some way ahead....He is always coming nearer but never makes up to it.

(3) The arrow in flight is at rest. For if everything is at rest when it occupies a space equal to itself, and what is in flight always occupies a space equal to itself, it cannot move.

(4) Half the time may be equal to double the time (a problem of rows of moving bodies).[a]

Brunet & Mieli (1), setting the Eleatic paradoxes in their position in the history of science,[b] indicate that they were, in fact, an anti-atomic, or rather anti-Pythagorean, *démarche*. Pythagorean discrete points in space and instants in time, or still worse, Leucippian-Democritean atoms, were pluralistic discontinuities against which Zeno set up his paradoxes. He wished to support a monistic continuity universe. Nothing can ever get anywhere, he said in effect, if it has to pass through an infinite number of points in a finite time. The fallacies involved, and the counter-propositions of the atomists, do not here concern us—what is of such great interest is that just about a century later than the time of Zeno, ground so similar was being gone over in the Mo-Ming discussions.[c]

It only remains to add that there may be a subsidiary meaning in the bird's shadow paradox (PC/25).[d] The statement that a shadow never moves (of itself) is found also in the *Mo Ching* (Ch/16), as will be seen when we cite the optical propositions; there I take it to mean that so long as the light-source and the object do not move their positions the shadow will never move. One should also remember the discourse in *Chuang Tzu* already quoted (p. 51) between Shadow and Penumbra, where the apparently dependent behaviour of the shadow is interpreted as perhaps the movement of an independent organism, either spontaneously rhythmic, or responding to a common cause. That concept may here be under attack.

(iii) *Universals and classification*

The Mo-Ming theories about classification are of course present in the paradoxes. There are several lesser statements and two greater ones. Paradox PC/12 (KT/1), that the fowl has three legs, is a restatement of the passage in Kungsun Lung's book where to the real legs of an animal are added the idea of the universal '*x*-type of legs in general'. PC/28 is similar. PC/14 probably means that both dogs and sheep are quadrupeds. LT/5 simply repeats the *Kungsun Lung Tzu* on white horses. PC/21 (LT/2) was understood by no one until Fêng Yu-Lan's demonstration that *chih* may mean universals; it is then easily recognised as a statement that the universals do not reach

[a] I use the version of Burnet (1), p. 367. They come from Aristotle, *Physics*, 239b, 5–33

[b] P. 128. See also, for bibliography, Cajori (1), Tannery (1), Diels-Freeman (1), p. 47, Freeman (1), p. 153.

[c] The dilemma of continuity and discontinuity is of course one of the fundamental themes of natural science in all ages. I do not know whether the 'wavicles' of modern physics have resolved it. Later (Sect. 26b) I hope to show that whereas the discontinuous atoms have dominated European thought, continuous waves have dominated that of the Chinese. It is noteworthy that in +2nd-century India, Nāgārjuna also propounded paradoxes like those of Zeno and Hui Shih purporting to prove the impossibility of motion. On such comparisons see further Ku Pao-Ku (1), pp. 129ff.

[d] Ku Pao-Ku (1), p. 123, concurs.

our perceptions, only material things can do this, and they, for their part, are infinite in number. To say that T-squares are not square and that compasses cannot make circles (PC/23) must surely mean that they cannot make squares and circles as perfect as the universals of these figures (cf. *Mo Ching*, Cs/70 above). Probably PC/24 means similarly that 'gimletness' cannot fit into handles.

The greater statements are HS/5 and HS/10. As regards 'little' and 'great' similarity-and-difference, opinions are rather varied. Fêng Yu-Lan assumed that the statement has reference to the differences seen by ordinary people between different things; as against the philosopher's realisation that in one way all things are different and in another way they are all similar. Hu Shih says that it embodies the idea of an essential and elemental unity underlying all apparent diversity and variation. Chang Ping-Lin (*1*) thought that it exposed the unreality of all possible classifications; Huang Fang-Kang (*2*) that it contained the Aristotelian idea of classification into a hierarchy of larger and smaller classes. All that can be said for certain is that it has something to do with classification.[a] The doctrine of 'the unity of similarity and difference' seems to be as characteristic of Hui Shih as 'the separateness of hardness and whiteness' is of Kungsun Lung.

'Love all things equally, the universe is one substance' (HS/10) has of course been recognised as the metaphysical side of the Mohist doctrine of universal love. But there was doubtless more to it than that.[b] The dictum links Mohism with Taoism closely. One remembers the Taoist doctrines that the sage covers everything impartially, that the Tao runs through everything in Nature, no matter how awful, disagreeable or trivial (pp. 47, 48), and that in the universe there is nothing really great or really small, nor should there be in human society either (p. 103). To this Huang Fang-Kang adds an element of 'Newtonian' universal attractive force (cf. pp. 40, 151 above, on attraction and repulsion).

(iv) *The role of the mind; epistemology*

Several of the paradoxes seem to refer to the mind's attainment of conceptual knowledge (*chih*[1]) by correlation of reflections (*lü*[2]). Fire is not 'hot' (PC/17) because 'heat' is added to the bundle of sense-perceptions which are always found to go together, by the activity of the mind (cf. *Mo Ching*, Ch/47). Eyes do not 'see' (PC/20) because the full process of seeing includes the correlation of experiences by the mind. Ideas without cogitation (LT/1) may, on the other hand, refer to Taoist recognition of unconscious mental activity, or to Taoist belief in the autonomic character of behaviour (cf. pp. 52 ff. above). Mind work is presumably also the meaning of HT/3.

[a] Recently Demiéville, reviewing Ku Pao-Ku (*1*), has sought to elucidate further this difficult text.
[b] Kuo Mo-Jo (*1*), p. 234, concurs. Note that Hui Shih used a slightly different term, not Mo Ti's *chien ai*, but *fan ai*, and he probably meant all natural things, not only all other human beings. The great artists of all ages, as well as the great naturalists, have followed his advice. Cf. pp. 270, 281, 368, 453, 471, 488, 581 below.

[1] 恕   [2] 慮

(v) *Potentiality and actuality*

This is a very obvious group, including the feathers (potentially) in the egg (PC/11, HT/5), the tails (formerly) attached to individual frogs (PC/16), and the puppy's growth into the dog (PC/27). One would probably not go far wrong in placing here also PC/14, if taken in an evolutionary sense (as in the transmutation passage of Chuang Tzu, p. 79); HT/2, since the abolition of intervening States might bring the western State Chhin and the eastern State Chhi into juxtaposition; and HT/4, which must refer to contemporary knowledge of sex-reversals in man and animals.[a] With this we move to the last groups.

(vi) *Natural wonders seemingly paradoxical*

I think the eggs of horses (PC/15) belong here. What was withheld from the wise and prudent Aristotle (and everyone else until the time of von Baer) was certainly not disclosed to Hui Shih and Tsou Yen. We cannot accept this as even the faintest anticipation of the discovery of the mammalian ovum. Nevertheless, the early mammalian foetus enwrapped in its membranes does roughly resemble an egg, and we can accept this paradox as a simple reminder that the processes of generation are roughly similar in all the higher animals. Of course, it might also have an evolutionary interpretation, like PC/14. The three other biological natural wonders seem to have verbal catches in them. The tortoise may be longer than the snake (PC/22) if you are talking about longevity; white dogs may be black (PC/29) if you are talking about their eyes and not their hair; the orphan colt, of course, never had a mother after it acquired the right to be called an orphan (PC/30, LT/6). These may all be warnings to specify more clearly what one is talking about.

Then as to PC/18, I suggest, instead of the usual explanation about echoes, that it may refer to volcanoes. Mountains may indeed issue from mouths in the earth. The ancient Chinese were living on the edge of the circum-Pacific earthquake and volcanic belt; active volcanoes may possibly have been known to them.[b]

In LT/4 we have a mechanical wonder. There are close parallels in the *Mo Ching* (Ch/24, Ch/52) (see below, Sect. 26c) where the properties of weights counterbalanced on ropes and pulleys are discussed.

Finally, the geometrical two-dimensional plane, which cannot form a pile because it has no thickness, and which yet can spread over a thousand *li* (HS/2), may be supposed to be brought in here as a mathematical wonder. Almost identical definitions are in the *Mo Ching* (Cs/19).

(vii) *An unclassified paradox*

The only paradox which seems of quite uncertain significance is PC/13, which says that the capital of the State of Chhu can contain the whole world. I have not met with any convincing exposition of its meaning. Perhaps Dubs (7) is right in saying that it asserts that compared with illimitable space, Ying and all China are equally small.

[a] E.g. *Mo Tzu*, ch. 19 (Mei (1), p. 113); *Lun Hêng*, ch. 7 (mistranslated by Forke (4), vol. 1, p. 327); *Shen Chien*, ch. 3. I am indebted to Mr Donald Leslie for these references. See on, p. 575.
[b] Cf. Sect. 23b below.

(e) LOGIC, FORMAL OR DIALECTICAL?

Paradoxes of Ming Chia type are sometimes met with in works generally considered Taoist. A striking conversation is found in the *Lieh Tzu* book,[a] where a philosopher named Hsia Ko,[1] who may or may not have been a real personage, answers questions propounded by the emperor Thang[2] of the Shang dynasty in a way reminiscent of the Kantian antinomies.

Thang of the Shang asked Hsia Ko saying, 'In the beginning, were there already individual things?' Hsia Ko replied, 'If there were no things then, how could there be any now? If later generations should pretend that there were no things in our time, would they be right?' Thang said, 'Have things then no before and no after?' To which Hsia Ko answered, 'The ends and the origins of things have no precise limits. Origins might be considered ends, and ends origins. Who can draw an accurate distinction between these cycles? What lies beyond all things, and before all events, we cannot know.'

So Thang said, 'What about space? Are there limits to upwards and downwards, and to the eight directions?' Hsia Ko said he did not know, but on being pressed, answered, 'If there is emptiness, then it has no bounds. If there are things, then they have bounds. How can we know? But beyond infinity there must exist non-infinity, and within the unlimited again that which is not unlimited.[b] (It is this consideration)—that infinity must be succeeded by non-infinity, and the unlimited by the not-unlimited—that enables me to apprehend the infinity and unlimited extent of space, but does not allow me to conceive of its being finite and limited.'

So Thang continued his questions and asked, 'What is there beyond the Four Seas?' Hsia Ko answered, 'Just the same as what there is here in Chhi.' 'How can you prove that?' said Thang. 'When travelling eastward I came to the land of Ying, and found that the people were quite the same as here. Inquiring about what was further east, I found that it was also the same. Travelling westward Pin was no different, and again beyond there was no difference. Thus I knew that the Four Seas, the Four Wildernesses, and the Four Uttermost Ends of the Earth are the same as where we ourselves live. The lesser is always enclosed by a greater, without ever reaching an end. Heaven and earth, which enclose the ten thousand things, are themselves enclosed in some outer shell, which must be infinite. How do we know that there is not some outer universe of which our own is but a part? These are questions which we cannot answer.[c] (In any case) Heaven and earth are material things and therefore imperfect.[d]

[a] Ch. 5, p. 1a.

[b] R. Wilhelm (4) adopts a different interpretation according to which Hsia Ko is speaking here of the infinitely small (atoms); this is seductive, but does not seem quite justified by the text.

[c] Later, in Section 20 on astronomy, we shall see how these ideas, carried down by the *Hsüan Yeh* doctrine, influenced all later Chinese scientific thought, and prevented it from undergoing any Aristotelian-Ptolemaic ossification.

[d] Tr. auct.; adjuv. Wieger (7), p. 131; Forke (6), p. 46; R. Wilhelm (4), p. 48; L. Giles (4), p. 82. To explain this imperfection, the text here goes off into mythological legends. Wilhelm says that this shows how insufficiently serious-minded the Taoists were, but it seems fairly obvious that the texts were conflated from two different sources.

[1] 夏革   [2] 湯

Here, as R. Wilhelm pointed out, the first and second questions correspond to the first antinomy of Kant, and the third to the third. The emphasis on infinity is distinctly Taoist, but the manner of treating it is akin to the methods of the school of Logicians.

It will be seen from the foregoing that the work of the Mo Chia (in its later phase) and the Ming Chia is of central importance for the study of the development of scientific thought in China. The thinkers of these schools attempted to lay foundations upon which the world of the natural sciences could have been built. Perhaps the most significant thing about them is that they show an unmistakable tendency towards dialectical rather than Aristotelian logic, expressing it in paradox and antinomy, conscious of entailed contradiction and kinetic reality. In this they strongly reinforced the tendencies which were characteristic of Taoism (cf. pp. 57, 77, 103 above), just as later on all these indigenous logical trends were to be reinforced by some of the schools of Buddhist philosophy (cf. pp. 423 ff. below).

At a later stage (Sect. 49) we shall enquire how far the differences of linguistic structure between Chinese and the Indo-European languages had influence on the differences between Chinese and Western logical formulations. It has been thought[a] that the subject-predicate proposition, and hence the Aristotelian identity-difference logic, is less easily expressible in Chinese. The distinction between being, or substance as such, and its attributes, is said to emerge less clearly; words like *shih*[1] and *yu*[2] conveying less sharp a conception of being than that which becoming enjoys in words such as *wei*[3] and *chhêng*.[4] Relation (*lien*[5]) was probably more fundamental in all Chinese thought than substance. Chang Tung-Sun cites a famous chapter of the *Tao Tê Ching*:[b]

Existence and non-existence mutually generate each other, the difficult and the easy complete each other, the long and the short demonstrate (*chiao*[6]) each other, high and low explain (*chhiung*[7]) each other, instrument and voice harmonise with each other, before and after follow each other.

as typical[c] of the Chinese tendency to dialectical logic, or, as he significantly calls it, 'correlative' logic.[d] 'The meaning of a term', he says, 'is completed only by its opposite.'[e] At any rate, Chinese thought, always concerned with relation, preferred

---

[a] See, for example, Chang Tung-Sun (1, 1, 4). Cf. p. 478 below.

[b] Ch. 2, tr. auct. adjuv. Chang Tung-Sun (1); Waley (4); Duyvendak (18).

[c] Compare the following from Wang Fu's[8] +2nd-century *Chhien Fu Lun*:[9] 'Poverty is born from riches, weakness comes from strength, order engenders disorder, and security insecurity' (tr. Balazs, 1). Here one begins to sense the implication of a dialectical account of social change.

[d] There is some ground for believing that the structure of the Chinese language is essentially favourable to the types of thinking now being explored in modern combinatory logic.

[e] Cf. p. 466 below.

[1] 是   [2] 有   [3] 爲   [4] 成   [5] 連   [6] 較   [7] 傾
[8] 王符   [9] 潛夫論

to avoid the problems and pseudo-problems of substance, and thus persistently eluded all metaphysics. Where Western minds asked '*what* essentially is it?', Chinese minds asked '*how* is it related in its beginnings, functions, and endings with everything else, and *how* ought we to react to it?'

In the same place (Sect. 49) we shall also enquire whether the development of Chinese scientific thought was adversely affected by the fact that syllogistic logic was not explicitly formulated. Syllogistic reasoning is of course not infrequently implicit in ancient Chinese texts; the form is complete, for instance, in the *Kungsun Lung Tzu*.[a] On the other hand the Peripatetics of Europe perhaps confined too rigidly the processes of thought. Perhaps the legal and theological preoccupations of Roman and Byzantine culture led to an excessive concentration on inferences and conclusions at the expense of premises. Yet for the natural sciences, premises have always been the most important part. In any case, the moderns have been harsh in their judgments on Aristotelian logic.

In the 17th century, all the proponents of the 'new, or experimental, philosophy' attacked scholastic logic—Francis Bacon, Joseph Glanvill,[b] John Amos Komensky (Comenius),[c] Robert Boyle,[d] Thomas Sprat[e] and many others. In his biography of Bacon, William Rawley wrote:

Whilst he was commorant in the university, about 16 years of age, as his lordship hath been pleased to impart unto myself, he first fell into the dislike of the philosophy of Aristotle—not for the worthlessness of the author, to whom he would ever ascribe all high attributes, but for the unfruitfulness of the way—being a philosophy (as his lordship used to say) only strong for disputations and contentions, but barren of the production of works for the benefit of the life of man; in which mind he continued until his dying day.

Later, Bacon constantly urged that Aristotelian logic was more hindrance than help. In the preface to the *Great Instauration*, he wrote:

As for those who have given the first place to Logic, supposing that the surest helps for the sciences were to be found in that, they have indeed most truly and excellently perceived that the human intellect left to its own course is not to be trusted; but the remedy is altogether too weak for the disease, nor is it without evil in itself. For the *Logic which is received*, though very properly applied to civil business and to those arts which rest in discourse and opinion, *is not nearly subtle enough to deal with Nature*; and in offering at what it cannot master, has done more to establish and perpetuate error than to open the way to truth.[f]

---

[a] Especially ch. 3; see Ku Pao-Ku (1), pp. 125 ff.
[b] In *Scepsis Scientifica; or, the Vanity of Dogmatising and Confident Opinion* (+1661).
[c] In *A Reformation of Schooles* (+1634, Eng. tr. by Samuel Hartlib, +1642).
[d] In *The Sceptical Chymist; or, Chymico-Physical Doubts & Paradoxes, touching the Spagyrists' Principles commonly call'd Hypostatical, as they are wont to be defended by the Generality of Alchymists* +1661).
[e] In *The History of the Royal Society of London, for the Improving of Natural Knowledge* (+1667).
[f] Italics mine.

And Thomas Sprat elaborated:

This very way of Disputing itself, and inferring of one thing from another alone, is not at all proper for the spreading of Knowledge. It serves admirably indeed, in those Arts, where the Connection between the Propositions is necessary, as in the Mathematicks, in which a long Train of Demonstrations may be truly collected from the certainty of the First Foundation; but in things of probability only, it seldom or never happens but that after some little Progress, the main Subject is not left, and the Contenders fall not into other Matters, that are nothing to the Purpose; for if but one Link in the whole Chain be loose, they wander far away and do not recover their first Ground again. In brief, Disputing is a very good Instrument to sharpen Men's Wits and to make them versatile and wary Defenders of those Principles which they already know; but it can never much augment the solid Substance of Science itself....[a]

In our own time, it was the considered judgment of Whitehead[b] that the popularity of the Aristotelian logic had been the greatest retarding force against which physics had had to contend. 'It is', he said, 'apart from the guardianship of mathematics, the fertile matrix of fallacies. It deals with propositional forms only adapted for the expression of high abstractions, the sort of abstractions usual in current conversation, where the presupposed background is ignored.'[c] Or again: 'It was a more superficial weapon than the scholastics deemed it. Automatically it kept in the background some of the more fundamental topics of thought, such as quantitative relations....'[d]

Putting the matter in another way, it provided the natural sciences with an inadequate tool for the handling of the greatest fact of Nature, so well appreciated by the Taoists, Change. The so-called laws of identity, contradiction, and the excluded middle, according to which X must be either A or not-A, and either B or not-B, were constantly being flouted by the fact that A was palpably turning into not-A as one watched, or else showed an infinite number of gradations between A and not-A, or else indeed was A from some points of view and not-A from others. The natural sciences were always in the position of having to say 'it is and yet it isn't'. Hence in due course the dialectical and many-valued logics of the post-Hegelian world. Hence the extraordinary interest of the traces of dialectical or dynamic logic in the ancient Chinese thinkers, including the Mohists whose writings we have been examining.[e] It may of course be said that Aristotelian logic was a necessary stage which European science had to go through. To this there can be no answer, for we shall never know whether, had environmental conditions in China been favourable for a development of the natural sciences, the Mohists or some other school would in their turn have formulated discrete syllogistic static logic, or whether it would have been possible for modern science to have arisen in Asia from more dialectical roots, or by the aid of some other system altogether.

The causes for the decay and disappearance of the two schools during the upheavals of the first unification of the empire remain unexplained. Presumably Chinese social

[a] *History*, p. 17. [b] (1), pp. 43, 66. [c] (2), p. 196.
[d] (2), p. 150. [e] Cf. Forke (13), p. 407.

life had the effect of polarising thought into the two moulds of Confucianism and Taoism. On the one hand the specific social aims of the literati precluded any close attention to logical problems. 'There is no reason', Hsün Tzu had already said, 'why problems of "hardness and whiteness", "likeness and unlikeness", "thickness or no thickness" should not be investigated, but the superior man does not discuss them; he stops at the limit of profitable discourse.'[a] Moreover, the Mohist ideal of universal love permeated Confucianism during the Han and after, modifying the Mencian principle of graded affection.[b] On the other hand, the Mohists, and hence Mo Ti himself, undoubtedly on account of their interest in war technology and scientific methodology, became incorporated into the Taoist tradition. There are, for example, legends about him in the *hsien*[1] literature. The *Mo Tzu* book was incorporated in the *Tao Tsang*.[c] The *San Kuo Chih* bibliography lists a *Mo Tzu Tan Fa*[2] (Alchemical Preparations of Mo Tzu). The *Sui* bibliography lists a *Wu Hsing Pien Hua Mo Tzu*[3] (Mo Tzu's Treatise on the Changes and Transformations of the Five Elements), and a *Mo Tzu Chen Nei Wu Hsing Chi Yao*[4] (Mo Tzu's Pillow Book of the Fundamental Actions of the Five Elements). We believe, however, that none of these Wei and Chin 'forgeries' are now extant.

It is usually said that the ideas of the Logicians were completely unknown to the people of the medieval period. But this seems to have been too confidently asserted, for what little we know of the 'Name-Principle' (*ming li*[5]) school in the Chin period suggests that logical discussion was continuing. We hear[d] of one logician called Hsieh Hsüan,[6] and elsewhere of an abstract argument between Yüeh Kuang[7] and a friend on coming-into-being and ceasing-to-be.[e] So also some of the circle of Wang Pi, including men such as Wang Tao[8] and Ouyang Chien,[9] maintained (in contrast to early Taoism) the thesis that 'words can completely express ideas'.[f] They were opposed by Yin Jung.[10] All this was in the +3rd and +4th centuries. Then Yen Chen-Chhing[11] (+709 to +785)[g] said, in a memorial inscription for a Taoist friend Chang Chih-Ho[12] (d. *c*. +780),[h] that he had written a book called *Chhung Hsü Pai Ma Fei Ma Chêng*[13] (Mystical Theses on Hardness, Whiteness and Horseness)—'though nobody knew much about it'. Moreover, the present preface and commentary of the *Kungsun Lung Tzu* were written in the Sung by Hsieh Hsi-Shen.[14]

---

[a] *Hsün Tzu*, ch. 2, tr. Dubs (8), p. 49.   [b] Dubs (14) has traced the steps of this.
[c] *TT* 1162.
[d] *Shih Shuo Hsin Yü*, ch. 4, p. 21*b*.
[e] P. 13*b*. Cf. Fêng Yu-Lan (1), vol. 2, p. 176; Ku Pao-Ku (1), p. 15, gives further names.
[f] See Fêng Yu-Lan (1), vol. 2, p. 185.
[g] We shall meet with this scholar again in Section 23 on geology.
[h] He is best known for his book *Yuan Chen Tzu*[15] (Book of the Original-Truth Master). His work on logic is listed in the bibliography of the *Hsin Thang Shu*, ch. 59, p. 3*b*.

---

[1] 仙	[2] 墨子丹法	[3] 五行變化墨子	[4] 墨子枕內五行紀要
[5] 名理	[6] 謝玄	[7] 樂廣	[8] 王導	[9] 歐陽建	[10] 殷融
[11] 顏真卿	[12] 張志和	[13] 沖虛白馬非馬證	[14] 謝希深
[15] 元眞子			

When one puts together the resemblances of the early Taoists to the pre-Socratics, and those of the Mohists and Logicians to the Eleatics and Peripatetics, and moreover, when one takes into account the enormous gaps known to exist in the ranks of the ancient Chinese writings which have come down to us,[a] one is left with the impression that there was little to choose between ancient European and ancient Chinese philosophy so far as the foundations of scientific thought were concerned, and, indeed, that in certain respects the advantage lay with the Chinese. If, then, these foundations became overgrown with the weeds of the rice-fields, and never received that superstructure of columns ornamented with gold and vermilion which they would have been capable of bearing, the fault is perhaps to be looked for in the factors of the environing intellectual climate of China. But the time has not yet come to speak of this.

Hui Shih used to deliver his views leaning against a dryandra tree....He diffused himself all over the world of things without satiety, till in the end he had only the reputation of being a skilful debater. Alas! Hui Shih, with all his talents, vast as they were, made nothing out; he pursued all subjects and never came back (with success). It was like trying to shout down an echo, or running a race with one's own shadow. Alas!

Thus Chuang Chou.[b]

[a] A concrete example may be given. The *Fa Yen* says (ch. 2, p. 6a), about +5, that the *Kungsun Lung Tzu* book extended to 'several tens of thousands of sophistic words'. Today it has but 2050.
[b] *Chuang Tzu*, chs. 2, 33 (tr. Legge (5), vol. 1, p. 186; vol. 2, p. 231).

12. THE *FA CHIA* (LEGALISTS)

IF THE student of the history of Chinese thought is often tempted to become impatient with Confucian sententiousness, he has only to read the writings of the Legalists to come back to Confucianism with open arms, and to realise something of that profound humanitarian resistance to tyranny which forms the background of the sacrificial liturgy of the Wên Miao. Had it been desirable to arrange these discussions of the ancient Chinese schools on the basis of a political spectrum, one should have treated of the Legalists before the Confucians, since they represent the extreme 'right' in political tendency just as the Taoists represent the extreme 'left'. It was natural, therefore, that they resembled the Confucians in being interested only in the governance of human society, and not in the processes of Nature. This would at first sight seem to make them of even less importance for the history of science than the Confucians, but in fact their connection is closer, since their beliefs raise in an acute form the problem of the relation between juridical law and natural law (in the sense in which the term is used in the natural sciences). It will be necessary in due course to examine the relations between these two conceptions as they developed in occidental civilisation, and to compare them with the parallel, but very different, course which events took in China.[a]

It has often been said that the peculiar glory of Chinese law lay in the fact that throughout its history (after the failure of the Legalists) it remained indissolubly connected with custom based on what were considered easily demonstrable ethical principles, and that enactments of positive law, with their codifications, were reduced to the absolute minimum. Yet perhaps this very aversion from codification and positive law, that is to say, the willed legislation of human rulers, was one of the factors which made the Chinese intellectual climate uncongenial to the development of systematised scientific thought. How this could be is the tale which we have to unfold.

Although the elaboration of China's first criminal codes goes back, as we shall see later (p. 522), to the −6th century, the rise of the school of the Legalists as such did not take place till the −4th. They flourished first in the north-eastern State of Chhi, and in the three succession States of Han, Wei and Chao (into which the former State of Chin had been divided after −403), but they reached their position of real dominance in Chhin during the −3rd century, where their policies helped to bring about that rise to power which enabled the last prince of Chhin to become the first emperor of a unified China (Piton, 1). We have already seen[b] how the draconic authoritarianism of the Chhin State and the short-lived Chhin dynasty brought a revulsion of feeling and led to the milder rule of the four Han centuries.

---

[a] Sect. 18, the last in the present volume.　　　　　　　　　　　　　　　　[b] Sect. 6.

The fundamental idea of the Legalists was that *li*,[1] the complex of customs, usages, ceremonies and compromises, paternalistically administered according to Confucian ideals, was inadequate for forceful and authoritarian government. Their watchword, therefore, was *fa*,[2a] positive law, particularly *hsien ting fa*,[3] 'laws fixed beforehand',[b] to which everyone in the State, from the ruler himself down to the lowest public slave, was bound to submit, subject to sanctions of the severest and cruellest kind. The lawgiving prince must surround himself with an aura of *wei*[4] (majesty) and *shih*[5] (authority, power, influence). This aspect was emphasised particularly by Shen Tao[6] (an older contemporary of Chuang Chou)[c] whose *floruit* is probably in the neighbourhood of −390. He must also possess the art (*shu*[7]) of statecraft, of conducting affairs and handling men. This aspect was emphasised by Shen Pu-Hai,[8] who was minister in Han State in −351 and died in −337.[d] The writings of these two men now exist only in the form of fragments. The central conception of *fa* or positive law was expounded with great clarity by Kungsun Yang[9] in a book, the *Shang Chün Shu*[10] (Book of the Lord Shang), which has come down to us.[e] Kungsun Yang was a descendant of the royal house of Wei who entered the service of the State of Chhin in −350 and was executed in −338.[f] The most scholarly and philosophical legalist was Han Fei,[11] whose life lay wholly in the following century (d. −233), and whose writings, in the form of the *Han Fei Tzu*[12] book, are available to us.[g] His biography in the *Shih Chi*[h] says that he studied under Hsün Chhing (see pp. 19, 26) with Li Ssu (later prime minister of the first Chhin emperor), who, however, connived at, or even arranged, his death in prison by poison when on an embassy from the State of Han to Chhin at the time when Chhin was subduing all the other States.

Names and particulars of many lesser figures of the Legalist school will be found in Duyvendak (3) and Liang Chhi-Chhao (1). Besides the books already mentioned, the *Kuan Tzu* contains many passages, and even whole chapters, of purely Fa Chia

---

a The ancient origin of this word is not without interest. Its old form was *fa*,[13] a character which incorporates the water radical with the word *chai*, meaning a kind of unicorn, and the sign for going away or being driven out (cf. p. 229). Granet (1), pp. 141 ff., describes an ancient magic rite or ordeal ceremony in which a bull was presented to the altar of the god of the soil, over which lustrations were sprinkled (hence the water radical). The contestants then read their oaths of innocence, but the guilty party was unable to finish, and was gored to death by the bull. Evil was thus driven out. The real meaning of the short form of the word was a mould or model, hence so many later *doubles entendres*.

b This phrase is in *Kuan Tzu*, ch. 55; and in the *Liu Thao*[14] (Six Quivers), a short military work containing ancient material, still used in the Sung. Manuscripts of the latter have been found at Tunhuang.

c Fêng Yu-Lan (1), vol. 1, pp. 153 ff., 318 ff.

d Fêng Yu-Lan (1), vol. 1, p. 319.

e Tr. Duyvendak (3). He is also called Shang Yang and Wei Yang.

f His biography, in ch. 68 of the *Shih Chi*, which is well worth reading, was translated by Duyvendak (3), pp. 8 ff., and by Pfizmaier (22). See also Fêng Yu-Lan (1), vol. 1, pp. 319 ff. Ruben (2) has compared him with Kauṭilya.

g Partial tr. W. K. Liao (1).

h Ch. 68, tr. Liao (1), vol. 1, p. xxvii (in part).

¹ 禮　² 法　³ 先定法　⁴ 威　　勢　　⁶ 慎到
⁷ 術　⁸ 申不害　⁹ 公孫鞅　¹⁰ 商君書　¹¹ 韓非
¹² 韓非子　¹³ 灋　¹⁴ 六韜

character, often rhymed; and there are certain other works of a very mixed composition which, if not wholly of Warring States period authorship, contain long passages from that time. Some of these, like the *Kuan Tzu*, embody transitional material linking the Taoists, the Logicians and the Legalists. Thus the *Kuei Ku Tzu*[1] (Book of the Devil Valley Master),[a] much of which probably dates from the −4th century, connects the Tao Chia and Fa Chia, using semi-naturalistic concepts not much found elsewhere, such as *phai*,[2] opening, and *ho*,[3] closing; or *fan*,[4] regression, and *fu*,[5] forward movement. Its 12th chapter (parallel with the 55th of Kuan Tzu) is definitely Legalist.[b] There is also the *Yin Wên Tzu*[6] (Book of Master Yin Wên)[c] which, though certainly not, as we have it today, the production of the early −4th century thinker of that name (a Mohist-Hedonist),[d] seems to contain Warring States material; it is a dull mixture of Taoist, Confucian, Logician and Legalist ideas.[e] Similarly, it is unlikely that the text of the *Têng Hsi Tzu*[7] book, as we have it today,[f] is contemporary with the jurist whose name it bears. Têng Hsi lived in the State of Chêng in the second half of the −6th century about the same time as Kungsun Chhiao (see on, p. 522), who is said to have composed the first penal code, which was cast on iron tripod cauldrons; Têng Hsi is supposed to have rewritten it on bamboo tablets.[g] He died in −501. We shall shortly recur to the *Têng Hsi Tzu* in speaking of Taoist-Legalist connections.

The central conception of *fa*, or positive law, enacted by the lawgiving prince without regard to considerations of accepted morality, or the goodwill of the people, appears everywhere in Shang Yang (Kungsun Yang was made prince of Shang) and Han Fei Tzu. Rules for rewards and punishments being made perfectly clear and definite, and published in every locality, the people will know how to behave. The law should, so to speak, apply itself, and not require the constant interference of the ruler. 'Law is the authoritative principle for the people, and is the basis of government (*Fa ling chê, min chih ming yeh; so chih chih pên yeh*[8]); it is what shapes the people' says Shang Yang.[h] If law is strong the country is strong, says Han Fei Tzu, if it is weak, the country will be weak (chapter 6, 'On having Regulations', *Yu Tu*[9]).[i] Punishments were to be deterrent in the highest degree. 'Punish severely the lightest

---

[a] Tr. Kimm (1). The better version by my friend Dr R. van Gulik, promised for many years, is believed to have been lost in manuscript during the war; a matter for great regret.

[b] Kuei Ku Tzu's political interests appear in the well-authenticated tradition which makes him the master of the two statesmen Su Chhin[10] and Chang I[11] who were responsible for the systems of alliances between States known as the Vertical and Horizontal Axes (*tsung hêng*[12]). Su Chhin's biography (ch. 69 of the *Shih Chi*) has been translated by Pfizmaier (23), and other passages concerning him by Margouliès (1), p. 13.

[c] Tr. Masson-Oursel & Chu Chia-Chien (1).

[d] Fêng Yu-Lan (1), vol. 1, p. 148.

[e] Escarra (1), p. 22; Forke (5).   [f] Tr. Forke (5); H. Wilhelm (2).

[g] Duyvendak (3), p. 69.

[h] Ch. 26, Duyvendak (3), p. 331.   [i] Liao (1), pp. 37 ff.

[1] 鬼谷子   [2] 捭   [3] 闔   [4] 反   [5] 覆   [6] 尹文子
[7] 鄧析子   [8] 法令者民之命也所治之本也   [9] 有度
[10] 蘇秦   [11] 張儀   [12] 縱橫

crimes, such was the law of Kungsun Yang', says Han Fei Tzu in his chapter 30, continuing, 'If small offences do not occur, great crimes will not follow, and thus people will commit no crimes and disorder will not arise.'[a] This idea, which occurs also in *Kuan Tzu*,[b] runs through all Legalist writings.[c] It was the Legalist version of the famous phrase in the *Shu Ching* (Historical Classic), *phi i chih phi*[1]—punishment to end punishment[d]—a phrase perhaps almost as devoid of justification as the 'war to end war' of our own time.

Much is said of the draconic nature of the punishments recommended by the Fa Chia. It should be made worse for the people to fall into the hands of the police of their own State than to fight the forces of an enemy State in battle.[e] The timorous should be put to death in the manner they most hate. Strictness in application of penalties should have no exceptions. A principle similar to the later *pao chia*[2] system was employed; men serving in the army were divided into squads of five men each, and if one of these were killed, the other four were beheaded for allowing it to happen.[f] Ranks depended on the number of the enemy slain. There was an elaborate system of delation and denunciation; omission to denounce a culprit was punished by the friend being sawn in two, and other tortures were employed.[g] The kind of thing of which the Legalists approved is shown in the story quoted by Han Fei Tzu[h] of Prince Chao of the State of Han.[i] The prince having got drunk and fallen asleep was exposed to cold, whereupon the crown-keeper put a coat over him. When he awoke he asked who had covered him, and on being informed, punished the coat-keeper but put the crown-keeper to death, on the principle that transgression of the duties of an office was worse than mere negligence.

The Legalists were conscious of this conflict between theoretically constructed positive law on the one hand, and ethics and equity, and even what one might call human common sense, on the other. Han Fei says:

Severe penalties are what the people fear, heavy punishments are what the people hate. Accordingly the sage promulgates what they fear in order to forbid the practice of wickedness, and establishes what they hate in order to prevent villainous acts. Thus the State is safe and no outrage can occur. From this I know well that benevolence, righteousness, love and favour are not worth adopting, while severe punishment and heavy penalties can maintain the State in order.[j]

[a] Liao (1), p. 295; Duyvendak (3), p. 60.
[b] Ch. 3.   [c] E.g. *Shang Chün Shu*, chs. 5, 13.
[d] Ch. 41 (Chün Chhen), Medhurst (1), p. 294; Legge (1), p. 233.
[e] *Shang Chün Shu*, ch. 5; Duyvendak (3), p. 210.
[f] *Shang Chün Shu*, ch. 19; Duyvendak (3), pp. 296 and 58. *Han Fei Tzu* also mentions this, with the implication that it applied to civil as well as military life (ch. 13, Liao (1), p. 115; and ch. 43).
[g] Duyvendak (3), p. 60. I believe that these punishments have to be taken seriously; in the later codes operated under Confucian aegis many punishments were retained only 'on paper', and there was much mitigation of their severity in practice, but the Legalists lived in a still comparatively primitive and barbarous age. Besides, cruelty was part of their system.
[h] Ch. 7 (Liao (1), p. 49).   [i] Reigned from −358 to −333.
[j] Ch. 14, tr. Liao (1), p. 128.

[1] 辟以止辟   [2] 保甲

Shang Yang, says Duyvendak, [a] is completely and consciously *amoral*. His great fear is that the people should become interested in the traditional virtues, and thereby set up other standards of conduct than those established by law. Virtue is not 'good-ness' or 'benevolence' but obedience to the law as fixed by the State, whatever it may seem good to the ruler that it should be. Hence the Legalist doctrine of the Six Parasitic Functions (*liu shih kuan*;[1] lit. 'Six Lice'). In its oldest formulation[b] these are named as care for old age, living on others (without employment), beauty, love, ambition and virtuous conduct. The things which sap the authoritarian State are extended in other lists[c] to include further—the study of the Odes and History Classics, the Rites, Music, filial piety, brotherly duty, moral culture, sincerity and faith, chastity and integrity, benevolence and righteousness, criticism of the army, and being ashamed to fight—all, except the two last perhaps, implying direct hits at the Confucian system of morality. Han Fei Tzu, for his part, comes forward with a parallel list, that of the Five Gnawing Worms (*wu tou*[2]) which destroy the State:[d] (1) the Confucian scholar praising the ancient sage-kings and discussing benevolence and righteousness; (2) the clever talker (or sophist; a hit at the Ming Chia?) using events to his private advantage and falsifying words; (3) the soldier of fortune collecting troops of adherents; (4) the merchant and artisan accumulating wealth; and (5) the official thinking only of personal interest.

The open conflict between Legalist law and Confucian ethics illustrates itself in concrete form in the debate as to whether a son should conceal his father's crime, or denounce it and give evidence against him. It had started already in the −6th century, for Confucius had decisively given his opinion that filial piety should prevail against State law.[e] Han Fei Tzu, however, argued with great insistence in the opposite sense, as a Legalist was bound to do.[f] With the ultimate overthrow of the Legalists, the orthodox Confucian view was transmitted to posterity in the *Hsiao Ching* (Filial Piety Classic).[g]

The complete rupture with traditional ethical concepts shows itself also in the positive recommendation of Shang Yang that officials should be chosen for their ruthlessness:[h] 'If virtuous officials are employed by the prince, the people will love their own relations; but if wicked officials are employed, the people will love the statutes (*Wang yung shan tsê min chhin chhi chhin: yung chien tsê min chhin chhi chih*[3])....In the former case the people will be stronger than the law; in the latter, the law will be stronger than the people.'

---

[a] (3), p. 85. The point is also well emphasised by Waley (6).
[b] *Shang Chün Shu*, ch. 4 (Duyvendak (3), p. 197).
[c] See Duyvendak (3), p. 85 and pp. 191, 197, 199, 256.
[d] Ch. 49 (tr. Escarra (1), pp. 31 ff.).
[e] *Lun Yü*, XIII, xviii. See also *Mêng Tzu*, VII (1), xxxv.
[f] *Han Fei Tzu*, ch. 49 (tr. Duyvendak (3), p. 115).
[g] We have already noticed (Sect. 7b) the parallel between this discussion and the *Euthyphro* dialogue of Plato.
[h] *Shang Chün Shu*, ch. 5 (tr. Duyvendak (3), p. 207).

[1] 六蝨官   [2] 五蠹   [3] 王用善則民親其親用姦則民親其制

It was quite in accordance with the general outlook of the Legalists that all their writings glorify first war and then to a lesser extent agriculture;[a] they had no use either for scholars or for merchants and traders. The third chapter of the *Shang Chün Shu* has for its title 'Agriculture and War', outside which there were no occupations (save that of Legalist administrator, needless to say) meriting any rewards. 'War is a thing that people hate, but he who succeeds in making people delight in war, attains supremacy.'[b] Trade should be hampered as much as possible by heavy tolls, merchants should be repressed by sumptuary laws, wine and meat should be heavily taxed, and trade in grain forbidden.[c]

There is one feature in Legalism which is of particular interest for the historian of science, namely, its tendency towards the quantitative. The word *shu*,[1] which often appears, means not only number but quantitative degree, and even statistical method. Already in the oldest parts of the *Shang Chün Shu*, says Duyvendak, there is a preference for expressing everything in numerical figures, points, units, degrees of penalties, numbers of granaries, amounts of available fodder, etc.[d] A later part of the same book says:

Rewards exalt and punishments degrade, but if the superiors have no knowledge of their method, it is as bad as if they had no method at all. But the method for right knowledge is power (*shih*[2]) and quantitative exactness (*shu*[1]). Therefore the early kings did not rely on their strength but on their power (*shih*[2]); they did not rely on their beliefs but on their figures (*shu*[1]). Now, for example, a floating seed of the *phêng*[3] plant,[e] meeting a whirlwind, may be carried a thousand li, because it rides on the power of the wind. If, in measuring an abyss, you know that it is a thousand fathoms deep, it is owing to the figures which you have found by dropping a lead line. So by depending on the power of a thing, you will reach your objective, however distant it may be, and by looking at the proper figures, you will find out the depth, however deep it may be....[f]

Again, chapter 14 condemns what it calls 'reliance on private appraisal' and speaks of the folly of trying to weigh things without standard scales, or forming an opinion about lengths in the absence of accepted units such as feet and inches.[g] Other schools, such as the Mohists, had made a good deal of rhetorical play with 'models' and 'measures', but this quantitative element in the Legalists was connected, I would suggest, with the discovery which they made that positive law, divorced from all ethical considerations, enabled them, and the rulers whom they advised, to achieve enhanced efficiency by strict regulation of weights, measures and dimensions. Hu

---

[a] Duyvendak (3), pp. 48, 83, 185.
[b] Ch. 18 (Duyvendak (3), p. 286).
[c] Duyvendak (3), pp. 49, 86, 177, 204, 313.
[d] (3), pp. 96 ff., 205, 207, 211, 266.
[e] *Erigeron kamtschaticum* (B II, 435). See Stuart (1), p. 164.
[f] Ch. 24, tr. Duyvendak (3), p. 318, slightly mod.
[g] Duyvendak (3), p. 262. Cf. *Han Fei Tzu*, chs. 6, 14 (Liao (1), pp. 45, 129).

[1] 數   [2] 勢   [3] 蓬

Shih,[a] indeed, points out that perhaps 'standard' was the oldest meaning of the word *fa*, since in *Kuan Tzu*, chapter 6, it is defined as including measures of lengths, weights, volumes of solids and liquids, T-squares and compasses. Hence possibly the significance of the bas-reliefs so often reproduced, in which the mythical ruler Fu-Hsi and his consort or sister Nü-Kua are represented as holding a T-square and a pair of compasses respectively (cf. Wu Liang tomb shrine).[b] It is obvious that natural law (in the juristic sense), or any other form of law dependent on ethical demonstrability, cannot regulate things ethically indifferent; it can recognise that parricide is 'unnatural' because never found in the *mores* of any known society, but it cannot justify the enforcement of an arbitrary rule that all chariot-wheels shall have a gauge of 4 ft. 8½ in., or that a given character shall be written in such and such a way, or that (in our own history) wool should not be exported from 16th-century England. Hence the significance of the famous action of the first emperor, Chhin Shih Huang Ti, on attaining power by the help of the Legalists in −221, when, as the *Shih Chi*[c] records: 'He unified the laws and rules (fixing the weight of) the *shih* (or *tan*; picul, about 133 lb.), and the lengths of the *chang* (about 10 ft.) and the *chhih* (about 10 in.). He standardised the gauge of chariot-wheels. He standardised the orthography of the characters (*I fa tu hêng tan chang chhih; chhê thung kuei; shu thung wên tzu*[1]).'[d] Duyvendak adduces[e] reasons for thinking that these applications of physico-mathematical thinking go back to Li Khuei,[2] whom we shall meet again as the author of an important, but long-lost, penal code (p. 523), and who was minister at the court of the State of Wei in the period −424 to −387.[f]

It is in connection with mathematics, geometry and metrology that we come upon the fundamental philosophical flaw in Legalist thinking. In their passion for uniformisation, in their reduction of complex human personal relations to formulae of geometrical simplicity, they made themselves the representatives of mechanistic materialism, and fatally failed to take account of the levels of organisation in the universe. The *Yin Wên Tzu* says:[g]

The ten thousand events are all gathered into a unity, the hundred measures all conform to a law. To be gathered into a unity is the height (lit. the very solstice) of simplicity. To conform to a law is the height (lit. the very pole-star) of facility.

---

[a] (2), p. 174.   [b] Cf. Fig. 28 in Vol. 1.

[c] Ch. 6, p. 13 b, Chavannes (1), vol. 2, p. 135.

[d] The last phrase refers, of course, to the orthographic standardisation of Li Ssu, which we have already noted (Sect. 6a).

[e] (3), pp. 43, 97. Some kind of standardisation of weights and measures certainly took place in the feudal States period, as we know from the *Yüeh Ling*, which says that it was done at the spring equinox (*Li Chi*, ch. 6, p. 52b (Legge (7), vol. 1, p. 260), *Lü Shih Chhun Chhiu*, ch. 6 (vol. 1, p. 12), tr. R. Wilhelm (3), p. 15). Moreover, in Chhin State it had been done at least since −347, when Shang Yang was in power (*Shih Chi*, ch. 68, p. 5a), cf. Duyvendak (3), p. 19.

[f] But, on the other hand, since Erkes (9), the mention of the standardisation of the gauge of chariot-wheels in the *Chung Yung*, ch. 28, has been generally regarded as evidence that that part of that work cannot be older than the Chhin dynasty.   [g] P. 3a (tr. Escarra & Germain, p. 24; eng. auct. mod.).

[1] 一法度衡石丈尺車同軌書同文字   [2] 李悝

The *Shang Chün Shu* elaborates:[a]

The former kings hung up balances with standard weights and fixed the lengths of the foot and the inch. Still today these are followed as models (*fa*[1])[b] because the divisions are clear. No (practical) merchant would proceed by dismissing standard scales and then deciding about the weights (of things), nor would he abolish feet and inches and then form opinions about the lengths (of things). Such (conclusions) would have no force (*wei chhi pu pi yeh*[2]). Turning one's back on models and measures (*fa tu*[3]), depending upon private conviction (*ssu i*[4]), takes away all force and certainty. Without a model, only a Yao could judge knowledge and ability, worth or its opposite. But the world does not consist exclusively of men like Yao. This was why the former kings understood that no reliance could be placed on individual opinions or biased approval; this is why they set up models and made distinctions clear. Those who fulfilled the standard were rewarded; those who harmed the public interest were put to death.

The whole argument, of course, used though it was again and again, depended on a false analogy, namely, that human conduct and human emotions could be measured as quantitatively as a picul of salt or an ell of cloth. Liang Chhi-Chhao (2), in his discussion of the Legalists, saw this extremely clearly.[c] The certainty and predictability of low-level phenomena cannot be found in the realms of 'free-will' at the higher levels. And he characterised the Legalist school as mechanistic (*chi hsieh chu-i*[5]), while the Confucians instinctively made allowance for the true organic (*sêng chi thi*[6]) character of man and of society.[d]

Mention has already been made of the connections between the Legalists and other schools, but it is noteworthy that they had a particular habit of distorting Taoist ideas for their own purposes. As regards the history of society, they rejected, indeed, the Taoist glorification of primitive pre-feudal collectivism, and naturally aligned themselves with the Mohist conception of primitive society as a *bellum omnium contra omnes*.[e] But in *Han Fei Tzu* we find the Taoist ideal of self-contained villages (cf. p. 100) curiously transmuted into a model of rural society on a Legalist pattern.[f] Similarly, the Taoist technical term *phu*,[7] which we have already identified (p. 114) as indicating the solidarity of pre-feudal tribal collectivism, was taken over by the Legalists to signify the simplicity and ignorance of a Spartan people occupied purely with war and agriculture.[g] The expression *wu wei*[8] meant for them, not the avoidance

---

[a] Ch. 14, p. 1*a* (tr. auct. adjuv. Escarra & Germain, p. 38; Duyvendak (3), p. 262).

[b] A pun on the word *fa*, meaning also, as it did and does, law.

[c] Esp. pp. 58, 61, 63, 64, 66.

[d] This is of considerable interest. Later (pp. 286, 474) we shall study the *philosophia perennis* of China as essentially a philosophy of organism. But not all modern Chinese philosophers have been fully conscious of the fact.

[e] Duyvendak (3), pp. 102 ff., adducing passages from *Shang Chün Shu*, chs. 7 and 23; *Kuan Tzu*, ch. 31; and *Han Fei Tzu*, ch. 49.

[f] Ch. 6, tr. Liao (1), p. 41. [g] Duyvendak (3), p. 86.

[1] 法 [2] 爲其不必也 [3] 法度 [4] 私議 [5] 機械主義
[6] 生機體 [7] 樸 [8] 無爲

of any activity contrary to Nature, but the absence of governing activity on the part of a ruler who has enacted sufficient positive law to allow of government being carried on 'automatically', and to ensure that this shall continue to happen even if his successors prove incompetent.[a] And the Taoist phrase 'Heaven and Earth are not benevolent' was only too easily given a Legalist meaning, as in the 1st chapter of *Têng Hsi Tzu*.[b] Thus the Tao of the Legalists, like that of the Confucians, was a Tao of human society, but unlike theirs it was not a universal ethical principle. It was the motive power of an aggressive authoritarian unit within the society of mankind, aiming at universal dominion. To the Confucian doctrine of 'government by (human-hearted) man' (*jen chih chu-i*[1]) it opposed a harsh and rigid 'government by laws' (*fa chih chu-i*[2]).

The implementation of all these ideas on the scale which the success of the State of Chhin made possible could hardly have failed to leave lasting marks on the structure of Chinese civilisation. But perhaps even the Legalists hardly realised how deep these were to prove. For it was the principles of Legalism which reflected that recrystallisation of Chinese society in which feudalism passed over into a new state of semi-stability—feudal bureaucratism.

As one sees it from the Legalist point of view, the change took the form of a frontal attack on the privileged and powerful feudal lords and their families. There was to be a levelling process in society. The laws were to be applied to all equally, and rank or reward was to be attained only on the grounds of merit in war or agriculture.[c] The severe restriction of its inheritance would lead to a situation in which all the intermediate and smaller feudal lords had disappeared, leaving the prince (and later the emperor) to govern by means of an enormous bureaucracy of officials. As we have already seen (Sect. 6*b*), and as must be set forth more fully later on (Sect. 48), this was what actually happened. The 'carrière ouverte aux talents' began with the decline of Chinese feudalism. Here are some plain statements:

Shang Chün Shu, chapter 17:

What I mean by the unification of punishments is that they should recognise no social distinctions (*So wei i hsing chê hsing wu têng*[3]). From ministers of state and generals, down to officials and ordinary folk, whosoever does not obey the king's commands, violates the interdicts of the State, or rebels against the statutes fixed by the ruler, should be guilty of death and should not be pardoned.[d]

Han Fei Tzu reaffirms it (chapter 6):

The law cannot fawn on the noble, just as the (carpenter's) string (when stretched) cannot yield to the crooked (*fa pu o kuei, shêng pu nao chhü*[4]). Whatever the law applies to, the wise

---

[a] Duyvendak (3), pp. 88, 99.   [b] H. Wilhelm (2), p. 59; Forke (5), p. 38.
[c] Duyvendak (3), pp. 82 ff., 91.   [d] Tr. Duyvendak (3), p. 278, mod.

[1] 人治主義   [2] 法治主義   [3] 所謂一刑者刑無等
[4] 法不阿貴繩不撓曲

cannot reject nor the bold defy. Punishment for faults must never skip ministers, nor rewards for good actions fail to reach commoners (*Hsing kuo pu pi ta chhen, shang shan pu i phi fu*[1]).[a]

The fundamental principle of bureaucratism is stated, indeed clearly defined, by Shang Yang (*Shang Chün Shu*, chapter 17):

Neither in high nor in low offices should there be an automatic hereditary succession to the offices, ranks, lands, or emoluments of officials (*Wu kuei chien, shih-hsi chhi kuan chang chih kuan-chio thien-lu*[2]).[b]

And finally the strategy of transition from feudalism to bureaucratism is subtly put by the same writer (*Shang Chün Shu*, chapter 4):

To remove the strong by means of a strong (people), brings weakness. To remove the strong by means of a weak (people), brings strength (*I chhiang chhü chhiang chê jo; i jo chhü chhiang chê chhiang*[3]).[c]

In other words, to overthrow the powerful feudal lords by strengthening the people would have been a Taoist measure, leading to a return of ancient collectivism, and it would weaken the ruler; but to overthrow the feudal lords while maintaining the people in a condition of weakness would strengthen the ruler. Thus was epitomised the way which led from Chou feudalism to the 'Lonely One' and his vast civil service of every dynasty after the Chhin. In a later context (Sects. 28*f* and 48) we shall consider the view that the concrete reality in the field of production which gave rise to these changes was the ever-growing importance of works of hydraulic engineering, for irrigation, water conservation, flood protection and tax-grain transportation always tended to transcend the boundaries of individual feudal domains; but this kind of underlying factor can only be touched upon here. Reference will also have to be made again to the land reforms of the Legalists[d] in which the ancient *ching thien*[4] (well and nine fields) system of feudal land tenure was abolished, and irregular fields with balks and headlands (*chhien mo*[5]) were laid out, the people acquiring the right to buy and sell land (Sects. 28*f*, 41 and 48).

We can now approach the significance of all this for the historian of science. It has a profound (though perhaps not at first sight obvious) bearing on the history of the gradual differentiation of the concepts of juridical law and the laws of Nature (in the sense of natural science), and on the comparative history of this development in China as opposed to Europe. The influence of the Legalist school did not, of course, die away very soon after the assumption of power by the Han dynasty, but there was

---

[a] Tr. Liao (1), p. 45.
[b] Tr. Duyvendak (3), p. 279.
[c] Tr. Duyvendak (3), p. 196.
[d] Cf. Duyvendak (3), pp. 44 ff.

[1] 刑過不避大臣賞善不遺匹夫
[2] 無貴賤尸襲其官長之官爵田祿
[3] 以強去強者弱以弱去強者強
[4] 井田　　　　[5] 阡陌

a steady replacement of Legalist by Confucian ideals lasting more than two centuries.[a] Slowly but surely Legalism was rejected by the Chinese people. As political stability was more and more achieved in the unified empire of the Han, the sense of strain of the Warring States period disappeared. Correspondingly the progressive liquidation of the great feudal houses removed one of the main obstacles against which the Legalists had contended. But the greatest factor was the reversion to li,[1] custom and usage based demonstrably on ethics, from fa,[2] positive law. Since the place of positive law could not be reduced beyond a certain minimum, the gulf between law and ethics was bridged by restricting codified law, according to ancient Chinese tendencies, to penal and criminal law. Law again became, as Duyvendak says, firmly embedded in ethics, and successive emperors, down to our own times, justified their mandates by invoking natural law (in the juristic sense), i.e. norms of behaviour universally considered moral—in fact, li[1]—and not positive laws. The Legalists had wanted to make law without any reference to the people's conceptions of right and wrong, so that it would act in an automatic mechanistic way; in the Chinese cultural milieu this was bound to fail. Moreover, the Confucians were able to make some acute observations on the failings of Legalist positive law; thus Hsün Chhing had already said: 'If there are laws, but they are not discussed, then those cases for which the law (code) does not provide, will certainly be let pass wrongly (*Ku fa erh pu i, tsê fa so pu chih chê pi fei*[3]).'[b] Such positive law as was retained in medieval China, therefore, was but a reflex of customary natural law, without any intrinsic force of its own. The will of the imperial lawgiver did *not* suffice to 'make it right'. And as a minor corollary of this general state of affairs it followed that quantitative standards such as weights and measures, the gauge of wheels and the rule of the road, reverted to comparative chaos until modern times.

Now while we must admire the exalted place taken by juristic natural law in Chinese civilisation, the question arises as to what its effects could have been on the development of thought in natural science. It is quite clear that in Europe juristic natural law and the Laws of Nature of the natural sciences sprang from a common root. For later Greek philosophy and Hebrew monotheism alike, the rational creator deity, whether Zeus or Jahveh, had laid down a celestial code of law which all created things were bound to obey, in a fashion exactly parallel to the princely and imperial lawgivers on earth. Positive law of earthly States (however strongly emphasised) could therefore not run contrary to that wider body of natural law (in the juristic sense) which all men everywhere spontaneously obeyed when they acted according to their natures; and this natural law in its turn was but a part of that body of universal law which controlled the behaviour of animals, plants and the stars in their courses. In so far, then, as natural law came to be overwhelmingly dominant in China, and positive law reduced to the minimum, one would expect that so different a balance might have

[a] Dubs (3), pp. 341 ff.; Duyvendak (3), pp. 126 ff.
[b] *Hsün Tzu*, ch. 9, p. 3a, tr. Duyvendak (3), p. 129.

[1] 禮   [2] 法   [3] 故法而不議則法所不至者必廢

had important effects on the development of the formulation of the regularities of Nature in the natural sciences. The weaker the role of the earthly ruler as lawgiver for man, the more difficulty there might have been in conceiving of a divine ruler as lawgiver for Nature. No lawgiver, no law. Or conversely, the stronger the role of (juristic) natural law, the greater facility there might have been in conceiving of the Laws of Nature as a kind of inescapable natural *li* in which the whole non-human world concurred.[a] An order, though no ordainer. Such is the problem which will present itself to us at the conclusion of this volume.

That the Legalists knew they were up to something fundamentally new is shown by many passages; one need refer only to the speeches of Kungsun Yang in the discussion with Duke Hsiao of Chhin.[b] But I find it quite impossible to believe that they could have succeeded as they did if there had not been some technological basis for their new social theories. Literary historians have been discussing for centuries the 'miracle' of the rise of Chhin, but I believe that we should look rather in the arsenals of that State to find the concrete technical innovation which permitted a system so oppressive and tyrannical to achieve its aims so outstandingly. In a later place (Sect. 30, Military Technology) evidence may be forthcoming as to what this invention possibly was.

The phenomenon of the Legalists was part of that great revolution during which Chinese society passed out of its classical stage. When the wave receded, the intermediate feudal lords were no longer there, but nor was authoritarian positive law unrelated to ethics;[c] only the bureaucracy, administered by Confucians rooted in custom and compromise, remained as the permanent network of government in a society based on agriculture and hydraulic engineering.[d]

---

[a] Cf. the remarkable passage from Hsün Chhing, quoted above, p. 27.

[b] *Shang Chün Shu*, ch. 1 (Duyvendak (3), p. 171).

[c] This of course does not mean that codified positive law did not continue, but it was restricted to a minimum and always formulated and interpreted in strict accordance with customary morality (*li*). There also grew up a great body of administrative law, rules and regulations applying mainly to the bureaucracy, which were collected in the *Hui Yao* and *Hui Tien* of the successive dynasties, but this too was subordinate to the basic principles of *li*.

[d] Hsiao Ching-Fang expresses this when he says, 'The Legalists wished to make law omnipotent, but the only result was that the emperor became omnipotent' (p. 67). Granet (5), p. 462, compares the Legalist princes with Greek tyrants—a suggestion which could provoke some far-reaching thoughts.

13. THE FUNDAMENTAL IDEAS OF CHINESE SCIENCE

(a) INTRODUCTION

WE NOW approach a field of central importance for the history of scientific thought in China, namely, the fundamental ideas or theories which were worked out from the earliest times by indigenous naturalists. Here there are three principal subjects requiring discussion: first, the theory of the Five Elements (*wu hsing*[1]); secondly, that of the Two Fundamental Forces (Yin and Yang[2]) in the universe; and thirdly, the scientific, or rather proto-scientific, use of that elaborate symbolic structure, the Book of Changes (*I Ching*[3]). It will be necessary to discuss not only their nature and later significance, but also their historical origin, and here our presentation will have to differ considerably from the Chinese traditions formerly generally accepted, and taken over more or less uncritically by the early occidental sinologists, but now superseded by the results of modern research. We have thought it desirable, moreover, to preface these discussions by a short account of the way in which some of the Chinese words most important for scientific thought originated as the ideographic script developed.

This section involves in a way the whole world-outlook of the ancient Chinese. It is therefore impossible not to refer to three important occidental works which have already been devoted to this subject, the *World-Conception of the Chinese* by Forke, the *Sinism* of Creel, and *La Pensée Chinoise* of Marcel Granet. The first of these is a relatively matter-of-fact work, which laudably gives the Chinese texts of most of the many passages which it quotes, but it has the disadvantage that it was written before or during the first world war when much less was known about the dating and authenticity of Chinese books than now, so that ideas of the Chou and the Han, the Sung and the Yuan are mingled in considerable confusion. Forke, indeed, exerted himself to find parallels and differences between Chinese thought and that of other civilisations. But what we miss most in his book is a critical evaluation of the worth of the Chinese ideas from the standpoint of a mind trained in the natural sciences; this no one at that time was in a position to give. The same applies to Creel's book of a dozen years later, which, however, had the serious defect of assuming that the cosmism and phenomenalism[a] of the Han was ancient. Creel himself now recognises the important part played by the School of Naturalists in launching many of the ideas which later became essential components of the Chinese world-outlook.[b]

[a] See below, pp. 247, 377 ff.   [b] See Creel (4), p. 86.

[1] 五行   [2] 陰陽   [3] 易經

Granet's volume was a much more sophisticated production than either of these two books—indeed, in its way a work of genius.[a] Granet, who was basically a socio-logical analyst of myths, had drawn (1, 2) an unforgettable picture of the peasant society, the aenolithic proto-feudalism, of the early Chou period, the time of the *Shih Ching* folksongs. Then, in his *Pensée Chinoise*, approaching the ideas of a Huai Nan Tzu or a Wang Chhung from the angle of that mythology and folklore out of which they had developed, he unfolded a vast panorama of Chinese thought, passing from facts of social life to concepts of time and space, from divination practices and magic squares to the theories of the elements, from the macrocosm to the microcosm and back again. The work is illuminated by flashes of deep insight, and Granet's most fundamental estimates of the specific characteristics of ancient Chinese thought are, I believe, correct—it will not be possible to complete this section without quoting from some of them. But he passes from one demonstration to another with so much assurance and clarity that one acquires the uneasy feeling of watching a conjuring performance, and ends with the conviction that none of the ancient Chinese naturalists can have been nearly so clear in their own minds about their own system as Granet was. We can hardly dare to take so godlike a way.

Granet would perhaps himself have admitted the existence of a measure of the subjective in his expositions. He himself says that 'however much one may approach China with a vivid imagination and critical spirit, she seems determined to show herself only through a literary and bookish veil'.[b] This is not the present writer's experience. The study of Chinese technology may go far if it starts from ancient techniques still being operated today. Material archaeology has already yielded brilliant finds, as in the case of the oracle-bones,[c] and we may expect marvellous results now that conditions in China permit excavation on a really generous scale.[d] And then Chinese science only comes to life when Chinese texts are read, as sinologists have rarely yet been able to read them, with the eyes of those who have been trained in the natural sciences. Perhaps the present book will prove that across the centuries and across the ideographic-alphabetic barrier, similar minds can still communicate.

---

[a] This is at last being recognised by sinologists, cf. the generous words of H. Franke (5), pp. 69ff.

[b] (5), p. 585.

[c] We shall see a striking example of this in the story of the discovery of the magnetic compass (Sect. 26*i* below).

[d] This is demonstrated by such excavations as that of the royal tombs near Chhêngtu in recent years (Fêng Han-Chi, 2). I have often noted promising sites for archaeological work when in China. Going up the Wei Valley from Sian to Paochi one passes to the right on the edge of the river terrace one enormous tumulus after another, the burial places of the Han and Sui emperors, too large, one would think, to have been rifled in past ages. Now (1953), important discoveries are constantly being made by the archaeological field teams of Academia Sinica, e.g. the Han ship model which will be discussed in Section 29.

(b) ETYMOLOGICAL ORIGINS OF SOME OF THE MOST IMPORTANT CHINESE SCIENTIFIC WORDS

Before proceeding further it may be of interest to glance at the processes whereby the Chinese acquired their stock of words without which no scientific communication could have gone on at all. This involves a short excursus into ideographic etymology. Owing to the discovery of the Anyang oracle-bones (cf. the discussion in Sect. 5b), we now have an abundance of information about the earliest known forms (−2nd millennium) of the Chinese characters, and these, together with others taken from the Shang and Chou inscriptions on bronze vessels, have yielded a copious graphic vocabulary, only part of which has yet been identified with the elements of the later stylised script still in use today. A very large number of identifications are however generally accepted, and we can therefore choose from among them a selection of ideographs which throw light on the origin of the stock of Chinese scientific terms.[a]

While this seems worth doing, it must be understood that these ancient etymologies probably had little influence on the thinking of the exponents of the proto-science of the Chhin and Han, or the scientific men of the Sung; on the mind of a magician like Luan Ta, or a sober pharmaceutical botanist like Thang Shen-Wei. Throughout Chinese recorded history many of these etymologies remained unknown, even to Hsü Shen, the +2nd-century father of Chinese lexicography, whose *Shuo Wên Chieh Tzu*[b] (Analytical Dictionary of Characters) was the predecessor of so vast a train of dictionaries and encyclopaedias. So far as we know, Hsü Shen never saw a bone inscription of the Shang period. He regularly gave, however, the 'lesser curly' (*hsiao chuan*[1]) or 'seal' forms of the characters, and it was only by comparing these with the forms found in bone and bronze inscriptions that scholars were enabled to decipher the latter. Hsü Shen misinterpreted many characters, but was right as to many more. Some of his ideas which were formerly thought to be absurd or fantastic have been confirmed by study of the bone inscriptions.

It is therefore not because the conceptual origins of Chinese characters used in scientific thinking had much effect on that thinking itself, that we take the opportunity here to glance at how they came to be formed. It is rather because the origins of a specifically ideographic scientific vocabulary cannot but be of interest as an aspect of the history of proto-science in general. That scholars of today know much more about the writing, and the thought behind it, of the Shang period, the formative time of Chinese orthography, than people did who lived a few centuries later, is just one of the paradoxes of archaeological science.

[a] This sub-section owes a great deal to the generous collaboration of Dr Wu Shih-Chhang, formerly of National Central University, now at Oxford, who is deeply learned in the most ancient forms of the characters. Our warmest thanks are here offered to him.

[b] Cf. Sect. 2 (Vol. 1, p. 31).

[1] 小篆

As for the original meanings of the roots in the earliest common Indo-Aryan language, before the separation from Sanskrit of the various language-groups of Europe—roots which were afterwards to form the words essential for an Aristotle or a Newton—their reconstruction must presumably be even more conjectural than the pursuit of the Chinese characters to their original forms. For in the one case there is nothing to go upon except permutations and combinations of a limited number of alphabetical letters representing ancient sounds, while in the other, though the sounds can only be guessed, the graphs, essential pictographs or drawings, remain, and their meanings and mutual relations through thought associations can be traced.

Such at least would be the case if it were not for the process whereby homophones were borrowed. As we saw above,[a] there was a tendency, from very early times, to use one character with the sense which properly belonged to another of the same sound. This latter might have a different form, or perhaps had not yet been provided with a form. It is therefore sometimes very difficult to be sure whether certain patterns and combinations ever really had semantic significance. Such purely phonetic loan-words were at any rate well calculated to mislead the unwary etymologist of three thousand years later.

I had intended to give, for the sake of comparison, a list of the Indo-European etymologies of the words chosen to exemplify the Chinese processes of graph-building.[b] But on looking into the dictionary of Pokorny (1), which gives the conjectured roots of the Indo-European language before the division into Sanskrit on the one hand and the Latin-Germanic tongues on the other, I found that their most ancient meanings do not seem to differ much from those which their derivatives bear today, disguised though they are from the recognition of the uninitiated. Thus 'above', Lat. *super*, goes back to *uper-*, meaning the same thing; 'wind', Lat. *ventus*, comes from *ue*, to blow; 'winter', Lat. *hiems*, originates from *ghei*, also meaning winter, or snowy. It is interesting, though well known, that 'law', Lat. *lex*, comes from an ancient root *leg* meaning to collect or bind together. *Lumen* and *luna* are both from *leuq*, light or to lighten. We are not therefore much further on. Occasionally, some point of interest emerges; thus 'woman', Lat. *femina*, comes from *dhei*, sucking or that which is sucked, and there may be a parallel here with the ideographic oracle-bone representation (no. 54 in Table 11), which seems to emphasise the breasts in distinguishing the figure from a male. But on the whole, it would seem that the thought-processes by which the basic words were formed in the Indo-European language group lie too far back to be attainable by any study of the alphabetical roots. They can hardly have been without exception onomatopoeic. But with all due caution, one might perhaps say that the ideographic system has preserved for our inspection some of these thought-processes.

In Table 11 a certain number of characters, selected for their interest as fundamental scientific terms, have been brought together. The English word is followed by

[a] Sect. 2 (Vol. 1, p. 30).
[b] On the question raised in this paragraph I gratefully acknowledge the help of Prof. N. Jopson.

Table 11. *Ideographic Etymologies of some of the words important in scientific thinking*

No.	Word	Modern Chinese romanisation	Modern Chinese character	Ancient oracle-bone, bronze, or seal, form	K no.	Remarks	Reference
1 a	affirmatory final particle (equivalent to 'x *is* y')	*yeh*	也		4	(a) Drawing of a cobra-like serpent. The semantic link, if any, would have been 'affirmation of danger'. The word for serpent, *shê* 蛇, is certainly related to it. But this explanation is less widely accepted than the following. (b) Drawing of the female external genitalia, the vulva. The semantic link, if any, would have been 'gate of being', hence 'affirmation of Being', and all lesser affirmations of qualities and attributes contained therein. This explanation was never challenged throughout Chinese history in spite of centuries of Confucian prudery.	Jung Kêng / Hsü Shen
1 b	affirmatory verb-noun, to be, is, existence	*shih*	是		866c	Drawing of the sun with a foot and other strokes below it, probably composing the word *chêng* 正, 'correct, straight, fair and square'; not illusory. Thus 'that which exists under the sun'. Vision was here taken as representative of all the other means by which we collect sense-data.	Hsü Shen
2 a	negative particle, 'not'	*pu*	不		999	Drawing of a flower-head on a stalk with two drooping leaves; thus, as regards sense, a borrowed homophone. The traditional explanation (Hsü Shen) was that it was an abstract concept symbol, i.e. a bird soaring aloft and *not* allowing itself to be caught.	Lo Chen-Yü
2 b	negative verb-noun, not to be, is not, non-existence	*fei* (*fei*, to fly)	非		579	Traditionally explained as the lower part of the word *fei* 飛, to fly, itself an old drawing of a bird; therefore two wings (or perhaps birds) back to back, i.e. *not* facing each other. Therefore (if Hsü was right) an abstract concept symbol. There are, of course, a number of other words signifying the affirmative and the negative with various nuances (see Sect. 49).	Hsü Shen

	Meaning	Romanization	Character	Ancient form	Ref. no.	Explanation	Authority
3	different	i	異		954	A frontal and linear representation of a man with arms raised protecting his head or making a gesture of respect. The latter meaning is found in bronze inscriptions; perhaps the gesture had reference to the assumed effulgence of a noble interlocutor. If the character is not purely a borrowed homophone, social difference between lords and people may thus have led to the idea of 'otherness', strangeness, difference' in general. The head itself is drawn in an unusual exaggerated way, possibly to represent a mask.	Wu Ta-Chhêng Ting Fo-Yen
4	like, similar to	ju	如		94g	The 'woman' and 'mouth' radicals combined. A very early borrowed homophone or phonetic loan-word. No archaic significance.	Wu Shih-Chhang
5	if	jo	若		777	A person kneeling, perhaps gathering plants. Some have thought that submissiveness is implied, hence 'to be harmonious, to concur, complaisant', hence, by further extension, 'granted that, if...'. But it is more likely to be purely a borrowed homophone.	Shang Chhêng-Tsu Kuo Mo-Jo
6	change, permutation	i	易		850	Drawing of a lizard, the meaning being derived either from colour-changes (cf. the chameleon), or rapid shifts of position.	Liu Hsin-Yuan
7	change, especially gradual change, and change of form	pien	變		178o	Apparently not found in bone or bronze inscriptions, therefore of relatively late invention. The meaning of the drawing is uncertain, but it contains two hanks of silk and Hsü Shen said that it meant 'to bring into order', as in spinning or reeling. The radical, placed below, shows a hand holding a stick, signifying 'movement, action'. If the character is not purely a phonetic loan-word, it may have implied change from disorder to order (cf. p. 74 above).	Wu Shih-Chhang
8	change, especially sudden change, and change of substance	hua	化		19	Drawing of two knives, i.e. coins of knife-money (cf. Vol. 1, p. 247). Currency exchange would thus have given rise to one expression of the idea of change in general. Cf. no. 27.	Wu Shih-Chhang
9	origin, first	yuan	元		257	Drawing of a figure of a man in profile, with emphasis on the head, therefore first, beginning', the head being the most important part of the body. Since the ancient people doubtless knew that the head grows faster than other parts of the body during the embryonic life of vertebrates, and is relatively larger then than later, there may be an echo of primitive biological knowledge here (cf. Sect. 43).	Wu Shih-Chhang

Table 11 (continued)

No.	Word	Modern Chinese romanisation	Modern Chinese character	Ancient oracle-bone, bronze, or seal, form	K. no.	Remarks	Reference
10	cause, to rely on, following	yin	因		370	Drawing of a mat with woven pattern. Hence a basis, 'something to be relied on'; the meaning being extended from the static to the temporal. It may be noted that the same drawing occurs again in hsiu 宿, the resting-place for the night, a term of importance in astronomy (the lunar mansion, see Sect. 20e).	Thang Lan; Wu Shih-Chhang
11	cause, reason, fact	ku	故		49i	The left-hand side of the ancient bronze graph is the radical meaning 'old, ancient'; its significance is not exactly known, but it originates from a drawing of a shield stored in an open rack. The right-hand side shows the hand holding the stick, symbolising action. The general meaning is clearly 'precedent' or 'prior action'.	Wu Shih-Chhang
12	make, do, act	wei	爲		27	Drawing of an elephant, with a man's hand on its trunk, symbolising prehensility and dexterousness.	Lo Chen-Yü
13	begin	shih	始		976p, e', g', h'	Drawing of a foetus (upside down) and a woman. Closely related to thai 胎, womb, and embryo. Hence here it probably signified a female embryo, a 'beginning of beginnings'.	Hsü Shen
14	go, move	hsing	行		748	Diagram of a crossroads.	Lo Chen-Yü
15	go away, deprive, send away	chhü	去		642	Drawing of a rice-basket covered with a lid. A homophone borrowed for the present meaning.	Thang Lan
16	come to, reach, attain	chih	至		413	An arrow hitting its target, or the ground.	Lo Chen-Yü
17	stop	chih	止		961	Drawing of a human foot.	Sun I-Jang
18	end, finished, exhausted	chin	盡		381	Drawing of a hand cleaning out a vessel with a brush.	Lo Chen-Yü

No.	Meaning	Romanization	Modern form	Ancient form	Ref.	Description	Authority
19	true, the truth, real	chen	真	(graph)	375	Seal form of uncertain representational significance, but (as we know from all the related derivative words) almost surely implying 'full, filled up, solid'. Hence the derived meaning of truth as opposed to 'empty, unreal'. The drawing of a full sack standing on a stool, if that is what it is, would thus be an abstract concept symbol.	Tuan Yü-Tshai, Wu Shih-Chhang
20	above, to ascend, to hand up	shang	上	(graph)	726	Geometrical pictograph.	Lo Chen-Yü
21	below, to descend, to hand down	hsia	下	(graph)	35	Geometrical pictograph.	Lo Chen-Yü
22	centre	chung	中	(graph)	1007	A flagstaff with two pennants, one above the trapezoid or bushell (still used today on Chinese masts) and one below it.	Lo Chen-Yü
23	region, side, square, quarter	fang	方	(graph)	740	Drawing of a plough or ard. By extension, the ploughed area.	Hsü Chung-Shu
24	go in	ju	入	(graph)	695	Drawing of a wedge or arrow-head.	Hsü Shen
25	come out	chhu	出	(graph)	496	Drawing of a human foot, shown as leaving an enclosed space such as a cave or a house.	Wang Kuo-Wei
26	south	nan	南	(graph)	650	Drawing of a musical instrument of some kind, perhaps a bell. How it acquired its eventual meaning is not known.	Kuo Mo-Jo
27	north	pei	北	(graph)	909	Drawing of two men back to back. Presumably a borrowed homophone. Perhaps akin to no. 8.	Hsü Shen
28	west	hsi	西	(graph)	594	Believed to be a drawing of a bird's nest, or (Ting Shan) a net to catch birds. Again presumably a borrowed homophone. But the graph looks very like a bundle.	Hsü Shen, Wang Kuo-Wei

Table 11 (*continued*)

No.	Word	Modern Chinese romanisation	Modern Chinese character	Ancient oracle-bone, bronze, or seal form	K. no.	Remarks	Reference
29	east	*tung*	東		1175	Not, as the traditional explanation (Hsü Shen) had it, the rising sun seen through a tree, but rather a sack or bundle, which in some forms is shown being carried on a man's back. Unless the word is purely a borrowed homophone, it is curious that an azimuthal direction should be connected with a bundle. Hsü Chung-Shu regards *tung* as an archaic form of *nang* 囊, bag, and *tho* 橐, bellows. This calls to mind the later Taoist expression *chhing nang* 青囊 'blue bag' for the heavens, hence the universe. Perhaps the equator and the ecliptic were the cords which tied up this bag (cf. the Iranian 'leashes' which controlled the planets; de Menasce; Mazaheri). The *tho* was also the metal-lurgical bellows, with which Lao Tzu compares the universe (*TTC*, ch. 5). Moreover, Hsü Chung-Shu relates the custom still existing in colloquial speech of calling 'things', in general *tung-hsi* 東西, to this ancient *nang*—the heavens and earth and all that is therein. Ting Shan concurs, and Wu Shih-Chhang, though doubting the cosmic relevance of the usage, has noted interesting classical variants in the common speech of certain provinces.	Hsü Chung-Shu, Ting Shan
30	heaven	*thien*	天		361	Drawing of a human figure with a large head. The obvious conclusion that it represents a primitive anthropomorphic deity has been drawn by many modern scholars.	Hsü Shen
31	sun	*jih*	日		404	Pictograph.	Hsü Shen
32	moon	*yüeh*	月		306	Pictograph.	Hsü Shen
33	bright, brightness	*ming*	明		760	A combination of the two foregoing.	Hsü Shen
34	light	*kuang*	光		706	Drawing of a kneeling human figure, with fire on its head, perhaps a torchbearer.	Hsü Shen, Lo Chen-Yü

No.	meaning	reading	char.	archaic forms	ref.	note	authority
35	year	*sui*	歲		346	Drawing originally symbolic of a special sacrifice, probably annual.	Wu Shih-Chhang
36	spring	*chhun*	春		463	Drawing of a plant sprouting in spring, with branches still not strong enough to support themselves.	Yeh Yü-Sên, Tung Tso-Pin
37	summer	*hsia*	夏		36	Drawing of unknown significance. The right-hand top element represents a pig.	Yeh Yü-Sên, Thang Lan
38	autumn	*chhiu*	秋		1092	Drawing of a tortoise. The character later evolved, through a series of stages now identified, into a combination of 'grain' and 'fire'.	Thang Lan
39	winter	*tung*	冬		1002	Probably not a drawing of two pendent icicles (Hsü Shen), but of falling branches with fruit or leaves on them.	Yeh Yü-Sên, Tung Tso-Pin
40	wind	*fêng*	風		625	Borrowed homophone from a somewhat similarly written character depicting the phoenix, or more properly speaking the peacock (*Pavo cristatus*), with its aigrette. The phonetic on the right of the bone form is probably, and suitably, a sail.	Wang Kuo-Wei
41	rain	*yü*	雨		100	Pictograph of raindrops.	Hsü Shen
42	snow	*hsüeh*	雪		297	Pictograph of snowflakes.	Wu Shih-Chhang
43	lightning	*tien*	電		385	Hsü Shen thought that this was an attempt to draw something far-stretching which accompanies rain. He interpreted *shen* 申, one of the cyclical characters (cf. Sect. 20h) as a symbol for stretching. But elsewhere he interpreted *shen* as a pictograph of lightning, and this is now to be accepted. The zigzag flash is accompanied by drops of rain. In *shen* 神, deity, divinity, the lightning graph persists as phonetic, indicating that the lightning was regarded by the ancient Chinese with the same awe as the thunderbolts of Zeus, Thor and Indra were by others. It is thus all the more striking that the wielder of the lightning, if originally fully personified, did not keep his personality long in the Chinese mind.	Wu Shih-Chhang
		(*shen*)	申				

Table 11 (*continued*)

No.	Word	Modern Chinese romanisation	Modern Chinese character	Ancient oracle-bone, bronze, or seal, form	K no.	Remarks	Reference
44	thunder	lei	雷		577	To represent the noise of thunder there was added to the lightning-pictogram a drawing of wheels rumbling among the flashes. The round objects have also been taken to be drums, but this is less plausible.	Hsü Shen Tuan Yü-Tsai Wang Yün Wu Shih-Chhang
45	rainbow	hung	虹		1172j	Drawing of an amphisbaena-like animal in the heavens. The zoomorphic (rain-dragon?) aspect persists in the modern character as the 'insect' radical.	Kuo Mo-Jo
46	life, birth	sêng	生		812	Drawing of a plant rising out of the ground; symbolic of vegetal growth.	Hsü Shen
47	with, together, belonging to the same group as	thung	同		1176	Drawing of a vessel covered with its lid. Some bronze forms give the latter a handle. A vessel and its lid certainly belong together.	Lo Chen-Yü
48	group, class, category	lei	類		529	No early forms known. The semantic significance is uncertain. In some ancient texts (such as the *Shih Ching*) the word meant 'good'. The graph has 'head' and 'rice' as phonetic with 'dog'. The traditional explanation was (apparently) that dogs were dogs, although there were many breeds of dog looking rather unlike each other.	Hsü Shen
49	young	shao	少		1149	Drawing of four cereal grains, a concept symbol for paucity of grain, hence 'fewness' in general, and 'young', by extension of meaning.	Wu Shih-Chhang
50	old	lao	老		1055	A drawing of an old man leaning on a stick.	Yeh Yü-Sên
51	death	ssu	死		558	Drawing of a man kneeling beside bones or a skeleton. Cf. Chuang Tzu and the skull.	Lo Chen-Yü
52	man, human being	jen	人		388	Drawing of a male human being.	Hsü Shen
53	man	nan	男		649	Drawing of a field and a plough. The male tiller of the soil is implied.	Hsü Chung-Shu

54	woman	nü	女		94	Drawing of a female human being.	Hsü Shen
55	body	shen	身		386	Drawing of the body of a pregnant woman.	Wang Yün / Chu Chün-Shêng
56	blood	hsüeh	血		410	Drawing of a sacrificial vessel with its contents.	Lo Chen-Yü
57	self	chi	己		953	Borrowed homophone. The drawing probably represents the wound cord of a tethered arrow (cf. Sects. 28e, 30d).	Kuo Mo-Jo
58	male, ancestor	tsu	祖		46	Phallus, hence phallic-shaped ancestral tablet. Originally tsu 且. Connected with mu 牡, now used only for animals.	Kuo Mo-Jo / Karlgren (9) / Waley (7) / Hopkins (28, 29)
59	female, ancestor	pi	妣		566n	Female external genitalia. In another form phin 牝, now used only for animals.	Erkes (11)
60	ruler, duke, public, just	kung	公		1173	Said to be again the male generative organ, the glans penis being emphasised. The traditional view (Hsü Shen) was that the character was a combination of pa 八, =pei 背, and ssu 私, i.e. 'turning the back on private interest', but this is not convincing.	Hopkins (8)
61	lines, design, pattern, ornament, a pictogram, literature, civilian, civilised	wên	文		475	A human figure viewed frontally, showing tattoo-marks or painted designs on the body. Cf. Liu Hsien (1).	Wang Yün / Hopkins (8)
62	sunny, bright, the south side of a hill, the Yang force	yang	陽 易		720	Traditionally (Hsü Shen), the upper part of this character is the sun; while the lower part represents slanting sunbeams (Wu Shih-Chhang). Cognate forms (unless Hopkins (19) erred) suggest a drawing of a man holding up a perforated jade disc, the pi 璧. This was not only a ritual object, but also perhaps, as will later be seen, the most ancient of Chinese astronomical instruments (Sect. 20g).	Sun Hai-Po / Jung Kêng
63	shady, dark, the north side of a hill, the Yin force	yin	陰		460 651 x, y, z	Drawing of yin 云 clouds, combined with chin 今, as phonetic, and (as in the previous word), fou 阜, hill.	Tuan Yü-Tsai

Table 11 (*continued*)

No.	Word	Modern Chinese romanisation	Modern Chinese character	Ancient oracle-bone, bronze, or seal, form	K no.	Remarks	Reference
64	metal	*chin*	金	金	652	Perhaps a drawing of a mine shaft with a cover or a hill above, the dots indicating lumps of ore.	Wu Shih-Chhang
65	wood	*mu*	木	木	1212	Pictograph of a tree.	Hsü Shen
66	water	*shui*	水	水	576	Pictograph of running water.	Hsü Shen
67	fire	*huo*	火	火	353	Pictograph of flames (cf. 'Mr Therm' as an example of convergence).	Hsü Shen
68	earth	*thu*	土	土	62	Drawing of the phallic-shaped altar of the god of the soil.	Wang Kuo-Wei Kuo Mo-Jo Karlgren (9) Waley (7) Hopkins (29)
69	vapour, steam, subtle matter	*chhi*	氣	氣	517c	Pictograph of rising vapour. Cf. Gk. *pneuma* ($\pi\nu\epsilon\hat{v}\mu\alpha$). The 'rice' component was a late addition.	Hsü Shen
70	way (in which Nature works, or which Society ought to follow)	*tao*	道	道	1048	Picture of a head (symbolising a person), *heading* somewhere on a road, hence 'way', hence 'the right way'.	Hsü Shen
71	natural pattern, the veins in jade, to cut jade according to its natural markings; principle, order, organisation	*li*	理	理	978d	No bone or bronze forms known. The graph given is a suggested reconstruction only. 'Field' and 'earth' are certainly the phonetic, 'jade' is the radical (cf. p. 473 below). The character must have been invented relatively late.	Hsü Shen
72	natural regularity, rule, law	*tsê*	則	則	906 / 906	(*a*) Perhaps a drawing of a knife being used to carve a set of pictures, or inscribe a code of laws, upon a ritual cauldron of bronze or iron (cf. p. 559 below). (*b*) Alternatively, the reference may have been to the table-manners of aristocrats as the pattern or rule which others should follow. In this case the knife was an eating-knife and the pot a flesh-pot.	Hsü Shen Chu Chün-Shêng Wu Shih-Chhang

	Meaning		Character	Ref.	Description	Authority
73	measured division, limit, bound, law	*tu*	度	801	No bone or bronze forms known. Drawing of which the essential part is the hand below. The hand (and arm) was one of the most important standards of measurement in ancient times.	Tuan Yü-Tsai
74	method, model, to model, mould, law (especially human positive law)	*fa*	法 / 灋 (old form)	642 *k, l, m*	The original form of this word combined 'water' with 'to go away', and *chai*, a legendary one-horned bull or unicorn. This animal was supposed to gore the guilty party in an ordeal at law before the altar of the god of the soil. Evil was thus driven out (made to go away), if indeed the unicorn or some other animal did not afterward play the part of a scapegoat itself; see Granet (1), pp. 141 ff. As late as the Han time there are instances of offenders being sent to fight with wild beasts in arenas (e.g. *Chhien Han Shu*, ch. 54). The water component probably arose, not (as Hsü Shen thought) from the belief that 'the law should be as level as water', but from the lustrations, aspersions, libations, or sprinklings which accompanied the ancient ceremony. Perhaps a 'sink-or-swim' ordeal was also involved. On bullfights see Bishop (9).	Hsü Shen
75	rule, regulation, standard musical tone	*lü*	律	502	The left-hand element is half of the crossroads pictograph (no. 14 above), i.e. a street, so the semantic significance of the character was the public announcement of government orders or laws—since the right-hand element depicts a hand holding a writing-brush. Hence the meaning of standardisation. Somewhat later the character came to be connected with the musical tones of the standard pitch-pipes. No bone or bronze forms known. Another etymology is given on p. 551 below.	Hsü Shen, Wu Shih-Chhang
76	virtue, power, property, *mana*	*tê*	德	919 *k*	The drawing combines the left-hand side of the crossroads pictograph (no. 14 above) with the primitive anatomical representations of the eye and the heart. The two latter certainly refer to seeing and thinking respectively. The former certainly refers to the social matrix. Hence the original meaning of this word was probably closely analogous to that of *mana* and *virtus*; the 'magnetic' power possessed by a leader of men whether priest, prophet, warrior or king, who came, saw, reflected, and conquered. The word *virtus* also first had to do with *man* in the fullest sense (*vīra*, hero). Hence, by extension, the *mana* or numinous quality of certain inanimate objects. Later, the 'virtues' of herbs and stones. Or of the Tao.	Wu Ta-Chhêng

Table 11 (*continued*)

No.	Word	Modern Chinese romanisation	Modern Chinese character	Ancient oracle-bone, bronze, or seal, form	K. no.	Remarks	Reference
77	ceremonial, *mores*, ethical social behaviour, natural law (juristic)	*li*	禮		597	Drawing of a ritual vessel containing two pieces of jade. Combined (especially in writing later than the bone or bronze inscriptions) with the radical meaning 'sign, signify, show, inform, deity, divinity, religious'. Some have thought that this was a drawing of the long and short divining sticks laid out (Karlgren (1), no. 553). But others (Kuo Mo-Jo) believe, very justifiably, that it is a disguised form of the phallic symbol (cf. no. 68 above).	Wang Kuo-Wei
78	number, to count, to calculate	*shu*	數		123*r*	No bone or bronze forms of this character are known. Here the radical is placed, unusually, to the right, and signifies action (cf. nos. 7, 11 above). The phonetic to the left (*li*) has a female figure below and some queer head-dress above. But whatever it meant, it is irrelevant to the semantic connotation of the whole, which originally signified 'frequent', and so by extension came to be used for number, and sums in which numbers frequently reappeared. Therefore an abstract concept symbol.	Hsü Shen
79	art, device, mystery, technique, process	*shu*	術		497*d*	The drawing has a glutinous millet plant (*Panicum miliaceum*) in the centre, but it is acting purely as a phonetic. The radical is the crossroads pictograph (no. 14 above). Abundant references in writings of the Han and earlier show that the original meaning of the word was 'roads' or 'streets', and this usage continued down to the +5th century. Just as in English we speak of 'ways and means', so this word gradually acquired the specific connotation of 'right way of doing something', hence 'correct technique'. *Tao* underwent a parallel evolution from the concrete to the figurative.	Hsü Shen
80	to count, reckon, calculate, compute	*suan*	算		174	No bone or bronze form known. Though some written forms resemble an abacus, this probably dates them as Han. The older form showed a drawing like *wang*, king, which almost certainly depicts, not jade (as Hsü Shen thought), but a pattern of bamboo counting-rods (see Sect. 19*f*). The whole is thus topped by the bamboo radical.	Wu Shih-Chhang

the modern pronunciation of its Chinese equivalent, and the character itself, then comes its bone, bronze, or seal form,[a] with the numerical reference to Karlgren (1), followed by a brief explanation of the archaic significance, if any. The last column gives a reference to the names of the Chinese scholars who were the first to give the accepted interpretation.[b] For comparison, recourse may be had to the contributions of western sinologists.[c]

When one surveys the material assembled in the Table, one finds that the fundamental terms necessary for the beginnings of science were formed much as might be expected, given the ideographic principle. Among these characters[d] only two of the simplest (nos. 20 and 21) may be regarded as pure geometrical symbolism. Twenty-six are drawings of non-human natural objects, of which eleven are biological and fifteen inanimate or cosmological. The human body and its parts account for twenty-two, and (as would be natural among any primitive people) seven of these are connected with the sexual and generative functions. Human actions frequently occur. Five characters are based on paths and motion along them, and twenty-three on tools and techniques of various kinds, including ploughing, weaving, basketmaking, brushing, signalling and computing. Of these techniques, ritual actions, whether of offering sacrifices or of dancing, give five characters. The technological total, including communications, thus wins the contest with twenty-eight characters. The circumstances of social life as such account for six. At least eight are borrowed homophones, and three or four are symbols for abstract concepts.[e] One alone still defeats archaeological analysis. Doubtless this distribution would change somewhat if a larger number of characters were analysed,[f] but it will suffice for the present purpose to show how, from the daily round of primitive life, ideographs were developed which ultimately acquired quite abstract meanings. Thus was furnished the technical terminology for proto-scientific and scientific thinking and experimentation. These prefatory pages aimed at nothing more.

[a] Bone forms are given in the table wherever possible, and failing one of these, a bronze rather than a seal form.

[b] Since this subject is a study in itself, and peripheral to the main theme of the present book, exact bibliographical data are omitted.

[c] Such as Chalfant (1) and Wieger (1). L. C. Hopkins devoted a lifetime to the study of the bone and bronze characters, but his etymological theories were liable to be rather whimsical. What he thought about the characters dealt with may be found in the following papers: Hopkins (3, 5–8, 10–15, 19, 22, 26).

[d] When two possible explanations of a single character had to be given, both were counted in the reckoning which follows. Borrowed homophones are also listed twice.

[e] Negative (no. 2b), fullness (no. 19), fewness (no. 49), and frequency (no. 78).

[f] It might be well worth while to construct a similar table for all the operative grammatical and other words which have been assembled in the vocabulary of 'Basic' Chinese.

(c) THE SCHOOL OF NATURALISTS (*YIN-YANG CHIA*), TSOU YEN, AND THE ORIGIN AND DEVELOPMENT OF THE FIVE-ELEMENT THEORY

The moment has now come to explain the origin and meaning of the fundamental theories of the Two Forces in the universe (Yin and Yang), and the Five Elements. It would, strictly speaking, be proper to discuss the former before the latter, since theoretically they lay, as it were, at a deeper level in Nature, and were the most ultimate principles of which the ancient Chinese could conceive. But it so happens that we know a good deal more about the historical origin of the Five-Element theory than about that of the Yin and the Yang, and it will therefore be more convenient to deal with it first.[a] It goes back to a thinker of whom we have not yet had occasion to say much, though he may be considered the real founder of all Chinese scientific thought, namely, Tsou Yen.[1] The exact dates of his birth and death are not known, but he must be placed approximately between −350 and −270. If he was not the sole originator of the Five-Element theory, he systematised and stabilised ideas on the subject which had been floating about, especially in the eastern seaboard States of Chhi and Yen, for not more than a century at most before his time. These datings of course contradict the traditional view, which accepted as genuine the whole of the Hung Fan chapter of the *Shu Ching* (Historical Classic) (see below, p. 242), and so placed the origins of the theory in the early Chou time, but they are nevertheless those which modern research indicates to be correct.

Since it is incumbent on us to follow with close attention every surviving footstep of this figure so venerable for historians of science, I reproduce part of the 74th chapter of the *Shih Chi*, which Ssuma Chhien entitled the biographies of Mencius and of Hsün Chhing, though in fact most of it deals with Tsou Yen.

The State of Chhi had three scholars named Tsou. The first of these was Tsou Chi,[2] whose lute-playing affected King[b] Wei[3] so much that he rose to a position in the administration of the State, was enfeoffed as the Marquis Chhêng, and received the seal of minister. He lived prior to Mencius.[c]

The second was Tsou Yen,[1] who came after Mencius. He saw that the rulers were becoming ever more dissolute and were incapable of valuing virtue, through which alone they might incorporate in themselves (the principles in) the Ta Ya songs (of the *Shih Ching*) and diffuse them among the common people. So he examined deeply into the phenomena of *the increase and decrease of the Yin and the Yang* (*Yin Yang hsiao hsi*[4]), and wrote essays totalling more than 100,000 words about their strange permutations, and about the cycles of the great sages from beginning to end. His sayings were vast and far-reaching and not in

[a] A great discussion on the history of both these theories is contained in the second part of vol. 5 of *KSP*; many scholars, including Liang Chhi-Chhao and Ku Chieh-Kang, contributed to it.
[b] Of Chhi (r. −377 to −331).
[c] So he must have been born about −400 or a little before.

[1] 騶衍   [2] 騶忌   [3] 威   [4] 陰陽消息

accord with the accepted beliefs of the classics. First he had to examine small objects, and from these he drew conclusions (*thui*[1]) [a] about large ones, until he reached what was without limit. First he spoke about modern times, and from this went back to the time of Huang Ti.[b] The scholars all studied his arts. Moreover, he followed the great events in the rise and fall of ages, and by means of their omens and (an examination into their) systems, extended (*thui*[1]) his survey (still further) backwards to the time when the heavens and the earth had not yet been born, (in fact) to what was profound and abstruse and impossible to investigate.

He began by classifying China's notable mountains, great rivers and connecting valleys; its birds and beasts; the fruitfulness of its waters and soils, and its rare products;[c] and from this extended (*thui*[1]) his survey to what is beyond the seas, and men are unable to observe.

Then starting from the time of the separation of the heavens and the earth,[d] and coming down, *he made citations of the revolutions and transmutations of the Five Powers* (Virtues) (*wu tê*[2]), arranging them until each found its proper place and was confirmed (by history).

Tsou Yen maintained that what the Confucians call the 'Middle Kingdom' (i.e. China) holds a place in the whole world of but one part in eighty-one. China, he named the Spiritual Continent of the Red Region (*chhih hsien shen chou*[3])....

> There follows a paragraph which is considered to be the actual words of Tsou Yen. It is therefore reserved for a few pages further on.

All his arts were of this sort. Yet if we reduce them to fundamentals, they all rested on the virtues of human-heartedness, righteousness, restraint, frugality, and the practice of the association of ruler with subject, superior with inferior, and the six relationships. It was only the beginning (of his doctrines) which was exaggerated and unbalanced (*lan*[4]).[e]

Kings, dukes and great officials, when they first witnessed his arts, fearfully transformed themselves, but later were unable to practise them. Thus Master Tsou was highly regarded in Chhi. He travelled to Liang, where King Hui[5][f] went out to the suburbs of the city to welcome him, and acted towards him with all the punctilio of a host towards a guest. He went to Chao, where the Prince of Phing-Yuan,[6][g] walking on one side (of the road), personally brushed off the dust from his seat. He went to Yen, where King Chao[7][h] acted as his herald, (sweeping the road with a) broom, and asked to take the seat of a disciple so as to receive his instruction.

Here in a palace built for him at Chieh-Shih,[8][i] the King went personally to listen to his teaching. Here Tsou Yen wrote the *Chu Yün*[9] (The Mastery of Time's Mutations—a book now lost). In all his travels among the feudal lords he received honours of this sort.[j]

[a] Note the use of the word which the Mohists employed as a technical term (cf. p. 183); extension (induction?).

[b] The legendary Yellow Emperor. Significant because always a favourite Taoist hero.

[c] None of this has come down to us. But we do have, in the *Chi Ni Tzu* fragments (see below, p. 275), some inventories of natural products, collected perhaps with an alchemical purpose, which may be of the same date and the same school.

[d] Doubtless a reference to the centrifugal cosmogony, as in the *Lieh Tzu* book, see on, p. 372.

[e] This is just a piece of apologetic on the part of Ssuma Chhien who feels he has to pretend, at least, to vindicate Tsou Yen's Confucian orthodoxy. [f] Of Wei (r. −370 to −319).

[g] Died −252. His biography (ch. 76 of the *Shih Chi*) has been translated by Pfizmaier (26). Cf. p. 185. [h] R. −311 to −278.

[i] Somewhere on the coast of Hopei between modern Taku and Shanhaikuan.

[j] Ch. 74, pp. 1b–3a, tr. Bodde in Fêng Yu-Lan (1), vol. 1, p. 159, and Dubs (5), slightly mod.

[1] 推   [2] 五德   [3] 赤縣神州   [4] 濫   [5] 惠   [6] 平原君

[7] 昭   [8] 碣石   [9] 主運

Compare this with Confucius who nearly starved to death in Chhen and Tshai, or Mencius who was surrounded with difficulties in Chhi and Liang—what a difference! Wu Wang, conquering Chou[1] (the last Shang emperor) with the doctrines of human-heartedness and righteousness, became emperor, while on the other hand Pai I starved rather than eat the bread of the Chou.[2] Duke Ling of Wei asked Confucius about military matters, but he would not answer. King Hui of Liang, planning to attack Chao, asked Mencius about it, but he (turned the question aside, and advised) a peaceful excursion to Fên. Such examples show that these men did not trim their ideas to suit the desires of worldly rulers. Yet (carrying this too far) is like trying to put a square peg in a round hole. Some say that I Yin (condescended to) carry cauldrons about in order to encourage Thang to take the throne, and that Pai Li-Hsi once fed the cart-oxen, yet Duke Mu, by employing him, became hegemon of the feudal lords. If ruler and adviser can agree beforehand (i.e. if the ruler's confidence is once won), then rulers can be brought into the Great Tao. So although the words of Tsou Yen were not disciplined, it seems that they played the same role as the cauldrons and the oxen.

Starting from Tsou Yen there were the Chi-Hsia[3] Academicians such as Shunyu Khun,[4] Shen Tao,[5a] Huan Yuan,[6] Chieh Tzu,[7] Thien Phing[8b] and Tsou Shih.[9] They all wrote books dealing with State affairs in order to influence the rulers. One cannot mention them all.

[There follow two pages on Shunyu Khun and Shen Tao—here omitted.]

Tsou Shih was one of the Tsou family of Chhi. He accepted the arts of Tsou Yen, and wrote essays about them. These were much appreciated by the King of Chhi, who gave to Shunyu Khun and all the others the title of Ta Fu (Minister of State). He built mansions for them along a broad street, with high gates and large halls, in which they were lodged with every manifestation of respect. And the guests of the other feudal Kings and Dukes said that Chhi was able to attract all the great scholars of the world.

(For example) there was Hsün Chhing of Chao, who at the age of fifty first came to spread abroad his teachings in Chhi (as a Chi-Hsia Academician). The science of Tsou Yen was grandiose and his reasoning eloquent. The writings of Tsou Shih were accomplished but difficult to put into practice. As for Shunyu Khun, if one lived with him a long time one sometimes got good sayings. So the people of Chhi praised the two Tsous, saying, 'For talking about Nature there is (Tsou) Yen; for carving dragons (i.e. making literary embellishments on Tsou Yen's doctrines) there is (Tsou) Shih; and for pithy sayings there is (Shunyu) Khun.'[c]

This long passage from Ssuma Chhien is extremely instructive. It gives one the impression that the Yin-Yang Chia[10] (the School of the Yin-Yang Experts), as the followers of Tsou Yen were afterwards called, was of a character rather different from any of the schools which we have so far examined, though closest to the Taoists.[d] Unlike the Taoists, however, the Naturalists (if we may adopt this henceforth as a convenient and suitable term for the school) did not shun the life of courts and kings; on the contrary, it would seem that they confidently felt themselves to be in possession

---

[a] A legalist. Cf. Fêng Yu-Lan (1), vol. 1, p. 153.
[b] Alternatively Thien Phien. Cf. Fêng Yu-Lan (1), vol. 1, pp. 132, 157.
[c] Ch. 74, pp. 3a–5b, tr. auct.
[d] Their exact relation to the Taoists has been examined in an interesting paper by Hsieh Fu-Ya (1).

[1] 紂   [2] 周   [3] 稷下   [4] 淳于髡   [5] 慎到   [6] 環淵
[7] 接子   [8] 田駢   [9] 騶奭   [10] 陰陽家

of certain facts about the universe which rulers could neglect only at their peril. If Tsou Yen had had the 'know-how' of the atomic bomb in his possession he could hardly have faced the rulers of the States with a steadier eye. For a brief period, then, we see this proto-science attaining great social importance and prestige, and the parallel with our own times is not so far-fetched as it might seem, since evidence will in a moment be forthcoming that the 'arts' of the Naturalists were in all probability by no means mere verbal speculation.

The visits of Tsou Yen to the feudal courts are certainly historical, and there is no reason to doubt the statements of the *Shih Chi* as to his welcome.[a] As for the Chi-Hsia Academy,[b] located outside one of the gates of the capital city of Chhi, Tsou Yen seems to have been the oldest member of it, and perhaps, as a citizen of that State, had inspired King Hsüan to found it. We have already noted (Sect. 5c) its great historic interest as roughly contemporary with the famous academies of ancient Greece. Besides the sophist Shunyu Khun, and the other men, mostly Taoists, mentioned in the above passage, there were also to be found among its members Mohists such as Sung Hsing[1] (or Khêng[2]) and, for a time, the greatest of Confucians, Mencius himself. It is thought that Chuang Chou may also have been of their company. The Academicians, whose official title indicated purely advisory duties, wore a special flat cap, the Hua Shan cap, and presumably special robes which went with it. What would one not give for a verbatim record of their discussions! The close association of the schools in this way must have led to borrowings of technical terms; we have noted the use of the Mohist term for induction in Ssuma Chhien's account of Tsou Yen's methods.

A point of interest which emerges from this account is that its writer, while critical of Tsou Yen to some extent, seeks to rehabilitate him from the Confucian point of view. However fanciful his doctrines about Nature may have been, says Ssuma Chhien, they ended in the sound inculcation of the virtues of human-heartedness, righteousness, etc. And in a later paragraph he suggests that Tsou Yen's teachings about the world of Nature were no more than entertainments intended to arouse the interest of the feudal princes, and win their confidence, so that later on he could proceed to keep them in the right way of Confucian good behaviour. This suggests that by the time of Ssuma Chhien, the Naturalist school having died out as an organised group, all its practical arts had passed over to the Taoists, while its five-element theories had become common property which the Confucians shared equally with everyone else. It seems clear that Ssuma Chhien did not understand the interest which the Naturalists had had in Nature, and thus put forward apologetics where none were called for. Of course he also had to account for the fact that Confucius had been

---

[a] His visit to Yen is also described in ch. 34 (Chavannes (1), vol. 4, p. 145). His visit to Liang (the capital of Wei) is also described in ch. 44 (Chavannes (1), vol. 5, p. 158).

[b] Description in Duyvendak (3), pp. 73 ff.; Chavannes (1), vol. 5, pp. 258 ff., who translates the relevant passage in *Shih Chi*, ch. 46.

[1] 宋鈃   [2] 牼

a failure from the worldly point of view while Tsou Yen had great success. Hence Tsou Yen's doctrines had to be made out, in some way or other, to have been really Confucian.

What, then, was the political dynamite which the Naturalists thought that they had discovered, and of the importance of which they were able to persuade the feudal princes? It will be best to give it in the words of Tsou Yen himself. In Ma Kuo-Han's enormous collection of fragments, we find a few pages[a] where are brought together all that remains of the books which were written by the Master or his immediate disciples, and which were still known to the Han bibliographers as the *Tsou Tzu*[1] (Book of Master Tsou) and the *Tsou Tzu Chung Shih*[2] (Master Tsou's Book on Coming into Being and Passing Away). Since they have not all hitherto been available in a Western language, we give them in full, with a few comments interspersed.

(1) What the Confucians call the Middle Kingdom (i.e. China) holds a place in the whole world of but one part in eighty-one. China is called the Spiritual Continent of the Red Region, and within it there are the nine provinces (*chou*[3]) which were those laid out by Yü the Great.[b] But these cannot be numbered among the real continents (i.e. *chou*[3] in its broader sense). The Middle Kingdom is only one of a total of nine continents, and these are the real Nine Chou.[3] Around each of these is a small encircling sea, so that men and beasts cannot pass from one to the other. But these Nine Chou form one division and make up one Great Continent. There are (again) nine such Great Continents, and around their outer edge is a vast ocean (*ta ying hai*[4]) which encompasses them and stretches to the bounds where the heavens and the earth meet.[c]

> Some have been tempted to see in this world-view, bold indeed for the −4th century, a direct influence of foreign thought, especially Indian (e.g. Conrady, 1). But perhaps it was no more than a resolute conviction, doubtless based on culture-contacts unknown to us, that the Chinese *oikoumene* was not the centre of the universe, and that there were other cultures in existence. The nine continents are often mentioned later, as in *Huai Nan Tzu*.[d]

(2) In Spring, fire should be kindled by twirling the fire-drill in elm and willow wood. In Summer, the wood of the jujube-tree (Chinese date) and the apricot tree should be used. In Autumn the oak and the *yu*[5]-tree[e] should be used. In Winter one must make use of the wood of the *huai*[6]-tree[f] and the *than*[7]-tree.[g]

> These admonitions probably have regard to the varying nature of the woods, how far hygroscopic, etc.

---

[a] *YHSF*, ch. 77, pp. 16a ff.

[b] Legendary emperor and hydraulic engineer who mastered the floods (cf. pp. 117, 119 and Sect. 28f below).

[c] From *Shih Chi*, ch. 74, p. 2a. Note the contact of the peripheral sky with the rim-ocean, characteristic of the *Kai Thien* cosmology (see Sect. 20d below).   [d] Ch. 3, p. 10b.

[e] B II, 537, probably a kind of oak, perhaps *Quercus crispula*.

[f] B II, 546, *Sophora japonica*, which gives a yellow dye.

[g] B II, 540, probably *Caesalpinia* spp., yielding a fine-grained hard wood, or *Dalbergia hupeana*, R 381. Sandalwood, though often used for this term in Legge's translations, is not quite the right equivalent. This passage is taken from Chêng Hsüan's commentary on the *Chou Li*; Biot (1), vol. 2, p. 195.

¹ 騶子   ² 騶子終始   ³ 州   ⁴ 大瀛海   ⁵ 櫄   ⁶ 槐   ⁷ 檀

(3) Provisions for administration and education, recommendations of complicated rites and ceremonies, or alternatively of simplicity in them; all are remedies (suitable for particular ages). They may be practicable during some periods, but with the passage of time they may have to be abandoned. As circumstances change, so these things must change. Those who insist on adhering to particular arrangements and will not change with the times, will never attain perfection in the art of ruling.[a]

> Here we have clearly the flexibility of Taoism and its appreciation of long-term social changes, as opposed, for example, to Confucian orthodoxy (as we have seen, in Hsün Tzu, p. 83) which denied change and adhered with intense conservatism to ancient customs.

(4) The State of Chhi sent Tsou Yen to the Prince of Phing-Yuan at Chao, where Kungsun Lung (the logician) and his pupil Chhiwu Tzu[1] were discussing about whether a 'white horse is a horse, or not', and so on. They asked Master Tsou about it. But Master Tsou refused to discuss it, saying: 'To speak about the Five Conquerors[b] and the Three Completions (*wu shêng san chih*[2])[c] is the type of discussion proper to the mundane world. But distinguishing different kinds or species (categories) to prevent them hurting (i.e. overlapping with) each other, showing up so-called heretical ideas in order to prevent their getting confused with so-called true doctrine, expressing thoughts by means of "universals" (*chih*[3])[d] and particulars, finding artful words with which to contradict others, making clever analogies (*pi*[4])[e] to upset each other's ideas, and stimulating people to argue until they themselves no longer know what they should think—all these are harmful to the Great Tao. And this jousting with words cannot but bring harm to the lords also.'[f]

> This passage, if, as probably, authentic, is of the greatest interest, since it confirms all that has been said above about the disinclination of the schools which were interested in Nature to collaborate in the efforts of the Mo-Ming thinkers towards building up a logic suitable for science. Tsou Yen cannot see how logic is going to benefit the Taoists and the Naturalists.

(5) I stood at the city of Min[5] and looked towards the capital of Sung.[6][g]
> Significance obscure; perhaps strategic?

(6) The four corners are not quiet.[h]
> This sentence seems to refer to political dangers, but may equally well be a statement concerning the motion of the earth. Cf. below, Sect. 20*d*, concerning the theory of the 'four displacements'.

[a] From *Chhien Han Shu*, ch. 64B, p. 1*a*, biography of Yen An.
[b] Undoubtedly the Five Elements.
[c] The meaning of this is uncertain. I suggest the three positions of Yin and Yang; (*a*) when Yin is at its position of completest dominance; (*b*) when Yang is at its position of completest dominance; and (*c*) when they are precisely equally balanced. In other words, the top and bottom of a wave curve, and the cross-over point. Cf. Sect. 26*b*.
[d] A very important technical term of the School of Logicians, cf. p. 185 above.
[e] Another technical term, this time of the Mohist logicians, cf. p. 183 above.
[f] From the lost *Pieh Lu*[7] (Bibliographical Companion) of Liu Hsiang, of which a fragment survives here in the form of a commentary to ch. 76, p. 5*b*, of the *Shih Chi*, i.e. the biography of the Prince of Phing-Yuan.
[g] From the *Shui Ching Chu*[8] (Commentary on the Water Classic), ch. 8, p. 20*a*.
[h] From the *Wên Hsüan*[9] (Collection of Literature) ed. Hsiao Thung,[10] *c.* +530, ch. 6, p. 3*b*.

[1] 綦毋子   [2] 五勝三至   [3] 指   [4] 辟   [5] 緡   [6] 宋
[7] 別錄   [8] 水經注   [9] 文選   [10] 蕭統

(7) The Five Elements dominate alternately. (Successive emperors choose the colour of their) official vestments following the directions (so that the colour may agree with the dominant element).[a]

(8) Each of the Five Virtues (Elements) is followed by the one it cannot conquer. The dynasty of Shun [b] ruled by the virtue of Earth, the Hsia dynasty ruled by the virtue of Wood, the Shang dynasty ruled by the virtue of Metal, and the Chou dynasty ruled by the virtue of Fire.[c]

(9) When some new dynasty is going to arise, Heaven exhibits auspicious signs to the people. During the rise of Huang Ti (the Yellow Emperor) large earth-worms and large ants appeared. He said, 'This indicates that the element Earth is in the ascendant, so our colour must be yellow, and our affairs must be placed under the sign of Earth.' During the rise of Yü the Great, Heaven produced plants and trees which did not wither in autumn and winter. He said, 'This indicates that the element Wood is in the ascendant, so our colour must be green,[d] and our affairs must be placed under the sign of Wood.' During the rise of Thang the Victorious a metal sword appeared out of the water. He said, 'This indicates that the element Metal is in the ascendant, so our colour must be white, and our affairs must be placed under the sign of Metal.' During the rise of King Wên of the Chou, Heaven exhibited fire, and many red birds holding documents written in red flocked to the altar of the dynasty. He said, 'This indicates that the element Fire is in the ascendant, so our colour must be red, and our affairs must be placed under the sign of Fire. Following Fire, there will come Water. Heaven will show when the time comes for the *chhi* of Water to dominate. Then the colour will have to be black, and affairs will have to be placed under the sign of Water. And that dispensation will in turn come to an end, and at the appointed time, all will return once again to Earth. But when that time will be we do not know.'[e]

Here in these last three fragments we have the essence of the half-scientific, half-political doctrine with which the Naturalists were able to frighten the feudal lords. The conception of the five elements itself was essentially a naturalistic, scientific one, and we shall shortly look more closely at it from that point of view, inquiring what exactly it implied, but Tsou Yen evidently extended it to the dynastic world, believing that every ruler or ruling house ruled only 'by the virtue of' one of the elements in the series. This provided, in effect, a theory for the rise and fall of ruling houses, bringing human affairs and their history under the same 'law' (though, so far

---

[a] From Ju Shun's commentary on the *Shih Chi*, ch. 28, p. 11a; cf. Chavannes (1), vol. 3, p. 328 ff.
[b] Legendary emperor.
[c] This was why, when Chhin Shih Huang Ti came to the throne of the unified empire, the Chhin dynasty was considered to have conquered by the power of Water, and its heraldry was black. See the following fragment. This paragraph is from the *Wên Hsüan*, ch. 59, p. 9b,
[d] The word used is *chhing*.[1] The primary meaning of this is 'blue', but there was a good deal of fluctuation in the colour terms in ancient Chinese (as in other ancient languages—cf. Homer). When used technically as the correlate of the element Wood, we take *chhing* as green. We shall return to the subject in Section 43 on physiology and vision.
[e] In fact it was the Han, since by then the five elements had each had their turn. This paragraph from the *Lü Shih Chhun Chhiu*, ch. 63 (vol. 1, p. 122), has also been translated by Fêng Yu-Lan (1), vol. 1, p. 161. All the translations given here are ours. There follows a very obscure prophecy, not translated.

[1] 青

as we know, that key-word was never used in this connection) as the phenomena of non-human Nature. The mechanism of both was the unvarying uniformity which came to be known as Mutual Conquest (*hsiang shêng*[1]), or Cyclical Conquest, wood overcoming earth, metal overcoming wood, fire overcoming metal, water overcoming fire, and earth overcoming water, at which point the cycle commenced all over again. All changes in human history were thus considered manifestations of the same changes which could be observed at the lower, 'inorganic' levels, and from which indeed the very conception of the elements had been derived. One may conjecture that the reason why the feudal lords found Tsou Yen's doctrines difficult to put into practice was that though they might be convinced of the truth of the theory which he and his school expounded with such conviction, it was rather difficult to ascertain exactly the elements by virtue of which they were ruling, and hence to take the necessary precautions. Moreover, whatever precautions they took, the cyclical mutations of Nature would continue on their course, so that no ruling house could hope to stabilise its position in perpetuity. Tsou Yen's 'discovery' was, we can see, but proto-science, yet its sociological interest lies in the fact that it was so widely and deeply believed, with the result that the success of the doctrines of the Naturalists, and later on of the Han Confucians, when they inherited them, reminds us somewhat of the political importance achieved by the natural sciences in our own time. And it is not easy to point to any very striking parallel in ancient Greece to the position of the Chinese Naturalists, however important the work of the pre-Socratics, Peripatetics and Alexandrians may be conceded to be as the foundation of modern science.

Tsou Yen's doctrine became crystallised, perhaps some time in the late −3rd century, in a short treatise known as the *Wu Ti Tê*[2] (The Virtues by which the Five Emperors Ruled). This is known to have been used by Ssuma Chhien,[a] but is probably not the same as the piece with the same title later incorporated in the *Ta Tai Li Chi*[3] (Record of Rites edited by the elder Tai)[b] and the *Khung Tzu Chia Yü*[4] (Table-Talk of Confucius),[c] though these embody the ideas. We know also of an imperial counsellor, Chang Tshang[5] (d. −142), who may have been important as a transmitter of the ideas of the Naturalist school under the early Han emperors.[d]

Although no doubt largely speculation, it is probable that the influence of Tsou Yen and his school rested on something more than that, and there is considerable reason to suspect that their 'arts' included astronomy and calendrical science. Thus in chapter 26 of the *Shih Chi* (on the calendar) Ssuma Chhien says:

Then the feudal kingdoms plunged into mutual wars; there were attacks and counter-attacks, rivalries of powerful princes, expeditions to rescue lords in distress, unions, treaties

---

[a] *Shih Chi*, ch. 1, p. 13a; Chavannes (1), vol. 1, p. cxliii.
[b] Ch. 62, tr. R. Wilhelm (6), p. 281. The book is +1st century, not −1st, as used to be supposed.
[c] Ch. 23. This is a book of the +3rd century, but compiled from earlier sources.
[d] *Shih Chi*, ch. 96, p. 1a. Chang Tshang was a mathematician and we shall meet him again in Sect. 19.

[1] 相勝   [2] 五帝德   [3] 大戴禮記   [4] 孔子家語   [5] 張蒼

and treacheries—in such a time who could find leisure to think about such things (as the calendar)? Tsou Yen was the only one who attained knowledge about the transmutations of the Five Virtues (Powers), and who discoursed on the differences between coming-into-being and passing-away, in such a manner as to make himself illustrious among the feudal lords.[a]

Moreover, there is much evidence which connects the Naturalists with the beginnings of alchemy. We have seen that Tsou Yen made lists of natural products, probably minerals, chemical substances and plants.[b] And there are two important texts which point to the alchemical interests of the school. The *Shih Chi* says:

From the time of (Kings) Wei[1] and Hsüan[2] of the State of Chhi[c] the disciples of Master Tsou discussed and wrote about the cyclical succession of the Five Powers. When (the King of) Chhin became (the First) Emperor (in −221), people from Chhi sent in memorials (bringing these theories to his notice). And the First Emperor (Chhin Shih Huang Ti) chose them and gave them employment. Moreover from first to last Sung Wu-Chi,[3] Chêng Po-Chhiao,[4] Chhung Shang[5] and Hsienmên Kao[6] were all[d] people from (the State of) Yen who practised the method of (becoming) immortals by the use of magical techniques, so that their bodies would be etherealised and metamorphosed by some trans-mutation (*hsing chieh hsiao hua*[7]).[e] For this they relied upon their services to the gods and spirits.

Tsou Yen was famous among the feudal lords (for his doctrine) that the Yin and the Yang control the cyclical movements of destiny. The men who possessed magical techniques, and who lived along the sea-coast of Yen and Chhi, transmitted his arts, but without being able to understand them. From this time on one cannot count the constantly increasing number of those persons who performed deceptive wonders, flatteries, and illicit practices.[f]

Then beginning with (Kings) Wei and Hsüan (of Chhi) and (King) Chao of Yen, people were sent out into the ocean to search for (the fairy isles of) Phêng-Lai,[8] Fang-Chang,[9] and Ying-Chou.[10] These three divine (island) mountains were reported to be in the Sea of Po,[11] [g] not so distant from human (habitations), but the difficulty was that when they were almost reached, boats were blown away from them by the wind. Perhaps some succeeded in reaching (these islands). (At any rate, according to report) many immortals (*hsien*) live there, and the drug which will prevent death (*pu ssu chih yao*[12]) is found there. Their living creatures, both birds and beasts, are perfectly white, and their palaces and gate-towers are made of gold and silver. Before you have reached them, from a distance they look like clouds, but (it is said that) when you approach them, these three divine mountain-islands sink below the water, or

---

[a] P. 4*a*, tr. Chavannes (1), vol. 3, p. 328, eng. auct.

[b] See on, Sects. 25, 33, for the lists in the *Chi Ni Tzu*, which may be contemporary.

[c] The two reigns covered −377 to −312.

[d] It is doubtful whether any of these four were historical personages; all were mentioned by Han writers as former 'immortals'. Yet they may well have been magician-Naturalists of Yen contemporary with, or earlier than, Tsou Yen. We have already suggested (p. 133) that *hsien-mên* means shaman.

[e] Note the expressions used in relation to what was said above, p. 141.

[f] This hints at alchemy.

[g] The present Gulf of Chih-Li.

[1] 威	[2] 宣	[3] 宋毋忌	[4] 正伯僑	[5] 充尚	[6] 羨門高
[7] 形解銷化		[8] 蓬萊	[9] 方丈	[10] 瀛洲	[11] 勃海
[12] 不死之藥					

else a wind suddenly drives the ship away from them. So no one can really reach them. Yet none of the lords of this age would not be delighted to go there.[a]

It seems fair to conclude from this authentic and fascinating passage that Tsou Yen's Naturalists were not only at the origin of the semi-Confucian Han speculation about the five elements, but also in close contact, if not identical with, the magical technicians of the seaboard States who were afterwards so important at the court of Han Wu Ti, and whom we shall come across again and again, in relation, for example, to the history of chemistry and of magnetism.

The second passage, vital for Tsou Yen's function in this complex, occurs in the *Chhien Han Shu*[b] and concerns events of a rather later date. I shall reserve most of it for the section on Chemistry, since it concerns the attempt of the Han Confucian, Liu Hsiang,[1] then a young man, to make gold artificially in −60. But it reveals clearly the transmission of secret writings or perhaps oral traditions of the school of Naturalists to the circle surrounding Liu An, the Prince of Huai Nan (Huai Nan Tzu).[c]

Liu Hsiang presented to the throne (a memorial concerning) the matter of reviving the arts and techniques of the divine immortals. Now (the Prince of) Huai Nan had had in his pillow (for safe-keeping) certain writings entitled *Hung Pao Yuan Pi Shu*[2] (The Secret Book of the Precious Garden). These writings told about divine immortals and the art of inducing spiritual beings to make gold, together with Tsou Yen's technique for prolonging life by a method of repeated (transmutation) (*chhung tao*[3]). People of that age had not seen these writings, but Liu Tê,[4] the father of Liu Hsiang, had, in the time of the Emperor Wu, investigated the case of the (Prince of) Huai Nan, and (after his downfall) had secured his books....[d]

It is, of course, true that alchemical writings of the −2nd century could have been fathered on Tsou Yen, according to a common custom of alchemists in all ages,[e] but this is not a necessary supposition. It is much more probable that Chinese alchemy (older, as we may see, than that of any other part of the world) began in the School of Naturalists during the −4th century.[f]

[a] Ch. 28, pp. 10b–11b, tr. Dubs (5). Something has already been said about these, quite historical, expeditions to discover the isles of the eastern sea, and we shall meet them again in an unexpected, cartographic, connection (Sect. 22). The passage is also found in *Chhien Han Shu*, ch. 25A, pp. 10a–11a. Translated also by Chavannes (1), vol. 2, p. 152, and vol. 3, p. 435.

[b] Ch. 36, p. 6b.   [c] Cf. p. 83.

[d] Tr. Dubs (5). Actually it cannot have been Liu Tê who investigated the case of Huai Nan Tzu, for Liu Tê cannot have been born before −126, and the downfall of the Prince took place in −123. Presumably it was Liu Hsiang's grandfather, Liu Pi-Chiang (−164 to −85).

[e] We have already seen a probable example of this in the books attributed in later ages to Mo Tzu, p. 202.

[f] I do not feel that Dubs (5) is convincing in connecting this with Tsou Yen's geographical views which placed China in the south-eastern corner of the world, and thence deducing foreign influence. The question involves the earliest origin of the idea of the immortality-conferring drug, and will be discussed in Section 33 on alchemy and chemistry below.

[1] 劉向   [2] 鴻寶苑祕書   [3] 重道   [4] 劉德

An early −4th-century origin of the five-element theory is in agreement with epigraphic evidence. Chhen Mêng-Chia (*1*) has drawn attention to what is believed to be the earliest mention of the elements, in an inscription on a jade sword-handle which may be dated not long after −400, and which is thought to be of Chhi State provenance. The inscription is in the form of an epigram, somewhat reminiscent of what one would expect to find in a Chinese counterpart, if there were one, of the Greek Anthology. It runs as follows:

(When the) *chhi* of the elements (is) settled, condensation (i.e. corporeality) (is brought about); this condensation (acquires) a spirit; (after it has acquired) a spirit it comes down (i.e. is born); (after it has) come down it (becomes) fixed (i.e. complete in all its parts); (after it has) become fixed (it acquires) strength; with strength (comes) intelligence; with intelligence (comes) growth; growth (leads to) full stature; and with full stature (it becomes truly) a Man.

(Thus) Heaven supported him from above, Earth supported him from below; he who follows (the Tao of Heaven and Earth) shall live; he who violates (the Tao of Heaven and Earth) shall die.[a]

This beautiful epigram allows us to visualise perhaps the mentality of the warriors of Chhi and Yen (and later of the unified empire) who were convinced that following the teachings of the Naturalists they were 'on the side of the angels', or as we might say, 'on the side of the forces of history'. It was not to be the last time in history that an attempt at a scientific view of the world would strengthen the will and courage of the soldiers of specific human social organisms.

Another *locus classicus* for the five-element theory is the Hung Fan[1] (Great Plan) chapter of the *Shu Ching* (Historical Classic).[b] This canonical work, traditionally ascribed to the early centuries of the first millennium before our era, is now considered a patchwork (like so many other ancient texts) from pieces of very varying age. That portion at least of the Hung Fan which treats of the five elements must be regarded as a Chhin interpolation of −3rd century or at least not older than Tsou Yen.[c] The passage[d] begins by saying that the doctrine of the five elements was one of the parts of the ninefold Great Plan which Heaven withheld from Kun (cf. p. 117 above) and gave to Yü the Great. These parts are, with the exception of one other ('the harmonious use of the five dividers of time', *hsieh yung wu chi*[2]), which is clearly astronomical, all concerned with human and social qualities and relations.

---

[a] Tr. auct. The large number of bracketed phrases in this epigram is due to the fact that the Chinese contains no verbs at all.

[b] Ch. 24 (in the *Chou Shu* division).

[c] I follow Maspero here (*2*), p. 439. There is another passage in the *Shu Ching* where the five elements are spoken of, less interesting, but not unimportant, as grain seems to be added as a sixth element. It occurs in the Ta Yü Mo chapter (ch. 3, in the *Yü Shu* division; Medhurst (*1*), p. 44; Legge (*1*), p. 47). Cf. Forke (*6*), p. 227. But the chapter is considered a +4th-century interpolation.

[d] Tr. Medhurst (*1*), p. 198; Legge (*1*), p. 140.

[1] 洪範   [2] 協用五紀

They are termed 'invariable principles' (*i lun*[1]). The character *i*[2] (K 1237, *c, g*) is therefore of some interest as an early approximation to the idea of a natural law (in the scientific sense), but it is of very infrequent use in this connection. Though it can mean, as here, natural norm, rule or law, it derives from oracle-bone graphs showing a ritual vessel containing pork and rice, and garlanded with silk, being held up by two hands. It therefore presumably originated in connection with some liturgical rubric.

Now the description of the five elements here gives us a little insight into the manner in which the Naturalists conceived of them. The translation is difficult owing to the laconic character of the phrases. For example, the text says *shui yüeh jun hsia*,[3] lit. 'water called soak descend (or down, or below)'. So one does not know whether to write 'water is said to be that which soaks, etc.' or 'that which is used for soaking...' or 'that which has a soaking quality', and so on. Yet it will be seen that a meaning emerges. The text says:

K1237g

As for the five elements, the first is called Water, the second Fire, the third Wood, the fourth Metal, and the fifth Earth. Water (is that quality in Nature) which we describe as soaking and descending. Fire (is that quality in Nature) which we describe as blazing and uprising. Wood (is that quality in Nature) which permits of curved surfaces or straight edges. Metal (is that quality in Nature) which can follow (the form of a mould) and then become hard. Earth (is that quality in Nature) which permits of sowing, (growth), and reaping.

That which soaks, drips and descends causes saltiness. That which blazes, heats and rises up generates bitterness. That which permits of curved surfaces or straight edges gives sourness. That which can follow (the form of a mould) and then become hard, produces acridity. That which permits of sowing, (growth), and reaping, gives rise to sweetness.[a]

All this suggests that the conception of the elements was not so much one of a series of five sorts of fundamental matter (particles do not come into the question), as of five sorts of fundamental processes. Chinese thought here characteristically avoided substance and clung to relation. We might therefore construct a table somewhat as follows:

WATER	soaking, dripping, descending, (dissolving?)	liquidity, fluidity, solution	saltiness
FIRE	heating, burning, ascending	heat, combustion	bitterness
WOOD	accepting form by submitting to cutting and carving instruments	solidity involving workability	sourness
METAL	accepting form by moulding when in the liquid state, and the capacity of changing this form by re-melting and re-moulding	solidity involving congelation and re-congelation (mouldability)[b]	acridity
EARTH	producing edible vegetation	nutritivity	sweetness

[a] Tr. auct. adjuv. Karlgren (12), p. 30.

[b] The category of mouldability was one of which the Chinese took particular notice, as befitted a people who were the finest bronze-founders of antiquity, and who made cast iron thirteen centuries

[1] 彝倫　　[2] 彝　　[3] 水曰潤下

This gives first the natural property or process which struck the imagination of the Naturalists, then an approximate modern equivalent, and thirdly the corresponding taste which they associated with the products of the natural activity in question. In such a formulation there is, of course, always the danger of attributing to the ancients ideas more sophisticated than those they really had, yet the *Shu Ching* passage seems to suggest it.

On this view the five-element theory was an effort to reach a provisional classification of the basic properties of material things, properties, that is to say, which would only be manifested when they were undergoing change. It is often pointed out, therefore, that the term 'element' has never been satisfactory for *hsing*,[1] the very etymology of which, as we have just seen (Table 11, p. 222, no. 14) had from the beginning the implication of movement.[a] As Chhen Mêng-Chia says, the five 'elements' were five powerful forces in ever-flowing cyclical motion, and not passive motionless fundamental substances. Nevertheless, the term 'element' has for so long been used of the Wu Hsing that it is hardly possible to discard it. One remarkably interesting aspect of this *Shu Ching* passage is its association of the five elements with the five tastes.[b] Although this is generally considered to be part of that far-reaching system of correlation of the five elements with everything in the universe which it was possible to classify in fives (see below, p. 261), the present correlation cannot quite be written off in this way, since it strongly suggests the chemical interests of the Naturalists. The association of saltiness with water, while natural indeed to a coastal people, suggests primitive experiments and observations on solution and crystallisation. The association of bitterness with fire, while perhaps the least obvious of the five, may imply the use of heat in preparing decoctions of medicinal plants, which would be the bitterest substances likely to be known. There would also be a connection of 'hot' and bitter in spices. The association of sourness with wood can readily be explained, since wood, as vegetal, would be connected with all kinds of plant substances which become sour on decomposition. The alkali in plant ashes would also taste sour. The association of acridity with metal points directly to smelting operations, many of which would give off highly acrid fumes, e.g. sulphur dioxide. Lastly, the association of sweetness with earth would be due to the finding of honey in bees' nests in the earth, and to the general sweet taste of cereals.

---

before Europeans. Some have thought that the word *fa*,[2] which afterwards came to signify 'law', first meant a mould, for it is a combination of the graphs for 'water' and 'deprivation' or 'going away'. That which could be liquid like water and from which the liquidity would go away, was mouldable. Granet (5) often stressed the effect which these origins had upon the notion of law, which for the Chinese never lost the undertone of 'modelling' or 'imitating' according to a pattern, and never acquired the undertone of 'bonds' imposed by a binder (lawgiver, *ligare*).

[a] I am indebted to Dr Tsang Chhi-Mou for emphasising this in private correspondence.

[b] The whole discussion of the passage in Granet (5), pp. 168, 308, is worth reading. See also Myers' paper (1) on taste terms among primitive peoples.

[1] 行   [2] 法

(1) Comparison with Element Theories of other Peoples

Some comparison between these Chinese theories and the thinking of the ancient Greeks about elements cannot be avoided. The Greek elements seem to go back to the beginnings of the pre-Socratic school, since they were discussed by Anaximander (c. −560)[a] who distinguished four (the usual four, Earth, Fire, Air and Water) as well as a fifth, the Non-Limited (*apeiron*, ἄπειρον), which was a kind of substratum of the others. In the ancient Orphic formulations, however, there had been only three,[b] and this tradition was perpetuated by some thinkers, such as Ion of Chios (c. −430).[c] According to Pherecydes of Syros (c. −550)[d] the elements warred with one another (a remarkable parallel with the theory of Mutual Conquest), and both he and Empedocles (c. −450)[e] associated each element with a particular god. This, too, is found in Chinese thought, for several lists of supernatural beings each associated with one of the elements have survived, for example, in the *Chi Ni Tzu* book.[f] The elements were termed by Empedocles 'roots' (*rhizomata*, ῥιζώματα) and the familiar word *stoicheia* (στοιχεῖα) was first used by Plato (−428 to −348). Contrary to a common belief, this word seems to have had no connection with the idea of movement, but in its most primitive sense signified a small stationary upright post, in fact a gnomon. Thence it acquired the meaning of 'simple component'. Nevertheless, Aristotle (−384 to −322), Tsou Yen's elder contemporary, taking over the doctrine of the four 'primary sorts of matter' (*prota somata*, πρῶτα σώματα), gave it a decidedly dynamic twist by considering them as qualities. The *stoicheia* of Aristotle were no longer earth, fire, air and water, but rather the dry, the hot, the cold and the moist, qualities which became so familiar in later European science and medicine as long as the Aristotelian domination lasted.[g] The assumption of these qualities by inert primal matter (*hule*, ὕλη) gave it its form (*eidos*, εἶδος). These elements could, and constantly did, change into one another,[h] one quality in a given phenomenon being replaced by its contrary (*alloiosis*, ἀλλοίωσις). Aristotle distinguished various kinds of combination; his *synthesis* (σύνθεσις) was what we should now call a physical mixture, while his *mixis* (μῖξις) more closely, if somewhat vaguely, resembled our conception of a chemical compound. His *krasis* (κρᾶσις), a word which had been of great importance in the Hippocratic medical corpus, was a balanced mingling of liquids or solutions.

A good deal of interest attaches to the fifth element of the Greeks, but I am not clear that it forms any parallel to the Chinese conceptions. Philolaos of Tarentum (c. −430)[i] felt the need of a fifth element because he thought that there ought to be some connection between the elements and the five known figures of solid geometry. He called it *holkas* (ὁλκάς), the hull (as if of a ship), or vehicle, and perhaps thought of it in a way somewhat similar to the *apeiron* of Anaximander. Plato followed this up,

---

[a] Freeman (1), p. 56.
[b] Freeman (1), p. 6.
[c] Freeman (1), p. 206.
[d] Freeman (1), p. 39.
[e] Freeman (1) pp. 181 ff.
[f] Forke (13), p. 502.
[g] Cf. W. D. Ross (1); B & M, pp. 238 ff.
[h] As the Stoics also thought; Arnold (1), p. 180.
[i] Freeman (1), p. 222, 231; B & M, p. 197.

identifying the fifth with *aether* ($\alpha\grave{\iota}\theta\acute{\eta}\rho$), a subtler kind of air, and Aristotle relegated it to the substance of the heavenly bodies, thus banishing it from the sublunary world.[a]

In general, one may say that while there are certain similarities between the Greek and Chinese theories of the elements, the divergencies are still more striking, and it seems unnecessary to assume any transmission.

In this connection reference must be made to the determined effort of Chavannes (7) to prove that the Chinese theory of the elements was derived, with the duodenary animal cycle, from neighbouring Turkic or Hunnish peoples about the middle of the − 1st millennium. He made a great deal of the fact that when in − 205 the first Han emperor conquered the territories of the former State of Chhin, he found sacrifices customary to only four celestial emperors or gods (white, green, yellow, and red), whereupon he directed that an additional sacrifice should thenceforward be offered to the black celestial emperor.[b] The north-western element in Chinese civilisation was doubtless, as we have seen (Sect. 5*b*), Turanian and nomadic, and Chavannes' conclusion here was that a theory of (four) elements came in with it. The State of Chhin was notoriously 'barbaric' in culture. This view was combated vigorously by de Saussure (8, 10)[c] and Forke,[d] who probably had the right on their side, though many of their reasons were certainly wrong, since they partly relied upon a belief in datings for books such as the *I Ching* and the *Chou Li* much earlier than those accepted today. Nothing is said about elements in the *Shih Chi* passage which describes the first Han emperor's surprise and his new instructions, and the whole incident could just as well be explained by assuming that the general cosmic system of fives which had been worked out during the preceding few centuries in the eastern coastal States of Chhi and Yen by the predecessors and successors of Master Tsou, had filtered through somewhat imperfectly to the backward but warlike State in the west. To make a Turkish transmission of western element-doctrine convincing Ssuma Chhien should have said something here about the element air or wind, but of this we hear nothing.[e] Moreover, the division of the heavens into five celestial palaces assuredly goes back some centuries before Tsou Yen.[f]

---

[a] But the Manichaeans brought it back, firmly establishing five elements in their religious philosophy (cf. Cumont (3), p. 16; Bousset (1), p. 231), and including it as one of them. Water and fire they shared with Chinese and Greeks alike, and for the remaining two they had light and wind. Perhaps their Persian and Central Asian connections fixed the number five for them; they recognised also five kinds of plants, five kinds of animals, and so on.

[b] *Shih Chi*, ch. 28; tr. Chavannes (1), vol. 3, p. 449.

[c] Pp. 249 and 351 in de Saussure (1).   [d] (6), p. 242.

[e] More interest attaches to parallels with the five elements in Iran, and their relations with Indian thought, but these are still under investigation (Sheftelowitz, 1).

[f] Though not to the − 2nd millennium, as de Saussure and Forke believed.

(2) THE NATURALIST-CONFUCIAN SYNTHESIS IN THE HAN

This digression has led us away from Tsou Yen and his school. One may ask how its doctrines were transmitted to the Han people. Here a key-figure was Fu Shêng[1] (fl. *c*. −250 to −175),[a] a scholar from Shantung (the old State of Chhi) who must have been born not long after the death of Tsou Yen and who lived through the whole dynasty of Chhin.[b] He was the expert on the *Shu Ching*, who, according to a well-known story, repeated most of it from memory after the burning of the books by the First (Chhin) Emperor, but since this whole event may indeed be apocryphal, the tradition may only mean that he and the group round him drastically re-edited or reconstructed the Historical Classic; this must have been the time when the five-element theory became incorporated in it. Fragments from Fu Shêng's commentary on the Classic, the *Shang Shu Ta Chuan*,[2] certainly exist,[c] and some are probably embedded in the Wu Hsing Chih[3] (Five Element Chapter) of the *Chhien Han Shu*. Steadily becoming more political and less scientific, the five-element theories were handed down through a succession of scholars,[d] such as Ouyang Shêng,[4] Ouyang Kao,[5] Hsiahou Shih-Chhang[6] and Hsiahou Shêng.[7] Ching Fang,[8] who was really an *I Ching* specialist (see below), may also have had something to do with it. A number of different schools grew up. In the last quarter of the −1st century Liu Hsiang and his son Liu Hsin occupied themselves with the theory, and produced the (now lost) *Hung Fan Wu Hsing Chuan*[9] (or perhaps *Wu Hsing Chuan Shuo*;[10] the title is uncertain, but the book was a discussion of the theory arising from the Hung Fan text). By the beginning of the +1st century the material was ready to serve, as it did, for the elaboration of Pan Ku's chapter, already referred to, of the *Chhien Han Shu*, which Eberhard (6) exhaustively analyses. Pan was writing in the third quarter of the +1st century. By that time the essentials of the theory had become surrounded by an enormous accretion of omen- and portent-lore of all kinds. The theory known as 'Phenomenalism' (to be described in the next section, p. 378) had become stabilised; according to this, governmental or social irregularities would lead to dislocations of the five-element processes on earth and deviations from the proper course of events in the heavens. Thus was the tradition started which continued through all the subsequent dynastic histories in their 'Five Element Chapters'— the proto-science of the Naturalists had turned into the pseudo-science of the Phenomenalists.[e]

[a] G 599.   [b] Cf. Eberhard (6).

[c] See Wu Khang (1), p. 230.

[d] Biographies in ch. 75 of *Chhien Han Shu*; see Table II in Tsêng Chu-Sên (1) after p. 86.

[e] This is, of course, not to say that these chapters of the later dynastic histories do not contain much matter of scientific interest, especially records of sunspots, auroras, meteor showers, and eclipses, which we shall note in later Sections. They are none the less valuable because the motives which led to their observation and recording were not quite those of modern science.

[1] 伏勝   [2] 尚書大傳   [3] 五行志   [4] 歐陽生   [5] 歐陽高

[6] 夏侯始昌   [7] 夏侯勝   [8] 京房   [9] 洪範五行傳   [10] 五行傳說

In order to place this transformation correctly in relation to Chinese cultural history as a whole, it must be remembered that the proto-scientific and pseudo-scientific thinkers nearly all belonged to the 'New Text School' (*chin wên chia*[1]), while their opponents formed the 'Old Text School' (*ku wên chia*[2]).[a] This division had arisen because of the discovery, during the −2nd century, of a set of versions of the classics (*Shu Ching, Shih Ching, Tso Chuan, Chou Li*, etc.) which differed from the texts previously accepted, and which were written in the archaic script of the early (Western) Chou. This occurred during the destruction of the supposed house of Confucius in *c*. −135 when Prince Kung of Lu (Lu Kung Wang[3]) was enlarging his palace. Many subsequent centuries of scholarly debate ended in the conclusion that the story of the single discovery was a legend, and that at least some of the 'old versions' were probably forgeries, though the *Shu Ching* were ones not identical with the present 'Old Text' chapters, compiled with ancient fragments about +320. The situation is thus a little confusing to grasp, for while the members of the New Text school were textually on stronger ground they accepted all the superstitious exaggerations of phenomenalism and other pseudo-sciences[b]; and while the members of the Old Text School put their faith in false documents, they were mostly nevertheless rationalists. Generally speaking, the New Text school was dominant in the Early Han, and its opponents in the Later. Most of the men just mentioned belonged to the New Text school, like Liu Hsiang whose son Liu Hsin, however, led the opposition. Imperial princes supported it, such as Liu Tshang,[4] Prince of Tung-Phing, and its greatest thinker was Tung Chung-Shu. But there were scientific minds on both sides, for besides famous scholars such as Khung An-Kuo,[6] Mao Hêng,[6] Mao Chhang[7] and Wang Huang,[8] the Old Text school included astronomers like Chia Khuei[9] as well as mutationists like Yang Hsiung,[10] and prepared the way for the greatest sceptic of all, Wang Chhung.[11]

The literature of the five-element theories which remains from just before and during the Han time is very large (and also tedious, fanciful and repetitive). But to show the kind of way in which the scholars were thinking and talking, I reproduce two extracts, one from the *Kuan Tzu* book (into which it must have been interpolated during the −3rd or −2nd century), and one from the *Chhun Chhiu Fan Lu* of Tung Chung-Shu. The *Kuan Tzu*[c] visualises the periodical dominance of each of the elements in turn during the cycle of the year, saying:

When we see the cyclical sign *chia-tzu*[12] arrive, the element Wood begins its reign. If the emperor does not bestow favours and grant rewards, but rather allows great cutting,

---

[a] Cf. Fêng Yu-Lan (1), vol. 2, pp. 7 ff., 133 ff., 673 ff.; Wu Khang (1), p. 186; Wu Shih-Chhang (1).
[b] Containing, of course, the kernel of truth of organic philosophy, see on, pp. 280, 526.
[c] Ch. 41, entitled 'The Five Elements'. Related material is in the preceding chapter on 'The Four Seasons', partially translated by Hughes (1), p. 215; Than Po-Fu *et al*. p. 88. A paraphrase of the passage quoted is given in *Huai Nan Tzu*, ch. 3, p. 8*b*.

[1] 今文家	[2] 古文家	[3] 魯恭王	[4] 劉蒼	[5] 孔安國
[6] 毛亨	[7] 毛萇	[8] 王璜	[9] 賈逵	[10] 揚雄
[11] 王充	[12] 甲子			

destroying and wounding, then he will be in danger. Should he not die, then the heir-apparent will be in danger, and some one of his family or consort will die, or else his eldest son will lose his life. (For spring is a time for growth, not destruction.) After seventy-two days this period is over.

When we see the cyclical sign *ping-tzu*[1] arrive, the element Fire begins its reign. If the emperor now takes hurried and hasty measures, epidemics will be caused by drought, plants will die, and the people perish. After seventy-two days this period is over.

When we see the cyclical sign *wu-tzu*[2] arrive, the element Earth begins its reign. If the emperor now builds palaces or constructs pavilions, his life will be in danger, and if city-walls are built (at this time) his ministers will die. (For the people should not be taken away from their harvesting.) After seventy-two days this period is over.

When we see the cyclical sign *keng-tzu*[3] arrive, the element Metal begins its reign. If the emperor attacks the mountains (by mining operations) and causes rocks to be pounded (for metallurgy) his troops will be defeated in war, his soldiers die, and he will lose his throne. After seventy-two days this period is over.

When we see the cyclical sign *jen-tzu*[4] arrive, the element Water begins its reign. If the emperor now (allows) the dykes to be cut, and sets the great floods in motion, his empress or great ladies will die, birds' eggs will be found to be addled, the young of hairy animals will miscarry, and pregnant women will have abortions. After seventy-two days this period is over.[a]

In this passage the connection between the five-element theory and prognostication and divination is very obvious. That which I take from Tung Chung-Shu shows it also. His chapter 'On the Five Elements', which must date from about −135, runs thus:

Heaven has five elements, first Wood, second Fire, third Earth, fourth Metal, and fifth Water. Wood comes first in the cycle of the five elements and water comes last, earth being in the middle. This is the order which Heaven has made. Wood produces fire, fire produces earth (i.e. as ashes), earth produces metal (i.e. as ores), metal produces water,[b] and water produces wood (for woody plants require water). This is their 'father-and-son' relation. Wood dwells on the left, metal on the right, fire in front and water behind, with earth in the centre. This, too, is their father-and-son order, each receiving from the other in its turn. Thus it is that wood receives from water, fire from wood, and so on. As transmitters they are fathers, as receivers they are sons. There is an unvarying dependence of the sons on the fathers, and a direction from the fathers to the sons. Such is the Tao of Heaven.

This being so, wood having produced fire nourishes it, while metal having died is stored up in water. Fire delights in wood, and through the operation of the Yang is nourished by it. Water, having conquered metal, through the operation of the Yin buries it. Earth, in its service to Heaven, 'uses all its loyalty'. Thus it is that the five elements correspond to the actions of filial sons and loyal ministers.[c] Putting the five elements into words (like this) they really seem to be five kinds of action, do they not?

[a] Tr. Forke (6), p. 259.
[b] Either because molten metal was considered aqueous, or more probably because of the ritual practice of collecting dew on metal mirrors exposed at night time (see on, Sect. 26 g).
[c] Note how fully Naturalism has become absorbed by Confucianism.

[1] 丙子　　　[2] 戊子　　　[3] 庚子　　　[4] 壬子

The fact that definite propositions can be made about them means that sage men can get to know them, and thereby increase their own loving-kindness and decrease their severity, lay stress on the nourishing of life, and take care about the funeral offices for the dead, in this way being in keeping with Heaven's ordinances. Thus as a son welcomes the completion of his years (of nurture), so fire delights in wood, and as (the time comes when) the son buries his father, so (the time comes when) water conquers metal. Also the service of one's sovereign is like the reverent service Earth renders to Heaven. Thus one can say that there are men (in tune with) the elements, and just as the five elements follow one another in orderly succession, so there are officials (in tune with) the elements, taking advantage to the utmost of their several capacities.

Thus wood has its place in the east and has authority over the *chhi* of spring. Fire has its place in the south, and has authority over the *chhi* of summer. Metal has its place in the west, and has authority over the *chhi* of autumn. Water has its place in the north, and has authority over the *chhi* of winter. This being so, wood takes charge of life-giving, and metal of death-dealing; fire of heat and water of cold. Men have no choice but to go by this succession; officials have no choice but to operate according to these powers. For such are the calculations of Heaven.

Earth has its place at the centre and is (as it were) the rich soil of Heaven. Earth is Heaven's thighs and arms, its virtue so prolific, so lovely to view, that it cannot be told at one time of telling. In fact earth is that which brings these five elements and four seasons all together. Metal, wood, water, and fire each have their own offices, yet if they did not rely on earth in the centre, they would all collapse. In similar fashion there is a reliance of sourness, saltiness and bitterness on sweetness. Without that (basic) tastiness, the others could not achieve 'flavour'. The sweet (i.e. edible) is the root of the five tastes. Thus earth is the controller of the five elements, and its *chhi* is their unifying principle, just as the existence of sweetness among the five tastes cannot but make them what they are. This being so, among the actions of sage men, there is nothing equal in honour to loyalty, that fidelity which I have described as being the characteristic virtue of Earth....[a]

The cyclical recurrence of the elements according to the order of their mutual production through the seasons of the year became in later centuries very much stylised.[b] In each of the five seasons of the year (the sixth month being considered separately between summer and autumn), the five elements would each be in one or other of the following phases; 'helping' (*hsiang*[1]), 'flourishing' (*wang*[2]), 'retiring' (*hsiu*[3]), 'undergoing imprisonment' (*chhiu*[4]) and 'dying' (*ssu*[5]). Later there were twelve phases corresponding to the twelve months of the year, which each of the five elements occupied in turn: (1) to receive breath (*shou chhi*[6]); (2) to be in the womb (*thai*[7]); (3) to be nourished (*yang*[8]); (4) to be born (*sêng*[9]); (5) bathed (*mu yü*[10]); (6) to assume the cap and girdle (*kuan tai*[11]); (7) to become an official (*lin kuan*[12]); (8) to flourish (*wang*[2]); (9) to become weak (*shuai*[13]); (10) to become ill (*ping*[14]);

---

[a] *Chhun Chhiu Fan Lu*, ch. 42, tr. Hughes (1), p. 294.
[b] One of the earliest statements of this is in *Huai Nan Tzu*, ch. 3, p. 14*b*.

[1] 相   [2] 旺   [3] 休   [4] 囚   [5] 死   [6] 受氣   [7] 胎
[8] 養   [9] 生   [10] 沐浴   [11] 冠帶   [12] 臨官   [13] 衰
[14] 病

(11) to die (*ssu*[1]); and finally (12) to be buried (*tsang*[2]). These elaborations were much used in fate-calculations.

But if the five-element theories were thus incorporated in Han political thinking, the conventional and orthodox Confucians of the Han time rejected Tsou Yen and all his works. This is shown by an interesting passage which constitutes chapter 53 of the *Yen Thieh Lun*[3] (Discourses on Salt and Iron), written by Huan Khuan[4] about −80 as the supposedly verbatim account of a conference between officials and Confucian scholars which took place in the previous year. The *Yen Thieh Lun* is a document of extraordinary interest, which will be carefully examined in the closing sections of the present book; here it is relevant because in the chapter in question the official group appeals to the memory of Tsou Yen as a man of the widest conceptions and profoundest knowledge, while the scholars decry him and state with perfect frankness the vulgar Confucianism which could see no value in science. From this point of view the passage deserves to be considered capital for the history of science in China. Here is the chapter ('Tsou Lun',[5] Discussion about Master Tsou):

The Lord Grand Secretary said: 'Master Tsou was sick of the later Confucians and Mohists who did not understand the vastness of Heaven and Earth, and the Tao of the universe, broad and bright. Knowing only one part, they thought they could talk about all nine parts; knowing only one corner of the world they thought that they understood the whole of it. They thought that they could determine heights without a water-level, and tell the difference between straight lines and arcs without using squares and compasses. But Tsou Yen was able to make inferences about the cycles of the great Sages from beginning to end, giving examples (from history) to kings, feudal lords, and illustrious scholars. Classifying China's famous mountains and connecting valleys, he pushed on to attain a knowledge of what was beyond the seas....'

> He has drifted into a verbal quotation from chapter 74 of the then recently written *Shih Chi*, which he continues, giving the account of Tsou Yen's geographical opinions about China not being the centre of the world, and about the Nine Great Continents (pp. 233, 236 above).

'(True), the heights of mountains, rivers and marshes (had also been) recorded in the Yü Kung ('Tribute of Yü' chapter in the *Shu Ching*) but Yü did not realise the far-reachingness of the Great Tao. This was why Chhin (Shih Huang Ti—who came after Master Tsou) wanted to reach the Nine Continents, and to attain to the Great Ying Ocean, to drive away the barbarians, and to be lord over the ten thousand countries. Ordinary scholars do nothing but worry over the affairs of their own small lands, and never go beyond their own villages and districts, so they have no idea of the meaning of the great world of the Empire.'

The scholars answered, 'Yao appointed Yü to be Minister of Works and to control the waters and the land. Following the natural course of the mountains, he marked out the heights with wooden posts, and delimited the Nine Provinces. But Tsou Yen was no sage; with strange and deceptive teachings he enchanted the six feudal kings, and so got them to

---

[1] 死   [2] 葬   [3] 鹽鐵論   [4] 桓寬   [5] 騶論

accept his ideas. This is what the *Chhun Chhiu* calls "the bewilderment of the feudal kings by one common fellow". Confucius said, "People do not know how to manage human affairs; how should they know about the affairs of the gods and spirits?" Those who have not yet attained to a knowledge of that which is near at hand, how should they know about the Great Ying Ocean? Therefore the *chün-tzu* should have nothing to do with things which are of no practical use. What is not concerned with government matters he should not investigate. The (legendary) Three Emperors believed in the Tao of the (Confucian) classics, and their bright virtue spread throughout the world. But the kings of the warring states believed in the seductive teachings (of people like Master Tsou), and they were conquered and destroyed. Moreover, when Chhin (Shih Huang Ti), having eaten up the known world, still wanted to grasp the ten thousand countries, he lost in the end even his own thirty-six commanderies—he wanted to reach the Great Ying Ocean, but instead lost even his own provinces and districts. If we thoroughly understand that these things are so, we had better restrict ourselves to modest planning.'[a]

Here we must not anticipate the social and economic significance of the *Yen Thieh Lun*; it need only be said that its discussions contrast Confucian conservatism with the new bureaucratic state organisation. The interest of the passage lies in the fact that it reveals how the ruling house of Chhin was influenced by the school of the Naturalists, and how the later Confucians, while unable to resist the political aspect of the five-element theories, rejected the other scientific components of Tsou Yen's teachings. Meanwhile its alchemical and pharmaceutical components had been absorbed into the Taoist complex, and had become for the Confucians quite heterodox.

In the bibliography of the *Chhien Han Shu*, no less than twenty-one books are assigned to the Naturalists' school (Yin-Yang Chia), but all were afterwards lost. The two books of Tsou Yen himself, mentioned above, are both there, as also one by Tsou Shih; and another book is by Chang Tshang (cf. above, p. 239). Two members of the Kungsun family, otherwise unknown, are represented. Taoist influence is suggested by the appearance of the Yellow Emperor, Huang Ti, in one of the titles.[b] The bibliographer comments:

The teaching of the Yin-Yang school began with the old official astronomers Hsi and Ho.[c] (The school) respectfully followed luminous Heaven, the successive symbols, the sun and moon, the stars and constellations, and the division of times and seasons for the people. Herein lay the good points of the school. But those who took this teaching too strictly and literally were bound by numerous restrictions and trivial prohibitions; they tended to give up reliance on human effort, and rely instead on the gods and spirits.[d]

[a] Tr. auct.

[b] It would be an interesting research to try to throw some light on the writers of these books, and their contents, from other ancient books and fragments. Perhaps this has been done, but I have not come across it. Cf. Forke (13), p. 506.

[c] Legendary astronomical officials mentioned in the *Shu Ching* (see Sect. 20c). The bibliographer had a theory that each one of the philosophical schools originated in one of the government ministries, but this was, of course, nonsense; see Hu Shih (6), Fêng Yu-Lan (4).

[d] Ch. 30, p. 22a, tr. auct.

In later history the five-element theory became more and more bound up with pseudo-sciences such as fate-calculation, which we shall describe a little further in the following Section. Thao Hung-Ching wrote books of this kind in the Liang dynasty (end of the +5th and beginning of the +6th century), and there are several fragments of about this date in Ma Kuo-Han's collection. The most important medieval book on the five elements, however, was the *Wu Hsing Ta I*[1] (Main Principles of the Five Elements) written by Hsiao Chi[2] and presented to the emperor of the Sui dynasty in +594. This deals more with scientific matters and less with fate-calculation than any of the subsequent books.[a] In the Thang there were Lü Tshai,[3] Li Hsü-Chung[4] and many others, some of whom will be referred to later.[b] By this time the five-element theories had become a universal commonplace of Chinese thought.

(d) ENUMERATION ORDERS AND SYMBOLIC CORRELATIONS

In considering the five-element theory as it was stabilised in Han time and for all later ages there are two aspects which particularly merit attention. These are (a) the Enumeration Orders, and (b) the Symbolic Correlations.

(1) THE ENUMERATION ORDERS AND THEIR COMBINATIONS

By the Enumeration Orders I mean the orders in which the five elements were named in the various ancient and medieval presentations of the subject. These orders were far from being always the same. We may distinguish the four most important ones as follows:[c]

(i) The Cosmogonic Order (*sêng hsü*[5])	*w*	*F*	*W*	*M*	*E*
(ii) The Mutual Production Order (*hsiang sêng*[6])	*W*	*F*	*E*	*M*	*w*
(iii) The Mutual Conquest Order (*hsiang shêng*[7])	*W*	*M*	*F*	*w*	*E*
(iv) The 'Modern' Order	*M*	*W*	*w*	*F*	*E*

These orders and their significance have been thoroughly analysed by Eberhard,[d] who laid under contribution a large number of texts dating from all periods down to the end of the Later Han.[e] Before considering the various orders separately, it should be noted that Eberhard arranged the elements in all the theoretically possible combinations and permutations. The total number of sequences so formed is thirty-six.

---

[a] See Chao Wei-Pang (1).   [b] See pp. 352, 358.

[c] For convenience we let *W* stand for wood, and *w* for water here. The other capital letters are self-explanatory.

[d] (6), pp. 41 ff.

[e] The medieval discussions (from Chin to Yuan) on this subject have not yet been analysed; we can only indicate a few useful sources. One would be the *Lü Chai Shih Erh Pien*[8] of +1205 (ch. 1, pp. 6 ff.).

[1] 五行大義   [2] 蕭吉   [3] 呂才   [4] 李虛中   [5] 生序
[6] 相生   [7] 相勝   [8] 履齋示兒編

He divided them into two groups of eighteen each, according to whether they proceeded in what might be called a 'clockwise' or an 'anti-clockwise' direction; these he identified with a solar and a lunar formulation respectively. In order to understand this one must remember that the five elements (or rather four of them) were from an early stage correlated with the points of the compass (cf. the passage from Tung Chung-Shu just quoted). Now the sun rises in the east, stands at midday in the south, sets in the west, and at night is considered to be in the north; the solar cycle will therefore be $E\ S\ W\ N$ (or, translated into terms of the elements, $W\ F\ M\ w$). We may note that this is akin to the Mutual Production Order. But another heavenly movement is the lunar cycle of 29·5306 days; the new moon appears in the west, the full moon rides in the south, the waning moon in the east, and to the north there is no visible moon. This order will therefore be $W\ S\ E\ N$ (or, translated into terms of the elements, $M\ F\ W\ w$). It does not occur in any of the four principal orders listed above, but not a few instances of it can be found in the various texts. This lunar order may also be considered a solar order if the annual, not the daily, movement of the sun is considered. For the points of rising and setting of the sun move $E\ N\ W\ S$. After rising due east at the spring equinox it rises more northerly as summer goes on, sinks then due west at the autumnal equinox, and then sets more southerly until the winter solstice is reached. This motion ($W\ w\ M\ F$ in terms of the elements) becomes therefore a special case of the lunar sequences.

Eberhard's statistical calculations show that of the eighteen possible 'solar' sequences no less than eleven are represented by one or another of the texts studied, while, on the other hand, only five of the eighteen possible 'lunar' sequences occur. The number of books searched by him for this purpose was about thirty. This preference is worth noting. It should be added that there is no Chinese authority for Eberhard's astronomical interpretation of the two different classes of sequences. It is hard to know how much significance should be attached to some of the rarer variants, some of which might have been simply random statements, but it is certain that the four principal orders to which we must now turn were very consciously intended.

(i) *The Cosmogonic Order* (*sêng hsü*[1]), $w\ F\ W\ M\ E$ (Eberhard's B₄ sequence)

About this there is not very much to say. It was the evolutionary order in which the elements were supposed to have come into being. In the *Chhien Han Shu* it occurs[a] as part of a discussion of a passage in the *Tso Chuan*[b] about the five elements, which was presumably really a −3rd-century interpolation. Only three of the texts give it, but these include one with great prestige, the Hung Fan chapter of the *Shu Ching* quoted above (p. 243); the others are the *Chi Chung Chou Shu*[2] (or *I Chou Shu*, a book of somewhat dubious authenticity; and the (probably Thang) *Kuan Yin Tzu* (cf. p. 443).

[a] Ch. 27A, p. 10a.   [b] Duke Chao, 9th year (tr. Couvreur (1), vol. 3, p. 166).

[1] 生序   [2] 汲冢周書

A point which should not be lost here is the significance of the fact that the series begins with water. There may have been, therefore, some ancient forgotten Chinese thinker who was the counterpart of Thales and who would have inspired the 'water-chapter' of the *Kuan Tzu* book, of which a translation has been given in the section on the Taoists (p. 42 above). A certain emphasis on water as the primal substance keeps on turning up, as was there said, with references, all through the history of Chinese thought. To these may here be added the curious thought of the Thang writer Wang Shih-Yuan[1] (*c.* +745), who in his *Khang Tshang Tzu*[2] describes a theory of 'moulting' or 'disrobing'. When earth (*thu*[3]) disrobes (*tho*[4]) we see only water (*shui*[5]); when water disrobes there is nothing left but 'empty *chhi*', then that unveils itself as naked Emptiness (*hsü*[6]), and after a last disrobing the entity ultimately appearing is the Tao.[a] In the Section on the Neo-Confucians of the +12th century we shall also find (p. 463 below) that they gave a primacy to water and fire (in a highly pre-Socratic manner), considering the other three elements as secondary.[b]

(ii) *The Mutual Production Order* (*hsiang sêng*[7]), *W F E M w* (Eberhard's A₁ sequence)

This is the order in which the five elements were supposed to produce each other. We have just met with it in the passage from the −2nd-century *Chhun Chhiu Fan Lu* (String of Pearls on the Spring and Autumn Annals) of Tung Chung-Shu. *W* produces *F* (by being consumed as fuel), *F* produces *E* (by giving rise to ashes),[c] *E* produces *M* (by fostering the growth of metallic ores within its rocks), *M* produces *w* (by attracting or secreting sacred dew when metal mirrors were exposed at night, or else by its property of liquefying),[d] and *w* produces *W* (by entering into the substance of plants), thus completing the cycle. This sequence occurs no less than thirteen times in the texts examined; a frequency more than twice as great as that of the occurrence of any other series. Among these texts may be mentioned the *Kuan Tzu*, the *Huai Nan Tzu* and the *Lun Hêng*.

From this order and from the next derive two very interesting secondary principles, as we shall shortly see.

One should note that this order describes the sequence in which the elements come into being at the successive seasons of the year, starting with spring (earth being left out on account of its central position). We may find it thus clearly stated in both the excerpts given above (from the *Kuan Tzu* book and from the *Chhun Chhiu Fan Lu*, pp. 248, 250).

[a] Cf. Forke (12), p. 318. One cannot help wondering whether Wang Shih-Yuan can have had an intuitive appreciation of the 'envelopes' of modern organismic science. While he could, of course, have known nothing of what the microscope was later to reveal, it was already quite obvious that living bodies embodied discrete organs and were themselves embodied in social groups. Ch. 1, p. 2*a*.

[b] This tendency begins in the Thang with the unknown Taoist who wrote the *Kuan Yin Tzu*, ch. 1, pp. 9*a*, 12*a*, *b*.

[c] Or, as Granet (5), p. 308, ingeniously suggests, by burning off trees and undergrowth so that cultivation of the earth can be carried on (cf. the *milpa* methods of the Mayas, and Section 41 on agriculture).   [d] Cf. Pliny, *Hist. Nat.* XXXIV, 146 (Bailey (1), vol. 2, pp. 59, 188).

[1] 王士元   [2] 亢倉子   [3] 土   [4] 蛻   [5] 水   [6] 虛
[7] 相生

(iii) *The Mutual Conquest Order* (*hsiang shêng*[1]), *W M F w E* (Eberhard's A$_4$ sequence)

This order, which describes the series in which each element was supposed to conquer its predecessor in turn, may be considered the most venerable of the four, since it is the one associated with the teaching of Master Tsou himself. We have already seen it in the last three paragraphs of the remaining fragment of the *Tsou Tzu* book (p. 238). Starting with the last link in the cycle, *W* conquers *E* (because, presumably, it can, when in the form of spades, dig it up and make shapes of it), *M* conquers *W* (because it can cut it and carve it), *F* conquers *M* (because it can melt and even volatilise it), *w* conquers *F* (because it can extinguish it), and *E* conquers *w*, thus completing the cycle (because it can dam it up and constrain it—a metaphor very natural for a people whose life depended so much upon hydraulic engineering and irrigation as did that of the Chinese).

This order occurs six times in Eberhard's texts, two of them being authorities as important as the *Huai Nan Tzu* and the *Lun Hêng*. It was obviously the most important from the political point of view, since it was put forward as an explanation of centuries of history, with the implication that it would continue to describe the saecular mutations of the elements, and hence could be used for prediction.

In the few words above on the parallels between the Chinese five-element theories and the Greek doctrines of elements I alluded to the fact that Pherecydes of Syros (−6th century) considered the elements, to be at war with one another. But an even closer parallel is found in the fragments of Heracleitus of Ephesus (*c.* −500), where he says that one thing lives the 'death' of another; Fire lives the death of Air, Air lives the death of Fire, and so on.[a] Yet it seems to me that the idea of successive mutual conquests as phenomena succeed one another in the eternal round of Nature is such an obvious one that there is little need to cast about for some evidence of transmission at such an early date.

(iv) *The 'Modern' Order*, *M W w F E* (Eberhard's D$_4$ sequence)

This is the most obscure of the four enumeration orders, for although its significance is not at all apparent, it is that which has come down into modern Chinese colloquial speech, where everyone learns of the 'Chin Mu Shui Huo Thu' even in nursery rhymes. Contrary to the expositions of Wieger[b] and Forke,[c] it cannot be explained as a Mutual Conquest Order. Eberhard (6) suggests, ingeniously but not very hopefully, that it may derive from an ancient mnemonic rhyme. It is not infrequent in ancient texts, occurring six times on his count, the occurrences including such important works as the *Kuo Yü*, the *Pai Hu Thung Tê Lun* and the *Huai Nan Tzu*. It is not there explained, and we have no explanation for it, nor is it involved in the interesting secondary principles, to which we must now turn.

---

[a] Freeman (1), p. 124; Diels-Freeman (1), p. 30. An echo of this got into ancient Hebrew folklore (Ginzberg (1), p. 93).   [b] (2), p. 31.

[c] (6), p. 120 of the German edition. The English edition, p. 291, does not retain this.

[1] 相 勝

(v) *Rates of Change; the Principles of Control and Masking*

In order to fix in the mind the four orders, they may be represented diagrammatically as follows:

(i) Cosmogonic Order $\qquad w \to F \to W \to M \to E$ $\qquad\qquad\qquad$ (B$_4$)

(ii) Mutual Production Order $\qquad\qquad\qquad\qquad\qquad\qquad\qquad\qquad$ (A$_1$)

$$
\begin{array}{ccc}
 & F & \\
 \swarrow & & \searrow \\
E & & W \\
 \searrow & & \nearrow \\
 & M \to w &
\end{array}
$$

(iii) Mutual Conquest Order $\qquad\qquad\qquad\qquad\qquad\qquad\qquad\qquad$ (A$_4$)

$$
\begin{array}{ccc}
 & E & \\
 \nearrow & & \searrow \\
W & & w \\
 \nwarrow & & \swarrow \\
 & M \leftarrow F &
\end{array}
$$

(iv) 'Modern' Order $\qquad M—W—w—F—E$ $\qquad\qquad\qquad$ (D$_4$)

Now from the second and third of these, two further principles were deduced, which may be called:

(*a*) The Principle of Control (*hsiang chih*[1]); and
(*b*) The Principle of Masking (*hsiang hua*[2]).

In the first system, which derives solely from a consideration of the Mutual Conquest Order, a given process of destruction is said to be 'controlled' by the element which destroys the destroyer. For example:

W destroys (conquers) E, but M controls the process.

M destroys (conquers) W, but F controls the process.

F destroys (conquers) M, but w controls the process.

w destroys (conquers) F, but E controls the process.

E destroys (conquers) w, but W controls the process.

It is true that this idea was employed in fate-calculation, as described by Chao Wei-Pang, at least as much as for the explanation of natural phenomena. Nevertheless, it must be pointed out that in drawing these conclusions the Chinese were following perfectly logical paths of thought which in our own time have been found applicable in numerous fields of experimental science, for instance, the kinetics of enzyme action, or the oecological balance of animal species. Thus the digestion of an oxidase by a protease and the consequent inhibition of the reaction which it would otherwise have catalysed would be an excellent example of the 'principle of control' worked out by

[1] 相 制　　　[2] 相 化

men such as Hsiao Chi in the +6th century.[a] Or again, the 'food chains' in natural oecological communities must obviously depend on the relative abundance of the various species which prey on one another in a sequence based on their sizes and habits. A factor which increases the abundance of a certain bird will indirectly benefit a population of aphids because of the thinning effect which it will have on the coccinnellid beetles ('ladybirds') which eat the aphids but are themselves eaten by the birds. Modern economic entomology is full of such examples.[b] The Chinese 'principle of control' could of course be criticised by pointing out that it involved an infinite regress, since if the mutual conquest order was a cyclical one no process strictly speaking could ever take place at all, but this objection could be but formal, since it was not to be supposed that all the elements were everywhere present effectively at one time.

When the principle of control was related to the Mutual Production, as well as to the Mutual Conquest order, the corollary followed that the 'controlling' element is always that one produced by the destroyed element. Thus W conquers E in a process which is controlled by M, but M is the product of E. This argument was used, following all too human Confucianist interpretations such as we have seen in Tung Chung-Shu above, to prove that a son had the right to take revenge on the enemy of his father. Nevertheless, there was a germ of dialectical thinking here. The idea of something which acts upon something else and destroys it, but in so doing is affected in such a way as to bring about later on its own change or destruction, has become rather familiar to us. Modern science is finding that not all agents act, as it were, with impunity. The conception of contractile enzymes, now coming into use as the result of experiments on the principal contracting protein of muscle tissue, which seems to be at the same time one of the most important enzymes connected with the breakdown of those phosphorus compounds which transfer energy to the muscle machinery, suggests that in carrying out their catalytic functions the enzyme-proteins in living cells may themselves suffer biologically important configuration changes. And, at a simpler level, innumerable reactions come to a stop because of the accumulation of the reaction-products, according to the Law of Mass Action.

The second principle, the 'principle of masking' (the character used simply means 'change', but this translation makes its meaning plainer), depends clearly on both the Mutual Conquest and the Mutual Production orders. It refers to the masking of a process of change by some other process which produces more of the substrate, or produces it faster than it can be destroyed by the primary process.

[a] Or, to take another example from current biochemical research, there is an enzyme, phosphorylase, which breaks down glycogen to hexose molecules, esterifying them with phosphate as it does so. But there is another enzyme which breaks the phosphorylase into two portions, thereby inactivating it, and making its enzymic function impossible (Keller & Cori).

[b] It is remarkable, in this connection, that what is probably the oldest record we have of an entomological method of pest control is Chinese, and goes back to the +3rd century. While it would perhaps be going too far to suggest that theoretical thinking of the kind described above could have been associated with this invention, the possibility can hardly be altogether excluded. See on, Sect. 42.

Thus:

W destroys (conquers) E, but F masks the process.

F destroys (conquers) M, but E masks the process.

E destroys (conquers) w, but M masks the process.

M destroys (conquers) W, but w masks the process.

w destroys (conquers) F, but W masks the process.

Here again, both for enzyme kinetics and for oecological analysis, examples spring to the mind. Larger carnivores may devour the lemmings of Norway, but although they may continue to do so at maximal speed, their efforts will rapidly be overtaken in the years when those still mysterious factors operate which so enormously increase the lemming population. Examples of competing processes which would illustrate this quite simple but perfectly justified deduction from the dual cycles of production and destruction respectively, could certainly be found from every branch of modern science.[a]

It is important to note that in both these principles there lurks a strongly quantitative element. The conclusions depend on quantities, speeds and rates. Thus to take two examples, allowing for the abstract nature of the thought, which does not readily go into visual images; on the first principle M destroys W (by cutting it up into fragments), but F 'controls' the process (by melting the metal faster than it can cut up the wood); on the second principle w 'masks' the process (by producing wood faster than the metal can cut it up). I suspect that these elaborations may have been worked out in answer to the very obvious criticisms which the simple enumeration orders themselves must have invited. It is quite a mistake to imagine that the early Chinese thinkers were content with these formulations. In this particular case we are fortunate in that we have, preserved in the *Mo Ching*, a fragment of the criticisms which the later Mohists levelled against the Naturalists. The following passage would probably date from about the time of the death of Tsou Yen (-270).

Ch. 43/275/4.35.[b] *The Five Elements*

C. The five elements do *not* perpetually overcome one another. The reason is given under 'quantity' (i^1).

CS. The five are metal, water, earth, wood, and fire.[c] Quite apart (from any cycle) fire naturally melts metal, if there is enough fire. Or metal may pulverise a burning fire to

[a] The reader may like to follow this further in current standard scientific works. The cycles of world change, both inorganic and organic, are discussed in the classical treatise of Lotka, while among many good books on animal and plant oecology, those of Shelford and Elton may be mentioned. The general principles of the chemistry of life and the speeds of reactions in living tissues may be approached through Baldwin (1), Tracey (1) and Fruton & Simmonds. An introduction to the understanding of enzymes and their role may be gained from Bacon (1) and Tracey (2).

[b] For the explanation of this numbering-system for the Mohist Canon, see above, p. 172.

[c] Note that this sequence does not correspond with any of the four principal Orders. It is D_1 in Eberhard's classification, though he attributes the closely allied sequence A_2 to Mo Tzu. The point is curious, but in view of the great corruptions of this text, too much weight cannot be attributed to it.

[1] 宜

cinders, if there is enough metal. Metal will store water (but does not produce it). Fire attaches itself to wood (but is not produced from it). We should recognise that the different things, such as (mountain-) elks (R 365) or (river-) fishes, all have their own specific merits.[a]

This attack on Master Tsou's Mutual Conquest theory seems remarkably like a riposte to the supercilious attitude which he apparently took to the logical investigations of the Mo-Ming schools (p. 237 above). It is extremely interesting as demonstrating the quantitative element in Mohist scientific thinking, natural enough in view of what they did in physics (Sect. 26 below), and generating various echoes in the long history of Chinese thought. As we shall shortly see, there were other critical attacks against other aspects of the five-element theory. Here I shall only add a passage from the *Wên Tzu*[1] book, a work of very uncertain date and authorship, possibly late Han. In it we find:

Metal may overcome wood, but with one axe a man cannot cut down a whole forest. Earth may overcome water, but with a single handful, one cannot dam up a river. Water may overcome fire, but with no more than a cup of it one cannot put out a large conflagration.[b]

Parallel passages are to be found in the *Pao Phu Tzu* book (+4th century)[c] and in *Chin Lou Tzu* (+550).[d] The *Kungsun Lung Tzu* also touches on the five-element theory in the course of a quasi-quantitative argument about colours.[e]

We can catch a glimpse of some of the technical terms which were used by the school of naturalists surrounding Liu An (Huai Nan Tzu) about −130 from a passage in the book which bears the name of the prince.[f] The generating element is called *mu*[2] (Generator) and the product *tzu*[3] (Offspring). When the Offspring produces the Generator, the process is called 'righteousness' (*i*[4]); for example, *w* producing *W*, though it is formed from *W* through *F*, *E* and *M*. When the Generator produces the Offspring, as in all stages of the Mutual Production Order, the process is called 'fostering' (*pao*[5]). When the Offspring and Generator are 'mutually' obtained (*F* and *E*, for example, both arising from *W*, the former directly, the latter indirectly), the process is called 'special effort' (*chuan*[6]). When the Generator overcomes the Offspring, the process is called 'control' (*chih*[7]); for instance *M* conquers *W*, though generating *W* through *w*). When the Offspring overcomes the Generator, the process is called 'surrounding' (*khun*[8]); for instance *M* conquers *W*, though being formed from *W* through *F* and *E*.

Notable here once again is the inability of these thinkers to coin new technical terms, the first, second, third, fourth and seventh being adopted without change from

---

[a] Tr. auct.
[b] Ch. 6, p. 11*b*, tr. auct.; cf. Forke (13), p. 341.
[c] Wai Phien, ch. 1.   [d] Ch. 4, p. 22*b*.
[e] Ch. 4; see Ku Pao-Ku (1), pp. 50 ff.; Perleberg (1), p. 107.
[f] Ch. 3, p. 15*b* of *Huai Nan Tzu*.

[1] 文子   [2] 母   [3] 子   [4] 義   [5] 保   [6] 專   [7] 制
[8] 困

obvious human relationships.[a] More interesting scientifically is the fact that in all these arguments about the elements, one is always formed from one, and not one from two or more; the thought is therefore not yet chemical. It was rather the polar concept of Yin and Yang which could, and did, lead to the idea of chemical reaction.[b]

(2) THE SYMBOLIC CORRELATIONS AND THE SCHOOLS
WHICH EVOLVED THEM

We turn now to the symbolic correlations. As was hinted above, the five elements gradually came to be associated with every conceivable category of things in the universe which it was possible to classify in fives. Table 12 sets these forth.[c] Such correspondences were the commonplaces of thought in growing measure from the Chhin dynasty onwards, and may be found in varying degrees of completeness in most of the ancient texts.

Some of these correlations were a natural and harmless outcome of the basic hypothesis itself. The association of the elements with the seasons was obvious enough, and it had been on their association with the cardinal points that the various sequences had been built up. What could have been more unavoidable than to link fire with summer and the south? This must have been of considerable antiquity, since we find fire (i.e. heat, and the grain ripened by it) in the autumn harvest character (no. 38 in Table 11), and its existence in the character for south (no. 26) is possible. Then the tastes (and probably also the smells, though the relation is not so clear) strongly suggest primitive chemistry, as we have seen above (p. 244). The colours invite much speculation. Since the cradle of Chinese civilisation was the land of yellow loess soil in the upper Yellow River basin (modern Shansi and Shensi, cf. Sects. 4, 5), it is quite plausible to suppose that for the centre that colour imposed itself. Then white in the west would stand for the perpetual snows of the Tibetan massif, with green (or blue) in the east for the fertile plains or the seemingly infinite ocean.[d] Finally, red in the south may have taken its origin from the red soil of Szechuan, the region which lies just south of Shensi and Shansi; there are, moreover, large areas also of red soil in Yunnan and towards Indo-China.[e] But with growing complexity came growing artificiality and arbitrariness.

---

[a] Attention has already been directed (p. 43) to this weakness in ancient Chinese science.

[b] See on, p. 278.

[c] Cf. Granet (5), pp. 375 ff.; Forke (4), vol. 2, pp. 431 ff.; (6), p. 240; Mayers (1), p. 332, (4). The Chinese term for the category is given at the head of each column.

[d] These suggestions derive from a conversation with Dr Ong Wên-Hao, at that time Minister of Economics and National Resources, and from Ku Chieh-Kang (1).

[e] Whatever may be the specific causes which in China led to these identifications, it is certain that the principle of identifying colours with the directions of space is found also in far-distant cultures. It is clear from Soustelle (1), pp. 12, 30, 56, 73 ff. and figs. 4a, 5a, b, that the Aztecs associated black with the north, red with the east, blue with the south and white with the west. Spinden (1) also refers to this system (p. 126), and it may be found in the history of de Sahagun (1). For the Mayas there are similar indications in the history of de Landa; cf. also Morley (1); and Recinos, Goetz & Morley.

Table 12. *The symbolic correlations*

Elements hsing 行	Seasons shih 時	Cardinal points fang 方	Tastes wei 味	Smells chhou 臭	Stems (denary cyclical signs) kan 干	Branches (duodenary cyclical signs) and the animals pertaining to them chih 支	Numbers shu 數
WOOD	spring	east	sour	goatish	chia i 甲乙	yin 寅 (tiger) and mao 卯 (hare)	8
FIRE	summer	south	bitter	burning	ping ting 丙丁	wu 午 (horse) and ssu 巳 (serpent)	7
EARTH	—a	centre	sweet	fragrant	wu chi 戊己	hsü 戌 (dog), chhou 丑 (ox), wei 未 (sheep) and chhen 辰 (dragon)	5
METAL	autumn	west	acrid	rank	kêng hsin 庚辛	yu 酉 (cock) and shen 申 (monkey)	9
WATER	winter	north	salt	rotten	jen kuei 壬癸	hai 亥 (boar) and tzu 子 (rat)	6

Elements hsing 行	Musical notes yin 音	Hsiu hsiu 宿	Star-palaces kung 宮	Heavenly bodies chhen 辰	Planets hsing 星	Weather chhi 氣	States kuo 國
WOOD	chio 角	1–7	Azure Dragon	stars	Jupiter	wind	Chhi
FIRE	chih 徵	22–28	Vermilion Bird	sun	Mars	heat	Chhu
EARTH	kung 宮	—	Yellow Dragon	earth	Saturn	thunder	Chou
METAL	shang 商	15–21	White Tiger	hsiu constellations	Venus	cold	Chhin
WATER	yü 羽	8–14	Sombre Warrior	moon	Mercury	rain	Yen

Table 12 (continued)

Elements hsing 行	Rulers[b] ti 帝	Yin-Yang 陰陽	Human psycho-physical functions shih 事	Styles of government chêng 政	Ministries pu 部	Colours ssu 色	Instruments chhi 器
WOOD	Yü the Great [Hsia]	Yin in Yang or lesser Yang	demeanour	relaxed	Agriculture	green	compasses
FIRE	Wên Wang [Chou]	Yang or greater Yang	vision	enlightened	War	red	weights & measures
EARTH	Huang Ti [pre-dyn.]	Equal balance	thought	careful	the Capital	yellow	plumblines
METAL	Thang the Victorious [Shang]	Yang in Yin or lesser Yin	speech	energetic	Justice	white	T-squares
WATER	Chhin Shih Huang Ti [Chhin]	Yin or greater Yin	hearing	quiet	Works	black	balances

Elements hsing 行	Classes of living animals chhung 蟲	Domestic animals shêng 牲	'Grains' ku 穀	Sacrifices[c] ssu 祀	Viscera tsang 臟	Parts of the body thi 體	Sense-organs kuan 官	Affective states chih 志
WOOD	scaly (fishes)	sheep	wheat	inner door	spleen	muscles	eye	anger
FIRE	feathered (birds)	fowl	beans	hearth	lungs	pulse (blood)	tongue	joy
EARTH	naked (man)	ox	panicled millet	inner court	heart	flesh	mouth	desire
METAL	hairy (mammals)	dog	hemp	outer door	kidney	skin and hair	nose	sorrow
WATER	shell-covered (invertebrates)	pig	millet	well	liver	bones (marrow)	ear	fear

[a] As we have just seen, the sixth month was sometimes supposed to be under the sign of Earth.
[b] There are many variants of this list; I give the names which appear in the fragment from Tsou Yen himself on p. 238 above, adding that of the First (Chhin) Emperor who believed his sway to be under the sign of water.
[c] We have already noticed (p. 245) that certain gods and spirits, of which little is known, were connected with the five elements. I omit them as lacking scientific interest (cf. Forke (6), p. 233).

One must realise that the correspondences as shown in the table are only a few out of very many more. Eberhard (6) lists over a hundred of them, giving references to the texts. Moreover, they are full of discrepancies and may be stated in different ways in the same text. In a valuable discussion, he distinguishes several circles of scholars who each contributed their part to the large edifice or network which finally resulted. First there was the Astronomical Group. Significantly enough it was, like the Naturalists, associated with the State of Chhi. There is some evidence that it may have gone back to the time of the *Shih Ching* folksongs (perhaps −9th century),[a] but by the −4th century it produced one of the greatest astronomers in Chinese history, Kan Tê[1] (whom we shall meet with again in the section on astronomy), and in the −1st there was an important astrologer of the same family,[b] Kan Chung-Kho.[2] This astronomical group was certainly responsible for the correlations between the elements and the denary and duodenary cyclical signs (given, as Stems and Branches, in Table 12), the elements and the *hsiu*[3] ('mansions', the equatorial divisions of the celestial sphere), the elements and the planets, and between the elements and the feudal States (for astrological reasons).

Secondly, there come three groups which all seem to derive directly from Tsou Yen and therefore to deserve the name of the Naturalist Groups. Eberhard distinguishes them as the Emperor-series Group, the Yin-Yang Group and the Hung Fan Group.

It is quite clear that the emperor-series group was connected with Tsou Yen, since his identification of the successive (legendary) emperors with the powers of the elements was what had given him his great political importance. The question of the later developments of the theory is extremely complex, and there are exhaustive studies on it by Ku Chieh-Kang (6) and Haloun (3) as well as Eberhard (6). It is interesting to note that the Han dynasty was not at all sure from what element it gained its authority. In the early −2nd century the view of Chang Tshang prevailed, that water was still dominant, since the Chhin had ruled too short a time to exhaust its virtue. But in −165 Chia I urged that the dominant element was earth, and eventually in −104 a change was made which lasted until the end of the first phase of the dynasty.

The Yin-Yang group is very obscure, and its members hardly distinguishable from other Naturalists. Tsou Yen, as we have seen, himself discussed the Yin and Yang. The only Han or pre-Han texts in which the correlation of these with the elements occurs (see Table) are the *Kuan Tzu* (chapter 40) and the *Pai Hu Thung Tê Lun* (chapter 2) of Pan Ku. As we shall later see, however, it had an influence on subsequent biological thinking (p. 334).

The third of these groups has been called the Hung Fan group, i.e. those Naturalists who studied (perhaps even invented) the passages concerning the five elements in the *Shu Ching*. Here again the interest tended to the human, social and political. Correlations were made with the psycho-physical functions of man (a viewpoint which is very prominent in the five-element chapters of the *Chhien Han Shu*), the

---

[a] Eberhard (6), p. 65.   [b] Cf. Tsêng Chu-Sên (1), p. 124.

[1] 甘德   [2] 甘忠可   [3] 宿

different styles of government, the Ministries, forms of morality, and so on. The personalities who would have been connected with this group were above all Fu Shêng and his successors, together with Tung Chung-Shu.

This concludes the trends which can be closely connected with the Naturalists.

Lastly we have two groups of considerable scientific interest. The Yüeh Ling Group was primarily agricultural, and the Su Wên Group primarily medical. The *Yüeh Ling*[1] (Monthly Ordinances) is a long section of the *Li Chi* (Record of Rites),[a] where it replaces the shorter *Hsia Hsiao Chêng*[2] (Lesser Annuary of the Hsia Dynasty) found in the *Ta Tai Li Chi* (Record of Rites arranged by the elder Tai).[b] The *Yüeh Ling* is also found complete in the *Lü Shih Chhun Chhiu*, and in large part in other books such as *Huai Nan Tzu*. The correlations with the five elements for which these agriculturalists were responsible concerned the seasons of the year (with or without a central element), possibly the colours, certainly the classes of living animals, the domestic animals, the grains, the weather and probably the sacrifices and the small gods to whom they were offered. It is striking that rice does not appear among their list of grains, though it does so in a parallel correlation belonging to the medical group; presumably the former originated in north China, or at an earlier date. No names of personalities are associated with this school.

The Medical Group, called after the most ancient surviving Chinese medical text, the *Huang Ti Su Wên Nei Ching*[3] (Pure Questions of the Yellow Emperor; Canon of Internal Medicine), was responsible for the physiological correlations. The date of this text is very uncertain, but the bulk of it must be at least early Han, and some of it may be from the Warring States period. There are associations between the elements and the viscera, the parts of the body, the sense-organs, and the affective states of mind. No names of personalities connected with this medical group have come down to us.

Thus was established the far-reaching system of symbolic correlations.[c]

(3) CONTEMPORARY CRITICISM AND LATER ACCEPTANCE

It would be a great mistake to imagine that it did not receive severe criticism. We saw above (p. 259) that in the −3rd century the mutual conquest theory was attacked by the later Mohists. We are now in a position to appreciate the demonstration, given by Wang Chhung[4] in his *Lun Hêng*[5] (Discourses Weighed in the Balance) of the late +1st century, of the absurdities to which the symbolic correlations led. In his Wu Shih[6] (Things and their Mutual Influences) chapter he speaks as follows:

The body of a man harbours the *chhi* of the Five Elements, and therefore (so it is said) he practises the Five Virtues, which are the Tao of the elements. So long as he has the

---

[a] Tr. Legge (7), vol. 1, pp. 249 ff. [b] Tr. R. Wilhelm (6), pp. 233 ff.
[c] The origin of the correlation between the elements and the numbers, and between the elements and the musical notes, remains quite obscure. On European parallels see p. 296 below.

[1] 月令 [2] 夏小正 [3] 黃帝素問內經 [4] 王充 [5] 論衡
[6] 物勢

five viscera within his body, the *chhi* of the five elements are in order. Yet according to the theory (which associates different animals with each of the five elements), animals prey upon and destroy one another because they embody the several *chhi* of the five elements; therefore the body of a man with the five viscera within it ought to be the scene of internecine strife, and the heart of a man living a righteous life be lacerated with discord. But where is there any proof that the elements do fight and harm each other, or that the animals overcome one another in accordance with this?

The sign *yin*[1] corresponds to wood, and its proper animal is the tiger.[a] *Hsü*[2] corresponds to earth, and its animal is the dog. *Chhou*[3] and *wei*[4] likewise correspond to earth, *chhou* having as animal the ox, and *wei* having the sheep. Now wood conquers earth, therefore the tiger overcomes the dog, ox and sheep. Again, *hai*[5] goes with water, its animal being the boar. *Ssu*[6] goes with fire, having the serpent as its animal. *Tzu*[7] also signifies water, its animal being the rat. *Wu*[8] conversely, goes with fire, and its animal manifestation is the horse. Now water conquers fire, therefore the boar devours the serpent, and horses, if they eat rats (are injured by) a swelling of their bellies. (So run the usual arguments.)

However, when we go into the matter more thoroughly, we find that in fact it very often happens that animals do not overpower one another as they ought to do on these theories. The horse is connected with *wu* (fire), the rat with *tzu* (water). If water really conquers fire, (it would be much more convincing if) rats normally attacked horses and drove them away. Then the cock is connected with *yu*[9] (metal), and the hare with *mao*[10] (wood). If metal really conquers wood, why do cocks not devour hares? Or again, *hai* stands for the boar (and water), *wei* for the sheep (and earth), and *chhou* for the ox (also earth). If earth really conquers water, why do oxen and sheep not run after boars and kill them? Furthermore, *ssu* corresponds with the serpent and fire, *shen*[11] with the monkey and metal. If fire really conquers metal, why do serpents not eat monkeys? (On the other hand) monkeys are certainly afraid of rats, and are liable to be bitten by dogs, (yet this is equivalent to) water and earth conquering metal (—which is not in accordance with theory)....[b]

So important was the Chinese sceptical tradition (which will be examined in the following Section) that Wang Chhung was doubtless not the only critic of the five-element theories. In the beginning they were helpful, so far as I can see, rather than harmful, to scientific thought in China, and certainly no worse than the Aristotelian theory of the elements which dominated European medieval thinking. Of course the more elaborate and fanciful the symbolic correlations became, the further away from observation of Nature the whole system tended. By the time of the Sung (+11th century) it was probably having a definitely deleterious effect on the great scientific movement which then developed.

In order to illustrate this I will quote from the *Mêng Chhi Pi Than* (Dream Pool Essays) of Shen Kua (+1086). The example is telling because, as the reader who runs

---

[a] In this paragraph Wang Chhung must have had in mind the system (new in his time) by which a series of twelve animals was associated with the duodenary cyclical signs and so applied to hours, days, years and compass directions, in association with the five elements. We shall come back to this animal series in Section 20 on astronomy.

[b] Ch. 14, tr. Forke (4), vol. 1, p. 105; Chavannes (7), p. 31, mod.

| [1] 寅 | [2] 戌 | [3] 丑 | [4] 未 | [5] 亥 | [6] 巳 | [7] 子 |
| [8] 午 | [9] 酉 | [10] 卯 | [11] 申 | | | |

through the whole of this book will appreciate, Shen Kua was one of the most widely interested scientific minds which China produced in any age. He is speaking here about transformations of the elements.

In the Chhien Shan district of Hsinchow there is a bitter spring which forms a rivulet at the bottom of a gorge. When its water is heated it becomes *tan fan*[1] (bitter alum, lit. gall-alum; probably impure copper sulphate, RP 87). When this is heated it gives copper. If this 'alum' is heated for a long time in an iron pan the pan is changed to copper. Thus Water can be transformed into Metal—an extraordinary change of substance.

According to the *Su Wên* (the medical classic) there are five elements in the sky, and five elements on the earth. The *chhi* of earth, when in the sky, is moisture. Earth (we know) produces metal and stone (as ores in the mountains), and here we see that Water can also produce metal and stone. These instances are therefore proofs that the principles of the *Su Wên* are right.

Take another example. In certain caves, where water keeps dropping, stalactites (*chung ju*[2])[a] are formed in abundance. Or, at the vernal and autumnal equinoxes, water taken from certain wells forms 'stone flowers' (*shih hua*[3])[b] (on evaporation). Or from (certain) strong brines (*ta lu*[4]) the Yin-essence stone (*yin ching shih*[5])[c] is formed, which is always moist (hygroscopic). All these are concretions changed and converted from Water.

Similarly, the *chhi* of Wood, when in the sky, is Wind. Now wood can produce fire, and wind can foster it. Such is the nature of the five elements.[d]

This passage seems clearly to suggest that Shen Kua was prevented by a too uncritical acceptance of the five-element theory from attaining an understanding of the nature of solution and mixture. Yet we cannot place such an + 11th-century mind in the right perspective without tracing the parallel development of thought in Europe. The observation of the precipitation of metallic copper in powdery or solid form by iron, with the formation of iron sulphate, described in the opening paragraph, was an excellent one, and perhaps the earliest in any language (for the Plinian reference[e] is obscure and uncertain). T. T. Read (4, 8) mentions that in our own times a process for the winning of copper from mine waters by precipitation with scrap iron was developed at Butte, Montana, in ignorance of the fact that it had been known in Moorish Spain and at least from the + 13th century in China. Basil Valentine, in his *Currus Triumphalis Antimonii*, noted the power of iron to precipitate copper from 'an acrid ley in Hungary',[f] an effect which Paracelsus in the + 16th century believed

[a] RP 63. Lit. 'bell-milk' or 'suckling bell', so called because the dripping lime-laden water seemed like milk, and the shape of the concretion formed was roughly similar to the convex clay moulds used in the casting of bells.

[b] RP 65*b*. One cannot say what crystalline substance or deposit this was.

[c] RP 120, 126. Probably mixtures of ammonium chloride, calcium and sodium sulphates and sodium chloride, certainly hygroscopic, which for centuries have appeared in Chinese markets.

[d] Ch. 25, para. 6 (p. 4*b*), tr. auct.

[e] Pliny, *Hist. Nat.* XXXIV, 149 (Bailey (1), vol. 2, pp. 61, 188).

[f] There is, of course, much doubt as to the date of 'Basil Valentine'. The book is more probably of the early + 17th century than of the + 15th which it purports to be, though doubtless based on earlier material (see J. Read (1), p. 136; v. Lippmann (1), p. 640).

[1] 膽礬   [2] 鍾乳   [3] 石花   [4] 大滷   [5] 陰精石

demonstrated the transmutation of metals, as also Stisser as late as +1690.[a] Van Helmont surmised that the copper was in the solution beforehand, and Robert Boyle proved it in his *Treatise on the Mechanical Causes of Chemical Precipitation* of +1675. It would therefore be somewhat unjust to censure Shen Kua for accepting a transmutation which did not receive its true explanation for six centuries after his death. The question really is to what extent the true explanation of such phenomena was delayed by prolonged uncritical acceptance of blanket theories such as that of the five elements, and on this count it would be difficult to acquit them.

(4) 'PYTHAGOREAN' NUMEROLOGY; TSÊNG SHEN

Before we can come to an assessment of the whole system of thought, it will be desirable to give two specimens of ancient Chinese naturalistic thinking, both of which are contained in the *Ta Tai Li Chi*[1] (Record of Rites of the elder Tai). According to tradition this was put together from older material by Tai Tê[2] sometime between −73 and −49, about the same time as the Younger Tai, Tai Shêng,[3] his cousin, prepared a similar compilation, originally known as the *Hsiao Tai Li Chi*[4] (Record of Rites of the Younger Tai), but subsequently incorporated into the Confucian Canon as the *Li Chi*. It is now known, however, from the work of Tsuda and Hung, that 'Tai Tê's' compilation was made, not in the −1st century, but between +80 and +105, probably by a group under the leadership of Tshao Pao.[5] The two versions are thought to have represented the views of two ritualist schools. The portions of the *Ta Tai Li Chi* which I wish to give here probably date from about the same time as Huai Nan Tzu (−2nd century).[b] The first, which constitutes the 58th chapter, bears the title 'Thien Yuan'[6] (The Roundness of Heaven).

Shanchü Li[7] asked Tsêng Tzu[8][c] saying, 'It is said that Heaven is round and Earth square, is that really so?' He replied, 'What have you yourself heard about it?' Shanchü Li said,

---

[a] Roscoe & Schorlemmer (1), vol. 2, p. 413.

[b] Forke's view (13), p. 147, that this material goes back to the generation after Confucius, can no longer be accepted.

[c] Tsêng Shen, one of the most famous pupils of Confucius. Here he might be just a mouthpiece, but there are reasons for thinking that he, and his disciples in turn, were more interested in natural phenomena and the beginnings of natural science than any other of the Confucian groups. There is a story (*Mêng Tzu*, IV (2), xxxi) that when the city in which Tsêng Shen was living was attacked by marauders, he left, saying to the caretaker: 'Do not allow any persons to lodge in my house, lest they break and injure the plants and trees.' The discussion following turns on the question whether Tsêng, in his position as an invited teacher, had been wrong in leaving or not, but for us the important thing is the hint that Master Tsêng's garden may have been a kind of botanic garden. This would at least agree with the interest in natural history attributed to him here. It should also be remembered that Tsêng Shen was the traditionally accepted author of the *Ta Hsüeh*, which contains the famous phrase 'the extension of knowledge comes from the investigation of things' (cf. Vol. 1, p. 48). Though this book is now attributed to Yochêng Kho (c. −260), some small portions of text may have come down from Tsêng Shen.

[1] 大戴禮記	[2] 戴德	[3] 戴聖	[4] 小戴禮記	[5] 曹褒
[6] 天圓	[7] 單居離	[8] 曾子		

'Your disciple does not understand these things; that is why I dare to ask you about them.'

Tsêng Tzu said, 'That to which Heaven gives birth has its head on the upper side; that to which Earth gives birth has its head on the under side. The former is called round, the latter is called square. If heaven were really round and the earth really square the four corners of the earth would not be properly covered. Come nearer and I will tell you what I learnt from the Master. He said that the Tao of heaven was round and that of the earth square. The square is dark and the round bright. The bright radiates (thu[1]) chhi, therefore there is light outside it. The dark imbibes (han[2]) chhi, therefore there is light within it. Thus it is that fire and the sun have an external brightness, while metal and water have an internal brightness. That which irradiates is active (shih[3]), that which imbibes radiation is reactive (hua[4]). Thus the Yang is active and the Yin reactive.

The seminal essence (ching[5]) of the Yang is called shen.[6] The germinal essence of the Yin is called ling.[7a] The shen and the ling (vital forces) are the root of all living creatures; and the ancestors of (such high developments as) rites and music, human-heartedness and righteousness; and the makers of good and evil, as well as of social order and disorder.

When the Yin and the Yang keep precisely to their proper positions, then there is quiet and peace. But if (the balance) leans to one side, then there is wind, if they clash there is thunder, if they cross each other's path there is lightning, if they are in confusion there is fog and clouds, if they are at harmony there is rain. If the Yang force conquers, clouds and rain result;[b] if the Yin force conquers, ice and frost are formed. Absolute supremacy of the Yang leads to hail, the converse leads to sleet. These are the changes of the two fundamental forces.

Hairy animals acquire their coats before coming into the world, feathered ones similarly first acquire their feathers. Both are born by the power of Yang. Animals with carapaces and scales on their bodies likewise come into the world with them; they are born by the power of Yin. Man alone comes naked into the world; (this is because) he has the (balanced) essences of both Yang and Yin.

The essence (or most representative example) of hairy animals is the unicorn, that of feathered ones is the phoenix (or pheasant); that of the carapace-animals is the tortoise, and that of the scaly ones is the dragon. That of the naked ones is the Sage.

The dragon cannot rise up without wind;[c] the tortoise cannot foretell the future without (the application of) fire. These are examples of the action of the Yang on the Yin.

These four (numinous) animals are the aids of the spirit of the Sage. Thereby the sage can be the master of heaven and earth, the master of the mountains and rivers, the master of the gods and spirits, and the master of the sacrifices in the ancestral temple. The sage marks carefully the Numbers of the sun and moon, so that he can observe the motions of the stars and the constellations, and thence arrange the four seasons in order according to their progressions and retrogradations. This is called the "calendar" (li[8]).

---

[a] R. Wilhelm (6) translates these by 'Geist' and 'Seele' respectively. Others would have put *animus* and *anima*. Such renderings seem to assume too much, so it is safer to leave the words as technical terms.

[b] One may note this phrase in connection with its age-old use in China as a poetical expression for sexual intercourse.

[c] For comments on this and later speculations tending towards the theory of flight, see Sect. 27j.

[1] 吐   [2] 含   [3] 施   [4] 化   [5] 精   [6] 神   [7] 靈
[8] 曆

The sages invented the twelve musical tubes, so as to provide standards for the eight notes, high and low, clear or blurred. These are called the "pitchpipes" (lü[1]).

The pitchpipes are in the domain of the Yin but they govern Yang proceedings. The calendar comes from the domain of the Yang, but it governs Yin proceedings. The pitchpipes and the calendar give each other a mutual order, so closely that one could not insert a hair between them.[a]

The sages established the five (kinds of) rites, in order to give the people a visible (standard). They ordained the five (degrees of) mourning, in order to distinguish between (what was due) to nearer and further relatives. They made music for the five-holed pipe, in order to encourage the *chhi* of the people. They put together (in various combinations) the five tastes, so as to observe the preferences of men. They established the proper places for the five colours, gave names to the five grains, and decided upon the relative standing of the five sacrificial animals....[b]

And all this is what is meant by saying that the root of all living creatures was also the origin of rites and music, and the maker of good and evil, as well as of social order and disorder.'[c]

The passage has a Pythagorean flavour,[d] and thus serves well as a prelude to the next one. After the geometrical opening there is a striking paragraph about the radiant energy which emanates (lit. spits forth) from the Yang, and is received (lit. tasted in the mouth) by the Yin. There follows a rather detailed meteorological development of the theme, and the speaker then passes to the biological world, where the animals are divided into Yang and Yin classes. The sudden and at first surprising remark that the sage is the chief representative of the naked animals gives the key to the whole passage. It is really a supreme statement of the truth stated by Granet[e] that Chinese thought refused to separate Man from Nature, or individual man from social man. This comes out first in the hint that the human microcosm carries its head, round like heaven, upwards; then in the firm statement (which could not be bettered by the most convinced modern exponent of evolutionary naturalism) that the basic forces seen at work in the lowest creatures[f] are the same as those which will at higher levels develop the highest manifestations of human social and ethical life; then in the appearance of the sage, who is set against the background of all Nature, and by virtue of his deep connection with it is able to be its master and ruler, even commanding the gods and spirits (who are seen as immanent natural forces, not as transcendent superhuman beings); and finally in the picture of human social organisation as the product of Nature, though indeed its highest product.

[a] This conviction goes back at least to the time of Liu Hsin (−50 to +22) and probably to the calendar-maker Têng Phing (fl. −104). Cf. p. 286 below.

[b] Several liturgiological sentences about sacrifices omitted.

[c] Tr. R. Wilhelm (6), p. 127, eng. auct.

[d] For a more detailed analysis of this parallelism, see Fêng Yu-Lan (1), vol. 2, pp. 93 ff.

[e] (5), pp. 338, 415. Cf. pp. 191, 196 above; 281, 368, 453, 488 below.

[f] Note that the word used for living creatures in the text is *chhung*,[2] a character which is simply the 'insect' radical three times repeated.

[1] 律   [2] 蟲

The second passage forms chapter 81 of the *Ta Tai Li Chi*.[a] It is mainly biological, but the Pythagorean flavour is more marked. We see the development of what may perhaps be called a 'numerology', a playing with numbers in which things are related which today we know do not have any simple relation with one another. The chapter is entitled 'I Pên Ming'[1] (The Metamorphoses of Life). It says:

The Master said, '(The Principle of) Change has brought into existence men, birds, animals, and all the varieties of creeping things, some living solitary, some in pairs, some flying and some running on the ground. And no one knows how things seem to each of them. And he alone who profoundly scrutinises the virtue of the Tao can grasp their basis and their origin.

Heaven is 1, Earth is 2, Man is 3. 3×3 makes 9. 9×9 makes 81. 1 governs the sun. The sun's number is 10. Therefore man is born in the tenth month of development.

8×9 makes 72. Here an even number follows after an odd one. Odd numbers govern time. Time governs the moon. The moon governs the horse. Therefore the horse has a gestation period of 11 months.

7×9 makes 63. 3 governs the Great Bear (Northern Dipper). This constellation governs the dog. Therefore the dog is born after only 3 months.

6×9 makes 54. 4 governs the seasons. The seasons govern the pig. Therefore the gestation time of the pig is 4 months.

5×9 makes 45. 5 governs the musical notes. The notes govern the monkey. Therefore the monkey is born after 5 months' development.

4×9 makes 36. 6 governs the pitchpipes. The pitchpipes govern the deer; therefore it remains 6 months in the womb.

3×9 makes 27. 7 governs the stars. The stars govern the tiger. Therefore the tiger is born in the 7th month.

2×9 makes 18. 8 governs the wind. The wind governs insects. This is why insects undergo changes in the 8th month of the year. And so it goes with all living things, each according to their kind.

Now birds and fishes are born under the sign of the Yin, but they belong to the Yang. This is why birds and fishes both lay eggs. Fishes swim in the waters, birds fly among the clouds. But in winter, the swallows and starlings go down into the sea and change into mussels.[b]

The habits of the various classes of animals are very different. Thus silkworms eat but do not drink, while cicadas drink but do not eat, and ephemeral gnats and flies do neither. Animals with scales and carapaces eat during the summer, and in winter hibernate. Animals with beaks (birds) have 8 openings of the body and lay eggs. Animals which masticate (mammals) have 9 openings of the body and nourish their young in wombs. Quadrupeds have neither feathers nor wings. Horned animals have no incisor teeth in their mouths. Animals which have neither horns nor incisor teeth are fat (pigs). Animals which have no feathers and no molar teeth are also fat (sheep). Animals born by day take after their paternal parents, those born by night take after their maternal ones. [When the Yin component prevails the offspring is female, when the Yang prevails it is male.][c]

---

[a] There is a parallel passage in *Khung Tzu Chia Yu*, ch. 25, and the numerological masterpiece with which the chapter opens occurs also in *Huai Nan Tzu*, ch. 4, pp. 6b, 7a (tr. Erkes (1), p. 61).

[b] We shall study this and other ancient beliefs on metamorphosis in Section 39 on zoology.

[c] The sentence in square brackets is added in the parallel passage in *Huai Nan Tzu*.

[1] 易 本 命

As to the earth, the east-west direction is the weft and the north-south direction the warp. In the mountains virtues accumulate,[a] and the rivers bring profit.[b] Heights correspond to life, depths to death. Mountains and hills are male, gorges and valleys female.[c]

Mussels, tortoises, and pearl (-oysters) wax and wane according to (the phases of) the moon. [d]

Men who live in places where the earth is solid, grow fat, those who live on loose soil are tall, those who live on sandy soil are thin. Fine-looking men are produced from places where the earth is fertile, but a poor soil breeds ugly ones.[e]

Animals which live in the water swim well and can endure cold, those which live in the earth (lit. eating the earth) have no hearts and do not breathe (e.g. worms). Those that eat wood (-y plants) are strong and wild (e.g. bears); those that eat grass are good runners and voiceless (e.g. deer); those that eat mulberry-leaves spin silk and turn into moths; those that eat flesh are fierce and bold (e.g. tigers); those that eat cereal grains are wise and ingenious (man); those that live on *chhi* (air) are illuminated and long-living (Taoist immortals); those that eat nothing at all are deathless and spiritlike.

Of feathered animals there are 360 kinds[f] and the phoenix is their headman; of hairy animals 360 kinds and the unicorn is their headman; of animals with carapaces 360 kinds and the tortoise is their headman; of scaly animals 360 kinds and the dragon is their headman; and of naked animals 360 kinds, and the Sage is their headman. These are the beautiful things which Heaven and Earth (lit. the Donator and the Receptor) have produced, and the numbers of the animals among the ten thousand beings.[g]

If a human ruler likes to destroy nests and eggs, the phoenix will not rise. If he likes to drain the waters and take out all the fishes, the dragon will not come. If he likes to kill pregnant animals and murder their young, the unicorn will not appear. If he likes stopping the watercourses and filling up the valleys, the tortoise will not show itself.[h]

Thus the (real) king moves only in accordance with the Tao, and rests only in accordance with *Li*,[1] (the principles of things and the tendency of the universe). If he acts contrary to these, Heaven will not send him long life, evil omens will appear, the spirits will hide themselves, wind and rain will not come at their usual times, there will be storms, floods and droughts, the people will die, the harvest will not ripen and domestic animals will have no increase.'[i]

We see, therefore, that the Naturalists, or whoever it was who wrote these passages, having a keen interest in the world of living organisms, made many creditable observations. But it was all in a framework of number-mysticism, as the opening paragraph shows. Traces of this were already evident in the table of symbolic correlations.

[a] Presumably a reference to the growth of ores in the hills, of which the same chapter of *Huai Nan Tzu* speaks (see on, in the Mineralogical Section, and with reference to alchemy, Sects. 25, 33).
[b] Presumably a reference to irrigation.
[c] Here is one of the basic texts of geomancy; see on, pp. 359 ff.
[d] Cf. Vol. 1, p. 150, in relation to culture-contacts, and below, in Zoology, Sect. 39.
[e] Cf. the passage from *Kuan Tzu*, quoted on p. 45 above.
[f] There would be, of course, on account of the approximate number of days in the year.
[g] Wilhelm (6) and Granet (5), pp. 138, 326, insist on reading 10,000 as 11,520, since that is the numerical value of the total number of lines in the 64 hexagrams (see below), but I do not feel convinced that that was the origin of the phrase, which is perhaps better rendered vaguely by 'myriad'.
[h] Many have seen, not unreasonably, early efforts at nature-conservation in passages such as this.
[i] Tr. R. Wilhelm (6), p. 250, eng. auct.

[1] 理

The time has nearly come to attempt an evaluation of the system which we have been discussing, but before doing so it should be pointed out that though these numerological formulations began their career in the −3rd century or a little before, and though they were of great interest to the scholars of the Han,[a] they still retained all their fascination for many minds as late as the Sung. The significance of this will be better appreciated shortly. Thus Tshai Chhen[1] (+1167 to +1230),[b] who was a direct pupil of Chu Hsi himself (see on, pp. 472 ff.), engaged in elaborate numerological speculations.

If one follows the numbers (shu[2]) (of all things) then one can know their beginnings, if one traces them backwards then one can know how it is that they come to an end. Numbers and Things are not two separate entities, and beginnings and endings are not two separate points. If one knows the numbers, then one knows the things, and if one knows the beginnings then one knows the endings. Numbers and Things continue endlessly—how can one say what is a beginning and what is an ending?[c]

We need not reproduce examples of this kind of numerical symbolism, which had nothing in common with true mathematics, and followed closely the kind of model already reproduced from the *Ta Tai Li Chi*. The point of interest is that +12th-century minds of the Neo-Confucian school could still be fascinated by it.

(e) THE THEORY OF THE TWO FUNDAMENTAL FORCES

Up to the present point more has been said about the five elements and their symbolic correlations than about the Yin and the Yang because we know rather more about the historical origin of the former theory. As we have seen, the two fundamental forces are not mentioned in any of the surviving fragments of Tsou Yen, though his school was called the Yin-Yang Chia, and in the *Shih Chi* and other documents, discussion of them is definitely attributed to him. There can be very little doubt that the philosophical use of the terms began about the beginning of the −4th century, and that the passages in older texts which mention this use are interpolations made later than that time.

Etymologically the characters are certainly connected with darkness and light respectively. The character Yin (cf. Table 11, no. 63, p. 227) involves graphs for hill (-shadows) and clouds; the character Yang has slanting sunrays or a flag fluttering in the sunshine, if indeed it does not represent a person holding the perforated disc of jade which was the symbol of heaven, the source of all light, and which may have been originally (cf. Sect. 20g) the most ancient astronomical instrument. These ideas

[a] They are abundantly found, for instance, in *Huai Nan Tzu*, ch. 3.
[b] Forke (9), p. 274.
[c] *Sung Yuan Hsüeh An*, ch. 67, p. 15a, tr. Forke (9), p. 277, eng. auct.

[1] 蔡沉   [2] 數

correspond with the way in which the terms are used in the *Shih Ching* collection of ancient folksongs. Yin evokes, as Granet (5) says, the idea of cold and cloud, of rain, of femaleness, of that which is inside and dark, such as the underground chambers in which ice was conserved against the summer. Yang evokes the idea of sunshine and heat, of spring and summer months, of maleness, and may refer to the appearance of a male ritual dancer. It is agreed also that Yin meant the shady side of a mountain or a valley (north of the mountain and south of the valley), the 'hubac' side; while Yang meant the sunny side (south of the mountain and north of the valley), the 'adret' side.[a]

Those who have studied the first appearance of the words as philosophical terms[b] find the *locus classicus* in the fifth chapter of the fifth appendix of the *I Ching* (the Hsi Tzhu;[1] the 'Great Appendix'), where the statement is made 'One Yin and One Yang; that is the Tao!' (*I Yin i Yang chih wei Tao*[2]).[c] The general sense must be that there are only these two fundamental forces or operations in the universe, now one dominating, now the other, in a wave-like succession. This appendix would date (at the earliest) from the late Warring States period (early − 3rd century).[d]

Other early mentions are those in the *Mo Tzu*, the *Chuang Tzu* and the *Tao Tê Ching*. The Book of Master Mo refers to the Yin and Yang twice in the technical sense; in chapter 6 where it says that every living creature partakes of the nature of Heaven and Earth and of the harmony of the Yin and the Yang, and in chapter 27 where the virtue of the sage-kings is said to have brought the Yin and Yang, the rain and the dew, at timely seasons. In *Chuang Tzu* the words are common; one may find at least twenty passages in which they occur in the technical sense. They once appear in the *Tao Tê Ching* (chapter 42, quoted on p. 78) where living creatures are said to be surrounded by Yin and to envelop Yang, and that the harmony of their life processes depends upon a harmony of these two *chhi*. Translators have been cautious about giving the words here their full technical sense, on account of the (now abandoned) early dating of Lao Tzu, but I believe that they should have it.

In other places the general view is that the mentions are later interpolations, e.g. *Shu Ching* (Chou Kuan chapter);[e] *Tso Chuan* (half a dozen passages).[f] But in such books as the *Hsün Tzu*, the *Li Chi* and *Ta Tai Li Chi* and the *Huai Nan Tzu*, which date from the − 3rd to the + 1st centuries, there is no reason to suspect any change of

---

[a] Granet (2), p. 245, says he borrowed these terms from the 'terminologie alpestre' and that they derive from *ad opacum* and *ad rectum* respectively.

[b] E.g. Liang Chhi-Chao (*1, 4*); Rousselle (2); Conrady (4).

[c] There has been great division of opinion as to how this all too simple affirmation should be translated. We cannot now accept as very satisfactory the version of Legge (9), p. 355: 'The successive movement of the inactive and active operations constitutes what is called the course (of things).' Granet (5), p. 119, gives a long discussion of the matter.

[d] We shall say a few words below (p. 306) on the difficult question of the dates of the *I Ching*.

[e] Ch. 40; see Legge (1), p. 228; Medhurst (1), p. 289.

[f] The common view (e.g. Forke (6), p. 170) that the philosophical use of the words goes back to the − 2nd millennium is now quite untenable.

[1] 繫辭   [2] 一陰一陽之謂道

the text. One piece which may be of considerable antiquity is the fragment known as *Chi Jan*,[1] apparently a chapter from a lost book purporting to be of the − 5th century, which gives the words of a more or less historical character, Chi Ni Tzu.[2] It describes the conversations which he carried on with Kou Chien, King of the southern State of Yüeh.[a] The text certainly seems to represent a Naturalist tradition of southern coastal origin, and most probably it would be contemporaneous with Tsou Yen. The king, meditating an invasion of the neighbouring State of Wu, asked his adviser about it. Chi Ni Tzu declined to talk about military affairs, and urged the king instead to observe natural phenomena in order to increase agricultural productivity and so enrich his people.

Chi Ni Tzu said, 'You must observe the *chhi* of Heaven and Earth, trace the (activities of the) Yin and the Yang, and know the Ku-Hsü.[3][b] You must understand survival and death. Only then can you weigh up your enemy....'

The king replied, 'Your principles are excellent.' So he observed the phenomena of the heavens (*yang kuan thien wên*[4]), collected and investigated the constellations and their positions (*chi chha wei hsiu*[5]) and devoted himself to the calendar (*li hsiang ssu shih*[6]). Thus his country became rich. And he rejoiced, saying, 'If I become the leader of all the kings, it will be due to the good planning of Chi Ni Tzu.'[c]

In order to gain a glimpse of the way in which the Han Confucians argued about these matters, we may look at a part of the fifty-seventh chapter of Tung Chung-Shu's *Chhun Chhiu Fan Lu* (*c.* − 135).[d] He says:

When Heaven is about to make the Yin rain down, men fall sick; that is, there is a movement prior to the actual event. It is the Yin beginning its complementary response (*hsiang ying*[7]). Also when Heaven is about to make the Yin rain down, men feel sleepy. This is the *chhi* of the Yin. There is (moreover) melancholy which makes men feel sleepy, this is the effect of Yin on Yin; and there is delight which keeps men fully awake, this is the Yang attracting the Yang. At night (the Yin time) the waters (a Yin element) flood more, by several inches. When there is an east wind, (fermenting) wine froths up more. Sick men are very much worse at night. When dawn is about to break, the cocks all crow and jostle one another; the morning's *chhi* invigorates their *ching*.[8] Thus it is that Yang reinforces Yang, and Yin reinforces Yin, and accordingly the (manifestations of the) two *chhi*, whether Yang or Yin, can reinforce or diminish each other.

[a] This interesting material will later call for comment in several different connections (Sects. 18, 25, 33).

[b] This rare term is explained as meaning the gate of heaven and the door of the earth. In later times it became a term for lucky and unlucky in divination, and it is certainly connected with the relation between the denary and duodenary cyclical signs.

[c] *Chi Jan*, Fu Kuo ch. (How to make the Country Prosperous); preserved in the *Wu Yüeh Chhun Chhiu*,[9] ch. 9 (Spring and Autumn Annals of the States of Wu and Yüeh); *YHSF*, ch. 69, p. 27 *b*, tr. auct.

[d] More material of this kind will be found in the following Section, on Wang Chhung and the Sceptical Tradition.

[1] 計然　　　　[2] 計倪子　　　[3] 孤虛　　　[4] 仰觀天文　　　[5] 集察緯宿
[6] 歷象四時　　[7] 相應　　　[8] 精　　　[9] 吳越春秋

Heaven has the Yin and Yang, and so has man. When the Yin *chhi* of Heaven and Earth begins (to dominate), the Yin *chhi* of man responds by taking the lead also. Or if the Yin *chhi* of man begins to advance, the Yin *chhi* of Heaven and Earth must by rights respond to it by rising also. Their Tao is one. Those who are clear about this (know that) if rain is to come, then the Yin must be activated and its influence set to work. If the rain is to stop, then the Yang must be activated and its influence set to work. (In fact), there is no reason at all for assuming anything miraculous (lit. connected with spirits, *shen*[1]) about the causation and onset of rain, though (indeed) its rationale (*li*[2]) is profoundly mysterious.[a]

For Tung Chung-Shu, the Yin and the Yang were only the supreme examples of all the polar opposites or 'correlates' in the world, pairs for which he used the technical term *ho*,[3] as in his chapter 53.[b]

Although it more properly belongs to our discussion of the symbolic hexagrams of the *I Ching* (Book of Changes), each of which is composed of six lines, whole or broken, corresponding to the Yang and the Yin respectively, this is perhaps the place to refer to one of the later elaborations of the system. Each of the hexagrams was primarily Yin or primarily Yang, and by a judicious arrangement it was possible to derive all sixty-four of them in such a way as to produce alternating Yin and Yang ones by continual dichotomy. I reproduce a diagram (Fig. 41) from the *I Thu Ming Pien*[4] (Clarification of the Diagrams in the Book of Changes) by Hu Wei[5] (+1706),[c] in which one may see how, for instance, the original Yang half splits into two, one of which is Yin and one Yang; each of these again splits into two, one of which is Yin and one Yang. The process continues until the sixty-four hexagrams are formed, and could naturally go on *ad infinitum*. The Yin and Yang components never become fully separated, but at each stage, in any given fragment, only one of them is manifested. It cannot but interest the scientific mind because the path of thought thus trodden by the *I Ching* scholars was one to which we have become accustomed in modern scientific thinking, namely, it was a principle of segregation. There is an analogy with what we now know as the recessive and dominant factors in a genotype, only the latter of which appear outwardly and visibly by their manifestation in the phenotype. More broadly speaking, the process recalls the phenomena seen in the morphogenesis of many animals (e.g. echinoderms, fishes, amphibia),[d] for which it has been necessary to develop the conception of morphogenetic fields. Here, then, is another example parallel with what was said above about the supposed interactions of the five elements leading to paths of thought which have, in our time, attained what one might call 'valid application' to Nature. In this case there may be an analogy not only with modern genetics and embryology but also perhaps with chemistry, in that successive steps of purification will lead only gradually to the separation of substances. In so far

---

[a] Tr. auct. with D. Leslie, adjuv. Hughes (1), p. 306.
[b] Cf. Fêng Yu-Lan (1), vol. 2, p. 42.
[c] It is based on the original chart of Shao Yung (+1011 to +1077) in the *Sung Yuan Hsüeh An*, ch. 10; a simplified form of which was given by Tshai Chhen (+1167 to +1230).
[d] Needham (12), pp. 127 ff., 271 ff., 477 ff., 656 ff.

[1] 神   [2] 理   [3] 合   [4] 易圖明辨   [5] 胡渭

PLATE XVI

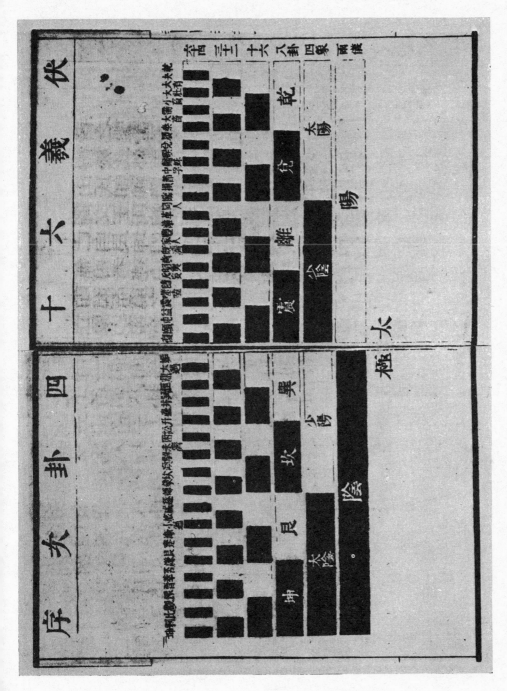

Fig. 41. Segregation Table of the symbols of the Book of Changes (*Fu-Hsi Liu-shih-ssu Kua Tzhu Hsü*), from Chu Hsi's *Chou I Pên I Thu Shuo* (+12th cent.), reproduced in Hu Wei's *I Thu Ming Pien*, ch. 7, pp. 2b, 3a, and elsewhere. Yin and Yang separate, but each contains half of its opposite in a 'recessive' state, as is seen when the second division occurs. There is no logical end to the process but here it is not followed beyond the stage of the 64 hexagrams.

as the *I Ching* scholars intuited that no matter how long the purification of material substance might be carried on there would still remain the positive and the negative combined together, even though in appearance one or the other might dominate, their thought was quite close, after all, to the perspectives of modern science. Indeed, their thinking here was 'field' thinking, though perhaps few of them could consciously have pointed to the fact that the north and south poles of a magnet are reproduced no matter how much the magnet may be divided into smaller magnets.[a] The point I am trying to make is that some elements of the structure of the world as modern science knows it were prefigured in their speculations. If these were divorced from the perfected study of Nature in experiment and mathematically formulated hypothesis, they were by no means unreasonable.

One thought which arises in the mind when examining the diagram is that *if* there was any undertone of attribution of good and evil to the Yang and Yin respectively, then the formulation is rather Manichaean. Mani's Persian followers (cf. Burkitt, 1) believed that man's duty was to sort out the good from the evil constituents in the mixture-universe, but that it was a task which perhaps would never be achieved. Elsewhere (Sect. 7b) something has already been said about the suggestions which have been made that the Yin-Yang theory owed its origin to stimulation from Persian religious dualism. The chief difficulty in believing this is that undertones of good and evil were in fact *not* present in the Chinese formulations of Yin-Yang theory. On the contrary, it was only by the attainment and maintenance of a real balance between the two equal forces that happiness, health or good order could be achieved. However, the attempt to derive the Yin-Yang of China from the dualism of Persia (e.g. Zoroastrianism) still continues (cf. P. Schmidt, 1). It is hardly possible to evaluate it until more is known of the dualistic myths and cosmologies of Iran and India (Przyłuski (2); Sheftelowitz) and their possible relation with Mesopotamian origins. It is too early to conclude with Rey[b] that the Chinese world-picture had no success outside the land of its origin. There is now indeed a tendency to return to a former view, which on the contrary derived Iranian dualism from Chinese Yin-Yang sources (de Menasce (1); Mazaheri). This, however, is based mostly upon the work of de Saussure (18, 19), valuable in other respects, but which in a matter such as this suffers from an often gross over-estimation of the antiquity of ancient Chinese texts.[c] In any case, the immense success which the theory met with in China testifies, as Bodde has said,[d] to the Chinese tendency to find in all things an underlying harmony and unity rather than struggle and chaos.

I am not sure that here again we do not have to deal with ideas of such simplicity that they might easily have arisen independently in several civilisations. There must be some weight in the connection made by Granet (1, 2) between the Yin-Yang theory and the social manifestations of sex-differences in early Chinese society, the seasonal festivals where the young people chose their mates and danced in ceremonial formations

[a] But magnetism is a Chinese science, see Sect. 26i.   [b] (1), vol. 1, p. 412.
[c] He frequently did not hesitate to draw conclusions about the −3rd millennium.
[d] (7), p. 22; (14).

which symbolised the eternal and profound duality in Nature. Moreover, what is not so often referred to in this connection is that one can find elements of this dualism, though of course in rudimentary form as compared with China, from one end of European history to the other. Freeman describes[a] the dualistic cosmology of the Pythagorean school (−5th century), embodied in a table of ten pairs of opposites.[b] On one side there was the limited, the odd, the one, the right, the male, the good, motion, light, square and straight. On the other side there was the unlimited, the even, the many, the left, the female, the bad, rest, darkness, oblong and curved.[c] All this is reminiscent of the Chinese system, but there is nothing to connect the two, unless we make the speculative assumption that some similar sort of polarity doctrine was originally Babylonian and spread thence in two directions.[d]

At the other end of the European story we have certain 17th-century thinkers who derived inspiration from the traditional Jewish mysticism of the Kabbalah, e.g. Robert Fludd (+1574 to +1637), whose thought has been closely analysed by Pagel (1). Fludd's *Medicina Catholica* pictured God as a chemist rather than a mathematician, with the world as his 'elaboratory'. In this world there was a series of polar opposites—on the one hand heat, movement, light, dilatation, attenuation; on the other hand cold, inertia, darkness, contraction, inspissation. To the sun, the father, the heart, the right eye and the blood corresponded the moon, the mother, the uterus, the left eye and mucus. It is especially interesting to find here the ancient opposition of condensation and dispersion, though it could have been of pre-Socratic origin more probably than Chinese. Nevertheless, one must admit that the alchemical interests of Fludd were no coincidence, and that the polarity of opposites (usually gold and mercury) runs through the whole of late medieval and 17th-century alchemy (cf. Muir (1); J. Read). Here we must not anticipate what will appear in Section 33 on alchemy, but if it is true (and all evidence points to it) that Chinese alchemy came to Europe through Muslim channels, then in a sense the Yin-Yang doctrine came with it, and Fludd inevitably acquired an indebtedness to Tsou Yen and Lao Tzu, even though he himself could never have been aware of the fact. And also there may be a sense in which all these ancient polarity theories lie buried in the foundations of the science of chemistry, since the reactivity of chemical substances depended, for the alchemists, on their position with regard to this polarity, and today we know that reactivity is only the outward and visible sign of the arrangements of those ultimate electrical charges, negative and positive, which make up what we call the material world.

---

[a] (1), pp. 81, 83, 136, 248.   [b] Cf. Aristotle, *Metaphys.* 1, 5.

[c] In Parmenides (c. −475) there was also a polarity-theory (Cornford (1), p. 219). Brightness, warmth, lightness (rarity), fire and maleness were said to exist. Darkness, cold, heaviness (density), earth and femaleness were said not to exist, i.e. to be the absence of the former qualities. Cf. Forke (6), p. 221. See further, pp. 296 ff. below.

[d] Loewenstein (1) has pointed out that the famous symbolic representation of Yin and Yang ☯ is similar to the indubitable swastika designs found on Chinese Neolithic pottery and also on Chou bronzes. There has been great divergence of opinion about the origin of the swastika, as may be seen from the literature which Loewenstein cites, but it is certainly Neolithic and almost certainly a dualistic fecundity symbol. Hence its connection with Yin and Yang. Perhaps it has something to do with the S-spiral designs so common on Yangshao pottery (cf. Vol. 1, p. 81).

(f) CORRELATIVE THINKING AND ITS SIGNIFICANCE; TUNG CHUNG-SHU

Let us recapitulate. The scientific or proto-scientific ideas of the Chinese involved two fundamental principles or forces in the universe, the Yin and the Yang, negative and positive projections of man's own sexual experience; and five 'elements' of which all process and all substance was composed. With these five elements were aligned and associated, in symbolic correlation, everything else in the universe which could be got into a fivefold arrangement. Around this central fivefold order was a larger region comprising all the classifiable things which would only go into some other order (fours, nines, twenty-eight),[a] and much ingenuity was shown in fitting the classifications together. Hence the number-mysticism or numerology, one of the main purposes of which was to relate the various numerical categories. What does it all mean?

Most European observers have written it off as pure superstition which prevented the rise of true scientific thinking among the Chinese. Not a few Chinese, especially natural scientists in modern times, have been inclined to adopt the same opinion. But their situation was somewhat different, since they had to deal with many thousands of traditional Chinese scholars who, unschooled in the modern scientific view of the world, still imagined that the ancient thought-system of China was a live issue as an alternative. Dying proto-scientific theories clung tenaciously to undying ethical philosophy. But our task is not concerned with the modernisation of Chinese society, which is quite capable of modernising itself; what we have to examine is whether in fact the ancient and traditional Chinese thought-system was merely superstition, or simply a variety of 'primitive thought', or whether perhaps it had something in it which was characteristic of the civilisation which produced it, and contributed some stimulus to other civilisations.

The first approach to the fivefold system of symbolic correlations was a sociological one. Durkheim & Mauss (1)[b] suggested that the numerical categories adopted had originally been based on the exogamous clan or phratry groups in primitive society. While others had thought that the exogamous groups were modelled on the categories, Durkheim & Mauss more plausibly proposed that it was the other way round. They had no difficulty in showing that for several cultures, such, for example, as the Amerindian Zuñis, there was a clear correspondence between the exogamous clans and the numerical categorisation; in the case of the Zuñis everything went in sevens

---

[a] Mayers (1) lists 317 such categories; he took them from the *Tu Shu Chi Shu Lüeh* (+1707) of Kung Mêng-Jen (already noted, Vol. 1, p. 50). But it gives a rather startling sidelight on Chinese thinking when we find that no less than eleven chapters of the *Thu Shu Chi Chhêng* encyclopaedia are consecrated to this subject in its *mathematical* section (*Li fa tien*, chs. 129–40). Bodde (5) has devoted a special paper to Chinese 'categorical thinking' in which he analyses the curious tabulation in the twentieth chapter of the *Chhien Han Shu*, where nearly 2000 historical and semi-legendary individuals were arranged in nine grades according to their virtue.

[b] Their Chinese evidence was based mainly on de Groot (2).

and there were seven clans.[a] But for the Chinese it was much more difficult to establish any such explanation, for the origins of the civilisation go too far back. Moreover, even if it could be established, it would not affect our estimate of the intellectual value of the more or less completed world-picture.

The approach of the analysts of magic was perhaps more interesting. Frazer, in his classical work, had stated two 'laws' and one general principle of magic. There was the 'law of similarity', according to which, for ancient magicians (and also those of modern primitive peoples), like produces like. There was the 'law of contiguity or contagion', according to which things which have been in contact but which are so no longer, continue to act upon one another. One can immediately see how the Chinese symbolic correlations would have worked in this respect, and one begins to visualise some of the motives which led to their establishment. Other scholars accepted and exemplified Frazer's theories of sympathetic magic, and also his general principle that while 'religion conciliates, magic constrains'. Some added further definitions, such as Hubert & Mauss (1, 2), who pointed out that magic involved primarily the isolated, solitary, operator, rather than the collectivity of religion. All agree, however, that 'magic has nourished science, and the earliest scientists were magicians.... Magic issues by a thousand fissures from the mystical life, from which it draws its strength, in order to mingle with the life of the laity and to serve them. It tends to the concrete, while religion tends to the abstract. It works in the same sense as techniques, industry, medicine, chemistry and so on. Magic was essentially an art of *doing* things.'[b] There is no need to labour the point, which has already been sufficiently well stressed in the Section on Taoism. The symbolic correlation system was exactly what the magicians needed for carrying on their operations. At that ancient stage of thought, how could they know what would conduce to the success of a technique and what would not? There had to be some way of choosing the conditions for the experiment, and naturally if one intended to do something which concerned water, it would obviously not help to wear red, which was the colour of fire. Of course the correlations were intuitive, not strictly rational. What else could they be?

A number of modern students—H. Wilhelm,[c] Eberhard (6), Jabłoński (1), and above all Granet (5)—have named the kind of thinking with which we have here to do, 'coordinative thinking' or 'associative thinking'. This intuitive-associative system has its own causality and its own logic.[d] It is not either superstition or primitive superstition, but a characteristic thought-form of its own. H. Wilhelm contrasts it with the 'subordinative' thinking characteristic of European science, which laid such emphasis on external causation. In coordinative thinking, conceptions are not subsumed under one another, but placed side by side in a *pattern*, and things influence one another not

---

[a] This parallel was made much of by Forke (4), App. 1; (6), in his account of the Chinese world-picture. Cf. also Haloun (3); Tsêng Chu-Sên (1), p. 76; Fei Hsiao-Thung (1). So also Soustelle (1) gives a table of Aztec correspondences (compass-points, colours, stages, winds, celestial bodies, birds, gods, years, etc.), p. 75. But there the relation with clans was not obvious.

[b] Hubert & Mauss (1).   [c] (1), esp. p. 35; (4), p. 45.

[d] Cf. what has been said earlier (pp. 52, 199) about Taoist and Mohist logic; also Granet (5), p. 336.

by acts of mechanical causation, but by a kind of 'inductance'. In the Section on Taoism (pp. 55, 71, 84) I spoke of the desire of the Taoist thinkers to understand the causes in Nature, but this cannot be interpreted in quite the same sense as would suit the thought of the naturalists of ancient Greece. The key-word in Chinese thought is *Order* and above all *Pattern* (and, if I may whisper it for the first time, *Organism*). The symbolic correlations or correspondences all formed part of one colossal pattern. Things behaved in particular ways not necessarily because of prior actions or impulsions of other things, but because their position in the ever-moving cyclical universe was such that they were endowed with intrinsic natures which made that behaviour inevitable for them.[a] If they did not behave in those particular ways they would lose their relational positions in the whole (which made them what they were), and turn into something other than themselves. They were thus parts in existential dependence upon the whole world-organism.[b] And they reacted upon one another not so much by mechanical impulsion or causation as by a kind of mysterious resonance.[c]

Nowhere are such conceptions better stated than in the fifty-seventh chapter of Tung Chung-Shu's −2nd-century *Chhun Chhiu Fan Lu*, which is entitled 'Thung Lei Hsiang Tung',[1] i.e. (in Hughes' excellent translation) 'Things of the Same Genus Energise Each Other'. We read:

If water is poured on level ground it will avoid the parts which are dry and move towards those that are wet. If (two) identical pieces of firewood are exposed to fire, the latter will avoid the damp and ignite the dry one. All things reject what is different (to themselves) and follow what is akin.[d] Thus it is that if (two) *chhi* are similar, they will coalesce;[e] if notes correspond, they resonate. The experimental proof (*yen*[2]) of this is extraordinarily clear. Try tuning musical instruments. The *kung*[3] note or the *shang*[4] note struck upon one lute will be answered by the *kung* or the *shang* notes from other stringed instruments. They sound by themselves. This is nothing miraculous (*shen*[5]), but the Five Notes being in relation; they are what they are according to the Numbers (*shu*[6]) (whereby the world is constructed).

(Similarly) lovely things summon others among the class of lovely things; repulsive things summon others among the class of repulsive things. This arises from the complementary

---

[a] Thus it came naturally to Yang Hsiung (*c.* −20) to say: 'All things are generated by intrinsic (impulses), (only) their withering and decay comes partly from without (*Wan wu chhüan yü yü nei, chhu lo yü wai*[7]).' Cf. p. 540 below.

[b] A living philosopher, Chang Tung-Sun, has said that 'the concept of all things forming one body has been a persistent tendency in Chinese thought from the beginning until now', (3), p. 117. In the +11th century Chhêng Hao said: 'The myriad patterns are all subsumed in the Great Pattern (*Wan li kuei yü i li yeh*[8])', *ECCS, Honan Chhêng shih I Shu*, ch. 14, p. 1a, ch. 15, p. 11a. See also pp. 12, 191, 196, 270 above; 368, 453, 471, 488, 581 below.

[c] Zimmer (1), in his profound study of Indian thought, came across traces of similar thinking. I find that he uses the expression 'organism of the universe' several times (pp. 14, 56). Perhaps the Chinese view of parts of the whole has a parallel in the Indian *sva-dharma* or 'intrinsic *dikaiosune*'. Cf. p. 304 below.

[d] *Pai wu chhü chhi so yü i, erh tshung chhi so yü thung.*[9] This statement is fundamental for the fivefold system of correlative thought. Things 'go' with each other according to definite rules.

[e] An anticipation of much in modern colloid chemistry and experimental morphology.

[1] 同類相動   [2] 驗   [3] 宮   [4] 商   [5] 神   [6] 數
[7] 萬物權輿於內徂落於外   [8] 萬理歸於一理也
[9] 百物去其所與異而從其所與同

way in which a thing of the same class responds (*lei chih hsiang ying erh chhi yeh*[1])—as for instance if a horse whinnies another horse whinnies in answer, and if a cow lows, another cow lows in response.

When a great ruler is about to arise auspicious omens first appear; when a ruler is about to be destroyed, there are baleful ones beforehand. Things indeed summon each other, like to like, a dragon bringing rain, a fan driving away heat, the place where an army has been being thick with thorns.[a] Things, whether lovely or repulsive, all have an origin. (If) they are taken to constitute destiny (it is because) no man knows where that origin is (*mei o chieh yu tshung lai; i wei ming; mo chih chhi chhu so*[2])....[b]

It is not only the two *chhi* of the Yin and the Yang which advance and retreat (*chin thui*[3])[c] according to their categories. Even the origins of the varied fortunes, good and bad, of men, behave in the same way. There is no happening that does not depend for its beginning upon something prior, to which it responds because (it belongs to the same) category, and so moves (*wu fei chi hsien chhi chih, erh wu i lei ying chih, erh tung chih yeh*[4])....

(As I said) when the note *kung* is struck forth from a lute, other *kung* strings (near by) reverberate of themselves in complementary (resonance); a case of comparable things being affected according to the classes to which they belong (*tzhu wu chih i lei tung chê yeh*[5]). They are moved by a sound which has no visible form, and when men can see no form accompanying motion and action, they describe the phenomenon as a 'spontaneous sounding' (*tzu ming*[6]). And wherever there is a mutual reaction (*hsiang tung*[7]) without anything visible (to account for it) they describe the phenomenon as 'spontaneously so' (*tzu jan*[8]). But in truth there is no (such thing as) 'spontaneously so' (in this sense). (I.e. every thing in the universe is attuned to certain other things, and changes as they change.) That there are (circumstances which) cause a man to become what in fact he is, we know. So also things do have a real causative (power), invisible though this may be....[d]

Now the classifiability of which Tung Chung-Shu is speaking is the capacity of the various things in the universe to go into the fivefold categorisation or others of various numerical values. It is extremely interesting that he takes the phenomenon of acoustic resonance as his demonstration experiment.[e] To those who could know nothing of sound-waves it must have been very convincing, and it proved his point that things in the universe which belonged to the same classes (e.g. east, wood, green, wind, wheat) resonated with, or energised, each other. This was not mere primitive undifferentiatedness, in which anything could affect anything else; it was part of a very closely knit universe in which only things of certain classes would affect other

---

[a] Allusion to *Tao Tê Ching*, ch. 30.

[b] This, as Hughes rightly says, is an important statement regarding causation. There follows the passage already quoted in connection with the Yin and Yang (p. 275).

[c] Yet another hint at wave-motion.

[d] Tr. auct. with D. Leslie, adjuv. Hughes (1), p. 305; Bodde, in Fêng Yu-Lan (1), vol. 2, p. 56.

[e] In this he follows an earlier statement couched in almost the same words—*Lü Shih Chhun Chhiu*, ch. 63 (vol. 1, p. 122), tr. R. Wilhelm (3), p. 161. Cf. *I Ching* (Wên Yen), R. Wilhelm (2), vol. 2, p. 11, Baynes tr., p. 15; *Chuang Tzu*, ch. 24, tr. Legge (5), vol. 2, p. 99.

[1] 類之相應而起也   [2] 美惡皆有從來以爲命莫知其處所   [3] 進退
[4] 無非已先起之而物以類應之而動者也   [5] 此物之以類動者也
[6] 自鳴   [7] 相動   [8] 自然

things of the same class. Wang Chhung says this in so many words in the + 1st century,[a] adding that it happens naturally, without purpose or striving (*Wu lei hsiang chih, fei yu wei yeh*[1]).[b] And thus causation was of a very special character, since it acted in a sort of stratified matrix and not at random. Inductance or resonance could be considered a kind of cue from one declining process indicating that it was time for the proper rising process to come upon the stage. Nothing was un-caused, but nothing was caused mechanically. The organic system in the prompter's book governed the whole. And the characters in the eternal dramatic cycle were, as has been said, in existential dependence upon the totality of the system, since if they failed in their cues they would cease to exist. But nothing ever did fail.

Later on[c] we shall have occasion to quote that famous sentence of Heracleitus: 'The Sun will not transgress his measures, otherwise the Erinyes, the bailiffs of Dike, will find him out.' Here a phenomenon of Nature could rebel, and could be forced into submission by the executive branch of a cosmic constitution. With great acuity, Misch[d] found the complementary passage to this in a text and its commentary from the *I Ching* (Book of Changes). Speaking of the top line in the first hexagram, *Chhien*, the text says 'The dragon exceeds its proper bounds; there will be occasion for repentance.'[e] Then the Wên Yen commentary explains:

This phrase 'exceeds its proper bounds' means that it knows now to advance but not to retire, how to survive but not how to be dissolved, how to obtain but not how to let go. He alone is the sage who, knowing progression and retrogression, coming into being and passing away, never loses his true nature. Truly he alone is the sage.

But the sage is only finding out what all natural bodies, celestial and terrestrial, spontaneously know and perform. Misch rightly maintained that Chinese thinkers in all the descriptions which they gave of the regularity of natural processes had in mind, not government by law, but the mutual adaptations of community life.[f] Harmony was regarded as the basic principle of a world-order 'spontaneous and organic'.[g] With this in mind, we can see in a new light the poetical philosophy of Hsün Tzu (cf. p. 27 above), who went so far as to exalt *li*[2] (good customs and traditional observances sanctioned by generally accepted morality) to the level of a universal cosmological principle. Not in human society only, but throughout the world of Nature, there was a give and take, a kind of mutual courtesy rather than strife among inanimate powers and processes, a finding of solutions by compromise,

---

[a] E.g. 'The *chhi* of like things intercommunicate, and their natures being mutually stimulated, respond (*Thung lei thung chhi, hsing hsiang kan tung*[3])'; *Lun Hêng*, ch. 10, tr. Forke (4), vol. 2, pp. 1 ff. and, better, Leslie (1). Cf. p. 304 below, on *kan* and *ying*.

[b] *Lun Hêng*, ch. 19 (Forke (4), vol. 2, p. 187). Cf. also particularly ch. 47.

[c] In Sect. 18 below, p. 533.   [d] (1), p. 196.

[e] Ch. 1, pp. 2b, 9a. Tr. R. Wilhelm (2), Baynes tr. vol. 2, p. 16; Legge (9), p. 417; mod.

[f] Pp. 122, 170, 206, 240.   [g] P. 210.

[1] 物類相致非有爲也   [2] 禮   [3] 同類通氣性相感動

an avoidance of mechanical force,[a] and an acceptance of the inevitability of birth and doom for every natural thing.

If this expresses something deeply true, as I believe it does, about the Chinese world-picture, of which the fivefold correlations were the abstract chart, then clearly the scholars of the Han and later times were not simply stuck in the mud of 'primitive thought' as such. We are all greatly indebted to Lévy-Bruhl for one of the most interesting analyses of primitive thought, and though we can accept much in his description of it, we shall have to conclude that he was far from justified in his belief that the Chinese and Indian world-pictures exemplified it. Lévy-Bruhl's account first aroused my interest because of his striking statement, 'For the primitive mind, everything is a miracle, or rather, nothing is; and therefore everything is credible, and there is nothing either impossible or absurd.'[b] I came across this just as I was noting, not without amusement, in a Taoist context (p. 443 below), the irritation shown by certain Christian scholars at the characteristic Chinese (really Taoist) attitude to miracles, namely, a readiness to accept them as a fact, but an incapacity to see that they proved anything, except that the magician must have possessed a peculiarly potent technique. Now the pre-logical mind, says Lévy-Bruhl, is insensitive both to logical and physical absurdity. Anything can be the 'cause' of anything else. If a steamship with one funnel more than usual calls at a small African town, and an epidemic follows, the appearance of the steamship is just as likely as anything else to be regarded as the cause. The selection of 'causes' at random from this undifferentiated magma of phenomena was called by Lévy-Bruhl the 'law of participation' in that the whole of the environment experienced by the primitive mind is laid under contribution, i.e. participates, in its explanations, without regard either for true causal connection or for the principle of contradiction.[c]

The point at which we have to diverge from Lévy-Bruhl's analysis is where he proceeds to describe coordinative or associative thinking as a variety of primitive thinking. Primitive in the chronological sense it may well be, but a mere department of 'participative' thought it surely is not. For once a system of categorisations such as the five-element system is established, then anything can by no means be the cause of anything else. It would seem truer to visualise that there were (at least) two ways of advance from primitive participative thought, one (the way taken by the Greeks) to refine the concepts of causation in such a manner as to lead to the Democritean

---

[a] In his strange opposition to Bacon, Locke and Newton, William Blake was asserting something very similar in the face of the 'industrial revolution'. For example, in *Jerusalem*:

'...cruel Works
Of many Wheels I view, wheel without wheel, with cogs tyrannic
Moving by compulsion each other, not as those in Eden, which,
Wheel within wheel, in freedom revolve in harmony and peace.' (1. 15.)
Cf. Bronowski (1), p. 87.   [b] (1), p. 377.

[c] I add one or two more touches of the brushwork of Lévy-Bruhl's picture. For the primitive mind, disease is never purely physical, death never natural. Every unusual phenomenon is a sign. Divination is an added perception, designed to discover mystical relations within the participating collectivity of man and Nature. Magic proceeds to utilise them.

account of natural phenomena;[a] the other, to systematise the universe of things and events into a pattern of structure, by which all the mutual influences of its parts were conditioned. On one world-view, if a particle of matter occupied a particular point in space-time, it was because another particle had pushed it there; on the other, it was because it was taking up its place in a field of force alongside other particles similarly responsive. Causation was thus not 'particulate' but 'circumambient.' Peering down the long avenues of time we can perhaps see the Newtonian universe at the end of the former view, and the Whiteheadian universe at the end of the latter. Yet so far as the development of modern natural science was concerned, the former was doubtless the latter's indispensable historical antecedent.

The idea that things which belonged to the same classes resonated with, or energised, each other, though so characteristic of Chinese thought, was not without parallels in Greece. Cornford (2) has detected these in what he calls the maxims of popular belief accepted by the philosophers from 'common sense' without scrutiny. Take Aristotle's three kinds of change. Movement in space was explained by asserting that like attracts like; growth, by asserting that like nourishes like; and change of quality, by asserting that like affects like. 'Democritus held that agent and patient must be the same or alike; for if different things act upon one another, it is only accidentally by virtue of some identical property.'[b] But there was also an opposite set of maxims that like things repelled one another—'Everything desires, not its like, but its contrary.'[c] All this has an evident relationship with the ideas of the pre-Socratics about 'love' and 'hatred' in natural phenomena, and it would be easy to see the origin of it in social practices, exogamy or endogamy, sympathetic magic, and so on. The point to be emphasised here is that while Greek thought moved away from these ancient ideas towards concepts of mechanical causation foreshadowing the complete break of the Renaissance, Chinese thought developed their organic aspect, visualising the universe as a hierarchy of parts and wholes suffused by a harmony of wills.

The primitive world-picture, says Lévy-Bruhl, is superseded as the definition and differentiation of concepts of beings and objects goes on. But if these concepts should crystallise at an intermediate stage, a civilisation may have to pay dearly for it. They will be thought adequate for reality when they are really not.

The system [he goes on] will claim to be self-sufficing, and then mental activity applied to these concepts will exert itself indefinitely without any contact with the reality which they claim to represent. Chinese scientific knowledge affords a striking example of this arrested development. It has produced immense encyclopaedias of astronomy, physics, chemistry, physiology, pathology, therapeutics and the like, and yet to our minds all this is nothing but balderdash. How can so much effort and skill have been expended in the long course of ages,

---

[a] The Aristotelian account had more in common with Chinese ideas, but it was too biological and modern science had to discard it in order to be born.

[b] Aristotle, *De Generatione et Corruptione*, 323 b 10.

[c] Plato, *Lysis*, 215 c.

and yet their product be absolutely nil? This is due, no doubt, to a variety of causes, but above all to the fact that the foundation of each of these so-called sciences rests upon crystallised concepts, concepts which have never really been submitted to the test of experience, and which contain scarcely anything beyond vague and unverifiable notions with mystical preconnections. The abstract, general form in which these concepts are clothed allows of a double process of analysis and synthesis which is apparently quite logical, and this process, always futile yet ever self-satisfied, is carried on to infinity. Those who are best acquainted with the Chinese mentality—like de Groot, for instance—almost despair of seeing it free itself from its shackles, and cease revolving on its own axis. Its habit of thought has become too rigid, and the need it has begotten is too imperious. It would be as difficult to put Europe out of conceit with her savants as to make China give up her physicians, doctors and *fêng-shui* professors.[a]

And Lévy-Bruhl adds a few similar strictures on Indian scientific thought into the bargain.

It would be hard to find a passage more misguided. By what right this eminent scholar, who could not read a single word of the encyclopaedias which he was condemning, dismissed the scientific and technological achievements of that civilisation to which his own owed so much, is not clear.[b] Obviously the historical effects of the numerous Chinese technological discoveries were uninfluenced by the quality of the world-picture of those who made them. Nor is the mass of empirical information contained in the despised encyclopaedias worth any the less because the world-picture of those who wrote them was not that which proved to be essential for the development of the science of Galileo and Newton. On the contrary, our proper conclusion seems to me to be that the conceptual framework of Chinese associative or coordinative thinking was essentially something different from that of European causal and 'legal' or nomothetic thinking. That it did not give rise to 17th-century theoretical science is no justification for calling it primitive. What remains to be seen is whether it was not related to a view of the world which modern science is now being obliged to incorporate into its own structure, namely, the philosophy of organism. And if so, the moment has come to ask the question, from what roots did the philosophy of organism arise?

I am anxious to get this point of divergence perfectly clear. Chinese coordinative thinking was *not* primitive thinking in the sense that it was an alogical or pre-logical chaos in which anything could be the cause of anything else, and where men's ideas were guided by the pure fancies of one or another medicine-man. It was a picture of an extremely and precisely ordered universe, in which things 'fitted', 'so exactly that you could not insert a hair between them' (cf. above, p. 270). But it was a universe in which this organisation came about, not because of fiats issued by a supreme

---

[a] P. 380.

[b] His reliance on de Groot (2), who was a specialist in the folk-customs and popular demonology of Amoy, could be paralleled by someone who would undertake to describe the world-outlook of the educated Englishman solely on the basis of the otherwise admirable accounts of British folklore by such writers as Cecil Sharp or Maud Gomme.

creator-lawgiver, which things must obey subject to sanctions imposable by angels attendant; nor because of the physical clash of innumerable billiard-balls in which the motion of the one was the physical cause of the impulsion of the other. It was an ordered harmony of wills without an ordainer; it was like the spontaneous yet ordered, in the sense of patterned, movements of dancers in a country dance of figures, none of whom are bound by law to do what they do, nor yet pushed by others coming behind, but cooperate in a voluntary harmony of wills.[a] 'No one was ever seen to command the four seasons', we shall read later (p. 561), 'yet they never swerve from their course.' And, however absurd may have been the conviction that dread evils would follow his failure,[b] the ritual of the emperor was the supreme manifestation of this belief in the oneness of the universal pattern. In the proper pavilion of the Ming Thang[1] or Bright House,[c] no less his dwelling-place than the temple of the universe, the emperor, clad in the robes of colour appropriate to the season, faced the proper direction, caused the musical notes appropriate to the time to be sounded, and carried out all the other ritual acts which signified the unity of heaven and earth in the cosmic pattern. Or to speak of scientific matters, if the moon stood in the mansion[d] of a certain equatorial constellation at a certain time, it did so not because anyone had ever ordered it to do so, even metaphorically, nor yet because it was obeying some mathematically expressible regularity depending upon such and such an isolatable cause—it did so because it was part of the pattern of the universal organism that it should do so, and for no other reason whatsoever.

The contrast between the two views of the universe, Chinese and modern, comes out very clearly in the use of numbers. Of course there were the Pythagoreans in Europe,[e] and much creditable mathematics, as a later section will show (Sect. 19), was done in China, but the correlative thinking of the Chinese involved quite naturally a number-mysticism—numerology, I have called it—which is just as distasteful to the modern scientific mind as the numerological fancies about the Great Pyramid. So far as I can see, this facet of correlative thinking contributed nothing to Chinese science, though its inhibitory effect was probably not very great either in view of all the other inhibitory influences. Bergaigne has excellently said:[f] 'Instead of the number depending on the actual (empirical) plurality of the objects perceived or pictured, it is, on the contrary, the objects whose plurality is defined by receiving its form from a mystical number decided upon (as if in a prepared framework) before-hand.' No one really interested in Chinese thought should fail to read the chapter of

---

[a] The dance metaphor comes so readily to the mind in considering Chinese organicism that the disappearance of dancing from late Chinese society seems very strange. But it remained full of vitality at least until the end of the middle ages. One of the most beautiful poems of the Han period ('The Dancers of Huainan', tr. Waley, 11) was written by Chang Hêng, one of its greatest scientists. And contemporary China has made a dazzling rediscovery of the dance.

[b] See below, pp. 378 ff.   [c] Cf. Granet (5), pp. 180 ff.; Soothill (5).
[d] The hsiu,[2] see below, Sect. 20e.
[e] Whose influence on Renaissance scientific thought was great. Cf. Sect. 19k below.
[f] (1), vol. 2, p. 156.

[1] 明堂   [2] 宿

Granet (5) on numerical symbolism.[a] 'The notion of the quantitative', he said, 'plays practically no role in the philosophical speculations of the (ancient) Chinese. Nevertheless number as such passionately interested their sages. But however great the arithmetical or geometrical knowledge of the corporations of surveyors, carpenters, architects, chariot-builders, and musicians, may have been, the sages never took any interest in it, except in so far as it facilitated (without ever being allowed to carry the sage away to uncontrollable consequences) what can only be called "numerical games". Numbers were manipulated as if they were symbols....'[b] And elsewhere: 'Numbers did not have the function of representing magnitudes, they served to adjust concrete dimensions to the proportions of the universe.'[c] Doubtless no criticism of ancient and medieval Chinese numerology can be too harsh. Yet I would suggest that both this and the more extravagant extensions of the symbolic correlations of the five elements were exaggerations of certain basic ideas as valid in their way, and as valuable for the future history of human thought, as those other basic ideas which gave rise, in the European middle ages, to extravagances such as the trials of animals by due process of law.[d]

For the ancient Chinese, time was not an abstract parameter, a succession of homogeneous moments, but was divided into concrete separate seasons and their subdivisions.[e] Space was not abstractly uniform and extended in all directions, but was divided into the regions, south, north, east, west and centre.[f] And they joined together in the tables of correspondences; the east was indissolubly connected with the spring and with wood, the south with summer and fire. When I read the words of Jabłoński (1), expounding the views of his master Granet, 'This idea of correspondence has great significance and replaces the idea of causality, for things are *connected* rather than caused', I vividly recalled the passage from Chuang Tzu already quoted (p. 52), where he compares the universe to the animal body. 'The hundred parts of the body are all complete in their places. Which should one prefer? Do you like them all equally? Are they all servants? Are they unable to control one another and need a ruler? Or do they become rulers and servants in turn? Is there any true ruler other than themselves?' The answers to Chuang Chou's rhetorical questions were certainly all intended to be no. Two centuries later Tung Chung-Shu repeated the thought when he wrote[g] that 'the constant course of Nature is that things in opposition to each other cannot both arise simultaneously (*Thien chih chhang Tao, hsiang fan chih wu yeh pu tê liang chhi*[1]). The Yin and Yang (for example) move parallel to each other, but not along the same road; they meet one another, and each in turn operates as the controller (*Ping hsing erh pu thung lu, chiao hui erh ko tai li*[2]). Such is their pattern (*Tzhu chhi wên*[3]).' The implication was that the universe itself is a vast

---

[a] Pp. 151 ff.   [b] P. 149.   [c] Pp. 273, 283.

[d] I shall return to this subject later on, cf. pp. 574 ff.

[e] Granet (5), p. 88; Hubert & Mauss (2), p. xxxi.

[f] Granet (5), p. 96. The ideas of the Mohists on space and time (see on, Sect. 26 c) were, of course, much more 'modern'.

[g] *Chhun Chhiu Fan Lu*, ch. 51 (tr. Bodde in Fêng Yu-Lan (1), vol. 2, p. 24, mod.).

[1] 天之常道相反之物也不得兩起   [2] 並行而不同路交會而各代理   [3] 此其文

organism, with now one and now another component taking the lead—spontaneous and uncreated it is, with all the parts of it cooperating in a mutual service which is perfect freedom, the larger and the smaller playing their parts according to their degree, 'neither afore nor after other'.[a]

In such a system causality is reticular and hierarchically fluctuating, not particulate and singly catenarian. By this I mean that the characteristic Chinese conception of causality in the world of Nature was something like that which the comparative physiologist has to form when he studies the nerve-net of coelenterates, or what has been called the 'endocrine orchestra' of mammals. In these phenomena it is not very easy to find out which element is taking the lead at any given time. The image of an orchestra evokes that of a conductor, but we still have no idea what the 'conductor' of the synergistic operations of the endocrine glands in the higher vertebrates may be. Moreover, it is now becoming probable that the higher nervous centres of mammals and man himself constitute a kind of reticular continuum or 'nerve-net' much more flexible in nature than the traditional conceptions of telephone wires and exchanges visualised (Danielli & Brown).[b] At one time one gland or nerve-centre may take the highest place in a hierarchy of causes and effects, at another time another, hence the phrase 'hierarchically fluctuating'.[c] All this is quite a different mode of thought from the simpler 'particulate' or 'billiard-ball' view of causality, in which the prior impact of one thing is the sole cause of the motion of another.[d] 'The conviction that the universe and each of the wholes composing it have a cyclical nature, undergoing alternations, so dominated (Chinese) thought that the idea of succession was always subordinated to that of interdependence. Thus retrospective explanations were not felt to involve any difficulty. Such and such a lord, in his lifetime, was not able to obtain the hegemony, because, after his death, human victims were sacrificed to him.'[e] Both facts were simply part of one timeless pattern.[f]

[a] Why should I not make use of numinous phrases from my own civilisation? The history of European thought contains some elements, after all, akin to that of China. There was Alcmaeon of Crotona with his *isonomia* (ἰσονομία), the democratic principle of balance in the humours of the body, opposed to *monarchia* (μοναρχία), and the forerunner of the *krasis* (κρᾶσις), or right mixture, of Hippocrates and Aristotle. There was the democratic element in all Christian thought. And perhaps the background of 'correlative thinking' may help to explain those vital and universal qualities of true democracy inherent in Chinese society which everyone who has lived in that country has experienced.

[b] This question is connected with the maintenance of steady states by closed sequences of dependence or feedbacks, now being studied by physiologists and communication engineers alike; cf. Tustin (1). This will come up again in connection with the south-pointing carriage (Sect. 27c below), the first cybernetic machine.

[c] Dr R. H. Shryock of Baltimore has pointed out to me that one might add here the example of historical causation at the sociological level. This would agree with the position of the Chinese as having possessed greater historical sense than that of any other ancient civilisation.

[d] On the high abstraction of causal chains see Hanson (1). Cf. Graham (1), p. 104.

[e] Granet (5), p. 330. Ssuma Chhien relates this (cf. Chavannes (1), vol. 2, p. 45) about Duke Mu of the State of Chhin.

[f] It would be right here to point out that this kind of retrospective causality has some similarity with the final cause of Aristotle. But it would be necessary to add that one of the greatest efforts of Renaissance science was directed (successfully) to ridding itself of final causes (e.g. in Francis Bacon). The final cause may be considered an anomaly in European thought, due to the individual genius of Aristotle.

Granet does not use the word 'pattern' because it has no exact equivalent in the French language, but that best expresses the conclusion of his thought.[a] I am convinced that his insight was sure when throughout his books (especially 5) he emphasised the concept of *Order* as at the basis of the Chinese world-picture.[b] Social and world order rested, not on an ideal of authority, but on a conception of rotational responsibility.[c] The Tao was the all-inclusive name for this order, an efficacious sum-total, a reactive neural medium; it was not a creator, for nothing is created in the world, and the world was not created.[d] The sum of wisdom consisted in adding to the number of intuited analogical correspondences in the repertory of correlations.[e] Chinese ideals involved neither God nor Law.[f] The uncreated universal organism, whose every part, by a compulsion internal to itself and arising out of its own nature, willingly performed its functions in the cyclical recurrences of the whole, was mirrored in human society by a universal ideal of mutual good understanding, a supple régime of interdependences and solidarities which could never be based on unconditional ordinances, in other words, on laws.[g] As is said in a fine passage in one of the Han apocrypha, the *Li Wei Chi Ming Chêng*:[1]

The movements of the rites accord with the *chhi* of Heaven and the *chhi* of Earth. When the four seasons are in mutual accord, when the Yin and Yang complement each other, when the sun and moon give forth their light (unimpeded by fogs or eclipses), and when superiors and inferiors are in intimate harmony with one another, then (all) things, (all) persons and (all) animals, are in accord with their own natures and functions (*ju chhi hsing ming*[2]).[h]

Thus the mechanical and the quantitative, the forced and the externally imposed, were all absent.[i] The notion of Order excluded the notion of Law.[j]

When I first read Granet's work on Chinese thought in Lanchow in 1943 I noted this: 'Instead of observing successions of phenomena, the (ancient) Chinese registered alternations of aspects. If two aspects seemed to them to be connected, it was not by

[a] I understand (from Dr E. Balazs) that Prof. Demiéville now uses the word 'ordonnancement' to translate Neo-Confucian *Li*.[3] See p. 476 below.
[b] E.g. p. 24. Tsêng Chu-Sên (1) has also recently given a good summary of it (pp. 71–82) and does not hesitate to apply to it the word 'holistic' (pp. 98, 137, 165). As Eitel (3) said more than seventy years ago, Chinese thought has always adhered to the view of Nature as one organic whole.
[c] (5), p. 145.   [d] P. 333.
[e] P. 375.   [f] P. 588.
[g] It may, of course, be said that the converse would be a truer statement, namely, that the Chinese conceptions of the world mirrored the characteristics of their society. To this I subscribe, but we must await the concluding sections of this book before we can explore its meaning (see also below in this Section, p. 337).
[h] *Ku Wei Shu*, ch. 18, p. 1*a*; tr. Bodde in Fêng Yu-Lan (1), vol. 2, p. 126, mod. Parallel passage in *Chin Shu*, ch. 11, p. 9*b*.
[i] Chinese music naturally shares the qualities of Chinese organic thought. Its watchword was always 'order without mechanical symmetry' (Dr Laurence Picken in a lecture, June 1954).
[j] Granet (5), pp. 589, 590.

[1] 禮緯稽命徵   [2] 如其性命   [3] 理

means of a cause and effect relationship, but rather 'paired' like the obverse and the reverse of something, or to use a metaphor from the Book of Changes, like echo and sound,[a] or shadow and light.'[b] In the margin I wrote 'A morphological view of the universe'. But I then had little conception of how true this was.

(1) ROOTS OF THE PHILOSOPHY OF ORGANISM

It was no part of Granet's purpose to consider what effect, if any, the Chinese organismic view of the world had at any time upon European thought. Having sketched a synthetic reconstruction of it in its ancient form his task was accomplished. But our curiosity demands further satisfaction. In a later Section (16) I shall seek to show that the greatest of all Chinese thinkers, Chu Hsi in the +12th century, developed a philosophy more akin to the philosophy of organism than to anything else in European thought. Behind him he had the full background of Chinese correlative coordinative thinking, and ahead of him he had—Gottfried Wilhelm Leibniz.

Here it is not possible to do more than mention the great movement of our time towards a rectification of the mechanical Newtonian universe by a better understanding of the meaning of natural organisation. Philosophically the greatest representative of this trend is undoubtedly Whitehead, but in its various ways, with varying acceptability of statement, it runs through all modern investigations in the methodology and the world-picture of the natural sciences—the numerous and remarkable developments of field physics, the biological formulations which have put an end to the sterile strife between mechanism and vitalism[c] while avoiding the obscurantism of the earlier 'Ganzheit' schools,[d] the Gestalt-psychology of Kohler; then on the philosophical level the emergent evolutionism of Lloyd Morgan and S. Alexander, the holism of Smuts, the realism of Sellars, and last but by no means least the dialectical materialism (with its levels of organisation) of Engels, Marx and their successors. Now if this thread is traced backwards, it leads through Hegel, Lotze, Schelling and Herder to Leibniz (as Whitehead constantly recognised), and then it seems to disappear.[e] But is that not perhaps in part because Leibniz had studied the doctrines of the Neo-Confucian school of Chu Hsi, as they were transmitted to him through the Jesuit translations and despatches?[f] And would it not be worth examining whether something of that

[a] See Legge (9), p. 369.

[b] Granet (5), p. 329. These examples did not really illustrate the point which Granet was trying to make, for the sound is prior to the echo, and the obstruction to the shadow. What he had in mind were patterns simultaneously appearing in a vast field of force, the dynamic structure of which we do not yet understand. C. G. Jung realised that the Chinese world-outlook had involved a causality principle other than that of Galilean-Newtonian science, and he spoke of it as 'synchronistic' (Wilhelm & Jung, p. 142).

[c] Woodger (1), v. Bertalanffy (1, 2), A. Meyer (1, 2), Needham (9, 10), Gerard (1).

[d] How burning this question still is among contemporary biologists may well be seen in some recent pages of Dalcq (e.g. (1), p. 125).

[e] Von Bertalanffy (2), p. 195, for example, finds no precursor other than Nicholas of Cusa.

[f] At the end of the present Section we shall see a remarkable instance of this.

originality which enabled him to make contributions radically new to European thought was Chinese in inspiration? It would hardly be wrong to say that Leibniz's monads were the first appearance of organisms upon the stage of occidental theorising. Whitehead[a] has pointed out that while Lucretius and Newton were able to explain what the world of atoms looks like to a surveying intellect, Leibniz alone tried to explain what it must be like to be an atom. His pre-established harmony (though couched in theist terms, as for a European milieu it had to be) seems strangely familiar to those who have become accustomed to the Chinese world-picture. That things should not react upon one another but all work together by a harmony of wills was no new idea for the Chinese; it was the foundation of their correlative thinking.[b]

If we may propose then as a hypothesis for further research that the philosophy of organism owes a great deal to Leibniz, and that his mind was stimulated by the Neo-Confucian version of Chinese correlativism,[c] several further points of interest follow. Whitehead (5, 6) has termed algebra the mathematical study of pattern. Could it be, therefore, only a coincidence that (as we shall later see) while geometry was characteristic of Greek,[d] algebra was characteristic of Chinese mathematics? From the Han onwards the whole effort of Chinese mathematicians could be summarised in one sentence; how to fit a particular problem into a certain pattern or model problem and solve it accordingly. During the Sung, alongside of the Neo-Confucians, a great school of Chinese algebraists grew up, who maintained their lead over the rest of the

[a] (2), p. 168.

[b] Look, for example, at Wang Chhung's discussion of the sage whose actions occur in parallel, as it were, with the virtue of Heaven and Earth—'how could he go before or after?'—*Lun Hêng*, ch. 12 (Forke (4), vol. 1, p. 134). In ch. 10, with regard to fate and destiny, Wang Chhung attributes systematically to the intrinsic developments of predetermined individual persons many effects which would ordinarily be regarded as due to their interactions with other persons. For example, a man destined to die young marries a girl whose fate it is to become a widow early (see the translation and commentary of Leslie, 1). Later, in Sect. 20*i*, we shall find this 'pre-established harmony' applied by Wang Chhung to explain the phenomena of eclipses. And then, a century later, the Buddhists added themselves to the great stream of Chinese thought. Some of their philosophers used formulations which would have interested Wang Chhung very much. For instance, the Mere Ideation School of the Thang (see below, pp. 405, 408) held that in the 'perfuming of seeds', cause and effect occur at one and the same time (cf. Chhen Jung-Chieh (4), p. 107).

[c] Graf (2) seems to be the one and only sinologist who has understood that the symbolic correlations were a preparation for Neo-Confucian organicism (vol. 1, p. 253).

[d] At first sight geometry would seem to be the study of pattern *par excellence*. But algebra deals with pattern in a still more abstract way, not limiting it to dimensions in space or to particular numerical values. In a famous lecture, Cornford (4) urged that the Greeks did not think in Galilean-Newtonian terms (any more than the Chinese); 'the ancients', he claimed, 'were not moderns in a state of infancy or adolescence' but something quite different. They liked to try to define the substance of things just as figures could be defined in deductive geometry. They sought for a 'timeless' truth in the classification of natural objects, and avoided the formulation of sequences of causes and effects. This is why such Aristotelian conceptions as the material and formal causes have no modern equivalents. Cornford illustrated his point by citing Aristotle's treatment of lunar eclipses, which were defined as 'attributes' or 'affections' of the moon, a property which characterised it just as other properties characterised triangles or polygons. Now if we compare this attitude with that of Wang Chhung (Sect. 20*i* below), who regarded eclipses as the result of an *internal* rhythm of the celestial bodies, we can sense the difference between the two ancient world-outlooks, neither of which was wholly that of modern science. For the Greeks what mattered was an ideal world of static form which remained when the world of crude reality was dissolved away. For the Chinese the real world was dynamic and ultimate, an organism made of an infinity of organisms, a rhythm harmonising an infinity of lesser rhythms.

world for a couple of centuries.[a] But there is a still more interesting speculation. After the time of Gilbert in the European 17th century the study of magnetism brought field physics into being, but it was not in Europe that the directive property of the magnet was first discovered, it was in China. On evidence which will be presented in the Section on physics (Sect. 26*i*) we are now able to say with fair confidence that by the + 1st century the Chinese were familiar with the south-pointing properties of pieces of magnetite made into short spoons and capable of turning about the axis of their bowls. One is at liberty to wonder whether it was only a coincidence that in a world where everything was connected with everything else, according to definite correlative rules, it should have occurred to the magician-experimenters as natural or possible that a piece of lodestone carved into the shape of the Northern Dipper should partake of its cosmic directivity? In a way, the whole idea of the Tao was the idea of a field of force. All things oriented themselves according to it, without having to be instructed to do so, and without the application of mechanical compulsion. The same idea springs to the mind, as will shortly be seen, in connection with the hexagrams of the *I Ching*, Yang and Yin, Chhien and Khun, acting as the positive and negative poles respectively of a cosmic field of force. Is it so surprising, therefore, that it should have been in China that men stumbled upon what was in very deed the field of force of their own planet?

(2) ELEMENT THEORIES AND EXPERIMENTAL SCIENCE IN CHINA AND EUROPE

Lastly we must consider for a moment the five-element theories from the more practical point of view as a help or a hindrance to the advance of the natural sciences. They may seem odd to the modern scientific mind approaching them without reflecting on the history of science in Europe. In the hands of the adepts they attained absurdities but these were no worse than medieval European theorising on elements, stars and humours. Looking back on all the foregoing, the five-element and Yin-Yang system is seen to have been not altogether unscientific. Anyone who is tempted to mock at the persistence of it should remember that the founding fathers of the Royal Society spent much of their valuable time in deadly combat with the stout upholders of the four-element theory of Aristotle, and other 'peripatetick fancies'.

For example, every chemist reads (or should read) the *Sceptical Chymist* of Robert Boyle, first published in + 1661. It was a book earnestly recommending, in the form of dialogues, the 'mechanical hypothesis of corpuscles' or atoms of elementary bodies (in our modern sense), as against the four elements of the Aristotelians on the one hand, and the *tria prima* (philosophical salt, sulphur and mercury) of the alchemical writers on the other. Towards the end of the fifth part the efforts of a Peripatetic to explain the combustion of a piece of green wood upon the theory of the four elements are rent limb from limb. Thus: 'He makes the sweat, as he calls it, of the

[a] Sarton (1), vol. 2, pp. 507, 755. All this is discussed in detail in Sect. 19 below.

green Wood to be Water, the smoak Aire, the shining Matter Fire, and the Ashes Earth; whereas, a few lines after, he will in each of these (nay, as I just now noted, in one distinct part of the Ashes), shew the Four Elements. So that...the former Analysis must be incompetent to prove that Number of Elements....' Elsewhere along the same front Joseph Glanvill was contending against the Aristotelian Henry Stubbe. And Marchamont Needham (who could be quite as rude as Stubbe), in his *Medela Medicinae* of +1665, cried, 'Away with the frigid Notion of the four elements, which Galen, out of Aristotle, makes to be the Principles of all mixt Bodies....' I need not dwell further on these 17th-century controversies, which are very well known. The only trouble about the Chinese five-element theories was that they went on too long. What was quite advanced for the +1st century was tolerable in the +11th, and did not become scandalous until the +18th. The question returns once again to the fact that Europe had a Renaissance, a Reformation, and great concomitant economic changes, while China did not.

One of the ironies of history is that the Jesuits were proud of introducing to China the correct doctrine of the four elements—just half a century before Europe gave it up for ever.[a]

(3) MACROCOSM AND MICROCOSM

While, as was said above, European organic naturalism in its modern form seems to begin with Leibniz, we must not forget that it had what was to some extent a pre-scientific forerunner in the famous doctrine of the macrocosm and the microcosm.[b] If anything in Europe was analogous to ancient and medieval Chinese thinking in terms of cosmic pattern or organism, it was this doctrine, though it never dominated Western ideas to the same degree.[c] Two analogies were involved: one postulated point-for-point correspondences between the body of man and the universe or cosmos as a whole, the other imagined similar correspondences between the human body and the society of the State. We must glance for a moment at these theories and compare them with Chinese parallels, asking if there was any difference of emphasis in the two civilisations.[d] We may call the larger theory the 'universe-analogy' and the smaller one the 'state-analogy'.[e]

Among the pre-Socratic fragments there is nothing very definite to be found, and it is not till the time of Plato and Aristotle (−4th century) that the ideas attain any

[a] See Trigault (1), Gallagher tr., pp. 99, 327, 447.

[b] I owe to a conversation with my friend Dr Owsei Temkin the realisation of the necessity for this subsection.

[c] Here may be mentioned a very peculiar kind of 'microcosm' in which East Asian people delighted—namely the miniature gardens brought to great perfection in China. We now have on the history of this subject an exhaustive monograph by Stein (2).

[d] The two outstanding monographs on the history of macrocosm-microcosm theories in Europe are those of A. Meyer (3) and Conger (1).

[e] On the place of analogy in scientific thought, see two interesting discussions by Temkin (1) and Arber (1)

prominence.[a] It may be said that Plato employed all the arguments[b] but never used the term 'microcosm', while Aristotle used the term at least once but was too empirical in his biology and too abstract in his cosmology, as Conger says, to care much for the idea. The first authentic occurrence of the term microcosm is in his *Physics*,[c] where he says, in the course of some arguments about motion, 'If this can happen in the living being, what hinders it from happening also in the All? For if it happens in the little world (it happens) also in the great, etc.' The Stoics continued what Plato had begun; most of them agreed that the world was an animate and rational being. Hence detailed correspondences with the being of man were inviting. Seneca, in his *Quaestiones Naturales* (c. +64), did not hesitate to draw them. Nature was, he believed, organised after the pattern of man's body, water-courses corresponding to veins, air-passages to arteries, geological substances to various kinds of flesh, and earthquakes to convulsions.[d]

This general outlook permeated late antiquity and the middle ages in Europe. It may be found everywhere. Philo Judaeus, a contemporary of Seneca, called man '*brachys kosmos*', βραχὺς κόσμος, the little world.[e] Manilius, the astronomical poet, gives the assignment of parts of the body to regions of the zodiac.[f] Galen in the +2nd century, though not emphasising the theory, alludes favourably to it.[g] Plotinus in the +3rd held extremely organicist views, though they were so saturated with supernaturalism as to have little influence on scientific thinking; there is much in the *Enneads* which suggests the conception of the universe as a hierarchy of wholes, those on one level being the parts of the wholes on the next.[h] Macrobius, about +400, said that certain philosophers called the world a large man and man a short (-lived) world.[i] While Clement of Alexandria accepted the universe-analogy, other early Christian fathers were inimical to it, but this opposition was only temporary, and one finds it in full swing in later patristic literature. It is interesting to note that the first works bearing the term microcosm in their titles were written within a few years of one another; both were of the +12th century, while one was Jewish and one Christian. The former was the *Sefer Olam Qaṭan* (Book of the Little World) by Joseph ben Zaddiq of Cordova (+1149),[j] and the latter, the *De Mundi Universitate Libri Duo, sive Megacosmus et Microcosmus*, of Bernard of Tours (c. +1150). If Bernard was inspired by the same tradition from which Joseph drew, it must in all probability

[a] There has been a persistent custom of regarding them as ultimately Babylonian (Bouché-Leclercq (1), p. 77; v. Lippmann (1), pp. 196, 666; M. Berthelot (1), p. 51) but textual evidence sufficient to pin it down is never given. The universe-analogy was very marked in ancient Indian writings, cf. *Ṛg Veda*, x, 90.

[b] The state-analogy is in *Republic*, 434, 441, 462, 580, and in *Laws*, 628, 636, 735, 829, 906, 945 and 964. The universe-analogy is of course in the *Timaeus*, esp. 35, 36.

[c] VIII, 2, 252 *b*.

[d] Clarke & Geikie tr., p. 126.

[e] *Quis Rer. Div. Haer.* XXIX–XXXI, 146–56.

[f] Bouché-Leclercq (1), p. 319. This idea flourished for centuries.

[g] *De Usu Partium*, III, x, 241.

[h] As is well known, Plotinus admired what he knew of Persian and Indian philosophy and wanted to go and study it in those countries.

[i] *Comment. in Somn. Scipionis*, II, xii, 11.   [j] Ed. Jellinek.

have been the +10th-century encyclopaedia of the 'Brethren of Sincerity' at Basra.[a] In this *Rasā'il Ikhwān al-Ṣafā'*[b] the detail of the correspondences drawn in the universe-analogy reached an apogee never equalled before or since[c] and far surpassing that of Seneca or other Hellenistic writers.

The universe-analogy was still vigorous in the 16th century.[d] There was never a more thorough-going and consistent supporter of it than Paracelsus, and it runs through all his alchemical and medical ideas.[e] His followers, such as Robert Fludd in *Medicina Catholica* of +1629,[f] elaborated the same lines of thought. What is striking about these 16th- and early 17th-century nature-philosophers is their sometimes close approximation to Chinese conceptions. When Fludd, speaking of polarity, sets up opposites such as the following:

> Heat—Movement—Light—Dilatation—Attenuation
> Cold—Inertia—Darkness—Contraction—Inspissation;

or

> Sun—Father—Heart—Right Eye—Sanguis vitalis,
> Moon—Mother—Uterus—Left Eye—Mucus,

he is talking like any Chinese exponent of the theory of Yin and Yang. When Giordano Bruno, regarding the universe as an organism composed of organisms, speaks of a sexual intercourse between the sun and the earth, whereby all living creatures are brought into being,[g] he is using an extremely characteristic Chinese metaphor of frequent occurrence.[h] Presumably, however, the origins of these presentations were 'Pythagorean'[i] and Neo-Platonic, rather than immediately oriental, since at this time recent Chinese influences could hardly be suspected.

Analogies for more than Yin and Yang polarity can be found in European thought. Even the symbolic correlations have at least traces there. When Agrippa of Nettesheim (+1486 to +1535), in his *De Occulta Philosophia*, compiled a correlative tabulation, it was strikingly similar to the venerable Chinese forms. He aligned the seven planets with the seven letters of the Name of God, the seven angels, seven birds, fish, animals, metals, stones, parts of the body, orifices of the head, etc., and did not forget the seven dwellings of the damned. Forke,[j] who laid much emphasis on this in his account of Chinese correlative thinking, was right enough in concluding that 'sixteenth-century Europeans were not a whit further advanced in the natural

---

[a] Already referred to above, p. 95.

[b] Especially the twenty-fifth and the thirty-third treatises in it. Cf. Flügel and Dieterici (1).

[c] Except perhaps in the nineteenth-century panpsychism of Fechner, of which Conger (1) gives an account (p. 88).

[d] The first occurrence of the word microcosm in the English language was about +1200, in Ormin's *Ormulum*: 'Mycrocossmos that nemnedd iss after Englisshe spaeche the little werelld.'

[e] Conger's epitome of them (pp. 56 ff.) is excellent.

[f] Admirably analysed by Pagel (1). Cf. p. 278 above.

[g] *De Immenso* (in *Opera Latine*, ed. Tocco & Vitelli), VI, i, p. 179.

[h] Cf. Forke (6), p. 68, with references to *I Ching, Lieh Tzu Li Chi*, etc.

[i] Cf. Aristotle, *Metaphys.* I, 5.

[j] (4), App. I, 6.

sciences than the Chinese philosophers of the Sung (+12th)'. The tradition runs on through Bruno (+1548 to +1600), who has tables of correspondences in his *De Imaginum Signorum et Idearum Compositione* of +1591, and Franciscus Patritius, whose *Nova De Universalis Philosophia* was almost contemporary (+1593).[a]

The question is, where did these correlative tabulations come from? There is no doubt that they were largely Arabic and Jewish. Philo Judaeus, fifteen centuries before Agrippa, had classified things in sevens.[b] In a great number of subsequent writers, but especially among the Jews and in Arabic works such as the *Rasā'il Ikhwān al-Ṣafā'*, there are 'Chinese' correlations—the parts of the body, the planets, the gods, the strings of the lyre, zodiacal constellations, seasons, elements, humours, letters of the alphabet, perform a complicated ballet in groups of fours and sevens. Though the Chinese category of fives is rarely, if ever, found, one cannot help wondering whether some inspiration from the −3rd-century Chinese School of Naturalists did not find its way through Indian contacts or over the Silk Road to Byzantium, Syria and other parts of the Near East.

It is here that the corpus of mystical Jewish writings, the Kabbalah, played an important part. Its origins are still extremely obscure; there seems to have been a connection with Gnosticism, Persian sufism, and conjectural influences from still farther east (Loewe, 1). The elements of the system go back to the −2nd century, but the earliest text (the *Sefer Yeṣirah*) dates only from the +6th and the first historical personage definitely connected with the Kabbalah (Aaron ben Samuel) died towards the end of the +9th. The chief text (the *Zohar*) is of the +10th. The system included a great deal of numerology and magico-mystical arranging of letters and numbers, many doctrines of demiurges and angels, and distinct similarities with Chinese thought in its lists of 'pairs' (*ziwwugh*, syzygies) of things, as if grouped in Yin and Yang categories.[c] Some references to metempsychosis might betray Buddhist, or at least Indian, influence, but other origins were certainly Greek, for Ptolemy and Proclus had associated the parts of the body, the senses, and the human psychological states, with the various planets.[d] The doctrine of the macrocosm and the microcosm naturally appears in the Kabbalah. Kabbalistic notions undoubtedly influenced that extraordinary man Raymond Lull (+1232 to +1316),[e] in whose works tables of

---

[a] In the following century, when the Chinese systems became known, they played their part in the controversies between the supporters of the 'ancients' and the 'moderns'. In +1690 William Temple, who argued for the superiority of the former, praised the Chinese civil service. In +1697, William Wotton, opposing him, relied heavily on discoveries which were Chinese (or in which the Chinese had participated), e.g. printing, mineral therapy, magnetic polarity, lunar influence on the tides, and fossils on hilltops—but attacked what he took to be the Chinese world-outlook, and especially ridiculed the symbolic correlations.

[b] *De Mund. Opif.* XXXV–XLIII, 104–28.

[c] Though more than a century old, the book of Franck (1) is still regarded as one of the best accounts of the Kabbalah. There are good studies by Scholem (1, 2). The relations of its system with correlative thinking can be clearly seen from the tables and diagrams in the interesting Latin exposition of +1677 edited by Knorr von Rosenroth & Franciscus Mercurius van Helmont. I am much indebted to my friend Dr Walter Pagel for some orientations concerning the Kabbalah.

[d] Details in Bouché-Leclercq (1).

[e] Lull's attempts at a mathematical logic were later referred to by Leibniz in his *De Arte Combinatoria*.

correspondences in the Chinese manner can be seen. Here was the immediate precursor of Agrippa of Nettesheim.

It may well turn out that the 'correlative thinking' of the 16th and early 17th centuries had more influence on scientific minds in the true dawn of modern science than has generally been allowed. This is indeed the theme which runs through all the brilliant contributions of Pagel on the 'dark side' of scientific discoverers such as J. B. van Helmont. Bruno, in abandoning geocentrism, did not give up the universe-analogy; he said that the sun in the megacosm corresponded to the heart in man.[a] It has now been shown, by Temkin (1), Pagel (4, 5, 6) and Curtis (1), that the discovery of the circulation of the blood by William Harvey[b] was at any rate partly inspired by the known relation of the sun to the meteorological water circulation cycle. We may also ask whether similar influences stimulated Leibniz in his elaboration of the first European philosophy of organic naturalism.[c] As L. Stein (1) pointed out, Bruno distinguished three 'minima' or irreducibles, God, the 'Monas monadum' in whom both greatest and least are one; the soul, which serves as an organising centre (significant idea) round which the body is formed; and the atom, which enters into the composition of all substances. But the source from which Leibniz got his term 'monad' is more probably Franciscus Mercurius van Helmont, since Leibniz mentions one of his works, *The Paradoxal Discourses of F. M. van Helmont concerning the Macrocosm and Microcosm, or the Greater and Lesser World, and their Union* (+1685). In any case the son of J. B. van Helmont was in the same tradition. We thus reach the conclusion that there may have been (if our surmise concerning the original Asian origin of the correlative thinking of European antiquity is justified) two channels leading to Leibniz, not only the Neo-Confucian material which the Jesuits translated (see below, pp. 496–505), but also far more ancient ideas which entered European thought through Jewish and Arabic intermediation more than a thousand years earlier.

In general, of course, correlative thinking, and the universe-analogy, failed to survive the triumph of the 'new, or experimental, philosophy'. Experiment, induction and the mathematisation of natural science swept all its primitive forms away, ushering in the modern world. Later, at the end of the Section on mathematics, we shall see how all old ideas of a spatially differentiated cosmos were driven out by the bold application of uniform geometrical Euclidean space to the whole universe. Any references which occur to the universe-analogy after the middle of the 17th century in scientific writings may be considered nothing more than rhetorical survivals.[d]

But we must turn back for a moment to the state-analogy. First utilised by Plato, it awoke to new life in the *Policraticus* of John of Salisbury (+1159),[e] which,

---

[a] *De Monade* (*Opera*, I, ii, p. 347, .

[b] *Exercitatio Anatomica de Motu Cordis et Sanguinis in Animalibus*, 1628.

[c] The first use of the term 'organism' in its modern sense in English occurs in Leibniz's older English contemporary, John Evelyn, who says in his *Sylva* (+1664) 'the organism, parts and functions of plants and trees'.

[d] Conger (1), p. 66.   [e] He was secretary to St Thomas of Canterbury.

according to Gierke (2), was the first elaborate attempt to draw correspondences between all the parts of the body and the organs of the State. The prince was the head, the senate was the heart; the eyes, ears and tongue were the frontier-guards, the army and the judiciary were the hands and arms, and the menial labourers were the feet. A theory so convenient for any governing class was not likely to be left un-cultivated; the only surprising thing is that its development was slow.[a] Mentioned by Shakespeare[b] in a famous passage in the opening scene of *Coriolanus*, it appears again, naturally, in the *Leviathan* of Hobbes (+ 1651), who added financial channels as arteries, money as blood, and counsellors as memory. In the 19th century thinkers such as Herbert Spencer and Walter Bagehot were fairly cautious in the use they made of it, but gross abuses of the state-analogy continue to occur in our own time.

Before looking at the other side of the picture, the parallels in Chinese thought, I should like to allude to the part played by the two analogies in the development of alchemy. This was on the whole beneficial; the statement of the *Tabula Smaragdina*[c] that 'That which is beneath is like that which is above'[d] was sound science. In later alchemy the sulphur ('of the philosophers') was thought to be the *materia prima*, from which all other substances could be derived, hence it was considered the true microcosm (Hitchcock, 1). We have already noted the essential part which the universe-analogy played in the systems of men such as Paracelsus and Fludd. It is curious to remember that the term 'microcosmic salt' (sodium ammonium hydrogen phosphate, $HNaNH_4PO_4$) lingered on long into modern chemistry; it was so called since it was first prepared from human urine in the early 17th century.

It is now time to re-examine the Chinese parallels. If definite statements of early date are not very common, this is because the universe-analogy was implicit in the whole world-outlook of the ancient Chinese.[e] The *Huai Nan Tzu*[f] (c. − 120) gives a very detailed statement of it, as also the *Chhun Chhiu Fan Lu*.[g] The *Li Chi* (Record of Rites), put together about − 50, says[h] that man is the heart and mind of heaven and earth and the manifestation of the five elements. The *I Ching*[i] likens heaven to the

---

[a] Its history has been written by Coker (1), and elsewhere (Needham, 15) I have had something to say about its social function.

[b] Who took it of course from Plutarch.

[c] See Steele & Singer (1). The origins of this ancient alchemical document are very obscure, but it probably emanates from Christian Egypt of the + 3rd century.

[d] This remained throughout the orthodox alchemical doctrine, and corresponded in some sense to the Taoist affirmation of the unity of Nature; it gave a point even to the worst extravagances of Para-celsus. When he said, for instance: 'He that knoweth the origin of thunder, winds, and storms, knoweth where colic and torsions come from.... He that knoweth what the planets' rust is, and what their fire, salt, and mercury, also knoweth how ulcers grow, and where they come from, as well as scabies, leprosy, etc.'—his words may be taken as a premonition of the modern unity of physics and chemistry which does indeed elucidate the phenomena of the human body no less than those of the heavens. Cit. Temkin (1).

[e] It was also extremely old in Indian thought (e.g. *Ṛg Veda*, x, 90, long before the pre-Socratics) and in Persia too (cf. Filliozat, 5).   [f] Ch. 3, p. 16*b* (tr. Chatley, 1); ch. 7.

[g] Esp. ch. 56 (tr. Bodde in Fêng Yu-Lan (1), vol. 2, p. 30).

[h] *Ku jen chê thien ti chih hsin, wu hsing chih tuan yeh*,[1] ch. 9, p. 62*b*.

[i] App. 8 (*Shuo Kua*), ch. 9.

[1] 故人者天地之心五行之端也

head and earth to the belly.[a] The whole theory of Phenomenalism, which we shall presently describe in relation to the sceptical attitude of Wang Chhung (see below, pp. 378–82), rested on a belief in a one-to-one correspondence between the ethics of human actions on earth and the parallel behaviour of the heavenly bodies. It was thus essentially anthropocentric. Its origins have been described in detail by Granet in his two chapters on the microcosm and the macrocosm,[b] where he elucidates its relations with the ancient astronomical theories.[c] All through the history of Chinese thought the universe-analogy goes on.[d] One can find it not only in the work of Tung Chung-Shu, but of Shao Yung[1] (+1011 to +1077), who nearly parallels the Basra Brethren of the previous century in his physiological-geological comparisons.[e] Wang Khuei,[2] writing about +1390, says:

The human body imitates (*fa*[3]) Heaven and Earth very distinctly and exactly. Just as Heaven and Earth have *ssu, wu, shen*, and *yu* (duodenary cyclical characters) in front and above, so the human heart and lung are located in front and above. In the heavens *hai, tzu, yin* and *mao* are below and behind, so the human kidney and liver are at the back and underneath. In addition the four limbs and the hundred bones all imitate the dispositions of heaven and earth. Thus human beings are the most spiritual of all living things.[f]

There can be no doubt of the prominence of the universe-analogy in Chinese thought. But did it have the same philosophical content as in Europe?

Before answering this question, a word must be said about the state-analogy in China. While the universe-analogy has been widely recognised, it does not seem to have been pointed out before that the Chinese had the state-analogy too. In the *Pao Phu Tzu*[4] (early +4th century) Ko Hung[5] says:

Thus the body of a man is the image of a State. The thorax and abdomen correspond to the palaces and offices. The four limbs correspond to the frontiers and boundaries. The divisions of the bones and sinews correspond to the functional distinctions of the hundred

---

a *Chhien wei shou, khun wei fu.*[6]
b (5), pp. 342, 361. It is in these chapters that he discusses the symbolic correlations.
c See below, Sect. 20*d.*
d Following *Huai Nan Tzu*, it was carried to fanciful lengths in the Thang and Sung books in the *Tao Tsang* (cf. Maspero (13), pp. 19, 34, 35, 36, 108, 118). Man's head was round like the heavens, his feet were square like the earth, his five viscera corresponded to the five elements, his twenty-four vertebrae to the 24 fortnights of the solar year, his 365 bones to the 365 days of the year, his 12 tracheal cartilages to the 12 months, his blood vessels to the rivers, etc. A list of the stars corresponding to the various parts of the body is given in the *Shang Chhing Tung-Chen Chiu Kung Tzu Fang Thu*[7] (Description of the Purple Chambers of the Nine Palaces of the Tung-Chen Heaven), a Sung book probably of the +12th century (*TT* 153).
e Forke (6), p. 122; (9), p. 34.
f *Li Hai Chi*, p. 20*a*, tr. auct.

¹ 邵雍   ² 王逵   ³ 法   ⁴ 抱朴子   ⁵ 葛洪
⁶ 乾爲首坤爲腹   ⁷ 上清洞眞九宮紫房圖

officials. The pores of the flesh correspond to the four thoroughfares.[a] The spirit corresponds to the prince. The blood corresponds to the ministers, and the *chhi* to the people. Thus we see that he who can govern his body can control a kingdom. Loving his people, he will bring peace to the country; nourishing his *chhi*, he will preserve his body. If the people are alienated the country is lost; if the *chhi* is exhausted the body dies.[b]

Ko Hung here stands between Plato and John of Salisbury. His words were not forgotten. They may be found copied, for example, in the *Huang Ti Chiu Ting Shen Tan Ching Chüeh*[1] (Explanation of the Yellow Emperor's Manual of the Nine-Vessel Magical Elixir), an alchemical compendium of Thang or Sung date.[c]

We may take it, then, that the developed forms of the universe-analogy and also the state-analogy were found in China as in Europe. It may not, therefore, be out of place to seek a common origin for their appearance in the two civilisations. Although, as has already been mentioned, Babylonian cuneiform texts do not seem to say much about them, R. Berthelot[d] has made the interesting suggestion that the whole conception of microcosm and macrocosm may have been derived from the methods of divination used in high antiquity, in which the future was foretold from the examination of the whole or part of a sacrificial animal. The Babylonians certainly did this, using the liver,[e] and the scapulimancy of the Shang Chinese[f] may be considered another form of it, while we have detailed sources of information about it in the writings of Latin authors, such as Cicero, Seneca and Pliny, concerning the haruspicy of the Etruscans, much of which the Romans took over.[g] In the theory of the *templum*,[h] a division into spaces, either of the expanse of the heavens or of the body or organ of a sacrificial animal, the divination depended on the appearance of 'signs' in one or other of the spatial divisions. The animal or its liver or intestines was thus acting as a 'microcosm'. Parallel with this viewpoint went the theory of the *saeclum* or recurring period, arising directly from the arbitrary resonance-periods of the periods of revolution of the celestial bodies.[i] Thus both space and time were divided up into separate parcels, prefiguring all later scientific divisions of space and time, and within the spatial sphere the small and the great were thought to mirror each other.

Was now the Chinese universe-analogy philosophically similar to the form which it took in Europe? I am strongly inclined to think not. Europe had the macrocosm-

---

[a] He is certainly thinking of the rectangular plan of many Chinese cities, where the ways from the four gates centre on the drum-tower.

[b] Ch. 18, p. 30*a*, tr. auct. Echoing an earlier version in *Chhun Chhiu Fan Lu*, chs. 22 and 78.

[c] *TT* 878.

[d] (1), especially pp. 24, 41, 118, 163 and 343.

[e] Lenormant (1, 2). Much evidence is brought forward by Piganiol (1) that the Etruscans were originally a people of Asia Minor and carried Babylonian-Chaldean culture with them in their migration to Italy. [f] This will be described in Section 14*a* below. Cf. Sect. 5*b* above.

[g] Bouché-Leclercq (2).

[h] Derived from the root *tem*, to separate or divide, e.g. *temenos*.

[i] We shall consider this later, in the discussion of calendrical science (Sect. 20*h*).

[1] 黃帝九鼎神丹經訣

microcosm doctrine, yes, and to that extent, a primitive form of organic naturalism, together with its minor counterpart, the state-analogy, but both were subject to what I shall call later on (Sect. 46) the characteristic European schizophrenia or split-personality. Europeans could only think in terms either of Democritean mechanical materialism or of Platonic theological spiritualism. A *deus* always had to be found for a *machina*. Animas, entelechies, souls, archaei, dance processionally through the history of European thinking. When the living animal organism, as apprehended in beasts, other men, and the self, was projected on to the universe, the chief anxiety of Europeans, dominated by the idea of a personal God or gods was to find the 'guiding principle'. One sees it again and again—in the world-soul animating the world-body in the *Timaeus*; or the leading principle, the *Hegemonikon* (ἡγεμονικόν) sought by the Stoics (who differed very much among themselves as to what it was);[a] or Seneca's summary statement that God is to the world as the soul to man;[b] repeated by Philo and Plotinus; and echoed by the *Pirké Rabbi Eliezer* in the +8th century.[c]

Yet this was exactly the path that Chinese philosophy had *not* taken. The classical statement of the organismic idea by Chuang Chou in the −4th century (cf. above, pp. 52, 288) had set the tone for later formulations, expressly avoiding the idea of any *spiritus rector*. The parts, in their organisational relations, whether of a living body or of the universe, were sufficient to account, by a kind of harmony of wills, for the observed phenomena.[d]

This conception is often clearly stated in the famous +3rd-century commentary of Hsiang Hsiu and Kuo Hsiang on the *Chuang Tzu*. For example, in chapter 6 someone asks the question, 'Who can associate in non-association and cooperate in non-cooperation?'[e] Hsiang and Kuo comment:

The hands and feet differ in their duties; the five viscera differ in their functions. They never associate with each other, yet the hundred parts (of the body) are held together with them in a common unity. Thus do they associate in non-association. They never (force themselves to) cooperate, and yet, both within and without, all complete one another. This is the way in which they cooperate in non-cooperation...Heaven and Earth are such a (living) body.[f]

The cooperation of the component parts of the organism is therefore not forced but absolutely spontaneous, even involuntary. In another passage the same commentators suggested that even if there were reluctance to cooperate, or positive 'anti-social' action, the self-regulating or cybernetic control of the world organism (as we might put it) was so powerful that all things would continue to work together for good.[g]

[a] Conger (1), p. 13.   [b] *Epist.* 65, 24.
[c] See Karppe (1), p. 135.

[d] In conversation with a distinguished European scholar on this, he said, 'The Chinese did not analyse their organisms then?' The point is that the body-soul antithesis was the wrong kind of analysis. But his reaction was typically western.   [e] Legge (5), vol. 1, p. 250.

[f] Ch. 6 (*Pu Chêng* ed. ch. 3A, p. 22a), tr. Bodde in Fêng Yu-Lan (1), vol. 2, p. 211, mod.

[g] This evokes both the early capitalist conviction that private advantage was the same as public benefit (probably at that time by no means unjustified), and also the conviction of modern marxists that progress can only be achieved through the struggle of contradictory theories and policies.

Commenting on the passage about relativity in the 17th chapter,[a] they said, concerning 'If we look at things from the point of view of the services they render...':

There are no two things under Heaven which do not have the mutual relationship of the 'self' and the 'other'. Both the 'self' and the 'other' equally desire to act for themselves, thus opposing each other as strongly as east and west. On the other hand, the 'self' and the 'other' at the same time have the mutual relationship of lips and teeth. The lips and the teeth never (deliberately) act for one another, yet 'when the lips are gone, the teeth feel cold'.[b] Therefore the action of the 'other' on its own behalf at the same time helps the 'self'. Thus though mutually opposed, they are incapable of mutual negation.[c]

As Fêng Yu-Lan says, this last conclusion is surprisingly reminiscent of the dialectic of Hegel.

Such was the current which came into European thought with Leibniz and which contributed to the widespread adoption of organic naturalism at the present day.[d] The controversies in biology between vitalism and mechanism, which continued to as late as 1930, were a direct inheritance of the European split-personality—either there was the machine plus an invisible mechanic or signalman, or there was the machine alone. The general recognition of the uselessness of these controversies, which has come with the understanding that an organism is not a machine at all, and neither needs an archaeus, nor can be fully 'reduced' to lower integrative levels, is of recent date.

Thang Chün-I, in an interesting essay (2) on ontological ideas in the philosophy of West and East, has put the matter in another way by saying that the Europeans tended to seek for reality outside or beyond phenomena, while the Chinese sought for it within them. Hence European philosophy began with Platonic and Aristotelian dualism, and only came late to systems such as those of Spinoza, Leibniz and Hegel. Chinese philosophy, on the contrary, began with the Taoist recognition of the eternal in the changing, went on to the organic naturalism of the Neo-Confucians,[e] and gave rise to the idealists of the Ming as a passing phase not in the strict line of succession.

The universe-analogy in medieval Europe[f] might therefore be said to have been vitiated by its position on the perennial battlefield between mechanical materialism and theological spiritualism.

Until the middle of the 17th century Chinese and European scientific theories were about on a par, and only thereafter did European thought begin to move ahead so rapidly. But though it marched under the banner of Cartesian-Newtonian mechanicism, that viewpoint could not permanently suffice for the needs of science—the time came when it was imperative to look upon physics as the study of the smaller organisms, and biology as the study of the larger organisms.[g] When that time came, Europe (or rather, by then, the world) was able to draw upon a mode of thinking very old, very wise, and not characteristically European at all.

[a] Legge (5), vol. 1, p. 380. Cf. p. 103 above. [b] This was a well-known proverb in ancient China.
[c] Ch. 17 (Pu Chêng ed. ch. 6B, p. 9b), tr. Bodde in Fêng Yu-Lan (1), vol. 2, p. 211, mod.
[d] See further, Sects. 16 and 18 below.
[e] On macrocosm and microcosm in Sung thought see Graf (2), vol. 1, e.g. pp. 37, 75.
[f] Whether or not its origin was Asian. [g] Whitehead (1), p. 150.

(g) THE SYSTEM OF THE BOOK OF CHANGES

Much has already been said about the importance in the Chinese world-view of action at a distance, in which the different kinds of things in the universe resonate with one another. In the +5th-century *Shih Shuo Hsin Yü* we find the following:

Mr Yin,[1] a native of Chinchow, once asked a (Taoist) monk, Chang Yeh-Yuan,[2] 'What is really the fundamental idea (*thi*[3]) of the Book of Changes (*I Ching*[4])?'

The latter answered, 'The fundamental idea of the *I Ching* can be expressed in one single word, Resonance (*kan*[5]).'

Mr Yin then said, 'We are told that when the Copper Mountain (Thung Shan[6]) collapsed in the west, the bell Ling Chung[7] responded (*ying*[8]), by resonance, in the east.[a] Would this be according to the principles of the *I Ching*?'

Chang Yeh-Yuan laughed and gave no answer to this question.[b]

Up to the present point the consideration given to the theories of the Five Elements and the Two Forces has indicated that they were a help rather than a hindrance to the development of scientific ideas in Chinese civilisation. Not until the 17th century, when the four Aristotelian elements were finally discarded in Europe, did these theories confer upon Chinese thought any degree of backwardness as compared with the world-picture of occidentals. But as to the third great component of Chinese scientific philosophy, the system of the Book of Changes (*I Ching*[4]), it will not be possible to form so favourable a judgement. Originating from what was probably a collection of peasant omen texts, and accumulating a mass of material used in the practices of divination, it ended up as an elaborate system of symbols and their explanations (not without a certain inner consistency and aesthetic force), having no close counterpart in the texts of any other civilisation. These symbols were supposed to mirror in some way all the processes of Nature, and Chinese medieval scientists were therefore continually tempted to rely on pseudo-explanations of natural phenomena obtained by simply referring the latter to the particular symbol to which they might be supposed to 'pertain'. Since each one of the symbols came, in the course of centuries, to have an abstract signification, such a reference was naturally alluring, and saved all necessity for further thought. It resembled to some extent the astrological pseudo-explanations of medieval Europe, but the abstractness of the symbolism gave it a deceptive profundity. We shall shortly see that the sixty-four symbols in the system provided a set of abstract conceptions capable of subsuming a large number of the events and processes which any investigation is bound to find in the phenomena of the natural world.

[a] Cf. Sect. 26*h* below.
[b] Tr. auct. The concepts of stimulus (*kan*) and response (*ying*) are basic in Chinese naturalism. Cf. *ECCS*; *I-Chhuan I Chuan*, ch. 3, p. 4*a*, expounded well by Graham (1), p. 124; also *I Shu*, ch. 15, p. 7*b*, and *Wai Shu*, ch. 12, p. 15*b*.

[1] 殷   [2] 張野遠   [3] 體   [4] 易經   [5] 感   [6] 銅山
[7] 靈鐘   [8] 應

The *I Ching* as we have it today is a very complex book, and it will first be necessary to give an idea of its contents. The symbols of which we have been speaking are all made up of sets of lines (*hsiao*[1]), some full or unbroken (*yang*[2] lines), others broken, i.e. in two pieces with a space between (*yin*[3] lines). These may have been connected respectively with the long and short sticks of ancient divination procedures.[a] By using all the possible permutations and combinations eight trigrams are formed and sixty-four hexagrams, all known as *kua*.[4] The *kua* are arranged in the book according to a definite order. Each *kua* is followed by a single paragraph of explanation; this is known as the *thuan*[5][b] and attributed traditionally to King Wên of the early Chou dynasty (*c.* − 1050).[c] A commentary follows, usually in six sentences; this is known as the 'Appended Judgments' (*hsi tzhu*[6] or *hsiao tzhu*[7]), and was traditionally attributed to Chou Kung (the Duke of Chou), another famous figure of the early Chou dynasty (*c.* − 1020).[d] Such is the canonical text, or Ching proper (*pên ching*[8]).

Still more complex are the commentaries and appendices which are known as the 'Ten Wings' (*Shih I*[9]). The first two of these constitute the 'Treatise on the Thuan' (*Thuan Chuan*[10]), traditionally supposed to have been written by Confucius, and there are two of them because the Ching is divided into two parts, the first dealing with thirty *kua* and the second with thirty-four.[e] The third and fourth constitute the 'Treatise on the Symbols' (*Hsiang Chuan*[11]), also supposed to have been written by Confucius, and also divided into two parts corresponding to the first and second halves of the Ching.[f] The fifth and sixth constitute the 'Commentary on the Appended Judgments' (*Hsi Tzhu Chuan*[12]), again divided into two parts, though these are not, as in the former appendices, associated specifically with the two parts of the Ching. This commentary has been known since the Early Han as 'The Great Appendix' (*Ta Chuan*[13]).[g] It deals with the basic trigrams from which the hexagrams are formed, as well as with the hexagrams themselves, some of which it interprets in terms of a theory of social evolution. The next portion, the seventh, is known as the 'Explanation

[a] Another view regards them as deriving from counting-rods used in ancient arithmetic; see below, Sect. 19*f*. *Hsiao*[1] is often pronounced *yao*.

[b] This character means (and is certainly a pictograph of) a running pig, but no explanation has survived of how or why it came to be used in the present sense (cf. Legge (9), p. 213).

[c] This is called 'Urteil' or 'Judgment' in the translations of Wilhelm (2).

[d] This is called the 'Linien' or 'Lines' statement in the translations of Wilhelm (2); repeated as 'Lines (*a*)' in vol. 2.

[e] They form Appendix I in Legge (9), but in Wilhelm (2) they are split up among the different *kua*, and given only in vol. 2 under the heading 'Commentary on the Decision' (*Kommentar zur Entscheidung*).

[f] These form Appendix II in Legge (9), but in Wilhelm (2) they are split up among the different *kua* and given in both volumes, under the heading 'The Image' (*Das Bild*). These wings are divided into 'Greater Hsiang' statements and 'Lesser Hsiang' statements; the former explain the hexagram from the meanings of its two trigram components, while the latter are isolated comments on the Hsiao Tzhu portion of the canonical text. Wilhelm (2) prints the latter as 'Lines (*b*)' in vol. 2 only, under each *kua* separately.

[g] It forms Appendix III in Legge (9), and Wilhelm (2) also prints it separately (vol. 1, p. 301 of the English edition), calling it the 'Great Treatise' or 'Great Commentary' (*Grosse Abhandlung*).

[1] 爻   [2] 陽   [3] 陰   [4] 卦   [5] 彖   [6] 繫辭   [7] 爻辭
[8] 本經   [9] 十翼   [10] 彖傳   [11] 象傳   [12] 繫辭傳   [13] 大傳

of the Sentences' (*Wên Yen*[1]) and deals only with the first two hexagrams, *Chhien* and *Khun*.[a] The eighth appendix, the 'Discourses on the Trigrams' (*Shuo Kua*[2]), is divided into eleven short chapters,[b] and forms the chief source for what the *I Ching* has to say about the symbolic correlations discussed in a previous section (p. 261 above). The ninth is a short 'Treatise on the Orderly Sequence of the Hexagrams' (*Hsü Kua*[3]).[c] The tenth, and last, is a rhyming 'Treatise on the Oppositions of the Hexagrams' (*Tsa Kua*[4]).[d] In many editions the *Thuan Chuan*, the *Hsiang Chuan* and the *Wên Yen* are divided up and placed in the text of the canon along with the *kua* to which they correspond, but the Great Appendix is always printed after the end of the canon, and often the last two 'wings' also.

Naturally the first thing we ask about the Book of Changes is its date. Unfortunately, this is one of the most disputed of sinological questions, and there has been great disagreement about it. We really know a good deal less now about the origin of the *I Ching* than Legge thought he did when he made his first translation of it in 1854, and published his final version, with a long introduction, in 1882. For he followed the traditional view almost in its entirety, whereas now no one would maintain that either King Wên or the Duke of Chou had anything to do with the book. Nevertheless, a conviction of the high antiquity of the *I Ching* was retained by many modern critical scholars, such as Ku Chieh-Kang and Hu Shih, who were willing to place the *Thuan Chuan* and *Hsiang Chuan* commentaries in the −6th century, and coming therefore from the time of Confucius, on which view the canon itself might go back to the −8th. Others, such as Lei Hai-Tsung, thought that the *Thuan Chuan* and the *Hsiang Chuan* themselves might go back to the −8th century.[e] An extreme view of the opposite sort was upheld by Kuo Mo-Jo, who maintained that not only the commentaries, but also the text of the canon itself, were written in the Warring States period (−3rd and −4th centuries), some material being added in the Han time.[f] For our present purpose it will perhaps be best to adopt the com-

---

[a] It is Appendix IV in Legge (9), but Wilhelm (2) prints it with the two *kua* to which it refers, under the title 'Commentary on the Words of the Text' (*Kommentar zu den Textworten*).

[b] Appendix V in Legge (9); Wilhelm (2) also keeps it separate from the text, printing it in vol. 1, p. 281 of the English edition as 'Discussion of the Trigrams' (*Besprechung der Zeichen*).

[c] Appendix VI in Legge (9), but Wilhelm (2) distributes it among the different *kua*, calling it 'Sequence of the Hexagrams' (*Die Reihenfolge*), in vol. 2 only.

[d] Appendix VII in Legge (9), but Wilhelm (2) distributes it among the different *kua*, calling it 'Miscellaneous Notes' (*Vermischte Zeichen*), in vol. 2 only. All these differences, which are extremely confusing, arise from the fact that Chinese editions of the *I Ching* are themselves very differently arranged, following the views of various different Chinese schools of thought, among which European sinologists also have been more or less consciously divided.

[e] A great debate on this subject, by a number of authors including Ku Chieh-Kang, Ma Hêng, Hu Shih, Chhien Mu, Li Ching-Chhih and others, will be found in *KSP*, vol. 3, pp. 1–308, cf. also *CIB*, 1938, vol. 3, pp. 67 ff.

[f] More recently, however, Kuo Mo-Jo has been inclined to believe, (4), pp. 81 ff., that much of the *I Ching* was written by Han Pei,[5] a shadowy figure of the −5th century, who is mentioned in the *Shih Chi*, ch. 67. There a list is given of mutationists who were supposed to have received the Book of Changes in direct succession from Confucius, and Han Pei is the second in this list. He would have lived a couple of generations earlier than Tsou Yen.

[1] 文言   [2] 說卦   [3] 序卦   [4] 雜卦   [5] 馯臂

promise position represented, for instance, by Li Ching-Chhih (*1, 2*), according to which the canonical text originated from omen compilations which might be as old as the −7th or −8th century[a] but did not reach its present form before the end of the Chou dynasty. The *Thuan Chuan* and the *Hsiang Chuan* commentaries would then have been written by Chhin and Han Confucians (strongly influenced by the School of Naturalists, and probably from the old States of Chhi and Lu). The *Hsi Tzhu Chuan* (the Great Appendix), and the *Wên Yen*, would be of the early Han, before the middle of the −1st century, though doubtless including some earlier material; while the last three appendices might well all be of the +1st century.

One of the obstacles to such a view which would at once occur to a traditional scholar, is that there is a famous passage in the *Lun Yü* (Conversations and Discourses of Confucius), where he says[b] that he would like to have many more years of life in order to devote himself to the study of the *I Ching*. But there is every reason to suspect this passage of being a later corruption.[c] The crucial fact is that there is no other mention of the *I Ching* in any reliable contemporary text before the −3rd century;[d] in comparison with this, the argument that the study of divination was not in accord with the known character of Confucius seems of secondary importance. On the other hand, the rulers of the State, and then of the Empire, of Chhin, were very interested in any kind of divination or magic (cf. Vol. 1, p. 101), so that the −3rd century would be a likely date for a divination-text to acquire importance. And indeed at this time the *I Ching* took the place of the lost *Yo Ching*[1] (Music Classic), which Hsün Tzu had mentioned, among the Confucian canonical books. A further point of interest here is that if we may judge from the *Chou Li* (Record of Rites of the Chou),[e] a Han compilation, the book which became the *I Ching* was not the only one of its class. There we find that the Grand Augur (Ta Pu[2]) was supposed to be in charge of the 'Three I':[3] first, the *Lien Shan*[4] (Manifestation of Change in the Mountains);[f] secondly, the *Kuei Tsang*[5] (Flow and Return to Womb and Tomb); and thirdly, the *Chou I* (Book of Changes of the Chou Dynasty). Little is known, however, of these other two systems, for only the third survived in full as the *I Ching*.[g] If the views here outlined about the dates of this book are accepted, it follows that all the *kua*-consultations in the *Tso Chuan*[h] must be regarded as interpolations.[i]

Much the best book in a Western language on the *I Ching* is, in my opinion, the

[a] Wu Shih-Chhang (*1*), however, points out that the stories in the *hsiao tzhu* concern Shang rulers and Shang-Chou relations, and that the phraseology resembles that of the oracle-bone inscriptions. These stories may thus be the oldest components.

[b] VII, xvi.   [c] See, for example, Dubs (*17*); Creel (*4*), p. 217.

[d] E.g. *Chung Yung, Mêng Tzu, Hsün Tzu.*

[e] Ch. 6, p. 22*b* (ch. 24, p. 4); tr. Biot (*1*), vol. 2, p. 70.

[f] Or 'Intermittent Eruptions of Mountains' (volcanoes?); as the *Kuang Shih Ming* thinks (ch. 2, pp. 13*a* ff.), but this is probably only late speculation.

[g] The fragments of the other two are in *YHSF*, ch. 1.

[h] Tabulation by Barde (*2*).

[i] As had long been held by Kuo Mo-Jo, e.g. (*4*), p. 79.

[1] 樂經   [2] 大卜   [3] 三易   [4] 連山   [5] 歸藏

small introduction by H. Wilhelm (4). Probably the best translation is that by his father, R. Wilhelm (2),[a] though in many ways that of Legge (9) is more useful.[b]

Waley (8), Li Ching-Chhih (2) and others have urged that the Book of Changes is essentially an amalgam of an omen- or 'peasant-interpretation'-text,[c] with subsequent divination texts of more sophisticated nature. The omens which caught the attention of the ancient Chinese farmers, as of all people in similar stages of culture everywhere, were: (a) subjective inexplicable sensations and involuntary movements; (b) unusual phenomena observed in plants and animals; and (c) unusual sidereal or meteorological phenomena. Waley illustrates the conflation of these two original sources as follows, choosing an example familiar to us:

Omen-text: A red sky at morning....
Divination-text: Unlucky. Unfavourable for seeing one's superiors.
Omen-text: A red sky at night....
Divination-text: Auspicious. Favourable for going to war.

and so on. In order to show how he brings out the ancient omen hidden behind the stilted translation of Legge we may give a couple of examples.

Kua no. 31, *Hsien*[1]

LEGGE (9)	WALEY (8)
Thuan: *Hsien* (indicates that, on the fulfilment of the conditions implied in it, there will be) free course and success. Its advantageousness (will depend on being) firm and correct, as in marrying a young woman. There will be good fortune. Hsiao Tzhu commentary: (1) The first line, divided, (shows) one moving his big toes.	A feeling (*kan*[2])[d] in the big toe,

[a] The editions of R. Wilhelm, especially the English translation by Baynes (quite sound in itself), constitute unfortunately a sinological maze, and belong to the Department of Utter Confusion. The arrangement which Wilhelm adopted was in the first place unnecessarily complicated and repetitive though it could have been saved by clearer explanations and a better use of Chinese characters. The publishers of the American version then made matters worse by using a series of type faces and sizes which in no way correspond with the relative importance of the numerous headings and subheadings differentiating the original material. Wilhelm seems to have been the only person concerned from first to last who knew what it was all about, and even he presented the late commentary material as an amorphous mass with no indication of the various authorships and their dates. All this is the more regrettable because the uniqueness of the work itself fully deserved the munificence of the Foundation which gave typographical beauty to the English translation.

[b] The translation of de Harlez (1) is not reliable and needs checking against the text and the versions of Legge and Wilhelm.

[c] Similar, perhaps, to some of those which have been discovered in the Pelliot collection of the Tunhuang manuscripts (nos. 2661 and 3105).

[d] In effect, the first sentence of the Thuan Chuan on this *kua* says, '*Hsien*[1] is here used in the sense of *kan*,[2] meaning (mutually) influencing'.

[1] 咸   [2] 感

(2) The second line, divided, (shows) one moving the calves of his leg. There will be evil. But if he abide (quiet in his place) there will be good fortune.

or in the calf,

(3) The third line, undivided, (shows) one moving his thighs, and keeping close hold of those whom he follows. Going forward (thus) will cause regret.

or in the thigh....

(4) The fourth line, undivided, (shows) that firm correctness which will lead to good fortune and prevent all occasion for repentance. If its subject be unsettled in his movements, (only) his friends will follow his purpose.

If you fidget and can't keep still, it means that a friend is following your thoughts.

Kua no. 39, *Chien*[1]

LEGGE (9)

WALEY (8)

Thuan: (In the state indicated by) *Chien*, advantage (will be found) in the south-west, and the contrary in the north-east. It will also be advantageous to meet with a great man. (In these circumstances) with firmness and correctness (there will be) good fortune.

Hsiao Tzhu commentary:

(1) (From the) first line, divided, (we learn that) advance (on the part of its subject) will lead to (greater) difficulties, while remaining stationary will afford ground for praise.

He who goes stumbling

shall come praised;

(5) The fifth line, undivided, shows its subject struggling with the greatest difficulties, while friends are coming to help him.

a great stumble means
a friend shall come.

Of course this analysis does not solve all the problems of so complicated a text as the *I Ching*. But it throws much light on the origins from peasant sayings, and the later overlay of divination prescriptions. At least four different forms of divination are at the bottom of the Book of Changes: (*a*) the peasant omen-interpretations; (*b*) the 'drawing by lot' of plant stalks, short and long, which gave the lines of the symbols; (*c*) the divination by marks on the heated carapaces of tortoises or shoulder-blades of mammals (see on, pp. 347 ff.); this gave, as we know from the oracle-bone inscriptions, much of the vocabulary of the interpolated clauses (Li Ching-Chhih, *1*); and (*d*) divination by tablets of some form or other (dice, dominoes), since the character *kua* originally meant a tablet.

[1] 蹇

If the Book of Changes had remained a strictly auguristic text, it would have become simply another of the numerous books on divination, about some of which a word will later be said. But the addition of the appendices (written, probably, as we have seen, by Naturalist scholars and diviners in or just before the Chhin and Han periods) gave it a higher cosmological and ethical status. To each of the *kua* was allotted an abstract significance. Thus, if we take the examples of the two *kua* which we have just been considering, no. 31 (*Hsien*) is explained as 'mutual influence' with the undertone of sexual union (for the lower part of the hexagram is 'male' and the upper part 'female'), and hence of 'reaction'. Similarly, no. 39 (*Chien*) is explained as denoting lameness, hence the arresting of movement or advance, and so 'retardation' or 'inhibition'. The more abstract the explanations became the more the system as a whole assumed the character of a *repository of concepts*, to which all concrete phenomena in Nature could be referred. It would be surprising, with no less than sixty-four of these, if a pseudo-explanation could not be found for almost any natural event.

Before proceeding further it will be desirable to see the *I Ching* system in tabulated form. Table 13 sets forth the trigrams and Table 14 the hexagrams of the Book of Changes. With consummate art, the last *kua* of the series is not 'Consummation' or 'Perfect Order', but 'Disorder, potentially capable of perfection and order'.[a] One is tempted to exclaim, with Sir Thomas Browne, 'All things began in Order, so shall they end, and so shall they begin again, according to the Ordainer of Order, and the mystical mathematicks of the City of Heaven.' But for the Chinese, this mystical Order had no Ordainer.

Glancing over the series one finds several points to notice. The doubles of the eight trigrams appear in their places, though not symmetrically scattered; two are at the beginning (nos. 1, 2), two in the middle (nos. 29, 30), and four near the end (nos. 51, 52; 57, 58). With the exception of four pairs (nos. 1, 2; 27, 28; 29, 30; 61, 62) all the rest are placed in couples as mirror-images with their axis of symmetry below the first of the pair and above the second. In six cases (nos. 5, 6; 7, 8; 11, 12; 13, 14; 35, 36; 63, 64) this implies that there is a simple inversion of the natural-object descriptive formula (e.g. E/H and H/E, see p. 315), but by the nature of the case the great majority of the pairs are mirror-images, not trigram-inversions.

Later broodings over the system separated the *kua* into eight 'Houses' of eight *kua* each; the details can readily be found in R. Wilhelm.[b] The significance of this idea, which derives all the *kua* from the original eight trigrams, probably relates to Yin and Yang symbolism. If one looks again at Table 13 one notes the solidity of *Chhien* (no. 1) as opposed to the cavity in *Khun* (no. 2), and one can hardly overlook a phallic significance in this, *Chhien* as the lance and *Khun* as the grail.[c] There are

---

[a] Cf. *Yü Chien*, ch. 1, p. 12a.

[b] (2), vol. 2, p. 263, Baynes tr. vol. 2, p. 373.

[c] Such interpretations are entirely in the style of ancient Chinese thought. In several of the *kua* in Table 14 it will be noted that the assemblage of lines was itself considered 'pictographic'. There are a number of detailed explanations of these *kua* in the Great Appendix (App. 5 and 6), linked with a descriptive account of social evolution, which is very interesting in itself (H. Wilhelm (4), pp. 105, 111).

similar cavities in the other three 'female' *kua*, *Sun*, *Li* and *Tui* (nos. 6, 7, 8). *Kên* (no. 5) has been thought of as an upturned bowl or mountain, and it stands for stability; *Chen* as a bowl or valley lying open to the thunder for which it stands—one would have expected it thus to be 'female', but doubtless the desire for symmetry was more compelling.

(i) From Omen Proverbs to Abstract Concepts

When we compare the data in Table 14 with the footnotes appended to it, we gain a comparatively clear idea of the way in which the system grew up. First there were the collections of ancient peasant-omens (about birds, insects, weather, subjective feelings, and the like); these were without doubt already in existence in the −6th century, when Confucius was living. Somehow or other these collections coalesced with the books of the professional diviners, books which preserved traditional lore relating to scapulimancy (see on, p. 347), divination by the milfoil sticks (p. 349), and other forms of prognostication. This process was probably well advanced by the late −4th century, the time of Tsou Yen. He was assuredly well learned in these arts, and so it came about that the School of Naturalists, and later the Confucians of Chhin and Han times, who inherited so many of their ideas, remodelled the text and added elaborate commentaries to it. Nor is there anything surprising in the fact that growing abstract conceptualisation of the *I Ching* symbols kept pace with the development of early science out of earlier magic. To Han scholars, really trying to take a naturalistic attitude to phenomena such as magnetism or the tides, it must have been the obvious thing to do. Yet really they would have been wiser to tie a millstone about the neck of the *I Ching* and cast it into the sea.

It is not easy to trace the exact steps in the process of conceptualisation, and the story of it would be well worth the careful attention of sinological research. One of the turning points seems to be the *Chou I Lüeh Li*[1] of Wang Pi[2] (+226 to +249),[a] in which he stated the basic principle that if, for example, the first *kua*, *Chhien*, signified firmness, it was quite unnecessary to bring in the symbol of the horse to explain it; or if the second *kua*, *Khun*, signified compliance, the symbol of the cow could be summarily dispensed with. What the *I Ching* meant to this brilliant mind of the +3rd century may be sensed from the following passage, taken from the opening of his book.

What is the Thuan[3] (explanation)?[b] It treats comprehensively the essence of a given hexagram, and explains the 'dominant factor' (*chu*[4])[c] from which it flows.

The many cannot rule the many. That which rules the many is the supremely solitary (*chih kua*[5]). [*Continued on p. 322.*]

[a] Cf. Fêng Yu-Lan (1), vol. 2, pp. 184ff., 187.
[b] The 'solution' or 'significance' of the symbolism; see p. 305.
[c] Lit. 'master'.

[1] 周易略例   [2] 王弼   [3] 彖   [4] 主   [5] 至寡

Explanation of Table 13. Significances of the trigrams in the Book of Changes

Col. 1: The assemblage of lines which form the *kua*.

Col. 2: Romanised name of the *kua*.

Col. 3: Chinese character of the *kua*.

Col. 4*a*: 'Sex' of the *kua*.

Col. 4*b*: Associated position in a 'family' (from ch. 10 of App. 8, the *Shuo Kua*).

Col. 5: Associated animal (taken mostly from ch. 8 of App. 8, but with other information added).

Col. 6: Associated natural object or 'emblem' (from ch. 11 of App. 8). This list is important since the hexagrams shown in Table 14 are usually described in these terms. For example, *kua* no. 39, *Chien*, consists of *Khan* (trigram no. 4) over *Kên* (trigram no. 5), i.e.

fresh-water (lake)
mountain

These are represented in Table 14 using the following abbreviations: *H* = heaven; *E* = earth; *T* = thunder; *Fw* = fresh-water (lake); *M* = mountain; *WW* = wind; *L* = lightning; *Sw* = sea-water (sea).

Col. 7: Associated element (the five elements here have to cover eight *kua*). The list, which betrays the association of most of the appendices with the School of Naturalists, comes from ch. 11 of App. 8.

Col. 8: Associated compass-point, according to the 'more ancient' *hsien-thien*[1] ('prior to Heaven') or *Fu-Hsi*[2] system (see on, Sect. 26*i*).

Col. 9: Associated compass-point, according to the 'later' *hou-thien*[3] ('posterior to Heaven') or *Wên Wang*[4] system, as given in ch. 5 of App. 8 (see on, Sect. 26*i*).

Col. 10: Associated season.

Col. 11: Associated time of day or night.

Col. 12: Associated type of human being (from ch. 11 of App. 8).

Col. 13: Associated colour (from ch. 11 of App. 8).

Col. 14: Associated part of the human body (from ch. 9 of App. 8).

Col. 15: Primary concept or 'virtue' of the *kua* (taken mostly from ch. 7 of App. 8).

Col. 16: Secondary abstract concept of the *kua*.

[1] 先天　　　[2] 伏羲　　　[3] 後天　　　[4] 文王

Table 13. *Significances of the trigrams in the Book of Changes*

1 kua	2	3	4a	4b	5	6	7	8	9	10	11	12	13	14	15	16
1	Chhien	乾	♂	father 陽	dragon, horse	heaven	metal	S	NW	late autumn	early night	king	deep red	head	Being, strength, force, roundness, expansiveness	Donator
2	Khun	坤	♀	mother 陰	mare, ox	earth	earth	N	SW	late summer, early autumn	afternoon	people	black	abdomen	Docility, nourishment of being, squareness, form, concretion	Receptor
3	Chen	震	♂	eldest son	galloping horse, or flying dragon	thunder	wood	NE	E	spring	morning	young men	dark yellow	foot	Movement, speed, roads, legumes and young green bamboo sprouts	Stimulation, excitation
4	Khan	坎	♂	second son	pig	moon and fresh water (lakes)	water	W	N	mid-winter	midnight	thieves	blood-red	ear	Danger, precipitousness, curving things, wheels, mental abnormality, abysses	Flowing motion (especially of water)
5	Kên	艮	♂	youngest son	dog, rat, and large-billed birds	mountain	wood	NW	NE	early spring	early morning	gate-keepers	—	hand and finger	Passes, gates, fruits, seeds	Maintenance of stationary position
6	Sun	巽	♀	eldest daughter	hen	wind	wood	SW	SE	late spring, early summer	morning	merchants	white	thigh	Slow steady work, growth of woods, vegetative force, mercantile talent	Penetration, mildness, continuous operation
7	Li	離	♀	second daughter	pheasant, toad, crab, snail, tortoise	lightning (and sun)	fire	E	S	summer	midday	amazons	—	eye	Weapons, dry trees, drought, brightnesses, catching adherence of fire and light	Deflagration, adherence
8	Tui	兌	♀	youngest daughter (concubine)	sheep	sea and sea water	water and metal	SE	W	mid-autumn	evening	enchantresses	—	mouth and tongue	Reflections and mirror-images, passing away	Serenity, joy

Explanation of Table 14. Significances of the hexagrams in the Book of Changes

Col. 1: The assemblage of lines which form the *kua*.

Col. 2: Romanised name of the *kua*.

Col. 3: Chinese character of the *kua*. It is thought that all the names derive from those characters which occurred most frequently in the prognostications from the *kua*.

Col. 4: Characterisation of the *kua* according to its two component trigrams named by their associated natural objects or 'emblems', e.g. no. 7, *E/Fw*, earth over fresh-water; or no. 21, *L/T*, lightning over thunder.

Col. 5: One or two of the more common lexicographical meanings of the character which constitutes the name of the *kua*.

Col. 6: Concrete or social significance of the *kua*.

Col. 7: Abstract significance of the *kua*. These meanings are derived mostly from the *Ching* text itself, and from the *Thuan Chuan* commentary.

Col. 8: Page number references to Legge (9).

Col. 9: Page number references to explanations in H. Wilhelm (4).

Col. 10: Page number references to explanations in R. Wilhelm (2), vol. 1 (German ed.).

Notes will be found on pp. 320, 321.

Table 14. *Significances of the Hexagrams in the Book of Changes*

1 kua	2	3	4	5	6	7	8 Legge	9 HW	10 RW
1	Chhien	乾	H/H	heaven, paternal, dry, male	Heaven, king, father, etc., ordering, controlling	*Donator*	57	—	1
2	Khun	坤	E/E	earth, maternal	Earth, people, mother, etc., supporting, containing, docile, subordinate	*Receptor*	59	—	6
3	Chun	屯	Fw/T	sprout	Initial difficulties, 'contre-démarrage'a	*Factors slowing the onset of a process*	62	—	10
4	Mêng	蒙	M/Fw	cover	Youthful inexperienceb	*Early stages of development*	64	—	14
5	Hsü	需	Fw/H	need, procrastinate	Cunctatory policyc	*Stopping, waiting*	67	—	17
6	Sung	訟	H/Fw	litigation	Strife, contention at lawd	*Opposition of processes*	69	—	20
7	Shih	師	E/Fw	army, general, teacher	Military affairse	*Organised action*	71	—	23
8	Pi	比	Fw/E	assemble	Union, concord	*Coherence*	73	—	26
9	Hsiao Hsü	小畜	WW/H	to rear	Creative force modified by mildness, taming	*Lesser inhibition*	76	—	29
10	Li	履	H/Sw	shoe, to tread	Hazardous success attained by circumspect behaviour, treading delicately	*Slow advance*	78	114	32
11	Thai	泰	E/H	prosperous	Geniality of spring, peace (in the Sung came to mean one of the progressive world periods)	*Upward progress*	81	—	34
12	Phi	否	H/E	bad	Beginning of autumn (in the Sung came to mean one of the retrogressive world periods)	*Stagnation, or retrogression*	83	—	38

Table 14 (continued)

	1 kua	2	3	4	5	6	7	8 Legge	9 HW	10 RW
13		Thung Jen	同人	H/L	lit. people together	Union, community	State of aggregation	86	—	40
14		Ta Yu	大有	L/H	lit. great having	Abundance of possessions, opulence	Greater abundance	88	—	43
15		Chhien	謙	E/M	humility	Hidden wealth, modesty	Highness in lowness	89	110	46
16		Yü	豫	T/E	pleased	Harmonious excitement, enthusiasm, satisfaction	Inspiration	91	—	49
17		Sui	隨	Sw/T	follow	Following	Succession	94	—	52
18		Ku	蠱	M/WW	virulent poison (cf. p. 136)	Troublesome work in a decaying society[f]	Corruption	95	—	55
19		Lin	臨	E/Sw	approach	Approach of authority	Approach	97	—	58
20		Kuan	觀	WW/E	to look	Contemplation, looking for omens,[g] letting influence radiate	View, vision	99	—	60
21		Shih Ho	噬嗑	L/T	gnawing;[h] sound of voices	Crowds, markets, and courts, criminal law	Biting and burning through	101	113	63
22		Pi	賁	M/L	bright	Ornamental	Ornament, pattern	103	—	66
23		Po	剝	M/E	to peel, flay	Falling, overthrowing, collapse, like a house held together only by its roof (kua pictographic)	Disaggregation, dispersion	105	—	69
24		Fu	復	E/T	return	Year's turning-point	Return	107	—	71

No.	Hexagram	Name		Code		Description				
25	☰☷	*Wu Wang*	无妄	H/T	not reckless, not false[i]	No recklessness, no insincerity, not guilt, yet difficulties	*Unexpectedness*	109	—	74
26		*Ta Hsü*	大畜	M/H	to rear	Creative force suppressed by something stationary and heavy	*Greater inhibition*	112	—	76
27		*I*	頤	M/T	jaws	Mouth (which the *kua* shows in pictographic form)	*Nutrition*	114	—	79
28		*Ta Kuo*	大過	Sw/WW	to overstep	Large excess, strangeness not necessarily unfavourable[j]	*Greater top-heaviness*	116	5, 118	82
29		*Khan*	坎	Fw/Fw	pit[k]	The edge of the ravine, danger and the reaction to it; below, the torrent of water	*Flowing motion*	118	—	84
30		*Li*	離	L/L	separate, apart	The meshes of a net (*kua* pictographic), catching adherence of fire and light	*Deflagration, adherence*	120	112	87
31		*Hsien*	咸	Sw/M	all (but here used for *kan*)[l]	Mutual influence, interweaving, wooing	*Reaction*	123	—	91
32		*Hêng*	恆	T/WW	constant[m]	Perseverance	*Duration*	125	—	93
33		*Thun*	遯	H/M	to hide oneself, conceal[n]	Withdrawal, retreat	*Regression (further advanced than no. 12)*	127	—	96
34		*Ta Chuang*	大壯	T/H	great strength[o]	Great strength	*Great power*	129	—	99
35		*Chin*	晉	L/E	to rise, advance[p]	Advance in feudal rank	*Rapid advance*	131	—	101
36		*Ming I*	明夷	E/L	intelligence repressed[q]	Lack of appreciation of the services of a good official	*Darkening, extinction of light*	134	4	104
37		*Chia Jen*	家人	WW/L	family people	Members of a family or household	*Relation*	136	—	106

Table 14 (*continued*)

1 kua	2	3	4	5	6	7	8 Legge	9 HW	10 RW
38	Khuei	暌	L/Sw	separated	Division and alienation	*Opposition*	139	—	110
39	Chien	蹇	Fw/M	lame^r	Lameness, inhibition	*Retardation*	141	—	112
40	Chieh	解	T/Fw	dissection, analysis	Unravelling	*Disaggregation, liberation*	144	108	115
41	Sun	損	M/Sw	spoil, hurt, subtract	Removal of excess, payment of taxes	*Diminution*	146	—	118
42	I	益	WW/T	benefit	Increase of resources, addition	*Increase, addition*	149	—	121
43	Kuai	夬	Sw/H	fork, settled, decision	Breakthrough, release of strain, 'détente'	*Eruption*	151	—	124
44	Kou	姤	H/WW	copulation	Advance to casual encounter, meeting, intercourse	*Reaction, fusion*	154	—	127
45	Tshui	萃	Sw/E	thicket, congregate	Process of collection, consolidation of people around a good ruler	*Condensation, conglomeration*	156	—	130
46	Shêng	升	E/WW	to rise	Career of a good official	*Ascent*	159	—	133
47	Khun	困	Sw/Fw	surrounded, distressed	Straitened, distress, bewilderment	*Enclosure, exhaustion*	161	—	135
48	Ching	井	Fw/WW	a well	Dependableness	*Source*	164	—	138
49	Ko	革	Sw/L	skins	Moulting of skins, hence change	*Revolution*	167	—	141

No.		Romanization	Character	Trigrams	Literal meaning	Description				
50	䷱	*Ting*	鼎	*L/WW*	tripod cauldron	Nourishment (of talents), (*kua* alleged pictographic)	*Vessel*	144	—	169
51	䷲	*Chen*	震	*T/T*	quake, rock, thunder	Moving exciting power	*Excitation*	148	—	172
52	䷳	*Kên*	艮	*M/M*	limit[s]	Stability, as of a mountain	*Immobility; maintenance of stationary position*	151	5	175
53	䷴	*Chien*	漸	*WW/M*	gradually tinge[t]	Slow and steady advance (like chemical changes produced by soaking—dyeing, retting, lixiviating)	*Development, slow and steady advance*	154	—	178
54	䷵	*Kuei Mei*	歸妹	*T/Sw*	lit. returning, younger sister	Marriage[u]	*Union*	157	—	180
55	䷶	*Fêng*	豐	*T/L*	abundance (good harvest)	Prosperity	*Lesser abundance*	160	—	183
56	䷷	*Lü*	旅	*L/M*	travel, travellers	Strangers, merchants	*Wandering*	163	—	187
57	䷸	*Sun*	巽	*WW/WW*	gentle	Penetration of wind	*Mildness, penetration*	165	—	189
58	䷹	*Tui*	兌	*Sw/Sw*	exchange	Sea, pleasure	*Serenity*	168	—	192
59	䷺	*Huan*	渙	*WW/Fw*	broad, swelling, irregular	Dispersion, alienation from good	*Dissolution*	170	—	194
60	䷻	*Chieh*	節	*Fw/Sw*	joints of bamboo	Term, section, regular division, regulation, meditation (on general phenomena), confinement, silence	*Regulated restriction*	173	107	197
61	䷼	*Chung Fu*	中孚	*WW/Sw*	lit. central; confidence	Inmost sincerity, kingly sway	*Truth*	176	105	199

Table 14 (continued)

1 kua	2	3	4	5	6	7	8 Legge	9 HW	10 RW
62	Hsiao Kuo	小過	T/M	to overstep slightly	Small excess	Lesser top-heaviness	201	—	180
63	Chi Chi	既濟	Fw/L	lit. end; up to the mark	Completion, successful accomplishment	Consummation, perfect order	204	—	183
64	Wei Chi	未濟	L/Fw	lit. not quite; not quite up to the mark v	Position when all is not yet completed nor successfully accomplished	Disorder, potentially capable of consummation, perfection, and order	207	—	187

a The *Ching* text for this *kua* contains an ancient peasant omen concerning the colour of the horse on which the bride arrives at her husband's house. A marriage is one of the things which have 'initial difficulties'.

b All the interpretations given diverge from what was the original sense of this *kua*. It concerned the Chinese equivalent of the Golden Bough. *Mêng*[1] was anciently another name for the dodder (*Cuscuta sinensis*; B II, 131, 181, 450, 451; III, 163), an epiphyte like the mistletoe. Its commoner names are *nü lo*[2] (women's net) and *thu ssu tzu*[3] (rabbit silk). The absence of roots in the dodder was a matter of great interest for the early Chinese naturalists, as we shall see later in connection with the question of action at a distance (and magnetism). Earlier still there can be no doubt that it was an important magical object. Parasitic plants have widely been considered sacred (Frazer, 1). Waley (8) shows that the phrase with which the *Ching* text opens, 'It was not I who sought the *mêng* boy, it was the *mêng* boy who sought me', was simply a spell for averting the evil consequences of tampering with the holy plant. The idea of 'youthful inexperience' arose from the ancient folk custom of thinking of this plant as a boy, and of course the ultimate abstract significance is still further removed.

c The significances here are based on a misunderstanding, of which we shall see further examples, due to the failure of ancient scribes to add radicals to their phonetics. The *kua* should be not *hsü*[4] but *ju*,[5] meaning some kind of creeping insect (the word never acquired a generic or specific use, cf. Sect. 39), concerning which there are five peasant-omens in the text.

d The contention is thought to have been about the division of war booty, for there is an omen in the text about prisoners laughing.

e The text has a bird-omen indicating that a parley will be successful.

f Here the significances built up on the basis of the *ku*-poison (cf. p. 136) were all somewhat misleading; what the ancient text had to do with was the taking of omens by observations of the behaviour of maggots in the flesh of animals sacrificed to ancestral spirits.

g The text refers to observations of whether sacrificial animals advance or retreat, and to observations of inspired utterances of boys undergoing initiation (cf. *Lun Hêng*, Forke (4), vol. 1, pp. 232, 237, 246; vol. 2, pp. 2, 3, 126, 162). Waley (8) even suggests that the original meaning of *tung*,[6] boy, was 'one who is ceremonially beaten', arguing from old forms of the character. The 'weird ditties of children' were still being attended to when Ricci gave his account of Chinese customs about +1600 (Trigault; Gallagher tr. p. 84).

h The basis of the text here is omens from objects found when eating food; all the rest is wide-ranging derivation.

i This is all a misunderstanding. *Wu-wang* was a single word, probably meaning a figure tied to a bull which was driven away from the village as a scapegoat. This is one of Waley's most beautiful identifications.

[1] 蒙   [2] 女羅   [3] 兔絲子   [4] 需   [5] 蠕   [6] 童

j The text contains a willow-omen.

k Pit is right; all the rest is thought-arabesques of late scholars. The ancient ceremony referred to is that of sacrificing to the moon in a pit (cf. *Li Chi*, ch. 34).

l As we saw above, *hsien*[1] has lost its heart radical, and ought to be *kan*.[2] The omen basis is that of tinglings in the limbs.

m Here the concept of duration refers to what Waley (8) calls the 'stabilising process' of ancient magicians. When once a favourable omen had been obtained it was necessary to perform a rite stabilising it (e.g. burying or locking up of objects). Since the character *hêng*[3] in its ancient form consists of the moon between two lines, the suggestion is that it was originally a rite performed at the first appearance of the new moon and directed to making a favourable condition of affairs last all through that lunar month. The lines may represent magic lines drawn round the omen in question, such as circumperambulations of new settlements. Such a view throws quite a new light on *Lun Yü*, XIII, xxii, where Confucius makes his famous remark, 'The people of the south have a saying, "It takes *hêng*[3] (usually translated as perseverance) to make even a soothsayer or a medicine-man (physician)." It's quite true. If you do not stabilise your virtue, disgrace will overtake you.' All this ends up in a perfectly abstract concept of duration.

n Here *thun*[4] is a mistake for *thun*,[5] has nothing to do with hiding, and refers to omens from movements of young pigs.

o This originates from omens about rams getting stuck in bushes.

p *Chin*[6] should be *chin*,[7] to insert. The text therefore probably had to do with magic increasing fecundity of domestic animals, insertion referring to coupling of the males and females.

q This again is a complete misunderstanding. Li Ching-Chhih showed that *ming-i* was the ancient name of a bird, and the omens concern it. The interpretations as given in the table are all later imaginations.

r As we saw above, this originates from omens about stumbling. Of course, for later scientific thinkers, a concept such as retardation was very useful.

s A complete misunderstanding. *Kên*[8] ought to be *khên*,[9] and the omen concerned the way in which rats gnawed the exposed bodies of sacrificial victims.

t Although, as we have seen (p. 57), ancient Chinese thinkers were impressed by slow chemical actions in solution, the most ancient form of this omen-text has probably nothing to do with that conception. The *kua* character can also mean 'to skim', and Li Ching-Chhih showed that it referred to the way in which the wild geese skimmed over natural objects such as rock-ledges and trees, thereby giving omens.

u Wedding-omens are at the bottom of this.

v In spite of the very high and philosophical interpretation of the final *kua*, analysis of the text of the *Ching* shows that it had once to do with omens taken from animals crossing streams. Thus Waley:

'If the little fox, when almost over the stream,
Wets its tail,
Your undertaking will completely fail.'

1 咸 2 感 3 恆 4 遯 5 豚 6 晉 7 搢 8 艮 9 齦

Motion cannot control motion. That which controls the motion of the world is absolutely one (*chên i*[1]).

Therefore in order that the many may all be equally sustained, the dominant factor must be entirely unitary. In order that all motions may be equally carried on, the origin of them cannot be dual.

Things do not struggle among themselves at random (*wu wu wang jan*[2]). They flow of necessity (*pi*[3]) from their principle (or principles) of order (*li*[4]). They are integrated by a root cause[a] (*thung chih yu tsung*[5]). They are gathered together by a single influence (*hui chih yu yuan*[6]). Thus things are complex (*fan*[7]) but not chaotic (*luan*[8]). There is multiplicity of them but not confusion (*huo*[9]).[b]

At first sight we are surprised to find the Aristotelian conception of an unmoved mover emerging from the phraseology of Wang Pi. If this was really his thought we might be tempted to attribute to him a visualisation of laws of Nature enacted by a transcendent lawgiver. But we may be sure that in speaking of the One he had in mind the immanent Tao. What he was trying to describe was perhaps a series of fields of force (as we might call them), contained in, but subsidiary to, the main field of force of the Tao, and each manifesting itself at different points in space and time. He believed that to each of these there corresponded one of the hexagrams of the *I Ching*, its sufficient characterisation being given in the *thuan* or 'explanation' attached to it in the Book of Changes. In this way man could know the most important 'dominant factors' or 'root causes' of things, and feel able to affirm with unshakable faith that though there was manifold complexity in the universe, there was no confusion.[c]

(2) A UNIVERSAL CONCEPT-REPOSITORY

I wish now to show that by means of the abstract significances attached to each of the sixty-four *kua*, Chinese scientists of medieval times had what amounted to a repository of concepts to which almost any natural phenomenon could be referred. This may be illustrated by a diagram such as that of Fig. 42. If we take two coordinates or parameters, one indicating time and the other space, we can then insert the various *kua* along these axes. From the point of origin outwards the number of time units is increasing, or, if we retrace our steps, diminishing; from the point of origin outwards spatial configurations are growing, or if we retrace our steps, shrinking. Certain concepts seem to require a position on a line joining the two coordinates, which we may think of as representing roughly motion.

---

[a] Lit. 'ancestor'.

[b] Tr. auct., adjuv. Petrov (eng. Mrs Wright), and Bodde, in Fêng Yu-Lan (1), vol. 2, p. 180.

[c] My attention was drawn to the importance of this passage by Dr A. F. Wright of Stanford University.

[1] 貞一   [2] 物无妄然   [3] 必   [4] 理   [5] 統之有宗
[6] 會之有元   [7] 繁   [8] 亂   [9] 惑

Upon this conceptual diagram it is possible to place all of the sixty-four *kua* except nineteen. Of these, thirteen may be said to represent ideas which are naturalistic but not spatio-temporal. They may be listed as follows:

No. 15. Highness in lowness.
No. 16. Inspiration.
No. 30. Deflagration (with the undertone of Adhesion, which might justify it for insertion on the diagram).
No. 31. Reaction, interweaving.
No. 44. Reaction, fusion.
No. 54. Reaction, union.
No. 34. Great power.
No. 36. Darkening.
No. 43. Decisive breakthrough.
No. 21. Biting or burning through.
No. 57. Mild penetration (as of airs and winds). Compare no. 53, which may have a similar sense with reference to aqueous penetration.
No. 58. Serenity (with the undertone of a calm sea).
No. 22. Ornament, pattern (which could be applied to animal and plant pattern).

In addition, there are three which come in a highly abstract category, that of truth and order. These are:

No. 61. Truth. No. 63. Order. No. 64. Potential Order.

At the other extreme there are three *kua* which seem irreducibly concrete and human. These are:

No. 20. View, vision. No. 27. Nutritivity. No. 25. Unexpectedness.

This demonstrates the extent to which the Chinese thinkers subsequent to the Han succeeded in getting away from the extremely human significances which the *kua* had all originally had.

We may thus classify the *kua* as in Table 15. The reason why the total comes to more than sixty-four is that it has seemed desirable to count some of the *kua* twice over, e.g. with application both to time's flow and to spatial growth and shrinkage.

Occidental students of the Book of Changes have sometimes been tempted to praise it. I do not speak now of the strange story of what Leibniz found in it (see on, p. 340), but of Western philosophical sinologists of later date, who had to accept it at its face value, without the advantage of modern researches on its real nature and origins.[a]

---

[a] Richard Wilhelm actually believed in it as a method of divination and encouraged this belief in others (cf. Wilhelm & Jung, p. 144); as one may well remember when using his translation. See the remarks of Graf (2), vol. 1, p. [8].

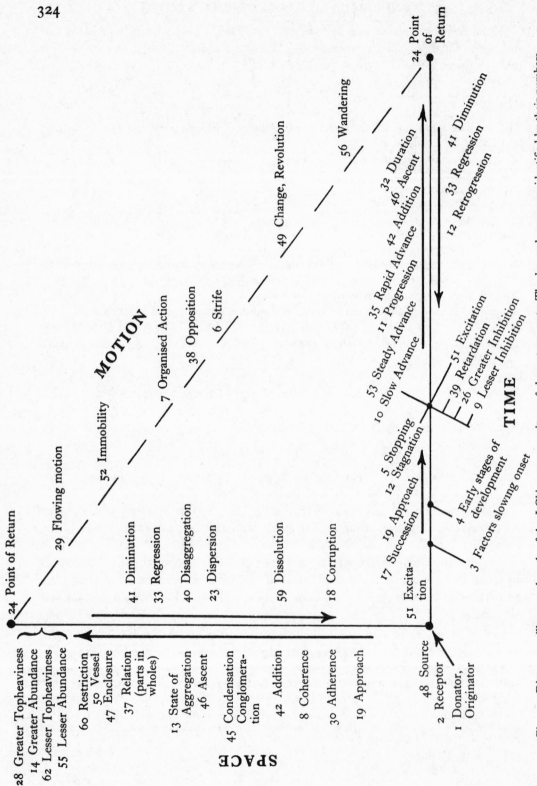

324

MOTION

29 Flowing motion
52 Immobility
7 Organised Action
38 Opposition
6 Strife
49 Change, Revolution
56 Wandering

24 Point of Return

SPACE

24 Point of Return
28 Greater Topheaviness
14 Greater Abundance
62 Lesser Topheaviness
55 Lesser Abundance
60 Restriction
50 Vessel
47 Enclosure
37 Relation (parts in wholes)
13 State of Aggregation
41 Diminution
33 Regression
40 Disaggregation
46 Ascent
23 Dispersion
45 Condensation Conglomera- tion
59 Dissolution
42 Addition
8 Coherence
18 Corruption
30 Adherence
19 Approach
51 Excita- tion
48 Source

TIME

24 Point of Return
41 Diminution
33 Regression
12 Retrogression
32 Duration
46 Ascent
42 Addition
35 Rapid Advance
11 Progression
53 Steady Advance
10 Slow Advance
51 Excitation
39 Retardation
26 Greater Inhibition
9 Lesser Inhibition
5 Stopping
12 Stagnation
19 Approach
17 Succession
4 Early stages of development
3 Factors slowing onset
2 Receptor
1 Donator, Originator

Fig. 42. Diagram to illustrate the role of the *I Ching* as a repository of abstract concepts. The *kua* or hexagrams, identified by their numbers, are arranged in relation to time, space, and motion.

Table 15. *Classification of the* kua *by categories*

			No. of *kua* in each category
Donator			1
Receptor and origin			2
Temporal	Duration	1	
	Forward motion	10	
	Stationary	2	
	Backward motion	3	
		16	16
Inhibition and retardation			4
Point of return			1
Spatial	Aggregating	6	
	Stationary	9	
	Disaggregating	6	
		21	21
Concepts involving motion		6	
Concept of immobility		1	
		7	7
Naturalistic but not spatio-temporal			13
Truth and Order			3
Irreducibly human			3
			71

More than sixty years ago Eitel (4) wrote:

There is underlying these diagrams a recognition of the truth that things are groups of relations.[a] The diagrams themselves are, to my mind, clearly ideal constructions, expressing real facts, and built up from the real elements of experience, though imperfect and fanciful. The diagrams are simply abstract types,[b] substituting an ideal process for that actually observed in Nature. They are formulae in which the multifarious phenomena are stripped of their variety, and reduced to unity and harmony. Causation is here represented as imminent change,[c] as the constant interaction of the bipolar power of Nature, which is never at rest, balanced or free, the mutually sustaining opposition of two forces which are essentially one energy, and in the activity of which divergence and direction are inherent.

One can only remark that it was very unfortunate that the 'ideal process' substituted for what was actually observed in Nature was an empty symbolism and not a series of

---

[a] Here he was hitting the nail on the head (cf. p. 199).
[b] In the theological sense.
[c] Did he mean to write 'immanent'?

mathematised hypotheses. When a little over thirty years ago Masson-Oursel said of the *I Ching* that 'it supposes a kind of translation of all natural phenomena into a mathematical language by means of a set of graphic symbols, germs of what Leibniz would have called a "universal character";[a] thus constituting a dictionary permitting men to read Nature like an open book, whether with intellectual or practical aims in view'—he was taking the name of mathematics in vain, as well as speaking of Nature in terms which a Pasteur, a Bohr or a Hopkins would never have dared to adopt.[b] For we are back again in that illusory realm of numerology, where number is not the empirical and quantitative handmaid of natural phenomena, but the categorical 'damsel of Nuremberg' in which they have to be made to fit.

Our judgement, it is safe to say, will not be similar to these, but it must be reserved until a few further stages in the argument have been completed. Of these the most important are the questions: (*a*) What, according to the *I Ching* Appendices, did the School of Naturalists and the scholars of the Han themselves think it was all about? and (*b*) How were the significances of the *kua* made use of by scientific writers during the succeeding centuries?

The answer to the first question seems to be that the idea of the Book of Changes as a repository of concepts was present from the −3rd century onwards. At the beginning of the 2nd chapter of the second part of the Great Appendix (App. 6) we read:

Anciently, when Pao Hsi (= Fu-Hsi) had come to the rule of all under Heaven, he looked up and contemplated (*kuan*[1])[c] the forms exhibited in the sky (the constellations), and he looked down, contemplating the processes (lit. methods; *fa*[2])[d] taking place on the earth. He contemplated the patterns (or ornamental appearances; *wên*[3])[e] of birds and beasts, and the properties of the various habitats and places. Near at hand, in his own body, he found things for consideration,[f] and the same at a distance, in events in general.[g] Thus he devised the eight trigrams, in order to enter into relations with the virtues of the bright Spirits, and to classify (*lei*[4]) the relations of the ten thousand things.[h]

The text then enters upon a list of the various inventions (nets, carts, boats, etc.) which the sages of old made. It is alleged that they got their ideas for these from contemplation of one or another of the *kua*. This chapter gives therefore a rather

---

   [a] He might have added John Wilkins (cf. pp. 344, 497 below). Cf. J. Cohen (1).

   [b] The comparison which H. Wilhelm (4), p. 75, makes between the Book of Changes and the Periodic Table of the chemical elements, is also an unfortunate one. The latter was almost wholly empirical, the former almost wholly arbitrary and imaginative.

   [c] Note the use of the word signifying the observation of omens.

   [d] The use of this word seems a little curious here, and suggests that it might have been borrowed from the Mohist logicians (cf. p. 173) to mean 'causes'.

   [e] One remembers that the most ancient forms of this character show a tattooed man (cf. p. 227).

   [f] Perhaps this is not so much physiological as a reminiscence of the 'tingling' omens (cf. p. 308).

   [g] Omens from birds, clouds, etc.

   [h] Tr. Legge (9), p. 382; R. Wilhelm (2), vol. 2, p. 251, mod.

¹ 觀   ² 法   ³ 文   ⁴ 類

interesting connection between ancient naturalistic thought and the beginnings of technologies, not, of course, because the sages really had anything to do with it, but because someone found it necessary to adduce reasons from the concept-repository for specific inventions. The traditions of inventors have already been mentioned elsewhere (Sect. 3d), and reference made to the interesting paper of Chhi Ssu-Ho (1) on the subject. Here are those given in the Great Appendix, part II, chapter 2:

Table 16. *Inventions mentioned in the Book of Changes*

Invention	Legendary sage	Attributed to		Explanation
		kua	no.	
Nets (and Textiles)	Fu-Hsi	*Li*	30	*kua* alleged pictographic
Ploughshares	Shen Nung	*I*	42	*WW/T*, both associated with wood (wooden ploughs)
Markets	Shen Nung	*Shih Ho*	21	*L* (i.e. sun)/movement on roads
Boats		*Huan*	59	wood/water
Carts		*Sui*	17	sprightliness/movement on roads
Gates		*Yü*	16	*kua* perhaps pictographic; movement/earth (walls)
Pestle and Mortar	Huang Ti,	*Hsiao Kuo*	62	wood/mountain (i.e. stone)
Bow and Arrow	Yao and Shun	*Khuei*	38	*L* (i.e. sun's rays like arrows)/passing away
Houses		*Ta Chuang*	34	*T* (i.e. inclement weather)/*H* (i.e. a space)
Coffins		*Ta Kuo*	28	serenity/wood
Quipu (knotted cords, as records)[a]		*Kuai*	43	speech/solidity (i.e. retention of things spoken)

It will be seen that the explanations are quite fanciful and arbitrary. Perhaps the claim that inventions stemmed from the Book of Changes was simply a device to add to its prestige.

We return to the general statements about the meaning of the *I Ching*. What the first part of the Great Appendix (App. 5) says is worth noting:

The *I* (*Ching*) was constructed in accordance with the measure (lit. water-level) of Heaven and Earth, therefore it is adjusted with perfect nicety to the Tao of Heaven and

---

[a] Cf. pp. 100, 445. This ancient method of recording has persisted in the Liu-Chhiu Islands until now (Simon, 1).

Earth. (Aided by the diagrams of the *I*) man can look up to observe (*kuan*[1]) the signs (*wên*[2]) in the heavens, and man can look down to study (*chha*[3]) the patterns (*li*[4]) on the earth. Thus he knows the causes (*ku*[5]) of darkness and light. Tracing things to their beginnings and following them to their end he knows what can be said about death and life. Noting how the seminal essence (*ching*[6]) and the *chhi* form all things, and how the wandering away of the soul (*hun*[7]) produces change, he understands the characteristics of the gods and spirits.

Between Man and Heaven-and-Earth there is a similarity, so there is no contradiction between him and them. His knowledge can embrace the whole world, and his Tao can set it all in order; thus he may avoid all error. He can act according to the nature of circumstances without being carried away in their current. He will rejoice in Heaven (because) he knows its ordinations; thus he will be free from anxiety.[a] He will rejoice in his estate, practising pure human-heartedness, and thus he will be able to attain love.

(The *I* gives) the moulds (*fan wei*[8])[b] of (all the) transformations of Heaven and Earth without any excess (or defect). All things are encircled by it so that nothing is left out. By it man can enter into relations with the Tao of day and night, and understand it. And thus his spirit is not bound down to any particular place. And the (principles of the) *I* (*Ching*) are not bound down to any particular corporeal manifestations...(ch. 4).

Production and reproduction is what may be called the principle of the (Book of) Changes...(ch. 5).

The sages were able to survey all the complex phenomena under the sky. They observed their forms and properties, and represented all things and their characteristics by emblematic diagrams (*hsiang*[9]). The sages also studied all the motive influences working under the sky. They contemplated their common actions and special natures in order to bring out the standard tendency of each one. Then they added their explanations (to each line of the diagrams), determining the good or evil indicated by it. These are called the *hsiao*.[10] (The diagrams) speak of the most complex phenomena in the world, but there is nothing distasteful in them. They speak of the subtlest movements in the world, and yet there is nothing of confusion in them...(ch. 8).[c]

What seems to show through these and other similar passages is the effort made by the School of Naturalists and the Han Confucians to erect the figures made by the long and short sticks into a comprehensive system of symbolism containing in some way all the basic principles of natural phenomena. Like the Taoists, they were looking for peace of mind through classification. Since, as we have seen, early science in China grew out from the same roots as magic, and since the arts of the Naturalists certainly included all forms of prognostication and *wu*[11] sorcery, this development was

---

[a] One cannot overlook here the clear statement of the search for ataraxy (cf. above, pp. 41, 63, below, p. 414).

[b] The metaphor is characteristically Chinese—that of metal casting. The *kua* would correspond to something invisible within Nature, like a set of moulds into which continuing creativity poured the molten *materia prima* of events and things. Cf. p. 243 above.

[c] Tr. Legge (9), R. Wilhelm (2), mod.

[1] 觀   [2] 文   [3] 察   [4] 理   [5] 故   [6] 精   [7] 魂
[8] 範圍   [9] 象   [10] 爻   [11] 巫

not unnatural. I shall shortly have a suggestion to make as to why it was pursued with such enthusiasm and persistence.[a] But first we must see to what uses it was put by later scholars who were trying to think scientifically about Nature.

(3) SIGNIFICANCE OF THE TRIGRAM AND HEXAGRAM SYMBOLS IN LATER CHINESE SCIENTIFIC THOUGHT

Here one can only choose a few examples from what is, of course, a vast literature. The first extension of the system of the *kua* seems to have taken place during the Early Han period, when they were brought into systematic association with sidereal movements and hence with the passage of time.[b] We know the names of certain men who were prominent in this, such as Mêng Hsi[1] and Ching Fang.[2] Hence arose the close connection of the *kua* with alchemy, where it was thought that the efficacy of chemical processes depended upon the exact time at which they were performed. But certain of the *kua* were also used, as we shall see, to symbolise chemical apparatus. Later on, the system of the *kua* was further extended to acoustics[c] and to speculations about the phenomena of living things, in biology, physiology and medicine. Many of these uses will be illustrated in the following pages. What should be kept particularly in mind in reading the passages quoted is the part which the *kua* were supposed to play; they come to be visualised not only as abstract formulations of all kinds of natural processes, but as invisible operators and causative factors.

One of the most important mutationists (*I Ching* specialists) of the Later Han period was Yü Fan[3] (+164 to +233), who in his commentary[d] gave a complete system of correlation between the eight trigrams, the movements of the sun and moon, the days of the month, and the ten 'stems' (denary cyclical characters). This need not be reproduced here since it is easily available.[e] The system was called the 'method of the contained stem' (*na chia*[4]).

---

[a] Mention should here be made of the *Thai Hsüan Ching*[5] (The Canon of the Great Mystery) written by Yang Hsiung[6] *c.* +10. This is a set of eighty-one tetragrams, the four lines of each being called *fang*,[7] *chou*,[8] *pu*[9] and *chia*[10] respectively from top downwards. There were broken lines of three, as well as of two, dashes. The tetragrams, each of which covered 4½ days of the year, were known as *shou*,[11] not *kua*. None of the names of these *shou* is identical with any of those of the *kua*. But this elaborate system did not seem to 'catch on'. Since tetragrams have been found on Chou bronzes (Schindler, 3), H. Wilhelm (4), p. 129, suggests that Yang Hsiung's work was based on a series of figures as old as the Book of Changes but of a different tradition—possibly one of those already referred to (p. 307). Alternatively W. Eberhard suggests (personal communication) that the *I Ching* hexagrams were connected with some pre-Chhin calendar system, while the tetragrams of Yang Hsiung were perhaps connected with the new calendar of Liu Hsin and Liu Hsiang. On Yang's system as a whole see Fêng Yu-Lan (1), vol. 2, pp. 139 ff.

[b] This is seen especially well in apocrypha such as the *I Wei Chi Lan Thu*,[12] described by Fêng Yu-Lan (1), vol. 2, pp. 106 ff.

[c] See Fêng Yu-Lan (1), vol. 2, pp. 118 ff.

[d] This no longer exists in full, but parts of it are given in the *Chou I Chi Chieh*[13] (Collected Commentaries on the *I Ching*), compiled by Li Ting-Tso[14] (fl. some time between +742 and +906).

[e] In the explanation of Fêng Yu-Lan (1), vol. 2, pp. 426 ff.; also Bodde (4), p. 116.

[1] 孟喜　　　[2] 京房　　　[3] 虞翻　　　[4] 納甲　　　[5] 太玄經　　　[6] 揚雄
[7] 方　　　[8] 州　　　[9] 部　　　[10] 家　　　[11] 首　　　[12] 易緯稽覽圖
[13] 周易集解　　　　　[14] 李鼎祚

Among Yü Fan's older contemporaries was the alchemist Wei Po-Yang,[1] who, as the author of the earliest extant Chinese alchemical book, will appear again prominently in the Section on chemistry.[a] His work, the (*Chou I*) *Tshan Thung Chhi*[2] (The Kinship of the Three), of +142, makes extensive use of the *kua*, and it is illuminating to quote side by side with it a few appropriate passages of the commentary on it written by the great Neo-Confucian philosopher, Chu Hsi, in +1197.[b] Chu Hsi starts out by saying[c] that it was not originally Wei Po-Yang's object to explain the *I Ching*, but that he made use of the Na Chia method in order to guide himself in the different times appropriate for adding reagents and withdrawing products. He goes on to say that the *kua Chhien* (no. 1) and *Khun* (no. 2) refer, among other things, to the chemical apparatus, while *Khan* (no. 29) and *Li* (no. 30) represent the chemical substances, and all the rest of the sixty *kua* are concerned with 'fire-times' (*huo hou*[3]), namely, the determination of the right moments for carrying out the chemical operations (and perhaps also the strength of the heating then to be employed).

WPY,[d] ch. 1. *Chhien* (no. 1) and *Khun* (no. 2) are the gateways of Change. They are the parents of all the *kua*.

CTC, p. 2*b*. ... The changes of Yin and Yang in connection with human beings refer to the Great Medicine of Golden Cinnabar. *Chhien* and *Khun* refer to the stove (*lu*[4]) and the reaction-vessels (*ting*[5]).

WPY, ch. 1. *Khan* (no. 29) and *Li* (no. 30) may be likened to the walls of a city, and their working is like that of the hub of a wheel which holds the axle in place.

CTC, p. 3*a*. They are likened to a city because of the places they occupy at the compass-points. They are likened to a wheel because their alternate exaltation and degradation (*shêng chiang*[6]) constitutes the principle of Change.[e]

WPY, ch. 1. The four male and female *kua* function like the bellows and the tuyau.

CTC, p. 3*a*. These *kua* are those in which the Yin and the Yang are combined, namely *Chen* (no. 51), *Tui* (no. 58), *Sun* (no. 57) and *Kên* (no. 52). The bellows (*tho*[7]), the piston (*pai*[8]), the bellows-bag (*nang*[9]) and the tuyau (*yo*[10]) are the tubular spaces (through which they work)[f]. ... (They also correspond to certain dates.)... The bellows should sometimes be worked slowly and sometimes rapidly (according to the degree of heating desired), just as the moon waxes and wanes.

[a] There is always a lingering doubt whether Wei Po-Yang was a truly historical person, but the dating of him and his book will be dealt with later on (Sect. 33). The full title of his book may perhaps best be translated: 'The Kinship of the Three; or, the Accordance (of the Book of Changes) with the Phenomena of Composite Things.'

[b] The *Tshan Thung Chhi Khao I*.[11]

[c] Pp. 1*b*, 2*a*, 3*b*.

[d] WPY = Wei Po-Yang. CTC = Chu Tzu's commentary. The former tr. Wu & Davis (1), mod., the latter tr. auct.

[e] Undoubtedly a reference to alternate oxidation and reduction, as of mercuric sulphide.

[f] There is of course a mystical undertone here, alluding to *Tao Tê Ching*, ch. 5, where the universe is compared to bellows, its use being its emptiness.

[1] 魏伯陽   [2] 周易參同契   [3] 火候   [4] 爐   [5] 鼎   [6] 升降
[7] 橐   [8] 鞴   [9] 囊   [10] 籥   [11] 參同契考異

WPY, ch. 2. The control of the Tao of Yin and Yang is like the work of a skilled driver, following his road as precisely as a carpenter works with his measures and inked plumb-lines.

CTC, p. 3*b*. ...The inked plumb-lines refer to the 'fire-times', calculated according to the sixty *kua*, as will later be explained.

WPY, ch. 3. At dawn *Chun* (no. 3) is at work, in the evening *Mêng* (no. 4) takes over control. Day and night each have a *kua* of their own, so we should use them according to their order.

WPY, ch. 3. The complete cycle runs from the moonless night to the night of the full moon. Then it is repeated all over again. There is a time to act and a time to refrain from action, according to the hour.

CTC, p. 4*a*. *Chi* (*-Chi*) (no. 63) and *Wei* (*-Chi*) (no. 64) are the *kua* of the last day of the lunar month. In the morning the former is suitable and in the evening the latter.

It is clear from this passage that in the alchemical tradition the two first trigrams and hexagrams were connected in some way with the apparatus employed, two others with the chemical substances, four more with the processes of heating, and all the rest with the times at which the experiments were to be carried out.[a] The hypostatisation of the *kua* comes out clearly in Wei Po-Yang's chapter 3, where it is actually said that certain *kua* are 'in control' at certain times. Besides the first four chapters already described, three other places in the *Tshan Thung Chhi* are particularly devoted to the *kua*; chapters 2 and 4, discussed by Fêng Yu-Lan (1) and translated by Bodde (4), in which the association of six trigrams with the phases of the lunar cycle are given; and chapter 18, where the cycle is described in different terms. Chapter 19 gives a diurnal cycle. These are listed in Table 17.

These cycles are worth working out on the diagram in Fig. 42, and its accompanying tables, since they throw some light on the mentality of their makers. In Table 17 the concepts of non-spatio-temporal character are bracketed.

Nothing could better illustrate the dialectical character of the correlative thinking embodied in the Book of Changes. No state of affairs is permanent, every vanquished entity will rise again, and every prosperous force carries within it the seeds of its own destruction.

In the above material the purely mystical element of alchemy (*nei tan*[1])[b] is more or less absent, but in the following passage from Chhen Hsien-Wei,[2] commenting in +1254 on the Fu[3] (Cauldrons, or Vessels) chapter of the *Kuan Yin Tzu*[4] book,[c] it is less easy to be sure that he is not referring to spiritual or psychological experiences. He may be talking about both this esoteric alchemy and the practical art at one and the same time. In any case, the *kua* are prominent.

---

[a] The keying of chemical operations to suitable sidereal times was of course a common feature also of later European alchemy. For example, Robert Norton in his *Ordinall of Alchimy* (+1477) says:
> 'The fifth Concord is knowne well of Clerks,
> Betweene the Sphere of Heaven and our suttill Werks';
and Elias Ashmole (*Theatrum Chemicum Britannicum*, +1652) comments: 'Our Author refers to the Rules of Astrologie for Electing a time wherein to begin the Philosophicall Worke....'

[b] As opposed to the 'exterior cinnabar' (*wai tan*[5]).

[c] *Wên Shih Chen Ching*, ch. 3, p. 1*b*. It is by an unknown Taoist of the Thang.

¹ 內丹　　　² 陳顯微　　　³ 釜　　　⁴ 關尹子　　　⁵ 外丹

Table 17. *Association of the* kua *with the lunar and diurnal cycles in the* Tshan Thung Chhi

THE CYCLE OF THE LUNAR MONTH, *Tshan Thung Chhi*, chs. 10 and 41

		no.	
(A)	*Fu*	24	Point of return (i.e. starting-point)
(B)	*Chen*	51	Excitation
(C)	*(Tui)*	58	Serenity (i.e. the process quietly at work)
(D)	*Chhien*	1	Donator (i.e. maximum of maleness, no moon)
(E)	*(Sun)*	57	Mild penetration
(F)	*Kên*	52	Immobility
(G)	*Khun*	2	Receptor (i.e. maximum of femaleness)
(A)	*Fu*	24	Point of return...

and the cycle recommences

THE DIURNAL CYCLE, *Tshan Thung Chhi*, ch. 42

		no.	
(A)	*Fu*	24	Point of return (i.e. starting-point)
(B)	*Thai*	11	Progression
(C)	*(Ta Chuang)*	34	Great power (i.e. acceleration of process)
(D)	*(Kuai)*	43	Decisive breakthrough
(E)	*Chhien*	1	Donator (i.e. maximum of maleness, noon)
(F)	*(Kou)*	44	Reaction
(G)	*Sui*	17	Succession
(H)	*Phi*	12	Stagnation
(I)	*(Kuan)*	20	Vision (?)
(J)	*Po*	23	Dispersion
(K)	*Khun*	2	Receptor (i.e. maximum of femaleness; midnight)
(A)	*Fu*	24	Point of return...

and the cycle recommences

...Now the sage, in this Seven Cauldrons chapter, has stated in detail the principles of Change. *Fu* is a pan or vessel in which things are changed by the action of water and fire. But very few of our modern scholars will not be astonished at his words. Some will think them heretical, some will think them false. But as Chuang Tzu said, you cannot talk to the blind about the beauties of literature, nor to the deaf about music, etc....

(For example) it is possible to make things pass through metal and stone. Now *Tui* (no. 58) is the *kua* for metal, and *Kên* (no. 52) is the *kua* for stone. But the *chhi* can penetrate them both, in the shape of mountains and marshes, and later, changes and transformations are accomplished....[a]

If you know how *Chhien* (no. 1) and *Khun* (no. 2) open and close, you will understand the principle of change. Then you will understand the intercourse between *Khan* (no. 29) and *Li* (no. 30), and (hence) the mutual antagonism of water and fire. The *chhi* penetrates

[a] Chhen Hsien-Wei is presumably thinking of subterranean water-channels, weathering, etc.

mountains and marshes, thunder and wind mutually fight—there is certainly a mechanism for all this.[a] What (I can recognise as) *Chen* (no. 51) and *Tui* (no. 58) in myself, is the lungs and liver in other people. If one could really enter into the spiritual aspect (symbolism?) of *Chen* and *Tui*, one would be able to see through their lungs and livers.[b]

A man's spirit and soul (*hun pho*[1]) are the refined essences of the Dragon and the Tiger (i.e. gold and mercury). If he can condense (*ning*[2]) the *chhi* of the *hun* and the *pho*, he can transform the Dragon and the Tiger within his viscera.[c]

Within the *Khan* (no. 29) there is the 'young lad' (*ying erh*[3]). Within the *Li* (no. 30) there is the 'beautiful girl' (*chha nü*[4]) (certainly mercury). If one can insert and fit (*tien*[5]) the solid reality of *Khan* into the emptiness of *Li*,[d] the 'young lad' and the 'beautiful girl' will see each other, and the shape of each will appear. This is the Tao.

Spiritual fire, shining within *Khan*, drives away the Yang which is inside the Yin. This Yang flies upwards and ascends, and at the 'original position of the spiritual fire' it meets with the Yin which is inside the Yang. These two capture each other, control each other, have intercourse with each other, and knot each other together.[e] It is like the taking hold of each other by the Golden Crow and the Rabbit (i.e. the conjunction of sun and moon), or the attraction of needles by lodestones. The two *chhi*, buttoning on to each other, and knotting each other together,[f] produce change and transformation. Sometimes the phenomena of the 'lad' and 'girl' appear, and sometimes the shapes of the Dragon and the Tiger. With numerous changes they fly about, rising, running and leaping, never quiet for a moment, and never coming out from the vessel and the stove. This is the time when the *Sun* (no. 57) wind should be blown to help the *Li* (no. 30) fire to bring about most fiercely the strongest transmutations. So will the true (cinnabar) medicine be condensed and aggregated. This is the Tao.

The two most important things are the observant mind (*kuan hsin*[6]) and the attracting spirit (*hsi shen*[7]); both helping the efficacy of the 'fire-times'. The meditation methods of the Buddhists seem to be valuable but are not really so. The Taoists who take deep breaths and swallow saliva are pursuing trifles and abandoning what really matters....[g]

From these excerpts it can be seen that the system of the *kua* was fully used by the alchemists in the Sung time and earlier. One should notice that both in Wei Po-Yang of the +3rd century, and in Chhen Hsien-Wei of the +13th, the *kua Khan* and *Li* both stand for the reacting chemical substances.[h] *Sun* is prominently connected with the ventilation of the furnace, and also (together with *Chen* and *Tui*) with the

[a] It might be a better reading to omit the 'wood' radical of this character, in which case Chhen Hsien-Wei is speaking of the 'germs' of things, as did Chuang Tzu (cf. p. 78). If so, his thought perhaps is that the *kua* represent in an abstract way these germs of things.

[b] I.e. if one really understood the symbolism one would understand the physiology (?).

[c] This expression has some connection with the respiratory exercises (p. 143).

[d] Frank sexual symbolism for the making of gold amalgam with mercury.

[e] The process described seems to be a vapour-solid reaction at the top of a reflux condenser system, similar to the *kerotakis* of Greek alchemy (cf. Sect. 33).

[f] Graphic similes for chemical combination.

[g] Tr. auct.

[h] I might add an instance of this from Shen Kua's *Mêng Chhi Pi Than* of +1086, *Pu* addendum, ch. 3, para. 13.

[1] 魂魄   [2] 凝   [3] 嬰兒   [4] 姹女   [5] 點   [6] 觀心
[7] 吸神

respiration of man—which was a perfectly correct parallel. But the vast majority of the *kua* stand for particular times.

It only remains to add an example or two from the biological field. In the *Li Hai Chi*[1] of Wang Khuei[2] (a book probably written in the early Ming, i.e. late +14th century) we find the following remarks about blood:

> The blood of man and animals is always red.[a] This is because it is Yin and belongs to watery things, which are under the aegis of the *kua Khan* (no. 29).[b] But the blood also harbours a Yang (component), and it is red too because of what it contains. The interaction of *Khan* with *Li* (no. 30) is what causes the motion of the *chhi* (of the blood). Now if the blood leaves the body for too long, it turns black, and if it be heated it also turns black; this is because it tends to return to its origin (i.e. the *kua Khun*, no. 2, earthiness).[c]

This is just like Wang Khuei, who noted many strange things of biochemical interest which no one else observed. But it shows the delusive nature of the *kua* system. The colour blood-red having been arbitrarily chosen in earlier centuries for association with *Khan*, it then becomes a fine and satisfying explanation for the red colour of blood to say that the *kua Khan* is controlling it. *Khan*'s partner, *Li*, played a similar role in explaining why there are some animals which have exoskeletons. We meet with this again and again. Thus Kao Ssu-Sun,[3] who wrote an excellent treatise on Crustacea about the year +1185, the *Hsieh Lüeh*,[4] says in his introduction: 'The Shuo Kua (appendix of the *I Ching*, App. 8) says that the *kua Li* (no. 30) controls (*wei*[5]) crabs. Khung Ying-Ta[6][d] explains this by saying that it is because they have their hard parts on the outside and their soft parts on the inside.'[e] Here the derivation is from a purely pictographic interpretation of *Li* as the seventh trigram (see Table 13), since it has a Yang line above and below, with a Yin line in the middle. According to this, *Khan* ought to stand for fishes, Sauropsida and mammals, but I have not seen this said in so many words; nevertheless, *Khan*'s animal is the pig, which is very soft outside. Everyone repeats the illuminating explanation of exoskeletons, even Li Shih-Chen at the end of the +16th century.[f]

Lastly, one physiological and one medical example. The *Li Hai Chi* says:

> The upper eyelid of human beings moves, and the lower one keeps still. This is because the symbolism of the *kua Kuan* (no. 20) embodies the idea of vision. Windy *Sun* (trigram no. 6) is moving above, and earthy *Khun* (trigram no. 2) is immobile below.
>
> Similarly, the human lower jaw moves while the upper one remains stationary. This is because the symbolism of the *kua I* (no. 27) embodies the idea of nutrition. Thundery *Chen* (trigram no. 3) is moving below, and mountainous *Kên* (trigram no. 5) is stationary above.

---

[a] He had not noticed the blue haemocyanin of Crustacea.
[b] See Table 13, the Trigrams, where blood-red is *Khan*'s colour.   [c] P. 8b, tr. auct.
[d] G 1055; commentator on the Book of Changes, Sui dynasty (+574 to +648).
[e] In *Shuo Fu*, ch. 36, p. 17b, tr. auct.
[f] *Pên Tshao Kang Mu*, ch. 45, p. 22a, in connection with crabs (cf. Read (5), no. 214, p. 33); and ch. 46, p. 28a, in connection with river-snails (*Paludina* spp.), cf. Read (5), p. 75.

[1] 蠡海集   [2] 王逵   [3] 高似孫   [4] 蟹略   [5] 為   [6] 孔穎達

Again, the eye is at the upper part of the head and its upper part moves; this is because the *chhi* of Heaven is active above. But the mouth is at the lower part of the head and its lower part moves; this is because the *chhi* of Earth is active below.[a]

The arthropod-vertebrate contrast comes out again in a medical context in the *Li Hai Chi*, where it is said:

The north of the (Yellow) River is the seat of *Khan* (no. 29), so the people up there have strong constitutions (*nei shih*[1]). The south of the (Yangtze) River is the seat of *Li* (no. 30), so the people have weak constitutions (*nei hsü*[2]). The former have the Yang inside and therefore need cold and purging medicines; the latter have the Yin inside and therefore need warm medicines and nourishing treatment.[b]

If this kind of argumentation tempts one to despair, one must remember that our European forefathers with their theological emblems and final causes were not much better off in the last decades of the +14th century, about the time of the foundation of the older Cambridge colleges. But as we read on, the devastating effects of the Book of Changes become more and more manifest.

Yet the interesting thing about these passages is the conception which must have been at the back of a great deal of *I Ching* thinking, namely, that of Heaven and Earth as one vast field of force, with *Chhien* and *Khun* as its two poles.[c] Of course Wang Khuei does not say so clearly, but he speaks as if it were quite natural that anterior-dorsal structures in the higher animals should orient themselves towards heaven, while posterior-ventral structures should orient themselves towards earth. The same idea seems to lie at the back of the cycles of 'fire-time' *kua* from Wei Po-Yang analysed a few pages above.

(4) THE BOOK OF CHANGES AS THE 'ADMINISTRATIVE APPROACH' TO NATURAL PHENOMENA; ITS RELATION TO ORGANISED BUREAUCRATIC SOCIETY AND TO THE PHILOSOPHY OF ORGANISM

The powerful hold which the essentially medieval system of the Book of Changes continued to exert upon Chinese minds even up to recent times is a matter of general knowledge. Everyone who has lived in China has known the profound attachment of old scholars to it. Legge (9) must have been speaking from personal experience when he wrote:

Chinese scholars and gentlemen who have got some acquaintance with 'western' science[d] are fond of saying that all the truths of electricity, light, heat, and other branches of 'Euro-

---

[a] P. 17*a*, tr. auct.   [b] P. 15*a*, tr. auct.   [c] H. Wilhelm (4), p. 41.

[d] The quotation-marks within this excerpt are mine, for science is and has always been, universal, a fact which is unaffected by the historical chance that the great upsurge of modern science which occurred in Europe in the 17th century had in due course to be transmitted eastwards. A thousand years earlier, the transmission had been in the opposite direction.

[1] 內實   [2] 內虛

pean' physics are in the eight trigrams. When asked how then they and their countrymen have been and are ignorant of those truths, they say that they have to learn them first from western books, and then, looking into the *I*, they see that they were all known to Confucius more than two thousand years ago. The vain assumption thus manifested is childish, and until the Chinese drop their hallucination about the *I* as containing all things that have ever been dreamt of in all philosophies, it will prove a stumbling-block to them, and keep them from entering upon the true path of science.

These words were written nearly a century ago, but now the pendulum has swung far in the opposite direction; indeed, the history of science in Asia suffers greatly from the fact that extremely few Chinese scientists have any time to spare for the examination of what they regard as the follies of their own medieval ages. But now the time has come to form our own judgment as to the role played by the Book of Changes in the development of Chinese scientific thought.

I fear that we shall have to say that while the five-element and two-force theories were favourable rather than inimical to the development of scientific thought in China,[a] the elaborated symbolic system of the Book of Changes was almost from the start a mischievous handicap. It tempted those who were interested in Nature to rest in explanations which were no explanations at all.[b] The Book of Changes was a system for *pigeon-holing novelty* and then doing nothing more about it. Its universal system of symbolism constituted a stupendous *filing-system*. It led to a stylisation of concepts almost analogous to the stylisations which have in some ages occurred in art forms, and which finally prevented painters from looking at Nature at all.[c] We may of course be prepared to admit that a filing-system for natural novelty can meet that need which, as I have pointed out above, was one of the greatest stimulatory factors of primitive science, namely, the need for at least *classing* phenomena, and placing them in some sort of relation with one another, in order to conquer the ever-recurring fear and dread which must have weighed so terribly on early men.[d] Any hypothesis would be better than none, but hypotheses which would take some of the terror out of disease and calamity there must at all costs be. At first sight it would seem that those who imagined the Democritean atoms were simply much more fortunate in their choice than those who thought they could seize the essence of all the moulding forces in the universe by means of a system of sixty-four linear diagrams. But the matter is not so simple as that.

There is a question here which refuses to be dismissed, namely, *why* did the universal symbolic system of the *I Ching*, to which Europe can offer nothing parallel,

---

[a] Fêng Yu-Lan (1), vol. 2, p. 131, concurs.

[b] Prof. H. H. Dubs has suggested to me that perhaps one of the reasons for the flowering of Chinese science in the Sung was that the Neo-Confucians, starting with Chou Tun-I, took much of the superstition out of the *I Ching*, and restored the *kua* to a purely symbolic use. But I fear that the concept of the *kua* as shadowy causative factors behind natural phenomena continued long after their time.

[c] This point arose in conversation with Mr M. Sullivan.

[d] As H. Wilhelm has pointed out, too (4), p. 24, the Book of Changes also embodies an optimistic psychology of attack in so far as the manifoldness of the universe was imagined to be made comprehensible by the *kua*.

grow up, and *why* did it show such extraordinary longevity and persistence? Could the answer have been given already in our description of it as a cosmic filing-system? Was the compelling power which it had in Chinese civilisation due to the fact that it was a view of the world basically congruent with the bureaucratic social order? Could one even describe it as the 'administrative approach' to natural phenomena? When Chinese scientific writers say that such and such a *kua* 'controls' such and such a time or phenomenon; when some natural object or event is said to be 'under the aegis of' such and such a *kua*, one is irresistibly reminded of the phrases familiar to all those who have served in government organisations—'a matter for your department', 'passed to you for appropriate action',[a] and so on. The Book of Changes might almost be said to have constituted an organisation for 'routing ideas through the right channels to the right departments'. Here, of course, it is not possible to give

Table 18. *Association of the* kua *with the administrative system in the* Chou Li

Pu	Associated concept	*kua*	Trigram no.	Hexagram no.
(1) General Administration	Heaven	*Chhien*	1	1
(2) Ministry of Education	Earth	*Khun*	2	2
(3) Ministry of Rites	Spring	*Chen*	3	51
(4) Executive	Summer	*Kên*	5	52
		Sun	6	57
		Li	7	30
(5) Ministry of Justice and Punishments	Autumn	*Tui*	8	58
(6) Ministry of Public Works	Winter	*Khan*	4	29

any description of Chinese bureaucratism, which must await the concluding sections of the book; the reader can only be asked at this stage to take it for granted that Chinese society was a bureaucratism (or perhaps a bureaucratic feudalism), i.e. a type of society unknown in Europe. The point to be made is that the system of the Book of Changes might be regarded as in a sense the heavenly counterpart of the bureaucracy upon earth, the reflection upon the world of Nature of the particular social order of the human civilisation which produced it.

This connection, moreover, was by no means unconscious in Chinese thought, if anyone had bothered to notice it. In the idealised administrative system elaborated by Han scholars and handed down to us as the *Chou Li* (Record of the Rites of Chou) each of the great ministries is associated with a season, and hence with a *kua* (Table 18). The descriptions in the *Chou Li* represent admittedly an ideal system which never exactly existed, but many of these ideas continued into later ages, as may be seen in the recent elaborate work on the administrative chapters of the official histories of the Thang dynasty by des Rotours (1).

[a] Or frequently, as here, non-action.

Such considerations lead us to what might be regarded as the dénouement of the present section. Perhaps the entire system of correlative organismic thinking was in one sense the mirror image of Chinese bureaucratic society. Not only the tremendous filing-system of the *I Ching*, but also the symbolic correlations in the stratified matrix world might so be described. Both human society and the picture of Nature involved a system of coordinates, a tabulation-framework,[a] a stratified matrix in which everything had its position, connected by the 'proper channels' with everything else. On the one hand there were the various Pu[1] or Ministries and departments of State (forming one dimension), and the Nine Ranks of officials (*chiu phin*[2])[b] (forming the other). Over against these there were the five elements or the eight trigrams or sixty-four hexagrams (forming one dimension), and all the ten thousand things divided among them and individually responsive to them (forming the other). One must of course avoid carrying such a comparison too far, for some of the most telling examples of the Chinese philosophy of organism (quoted on pp. 51 ff.) come from Chuang Tzu, who lived at least a couple of centuries before Chinese bureaucratism had got into its stride—nevertheless, one may say that the conditions for it were always there in Chinese society; there were constellations of career officials in each of the feudal States, and the concrete basis of bureaucratic power, hydraulic conservation works, had already begun to play an important role in Chuang Tzu's time. In making the obvious comparison between Taoist organicism and Democritean-Epicurean atomism can we consider it a mere coincidence that the former arose in a highly organised society where conservancy-dictated bureaucratism was dominant while the latter arose in a world of city-states and individual merchant-adventurers? I believe that we cannot, but the deep contrasts between European and Chinese society must be held over for the latter part of this book.

It would not, however, be anticipating too much what must there be said to point out that Chinese bureaucratism was fundamentally agrarian, and based upon agricultural production in a context of irrigation and water-control; as opposed to the maritime emphasis of the European city-states. Granet (5) was seeing another facet of Chinese society, therefore, when he underlined, in a famous passage, the agrarian and rustic elements in the Chinese world-picture:[c]

People like to talk about the gregarious instinct of the Chinese, and to attribute to them an anarchic temperament. In fact, their spirit of associativeness, and their individualism, are rustic and peasant qualities. Their idea of Order derives from a healthy country feeling for *good understanding*. The checkmate of the Legalists, the joint success of the Taoists and Confucians, proves it. This feeling, wounded by excessive administrative intrusions, equalitarian constraints, or abstract rules and regulations, always rested (allowing of course for individual variations) upon a kind of passion for autonomy, and upon a need, no less strong,

---

[a] Allusion is made elsewhere to the early appearance of coordinate-like tabulation frames in China (Sects. 2 (Vol. 1, p. 34) and 19*f*, *h*). Surely this was no coincidence.

[b] Cf. Mayers (1), p. 364.   [c] (5), p. 590

[1] 部   [2] 九品

for comradeship and friendship. State, Dogma, Law, were powerless as compared with Order. Order was conceived as a Peace which no abstract forms of obedience could establish, no abstract reasoning impose. To make this Peace reign everywhere, a taste for conciliation was necessary, involving an acute sense of compromise, spontaneous solidarities, and free hierarchies. Chinese logic was no rigid logic of subordination, but a supple logic of hierarchies, and its conception of Order never lost the concrete content of the ideas and emotions which gave it birth. Whether you call it the Tao, and see in it the principle of all autonomy and all harmony, or whether you symbolise it as Li, and see in Li the principle of all hierarchy and equitable partition, the idea of Order retains (in highly refined form, of course, yet never far from its rustic origins) the meaning that to understand and to induce understanding is to create Peace in oneself and around oneself. All Chinese wisdom arises from this. Its nuance may be more or less mystical or positivist, more or less naturalist or humanist, that does not matter much—in all the Schools we find the idea (expressed in concrete symbolism and none the less efficacious for that) that there is no difference between the principle of universal good understanding and that of universal intelligibility. All knowledge, all power, proceeds from the Li and the Tao. All acceptable rulers must be saints or sages. All authority rests on Reason.

Still broader consequences follow. Greek atomism and mathematics are doubtless rightly regarded as the foundations of the Cartesian-Newtonian science of the European 17th century. In the womb of modern capitalist society they gave birth to the 'modern' science of our immediate forefathers, Dalton, Huxley and the mechanical materialists. But science since their time has been obliged to become still more 'modern', to assimilate field physics, and to take account of parts of the universe, the enormously great and the enormously small, which transcend the range of sizes for which the Newtonian world-picture was constructed.[a] Deepening knowledge of biological phenomena, too, has necessitated a reformulation of scientific concepts in which the philosophy of organism has had a vital part to play. But the philosophy of organism was not, to begin with, a product of European thinking; we suspect that Leibniz may have been influenced by it in its systematic Neo-Confucian form. An unexpected vista thus opens before our eyes—the possibility that while the philosophy of fortuitous concourses of atoms, stemming from the society of European mercantile city-states, was essential for the construction of modern science in its 19th-century form; the philosophy of organism, essential for the construction of modern science in its present and coming form, stemmed from the bureaucratic society of ancient and medieval China. The new forms which science is taking today do not of course supersede the 'classical' system of Newtonian natural science; they are simply rendered necessary by the fact that science today has to deal with realms of the universe which that system did not envisage. All that our conclusion need be is that Chinese bureaucratism and the organicism which sprang from it may turn out to have been as necessary an element in the formation of the perfected world-view of natural science, as Greek mercantilism and the atomism to which it gave birth.

[a] Cf. the address by Niels Bohr at the Newton Tercentenary meetings.

Of course if these suggestions should be substantiated it would not be the only instance of a kind of oscillation in the application of fundamental ideas, as between Man and Nature. One thinks of the parallel of natural selection. As is generally known, Darwin obtained inspiration from Malthus, and applied with much validity to Nature what Malthus had somewhat unjustifiably applied to Man. Then later on the formulations of Darwin were brought back into human society and quite unjustifiably applied there. So in the present case a theoretical organicism which Leibniz and Whitehead applied to Nature had perhaps originated as a reflection in Nature of Asian bureaucratic society. It will be understood that none of these meditations justify in any way the position of the Book of Changes, or palliate the evil effects which it had on Chinese scientific thinking. The gigantic historical paradox remains that although Chinese civilisation could not spontaneously produce 'modern' natural science, natural science could not perfect itself without the characteristic philosophy of Chinese civilisation.

(5) ADDENDUM ON THE BOOK OF CHANGES AND THE BINARY ARITHMETIC OF LEIBNIZ

The mention of Leibniz brings us to a matter which is perhaps more curious than important in the history of science. Among his Chinese studies and discoveries was a mathematical interpretation of the diagrams in the Book of Changes, the significance of which is still somewhat disputed. This extraordinary story is best told in two rather inaccessible papers by H. Wilhelm (5) and Bernard-Maître (6).[a]

Our ordinary arithmetic has 10 as its base and the addition of a zero in the last integral place multiplies the number by 10. But this usage is purely arbitrary. Arithmetic could have been based on 12 instead of 10, in which case the third and the quarter would not have involved fractions of whole numbers. Some properties of numbers are fundamental to any system, while others depend upon the base arbitrarily chosen. For instance, the fact that adding odd numbers together gives the series of squares would be so whatever base had been chosen. But that all the multiples of 9 are figures which when added together give 9 (or a multiple of 9 less than the one in question) is not a fundamental property, and simply comes about because 9 is the penultimate number in the base series arbitrarily chosen. It occurred to Leibniz that an arithmetic to the base 2 would be possible and might be useful; in this 'binary' or 'dyadic' arithmetic a zero added to any number would have the power of multiplying it only by 2, just as in ordinary arithmetic it multiplies by 10. The numbers would therefore be represented in the following way:

1 = 1	6 = 110	11 = 1011	16 = 10000
2 = 10	7 = 111	12 = 1100	and so on
3 = 11	8 = 1000	13 = 1101	
4 = 100	9 = 1001	14 = 1110	
5 = 101	10 = 1010	15 = 1111	

[a] See also Vacca (8).

The first description of this system was given in a paper by Leibniz in +1679, 'De Progressione Dyadica'. The full publication appeared in the *Mémoires de l'Académie Royale des Sciences* for +1703 under the title 'Explication de l'Arithmétique Binaire' (Leibniz, 4), in which examples of addition, subtraction, multiplication and division in the binary system were given.[a] But the subtitle goes on to say '...qui se sert des seuls caracteres o et 1, avec des remarques sur son utilité et sur ce qu'elle donne le sens des anciennes figures chinoises de Fohy'. What had happened in the meantime?

What had happened was that Leibniz had come into contact with one of the Jesuit missionaries in China, Fr. Joachim Bouvet,[b] who was particularly interested in the Book of Changes, and with whom Leibniz carried on a long correspondence,[c] lasting from +1697 to +1702. The discovery that the *I Ching* hexagrams could be interpreted as another way of writing numbers according to the binary system, if the unbroken lines (Yang *hsiao*) were taken to represent 1, and the broken lines (Yin *hsiao*) to represent 0, seems to have been in the first place the idea of Bouvet rather than of Leibniz. Bouvet had brought the Book of Changes to Leibniz's attention in 1698, but it was not until Leibniz had sent him a table of his binary numerals in April 1701[d] that the identity with the hexagrams was realised, and in November of the same year Bouvet despatched to Leibniz two complete diagrams of the series. One of these was the 'segregation-table' which has already been reproduced as Fig. 41, and the other was a square and circular arrangement (shown in a folding plate in Legge, 9). Neither of these gives the *kua* in the so-called Wên Wang order in which they are arranged in the *I Ching* text, and according to which the charts in most editions show them. Known as the Fu-Hsi (*hsien thien*,[1] 'prior to heaven') system, Bouvet's form does not begin with *Chhien*, but with *Khun*, running 2, 23, 8, 20, 16, 35, 45, 12 and so on, in such a way that instead of having mirror-images or inversions next to one another, there is a methodical progression with a gradually increasing number of unbroken lines, exactly as required for Leibniz's notation. Thus *Khun* (no. 2) corresponds to 000000, *Po* (no. 23) to 000001, *Pi* (no. 8) to 000010, *Kuan* (no. 20) to 000011, *Yü* (no. 16) to 000100, and so on through the whole 64. Actually the Fu-Hsi order (which we shall have to discuss again in Section 26*i* because according to the two orders the compass-point associations differ) is not ancient at all, and cannot be traced further back than the Sung philosopher Shao Yung[2] and his *Huang Chi Ching*

---

[a] This paper is reproduced in full in Bernard-Maître (6).

[b] On Bouvet, see Dehergne (1); Pfister (1), pp. 433 ff.

[c] This correspondence is preserved in the Library at Hanover, and by one of the ironies of history it has as yet been published fully only in Japanese and Chinese. The Japanese scholar Gorai Kinzo copied it at Hanover and translated it into Japanese. It was then put into Chinese by Liu Pai-Min and appeared in Li Chêng-Kang's *Collection of Treatises on the I Ching* (1).

[d] He had done this because he attached religious and mystical significance to the binary arithmetic. 'All combinations arise from unity and nothing, which is like saying that God made everything from nothing, and that there were only two first principles, God and nothing.' Leibniz hoped that the Chinese might be induced to accept Christianity by such quasi-mathematical demonstrations.

[1] 先天　　[2] 邵雍

Shih Shu[1] (Book of the Sublime Principle which governs all things within the World) of about +1060. As H. Wilhelm (5) points out, Fr. Bouvet knew only of this arrangement because he was closely connected with Chhing dynasty court life, where Neo-Confucianism was still orthodox, and the new criticism of the *I Ching* by men such as Ku Yen-Wu,[2] Hu Wei,[3] Chang Erh-Chhi[4] and Wang Chhuan-Shan[5] was unknown. In a way, this was a fortunate chance.

Not unnaturally, Leibniz was amazed that he should find his binary notation employed for the series of numerals 63 to 0 in the hexagrams of the Book of Changes, which in his day were universally believed to go back to at least the −2nd millennium. He continued to descant on his joint discovery with Bouvet for the rest of his life, as, for instance, at the end of his long letter of +1716 on Chinese philosophy analysed below (p. 501), where the fourth section is entitled 'Des Caractères dont Fohi, Fondateur de l'Empire Chinois, s'est servi dans ses Ecrits, et de l'Arithmétique Binaire'.[a] And the discovery continued to arouse interest in the 18th century, as the publication by Haupt in 1753 testifies.

The real point of interest for the history of science, however, is what significance, if any, attaches to this story. 'The phenomenon', writes H. Wilhelm, 'that two speculative minds, six and a half centuries apart in time, living at opposite ends of the world, and starting from altogether different foundations, should have arrived at the same scheme of order is really astonishing. One cannot help feeling that the coincidence was not an accidental one, and that somehow both systems must rest upon the same natural basis.' Waley (9), who at that time (1921) accepted an extremely high antiquity for the hexagrams, suggested that Leibniz's discovery implied some understanding of the zero and of positional value by the Chinese long before −1000. In spite of Pelliot's criticism (15) of this, Olsvanger (1) (who continues to accept impossible legendary dates) retains, in his apparently independent rediscovery of binary arithmetic in the hexagrams, the idea that they embody an understanding of place-value and zero.[b]

Such suggestions should of course be discarded. The men who invented the hexagrams were simply concerned to form all the permutations and combinations possible from two basic elements, the sticks long and short. These once formed, it could have been obvious that several equally logical arrangements might be possible, and in fact two of them ultimately acquired great prominence, though others could be devised without difficulty. The chief defect in the attribution of mathematical significance to the hexagrams is that nothing was further from the thought of ancient *I Ching* experts than any kind of quantitative calculation, as Granet has sufficiently shown. In so far as the diviners worked with 'mutations' of the hexagrams, substi-

---

[a] Kortholt (1735), vol. 2, p. 488.

[b] Leibniz and Bouvet naturally assumed that while the ancient Chinese had had an understanding of binary arithmetic, it had long been forgotten.

[1] 皇極經世書   [2] 顧炎武   [3] 胡渭   [4] 張爾岐   [5] 王船山

tuting broken for unbroken lines and vice versa, they might be considered to have been executing simple binary arithmetical operations, but they certainly did so without realising it. One must surely ask of any invention, whether mathematical or mechanical, that it be made consciously and for us. If the *I Ching* diviners were unconscious of the binary arithmetic and made no use of it, the discovery of Leibniz and Bouvet has only the significance that the system of abstract order embodied in Shao Yung's version of the *I Ching* happened to be the same as the system of abstract order involved in the binary arithmetic. The belief of Leibniz and Bouvet that God had inspired Fu-Hsi to put it there need not detain us.

Recently Barde (1) has come forward with a more plausible suggestion. He thinks that the lines of the *kua* were connected not so much with long and short sticks for divination, as with the counting-rods which the Chinese certainly used from ancient times.[a] The symbols would thus have been derived from the procedures involved in the use of an arithmetic to the base 5, in which the weak or broken lines would have been rods having the value of 1, while the strong or unbroken lines would have been rods having the value of 5.[b] That arithmetics to the base 5 have existed among primitive peoples is a well-known fact of anthropology.[c] It may be not without significance that the first five Chinese numerals are, and were, rodlike; while in the Roman numerals there is a clear survival of arithmetic to the base 5, since 6 is 51, 7 is 52, etc. An ancient form of multiplication, before the construction of the multiplication-table to the base 10, would have needed the memorisation of certain numbers—25 (the sum of the first five odd numbers), 144 (the first six odd numbers each multiplied by 4), and 216 (the first six odd numbers each multiplied by 6). These are precisely the numbers which appear prominently in the Great Appendix of the *I Ching*.[d] If this is on the right track, the magical-divinatory symbols would have been a degeneration of a very ancient form of arithmetic. A corollary is that the hexagrams would have been primary, and the trigrams a later product of analytic thought; Barde has assembled sinological evidence to show that this was in fact the case.

It only remains to add that Olsvanger and Barde translate the hexagrams of the Wên Wang (*I Ching* text) block order into ordinary numerals, by way of the binary system or otherwise, and find a variety of magic squares. While it is probably true[e] that the discovery of the properties of magic squares occurred earlier in China than anywhere else, the magic squares obtained from the Book of Changes are rather complicated and it is hard to convince oneself that the Chinese mutationists ever had any such thought in mind when they arranged their hexagrams.

[a] See below, in Section 19*f* on mathematics.

[b] Alternatively the unbroken lines stood for odd numbers and the broken ones for even. But it is noteworthy that the Chinese abacus, which probably goes back to the early centuries of our era, has sliding balls of two values, 1 and 5, separated by a rail (see Sect. 19*f* below).

[c] They would have arisen very naturally from the use of only one hand instead of both.

[d] Ch. 9; tr. Legge (9), p. 365; R. Wilhelm (2), vol. 2, p. 236; eng. Baynes (1), vol. 1, p. 333. There were many fanciful explanations of these in Chinese literature, cf. *Yü Chien*, ch. 1, p. 3*a*.

[e] See below, in Section 19*d*.

A dozen years ago the subject might have been left at this point. But recent developments have shown that the binary or dyadic arithmetic of Leibniz is far from being a mere historical curiosity. It has been found to be, as Wiener points out in his important book on 'cybernetics' (the study of self-regulating systems whether animal or mechanical), the most suitable system for the great computing machines of the present day.[a] It has been found convenient to build them on a binary basis, using only 'on' or 'off' positions, whether of switches in electrical circuits or of thermionic valves,[b] and the type of algorithm followed is therefore the Boolean algebra of classes, which gives only the choice of 'yes' or 'no', of being either inside a class or outside it.[c] It is thus no coincidence that Leibniz, besides developing the binary arithmetic, was also the founder of modern mathematical logic and a pioneer in the construction of calculating machines.[d] As we may later see, Chinese influence was responsible, at least in part, for his conception of an algebraic or mathematical language, just as the system of order in the Book of Changes foreshadowed the binary arithmetic. In 1642 Blaise Pascal had constructed the first adding machine, but it was Leibniz who in 1671 conceived the first machine which should be able to multiply, though this was not carried out in the metal until the time of Thomas in France in 1820. The first conception of a universal calculating machine was that of Babbage in England in 1832, and its first realisation had to await the work of Aiken in America a little over a century later. It is not in the least surprising, says Wiener,[e] that the same intellectual impulse which brought about the development of mathematical logic led at the same time to the ideal or actual mechanisation of the processes of thought, for both were essentially devices intended to achieve the most perfect precision and accuracy by cutting out human prejudice and human frailty.

There is, moreover, a further perspective.[f] The computing machine of today, with its consecutive switching devices and its systems of feedbacks for automatic maintenance of a predetermined plan of operations, has been regarded as an almost ideal model of the animal central nervous system.[g] Obviously its input and output need not be in the form of numbers or diagrams, but might well be, respectively, the readings of artificial sense-organs such as light-sensitive cells, pH recorders, microphones, touch-switches, etc., on the one hand, and all kinds of effector servo-mechanisms, such as solenoids, on the other. This has been overlooked so long because physiologists and biochemists have tended to think in terms of energy sources and utilisation rather than in terms of signals, i.e. as power engineers rather than as communication engineers. It is becoming possible to visualise the future social implications of giant mechanisms of control which could render the entire functioning of a complex factory automatic. Nor is the effect likely to be in the industrial field alone,

---

[a] Wiener (1), pp. 10, 139. Some engineers, e.g. Pollard (1), recognise the Chinese ancestry. The application to fast electronic counting-circuits was first made by Wynn-Williams in 1932.

[b] Cf. Comrie (1); Bush & Caldwell (1); Lilley (1); Aiken & Hopper (1); Hartree (1); Berkeley (1).

[c] Wiener (1), p. 140.   [d] Cf. Michel (5).   [e] (1), p. 20.

[f] Noted elsewhere in the Chinese context only by Cassian in his introduction to Perleberg (1).

[g] Wiener (1), pp. 22, 36.

for it is pointed out that within the central nervous system of the higher living organisms the neurons themselves seem to act according to the principles of the binary arithmetic, namely, in their property familiar to physiologists as the 'all-or-none reaction'.[a] They are either at rest, or else when they 'fire' they do so in a manner almost independent of the nature and intensity of the stimulus. Of course this does not mean that graded responses are not often found in neurophysiological phenomena, but simply that there is ground for believing that these are the summation effects of populations of neurons each of which follows an all-or-none law. Here, then, we see how the binary arithmetic, stumbled upon by Shao Yung in his arrangement of the *I Ching* hexagrams and brought to consciousness by Leibniz, might be said in a very real sense to have been built into the mammalian nervous system long before it was found convenient for the great computing machines of modern man.

[a] Sherrington (1), p. 70; Wiener (1), p. 141.

14. THE PSEUDO-SCIENCES AND THE SCEPTICAL TRADITION

SUPERSTITIOUS PRACTICES flourished in China just as strongly as in all other ancient cultures.[a] Divination of the future, astrology, geomancy, physiognomy, the choice of lucky and unlucky days, and the lore of spirits and demons, were part of the common background of all Chinese thinkers, both ancient and medieval. The historian of science cannot simply dismiss these theories and practices, for they throw much light on ancient conceptions of the universe. Moreover, as has already been emphasised (pp. 83, 136, 240, 280), and as will be seen in striking examples later on (Sects. 22*f*, 26*i*, 32, 33), some of these magical practices led insensibly to important discoveries in the practical investigation of natural phenomena. Since magic and science both involve positive manual operations, the empirical element was never missing from Chinese 'proto-science'. On the other hand, scepticism was an essential part of that critical spirit which was the second requirement for the development of scientific thinking, and it is worthy of remark that this sceptical element also was never lacking from the traditions of Chinese thought.[b] The third element which would have been necessary for the unfolding of modern science in the purely Chinese milieu was the formation of mature hypotheses couched in mathematical terms and experimentally verifiable. These alone could supersede the primitive theories which have been described in the preceding Sections. But this was the only one of the three elements which never spontaneously arose.[c] The present Section will be devoted to the sceptical tradition[d] and its greatest representative, Wang Chhung, whose life fell in the +1st century. He typifies those men who, while remaining basically Confucian, were nevertheless attracted by the Taoist interest in Nature.

(*a*) DIVINATION

In order to understand what the sceptics were reacting against, it is necessary first to recount, in the briefest form, the principal types of pseudo-scientific belief in Chinese culture. They were 'techniques of destiny' (*shu shu*[1]), means of foretelling future events.

[a] Among the classics on this subject, especially with reference to Babylonian origins, are the books of Bouché-Leclercq (2) and of Lenormant (1, 2).

[b] Fêng Yu-Lan (1), tr. Bodde (4), pp. 122ff.; (1), vol. 2, p. 433, has also well contrasted the spirit of verification and precision which is found in Wang Chhung with the desire to gain control over the forces of Nature which is so marked in the later Taoists, especially the alchemists (see on, Sect. 33).

[c] See Section 19*k* below.

[d] The only paper known to me which covers in any way the field of this Section is the short sketch of Forke (10).

[1] 術數

(1) SCAPULIMANCY AND MILFOIL LOTS

First, as to divination. From the highest antiquity the Chinese had the conviction that it was possible to foretell the future, at least in so far as the affairs of princes and States were concerned, by processes of divination which gave a yes-or-no answer. The oldest technique was no doubt scapulimancy, the heating of tortoise carapaces or ox and deer shoulder-blades with red-hot metal, and the interpretation of the resulting cracks. The very word for divination (*chan*[1]) may be derived from an ancient pictogram of a scapula so treated (Hopkins, 21). It is to this technique, as we have already seen (Sect. 5*b*), that we owe most of the information which we have about Chinese society in the −2nd millennium, and all the information available concerning the most ancient forms of Chinese writing (Creel, 1, 2). The use of tortoise carapace and sternum was introduced after the mammalian shoulder-blades had long been current. The identity of the reptile which produced it is not quite certain, but the Chinese biological tradition (in the *Pên Tshao* series) was that it was the *shui kuei*,[2] an animal now identified with Reeves' terrapin.[a] Direct examination of large fragments indicates rather that it was a land-tortoise of a species now extinct, *Pseudocadia anyangensis* (Sowerby, 1). Some authorities believe that the carapaces and bones had to be imported from far to the south, outside the primary zone of Chinese culture. The consultation of the carapace or scapula, to obtain predictions either fortunate (*chi*[3]) or unfortunate (*hsiung*[4]), was termed *pu*;[5] or the 'resolution of doubts' (*chi i*[6]).[b] As a solvent for neuroses of indecision the method probably paid its way.

During the Chou period another procedure came to acquire an importance almost equal to that of scapulimancy, namely, the 'drawing of lots' by means of the dried stalks of a plant known as the Siberian milfoil[c] (*Achillea sibirica*),[d] called *shih*.[7] The technical term for consultation of the milfoil, corresponding to *pu*,[5] was *shih*.[8] It is interesting, in the light of the discussion of the shamanic component of Taoism in Section 10*h* to note that *wu*[9] (wizard) is one of the chief parts of this character. There is some possibility that it was a system of choosing long or short sticks in this method of divination which led to the arrangement of unbroken and broken lines in the trigrams and hexagrams of the *I Ching* (Book of Changes) which has just been discussed. But this work bears scapulimantic traces also (Wu Shih-Chhang, 1).

Most of the classical books (e.g. the *Li Chi* (Record of Rites), the *Chou Li* (Record of the Rites of Chou), the *Shu Ching* (Historical Classic), etc.) make mention of these two methods (Fig. 43). The milfoil was consulted mainly on affairs of lesser importance, the tortoise-shell for the greater, but frequently both were used. In this case matters became complicated, for the two methods naturally sometimes disagreed. On the

[a] R 199. *Emys*, or *Geoclemys*, *reevesii*.
[b] *Shu Ching*, ch. 24 (Hung Fan); Karlgren (12), p. 32.
[c] Or yarrow.   [d] R 1; B II, 428 and III, 71.

[1] 占   [2] 水龜   [3] 吉   [4] 凶   [5] 卜   [6] 稽疑   [7] 蓍
[8] 筮   [9] 巫

Fig. 43. A late Chhing representation of the legendary Emperor Shun and his ministers, including Yü the Great, consulting the oracles of the tortoise-shell and the milfoil. From *SCTS*, ch. 3, Ta Yü Mou.

basis of the Hung Fan chapter of the *Shu Ching*[a] and other texts, the following table may be made:[b]

Pro		Contra	
T, M	or	*T, M*	Definitely favourable or unfavourable as the case might be.
T or *M*		*M* or *T*	Milfoil valid for the immediate future, tortoise-shell valid for the further future.
T, M, P	or	*T, M, P*	Favourable or unfavourable as the case might be, in spite of the opinions of ministers and people.
T, M, m	or	*T, M, m*	Favourable or unfavourable as the case might be, in spite of the opinions of prince and people.
T, M, p	or	*T, M, p*	Favourable or unfavourable as the case might be, in spite of the opinions of prince and ministers.

KEY. T = tortoise-shell; M = milfoil; P = opinion of the Prince; m = opinion of the Ministers; p = opinion of the People.

Such a schematic presentation of course takes no account of the cases which could doubtless be cited of actions which took place contrary to it[c]—if the last category was ever acted upon, it must have meant a strange alliance of superstition and democracy.

During the Han and in later times the popularity of the tortoise-shell or scapula decreased.[d] Late encyclopaedias such as the *Thu Shu Chi Chhêng*[e] contain indeed a mass of information on scapulimancy, and all that was known about it was collected by Wang Wei-Tê[1] in his *Pu Shih Chêng Tsung Chhüan Shu*[2] of +1709, but nevertheless modern scholars have naturally had difficulty in interpreting the meanings of the cracks on the oracle-bones of the −2nd millennium. The milfoil, on the other hand, has descended continuously to the Taoist temples of the present day, where simple folk choose a stick from a box rattled by the attendant Tao-shih and are then given a future-foretelling paper corresponding to the number on the stick.

(2) USE OF THE SYMBOLS OF THE BOOK OF CHANGES

During the Warring States period (−4th and −3rd centuries) a third method of prognostication grew up, namely, the random selection of the trigrams of the *I Ching* and their combination and recombination.[f] As each one had come to stand for various more or less well-defined abstract ideas and broadly sketched natural processes (see

---

[a] Tr. Karlgren (12), p. 33.   [b] Wieger (2), p. 35.

[c] Cf. the examples collected in Wieger (2), pp. 67 ff. (the dates given are of course subject to all reservations).

[d] Nevertheless, these methods continued to be used by the thousands of diviners whose names are scattered through the official histories. The biography of one of them, Ssuma Chi-Chu[3] (ch. 127 of the *Shih Chi*), has been translated by Pfizmaier (36); he lived at Chhang-an and died about −170. Pfizmaier (56) also translated ch. 95 of the *Chin Shu*, which gives details of more than twenty famous diviners flourishing in the +3rd and +4th centuries.

[e] *I shu tien*, chs. 541–64.

[f] Significantly, the milfoil sticks were generally used for the selection.

[1] 王維德   [2] 卜筮正宗全書   [3] 司馬季主

Section 13g), it was not very difficult to draw conclusions as to what their fortuitous juxtapositions portended. In the *Tso Chuan* (Commentary on the Spring and Autumn Annals) as we have it today, there are many accounts of consultations using the *I Ching* symbols.[a] But though these annals purport to cover the period from −722 to −453 approximately, we know that they were expanded and retouched in the neighbourhood of −250, at which time the *I Ching* consultations were probably inserted, as well as many of the speeches.[b] The procedure was somewhat analogous to the 'sortes virgilianae' of medieval Europeans.

This method became exceedingly widespread during the Han dynasty, and books, the core of which may well be of that time, still survive, as, for example, the *I Lin*[1] of Chiao Kan[2] (fl. −85 to −40),[c] and the *I Chuan*[3] of his pupil Ching Fang[4] (fl. *c.* −51).[d] The *Thai Hsüan Ching*[5] (Canon of the Great Mystery) of Yang Hsiung[6] (−53 to +18)[e] is classified as belonging to this kind of literature. Thus by the time of Wang Chhung there were several important works expounding the system of prognostication by the trigrams and hexagrams. After his time, the ineradicable belief persisted, and Kuo Pho[7] (+276 to +324)—who had a hand in so many pseudo-sciences—published in the Chin dynasty his *I Tung Lin*[8] (Grottoes and Forests of the Book of Changes). Another outstanding work was the *Chhien Hsü*[9] (The Hidden Emptiness) of the great Sung scholar Ssuma Kuang[10] in the +11th century. Belief in the *I Ching* still persists. All these books, together with others, and with much additional miscellaneous material, are found in the +18th-century encyclopaedia, *Thu Shu Chi Chhêng*.[f]

Chatley (5), looking at the matter scientifically, has well said: 'There can be little doubt, when one studies the different forms of divination, that it was the ancient belief that any group of different units whose arrangement after a shuffling process was impossible to predict, would serve for purposes of prophecy. Unseen powers would be able to affect the slight variations of circumstance which determine the final configuration, while those initiated into the code explaining all the possible configurations were thereby able to interpret the will and knowledge of the unseen powers.' He adds that just as in occidental techniques of divination, the diviner was directed to concentrate his attention on the object to be known, presumably so that the spiritual influences could control the muscular and other elements in his shuffling process; so divination by the *I Ching kua* was preceded by the burning of incense and

---

[a] These have been collected by Wieger (2), pp. 115 ff., and Barde (2).

[b] As Dubs (7), p. 69, has pointed out, there is a strange silence about the *I Ching* in all other Chou writings before the −3rd century, so either it was too mysterious to be mentioned, or else it did not exist, and the latter view is the more probable. The general opinion is that the *I Ching* was from the beginning a book of divination. See pp. 304 ff. above.

[c] G 349.   [d] G 398.   [e] G 2379. See p. 329 above.

[f] *Ching chi tien*, chs. 95–110. But according to the *Ssu Khu Chhüan Shu Tsung Mu Thi Yao*, ch. 108, p. 6a, the works of Yang Hsiung and Ssuma Kuang were little used for divination.

[1] 易林   [2] 焦贛   [3] 易傳   [4] 京房   [5] 太玄經   [6] 揚雄
[7] 郭璞   [8] 易洞林   [9] 潛虛   [10] 司馬光

the recitation of prayers. Fifty sticks were then shuffled into two groups (or three), and the odd sticks counted out by cycles of eight, thus determining the complete or broken character of each of the six component lines (hsiao[1]) in the hexagram resulting. Methods of this kind have been used from antiquity up to the present time.

(3) ASTROLOGY

The next great department of pseudo-science which must be mentioned is astrology (hsing ming[2]). But like all the other systems of Chinese pseudo-scientific thought here mentioned, it has hardly been investigated at all by modern historians of science. There is nothing corresponding, for Chinese astrology, to the excellent treatises of Bouché-Leclercq (1), Boll (1), or Boll, Bezold & Gundel on Greek and ancient Mediterranean astrology;[a] and still less is there anything paralleling the exhaustive treatise of Thorndike (1) on magic and the pseudo-sciences in general. A few short and scattered papers are alone available.

Chinese astrology was bound from the outset to take a somewhat different course from its European-Mesopotamian counterpart, since the Chinese (as will be fully explained in Section 20 on astronomy) did not, in the earliest times, pay much attention to the heliacal risings and settings of stars which so interested the Babylonians, Egyptians and Greeks. They concentrated their attention rather on the circumpolar constellations which never rise and never set, but perform their apparent diurnal revolution around the pole-star in full view throughout the hours of darkness. These were divided, as we shall see, into twenty-eight 'mansions' (hsiu[3]), or radiating divisions separated by hour-circles.[b] These hsiu or hour-angle segments did not form, as is sometimes said, a zodiac, since the moon and the sun did not move among their defining stars, which were mostly equatorial or not ecliptic. Consequently, the Chinese astrologers gave less emphasis to what star or constellation was 'in the ascendant' at the time of any particular event on earth concerning which inquiries were made (cf. Fig. 44), and used a variety of other methods.

As Eisler (1) points out, in his recent survey of the astrological element in ancient astronomy, the outstanding characteristic of the oldest astrology is that it was never concerned with individual human beings (unless they were of royal blood), and always with prognostications concerning affairs of State, the chances of war, the prospects of the harvest, and so on. There is here a general parallel to the kinds of questions asked on the ancient Chinese oracle-bones. The thousands of cuneiform astrological tablets in the museums of Europe, forming the 'Reports of the Magicians and Astrologers of Nineveh and Babylon' (so runs the striking title of Thompson), show no ancient instance of a horoscope drawn up for an individual person. The earliest examples are from Hellenized Babylon in −176 (a birth-horoscope) and −169 (a conception-

---

[a] I would mention also the works of Cumont (1), Nilsson (1), and Thierens (1).

[b] Each hsiu was associated with a particular feudal state; the list is given in Huai Nan Tzu, ch. 3, p. 15a. Cf. Sect. 22d on the later 'fên yeh' system.

[1] 爻   [2] 星命   [3] 宿

Fig. 44. A Chinese horoscope of the +14th century. The nineteenth of a series of 39 sample horoscopes indicating all kinds of fortunes in life; here a person who is destined to achieve fame. The series constitutes the *Chêng shih Hsing An* (Astrological opinions of Mr Chêng) appended (as chs. 18 and 19) to the *Hsing Tsung*, a compendium of astrology attributed to Chang Kuo of the Thang (+8th cent.). From *TSCC, I shu tien*, ch. 584, *hui khao* 20, p. 19*b*. Favourable features of the horoscope are shown in the top right-hand box, unfavourable ones opposite on the left. Immediately underneath and at the bottom corners are shown the celestial influences governing 42 different aspects of life and health. Among them are included, besides the sun, moon and five planets (represented by their element names), Rahu and Ketu (the nodes of the moon's path), comets and vapours. The seventh or outermost ring of the disc itself gives constellation names, the fifth gives *hsiu*, and the first contains the twelve cyclical characters which are also compass points. Segment significances are defined by the fifth ring. They concern, counting counterclockwise from the right (at half-past two), fate (i.e. longevity), wealth, brothers, landed property, sons, servants, marriage and women, illness, travel, official position, happiness and bodily constitution. The order and nature of these twelve segments show at once that they are none other than the twelve houses or cusps (*loci, topoi*) of Hellenistic astrology as it was systematised in the time of men such as Sextus Empiricus (*c.* +170) and Firmicus Maternus (*c.* +335). The houses were so many immobile divisions of the celestial sphere, and horoscopes were cast according to the positions occupied by zodiacal constellations, planets, and certain stars at the time of the individual's birth (see Bouché-Leclercq (1), pp. 280 ff.; Eisler (1), p. 39). The word horoscope itself thus derives from the time-determining stars for the rising of which astronomers were accustomed to look in ancient Egypt and Babylonia. It can be seen that Chinese astrology included much, at any rate, which was common to all the peoples of the Old World. In the particular case here shown, two of the houses (the sixth and the twelfth) are inverted. But East Asian horoscopic houses could differ much more than this from the Greek order if we may judge from an Annamese diagram recorded by Huard & Durand (1), p. 67.

PLATE XVII

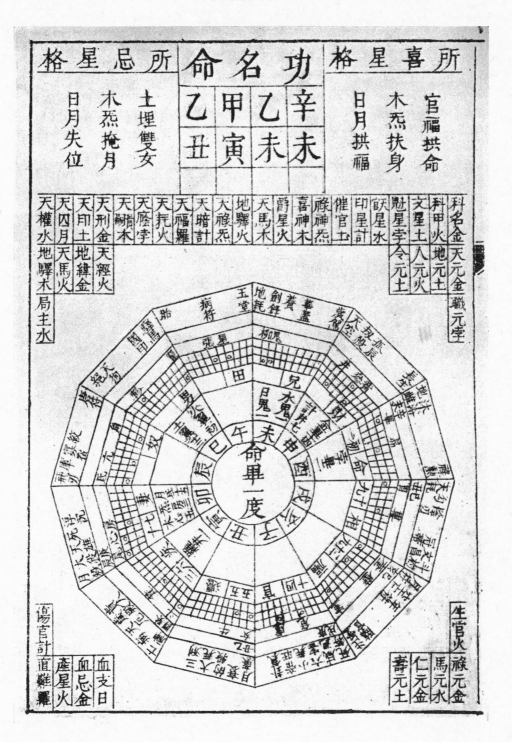

horoscope). It was said that a disciple of the Babylonian astronomer Berossus, who emigrated to Cos in about −280, was the originator of horoscopes; and Eisler[a] considers it safe to conclude that the application of celestial observations to the fates of individuals, the 'democratisation of astrology' as Pelseneer calls it,[b] was started by exiled Babylonian 'star-clerks' some time in the −2nd century.

In the preceding centuries there were certainly strong parallels between Babylonian and Chinese predictions, as was shown in a classical paper by Bezold (1).[c] He set side by side a number of statements made in cuneiform tablets, most of which came from the library of King Ashurbanipal (−7th cent., but were copies of texts from as far back as the −14th), and in the Thien Kuan Shu[1] chapter (ch. 27) of the *Shih Chi* (Historical Record) of Ssuma Chhien, written about −100, but undoubtedly containing astronomical and astrological traditions of much older date. For example:

(*a*) Cuneiform: If Mars, after it has retrograded, enters Scorpio, the King should not be negligent of his watch. On so unlucky a day, he should not venture outside his palace.

    Shih Chi:[d] If (the) fire-(planet)(Mars) forces its way into the *hsiu* Chio[2][e] then there will be fighting. If it is in the *hsiu* Fang[3][f] or the *hsiu* Hsin[4][g] this will be hateful to kings.

(*b*) Cuneiform: If Mars is in (name of constellation missing) to the left of Venus, there will be devastation in Akkad.

    Shih Chi:[h] When Ying-Huo[5] (Mars) follows Thai-Pai[6] (Venus), the army will be alarmed and despondent. When Mars separates altogether from Venus, the army will retreat.

(*c*) Cuneiform: If Mars stands in the house of the Moon (and there is an eclipse), the King will die, and his country will become small.

    Shih Chi:[i] If the Moon is eclipsed near Ta-Chio[7][j] this will bring hateful consequences to the Dispenser of Destinies (the Ruler).

(*d*) Cuneiform: If the Northern Fish (Mercury) comes near the Great Dog (Venus), the King will be mighty and his enemies will be overwhelmed.

    Shih Chi:[k] When Mercury appears in company with Venus to the east, and when they are both red and shoot forth rays, then foreign kingdoms will be vanquished and the soldiers of China will be victorious.[1]

---

[a] (1), p. 164.   [b] (1), p. 36.

[c] Bezold was an outstanding authority on Babylonian astronomy and astrology; cf. Bezold (2); Bezold & Boll (1); Boll & Bezold (1); Bezold, Kopff & Boll; etc.

[d] Ch. 27, p. 6*b*, tr. Chavannes (1), vol. 3, p. 346.   [e] α, ζ Virginis; Schlegel (5), p. 87.

[f] π and other stars in Scorpio; Schlegel (5), p. 113.

[g] Antares and σ Scorpionis; Schlegel (5), p. 138.

[h] Ch. 27, p. 20*a*, tr. Chavannes (1), vol. 3, p. 366.

[i] Ch. 27, p. 30*b*, tr. Chavannes (1), vol. 3, p. 388.

[j] Schlegel (5), p. 98. Ta-Chio is a single star (Arcturus), a paranatellon of the *hsiu* Khang.[8] *Paranatellontes asteres*, or 'corresponding stars', as the Greeks called them, are extra-zodiacal stars or constellations which rise, culminate and set at the same time as the zodiacal constellations (in Greece), or which culminate at the same time as a given *hsiu* (in China).

[k] Ch. 27, p. 27*b*, tr. Chavannes (1), vol. 3, p. 381.   [1] Eng. auct.

[1] 天官書   [2] 角   [3] 房   [4] 心   [5] 熒惑   [6] 太白

[7] 大角   [8] 亢

It would be tedious to give many examples of this kind, but I may add one or two from another source, the *Ku Wei Shu*.[a]

The Thien Chieh[1] (Heavenly Street)[b] lies between the *hsiu* Mao[2][c] and the *hsiu* Pi.[3][d] The sun, the moon, and the five planets go in and out (by this street of heaven). If Ying-Huo[4] (the planet Mars) stays in this street, and does not go through it, then the whole world will be in danger (of disorder).

The Chüan Shih[5] (Hanging Tongue)[e] governs rumours. If Ying-Huo[4] (Mars) stands near by it, there will be rebellions among the people, the prince will be injured by rumours, and robbers will arise.

These quotations simply illustrate the fact that the interests of the Chinese astronomers of the Chou period (and indeed of the early Han) were very similar to those of their still earlier Babylonian colleagues (Edkins, 3).[f] Prediction was based on: (*a*) the moon, its altitude, its conjunctions with planets, and with fixed stars and constellations, e.g. Gemini, Spica, Scorpio, etc.; (*b*) the sun, its zodiacal house or its *hsiu*, and its colour; and (*c*) the planets, especially their times of rising and setting, and their conjunctions, such as those of Saturn with Mars, Jupiter with Venus, and Mercury with Venus; also their positions with regard to the fixed stars and constellations. But many constellations known to the Babylonians were not recognised by the Chinese as such, and conversely there were many groups of fixed stars accepted by the Chinese which were not differentiated either by the Babylonians or the Greeks.[g] Bezold's contention therefore was that the system of prognostication, rather than specific astronomical knowledge (since the naming of stars and the drawing up of lists was going on in China independently), had passed from Mesopotamia to China during some period about the middle of the −1st millennium, or a little later. This seems quite a plausible view.[h]

---

[a] This is a late collection of apocryphal Han books containing prognostications, of which more will be said below, pp. 380, 382, 391. Here I quote ch. 7, p. 6*b* (tr. auct.). Many parallels in *Chin Shu*, ch. 11.

[b] Schlegel (5), p. 302; identical with the *hsiu* Pi,[6] α Andromedae and γ Pegasi.

[c] Schlegel (5), p. 351; the Pleiades.   [d] Schlegel (5), p. 365; the Hyades.

[e] Schlegel (5), p. 363; six stars in Perseus.

[f] Certain ancient fragments, unknown to Bezold and Edkins, call for new investigation, e.g. the *Sung Ssu-Hsing Tzu-Wei Shu*[7] by the early −5th-century astrologer Shih Tzu-Wei[8] (*YHSF*, ch. 77, p. 12*a*). Also the *Wu Tshan Tsa Pien Hsing Shu*,[9] perhaps of the −3rd century (*YHSF*, ch. 76, p. 57*a*).

[g] Details will be given in the Astronomical Section (20*f*).

[h] Some have been tempted to think that such influences may have been reinforced about Ssuma Chhien's own time (−2nd century) by theories coming from, or through, the Indian culture area. The bibliography of the *Chhien Han Shu* (ch. 30, p. 42*b*) contains the titles of no less than six books of astrology afterwards lost, all beginning with the words *Hai Chung* (lit. within the seas). One, for instance, was entitled *Hai Chung Hsing Chan Yen*[10] (Verifications of Hai-Chung Astrology); another *Hai Chung Wu Hsing Shun Ni*[11] (The Hai-Chung System of Planetary Progressions and Retrogradations), A third dealt with comets and rainbows, while others connected celestial happenings with specific

[1] 天街   [2] 昴   [3] 畢   [4] 熒惑   [5] 卷舌   [6] 壁
[7] 宋司星子韋書   [8] 史子韋   [9] 五殘雜變星書   [10] 海中星占驗
[11] 海中五星順逆

But now when did the application of star-lore to individual human fortune-telling come about in China? It seems to us today incredible that 'astral influences' could have been taken so seriously by so many millions of people throughout the generations both in China and the West, but Eisler[a] has clearly elucidated the plausibility of the idea. At first it was believed that stars were 'born' anew each time that they rose, and Heracleitus said that there was a new sun every day.[b] Meteors were thought of as souls descending to enter their appointed bodies—there are at least four examples of this conception in the *Chu Shu Chi Nien* (Annals of the Bamboo Books—perhaps −4th century). Solar and lunar influences on the seasons, so striking for primitive agricultural populations, were obvious,[c] and there was the familiar phenomenon of menstruation, which seemed to prove a direct effect. And it is here that we may see much significance in that strange belief to which we referred in the first volume of this work (Sect. 7a), namely, the conviction that marine invertebrate animals (such as molluscs or sea-urchins) grew fat and thin in response to the phases of the moon. The chapter of the *Lü Shih Chhun Chhiu* (Master Lü's Spring and Autumn Annals, c. −240), which contains the *locus classicus* on this subject, is largely concerned with all kinds of believed actions at a distance.[d] The effects of the moon on animals and plants, real or illusory, will be carefully examined in Section 39; here they are only mentioned to help us to understand the outlook of those who were prepared to extend to individuals the effects and influences which for a thousand years previously had been of admitted importance for affairs of State.

In the absence of adequate studies it is hard to say how far later Chinese astrology made use of methods paralleling those used in the West. There was the system of dividing the ninth (non-moving) sphere into eight sectors (later increased to twelve), and noting which stars and constellations were in which division at the time of birth or other event inquired about.[e] This was expounded by Manilius and Firmicus Maternus, and seems to have been evolved in the +2nd century. Then there were

States and their ministers. It is argued that if there were 'within-the-seas' books, there must also have been 'overseas' (*hai wai*) books and systems, and that these might have been Indian. But *hai chung* might also have referred to the magic islands of the Eastern Sea, such as Phêng-Lai (cf. p. 240), adopted by certain astrologers as the origin of their school. Again, as we shall see in Sect. 20f, the +2nd-century astronomer Chang Hêng spoke of a large number of stars which were taken account of by the *hai jen*, sailors, or sea-coast people.

[a] (1), pp. 41, 66, 140, 161, etc.   [b] Freeman (1), p. 112.

[c] Eisler (1) has brought out the full force of this in a passage which I cannot forbear from quoting: 'If people believe, and indeed know, their calendar (the change of their climatic seasons) to be determined by the position of the sun relative to certain stars, just appearing or disappearing before sunrise or after sunset; and if they know that their solar year is roughly divided into twelve months by the phases of the moon taking place in the neighbourhood of certain groups of fixed stars—it is natural that they should be driven to the conclusion that the periodic changes of weather (heat, cold, rain and storms) and the sprouting, fruiting and withering of all vegetation, are regulated by the apparent serpentine movement of the sun and moon past the milestones of their celestial journey, i.e. the various constellations appearing and disappearing in their wake or heralding their advance' (p. 154). The classical examples are the heralding of the annual flood of the Nile by the heliacal rising of Sirius, the sprouting of cereals in Mesopotamia by that of Spica, and the Italian grape-harvest by Vindemiatrix.

[d] Ch. 45 (vol. 1, p. 88).

[e] Eisler (1), p. 37. The animal zodiac signs were supposed to have stood in these places at the creation (theory of the *thema mundi*).

prognostications based on the decan-stars,[a] that is to say, those paranatellons the heliacal risings and settings of which can be used to determine the exact hour if the date is known, or the exact date if the time is known.[b] These were studied by the Egyptians as early as −2000. The Greeks called them *leitourgoi* ('stars on duty') or *theoi boulaioi* ('advisory gods'), and considered that every ten days one was sent as a messenger from those above to those below, and vice versa (i.e. setting and rising). Fanciful potencies were attributed to each of the decan-stars, and conclusions drawn respecting those who were born at the time of their rising. Thirdly, there was astrology based on the zodiacal constellations themselves, each of which, probably shortly after Aristotle's time, was associated with one of the four elements and with specific regions of the earth's surface. Fourthly, there was astrology based on the motions of the planets. Observations were made of their position relative to the zodiacal constellations, their declinations north or south of the ecliptic ('exaltations'), their conjunctions with each other ('sympathies and antipathies'), and the apparent loops and retrogradations of their orbits. From the pioneer studies which have been made on Chinese astrology (e.g. Chatley, 3, 5, 6), it seems that most of these methods were developed and used in China.[c]

In Wang Chhung's time, however, the application of astrology to individuals (horary, judicial, or genethliacal astrology) was only just beginning, and as we shall later see, it is interesting that this was almost the only one of the pseudo-sciences which he did not strongly attack. The first book is subsequent to his time, namely, the *Yü Chao Shen Ying Chen Ching*[1] (or *Yü Chao Ting Chen Ching*;[2] True Manual of Determinations by the Jade Shining Ones) attributed to Kuo Pho of the late +3rd century.[d] It is significant also that the first astrological expert whose biography is given in the relevant section of the *Thu Shu Chi Chhêng* encyclopaedia is Wei Ning[3] (fl. +550 to +589) of the Northern Chhi dynasty. By the time of the Thang so great an elaboration had taken place that a voluminous encyclopaedia could be produced, *Hsing Tsung*[4] (The Company of the Stars) by Chang Kuo;[5] it is dated +732. Another book of his, the *Hsing Ming Su Yuan*[6] (Astrology traced back to its Origins) is still extant. The great Yehlü Chhu-Tshai of the Liao dynasty (see Sects. 6*i*, 20*g*, 27*i*) also wrote on astrology,[e] and important works on it were still being produced at the end of

---

[a] The ten-day stars of modern nautical almanacs.

[b] Eisler (1), p. 99. This was the origin of the term 'horoscope' for such a star was a *horoskopos*, 'hour-pointer' or 'hour-observer'.

[c] In later periods there was of course contact with Iranian (cf. Ishida), Indian (cf. Geden; v. Negelein) and Muslim (cf. Nallino) astrology. Sogdian planetary astrology became particularly popular.

[d] It is more than probable, however, that the greater part of this production dates from the Sung, nearly a thousand years later, and may be from the hand of Chang Yung,[7] a writer of whom otherwise nothing is known.

[e] Like one of his clansmen, Yehlü Shun,[8] who may be identical with a man who was emperor of the Liao for one year (+1122), and who wrote *Hsing Ming Tsung Kua*[9] (General Descriptions of Stars and their Portents).

[1] 玉照神應眞經　　　[2] 玉照定眞經　　　[3] 魏寧　　　[4] 星宗　　　[5] 張果
[6] 星命溯源　　　　　[7] 張顒　　　　　　[8] 耶律純　　　　[9] 星命總括

the Ming.[a] All those mentioned here, and others also, are contained more or less *in extenso* in the *Thu Shu Chi Chhêng*.[b] Apart from this, there is the imperial astrological compendium, the *Chhin-Ting Hsieh Chi Pien Fang Shu*,[1] published thirteen years later (+1739).

(4) CHRONOMANCY; LUCKY AND UNLUCKY DAYS

Closely related to astrology was another system of beliefs, not peculiar to the Chinese, but cultivated by them, namely, the choosing of lucky and unlucky days (*hsüan tsê*[2]). Eisler (1) offers evidence that this goes back to Babylonia and Egypt; Herodotus, for instance, says that the Egyptians knew the gods who controlled each day, and what fate belonged to those born thereon.[c] Hence the expression *dies Aegyptiaci* in the Roman calendar. The idea is also in Hesiod. It seems without doubt to have been based originally on the phases of the moon, as is indicated by the fact that occidental books concerning it are called Selenodromia or Lunaria. As we shall see, this was one of the superstitions combated by Wang Chhung. But the largest works concerning it seem to be as late as the +17th century.[d] Until very recently, calendars produced in country towns always marked lucky and unlucky days, and not many years ago the Academia Sinica itself began to publish rural calendars in order to attack the super-stition and to impart elementary astronomical information.

(5) PROGNOSTICATION BY THE DENARY AND DUODENARY CYCLICAL CHARACTERS

A far more elaborate system of prognostication developed from the calendrical system, involving the use of the twelve horary characters (Branches, *chih*[3]) and the ten 'celestial Stems' (*kan*[4]). It was known simply as '(fate) calculation', *thui ming*.[5][e] It is the only Chinese pseudo-science known to me for which we have adequate modern investigation—I refer to the satisfying paper of Chao Wei-Pang (1). There is no doubt that at the time of Wang Chhung the system was in its infancy, though many of his attacks seem to be directed against ideas similar to it, as we shall see.

Elsewhere something is said about the origin of the twelve Branches and the ten Stems.[f] It was their combination into a recurring sexagenary cycle which gave rise to the Chinese calendrical system as we know it, and according to the general opinion (cf. Ku Yen-Wu's[6] *Jih Chih Lu*[7])[g] this did not take place before the time of Wang

---

[a] For its place in modern Chinese life see Doré (1), pt. 1, vol. 3, pp. 277 ff. On Japanese astrology there is a curious book by Severini (1).

[b] *TSCC, I shu tien*, chs. 565–592. [c] II, 82.

[d] *TSCC, I shu tien*, chs. 681–4, 687–701. Trigault (Gallagher tr. p. 548) has a graphic account of the difficulty which the early friends of the Jesuits had in breaking away from this deeply-rooted superstition.

[e] Alternatively *lu ming*,[8] because the point of greatest interest was how high the inquirer would be likely to go in the official bureaucracy.

[f] Sects. 5a, 20h. [g] Ch. 20, para. 2 (vol. 2, p. 29).

¹ 欽定協紀辨方書　　² 選擇　　³ 支　　⁴ 干　　⁵ 推命
⁶ 顧炎武　　　　　　⁷ 日知錄　　　　　　⁸ 祿命

Mang (+13), so far as years were concerned, though as applied to days it goes back to the time of the Shang oracle-bones. Thus the fate-calculators used the stem-branch combination of the day, month and year of birth as the basis of their conclusions.[a] Obviously this had distinct, though indirect, astrological connotations. Then came the identification of all the stem-branch combinations with one or other of the five elements;[b] this first occurs in the books of Kuan Lo[1] of the Three Kingdoms period (+3rd century), so that it probably grew up in the generations of the Later Han just succeeding Wang Chhung's time.

Kuan Lo is considered to have been the first of this school. 'My fate is with *yin*[2]',[c] he is supposed to have said, 'I was born at night during an eclipse of the moon. Heaven has fixed numbers, which can be known, though the common people do not know them.'[d] Another book[e] contains what is also thought to have been a saying of his: 'By the contained note (*na yin*[3]) one may judge one's fate.' This 'contained note' means simply the element which is associated with the particular stem-branch combination, and musical phraseology is used, since the notes on the standard bamboo pitch-pipes were each associated with an element.[f]

All books whose titles begin with the words *San Ming* (three kinds of fates) belong to this class. Thao Hung-Ching (+451 to +536), whom we have met before, and often shall again as Taoist, botanist and alchemist, wrote two, the *San Ming Chhao Lüeh*[4] and the *San Ming Li-Chhêng Suan Ching*,[5] but both are lost. From fragments of Lü Tshai[g] which remain it seems clear that the system had already reached its maximum development by the time of the Thang. The most famous fate-calculator of the Thang was Li Hsü-Chung,[6] who graduated in +795 and was an imperial censor about +820. His book *Li Hsü-Chung Ming Shu*,[7] still extant, is the oldest book on this subject which we have.[h] In the following century Hsü Tzu-Phing[8] made an important commentary on the *San Ming Hsiao Hsi Fu*[9] which had been written by an unknown author calling himself Lo Lu Tzu[10].[i] Finally, in the Ming, the pseudo-science was digested into an encyclopaedic work, the *San Ming Thung Hui*,[11] by Wan Min-Ying.[12]

There is no need to enter into the late elaborations of the system, but one may note that in the Sung the hour of the event in question was added to the day, month and

[a] In the Chou period, each stem and branch was associated with a particular feudal State; the list is in *Huai Nan Tzu*, ch. 3, p. 15b.
[b] Cf. above, Table 12.
[c] One of the duodenary branches.   [d] *San Kuo Chih*[13] (*Wei Shu*), ch. 29, p. 26b.
[e] The *Wu Hsing Ta I*[14] by Hsiao Chi[15] of the Sui (c. +600), ch. 3 (sect. 4), p. 18a.
[f] One can sense here a Chinese equivalent of the 'music of the spheres'.
[g] See below, p. 387.
[h] Though it is incomplete and has later interpolations.
[i] 'The Beadstring Master', perhaps because the days go round like a string of beads.

[1] 管輅   [2] 寅   [3] 納音   [4] 三命抄略   [5] 三命立成算經
[6] 李盧中   [7] 李盧中命書   [8] 徐子平   [9] 三命消息賦
[10] 珞琭子   [11] 三命通會   [12] 萬民英   [13] 三國志   [14] 五行大義
[15] 蕭吉

year, thus forming the 'four pillars' (*ssu chu*[1]). The whole development may be regarded as an offshoot of judicial astrology, possible only among a people who had a complex cyclical calendar returning to its starting-point at rather long intervals.[a]

(6) GEOMANCY (*FÊNG-SHUI*)

From divination depending on the heavens we now pass to divination depending on the earth. It was quite natural in the Chinese cosmological system that the latter should be considered as important as the former. The far-reaching pseudo-science of geomancy (*fêng-shui*,[2] lit. winds and waters) has received somewhat more attention from modern scholars than astrology,[b] but still nothing like as much as it deserves; later[c] we shall appreciate its great importance with relation to the discovery of the magnetic compass. It has been well defined by Chatley (7) as 'the art of adapting the residences of the living and the dead so as to cooperate and harmonise with the local currents of the cosmic breath'. If houses of the living and tombs of the dead were not properly adjusted, evil effects of most serious character would injure the inhabitants of the houses and the descendants of those whose bodies lay in the tombs, while conversely good siting would favour their wealth, health and happiness. Every place had its special topographical features which modified the local influence (*hsing shih*[3]) of the various *chhi* of Nature. The forms of hills and the directions of watercourses, being the outcome of the moulding influences of winds and waters, were the most important, but, in addition, the heights and forms of buildings, and the directions of roads and bridges, were potent factors. The force and nature of the invisible currents would be from hour to hour modified by the positions of the heavenly bodies, so that their aspects as seen from the locality in question had to be considered. While the choosing of sites was of prime importance, bad siting was not irremediable, as ditches and tunnels could be dug, or other measures taken, to alter the *fêng-shui* situation (Fig. 45).

This set of ideas is no doubt of high antiquity. In the chapter from the *Kuan Tzu* book quoted on p. 42, which may well contain material of the −4th century, we noted that the *chhi* of the earth flowed in vessels comparable with those in the body of man and animals. In Wang Chhung's time (*c.* +80), the system had developed sufficiently for him to argue against it,[d] as we shall see below. It is extremely probable that it was already well recognised by the beginning of the Han (*c.* −200). The *Shih Chi*[e] mentions a class of diviners called *khan yü chia*[4] (diviners by the canopy of Heaven

---

[a] Thus it would not have been possible in Europe, but in Maya civilisation it would have been not only possible but much more complicated than in China.

[b] Eitel (2); de Groot (2), vol. 3, pp. 935 ff.; Hubrig (1); and shorter accounts by Porter (2); Dukes (1). It must be added that the exposition of Eitel is often inaccurate and contains many ideas which are not acceptable today. Edkins (14) made a glossary of technical terms. Chinese geomancy was evidently entirely different from divination methods which passed under that name in the West (cf. Thorndike (1), vol. 2, pp. 110ff.) or in Arab Africa (cf. Maupoil, 1).

[c] Sect. 26*i*.   [d] Cf. Forke tr. (4), vol. 1, p. 531.

[e] Ch. 127, p. 7*b*; the comments of Chhu Shao-Sun on Ssuma Chi-Chu's biography.

[1] 四柱   [2] 風水   [3] 形勢   [4] 堪輿家

and the chariot of Earth). The bibliography of the *Chhien Han Shu* mentions two books with significant titles, the *Khan Yü Chin Kuei*[1] (Golden Box of Geomancy) and the *Kung Chai Ti Hsing*[2] (Terrestrial Conformations for Palaces and Houses)—both have long been lost. Then one of Wang Chhung's contemporaries, Wang Ching,[3] who was much occupied with astronomy and mathematics and who died in the year that the *Lun Hêng* was probably written (+83), seems clearly[a] to have studied geomancy (*khan yü*). His biography suggests that at this stage it may have had a certain connection with hydraulic engineering works and water-control.

The real consolidation of the system,[b] however, seems to have taken place in the Three Kingdoms period, when Kuan Lo (+209 to +256) probably wrote about it, although it is as yet impossible to say how much, if, indeed, any, of the *Kuan shih Ti Li Chih Mêng*[4] (Mr Kuan's Geomantic Indicator) which we still possess, is from his hand, or his time. In the +4th century Kuo Pho also wrote[c] on *fêng-shui*, but it is again very doubtful whether any of the present *Tsang Shu*[5] (Burial Book) ascribed to him, is his. In the +5th century (Liu Sung) there was Wang Wei,[6] whose *Huang Ti Chai Ching*[7] (The Yellow Emperor's House-Siting Manual) is still extant.[d] In the Thang there was the *Chhing Nang Ao Chih*[8] (Mysterious Principles of the Blue Bag, i.e. the Universe) ascribed to the famous geomancer Yang Yün-Sung;[9] and the series culminates, though it by no means ends, with the *Khan Yü Man Hsing*[10] (Agreeable Geomantic Aphorisms) by the eminent Yuan mathematician[e] Liu Chi[11] (+1311 to +1375).[f] But as to the beginning of the story, it may be significant that the biographies of *fêng-shui* experts given in the *Thu Shu Chi Chhêng* encyclopaedia include those of only three men prior to Kuo Pho.[g] The first was Chhu Li Tzu[12] (whose biography is in the *Shih Chi*)[h] of the late Warring States period (−3rd century), the second Chu Hsien-Thao[13] of Chhin, and the third a certain Chhing-Wu[14] (Blue Raven Master) placed some time in the Han, and said in some accounts to have been the author of a *Tsang Ching*.[15]

The two currents, Yang and Yin, in the earth's surface, were identified with the two symbols which apply to the eastern and western quarters of the sky, the Green Dragon (Chhing Lung[16]) of spring in the former case, the White Tiger (Pai Hu[17]) of autumn in the latter.[i] Each of these would be symbolised by configurations of the

---

[a] From his biography in ch. 106 of the *Hou Han Shu*.

[b] The only bibliographical catalogue of geomantic books seems to be that of Chhien Wên-Hsüan. Some useful pages on the history of this literature are contained in Wang Chen-To (5), pp. 110 ff.

[c] Sarton (1), vol. 1, p. 353; de Groot (2), vol. 3, p. 1001.

[d] *TT* 279.   [e] Sarton (1), vol. 3, p. 1536.

[f] These, and a number of other books, are printed more or less *in extenso* in *TSCC, I shu tien*, chs. 651–78. Kuo Pho's book is there written *Ching* not *Shu*.

[g] Apart from Kuan Lo.   [h] Ch. 71.

[i] As will later appear, there is also alchemical symbolism here. And sexual also, since the hills should be in mutual embrace (*kung pao*[18]).

[1] 堪輿金匱	[2] 宮宅地形	[3] 王景	[4] 管氏地理指蒙	[5] 葬書
[6] 王微	[7] 黃帝宅經	[8] 青囊奧旨	[9] 楊筠松	[10] 堪輿漫興
[11] 劉基	[12] 樗里子	[13] 朱仙桃	[14] 青烏	[15] 葬經
[16] 青龍	[17] 白虎	[18] 弓抱		

PLATE XVIII

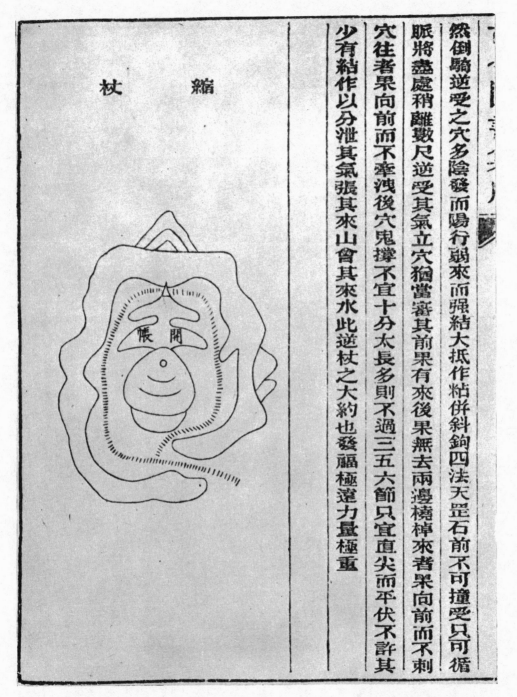

縮　杖

開帳

然倒騎逆受之穴多陰發而陽行弱來而強結大抵作粘併斜鈎四法天罡石前不可撞受只可循

賑將盡處稍離數尺逆受其氣立穴猶當審其前果有來後果無去兩邊橈棹來者果向前而不剌

穴往者果向前而不牽洩後穴鬼撐不宜十分太長多則不過三五六節只宜直尖而平伏不許其

少有結作以分泄其氣張其來山曾其來水此逆杖之大約也發福極遠力量極重

Fig. 45. Illustration from a work on geomancy (*fêng shui*), the *Shih-erh Chang Fa* (Method of the Twelve Chang), attributed to Yang Yün-Sung of the Thang (*c.* +880). From *TSCC, I shu tien*, ch. 666, *hui khao* 16, p. 2*b*. The chart shows a particular site for a tomb, towards the tip of a range of small hills separating two valleys with streams, the whole being enclosed by two further ranges of foothills. It is said that the higher these latter ranges are, the better, and that there should not be a 'tongue' or high ridge connecting the inner hills with the main massif (shown at the top as if in elevation). This kind of site is called '*so chang*' because the *chhi* of the mountain is 'condensed' around the tomb site. The relation of this kind of drawing to physiographic map-making (see Section 22*d*) is evident.

ground. The former ought always to be to the left, and the latter to the right, of any tomb or habitation, which should preferably be protected by them, as if in the crook of an elbow. But this was only the beginning of the complexity, since high and abrupt escarpments were considered Yang, and rounded elevations Yin. Such influences (*shan ling*[1]) had to be balanced, if possible, in the selection of the site, so as to obtain three-fifths Yang and two-fifths Yin. Needless to say, the trigrams and hexagrams, the sexagenary cycle of stems and branches, and the five elements, were woven into the reckoning. There was in general a strong preference for tortuous and winding roads, walls and structures, which seemed to fit into the landscape, rather than to dominate it; and a strong objection to straight lines and geometrical layouts.[a] Isolated boulders were also considered unlucky. In many ways *fêng-shui* was an advantage to the Chinese people, as when, for example, it advised planting trees and bamboos as windbreaks, and emphasised the value of flowing water adjacent to a house site. In other ways it developed into a grossly superstitious system. But all through, it embodied, I believe, a marked aesthetic component, which accounts for the great beauty of the siting of so many farms, houses and villages throughout China.[b]

There is now no doubt that the magnetic compass was first developed for *fêng-shui* purposes. Description of the compass as used by the geomancers must be deferred until Section 26 on physics; known as the 'dial-plate' (*lo-phan*[2]), it is marked not only with the compass points but also with *kua* (of the *I Ching*), stem-branch combinations and many other symbols.[c] Any anticipatory remarks in this place as to the period at which it was probably developed would spoil a truly remarkable story. We may only mention that the ancestor of the *lo-phan* or geomancer's compass was the diviner's board, the *shih*.[3] This consisted of two boards, the upper discoidal, corresponding to Heaven, and the lower square, corresponding to Earth. The Northern Dipper (Great Bear) was marked on the upper plate and both carried signs for the points of the compass. This diviner's board undoubtedly goes back to the −2nd, and probably to the −3rd century at least, and was at any rate coeval with the

[a] I have myself vividly experienced the effects of becoming accustomed to the Chinese point of view in these matters. In my youth I greatly admired the gardens and park of Versailles, but when many years later I visited it again after having become acquainted in the interval with the Summer Palace (I Ho Yuan) at Peking, it was with a feeling of desolation that one surveyed its geometrical arrangement, imprisoning and constraining Nature rather than flowing along with it. In this connection it may be mentioned that, as Lovejoy (3) has shown, the movement which in Europe sent the geometrical garden out of fashion in the 17th and 18th centuries, drew its inspiration, like other aspects of romanticism, from demonstrable Chinese sources. Sir William Temple, in his *Essay upon the Gardens of Epicurus* of 1685, introduced the idea of the 'picturesque' with a reference to a Chinese canon of taste—*sharawadgi*. This term long puzzled lexicographers until Y. Z. Chang (1) proposed that it is nothing but a corruption of the phrase *sa lo kuei chhi*[4]—'impressive and surprising because of its careless gracefulness'. Cf. Bald (1); Chhen Shou-Yi (2).

[b] In modern times the 'gentlemen of Ganchow' (*Ganchow hsien-sêng*), in Chiangsi, were particularly noted as expert in the art. Cf. the Toledan letters of Europe (Sarton (1), vol. 3, pp. 1110, 1113). The founder of their school was Yang Yün-Sung (Thang). The other chief school was that of Fukien, which made relatively more use of the compass.

[c] Cf. Eitel (2), de Groot (2). See Fig. 46.

[1] 山靈   [2] 羅盤   [3] 式   [4] 灑落瑰奇

Fig. 46. A late Chhing representation of the selection of a city site; the geomancer is consulting his magnetic compass. From *SCTS* (ch. 32, Shao Kao). The depiction of the use of the magnetic compass in an illustration of a Chou period text is of course an anachronism.

beginnings of *fêng-shui*, for which art it was obviously convenient, if not essential, to have an accurate indication of the points of the compass during the day in any weather. Evidence will later be presented that this diviner's board has a connection with the game of chess as well as with the origin of the magnetic compass, and that perhaps its earliest use was a form of divination carried out by casting pieces ('men'), like dice, on to it. We shall also find that one of the most crucial passages in all Chinese literature about the magnetic compass is to be found in Wang Chhung's own writings. During later centuries expertise with the *shih* declined (presumably as some form of the magnetic compass became known), but it was still discussed at length by the Buddhist monk I-Hsing[1] during the Thang, whose material forms the core of the Chhing collection *Liu Jen Lei Chi*[2] (Compendium of (Divination by) the Six Cardinal Points— i.e. N., S., E., W., above and below)[a] and its continuation, the *Liu Jen Li-Chhêng Ta Chhüan Chhien*.[3] Moreover, the Sung bibliography in the *Thung Chih Lüeh* (c. +1150) lists no less than twenty-two books on the use of the *shih*, but apparently all of them have perished.[b]

Lastly, we pass to the methods of divination which concerned, not heaven or earth, but specifically human things—physiognomy, oneiromancy and glyphomancy.

(7) PHYSIOGNOMY AND CHEIROMANCY

Physiognomy (*hsiang shu*[4]) was the belief that the fortune of the individual could be foretold by examination of his physical characteristics, his facial appearance, bodily form, and so on.[c] Sarton has shown[d] how large a part the belief in physiognomy played in the medieval occident, and it was also very prominent in Islamic culture (Mourad). Though there may have been some Indian influence,[e] there is no doubt of its antiquity in China, for Hsün Chhing (−3rd century) devotes a special chapter to combating it.[f] An affair of State treason in which it was involved occurred in +67 in the Later Han.[g] But the principal works on it were not written until comparatively late; for example, the *Thai-Chhing Shen Chien*[5] (probably of the Sung), and the +14th-century *Shen Hsiang Chhüan Pien*[6] of Yuan Chung-Chhê,[7] of which a short

---

[a] Contained in *TSCC, I shu tien*, chs. 717–44.

[b] It is interesting that one of them is attributed to Wu Tzu-Hsü,[8] a statesman of the Chou, who figures so prominently in one of Wang Chhung's greatest chapters. Having been unjustly done to death, he was supposed to have been thrown into the Chhien-thang river near Hangchow, and the anger of his spirit was supposed to cause the periodical tidal bore which occurs there; see Sect. 21*i*, Meteorology. There is much concerning Wu Tzu-Hsü and *fêng-shui* in the *Wu Yüeh Chhun Chhiu*[9] (Spring and Autumn Annals of the States of Wu and Yüeh), but this book was written in the Han, which only suggests again that it was then that the system grew up.

[c] A kind of 'phrenology' was also included.

[d] (1), vol. 3, pp. 270, 1232.   [e] See Chi Hsien-Lin (1).

[f] Ch. 5, tr. Dubs (8), pp. 67 ff.

[g] *TH*, p. 690.

[1] 一行   [2] 六壬類集   [3] 六壬立成大全鈐   [4] 相術
[5] 太清神鑑   [6] 神相全編   [7] 袁忠徹   [8] 伍子胥   [9] 吳越春秋

mention has been made by H. A. Giles.[a] For information on physiognomy in the recent Chinese past, Doré may be consulted.[b] One very interesting outcome of physiognomy and its offshoot, cheiromancy,[c] was the early discovery by the Chinese of the practicability of identification by finger-printing.[d]

(8) ONEIROMANCY

Oneiromancy (*chan mêng*[1]), or prognostication by dreams, was also practised in China, as in most ancient civilisations, though it can hardly be said to have taken a very important place there.[e] The *Chou Li* says that the interpretation of dreams was in the department of the Grand Augur (Ta Pu[2]), and mentions a special expert of lower grade (Chan Mêng[1]) who specialised in it.[f] Here again the chief book was late, the *Mêng Chan I Chih*[3] of Chhen Shih-Yuan,[4] published in +1562 (Ming). How far certain aspects of Chinese dream-interpretation might be considered, as Chinese themselves are sometimes inclined to think, anticipations of Freudian psychology, would be a subject worth investigation.[g]

(9) GLYPHOMANCY

Glyphomancy (*chhê tzu*[5]) is a very curious game, which could only have arisen in a culture with an ideographic language. It consisted in dissecting the written characters of personal and other names, with a view to making prognostications from them. Two chapters are devoted to it in the *Thu Shu Chi Chhêng* encyclopaedia,[h] and de Groot (3) has briefly explained the methods used. He doubted if it was older than the Thang. An allied superstition was 'automatic' or 'planchette' writing (*fu chi*[6, 7]); this was known and used in the late Sung, how much earlier we do not know.[i]

---

[a] (5), p. 178. This book, with other material, is to be found in *TSCC, I shu tien*, chs. 631–50. It was largely the work of his father Yuan Kung,[8] but much of the material is old and Chhen Thuan's name is also attached to it.

[b] Pt. 1, vol. 3, pp. 223 ff.

[c] On this, see Arlington (1). We have not, however, been able to ascertain how far developed cheiromancy, or 'palmistry', was an indigenous growth.

[d] See on, in Section 43 on anatomy.

[e] Cf. Wieger (2), pp. 73, 93, 117, etc. Cf. *Shu Ching*, ch. 17 (Yüeh Ming).

[f] Ch. 6, pp. 23*a*, 28*a* (ch. 24); tr. Biot (1), vol. 2, pp. 71, 82.

[g] Four chapters on dreams in the *Thai-Phing Yü Lan* encyclopaedia have been translated by Pfizmaier (84). Cf. *ECCS, Honan Chhêng shih I Shu*, ch. 2B, p. 4*a*, ch. 18, pp. 16*a* ff., 34*a*, ch. 23, p. 2*a*.

[h] *I shu tien*, chs. 747–8.

[i] See Chao Wei-Pang (2); Howell (1). First-hand description by Eitel (5).

[1] 占夢   [2] 大卜   [3] 夢占逸旨   [4] 陳士元   [5] 拆字
[6] 扶箕   [7] 扶乩   [8] 袁珙

(b) SCEPTICAL TRENDS IN CHOU AND EARLY HAN TIMES

The preceding pages will have sufficed to give a rough idea of the background of superstition against which the Chinese sceptics have to be set. But early though it arose in Chinese history, the beginning of the sceptical and rationalising tradition runs it very close. To illustrate this I quote from the *Tso Chuan*[a] a passage referring to events in −679:

Prince Li, having heard the story about the apparition of the two serpents, asked Shen Hsü[1] about them, saying, 'Do people still see apparitions of evil augury?' Shen Hsü replied, 'When a man fears something, his breath (*chhi*[2]), escaping, attracts an apparition relating to that which he fears. These apparitions have their principle in men. When men are without fault, no ominous apparitions appear. But when men throw away the rules of constant behaviour, they appear. Such is the way in which they are caused.'[b]

Although with an undertone of reference to the 'phenomenalism' to be discussed below (p. 378), i.e. the doctrine that moral faults give rise to natural calamities; the passage does also express the idea that ghosts and apparitions are of a subjective nature, and the projections of men's minds. Another passage, under date −540, attracts one's attention, in which a minister, Kungsun Chhiao,[c] argues that the health of a prince depends on his work, his journeys, food, joys and sorrows, and not on the spirits of the rivers and mountains or the stars.[d] Many of the −6th and −5th century statesmen appear to have taken strongly rationalist positions, especially regarding magic, sacrifices, prayers, etc., e.g. Yen Ying.[3]

When we come to the period of the philosophers we find many among them who expressed sceptical and rationalist views. Thus Hsün Chhing (−3rd century) says:

If (officials) pray for rain and get rain, why is that? I answer, there is no reason at all. If they do not pray for rain, they will nevertheless get it. When (officials) 'save the sun and moon from being eaten',[e] or when they pray for rain in a drought, or when they decide an important affair only after divination—this is not because they think that they will in this way get what they want, but only because it is the conventional thing to do. The prince

[a] Duke Chuang, 14th year.
[b] Tr. Couvreur (1), vol. 1, p. 160, eng. auct. The remarkably advanced tone of this passage may be appreciated by a glance at the most recent book on the psychology of hallucinations (Tyrrell).
[c] Cf. pp. 206, 522, and Forke (13), pp. 92, 96.
[d] Duke Chao, 1st year (Couvreur (1), vol. 3, p. 33). Cf. another passage under −643, Duke Hsi, 16th year (Couvreur (1), vol. 1, p. 311). Of course *Tso Chuan* passages can never be firmly attributed to the date which they purport to bear, on account of the drastic rewriting and interpolation which went on in the Han. But the rationalist tradition is undoubtedly very old. Yen Ying's speech on the comet which was alarming the Prince of Chhi (*Tso Chuan*, Duke Chao, 26th year; Couvreur (1), vol. 3, p. 416) is worth reading. Cf. Forke (13), pp. 82, 89.
[e] By beating gongs, etc. during an eclipse.

[1] 申繻   [2] 氣   [3] 晏嬰

thinks it is the conventional thing to do (*i wei wên* [1]), but the people think it supernatural (*i wei shen* [2]). He who thinks it is a matter of convention will be fortunate; he who thinks it is supernatural will be unfortunate.[a]

And a little earlier in the same chapter he has a splendid passage in the best Confucian vein. It begins with a statement of ataraxy like that of the Taoists, and continues by affirming that portents and presages mean very little compared to good or bad government.

> When stars fall or the sacred tree groans,[b] the people of the whole State are afraid. We ask, 'Why is it?' I answer: there is no (special) reason. It is due to an aberration of heaven and earth, to a mutation of the Yin and Yang. These are rare events. We may marvel at them but we should not fear them. For there is no age which has not experienced eclipses of the sun and moon, unseasonable rain or wind, or strange stars seen in groups. If the prince is illustrious and the government tranquil, although these events should all come together in one age, it would do no harm. If the prince is unenlightened and his government bent on evil, although not one of these strange events should occur, that would do him no good....
>
> But when human ominous signs come, then we should really be afraid. Using poor ploughs and thereby injuring the sowing, spoiling a crop by inadequate hoeing and weeding, losing the allegiance of the people by government bent on evil—when the fields are uncultivated and the harvest is bad, when the price of grain is high and the people are starving, when there are dead bodies on the roads—these are what I mean by human ominous signs....[c]

And in a previous connection (p. 27) we have seen exemplified in another striking passage the agnostic rationalism of Hsün Tzu, with respect to spirits and their believed responsibility in the causation of disease. We also just noted that one of his chapters (ch. 5) is entirely devoted to an attack on physiognomical fortune-telling.

Han Fei Tzu partakes of the same tradition. 'If the ruler believes in date-selecting, worships gods and demons, puts faith in divination, and likes luxurious feasts, then ruin is probable.'[d] And elsewhere he cites battles between States ending in the ruin of one, when both had been encouraged by the tortoise-shell and the milfoil.[e]

During the Han, Confucianism separated into two rather sharply contrasting currents. 'When Confucianism was established as a "State religion" in the −2nd century, it was not agnostic Confucianism, but theistic Mohism in a Confucian disguise. And when Taoism as a religion arose in the +2nd century, it was no longer the naturalism and atheism of Lao Tzu and Chuang Tzu, but theistic Mohism together with a thousand superstitious features from the religion of the common people' (Hu Shih, 4). Han Confucianism, moreover, adopted most of the proto-scientific and

---

[a] *Hsün Tzu*, ch. 17, p. 22*b*, tr. Dubs (8), p. 181, mod.

[b] For Western parallels to this particular presage, cf. Bouché-Leclercq (2), vol. 1, p. 177.

[c] *Hsün Tzu*, ch. 17, p. 21*a*, tr. Dubs (8), p. 179, mod. It is interesting that this passage was still being quoted fifteen centuries later, in the sceptical book *Pien Huo Pien* of +1348 (ch. 1, p. 15*b*); see below, p. 389.

[d] *Han Fei Tzu*, ch. 15 (Liao (1), vol. 1, p. 134).

[e] Ch. 19 (Liao (1), vol. 1, p. 156).

[1] 以爲文   [2] 以爲神

semi-magical theories of the school of Tsou Yen, the Yin-Yang dualism and the Five Elements, together with all kinds of divination practices and mantic portent-lore. Owing to the essentially moralistic character of Confucian thought, the older Taoist recognition of the ethical neutrality of science was abandoned, and the ideas of the Naturalists were preferred, confirming and making explicit a suspicion which the Chinese had long entertained, namely, that ethical regularity and cosmic regularity were one.[a] Ethical or ritual irregularities were believed to be directly connected with cosmic irregularities. This is the essence of what we may call 'phenomenalism'. Consideration of it may be deferred for a moment so that we can approach it through Wang Chhung's attack upon it.

The other current was, however, the continuation of the agnostic and sceptical tradition,[b] to which Confucius himself had given indirect sanction. The *Hou Han Shu* gives an interesting story, referring to +46, which is well worth quoting:

When Liu Khun[1] was prefect of Chiang-ling, his city was devastated by fire. But he prostrated himself before it, and it immediately went out. Later, when he became prefect of Hung-nung, the tigers (which had previously infested the place) swam across the Yellow River with their cubs on their backs and migrated elsewhere. The emperor heard about these things and wondered at them, and promoted Liu Khun Chief of the Personnel Department. The emperor said to him, 'Formerly at Chiang-ling you turned back the wind and extinguished the conflagration, and then at Hung-nung you sent the tigers north of the River; by what virtue did you thus manage affairs?' Liu Khun replied, 'It was all pure chance (*ou jan*[2]).' The courtiers to left and right could not restrain their smiles (to see a man losing such a fine opportunity of getting on in the world). But the emperor said, 'This reply is worthy of a really superior man! Let the annalists record it.'[c]

Of an elder generation was Huan Than[3] (−43 to +28),[d] a scholar who held that 'life is like the flame of a lamp, going out when the fuel is exhausted', and who rebuked the emperor Kuang Wu for his belief in various forms of divination. His *Hsin Lun*[4] (New Discussions), which shows a very sceptical attitude, now exists only in a form reconstituted from fragments, in Yen Kho-Chün's collection.[e] We shall meet with Huan Than again in the Astronomical Section (20), since he tried to measure temperature and humidity. We know that he was deeply interested in scientific questions, as it is recorded that he often discussed them with his friend Yang Hsiung (cf. Sect. 20*g*). He did not believe in any form of prognostication, but nevertheless accepted the phenomenalism of his time (see on, p. 380).

[a] This formulation is due to Creel (3). Cf. p. 247 above and pp. 378 ff. below.
[b] These two traditions (the agnostic-sceptical and the theistic-magical) were to some extent associated with the opposing sides in the Old Text versus New Text controversies, which continued all through the Han (particulars in Tsêng Chu-Sên (1), pp. 137 ff.). Cf. p. 248 above.
[c] Ch. 109A, p. 5*a*, repeated in *Tzu Chih Thung Chien*, ch. 43, p. 28*b*, *Thung Chien Kang Mu*, ch. 9, p. 103*a*; *TH*, vol. 1, p. 675 (tr. auct. adjuv. Wieger).
[d] G 844; Forke (12), pp. 100 ff.
[e] *CSHK*, Hou Han section, ch. 15.

[1] 劉昆   [2] 偶然   [3] 桓譚   [4] 新論

(c) THE SCEPTICAL PHILOSOPHY OF WANG CHHUNG

These preliminaries having been completed we can now appreciate the work of Wang Chhung,[1] one of the greatest men of his nation in any age, who has often, not altogether inappropriately, been called the Lucretius of China. His merit in the history of Chinese science is well appreciated by modern Chinese scientists and scholars, as is seen from articles such as those of Wang Chin (1) and Hu Shih (3). His biography in the *Hou Han Shu*[a] tells us that he was born in +27 (he was thus a younger contemporary of Liu Khun), and that he was an indefatigable student and writer. He was constantly in and out of official positions, his independence of mind rendering him a difficult colleague, and he never attained high rank in the bureaucracy. He died in +97, and his great work, the *Lun Hêng*[2] (Discourses weighed in the Balance),[b] must have been written in the years +82 and +83.

First, as to his conception of Nature. He fully accepted the fundamental Yin-Yang dualism and the theory of the Five Elements, though not uncritically.[c] He makes little use of the terms *Tao*[3] or *li*,[4] but adopts a very thorough determinism symbolised by the term *ming*,[5] fate or destiny, analogous to the *anangke* (ἀνάγκη) of the pre-Socratics. Like the Taoists he denies consciousness to Heaven and holds a naturalistic world-view in which *tzu jan*,[6] spontaneity, is the watchword (cf. p. 51). The principles of maleness and femaleness seemed to him to be right at the heart of Nature, since heaven was equated with Yang and earth with Yin;[d] it is interesting that exactly the same idea occurs in Lucretius (*pater aether* and *mater terra*).[e] Calamitous changes supervene if the Yin and the Yang are at variance,[f] and there is a wave-like succession of dominance, the Yang handing over to the Yin when it has reached its climax and vice versa[g] (see on, Sect. 26b). Like the 'universal law' of the Stoics (see on, p. 534), but with significant differences, Wang Chhung's view is that 'Heaven and Man have the same Tao (*Thien jen thung Tao*[7]) in which good and evil do not differ. If something is impossible in the Tao of man, we know that it could not come into effect under the Tao of heaven either.'[h]

Wang Chhung took over and elaborated the ideas of rarefaction and condensation

---

[a] Ch. 79.

[b] Complete translation by Forke (4). In the following pages, references to this may appear as 'tr.' only. It is preceded by an excellent discussion of Wang Chhung's world-outlook. If this is not available, Li Shih-I's paper (1) may be consulted.

[c] Above, p. 266, we gave a good example of his criticism of the five-element theories current in his time.

[d] Ch. 14 (tr. vol. 1, p. 104), ch. 32 (tr. vol. 1, p. 261), ch. 15 (tr. vol. 1, p. 322). We cite chapter numbers in the order in which they occur in the original text, and not in the order in which Forke (4) rearranged them (he has a table of correspondences in his vol. 2, p. 421).

[e] *De Rer. Nat.* I, 250–3; II, 991–5.

[f] Ch. 55 (tr. vol. 2, p. 16).

[g] Ch. 46 (tr. vol. 2, p. 344).

[h] Ch. 17 (tr. vol. 2, p. 157). Cf. p. 488 below.

[1] 王充   [2] 論衡   [3] 道   [4] 理   [5] 命   [6] 自然   [7] 天人同道

which we have already met among the Taoists (p. 40). Life arises from a condensation of the *chhi* of Yin and Yang.[a] Accordingly:

As water turns into ice, so the *chhi* crystallise to form the human body (*shui ning wei ping, chhi ning wei jen*[1]). The ice, melting, returns to water, and man, dying, returns to the state of a spirit (*fu shen*[2]). It is called spirit just as melted ice resumes the name of water. When we have a man before us we use a different name. Hence there are no proofs for the assertion that the dead possess consciousness, or that they can take a form and injure people.[b]

Elsewhere he says that all those who study the art of immortality and trust that there are means by which one can avoid dying, must fail as surely as ice cannot be prevented from melting.[c]

Following the train of thought which led ultimately to Sir Thomas Browne's remark 'Life is a pure flame and we live by an invisible sun within us', and to Mayow's candle-flame and Benedict's calorimeter nearly two thousand years later, Wang Chhung identified the vital spirit (*ching shen*[3]) of living things with the fiery Yang principle, and the wet tissues, the flesh and bones (*ku jou*[4]), with the aqueous Yin principle. This led him to a position somewhat reminiscent of the distinction made by Aristotle between form and matter, *hule* and *eidos* ($\H{\upsilon}\lambda\eta$ and $\epsilon\tilde{\iota}\delta\sigma$); except that for Wang Chhung it was possible (though extremely dangerous) for the two principles, especially the Yang, to exist and manifest themselves independently. In chapter 65 he says:

That by which man is born are the two *chhi*[d] of the Yin and the Yang. The Yin *chhi* produces his bones and flesh; the Yang *chhi* his vital spirit. As long as he is alive the Yin and the Yang *chhi* are in good order, hence bones and flesh are strong, and the vital force full of vigour; the former gives him muscular energy, and the latter consciousness. The former continues strong and robust, and the latter manifests the power of speech. While bones and flesh are entwined and linked together, they always remain visible and do not perish.[e]

But the fiery solar Yang *chhi* can appear independently, and that is the cause of ghosts and apparitions, as well as of the lightning flash.[f]

What people call unlucky or lucky omens, and ghosts and spirits, are all produced by the *chhi* of the Great Yang (i.e. the sun; acting alone). This solar *chhi* is identical with the *chhi* of Heaven. As Heaven can generate the body of man, it can also imitate his appearance.... When the Yang *chhi* is powerful, but devoid of the Yin, it can merely produce a semblance, but no body. Being nothing but the vital spirit without bones or flesh, it is vague and diffuse, and when it appears it is soon extinguished again.[g]

[a] Though not in the passage which I give here, Wang Chhung elsewhere (e.g. ch. 62, tr. vol. 1, p. 196) uses the same terms as the Warring States Taoists, e.g. *san*[5] for dispersion or rarefaction.

[b] Ch. 62 (tr. Forke (4), vol. 1, p. 192); the affirmation is repeated later in the same chapter (tr. vol. 1, p. 196).   [c] Ch. 24 (tr. vol. 1, p. 350).

[d] I need not again insist on the untranslatability of this word, which has connotations similar to the *pneuma* of the Greeks, and to our own conceptions of a vapour or a gas, but which also has something of radiant energy about it, like a radioactive emanation.

[e] Tr. Forke (4), vol. 1, p. 249.

[f] For Wang Chhung's splendid naturalistic account of lightning and thunder, see on, in Section 21 on meteorology.   [g] Tr. Forke (4), vol. 1, p. 249, mod.

[1] 水凝爲冰氣凝爲人   [2] 復神   [3] 精神   [4] 骨肉   [5] 散

In many other places in his book, Wang Chhung describes the poisonous and dangerous character of this pure 'form' emanating from the source of all fire and heat,[a] and he considers it the best explanation for all recorded cases of injuries due to supernormal manifestations.[b] For an example of the formless Yin material before it is vivified by the Yang *chhi* he goes straight to the undeveloped hen's egg. He says:

Before a hen's egg is incubated, there is a formless mass (*hung-jung*[1])[c] within the shell, which, on leaking out, is seen to be of an aqueous nature. But after a good hen has incubated the egg, the body of the chick is formed, and when it has been completed, the chick can pick the shell and kick (its way out). Now human death (is a return to) the time of the formless mass. So how could the *chhi* of this formlessness injure anybody?[d]

Here he is not far from the concepts of differentiation and de-differentiation, or decomposition. Lastly, he uses the analogy of a sack which collapses when there is no more rice or millet inside it, thus losing the form which it had temporarily possessed;[e] or a gourd which must change its shape on desiccation.[f]

In accordance with his naturalism, Wang Chhung saw much scope for chance and strife in Nature, as well as necessity. In order to show that it is unreasonable to insist on relating everything that happens to human beings to their known or alleged moral merits or demerits, he takes the example of smaller things in non-human Nature, emphasising the role of chance and accident in their fates. Thus he says:

Mole-crickets and ants creep on the ground. If a man lifts his foot and steps on them, crushed by his weight, they die at once, while those which are untouched continue alive and unhurt. Wild grass is burnt up by fires kindled by the friction of chariot-wheels. People think the tufts of grass which have not been consumed are happy, and call them 'lucky grass'. Nevertheless that an insect has not been trodden upon or grass not been reached by a brush-fire, is no proof of their excellence. The movement of the feet and the spread of the fire are purely accidental.

The same reasoning holds good for the breaking out of ulcers. When the free circulation of the humours is stopped, they coagulate and form a boil; as it begins to run, it becomes a sore, finally blood comes out and pus is discharged. Are these particular pores, where the ulcer breaks through, 'better' than others? No, it is just that the working of the good constitution has been checked in some places.[g]

He ends by an amusing story attributed to Confucius, who used to pass by the ruinous gate-tower of the Lu capital at great speed. When his disciples pointed out

---

[a] Especially in ch. 66, 'On Poison' (tr. vol. 1, pp. 298 ff.); and tr. vol. 1, p. 245.

[b] As against other theories of the origins of ghosts, described in ch. 65 (tr. vol. 1, pp. 239 ff.), such as the view that they are the spirits of the cyclical signs.

[c] Note that Wang Chhung avoids the use of the famous Taoist technical term *hun-tun* here.

[d] Ch. 62 (tr. vol. 1, p. 199).

[e] Ch. 62 (tr. vol. 1, p. 192); ch. 7 (tr. vol. 1, p. 329).

[f] Ch. 7 (tr. vol. 1, p. 329).

[g] Ch. 5 (tr. vol. 1, p. 151). On the pathological theory of stasis see above, Sect. 7*j* (Vol. 1, p. 219), and below, Sect. 44.

[1] 澒溶

that it had been liable to fall down for a very long time already, Confucius replied that that was just what he disliked about it. So he, at least, appreciated the relation of chance and necessity. We shall return to this shortly in connection with Wang Chhung's views on human destinies.

Not only had the universe not been made for man[a] but there was no evidence for design in any of it. This denial of creativity, attractive though it is to the modern scientific mind in some respects, had other less desirable consequences for science, which we shall analyse in Section 18. In Wang Chhung's 14th chapter we read:

Tilling, weeding and sowing are designed acts, but whether the seed grows up and ripens or not depends upon chance and spontaneous action (*ou tzu-jan*[1]). How do we know? If Heaven had produced its creatures on purpose, it ought to have taught them to love each other and not to prey upon and destroy one another. It might be objected that such is the nature of the five elements, that when Heaven created all things it imbued them with the *chhi* of the five elements and that these fight together and destroy one another. But then Heaven ought to have filled its creatures with the *chhi* of one element only, and taught them mutual love, not permitting the five elements to war against one another and mutually destroy each other.[b]

Forke rightly compares this criticism with that of Lucretius.[c]

(d) CENTRIFUGAL COSMOGONY

Closely related to what has already been said about Wang Chhung's theory of changing densities, of rarefaction and condensation, is the theory of a 'centrifugal' cosmogony which he sets forth.[d] This did not begin with him nor did it end with him. The idea of the earth having been formed by solidification of the centre of a gyrating mass is so similar to modern cosmological views about nebulae and the formation of planetary systems around stars (suns) that a study of its history is of much interest.

We have not been able to trace the idea in China definitely earlier than the *Huai Nan Tzu* book (*c.* −120), where in the astronomical chapter (ch. 3) one finds:

Before the heavens and the earth took shape, there was an abyss without form and void (*fêng-fêng i-i tung-tung shu-shu*[2]); hence the expression 'Supreme Light'. The Tao began with Emptiness and this Emptiness produced the universe. The universe produced *chhi*

---

[a] Cf. the story in *Lieh Tzu*, p. 55 above.
[b] Tr. Forke (4), vol. 1, p. 104. See also p. 266 above.
[c] *De Rer. Nat.* II, 177–81; V, 185–9.
[d] It would be better to call it 'centripetal', for in all presentations of the cosmogony the more solid matter comes to the centre and the lighter matter flies outwards, i.e. exactly contrary to what really happens in centrifugation. The analogy which the ancient thinkers actually had in mind was that of an eddy or whirlpool, in which heavy particles will be rounded up as a mass centrally.

[1] 偶自然 [2] 馮馮翼翼洞洞濁濁

(vital gaseous emanation), and this was like a stream swirling between banks.[a] The pure *chhi*, being tenuous and loosely dispersed, made the heavens; the heavy muddy *chhi*, being condensed and inert, made the earth. The pure and delicate *chhi* coming together and making a whole was an easy matter, but the condensation of the heavy and turbid material was difficult. Consequently the heavens were finished first, and the earth became solid later. The combined essences of heaven and earth became the Yin and the Yang, and four special forms of the Yin and the Yang made the four seasons, while the dispersed essence of the four seasons made all creatures....[b]

If reasons had not already been given for attributing to the Chinese thinkers observations on the phenomena of condensation and rarefaction at a time before the transmission of ideas on the subject from Greece was at all likely (cf. Sects. 7, 10), one would have found it hard to believe that this statement was not a reflection of the well-known cosmogonic theory of Anaximenes of Miletus (fl. −545).[c] As is well known, it was reiterated in various forms by Empedocles of Akragas (fl. −450),[d] Anaxagoras of Clazomenae, whose *floruit* is about the same,[e] and the atomists Leucippus and Democritus (fl. −425).[f] It was known as the vortex theory, and of course appears in Lucretius (−98 to −55).[g] Since the Silk Road was opened at the end of the −2nd century there was no reason why an echo of Lucretius' great presentation could not have reached Wang Chhung, yet for this there is no evidence.

If the passage in *Lieh Tzu*[h] is genuine, however, the Chinese expression of the idea can be put back in the −4th century.[i] Thus it runs:

We say that there was a great (Principle) of Change (*thai i*[1]), a great Origin (*thai chhu*[2]), a great Beginning (*thai shih*[3]) and a great Primordial Undifferentiatedness (*thai su*[4]). At the great Change, *chhi* was not yet manifest. At the great Origin, *chhi* began to exist. At the great Beginning came the beginning of form and shape. In the great Primordial Undifferentiatedness lay the beginning of matter. When *chhi*, shapes and matter, were still indistinguishably blended together, that state is called chaos (*hun-lun*[5]). All things were mixed in it, and had not yet been separated from one another....The purer and lighter (elements) (*chhing chhing chê*[6]), tending upwards, made the heavens; the grosser and heavier (elements) (*cho chung chê*[7]), tending downwards, made the earth....[j]

---

[a] The writer presumably had in mind here the deposition of silt on the convex sides of meanders as opposed to the scouring of the opposite sides. This is another instance of the interest which the ancient Chinese took in matters related to hydraulic engineering.

[b] Tr. Hughes in Fêng Yu-Lan (2), p. 112, mod.

[c] Rey (1), vol. 2, p. 94; Freeman (1), p. 66.   [d] Freeman (1), p. 187.

[e] Freeman (1), p. 268.   [f] Freeman (1), pp. 287, 304.

[g] *De Rer. Nat.* v, 439–49, 485–93.   [h] Ch. 1, p. 3 *b*.

[i] There is in Wieger (2), p. 144, a passage which could easily confuse the unwary. A statement of the centrifugal cosmogony occurs in a chapter of extracts from the *Li Chi* and other writings alleged to be those of the disciples of Confucius, but it is clearly a Sung (+13th century) commentary on ch. 9 (Li Yün), p. 66*a*, of that book.

[j] Tr. auct. adjuv. L. Giles (4), p. 20; Wieger (2), p. 272, (7), p. 69. It is repeated in a closely similar passage in one of the Han apocrypha, the *I Wei Chhien Tso Tu*,[8] ch. 1. Cf. *TPYL*, ch. 36, p. 5*b*.

[1] 太易   [2] 太初   [3] 太始   [4] 太素   [5] 渾淪   [6] 清輕者
[7] 濁重者   [8] 易緯乾鑿度

And here is Wang Chhung's version:

The *I Ching* commentators say that previous to the differentiation of the original *chhi* there was a chaotic mass (*hun-tun*[1]). And the Confucian books speak of a wild medley (*ming-hsing mêng-hung*[2]), and of the (two) *chhi* undifferentiated. When it came to separation and differentiation, the pure (elements) (*chhing chê*[3]) formed heaven, and the turbid ones (*cho chê*[4]) formed earth....[a]

It is particularly interesting that in none of these ancient texts is there a definite statement as to *what* rose up and *what* settled down; the Chinese language made it unnecessary to be precise, for the word *chê*,[5] which can mean stuff, things (or even more often people), came naturally to the writers.[b] The point is important for the history of atomism, for, as we shall see in a moment, Chu Hsi's final presentation of the theme, in the +12th century, comes very near to speaking of particles.

Another mention of the centrifugal cosmogony occurs not long after Wang Chhung's time, for, as Maspero[c] has pointed out, the Taoist book *Thai Shang San Thien Chêng Fa Ching*[6] (Exalted Classic of the True Law of the Three Heavens)[d] contains an account of the differentiation of nine *chhi*, the light ones mounting to make the sky and the heavy ones sinking to make the earth. This is considered to have been written before the +4th century. The idea is also expressed in the +4th-century *Pao Phu Tzu*[e] and in Thang books such as *Kuan Yin Tzu*[f] and *Wu Nêng Tzu*[g] But the definitive statement is that of the great Neo-Confucian Chu Hsi,[7] in the late +12th century. It is in chapter 49 of *Chu Tzu Chhüan Shu*:[8]

Heaven and earth were in the beginning nothing but the *chhi* of Yin and Yang. This single *chhi* was in motion, grinding to and fro (*mo lai mo chhü*[9]),[h] and after the grinding had become very rapid, there was squeezed out (*tso*[10]) a great quantity of sediment (*hsü to cha tzu*[11]). There being no way by which it could escape from within, it coagulated and formed an earth in the centre. The purest (elements) of *chhi* (*chhi chih chhing chê*[12])[i] became the sky, the sun and moon, and the stars, which are permanently revolving and turning round outside. The earth was in the centre motionless, but not 'below'.

[a] Ch. 31, tr. Forke (4), vol. 1, p. 252. Wang Chhung's contemporary, Pan Ku,[13] gives a similar but shorter statement in his *Pai Hu Thung Tê Lun*,[14] ch. 4, p. 1a.

[b] Even when the antecedent noun is *chhi*, as with Wang Chhung, the precision is not much enhanced as the concept is so vague.

[c] (7), p. 201; (13), p. 124.

[d] *TT* 1188.

[e] Ch. 1 (Feifel (1), p. 119).   [f] Forke (12), p. 356.

[g] Forke (6), p. 56.

[h] Forke (6), p. 106, translates 'by the grinding of the *particles* against each other a violent friction ensued', but this is not in the text.

[i] Forke again says 'particles', but the vague word used does not justify it. Le Gall (1), p. 120, avoided this in both sentences.

[1] 渾沌   [2] 溟滓濛涒   [3] 清者   [4] 濁者   [5] 者
[6] 太上三天正法經   [7] 朱熹   [8] 朱子全書   [9] 磨來磨去
[10] 捽   [11] 許多渣滓   [12] 氣之清者   [13] 班固   [14] 白虎通德論

Heaven moves and has moved unceasingly, turning round by day and by night. Earth, that bridge on which we stand, is in its centre. If Heaven was to stop only for an instant, the earth would collapse. But (in the beginning) the gyration of Heaven was so rapid that there was a great mass of sediment crystallising and coagulating in the middle. This sediment was the sediment of *chhi*, and it is the earth. Therefore it is said that the purer and lighter parts became the sky, and the grosser and more turbid ones earth.[a]

Thus we see that although the text trembles on the verge of saying that the sediment is formed of particles (atoms) made small by mutual friction, it does not actually do so.[b] The theme continues into the Ming in less interesting statements such as that of Yeh Tzu-Chhi in his *Tshao Mu Tzu*[1] of about +1378.

Martin (6) long ago made the interesting suggestion that these ideas, conveyed to Europe through Jesuit channels, might have influenced Descartes' theory of vortices (*tourbillons*) in the physical aether.

(e) WANG CHHUNG'S DENIAL OF ANTHROPOCENTRISM

I now return to Wang Chhung. What was his view of the position of man in this universe? First of all he made a frontal attack upon the Chinese State 'religion' by an uncompromising resistance to anthropocentrism of any kind.[c] Again and again he returns to the charge that man lives on the earth's surface like lice in the folds of a garment.[d] At the same time, he admits that among the 300 (or 360) naked creatures, man is the noblest and most intelligent.[e] But if fleas, he said, desirous of learning man's opinions, emitted sounds close to his ear, he would not even hear them; how absurd then it is to imagine that Heaven and Earth could understand the words of Man or acquaint themselves with his wishes.[f] This position once gained, the whole

[a] P. 19*a*, tr. auct.

[b] Wieger (4), p. 624, affirms that Greek pre-Socratic concepts reached the Chinese through the *Laṅkāvatāra Sūtra* (see p. 405) from India. This Buddhist text was translated into Chinese in +430. The names of Hindu heretical sects which he identifies with various pre-Socratic doctrines (Anaximander, Anaximenes, Thales, etc.) occur, however, not in the *Sūtra*, but in a commentary on it (Takakusu (1), vol. 32, pp. 156–8) which was translated in +520. Wieger's identification is highly dubious (Dr A. Waley, personal communication), and even if it were accepted it would obviously not account for the Chinese statements of the centrifugal cosmogony before the Christian era. It must, of course, be remembered that the Indians themselves had something similar to the centrifugal cosmogony. The three components (*guṇa*) into which, according to the Sāṃkhya philosophy, the primitive undifferentiated Nature-stuff (*prakṛti*) separated, were lightness, heaviness and motion. There is dispute about the connections of this philosophy with the beginnings of Buddhism (Jacobi (1); Keith (3); Garbe (2); Thomas (1), p. 91). No evidence, so far as I can see, has been adduced to show that Chinese thought in, say, the −3rd century was influenced by this philosophy, and the case resembles that of atomism, which has already been discussed (Sect. 7*b*, Vol. 1, p. 154).

[c] In this way his position in the history of Chinese thought may be termed 'Copernican'.

[d] I noted at least four places—ch. 14 (tr. vol. 1, p. 103), ch. 15 (tr. vol. 1, p. 322), ch. 43 (tr. vol. 1, p. 109), ch. 71 (tr. vol. 1, p. 183).

[e] Ch. 72 (tr. vol. 1, p. 528); ch. 38 (tr. vol. 2, p. 105).

[f] Tr. vol. 1, p. 183.

[1] 草木子

weight of Wang Chhung's attack on superstition was deployed. Heaven, being incorporeal, and Earth inert, can on no account be said to speak or act;[a] they cannot be affected by anything which man does;[b] they do not listen to prayers;[c] they do not reply to questions.[d] Hence was swept away the whole basis of the systems of divination which were described at the beginning of this section.[e]

What remained of superstition after Wang Chhung's denial of anthropocentrism had done its work, he then attacked either by demonstrating the statistical absurdity of some beliefs, or the sheer unreasonableness of others. The thousands of prisoners in the gaols, or all the inhabitants of the city of Li-yang, which was flooded during a single night and sank to the bottom of a lake, cannot all have chosen unlucky days for their business; nor could the choice of auspicious ones account for all the scholars who attain high official rank.[f] As for sacrifices to the ghosts and spirits, the whole thing is complete nonsense. In order to give the feel of one of Wang Chhung's diatribes, part of the one on this subject may be chosen. In chapter 75 he says:

The world places confidence in sacrifices, trusting that they procure happiness; and likewise it approves of exorcisms, fancying that these remove evil. The first ceremony performed at exorcising is the setting out of a sacrifice, which we may compare with the entertainment of guests among living men; but after the savoury food has been hospitably set out for the spirits and they have eaten of it, they are chased away with swords and sticks. If the spirits were conscious of such treatment they would surely stand their ground, accept the fight, and refuse to go; and if they were susceptible of indignation, they would cause misfortune. But if they have no consciousness they cannot possibly effect any evil. Accordingly exorcising is lost labour, and no harm is caused by its omission.

Besides, it is disputed whether spirits have a material form (*hsing hsiang*[1]). If they have, it must be like that of living men. But anything with the form of living men must be capable of feeling indignation, and exorcism would therefore cause harm rather than good. And if they have no material form, driving them away is like (trying to) drive out vapour and clouds, which cannot be done.

And since it cannot (even) be ascertained whether the spirits have a material form, we are not at all in a position to guess their feelings. For what purpose do they gather in human dwellings anyway? If disposed to killing and injury, they will, when exorcised, simply abscond and hide, but return as soon as the chase is over. And if they occupy our homes without nefarious purposes, they will not be harmful, even though not expelled.

[a] Ch. 71 (tr. vol. 1, p. 183).
[b] Ch. 43 (tr. vol. 1, p. 110).　　　　　　　　　　[c] Ch. 43 (tr. vol. 1, p. 113).
[d] Ch. 71 (tr. vol. 1, p. 184). As Leslie (1) has pointed out, this general position of Wang Chhung constantly involved him in the affirmation that small causes cannot produce big effects. This was contrary to a favourite Taoist point of view (cf. the passage from Than Chhiao, p. 451 below), and here the Taoists were at least as right as Wang Chhung.
[e] This does not mean that Wang Chhung rejected all the divination procedures used in his time, but he regarded the results (more or less reliable) as phenomena occurring wholly within the natural order and at least as likely to deceive as to enlighten.
[f] Ch. 72, the whole of which is devoted to a slashing attack on the principle of lucky and unlucky days (tr. vol. 1, pp. 525 ff.).

[1] 形象

There follows an analogy between the return of the spirits after exorcism, and the crowd pressing round an official procession, which gathers again as soon as the lictors move on; or birds coming back to eat corn in a farmyard each time after being driven away.

Decaying generations cherish a belief in ghosts. Foolish men seek relief in exorcism. When the Chou dynasty rulers were going to ruin, sacrifice and exorcism were believed in, and peace of mind and spiritual assistance were thus sought. The foolish rulers, whose minds were misled, forgot about the importance of their own behaviour, and the fewer were their good actions the more unstable their thrones became. The conclusion is that man has his happiness in his own hands, and that the spirits have nothing to do with it. It depends on his virtues and not on sacrifices.[a]

This rises to a prophetic level reminiscent of Isaiah, and perhaps it is rather characteristic of Chinese civilisation that if we had to look anywhere for an analogue of the moral force of the Hebrew prophets, it would be found among the most atheist and agnostic of the Confucian rationalists.

Another set of ideas against which Wang Chhung directed destructive criticism was the Taoist belief in the possibility of the attainment of material immortality by the aid of techniques.[b] He compares Taoist longevity practices with biological metamorphoses[c] which nevertheless do not stop the quail or the crab from being eaten in the end. Moreover, the life-span of those animals (insects) which have the most complete of metamorphoses does not compare favourably with that of animals which do not metamorphose.[d] Ataraxy may be considered helpful to longevity, but plants and herbs, though quite dispassionate, often live only for one year.[e] Living organisms can be injured by excessive ventilation, so what is the point of respiratory exercises? Rivers acquire turbidity as they flow through the land, so what is the use of trying to increase the circulation by gymnastics? (i.e. the blood stream will be purest if it is not interfered with).[f] Wang Chhung has a single passing mention of alchemy, saying that 'one hears that the Taoists eat the essence of gold and jade' (*chin yü chih ching*[1]), but he may have known more about it than appears, for a page or two previously he speaks of the 'yellow and the white' (a well-known esoteric term for alchemy), and turns the tables on the alchemists by saying that surely yellow is the sign of mature ripeness and impending decay in plants, and white the symbol of white-haired old age. 'Yellow and white are like the frying of meat and the cooking of fish', i.e. all chemical changes are irreversible.[g]

[a] Tr. Forke (4), vol. 1, pp. 532, 534; de Groot (2), vol. 6, p. 934, mod. Cf. Giles (12), pp. 96 ff.
[b] See esp. ch. 7 (tr. vol. 1, pp. 325 ff.) and ch. 24 (entitled 'Empty Taoist Nonsense', tr. vol. 1, pp. 332 ff.).
[c] Tr. vol. 1, p. 326. Wang Chhung's zoology was rather uncritical. By metamorphosis he included true metamorphosis (as of insects; cicada, silkworm, mole-cricket, etc.), moulting of skin (as in snakes and other reptiles), and a number of fancied transformations such as those of frogs into quails and sparrows into clams (cf. Sects. 7h, 10i, 16a, 39), which were generally thought to occur in his time.
[d] Tr. vol. 1, p. 327.   [e] Tr. vol. 1, p. 347.
[f] Tr. vol. 1, pp. 348, 349.   [g] Tr. vol. 1, pp. 337, 339.

[1] 金玉之精

But it seems that while, as might be expected of a semi-Confucian, Wang Chhung was for most of his life in favour of Taoist naturalism but against Taoist experimentalism,[a] he changed his mind as he grew older. His biography says that at the very end of his life he wrote a book on 'nourishing the vital spirit', *Yang Hsing Shu*,[1] but it has not survived.[b]

In general, however, Wang Chhung remained a thorough enemy to all forms of 'supernaturalism'.[c] He combated with particular force the vast body of legend in which his contemporaries universally believed; for example, the many stories of intercourse with dragons, supernatural births, etc.[d] Many of his arguments are of course quite sound biologically, but one gets the impression that he is using a steam-hammer to crack a nut until one remembers how seriously his contemporaries took these traditional marvels. Later, in Section 21 on meteorology, I shall give a long passage in connection with the history of tidal theory, which shows to perfection how he would take a legendary explanation and proceed to tear it limb from limb. The nearer the legendary explanation came to facts which Wang Chhung felt were susceptible of proper scientific explanations, the harder he worked to destroy it—this appears in his chapters which deal with astronomical subjects, which we shall examine briefly in the appropriate Section (20).

Half obscured by the dust of battle which rises from the pages of the *Lun Hêng* are the names for the various schools of magicians which existed at the time. There were the 'horoscopists and seers' (*kung chi shê shih chê*[2]) who selected lucky and unlucky days;[e] the 'geomancers' (*chan shê shih chê*[3]) and the 'soothsayers' (*kung chi chih chia*[4]) who attended to the siting of houses;[f] and the 'meteorologists' (*hou chhi pien chê*[5]) who watched for portents.[g] Wang Chhung has less criticism of the 'physicists' (*chi Tao chih chia*[6]) who are mentioned in connection with amber and the magnet, cast burning mirrors and attempt to foretell fates by the five elements and the denary and duodenary cyclical signs.[h] But the school which he most hated was that of the 'phenomenalists'[i] (*pien fu chih chia*;[7] the 'change-and-reversion school'), no doubt on account of their great political importance. Their ideas were to him what *religio* was for Lucretius.

---

[a] He commended the *wu wei* administration of Tshao Shen and Chi Yen (cf. p. 70); tr. vol. 1, p. 94.

[b] This is what Forke (4) translates so oddly as 'Macrobiotics'.

[c] I use quotation marks here because it should be remembered that for the characteristic and instinctive Chinese world-view in all ages there could be nothing supernatural *sensu stricto*. Invisible principles, spirits, gods and demons, queer manifestations, were all just as much part of Nature as man himself, though rarely met with and hard to investigate.

[d] Ch. 15 (tr. vol. 1, pp. 318 ff.).   [e] Ch. 72 (tr. vol. 1, p. 525).

[f] Ch. 72 (tr. vol. 1, p. 531).   [g] Ch. 61 (tr. vol. 2, p. 275).

[h] Ch. 47 (tr. vol. 2, p. 349), ch. 74 (tr. vol. 2, p. 413).

[i] We apologise for continuing the use of this term, which derives from Forke, and which has, of course, nothing whatever to do with modern meanings of the same word in 19th-century or contemporary philosophical discussions. We have not been able to think of a better one.

---

[1] 養性書   [2] 工伎射事者   [3] 占射事者   [4] 工技之家

[5] 候氣變者   [6] 伎道之家   [7] 變復之家

(f) THE PHENOMENALISTS AND WANG CHHUNG'S STRUGGLE AGAINST THEM

We enter this subject by an unexpected door, namely, Wang Chhung's position regarding the primitive collectivism praised by the Taoists. He begins by speaking in a very Taoist way on the matter,[a] but quickly diverges to attack the phenomenalists. The passage is in chapter 54:

Ceremonies originate from want of loyalty and good faith, and were the beginning of confusion. On this score people find fault with one another, which leads to mutual reproof (of superiors and inferiors). At the time of the Three Rulers people sat down informally (without attending to precedence) and walked about at their ease. They worked themselves instead of using horses and oxen. Simple virtue was the order of the day, and the people were unsophisticated and ignorant (of social distinctions). Minds acquainted with 'know-ledge' and 'cleverness' had not then developed.

Originally there were no calamities or omens, or if there were, they were not considered as reprimands (from Heaven). Why? Because at that time people were simple and un-sophisticated, and did not restrain or reproach one another. Later ages have gradually declined—superiors and inferiors contradict one another, and calamities and omens constantly occur. Hence the hypothesis of reprimands (from Heaven) has been invented. Yet the Heaven of today is the same Heaven as of old—it is not that Heaven anciently was kind, and now is harsh. The hypothesis of Heavenly Reprimands (*Thien chhien kao chih yen*[1]) has been put forward in modern times, as a surmise made by men from their own (subjective) feelings.[b]

This was a frontal attack on a very deep-rooted and powerful group of beliefs and people. As has already been indicated (p. 247) some of the Han Confucians developed the ideas of the School of Naturalists so as to produce a system in which any ethical irregularity actually *caused* cosmic irregularities.[c] The *Huai Nan Tzu* book has a firm statement on the subject.[d] One of the chief representatives of this strange synthesis was Tung Chung-Shu (−179 to −104), whose *Chhun Chhiu Fan Lu* (String of Pearls on the Spring and Autumn Annals) has already been referred to in other connections.[e] Several chapters of this work set forth the theory, notably chapter 44,[f]

---

[a] I doubt if Wang Chhung took this Taoist doctrine very seriously, for he has no less than five chapters (56–60) praising the Han dynasty and his own time, as much better than the Chhin and Warring States periods, at any rate.

[b] Tr. Forke (4), vol. 1, p. 100, mod.

[c] So I write, but the conception was often more subtle than the words would imply. It was as if ethical irregularities were disturbances in the cosmic pattern at one point which were bound to induce other (physical) disturbances elsewhere, not by direct action but by a kind of shock signalled through the vast ramifications of one organic whole. It was, as we might say, a matter of communication engineering, not of mechanical power.   [d] Ch. 3, p. 2a, b.

[e] Pp. 249, 281. See the excellent article on him by Yao Shan-Yu (4). Tung Chung-Shu was by no means the only writer of note on this; there was, for instance, Yang Hsiung[2] with his *Thai Hsüan Ching* (Canon of the Great Mystery); cf. Forke (7), and p. 329 above.

[f] Tr. d'Hormon *et al.*

[1] 天譴告之言   [2] 讖緯

entitled Wang Tao Thung San[1] (That the Action of the Prince puts into Communication the Three Agents of the Universe), and chapter 64,[a] entitled Wu Hsing Wu Shih[2] (The Five Elements and the Five Daily Affairs). It would be tedious and unnecessary to follow out the details, built up as they were on an entirely imaginary basis, but chapter 64 maintains, for example, that if the emperor and his ministers do not practise *li*,[3] the rites and usages, there will be excessive gales and trees will not grow properly; if the emperor's speech is not in accord with reason, metals will not be malleable[b] and there will be terrible thunderstorms;[c] if the emperor's audiences fail to be discriminating, there will be rainstorms and floods, etc.[d] Not only the emperor but the whole bureaucracy was involved in this, faults of local officials producing local effects. Another aspect of phenomenalism was a kind of inverted astrology, perturbations of the planets' motions being ascribed to governmental irregularities; this may be well seen in a fragment of the *Wên Tzu* book, probably of Han date.[e] It comes out clearly also in a statement of Yang Hsiung,[4] in his *Fa Yen*[5] (Model Discourses) of about +5:

Someone asked whether a sage could make divination. (Yang Hsiung) replied that a sage could certainly make divination about Heaven and Earth. If that is so, continued the questioner, what is the difference between the sage and the astrologer (*shih*[6])?[f]

(Yang Hsiung) replied, 'The astrologer foretells what the effects of heavenly phenomena will be on man; the sage foretells what the effects of man's actions will be on the heavens.'[g]

Hu Shih[h] has chosen from Tung Chung-Shu a sentence which sums it up: 'The action of man, when it reaches the highest level of goodness or evil, will flow into the universal course of heaven and earth, and cause responsive reverberations in their manifestations.' This 'moral reactivity' of Nature was of course not unknown in the history of ideas in the occident,[i] but the Chinese seem to have carried it further and given it more persistence, automatism and logical schematisation than anyone else.

The flood of literature which grew up devoted to the detection and interpretation of the meaning of all abnormal or catastrophic phenomena in the skies or on earth has left behind a mass of débris in every one of the dynastic histories. Some of their

---

[a] Tr. Hughes (1), p. 308; cf. Fêng Yu-Lan (6), p. 124.

[b] The general idea was that objects or substances which manifested *par excellence* the qualities of the five elements, would lose their normal properties (cf. Pfizmaier, 58).

[c] Death by lightning was naturally considered a particularly striking example of a 'heavenly reprimand', but I defer Wang Chhung's eloquent passages on the nature of thunder and lightning until Sect. 21 on meteorology.

[d] Han interpolations carried this into books such as the *Kuan Tzu*, e.g. ch. 40.

[e] See Forke (13), p. 351.

[f] Lit. State Astrologer, or astrological official.

[g] Ch. 6, p. 9*b*, tr. auct.; cf. Forke (12), p. 95.

[h] (3), p. 44.

[i] See Cornford (1), p. 5, quoting Hesiod and Sophocles; and also p. 55. Patai (1) describes remarkable Jewish parallels. Any plague, flood, or 'visitation' from God, would come in an analogous category (see below, p. 575).

[1] 王道通三   [2] 五行五事   [3] 禮   [4] 揚雄   [5] 法言   [6] 史

longest sections (the 'Five Element' chapters) deal with these matters.[a] In Wang Chhung's time the Classics were being searched for material which would fit into the new theories, and when that proved insufficient, entirely new texts were invented which were called Wei Shu[1] ('Weft Books') by analogy with the Ching,[2] the Classics themselves (the original meaning of the word *ching* had been 'warp').[b] The authority of this class of apocryphal literature, says Hu Shih, became so exalted that throughout the +1st and +2nd centuries many important State policies, such as calendar reforms or the selection of heirs to the throne, were decided solely by reliance on these invented books.[c] The only good account of them in a European language is that of Tsêng Chu-Sên (1), though Bruce (3) has written briefly on the eight Wei books associated with the Book of Changes, which alone of this whole literature have come down to us in more than fragmentary form.[d] Much work on the *chhan-wei*[3] literature, as it came to be called, is appearing in Chinese by Chhen Phan, and some of it is available in the form of a French résumé by Kaltenmark (1).

To all this Wang Chhung led the opposition. He brought forward every argument he could think of to assert (*a*) that excessive seasonal heat and cold do *not* depend on the ruler's joy and anger;[e] (*b*) that plagues of tigers[f] and of grain-eating insects[g] are *not* due to the wickedness of secretaries and minor officials; (*c*) that natural calamities and unlucky events are *not* manifestations of Heaven's anger;[h] and (*d*) that hard winters are not due to cruelties and oppressions.[i] In his main chapter on this subject[j] he ends by saying:[k]

The heart of high Heaven is in the bosom of the Sages. When Heaven reprimands it does so through the mouths of the Sages. Yet people do not believe their words. They trust in the *chhi* of calamitous events, and try to make out Heaven's meaning therefrom. How far is this away (from the truth)![l]

[a] That of the *Chhien Han Shu* has been elaborately analysed by Eberhard (6) and Bielenstein (1). Those of the (Liu) *Sung Shu* (+4th and +5th centuries), chs. 30–4, have been translated by Pfizmaier (58) and those of the *Hsin Thang Shu* (+7th to +9th centuries), chs. 34–6, 88, 89, by Pfizmaier (67).

[b] The *Lu Thu*,[4] a prognosticatory book presented to Chhin Shih Huang Ti by the magician Master Lu[5] in −215, has often been regarded as the earliest of the kind. The Chhan books began to flourish only after about −40.

[c] These forgeries almost invite comparison with the place taken by the Forged Decretals in European history. From +862 onwards the Papacy made use of a large number of decrees which purported to have been issued by Popes between the time of the apostles and that of the first genuine decrees (+385), but which were +9th-century forgeries. A good account of these, and other similar inventions which built up the papal power, will be found in A. Robertson (sen.) (1), pp. 236 ff. It seems rather characteristic of the two civilisations that the great forgeries of Europe should have been legal and administrative, while those of China should have been pseudo-scientific.

[d] Four of them will be found in Sun Chio's *Ku Wei Shu*, chs. 14–16.

[e] Ch. 41 (tr. vol. 1, pp. 278 ff.).

[f] Ch. 48 (tr. vol. 2, pp. 357 ff.).

[g] Ch. 49 (tr. vol. 2, pp. 363 ff.).

[h] Ch. 55 (tr. vol. 2, pp. 16 ff.).

[j] Ch. 42 (tr. vol. 1, pp. 119 ff.).

[l] Tr. vol. 1, p. 129.

[i] Ch. 43 (tr. vol. 1, pp. 109 ff.).

[k] With another touch of Hebrew prophecy.

[1] 緯書　　　[2] 經　　　[3] 讖緯　　　[4] 錄圖　　　[5] 盧

And in chapter 53,[a] after giving (in his usual logical way) instances of proverbially benevolent rulers in whose time calamities occurred, he turns the tables on the phenomenalists by saying that instead of natural calamities depending upon human virtues, human virtue depends on natural calamities, for example, in famine:

What are the causes of disorder? Are they not the prevalence of robbery, fighting and bloodshed, the disregard of moral obligations by the people, and their rebellion against their rulers? All these difficulties arise from want of grain and other foods, for the people are unable to bear hunger and cold (beyond a certain limit). When hunger and cold combine, there are few who will not violate the laws; but when they enjoy both warmth and food, there are few who will not behave properly.[b]

The recognition of this elementary fact was already old, if anyone had cared to notice it; it may be found, for instance, in the first chapter of the *Têng Hsi Tzu* book.[c] Fundamentally Wang Chhung falls back on maintaining (quite correctly) that all the assumed coincidences were due to pure chance—thus in chapter 41:

The setting in of torrid and frigid weather does not depend on any governmental actions, but heat and cold may chance to be (*tsao*[1]) coincident with rewards and punishments, and it is for this reason that the phenomenalists (*pien fu chih chia*[2]) (falsely) describe them as having such a connection.[d]

Here we meet again with the idea of the pre-established harmony.[e] The phenomenalists thought that they had detected invariable manifestations of it; Wang Chhung was convinced that they were wrong, though he allowed the possibility of occasional 'chance' coincidences within its framework.

As Leslie (2) has pointed out, Wang Chhung's denial of phenomenalism involved him in another position more seriously aberrant from the main Chinese tradition, namely, the denial of 'action at a distance'.[f] This naturally followed from his attack upon anthropocentrism, but he often returns to the charge. That the dragon can cause rain to fall he admits, but only within a radius of 100 li;[g] there may be telepathy but not beyond a limited distance;[h] the death far away of three successive fiancés of Wang Mang's aunt cannot possibly have been due to any baleful influence radiating from her.[i] Yet Wang Chhung himself felt obliged to uphold certain forms of action at a

---

[a] Tr. vol. 2, pp. 9 ff. [b] Tr. vol. 2, p. 12.

[c] H. Wilhelm (2). And there are powerful statements of it also in *Kuan Tzu*, ch. 1, p. 1 *a* (cf. Forke (13), p. 76); and much later on, in the Thang *Hua Shu*, ch. 5, p. 26*b* (cf. Forke (12), p. 346).

[d] Tr. Forke (4), vol. 1, p. 281, mod. See also vol. 1, pp. 127, 128, 283; vol. 2, p. 357.

[e] P. 292 above.

[f] Indeed, this doctrine, which was fundamental for the organic quality of the ancient and medieval Chinese world-view, persisted in spite of Wang Chhung, and led to many valuable features of early Chinese science, such as the discovery of magnetic polarity. [g] Ch. 41 (tr. vol. 1, p. 280).

[h] Ch. 19 (tr. vol. 2, p. 189). [i] Ch. 10 (tr. vol. 2, p. 6) and ch. 11 (tr. vol. 1, p. 306).

[1] 遭 [2] 變復之家

distance, especially the influences emanating from the stars and constituting an important part of the endowment of individual human beings.[a] Nature could thus affect men, but it was really too much to be asked to believe that the puny doings of men could affect Nature. And if influences there were which radiated from individual beings, then their range was but short.

Wang Chhung's protest, alas, was not very efficacious,[b] and the *Chhan-Wei Shu*[1] (Apocryphal Treatises on Prognostications) continued in favour until as late as the Thang. The sinological study of this corpus has hardly as yet begun,[c] and it is likely to prove tedious, though one must recognise that phenomenalism was responsible for many of the very old series of systematic observations of comets, meteor-showers, sun-spots, etc., to which we shall devote attention in the relevant Sections. The *Chhan-Wei* books were put together in modern times into collections such as the *Ku Wei Shu*[2] by Sun Chio[3] of the Ming, and the *Chhi Wei*[4] by Chao Tsai-Han[5] in 1804. They are full of material concerning auspicious and ominous signs, prophecies alleged to have foretold historical events, and astrological matters. But as time went on, and as one rebellion after another claimed justification from these apocryphal books, governmental bureaucratic circles began to frown upon them, and the Sui emperor Yang Ti (+605 to +617) ordered them to be burnt whenever they could be got hold of.[d] It was not that the beliefs had lost their interest, but the government felt it inexpedient that the common people should engage in arguments about the relations between the emperor and Heaven. This prohibition had close parallels in European history (cf. p. 537). Eventually it had its effect, and with the rationalist influence of the Neo-Confucians in the Sung, this particular form of superstition largely died out.

(g) WANG CHHUNG AND HUMAN DESTINY

So far we have seen only the negative side of Wang Chhung. His positive teaching, so far as human affairs were concerned, dealt mainly with human fate (*thien ming*[6]). His conception of destiny, says Forke,[e] was not an inexorable decree of Heaven laid down for each individual, not the *heimarmene* (εἱμαρμενη, that which falls to one's lot) of the Greeks, or the Roman *dira necessitas*, nor quite the same thing as the pre-destination of patristic writers; but depended on (*a*) the spiritual essence (*ching shen*[7])

---

[a] See immediately below. He himself was also one of the first to state (ch. 16) the moon's relation to the tides, and he accepted (chs. 10 and 32) the classical case of the moon's influence on shellfish (see Sects. 7*a* and 39).

[b] Even though strongly supported by the astronomer Chang Hêng in the following century.

[c] Cf. Nagasawa (1), p. 135.

[d] Roman parallels in Cramer (1).

[e] (4), vol. 1, p. 26.

[1] 讖緯書   [2] 古微書   [3] 孫瑴   [4] 七緯   [5] 趙在翰
[6] 天命   [7] 精神

with which each individual human constitution (*hsing*[1]) was endowed,[a] (*b*) specific influences emanating from the stars (*tê hsing chih ching*[2]),[b] and (*c*) the effects of chance.

Let us take these three factors in turn. By constitution Wang Chhung meant not only mental endowment but something very physical; thus he urges that the legendary supernumerary nipples of Wên Wang,[c] and the vision anomalies of Wu Wang, were already present long before birth.[d] 'The fate of individuals is inherent in their bodies, just as with birds the distinction between cocks and hens exists already within the egg-shell.'[e] His emphasis on genetic inheritance, as against environmental influences, is here most interesting. He was touching on the determination problem so basic in modern experimental embryology two thousand years later. 'As the shape of a vessel, once completed (i.e. fired in the potter's kiln), cannot be made smaller or larger, so the duration of the corporeal frame having once been settled, cannot be shortened or prolonged.'[f] 'Man may be thought of as having been moulded and baked in the furnace of Heaven and Earth—how could he still undergo a change after his shape has been fixed?'[g] 'A boy of fifteen is like silk, his gradual changes into good or bad resembling the dyeing of boiled silk with indigo or vermilion colour, which makes it blue or red.'[h] This kind of argument had led Hsün Chhing three centuries previously to emphasise the role of education, but Wang Chhung, much more predestinarian, placed, as it were, the moment of determination further back in the individual ontogeny, and generally looked upon it as already past at birth.[i] 'The deaf and the dumb', he said, 'the crippled and the blind, (are people whose) *chhi* met with harm in the uterus, so that they received a warped nature.'[j] However, he thought it was possible to 'instruct' embryos while still in the womb,[k] if the mother followed the rules of *li*.[3]

In the light of the discussion of astrology at the beginning of this section, Wang Chhung's second (astral) factor is of much moment. He lived just at the time when astrology was spreading out to the masses of the people from its former restricted function at the courts of the rulers. He was of course well aware of the names and

[a] Ch. 12 (tr. vol. 1, pp. 130 ff.); and ch. 4 (tr. vol. 1, pp. 313 ff.).

[b] Forke qualifies this component as 'supernatural', but the term is surely incorrect. In Wang Chhung's universe there was, by definition, nothing supernatural. Cf. ch. 6 (tr. vol. 1, pp. 136 ff.).

[c] One of the early sage-kings at the beginning of the Chou dynasty. On the biology of supernumerary mammae see Speert (1).

[d] Tr. vol. 1, p. 131. The reference to sheep foetuses in Forke's translation here is an error.

[e] Tr. vol. 1, p. 132.

[f] Tr. vol. 1, p. 325.

[g] Tr. vol. 1, p. 330.

[h] Tr. vol. 1, p. 374.

[i] Wang Chhung's attitude on the 'human nature' problem has already been referred to in Sect. 9*d*.

[j] Ch. 6, tr. Leslie, cf. Forke (4), vol. 1, p. 141.

[k] The idea of the possibility of 'maternal impressions' influencing the development of the foetus has a very long history; cf. Needham (2), pp. 11, 193. There is a parallel here in the Pali *Kathavatthu*, XXII, 4 (tr. Aung & Rhys Davids (1), p. 360), and the idea of *thai chiao*[4] often appears in Chinese literature, e.g. the partly −2nd-century *Hsin Shu* of Chia I, ch. 55, and the +6th-century *Yen shih Chia Hsün*, ch. 1, p. 2*b*.

[1] 性　　　[2] 得星之精　　　[3] 禮　　　[4] 胎敎

duties of the astrological officials, the *shih kuan*[1] (or simply *shih*),[a] or the *fêng hsiang shih*[2] and *pao chang shih*[3] of the *Chou Li*[b] who kept watch on the palace towers and recorded celestial events in old Chaldean style. On occasion, he combated traditional accounts of the behaviour of heavenly bodies which he was convinced were absurd, such as the story that on the occasion of a certain battle the sun regressed through three *hsiu*, and another to the effect that Mars passed rapidly through three *hsiu* when a certain feudal lord had uttered some particularly excellent maxims.[c] But he nevertheless believed that among the most important of all influences acting upon men during the formative period of their lives were those of the stars. In chapter 6 he says:

As regards the transmission of wealth and honour, that depends on the *chhi* which the nature obtains; it receives an essence (*ching*[4]) emanating from the stars. Their hosts are in heaven, and heaven has their signs (*hsiang*[5]). If a man receives (at his birth?) a heavenly sign implying wealth and honour, he will obtain wealth and honour. If a man receives (at his birth?) a heavenly sign implying poverty and misery, he will become poor and miserable. Therefore it is said that (all dispositions depend on) Heaven. But how can this be? Heaven has its hundreds of officials and its multitudes of stars. Heaven sends forth its *chhi*, and the stars send forth their essences (*ching*[4]), and the essences are in the midst of the *chhi*. Men imbibe this *chhi* and are born. As long as they cherish it they grow. If they obtain a sort which means honour, they will be men of rank, if not, they will be common people. Their position will be higher or lower, and their wealth greater or lesser, according to the position of the stars concerned, whether more honourable or less honourable, smaller or greater. Heaven has its hundreds of officials and its multitudes of stars; so also we have on earth the essences of the myriad common people, of the Five Emperors and the Three Rulers. Heaven has its Wang Liang or Tsao Fu,[d] and on earth there have also been such men. It was because the latter were endowed with the *chhi* of their heavenly counterparts that they became skilled in chariot-driving.[e]

This is at any rate quite unequivocal, and paradoxically it may well be the first statement in Chinese literature of individual astrology. The paradox lies in the probability that it was precisely Wang Chhung's scientific naturalism which pushed him into this theory, as a means of escaping from the arbitrary endowments of local gods and spirits and other 'supernatural' agencies. The stars were at least regular in their motions. We should not forget the argument of Burkitt (2) that Western gnostic astrology was in many ways an attempt to fit religion to the science of its time. And Wang Chhung's attitude must be seen as part of his general cosmology in which the *chhi* of Yang and Yin were associated with the sun and moon respectively.[f] It may

[a] *Lun Hêng*, ch. 52 (tr. vol. 1, p. 319); ch. 68 (tr. vol. 2, p. 376).
[b] Chs. 17 and 26; Biot (1), vol. 1, p. 413, vol. 2, p. 112, and vol. 1, p. 414, vol. 2, p. 113, respectively.
[c] Tr. vol. 2, p. 174.
[d] Two famous charioteers of old. The asterism here referred to corresponds to five stars in Cassiopeia (Schlegel (5), p. 329).
[e] Tr. auct. adjuv. Forke (4), vol. 1, p. 138, mod.
[f] Cf. tr. vol. 1, p. 241.

[1] 史官   [2] 馮相氏   [3] 保章氏   [4] 精   [5] 象

turn out to be one of the paradoxes of the history of science in China that individual judicial astrology was founded by the greatest sceptic of them all.

The last of the three components was Chance. This Wang Chhung tried to analyse further,[a] distinguishing between the effects of time (*shih*[1]); contingencies (*tsao*,[2] lit. meetings), e.g. general calamities in which many people perish at the same time, irrespective of what their fates might otherwise have been; luck or fortune (*hsing*[3]), e.g. the arrival of a general amnesty after a man has been imprisoned; and incidents (*ou*[4]), e.g. chance encounters with men who have high official posts in their gift, etc.[b] He hardly succeeds in establishing the definitions of these various forms of chance. His contemporaries distinguished between three types of joint destined and willed results—*chêng*[5], i.e. the natural fate not interfered with by the will; *sui*,[6] adjuvant, i.e. an evil will aiding an evil fate or vice versa; and *tsao*,[2] contrary, i.e. an evil will acting against a good fate or vice versa. But Wang Chhung did not agree with them, and believed in a strict pre-determined pattern in which the will of the individual could accomplish little or nothing.[c]

From all this it followed that Wang Chhung accepted not only the individual astrology which was developing in his time (if, indeed, he was not himself the patron of it), but also most of the physiognomy (or 'anthroposcopy', as Forke calls it), since naturally the type of constitution possessed by a person would be expected to have its outward and visible signs.[d] He has a special chapter devoted to this subject, in which he gives many examples of physical characteristics which he thought could be justifiably connected with inward constitution and consequent destiny. He felt himself to be the first also to correlate physiognomical signs with character.

'In eulogy we say' (to use a stock phrase of the dynastic histories) that Wang Chhung was one of the greatest figures of his age from the point of view of the history of scientific thought.[e] Hu Shih has drawn attention to Wang Chhung's words (in chapters 61 and 84):[f]

One sentence is enough to sum up my book—it hates falsehood (*chi hsü wang*[7]). Right is made to appear wrong, and falsehood is regarded as truth. How can I remain silent? When

---

[a] His idea of it has some similarity with those involved in our mathematically expressible 'laws of chance'; he often uses the word *shu* (numbers) in this connection, e.g. *shih ou chih shu*[8] at the beginning of ch. 10. But he did not think in 'statistical' terms; what he visualised was the operation of a vast unseen loom weaving automatically a pattern determined from the beginning, and chance was part of its mechanism.

[b] Ch. 6 (tr. vol. 1, p. 142); ch. 10 (tr. vol. 2, pp. 1 ff.).

[c] Ch. 6 (tr. vol. 1, p. 156) and ch. 20 (tr. vol. 1, pp. 156ff.).

[d] Ch. 11 (tr. vol. 1, pp. 304 ff.); ch. 50 (tr. vol. 1, pp. 359 ff.).

[e] It is curious that the *Lun Hêng*, unlike most Han and pre-Han books, was never extensively commented upon. Perhaps Wang Chhung was too extreme for the general run of temporising scholars, in which again he parallels Lucretius. Sir Thomas Browne in the 17th century, writing to his son, advised him not to pay too much attention to Lucretius because ''tis no credit to be punctually versed in him'. On the other hand, Wang Chhung's style is more explanatory and repetitive than that of any other ancient Chinese writer.   [f] Tr. vol. 2, p. 280; vol. 1, p. 89.

[1] 時   [2] 遭   [3] 幸   [4] 偶   [5] 正   [6] 隨   [7] 疾虛妄

[8] 適偶之數

I read current books of this kind, when I see the truth overshadowed by falsehood, my heart beats violently and the pen trembles in my hand. How can I be silent? When I criticise them, I study them, check them against facts, and show up their falsehood by appealing to evidence.[a]

But unfortunately the main value of Wang Chhung's work was negative and destructive. If only he could have devised some hypotheses more fruitful for science and technology than the Yin and Yang dualism and the five elements, his services to Chinese thought would have been greater still.

We do not hear of a group of immediate disciples, but we know that Tshai Yung[1] (+133 to +192)[b] and Wang Lang[2] (d. +228)[c] both greatly prized Wang Chhung's book. Probably it also affected Hsün Yüeh[3] (+148 to +209),[d] who vigorously opposed Taoist superstitions. In this connection it is significant to find that the *Jen Wu Chih*[4] (Study of Human Abilities), written by Liu Shao[5] about +235, and the most important book on the psychology of character in old Chinese literature,[e] has nothing whatever to say about physiognomy. It is based entirely on a rationalistic observation of psychological traits and their effects in human affairs. Then in the Chin, there was Phei Wei[6] (+267 to +300) who continued the tradition of the Han sceptics. His works remain to us, however, only in the form of fragments collected by Yen Kho-Chün.[f]

(h) THE SCEPTICAL TRADITION IN LATER CENTURIES

Throughout subsequent Chinese history the sceptical rationalist tradition runs on. Indeed, it stands out as one of the great achievements of the culture, when one compares it with the rabble of religious and magical writings dominant in some other civilisations. The ridicule of spirits became a commonplace of Confucianism. A few centuries after the death of Wang Chhung the growing power of Buddhism brought great reinforcements to the side of superstition, but Confucians were never lacking to oppose it—thus the histories record for +484 a great debate before the Prince of Ching-Ling in which Fan Chen[7] attacked the doctrine of *karma* (the explanation of the good or evil fortune of this life in terms of good or evil actions performed in previous incarnations).[g] Wang Chhung had already challenged the belief in immortality,[h] Fan

---

[a] Tr. Hu Shih (3), p. 46.　　　　[b] G 1986.　　　　[c] G 2195.

[d] G 811; Forke (12), p. 135; biography by Busch (1).

[e] Forke (12), p. 196; tr. Shyrock (2).

[f] *CSHK*, Chin section, ch. 33, p. 3a. Forke (12), p. 226.

[g] *Liang Shu*, ch. 48, p. 7a. Hou Wai-Lu & Chi Hsüan-Ping (1) have recently given a general account of Fan Chen's philosophical materialism.

[h] Forke (8) has assembled his arguments on this subject and contrasted them with those of Plato, much to the latter's disadvantage.

[1] 蔡邕　　　[2] 王郎　　　[3] 荀悅　　　[4] 人物志　　　[5] 劉邵　　　[6] 裴頠

[7] 范縝

Chen now said: 'The spirit is to the body what the sharpness is to the knife. We have never heard that after the knife has been destroyed the sharpness can persist (*Shen chih yü hsing, yu li chih yü tao. Wei wên tao mei erh li tshun*[1])'.[a] Fan's views were embodied in an essay entitled *Shen Mieh Lun*[2] (On the Destructibility of the Soul)[b] which so alarmed the Buddhists that more than seventy refutations of it were written;[c] Hu Shih (4) says that the best was that of Shen Yo,[3] who argued that the knife can be recast into a dagger, which may be of quite different shape, but in which the sharpness will be the same, i.e. 'reincarnated'. It does not seem very convincing to us.

Other records for the years around +631 describe many encounters in which the Confucian scholar Fu Yi[4] worsted Buddhist thaumaturgists.[d] His contemporary, Lü Tshai,[5] a figure reminiscent in many ways of Wang Chhung, was ordered by the emperor in +632 to edit the existing books of divination and of Yin-Yang and five-element theory, which he did, adding sceptical prefaces to each section.[e] His works seem to be lost, but some extracts preserved in the *Pien Huo Pien*[f] (see p. 389) show that he used similar arguments to those of Wang Chhung, for example, that catastrophes in which hundreds of people perish at one time make nonsense of the doctrine of individual destinies. Stories directed against superstition of all kinds now become more and more frequent. The *Pien I Chih*[6] (Notes and Queries on Doubtful Matters) written by Lu Chhang-Yuan[7] of the Thang, for instance, has a story of a temple where, it was believed, the body of a Taoist nun remained for centuries without decomposition. But at last a group of wild young men burst in after a feast, and found nothing in the coffin but mouldering bones.[g] This event is dated about +770. A little later Liu Tsung-Yuan the poet argued strongly against phenomenalism.[h] Then in +819 came the celebrated incident of the protest made by Han Yü[8] against the official reception by the emperor of a relic of the Buddha.[i]

In the Sung, Chhu Yung[9] gave, in his *Chhü I Shuo Tsuan*[10] (Discussions on the Dispersion of Doubts), a book not otherwise remarkable for its scepticism,[j] a subjectivist interpretation, along the lines of auto-suggestion, of the effectiveness of Taoist charms and talismans. Hu An-Kuo,[11] in one of his historical commentaries,[k] enunciates

---

[a] *Thung Chien Kang Mu*, ch. 28, p. 1b; tr. Wieger (1), p. 1155; eng. auct.

[b] In *TSCC, Jen shih tien*, ch. 23; translation by Balazs (3).

[c] But there were many who agreed with Fan Chen, e.g. the Juan Hsüan-Tzu[12] mentioned in *Shih Shuo Hsin Yü*, ch. 2A, p. 11a.

[d] Wright (3); *TH*, p. 1344. [e] *TH*, p. 1345.

[f] Ch. 3, p. 3b. [g] Ch. 34, p. 20a.

[h] *Yü Chien*, ch. 4, p. 1b.

[i] Full details and translation of the memorial in Dubs (16).

[j] But which has an importance in the history of magnetism; see Sect. 26i.

[k] See Forke (9), p. 121. The passage is translated by Wieger, *TH*, p. 1430, who pours sarcasm on it most unjustifiably.

---

[1] 神之於形猶利之於刀未聞刀沒而利存　　　[2] 神滅論　　　[3] 沈約
[4] 傅奕　　　[5] 呂才　　　[6] 辯疑志　　　[7] 陸長源　　　[8] 韓愈　　　[9] 儲泳
[10] 袪疑說纂　　　[11] 胡安國　　　[12] 阮宣子

the doctrine that while in the long run Heaven rewards the good and punishes the evil, this is a statistical process, and not valid for individual persons and events. Another passage (in a Ming commentary)[a] shows how the ancient Chhan-Wei ideas were rationalised by considering omens not as signs of inevitable evil, but as a kind of celestial semaphore indicating what actions men have already done, and hence what they are likely to continue to do. In this way the admonitory character of 'heavenly reprimands' was weakened—'If a tree is planted, Heaven helps it to grow; if it is cut down Heaven helps it to rot.'[b] A staunch sceptic of this time was Shih Chieh[1] (+1005 to +1045).[c] In his *Shih-Tshu Lai Chi*[2] he wrote:

I believe that there are three illusory things in this world, immortals (*hsien*[3]), the alchemical art, and Buddheity. These three things lead all men astray, and many would willingly give up their lives to obtain them. But I believe that there exists nothing of the sort, and I have good grounds for saying so. If there were any one man in the world who had obtained them, no one would be more honoured than he. Then no one would strive without accomplishment or pray without response....Chhin Shih Huang Ti wished to become immortal, Han Wu Ti wished to make gold, and Liang Wu Ti wished to become a Buddha, and they spent themselves in these aims. But Chhin Shih Huang Ti died on a far journey, Liang Wu Ti starved (himself) to death, and Han Wu Ti never obtained any gold....[d]

But the whole of the Sung Neo-Confucian school had an intensely naturalistic and sceptical tendency, as we shall find in the special Section devoted to them.[e]

The Yuan saw the appearance of a distinctly scientific mind, Liu Chi[4] (+1311 to +1375),[f] who sought to explain natural phenomena by purely natural causes. Taking up Wang Chhung's arguments about lightning, he strenuously combated the idea that death by lightning is anything but mere chance. In his book, the *Yu Li Tzu*,[5] he compares death to the pouring back of a cup of water into the sea; the *chhi* returns to the universal mass of *chhi*. There is also an interesting conversation between two imaginary characters, Chhu Nan-Kung and Hsiaoliao Tzu-Yün. Thus it runs:

Chhu Nan-Kung asked Hsiaoliao Tzu-Yün, 'If Heaven has a boundary, what things could be outside it? Yet Heaven must have a boundary, for all things which have form (*hsing*[6]) must have boundaries (*chi*[7]); according to all general principles (*li*[8]) and influences (*shih*[9]) which we know.'
Hsiaoliao Tzu-Yün replied, 'About those things which are outside the six cardinal points the sages did not speak.'

[a] Tr. Wieger, *TH*, p. 1695. [b] *Chung Yung*, XVII, 3 (Legge (2), p. 263).
[c] Forke (9), p. 8. [d] Ch. 2, p. 48*b*, tr. Forke (9), eng. auct.
[e] Traditional Confucian scepticism may be traced in a thousand examples of popular literature. I will only cite as one reference the tale no. 17 in Eberhard (5), p. 75.
[f] Forke (9), p. 306; G 1282; Sarton (1), vol. 3, p. 1536. This is a good instance of the deficiencies of Giles' *Biographical Dictionary*. No one would ever suspect from his entry there that Liu Chi was of high scientific attainments, an astronomer and mathematician, though of course also an astrologer (like Kepler later).

[1] 石介   [2] 石祖徠集   [3] 仙   [4] 劉基   [5] 郁離子   [6] 形
[7] 極   [8] 理   [9] 勢

Chhu Nan-Kung laughed and said, 'As the sages did not know anything about them, of course they could not speak of them. But the sages followed the motions of the heavens with the help of astronomy and calendrical science. They examined the constellations by the use of instruments. They checked the quantitative changes of the heavens using calculations. Heaven's principles they elucidated by the assistance of the *I Ching*. Everything which the ear can hear, the eye can see, or the mind can think, the sages investigated, leaving not the minutest matter in darkness—except what Heaven obstinately conceals, and for that man has no methods whereby he can reveal it. That is the point. If you had said, "They did not know" instead of, "They did not speak about it", you would have been quite right.'[a]

Liu Chi's writings and life should be more closely studied. We shall meet him again.[b]

One of his contemporaries, Hsieh Ying-Fang[1] (fl. +1340 to +1360),[c] made a whole collection of anti-superstitious material in his *Pien Huo Pien*[2] (Disputations on Doubtful Matters). In this will be found quotations relating to attacks by Confucian scholars on praying for longevity,[d] burning paper money for the dead,[e] the Buddhist ideas of immortality and hell,[f] metempsychosis or reincarnation,[g] divination from the *I Ching*,[h] astrology,[i] lucky and unlucky days,[j] geomancy,[k] physiognomy,[l] and the like. Famous passages from Hsün Chhing (see above, p. 27) and on Hsimên Pao (p. 137) are quoted, and texts in which the abolition of Buddhist temples and images was recommended by Confucian scholars,[m] who claimed that Heaven had not favoured those dynasties which supported Buddhism.[n]

In the early Ming the most outstanding sceptic was perhaps Tshao Tuan[3] (+1376 to +1434),[o] who was a great admirer of the *Pien Huo Pien*. His *Yeh Hsing Chu*[4] (Candle in the Night) was written for his own father, who was an adherent of Buddhism. In the late Ming the greatest name was that of Wang Chhuan-Shan[5] (+1619 to +1692),[p] whose acquaintance we shall make in Section 17*c* below. By his time the two sides in the perennial controversy had become quite clearly defined.[q] Confucian scepticism and Taoist empirical pseudo-science had become well-worn ground, and such was the position when the 'new, or experimental, philosophy' reached China with the Jesuits in the early years of the 17th century.

[a] *Yu Li Tzu*, p. 4*a*, tr. Forke (9), eng. auct. Cf. pp. 8, 198 above.
[b] We have already met him (p. 360) as the author of a book on geomancy. This was evidently his favourite among the pseudo-sciences, just as Wang Chhung accepted astrology and physiognomy.
[c] G 746. [d] Ch. 1, p. 2*a*.
[e] Ch. 1, p. 13*a*, *b*. [f] Ch. 2, p. 8*a*.
[g] Ch. 1, p. 16*b*. [h] Ch. 2, pp. 4, 5.
[i] Ch. 3, p. 5*b*. [j] Ch. 3, p. 7*a*.
[k] Ch. 2, p. 10*a*. [l] Ch. 3, p. 12*a*.
[m] Ch. 1, pp. 13*a*, 18*a*. [n] Ch. 2, p. 7*b*.
[o] Forke (9), p. 347; G 2015.
[p] Forke (9), p. 484.
[q] In the final phase there was nevertheless a distinct tendency to 'see both sides of the question'. It is very significant that the *Thu Shu Chi Chhêng* encyclopaedia of +1726 is careful to include Hsün Chhing's chapter against physiognomy, and to print, side by side with an immense mass of detailed material on *fêng-shui*, lucky and unlucky days, fate-calculations, etc., the relevant diatribes of Wang Chhung against them. Nor was Mo Ti's chapter against fatalism forgotten.

[1] 謝應芳 [2] 辯惑編 [3] 曹端 [4] 夜行燭 [5] 王船山

(*i*) CHINESE HUMANISTIC STUDIES AS THE CROWNING ACHIEVEMENT OF THE SCEPTICAL TRADITION

As a kind of appendix to this section I propose to add something on the development of humanistic studies, textual criticism, and archaeology in China. These were fields of scholarly activity in which the sceptical tradition could find full outlet, with fruitful results, since the material for empirical interest was ready to hand, and the progress of research not inhibited by the failure to mathematise scientific hypotheses and to test them by experiment. Thus it came about that China was the very home of the humanistic sciences, the 'Geisteswissenschaften', maintained at a higher level over a longer continuous period than in any other civilisation.

A glance at Sandys' well-known history of classical scholarship (in Europe)[a] suffices to remind one that although the critical study and dating of ancient texts rose to a high level among the Alexandrians, as, for instance, Aristophanes of Byzantium (*c.* −195) and Aristarchus of Samothrace (*c.* −150)—contemporaries of the early Han scholars—little trace of it survived into Byzantine or medieval civilisation, where manuscripts were simply copied and collected.[b] Even the earlier phases of the Revival of Learning, from the beginning of the 14th century, did not rediscover the methods of scientific textual criticism. The Italian humanists, says Sandys, were concerned mainly with imitation and reproduction of classical models, while the French 'poly-histors' of the 16th century were mainly marked by vast and industrious erudition. Not until the age of Bentley (1662 to 1742) and his successors did modern scientific textual criticism arise; and perhaps it was no coincidence that this occurred in step with the triumphs of the 'new, or experimental, philosophy' in which the methods of modern natural science were for the first time applied.

At an earlier stage[c] we had occasion to note the striking contrast drawn by Hu Shih (1) between the use of the scientific method in the 17th century in Europe and China. While Galileo, Harvey and Newton were applying it to natural phenomena in Europe, Ku Yen-Wu and Yen Jo-Chü were applying it to philological studies in China. But unlike the European natural scientists, who were discovering the world anew with a new method, the Chinese philologists were continuing a tradition of literary scholarship already very old. It may be said perhaps to have begun in the Han dynasty, when scholars applied some of that scepticism, which Wang Chhung represented, to the examination of ancient texts. Such studies were encouraged for

---

[a] I have not found any Chinese work which could be considered the counterpart of that of Sandys. The opinions of eminent scholars of the various dynasties on the dates and authenticities of ancient works have, of course, been collected together (see below). Textual emendations also naturally form part of the extremely elaborate commentaries with which it has been customary for centuries past to furnish Chinese books.

[b] The course of this field of study, it is interesting to note, closely resembles that of geography and quantitative cartography (see below, Sect. 22*d*), where also there was an almost complete loss of the Greek tradition, reducing the level of Europeans in such questions far below that of their Chinese contemporaries.

[c] In Section 6*j* (Vol. 1, p. 146 above).

reasons of State, and elsewhere (in the Introduction)[a] mention was made of the two famous assemblies, in which scholars discussed the reliability and meaning of the versions of the classical books—the Shih Chhü Conference of −51, and the Pai Hu Kuan Conference of +79. The *Hsi Ching Tsa Chi* (Miscellaneous Records of the Western Capital)[b] preserves for us a discussion between Han scholars as to the date of the ancient dictionary *Erh Ya*,[1] some of them expressing the gravest doubts that it could really go back, as supposed,[c] to the period of Chou Kung (−11th century).[d]

But the real flowering time of Chinese critical humanism came during the Sung dynasty (+10th to +13th centuries), contemporaneous, significantly enough, with a peak of activity in all branches of the natural sciences and the technologies (see pp. 493 ff. below), and the rise of that great philosophical achievement of the scientific view of the world, Neo-Confucianism (see Section 16).[e] The humanistic and the philosophical movements started almost simultaneously at the end of the +10th century—a time when Europe had nothing to show even remotely comparable.

Perhaps one of the most important triggers which set the movement going was dissatisfaction with the Chhan-Wei apocryphal books (see pp. 380 ff. above), large portions of which had become embedded in the commentaries currently accepted. Sun Fu[2] (+992 to +1057) launched an attack on these, but once the examination of books with the new critical methods got under way, it was not long before the ancient texts themselves, as well as the commentaries on them, came under fire. Sun Fu found many discrepancies in the *Tso Chʻan* (Commentary on the Spring and Autumn Annals), and similar doubts were extended by the famous Sung Confucian Ouyang Hsiu[3] (+1007 to +1072) to the classic Mao commentary on the *Shih Ching* (Book of Odes). Very soon the text of the ancient folk-songs themselves was sceptically examined. By the middle of the +11th century it was widely (and quite rightly) doubted whether any of the *I Ching* (Book of Changes) came from the hand of Confucius, and (again with justice, though this did not go far enough) the *Chou Li* (Record of the Rites of the Chou dynasty) brought down to the Warring States period.[f] Chêng Chhiao[4] (+1104 to +1162), the great author of the *Thung Chih* history and a contemporary of Chu Hsi (China's Aquinas), denied the high antiquity of the *I Ching*, the *Erh Ya*, and other classics, and in his *Shih Ku Wên*[5] (The Stone Drum Inscriptions) maintained that these writings, traditionally ascribed to the beginning of the Chou dynasty (−11th century), dated only from the second half of the −3rd. Wu Yü[6] (d. +1155) attacked the *ku wên*[7] text of the *Shu Ching* (Historical Classic) as falsified,

---

[a] Vol. 1, p. 105 above.   [b] Ch. 3, p. 2b.
[c] It is now considered to have been put together in the Chhin and early Han periods, i.e. −3rd and −2nd centuries. Cf. Sarton (1), vol. 1, p. 110.
[d] On the general aspects of the critical scholarship of the Han period and its immediate successors see Nagasawa (Feifel) (1), pp. 116 ff.
[e] The connection has also been noted by Chhen Jung-Chieh (1), p. 261.
[f] The common tradition had been that it went back to the beginning of the first millennium before our era.

[1] 爾雅   [2] 孫復   [3] 歐陽修   [4] 鄭樵   [5] 石鼓文   [6] 吳棫
[7] 古文

and Chu Hsi must have agreed with him since he wrote no commentary on it. In the next generation the process was extended to Taoist as well as Confucian books— Yeh Shih[1] (+1150 to +1223) was almost able to demonstrate (what we now know to be true) that the *Kuan Tzu* book had nothing whatever to do with the Chou statesman Kuan I-Wu, and threw grave doubts on the supposed connection of Tzu-Ssu, Confucius' grandson, with the *Chung Yung* (Doctrine of the Mean).[a] In the Yuan dynasty, the movement continued with little loss of impetus. Sung Lien[2] (+1310 to +1381) published in +1358 his *Chu Tzu Pien*[3] (Discussions on the Writings of the Ancient Philosophers), in which he weighed up the authenticity of more than fifty philosophical books. He was not by any means the first to write such a work, having had at least ten predecessors, but his summary was considered one of the best.

From critical emendation it was an easy step to far-reaching rewriting and re-organisation of the classics, and the Sung scholars are now thought to have carried this somewhat too far. The most famous instance is Chu Hsi's rearrangement of the *Ta Hsüeh* (Great Learning), but there are many other examples, e.g. Liu Chhang's[4] (fl. +1060) versions of the *Shu Ching* and the *Chhun Chhiu*, and Wang Po's[5] (+1197 to +1274) changes in the *Shih Ching*. In the Yuan, Wu Chhêng[6] (+1249 to +1333) revised the *Li Chi* (Book of Rites).

It would be easy to multiply examples, but for the purpose of this book sufficient has been said. Naturally there were compendia in which the best opinions on the dates and authenticities of ancient books were collected together,[b] and one of the most famous of these was the *Chün-Chai Tu Shu Chih*[7] of Chhao Kung-Wu[8] which appeared about +1175. This was greatly relied upon by the compilers of the Imperial Manuscript Library catalogue, the *Ssu Khu Chhüan Shu Tsung Mu Thi Yao* (see Sect. 3b) of +1782.

It is interesting that after the fall of the Mongols and throughout the Ming dynasty, the movement was largely in abeyance.[c] Perhaps the nationalism arising from the power of an indigenous dynasty militated against criticism of the semi-sacred texts of the nation; in any case the creative literature of novels and drama was more in the foreground. But when two centuries had elapsed, at the end of the 16th century, and just before the rise of the Chhing, sceptical and critical historical analysis began again with intensified force. Although this coincided with the coming of the Jesuits to China, and also with the great rise of science, both natural and humanistic, in Europe, there seems no reason for thinking that it was in any way influenced by those events.[d]

---

[a] This, however, even today, can hardly be said to be disproved; cf. Hughes (2), Wu Shih-Chhang (1).

[b] Modern opinions on this important subject will be found in the series of volumes in *KSP*, and in a book of Liang Chhi-Chhao's (3).   [c] For certain exceptions see Hummel (5).

[d] Such influence is sometimes suggested, as by Creel (6), p. 219. But the founder of the School, Chhen[9] Ti the phonologist (+1540 to +1617), was at the height of his powers two or three decades before there were any Jesuits in China. Moreover, there seems to be no concrete evidence for Jesuit stimulation of its greatest figures such as Ku Yen-Wu.

[1] 葉適   [2] 宋濂   [3] 諸子辨   [4] 劉敞   [5] 王柏   [6] 吳澄

[7] 郡齋讀書志   [8] 晁公武   [9] 陳第

The new school of textual criticism was known as that of *Khao Chêng Hsüeh*,[1] and its chief representative was Ku Yen-Wu[2] (+1613 to +1682), one of the two men referred to in the passage from Hu Shih which has already been quoted (vol. 1, p. 146). Its achievements were not at all inferior to the criticism of the age of Bentley in Europe, and it started about a hundred years earlier.[a]

As an example of its work, one may see what Yen Jo-Chü[3] (+1636 to +1704) did to the *Shu Ching*. He was able to prove that many chapters of the *Shang Shu*, together with the alleged −2nd-century commentary of Khung An-Kuo[4] upon them, were forgeries of the Eastern Chin dynasty (+4th century). Thereby he gave, as Nagasawa puts it, a fatal wound to a venerable classic which had been considered as sacred by successive generations of Confucian scholars for more than a thousand years, and upon which innumerable commentaries had been written. The destructive criticism of Hu Wei,[5] who showed that the Ho Thu[6] and the Lo Shu[7] (+1633 to +1714) (ancient mathematical magic squares) had nothing to do with the *I Ching*, but had been attached to it by the Wu Tai Taoist, Chhen Thuan,[8] (+906 to +989) comes in the same category. An important book was that of Yao Chi-Hêng,[9] (+1647 to +1715?) the *Ku Chin Wei Shu Khao*[10] (Investigation into Forged Books, New and Old). And there were the brothers Wan Ssu-Ta[11] (+1633 to +1683) and Wan Ssu-Thung[12] (+1643 to +1702), whose opinions about the dates of the *Chou Li* and the *Li Chi* were not far from those now accepted.[b] Afterwards these studies, ever growing, separated into various schools, such as the Wu[13] group, headed by Hui Tung[14] (+1697 to +1758), and the Huan[15] group, founded by Tai Chen[16] (+1723 to +1777), whom we shall meet with again in connection with the resurgence of materialist philosophy in the Chinese 18th century. In this way, the movement started at the beginning of the 17th century was carried forward through such men as Tshiu Shu[17] (+1740 to +1816) into the humanistic studies of the present day. This in itself is not surprising; the point of interest is that a continuity of sceptical and critical philology can be traced back, through a period of brilliance in the Sung (when Europe was sunk in uncritical traditionalism), to the first beginnings of criticism in the Han (corresponding roughly to the achievements of the Alexandrians in Europe).

Was it not entirely natural that this tradition should be accompanied by the rapid growth of archaeology? Indeed, the Chinese were, as Sarton has said, 'born archaeologists'. Of the earliest archaeological efforts one may read in the monograph of Wei Chü-Hsien,[c] but like all the other sciences mentioned, it took a leap forward

---

[a] Cf. Hu Shih (5). In its later phases it blended with the School of Han Learning (*Han hsüeh*[18]), which opposed or reinterpreted Neo-Confucianism.

[b] I.e. that they are, in the main, works of the Han, not of the Chou.

[c] *Chung-Kuo Khao-Ku-Hsüeh Shih*, which would be well worth translating into a Western language

[1] 考證學	[2] 顧炎武	[3] 閻若璩	[4] 孔安國	[5] 胡渭
[6] 河圖	[7] 洛書	[8] 陳摶	[9] 姚際恆	[10] 古今僞書考
[11] 萬斯大	[12] 萬斯同	[13] 吳	[14] 惠棟	[15] 皖	[16] 戴震
[17] 崔述	[18] 漢學			

to a thoroughly scientific level in the Sung dynasty; this is described in a valuable paper by Wang Kuo-Wei (1).

In the middle of the +11th century Ouyang Hsiu wrote his *Chi Ku Lu*[1] (Collection of Ancient Inscriptions), perhaps the earliest work on epigraphy in any language. The interest was practical as well as theoretical, for an early +12th-century book[a] records that Ssuma Chhih,[2] when Governor of Fênghsiang, took particular care to preserve the stone drums mentioned above by building a hall over them to protect them from the weather. Excavations were also undertaken.[b] In +1134 Têng Ming-Shih[3] compiled a treatise on the origins of the clan and family names.[c] In +1149 Hung Tsun[4] published the *Chhüan Chih*[5] (Treatise on Coinage), which Sarton considers the first independent work on numismatics in any language.[d] In the opening year of the century, when the emperor Hui Tsung had come to the throne, he had embarked upon the formation of an archaeological museum; the catalogue of this was issued by Wang Fu[6] in +1111 as the *Po Ku Thu Lu*[7] (Illustrated Record of Ancient Objects). An earlier work of the same kind had been the *Khao Ku Thu*[8] of Lu Ta-Lin[9] published in +1092.[e] The archaeological passion in these works far outshone anything which Europe could show until a much later time. 'In these books', says Li Chi, 'a system was created for recording and reproducing antiquities, which, except for minor details due to improvements in modern printing, has been taken as a model for all treatises on antiques until the present day. It may not be possible to test the accuracy of their measurements or reproductions, but their desire to be accurate is more than obvious, and the ingenuity and correctness of most of their identifications have been confirmed by modern criticism.' About the same time the epigraphic work commenced by Ouyang Hsiu was continued in the great collection of Han inscriptions made by Hung Kua,[10] the *Li Shih*,[11] which appeared from +1167 to +1181.

During the Yuan dynasty, the archaeological movement continued parallel with that of the sceptical humanists, and about +1307 Wuchhiu Yen[12] produced the first

---

[a] The *Ching-Khang Hsiang Su Tsa Chi*[13] of Huang Chao-Ying,[14] ch. 6, p. 3*b*.

[b] Cf. Laufer (8), p. 21.

[c] No similar treatise was produced in Europe till much later, but Sarton (1), vol. 2, p. 140, records two in Arabic and Japanese literature respectively, both somewhat antecedent to Têng. His book was entitled *Ku Chin Hsing Shih Shu Pien Chêng*.[15]

[d] (1), vol. 2, pp. 140, 262. He says 'independent', since accounts of currencies were included as a matter of course in all Chinese dynastic histories from the time of the Han onwards. For European numismatics we have to await the late +16th century.

[e] The famous work on jade, the *Ku Yü Thu Phu*[16] (Illustrated Record of Ancient Jades), apparently prefaced by Lung Ta-Yuan[17] in +1176, and so much used by Laufer (8), has been shown to be a forgery of the 18th century (cf. Hansford, 1). Another spurious collection is the *Li Tai Ming Tzhu Thu Phu*[18] (Famous Ceramic Pieces of all Ages), allegedly Ming but really 18th century, and perhaps from the same hands (Pelliot, 37). But these very fabrications bear witness to the existence of an archaeological 'public', and wide interest among educated non-specialists.

---

[1] 集古錄　　　[2] 司馬池　　　[3] 鄧名世　　　[4] 洪遵　　　[5] 泉志

[6] 王黼　　　[7] 博古圖錄　　　[8] 考古圖　　　[9] 呂大臨　　　[10] 洪适

[11] 隸釋　　　[12] 吾邱衍　　　[13] 靖康緗素雜記　　　[14] 黃朝英

[15] 古今姓氏書辨證　　　[16] 古玉圖譜　　　[17] 龍大淵　　　[18] 歷代名磁圖譜

treatise on sphragistics (seal inscriptions) in any language, the *Hsüeh Ku Pien*[1] (On our Knowledge of Ancient Objects). The archaeology of jade was studied further by Chu Tê-Jun[2] in his *Ku Yü Thu*[3] of +1341. During the Ming, the two traditions declined together, but as soon as the 17th century opened, archaeology arose once more to continue as one of the glories of Chinese scholarship. They have never flourished better than at the present time.

Hu Shih (6) tells us that in looking over some of his father's unpublished writings some years ago, he found a volume of notes made when he had been a student at the Lung-Mên Academy at Shanghai about 1875. These were written on notebooks printed by the college for the use of its students. On the top of every page was printed a motto reading, in part: 'The student must first learn to approach the subject in a spirit of doubt....The philosopher Chang Tsai used to say, "If you can doubt at points where other people feel no impulse to doubt, then you are making progress."' Such indeed has been the spirit of those Chinese thinkers who have kept the torch of intellectual freedom burning throughout the ages.

The object of these few notes has been to show that the tradition of Chinese sceptical rationalism was not merely empty and theoretical, nor was it just unconventional and destructive. Interest in the world of non-human Nature the Confucians did not share with the Taoists, admittedly, but within the domain of human life and thought a field was open for the application of the scientific method, with all its rigour, in so far as it can ever be applied where experimentation is impossible. Encouraged by the world-outlook of bureaucratic literature-dominated society, the scholars threw all their energies into history, philology and archaeology. What came out of it was not that terrifying strength which has transformed the world of space and time, but rather a vast edifice of knowledge of the past of a people, an edifice to which its European counterpart has only been for the past two centuries even comparable.

On 9 July 1704, Yen Jo-Chü, one of the greatest builders of this edifice, lay dying in Peking, of a disease which a few milligrams of one of the drugs which modern natural science would later discover, could have cured. The picture exemplifies both the nobility and weakness of medieval Chinese humanism.

[1] 學古編　　[2] 朱德潤　　[3] 古玉圖

15. BUDDHIST THOUGHT

IF MACAULAY'S NEW ZEALANDER, or (had Amerindian civilisation been allowed to attain full development) some future Aztec historian, were proposing to describe the rise and flowering of the sciences and technologies of Europe, he would certainly have to devote a chapter to the effects which specifically Christian ideas might have had in bringing about this development. It is interesting to speculate on what he would say. He might refer to the importance of the concept of a personal creator deity, or to the reality of the time-process (since the incarnation occurred at a definite point in time), or to ideas of a 'democratic' character which lent value to the soul of (and hence perhaps to the observations of Nature made by) every individual human being. Some of these points will probably call for comment in the concluding sections of this book. Here we are faced with a similar problem, namely, that of deciding what effects Buddhism[a] had, when introduced into China, upon the development of scientific thought. Since, so far as I have been able to see, these effects were very largely inhibitory, it might be possible to dismiss the question rather shortly were it not for the fact that inhibitory processes concern us as much as adjuvant ones. After all, one of the most interesting parts of the subject to which this book is devoted is the question why modern science and technology did not spontaneously develop in East Asia.

(a) GENERAL CHARACTERISTICS

The study of Buddhism is apt to be unsatisfying to natural scientist and sinologist alike.[b] There seems to be some lack of agreement as to what the primitive doctrine really was, and the dating of the most important texts seems to be so vague as to cause much uncertainty concerning the general history of the ideas. There was often no clearly defined orthodoxy, and there have been many varying and often incompatible, but apparently equally weighty, opinions as to how Mahāyāna Buddhism (the form prevailing in China) differed from the earlier Hīnayāna Buddhism.[c] 'It is

---

[a] The general term for Buddhists in medieval Chinese texts is Shih Chia,[1] after the first syllable of the transliteration of the clan name of Gautama Buddha, Sākyamuni, i.e. *Shih-ka-mou-ni*.[2] The first of these syllables served in China as the 'family name' of all monks, hence the two characters of their names are hyphenated throughout this book.

[b] I may mention here the guides which I have found most useful: the books of Oldenberg (1), Rosenberg (1), D. T. Suzuki (1), Keith (1), Rhys Davids (1, 2), Takakusu (1, 2) and, above all, E. J. Thomas (1). Wieger (2) has given a general survey of modern Chinese Buddhist doctrine, according to his usual style, and elsewhere (9) has translated many excerpts from the sūtras. The excellent survey of Conze (1) and the anthology of Conze *et al.* reached us only after this Section was written.

[c] In this Section we depart from our usual practice of invariably giving romanisations in the text when Chinese characters are supplied in footnotes. Many Chinese Buddhist terms are simply transliterations from the Sanskrit, and the syllables have phonetic values not quite the same as in normal

[1] 釋家   [2] 釋迦牟尼

still disputed', writes Thomas,[a] 'whether original Buddhism was "nothing but vulgar magic and thaumaturgy coupled with hypnotic practices", or whether Buddha was a "follower of some philosophic system in the genre of Patañjali's",[b] to take two extreme views.' All that is certain is that 'the philosophic system came to exist, with theories of the nature of the individual, his career according to a law of causation, and the doctrine of his final destiny; and then with the Mahāyāna movement a transformation of all these problems through a new theory of reality and a conception of the Enlightened One which made him indistinguishable from the highest conceptions of Hindu deity.'

At the outset it may be desirable to give, in the form of a brief table, a summary of the principal dates involved (Table 19). One of the most striking facts about the chronology is that the earliest written traditions which have come down to us do not date back before the +4th century, when the *Dīpavaṃsa* (History of the Island of Ceylon) was written, to be followed a century later by the *Mahāvaṃsa* (The Great Chronicle).[c] There is thus nothing in any way equivalent to the *Spring and Autumn Annals*, or even to the *Shih Chi*. We are far more certainly informed about the life and times of Confucius than we are about the beginnings of Buddhism, though these include a period considerably later. However, the Canon was growing up from the −1st century onwards, particularly the books known as the *Abhidharma* (*tui fa*[1]) or discussions of various philosophical aspects of the faith.[d]

A fact of much importance is that the Buddhists split into sects long before there were any written records at all. On the one hand there were the Sthaviravādins[2] (or Theravādins, lit. Elders, *shang tso pu*[2]), and on the other the Sarvāstivādins (*i chhieh yu pu*[3]). The latter derived their name from their metaphysical realism, and yet it was they who ultimately gave rise to the Mahāyāna sects, with all the idealist philosophy associated with them.[d]

All sects and schools, however, were united on certain fundamentals. The theory of *karma* was pre-Buddhist, for a transmigration or metempsychosis of the soul, which would experience happiness or misery in successive rebirths, is to be found in the

use; these we give simply in the form of the characters. For example, *Budh* was transliterated by the character now pronounced *Fo*,[4] for in the +3rd century it had a terminal consonant and reproduced the foreign sound fairly well. But when the idea was fully translated into Chinese we give, as usual, romanisations as well as characters. We start from this point onwards, although it should of course be understood that the Chinese transliterations or translations were in many cases adopted centuries later than the beginnings of which we are now speaking. Reliance has been placed on the dictionary of Soothill & Hodous.

[a] (1), p. 57.
[b] The +5th-century yoga master of course, not the −2nd-century grammarian of the same name.
[c] Thomas (1), p. 7. There are also the *Avadānas* (parables, Thomas (1), p. 279), but they do not seem to be any earlier.
[d] Thomas (1), p. 158.
[e] Thomas (1), p. 169.

---

[1] 對法　　　[2] 上座部　　　[3] 一切有部　　　[4] 佛

Table 19. *Chronology of the rise of Buddhism*

−563 to −483	Life of the founder of the religion, Gautama Siddhārtha,[1] prince of a small country in northern India, Kapilavasthu. (But some authorities place it a century later.)
−483	First Council at Rājagaha.
−338	Second Council at Vesālī.
−321	Maurya Empire founded by Chandragupta (cf. the unification of China by Chhin Shih Huang Ti some ninety years later).
−269 to −237	Reign of Aśoka (Wu-Yu Wang[2]). This is the earliest time from which any epigraphic evidence bearing on Buddhism exists.
−247	Third Council at Pātaliputra.
−246	Mission of Mahindra to Ceylon.
−2nd cent.	Beginnings of Mahāyāna doctrines, continued under the Kushāna kings in the −1st century.
+65	The first date at which we can place the appearance of Buddhist monks and laymen in China. They formed a community at Phêng-chhêng (modern Hsüchow in Chiangsu province) under the protection of a Han prince, Liu Ying,[3] who was also a patron of Taoism. A letter to him from the emperor mentions them (*Hou Han Shu*, ch. 72, p. 6a). See Maspero (12), p. 204, (13), p. 186, (19, 20). The work of O. Franke (5) and Maspero (5) has shown that the story of the sending out of ambassadors by the emperor Han Ming Ti (+58 to +75), as the result of a dream, and their subsequent return with books, images, and Buddhist monks in person, is nothing but a pious legend fabricated at the beginning of the +3rd century.
+78	Accession of Kaniṣka.[4]
+100	Council of Sarvāstivādins under Kaniṣka.
+2nd cent.	Rise of the dialectical Mādhyamika School of Nāgārjuna.[a]
+148	Arrival of the Parthian Buddhist An Chhing.[5] Among other missionaries of the late +2nd century Chu Shuo-Fo[6] the Indian, and Chih-Chhan[7] the Yüeh-chih, may be remembered.[b] From this time onward, a vast work of translation of texts went on.
+5th cent.	Rise of the idealist Yogācāra School of Vasubandhu[a] and Asaṅga.[a]
+6th cent.	Rise of Dignāga's School of Logic (Chhen-Na[8]).[c]
+7th cent.	Śāntideva, Dharmakīrti, and the rise of the Tantric Schools.

a Chinese equivalents will be given below when the schools are discussed.
b The best exposition of the Indian missions to China is that of Bagchi (2).
c See Tucci (1, 2).

[1] 瞿曇悉達   [2] 無憂王   [3] 劉英   [4] 迦膩色伽   [5] 安清
[6] 竺朔佛   [7] 支讖   [8] 陳那

Upanishads.[a] But Buddhist *karma*[1] (*tso*[2] or *yeh*[3]) differed from this (and here was the ethical insight of the founder) in that the happiness or misery was regarded as being based only on moral or ethical grounds, and not on whether ritual or sacrificial acts had been performed. Good actions were therefore the inescapable cause (*yin*[4]) of happiness, and bad actions of misery (*shan yin lo kuo; o yin khu kuo*[5]), and this would certainly show itself in future existences if not in the present one. The Jains and other ascetic sects in India had always tried to reduce or improve the *karma* of the individual by ascetic practices, often carried to an extreme, but all the legends of Gautama's life agree that these he decisively rejected. His doctrine was embodied in the basic 'Four Noble Truths'; (1) suffering exists; (2) its cause is thirst (*tṛṣṇā*), craving, or desire; (3) there is an overcoming of suffering (*nirodha, nirvāṇa*); (4) by means of the self-training of the 'Noble Eightfold Path', which included all kinds of psychological and mortificatory exercises short of extreme asceticism. These things constituted the *dharma*[6] (*fa*,[7] or 'law').[b]

Buddhist thought perpetually revolved round the notion of retribution, abstracting itself as ethical causality. The essence of the *dharma* was considered to be *pratītya-samutpāda* (*yuan chhi fa*[8]), the chain of causation. Further analysed, and considered apart from any particular chain of rebirths and their fates, it fell into the form of a famous sorites one classical presentation of which is found in the *Lalitavistara Sūtra* of the +1st or +2nd century.[c] This is the cycle of the Twelve *Nidānas* (*shih-erh yin yuan*[9]) (Table 20).

The expression 'cycle' may well be used, for in all Buddhist preaching and iconography there was from early times the tendency to use the symbol of the wheel. From this vicious circle Buddhism aimed to set men free. In general, the Hīnayāna schools emphasised primarily the salvation of the individual, while Mahāyāna emphasised primarily his actions in effecting the salvation of others, but as long as the original impetus lasted in any recognisable form, *liberation* from the world of phenomena was central to it.

The internal part of this scheme showed, it may be said, a certain primitive appreciation of the sensory and motor aspects of the human nervous system, but the beginning and end of it are little more than a series of *non-sequiturs* which demand the eye of religious faith for their acceptance. This was not all that the Buddhists had to

---

[a] Thomas (1), pp. 12, 110.

[b] See Stcherbatsky (4). Its four parts were epitomised in the four words *khu*,[10] *chi*,[11] *mieh*[12] and *tao*[13]

[c] This has several titles in translation: the *Shen Thung Yu Hsi Ching*[14] (Extended Account of the Sports of the Bodhisattva, i.e. the Buddha before his enlightenment); the *Fang Kuang Ta Chuang Yen Ching*;[15] and the *Phu Yao Ching*[16] (N 159, 160). Note the manner in which we refer here and hence-forward to Nanjio's *Catalogue of the Buddhist Tripiṭaka or Canon* (including Ross' *Index* to it). TW 186, 187 indicates the numbers in the more recent catalogue of Takakusu & Watanabe. On the twelve *Nidānas* see Oltramare (1).

---

[1] 羯磨	[2] 作	[3] 業	[4] 因	[5] 善因樂果惡因苦果
[6] 曇摩	[7] 法	[8] 緣起法	[9] 十二因緣	[10] 苦 [11] 集
[12] 滅	[13] 道	[14] 神通遊戲經	[15] 方廣大莊嚴經	[16] 普曜經

say on physiology, however. They analysed the body, mind, and also soul or spirit, if any, into five *skandhas* (bundles, *yün*[1]) or 'faggots' of elements, which were attached together at birth and scattered at death. Four of these were 'immaterial', grouped under *nāma*; these included the *saṃskāra*, *vijñāna*, *sparśa* and *vedanā* of the table below;[a] and one was 'material', the *rūpa* of the table (*sê*[2]).[b] In this were incorporated the four elements[c] (*ssu ta*[3]): earth (*thu*[4] or *ti*[5]) with the nature of solidity (*chien*[6]);

Table 20. *The cycle of the Twelve* Nidānas

IGNORANCE (*avidyā; wu ming* 無 明)	causes the appearance of	the AGGREGATES (*saṃskāra; hsing,* 行) (these are considered to mean manifestations of the will)
the AGGREGATES (*saṃskāra*)	cause the appearance of	CONSCIOUSNESS (*vijñāna; shih,* 識)
CONSCIOUSNESS (*vijñāna*)	causes the appearance of	MIND AND BODY (*nāmarūpa; ming sê,* 名 色)
MIND AND BODY (*nāmarūpa*) (lit. name and form)	cause the appearance of	the SIX SENSE-ORGANS (*ṣaḍāyatana; liu ju,* 六 入)
the SIX SENSE ORGANS (*ṣaḍāyatana*)	cause the appearance of	CONTACT (*sparśa; chhu,* 觸)
CONTACT (*sparśa*)	causes the appearance of	SENSATION (*vedanā; shou,* 受)
SENSATION (*vedanā*)	causes the appearance of	CRAVING[d] (*tṛṣṇā; ai,* 愛)
CRAVING (*tṛṣṇā*)	causes the appearance of	GRASPING[d] (*upādāna; chhü,* 取)
GRASPING (*upādāna*)	causes the appearance of	COMING INTO EXISTENCE (*bhava; yu,* 有)
COMING INTO EXISTENCE (*bhava*)	causes the appearance of	BIRTH (*jāti; sêng,* 生)
BIRTH (*jāti*)	causes the appearance of	OLD AGE, SICKNESS, DEATH AND ALL MISERIES (*jarāmaraṇa; lao ssu,* 老 死)
OLD AGE, ETC., AND ALL MISERIES (*jarāmaraṇa*)	cause the appearance of	IGNORANCE (*avidyā; wu ming,* 無 明)

[a] Thomas (1), p. 97.

[b] One cannot get the feel of this unless one remembers that in Chinese *sê*, the literal meaning of which is 'colour', stands for colour in the widest sense, 'the lust of the eye and the pride of life', and also sex-relations in general.

[c] What connection the Indian doctrine had with Aristotelianism I do not know, but the great differences between the Buddhist four elements and the typically Chinese five elements should be noted. Earth, water and fire were in common, but wind replaced metal and wood. The Indian system seems identical with the Greek.

[d] One can easily see how such doctrines could quickly become confused with Taoist *jang* (cf. p. 61) or 'yieldingness'. These two *nidānas* were often personified as Māra[7] the Tempter, so prominent on the Tunhuang frescoes. For a recent introductory account of the Tunhuang site see Vincent (1).

[1] 蘊   [2] 色   [3] 四大   [4] 土   [5] 地   [6] 堅   [7] 寬羅

water (*shui*[1]) with the nature of fluidity (*shih*[2]); fire (*huo*[3]) with the quality of heat (*nuan*[4]); and wind (*fêng*[5]) with the quality of motion (*tung*[6]). Although so reminiscent of Aristotle and Galen, it does not seem that this classification ever had any notable effect on Chinese scientific thought.

In the pre-Buddhist times of the Vedas and the Upanishads there had been a relatively naïve belief in the existence of an individual soul, and Indian idealist metaphysics had taken its first origin in the famous 'discovery' of the 'identity of the *ātman* and the *brahman*', the union of the individual soul with the universe or with God. The Buddhists, however, strenuously denied the existence of an *ātman* (*shen wo*[7]), while at the same time they maintained that the constituents of the individual (*skandhas, yün*) continued in subsequent incarnations until they were finally disposed of, if and when the individual attained the status of an *arhat* (*lo-han*[8]). This was the same process as 'entering *nirvāṇa*' (*nieh-phan*[9]). It was absolute release, the deliverance from the load of evil *karma*, the *lysis* of the *skandhas* (*chieh tho*[10]; lit. dissection and disrobing);[a] and the meaning of the Sanskrit word was the blowing-out of a flame or its dying away for lack of further fuel.[b] The theory of the permanence (*śāśvatā*) of the *ātman* was attacked as a heresy (*ātmavāda; wo shuo*[11]) and in the *Āgamas* (doctrinal sūtras going back to the −3rd century, though not written down till much later) is considered a form of *upādāna* (grasping, *chhü*[12]). Thus one had the 'heresy' of *ātman*'-ism' (*ātmadṛṣṭi*), or the 'misconception' of the *ātman* (*ātmagrāha*), while the true doctrine was that of the non-existence of the *ātman* (*nirātmavāda; wu wo shuo*[13]).[c] On the other hand, the Buddhists also attacked the opposite, materialist, theory that at death the individual was annihilated (*ucchedavāda; tuan mieh chien*[14]).[d] Some Buddhist schools preferred to introduce a new category, that of individuality as such (*pudgala*[15]), but this did not become orthodox.[e] Yet the individual did transmigrate, loaded with the *karma* of its past actions, and as the *gandharva* (the being to be reborn) entered into the embryo or the womb (*garbha, thai*[16]). 'Consciousness was not something permanent which existed unchanged from birth to birth, but simply one form which the individual assumed at certain stages of his existence.'[f] Hence

---

[a] We have already seen this expression used, in part, in Thang scientific philosophy, with purely naturalistic meaning (p. 255). Cf. p. 463 below.

[b] Thomas (1), pp. 119 ff. I cannot help remarking that though the Indian idea was poetical enough, the Chinese transliterators used a word, *nieh*, which properly means 'slimy black mud', doubtless to give the idea of absorption into the primal chaos (cf. the Taoist ideas on this, p. 115, and in Neo-Confucianism, p. 486). Of course the word quickly took on highly numinous implications in China, but the choice was rather characteristic.

[c] Later we shall note numerous examples of the influence of this idea on Taoist thinkers, and we have already seen how it linked up with ancient Chinese mechanistic conceptions (cf. the *Kuan Yin Tzu* passage cited on p. 54).

[d] In the preceding section we have just seen how strong a reaction was called forth on the part of the Buddhists by Confucian sceptics who adopted such views (pp. 387, 410, 414). Cf. Bodde's remark in Fêng Yu-Lan (1), vol. 2, p. 286n.

[e] Thomas (1), p. 100.   [f] Thomas (1), p. 105.

[1] 水	[2] 濕	[3] 火	[4] 煖	[5] 風	[6] 動	[7] 神我
[8] 羅漢	[9] 湟槃	[10] 解脫	[11] 我說	[12] 取	[13] 無我說
[14] 斷滅見	[15] 補伽羅	[16] 胎

arose the interest of the Buddhists in embryology, quite parallel with that of 17th-century Christian theologians, as we shall examine in the appropriate section.

As for the method of the universe in allocating rebirths, depending on the merit acquired or evil to be expiated, there was a series of 'careers' (*gati; chhü*[1]). Man could be reborn as a god (*thien shen*[2]), a man, a hungry ghost (*preta; o kuei*[3]), an animal (*chhu*[4]) or in hell (*yü*[5]).

In order to round out the picture of Buddhism before its entry into China, it must be remembered that in its earliest form it was a doctrine intended for mildly ascetic hermit monks, living in community (*vihāra; yuan*[6]), and for them alone were the rules of discipline (*vinaya; lü*[7]) formulated. Only later was the religion extended to house-holders and others in ordinary life. Thomas has pointed out that perhaps it was this fact which rendered the disappearance of Buddhism from India comparatively easy. Once the educated monk and his community disappeared there was no essential principle to distinguish the Buddhist layman from the Hindu. For though the importance of caste had always been denied by Buddhism, it had never been condemned and fought as a practice of the laity.[a] As for the life of the monasteries, it is certain that the early Buddhists took over current yoga[8] practices, including meditation techniques of self-hypnotisation (*samādhi; ting*[9] or *san-mei*[10]) and the deep insight (*jñāna; hui*[11] or *chih*[12]) which was felt to be produced thereby. It was also undoubtedly believed that by such means 'supernatural' powers (*ṛddhi; shu*[13]) could be acquired, e.g. materialisation of emanation-forms or multiplied emanation-forms of an individual, levitation, telepathy, the rendering of human bodies transparent and minds readable, invisibility, control of the thermo-regulatory functions of the body, and of other functions normally autonomic, etc. There is undoubtedly a basis of fact in these physiological games, the investigation of which is a worthwhile study,[b] but they can hardly have served any more useful purpose than to impress ancient and medieval princes and people, who were highly partial to them. As regards the gods, their worship was tolerated by Buddhism, but they were not considered the basis of morality, nor the bestowers of lasting happiness. Later Buddhism embarked upon the incorporation of all the former gods of territories which it newly conquered, enrolling them as protectors of the faith on a grand scale, sometimes with the effect, as in Tibet perhaps, of obscuring almost entirely what the original doctrine had been.

Yet Buddhism never lost the character of its primary refusal to give answers to questions which it considered unnecessary since concerned with things unknowable. A list of undetermined questions runs like a creed, it has been said, through all Buddhist history. These were: (1) whether the universe is eternal or not; (2) whether or not it is finite; (3) whether the vital principle (*jīva; shou*[14] or *yu ming*[15]) is the same as

---

[a] This was not, of course, an issue for Buddhism in Central Asia, Tibet and China.
[b] See the remarks on p. 144.

[1] 趣   [2] 天神   [3] 餓鬼   [4] 畜   [5] 獄   [6] 園   [7] 律
[8] 瑜伽   [9] 定   [10] 三昧   [11] 慧   [12] 智   [13] 術   [14] 壽
[15] 有命

the tangible body or not; (4) whether after death a *tathāgata*[1] (Buddha; *ju lai*[2]) exists or not. Perhaps this was another feature which made it inimical to scientific speculation.

(b) THE LESSER AND THE GREATER CAREERS

We are now in a position to consider the two forms into which Buddhism crystallised, the so-called Hīnayāna form (primitive and 'protestant', one is tempted to call it), and the so-called Mahāyāna form (developed and 'catholic'—though such a comparison can be taken only with the lightest touch). The Hīnayāna consisted of eighteen schools, but of three of them only do we have any detailed information. These were the Sthaviravādins[3] (also called Theravādins), the Sarvāstivādins,[4] and the Mahāsaṃghikas.[5,6] The written Canon of the Sthaviravādins is preserved in Pali.[a] Theravāda Buddhism has survived in Ceylon, Burma, Siam and other parts of south-east Asia. The Mahāsaṃghikas, intermediate between Hīnayāna and Mahāyāna, developed the docetic doctrine that the physical body of the historical Buddha had never been more than a false apparition. Some of their works exist in Sanskrit and others in Chinese translations. The Sanskrit Canon of the Sarvāstivādins is largely preserved in Tibetan and Chinese translations.

From the Sarvāstivādin and Mahāsaṃghika schools developed the teaching of the 'greater career' (*mahāyāna; ta chhêng*[7]).[b] The Hīnayāna advocated individual progress to *arhat*-ship. Their opponents described this as the 'disciple's career' (*śrāvaka-yāna*[8]) or 'lesser career' (*hīnayāna; hsiao chhêng*[9]). Alternatively the Hīnayāna envisaged the goal of a *pratyeka-buddha* (*yuan chio*,[10] i.e. 'riddle-reasoning enlightened one'; or later *tu chio*,[11] 'solitary enlightened one'), i.e. a Buddha who does not preach. On the new view, full Buddhahood, as distinct from the 'selfish' ideals of the Hīnayāna, was to be the goal for all, monks and laymen alike; it was claimed that Buddha had worked for the salvation of everyone, and that all his followers should do so too, by deliberate submission, if necessary, to a further series of rebirths, thus postponing the individual's attainment of *nirvāṇa*. Such a sacrifice is very reminiscent of the altruistic doctrines of certain other religious faiths, notably Christianity, and there have not been wanting those who have seen some mutual influence.[c] It must be emphasised, however, that nothing is really known about the origin of the Mahāyāna ideas; all that is certain is that they were already highly developed by the +2nd century, just in time to be conveyed into China as what was perhaps the most 'modern' and attractive presentation of Buddhism. On the new views, the world was full of *bodhisattvas*,[12] half-mythical beings and their reincarnations, who under-

[a] Owing to the fact that the Pali Canon is much more complete than that in Sanskrit, and dealt with a form of Buddhism historically earlier than that represented in Sanskrit texts, it exerted at first too great an influence on modern scholars, who supposed that it had been written down earlier, and was more reliable, which is not the case. Practically all the Pali Canon has now been translated.

[b] *Chhêng* means riding, and is the classifier of vehicles (cf. Vol. 1, p. 39), hence the common, but unsatisfactory, English term 'Greater' and 'Lesser' Vehicle.

[c] Reichelt (1), pp. 99ff.; Anesaki (1); de Lubac (2); Keith (5), pp. 601ff., variously for or against.

[1] 陀多竭多　　　[2] 如來　　　[3] 悉替耶部　　　[4] 薩婆多部　　　[5] 摩訶僧祇部
[6] 大衆部　　[7] 大乘　　[8] 聲聞乘　　[9] 小乘　　[10] 緣覺　　[11] 獨覺　　[12] 菩薩

took to save the world; to these it was recommended that definite religious worship (*bhakti*) should be accorded. *Nirvāṇa* was now played down, and even the ideal of *arhat*-ship said to be illusory, perhaps with the very sound idea that self-culture, or self-culture alone, would never attain the salvation of the self; only the effort to save others could lead to the salvation of the self. Here was a ready-made paradox quite to the Taoist taste, and one which they would evidently find no difficulty in appropriating.

The great document of this new system is the *Saddharma-puṇḍarīka Sūtra*,[1] dated about +200 (*Miao Fa Lien Hua Ching*[2]—The Lotus of the Wonderful Law).[a] It praises the *bodhisattvas*, promises the completeness of Buddhahood with its omni-science to all, and includes much about the cyclical world-catastrophes which were supposed to occur.[b] But perhaps the most beautiful expression of the system is the +7th-century poem of Śāntideva, the *Bodhicaryāvatāra*,[c] which speaks of love for all beings, and the burning desire of the Buddhist to tranquillise their pains. As time went on the Buddhist laity knew almost nothing but *bodhisattva* worship, and the Four Truths were taught only to monks.

But 'the real break', says Thomas,[d] 'came with the teaching of the Void'. Till then, the newer ideas could be looked on as a not illogical amplification of the older. But *śūnyavāda*,[3] or the doctrine of the total unreality (*hsü*[4]) of the world of experience, jolted Buddhism violently on to a new course. Hitherto the cycle of rebirths had been thought of as proceeding in a real universe, but now everything was pictured as a delusive shadow-play, and release into *nirvāṇa* as release from the necessity of having to watch it. This was the work of the *Mādhyamika* school (*wu hsiang khung chiao*[5]), which probably started in the −1st century, but was systematised by its greatest figure, Nāgārjuna[6] (Lung-Shu[7]) (fl. *c.* +120), who lived just after the introduction of Buddhism into China.[e] The principal document is the (*Mahā*)-*Prajñāpāramitā Sūtra*[8] (The Perfection of Wisdom),[f] with Nāgārjuna's commentary, the *Ta Chih Tu Lun*.[9] The famous *Vajracchedikā Sūtra* (Diamond-Cutter; *Chin Kang Ching*[10])[g] is a condensation of this:[h]

> As stars, as faults of vision, as a lamp,
> As Māyā (deception, illusion), as hoarfrost, or a bubble,
> As dream, or as the lightning flash,
> So should one look on relative things...

[a] N 134ff. TW 262ff. Tr. Soothill (3). In view of the Chinese use of the words *ching*[11] and *wei*[12] it is interesting that *sūtra* also meant thread of warp or weft.

[b] See on, p. 485, for the place which these took in Neo-Confucian thought.

[c] Tr. de la Vallée Poussin (6) and Barnett (2).   [d] (1), p. 201.

[e] His 'Middle Way' has been translated by Walleser (2).

[f] N 19, 20, 935, TW 220; tr. Lamotte (1); Conze (4). See also Vidyabhusana (1).

[g] N 10–15, TW 235ff.

[h] One of six translations was made by Kumārajīva[13] (d. +412). His biography (in ch. 95 of the *Chin Shu*) has been translated by Pfizmaier (56).

[1] 薩達喇摩奔荼利迦經   [2] 妙法蓮華經   [3] 舜若多見   [4] 虛
[5] 無相空敎   [6] 那伽閼剌樹那   [7] 龍樹   [8] 大般若波羅蜜多經
[9] 大智度論   [10] 金剛經   [11] 經   [12] 緯   [13] 鳩摩羅什

it says.[a] Everything is in perpetual change, not for one moment the same, and therefore not real.[b] Form, feelings, all the *skandhas*, are nothing but delusion (*māyā;*[1] *mi,*[2] *huan*[3] or *huan ching*[4]).[c] It is 'darkness' not to realise this (*mei yü-chou chih chen li wei mi*[5]). There are no individual permanent entities (*wu chhang*[6]), or their masters (*chu-tsai*[7]). Selfhood too, therefore, is delusion, and thus the doctrine of non-self (*wu wo*[8]) was turned against the whole *hīnayāna* scheme of salvation by attainment of *arhat*-ship. For in order to attain individual salvation, any aspirant would have to assume the existence of some sort of individuality continuing sufficiently long to be ultimately liberated, and this would be to fall into the heresy of 'self-ism' (*satkāya-dṛṣṭi*). Nirvāṇa thus became a kind of noumenal Absolute, of which nothing whatever could be predicated, and which could be attained only by mystical ecstasy. Moreover, Nāgārjuna's logical school made a destructive analysis of the central principle of causation.[d] The doctrine of Mahāyāna thinkers was popularised in works such as the *Suvarṇa-prabhāsa Sūtra* (*Chin Kuang Ming Tsui Shêng Wang Ching*[9])[e] translated into Chinese c. +415. It would seem impossible to overestimate the importance which the doctrine of *māyā* (illusion) had in Chinese Buddhism; it was this which perhaps most of all made it irreconcilable with Taoism and Confucianism, and helped to inhibit the development of Chinese science.

Of course it was one thing to declare the visible universe an illusion, and another to take the further step of asserting metaphysical subjective idealism. An illusion may be experienced by only one observer, hence it was inevitable that some should go on to say that the whole universe is a creation of the observing mind, whether of the individual human being, or of the Buddhas and *bodhisattvas*. This 'doctrine of Mere Ideation' (*vijñapti-mātra; chih shih shuo*[10]) appeared with the *Laṅkāvatāra Sūtra*[11] (The Entrance of the Good Doctrine into Laṅkā)[f] of perhaps the +3rd century, and translated into Chinese in +430 and +433. More psychological than Nāgārjuna's logic it was markedly hostile to the *hīnayāna* people and to other Indian philosophical schools (such as the Nyāya;[12] *yin ming lun tsung*[13]—and the Sāṃkhya[14]),[g] which it attacked as heretical (*tīrthakara; wai tao*[15]). The world was nothing but mind (*cittamātra; wei shih*[16] or *wei hsin*[17]), and the individual's mind (*sensus communis* correlating sense-perceptions—*manas*[18], *i*[19]) was conceived of as part of the universal mind (*tathāgata-garbha*,[20] the womb or embryo of Buddhahood; *ju lai tsang*[21]). Or

[a] V. 84; Thomas (1), p. 214; Conze (4), pp. 19, 97.
[b] The Taoists were in complete agreement with this premise but not with its conclusion.
[c] Māyā was of course a very ancient conception in India (see Keith (5), p. 531).
[d] Thomas (1), p. 220. See p. 423 below.   [e] N 127, 130, TW 663ff.
[f] N 175-7, TW 670ff. Eng. tr. D. T. Suzuki (2) and commentary (3).
[g] Cf. Garbe (2); Berriedale Keith (3).

[1] 慶也 [2] 迷 [3] 幻 [4] 幻境 [5] 昧宇宙之眞理謂迷
[6] 無常 [7] 主宰 [8] 無我 [9] 金光明最勝王經 [10] 只識說
[11] 楞伽阿跋多羅寶經 [12] 那雅 [13] 因明論宗 [14] 三彌叉
[15] 外道 [16] 唯識 [17] 唯心 [18] 末那 [19] 意 [20] 陀多竭多胎
[21] 如來藏

rather it was to be compared with transient waves (*pho*[1]) coming and going on the surface of the universal 'store-consciousness' (*ālaya-vijñāna;*[2] *chen ju*[3]).[a] It was considered that this world-picture could hardly be proved by logical argument, but that it should be accepted by a kind of 'conversion' or 'revulsion' of feeling (*parāvṛtti; chuan i,*[4] *chhêng fo kuo*[5] or perhaps *chhan*[6]). These doctrines came to their climax in the school of Yogācāra[7] led by Vasubandhu[8] (Thien-Chhin[9])[b] and Asaṅga (Wu-Chu[10]) in the second half of the +5th century. Hsüan-Chuang wrote a large work on them.[c] Out of them developed the so-called *trikāya* (*san shen*[11]) theory of the three bodies of Buddha, the *dharmakāya* or non-material body (*fa shen*[12]), the *sambhogakāya* or appearance body for preaching (*pao shen*[13]), and the *nirmāṇakāya* or transformation body (*hua shen*[14]); the whole approaching a worked-out theory of incarnations. The two schools of the Mādhyamika and the Yogācāra retained till the end their position as the leading divisions of Buddhist philosophy.

(c) THE BUDDHIST EVANGELISATION OF CHINA

From the middle of the +2nd century onwards Buddhist texts poured into China in an unceasing stream, reaching a maximal influx perhaps in the +5th. Many Indian monks spent their lives in China translating them in collaboration with Chinese scholars.[d] It was quite impossible for the Chinese (or later the Japanese) to recognise any chronological sequence in the mass of works which they received, and purely artificial classifications were therefore set up. Since Chinese Buddhism is commonly spoken of as Mahāyāna, it is often forgotten that the Chinese received and treasured a very large number of Hīnayāna writings too. The theory of five periods in Gautama's preaching thus arose, and he was imagined to have given forth the most diverse doctrines as part of a complicated preaching plan. The first was represented by the *Buddha-avataṃsaka Sūtra* (The Adornment of Buddha—*Hua Yen Ching*[15]),[e] a distinctly Yogācāra document. The second embodied the traditions (*Āgamas*[16]) of the Sarvāsti-vādins, purely *hīnayāna*. The third, fourth and fifth were all *mahāyāna*, but in the wrong chronological order, starting with the *Laṅkāvatāra* and *Suvarṇa-prabhāsa*

---

[a] Thomas (1), p. 240, remarks that this no doubt arose from observations on the relations of the subconscious to the conscious mind. In the late +4th century Chu Tao-Sêng[17] equated it with *li*,[18] a word which was being used (as by Ko Hung, cf. pp. 438, 477) for the principles of natural things (Thang Yung-Thung, 1).

[b] Vasubandhu's *Viṃśatikā* (*Wei Shih Erh-shih Lun;*[19] Treatise in Twenty Stanzas on Mere Ideation) has been translated from the Chinese by Hamilton (1). This is full of arguments about atoms. Chu Pao-Chhang (1) has detected Whiteheadian ideas in it. TW 1588 ff.

[c] The *Chhêng Wei Shih Lun* (Completion of the Doctrine of Mere Ideation). It was a conflation of translated texts and commentaries. Tr. de la Vallée Poussin (3). Cf. Fêng Yu-Lan (1), vol. 2, pp. 299 ff., 319, 330. TW 1585.

[d] Bagchi (1) gives a brief account of the more important of them, with biographical details.

[e] TW 278, 279.

[1] 波   [2] 阿賴耶識   [3] 眞如   [4] 轉依   [5] 成佛果   [6] 懺
[7] 瑜伽阿闍梨   [8] 婆藪盤豆   [9] 天親   [10] 無著   [11] 三身
[12] 法身   [13] 報身   [14] 化身   [15] 華嚴經   [16] 阿含
[17] 竺道生   [18] 理   [19] 唯識二十論

Sūtras, going on to the *Prajñāpāramitā Sūtra*, and ending with the *Saddharma-puṇḍarīka Sūtra*. The five periods were thus (1) *hua yen*;[1] (2) *a han*;[2] (3) *fang têng*[3] (mixed); (4) *phan jo*;[4] and (5) *fa hua*.[5] This remarkable theory was first set forth by the founder of the Thien Thai[6] school, Chih-I[7] (d.+597),[a] and his successor, Tu Shun[8] (d. +640). It has significance for the history of religion, not of science.[b]

Another characteristic product of Chinese Buddhism was the Chhan[9] (*dhyāna*) method or way, a mysticism of purest quality supposed to have been founded by the Indian Bodhidharma[10] (Ta-Mo[11])[c] who died *c*. +475. Rejecting all sūtras (*ching*[12]) and *śāstras* (*lun*[13]), it eschewed philosophy and relied entirely on mystical faith, with intense and prolonged contemplation.[d] Of immense cultural and artistic influence, it can have been only one further factor inimical to science.

The last type of Chinese Buddhism which need be mentioned[e] is the so-called Pure Land sect (*ching thu tsung*[14]), which believed that through devotional practices individuals could hope to be reborn in a pure or happy land (*sukhāvatī*) somewhere in the Far West where they would be able to listen to particularly efficacious preaching concerning *nirvāṇa*. Ultimately the concept of *nirvāṇa* dropped out and only the pure land remained. The idea was actually an old one, found in Pali writings, but came to its flowering only in China and Japan, where many Buddhists still today pray to Amitābha or Amida Buddha,[15] or to Avalokiteśvara (Kuan Yin[16]—originally a male deity, but transformed in some curious way into a woman holding a child and closely resembling the Mary statues of Christianity), for entry to the pure land.[f] There is some reason for thinking that the idea of the Pure Land may owe something to the paradises imagined by the Taoists—we saw a good example in *Lieh Tzu*.[g] It began, at any rate, quite early, in the +4th century, with Hui-Yuan,[17] a famous monk.

It was not long before numerous indigenous philosophical schools began to emerge.[h] The chief source for our knowledge of them is the *Chung Kuan Lun Su*[18] (Commentary on the *Mādhyamika Śāstra*)[i] written by the monk Chi-Tsang[19] (+549 to +623). As can be seen from the names of some of them, such as the Schools of Original Non-Being (*Pên Wu Tsung*[20]), Stored Impressions (*Shih Han Tsung*[21]) and Phenomenal

[a] G 376.

[b] See further Wieger (2), pp. 351 ff., 392 ff.; Fêng Yu-Lan (1), vol. 2, p. 284.

[c] G 14.

[d] See Ui (1); D. T. Suzuki (4); Blofeld (2).

[e] On the sects of Chinese Buddhism consult Takakusu (2); Forke (12), p. 191; and Blofeld (1)

[f] See Thomas (1), p. 254; Reichelt (1), pp. 112 ff.

[g] Ch. 5, tr. R. Wilhelm (4), p. 53, see p. 142 above. The Buddhists in China must have borrowed a good deal from earlier Chinese sources; I note, for example, that the story of the man who was afraid the sky would fall down, which we read on p. 41 from *Lieh Tzu*, ch. 1, occurs again in the *Kāśyapa-parivarta Sūtra* (N 805, TW 350 ff.), cit. Suzuki (1), p. 386.

[h] See the description of them by Fêng Yu-Lan (1), vol. 2, pp. 243 ff. Cf. Ware (4).

[i] TW 1824.

¹ 華嚴   ² 阿含   ³ 方等   ⁴ 般若   ⁵ 法華   ⁶ 天台
⁷ 智顗   ⁸ 杜順   ⁹ 禪   ¹⁰ 菩提達摩   ¹¹ 達摩   ¹² 經
¹³ 論   ¹⁴ 淨土宗   ¹⁵ 阿彌陀佛   ¹⁶ 觀音   ¹⁷ 慧遠
¹⁸ 中觀論疏   ¹⁹ 吉藏   ²⁰ 本無宗   ²¹ 識含宗

Illusion (*Huan Hua Tsung*[1]), they elaborated various forms of metaphysical idealism. Within the illusory realm of existence, however, there might be different forms of matter. The School of Matter as Such (*Chi Sê Tsung*[2]) distinguished between fine or impalpable matter and the coarse matter of familiar objects; this was one of the many unsuccessful attempts to introduce Indian atomism into China.

Buddhist thought was always partial to such quasi-physical speculations. In Hsüan-Chuang's philosophy, the 'store consciousness' mentioned above was also a 'seed consciousness', for it contained in itself the seeds or germs (*bījas*; *chung-tzu*[3]) of all things. Perhaps there was a Stoic echo here. The development of these seeds was stimulated by influences from the seven other kinds of consciousness; this was called 'perfuming' (*vāsanā*; *hsün hsi*[4]). Someone had noticed the extremely small amounts of certain highly odoriferous substances which are enough to make their presence felt.[a] The elaborations of this theme would be worth the attention of those who study the history of the idea of 'action at a distance' in the West, though for that the whole of Chinese organicism is relevant.[b]

What the effect of all this was on Chinese thought can to some extent be imagined, but although a certain amount of study has been made of its repercussions,[c] there still lacks, perhaps in any language, a thorough analysis of the influence of Buddhism on Chinese philosophical and scientific thought.[d]

The first Buddhists arrived in China about the middle of the +1st century (though the affair was not as the legendary accounts have it). At the end of the +2nd century the interesting little book *Li Huo*[5] (The Resolution of Doubts) was written by a layman whose family name was Mou,[e] and who has come down in history with the usual terminal particle of philosophers as Mou Tzu.[6] He had lived for some time in Indo-China where he had become acquainted with Buddhism, and Pelliot (14), in the preface to his translation of the work, accepts a date for it close to +192. It is a dialogue reminiscent of the *Milindapañha*[7] (the Graeco-Buddhist book in which the Bactrian king Menander asks questions of the monk Nāgasena)[f] which contains nothing of particular interest for the history of science. Nevertheless, Mou Tzu is extremely polite to Confucianism and Taoism, seeking to justify Buddhism by quotations from the indigenous Chinese classics. Already at this early time it is interesting to see that his interlocutor complains bitterly of the vast mass of sūtras and other Buddhist literature, which seems to him quite unnecessary, but Mou Tzu justifies it by saying that the Buddhist books deal with the infinitely great and the

[a] Cf. the 'wafts' so frequently seen depicted in the Tunhuang frescoes.
[b] Cf. especially p. 381 above.
[c] For example, the relevant chapters in Fêng Yu-Lan (1), and Hu Shih (4, 7).
[d] As to the process of translation whereby the Chinese Tripiṭaka or Buddhist patrology came into existence, a convenient orienting summary is found in Sarton (1), vol. 3, pp. 466 ff.
[e] His given name is not certain, perhaps Po.[8]
[f] TW 1670.

[1] 幻化宗   [2] 卽色宗   [3] 種子   [4] 薰習   [5] 理惑   [6] 牟子
[7] 那先比丘經   [8] 博

infinitely small, the infinitely old (before the formation of the existing world) and the far future (when this and succeeding worlds shall have passed away and others shall have been formed). Sceptical Confucianism comes out in the complaint about the improbability of the bodily abnormalities of Gautama Buddha, but the interlocutor is 'confuted' by quotations from classical legends which the Confucians were supposed to accept (e.g. the supernumerary nipples of Wên Wang). So also in the +3rd century, Buddhists such as Fa-Ya[1] and Khang Fa-Lang[2] used Taoist technical terms in their expositions; this was known as 'explaining by analogy' (ko i[3]).[a]

The following centuries were full of minor thinkers who sought to combine Buddhism with Confucianism and Taoism. Among these syncretists were Sun Chho[4] (+310 to +368),[b] Chang Jung[5] (+420 to +497)[c] and Chou Yung[6] (+465 to +498).[d] Others, such as Ku Huan[7] (+430 to +493),[e] whose book, the *I Hsia Lun*[8] (Discourse on the Barbarians and the Chinese), appeared in +467, while admitting a close similarity between Buddhism and Taoism, considered that the former was suitable for Indians but not for Chinese, and that the latter should therefore be supported. Many rejoinders to this were written from the Buddhist side. Hsiao Tzu-Hsien,[9] however, hit the nail on the head when he said,[f] 'For Confucius and for Lao Tzu the regulating of the things in this world is the main objective, but for the Buddhists the objective is to escape from this world (*Khung Lao chih shih wei pên; Shih shih chhu shih wei tsung*[10]).' Ku Huan adopted the widespread identification of the immortality of the Taoist *hsien*[11] with the *nirvāṇa* of the Buddhists; but again his opponents, as reported by Hsiao Tzu-Hsien, pointed out with admirable clarity that Taoist immortality was of a materialist character, while Buddhist liberation from desire was extinction even of the spirit. 'In the transmutation (of a person) into a *hsien*[11] the transformation of the (bodily) form is the main thing; but for *nirvāṇa* the first necessity is the refining of the spirit (*Hsien hua i pien hsing wei shang; Ni-huan i thao shen wei hsien*[12]).'[g] However, the movement continued, with names such as Mêng Ching-I[13] (end +5th) and Liu Chou[14] (+519 to +570).[h]

[a] On this see Thang Yung-Thung (2). The problem was exactly the same as that presented 1500 years later when the vocabulary of modern science had to be incorporated into the Chinese language. Transliterate, and explain the ugly compound resulting? Or employ an already existing Chinese term and distort the meaning? We shall return to this dilemma in Sect. 49.

[b] Forke (12), p. 229; Fêng Yu-Lan (1), vol. 2, p. 240.   [c] Forke (12), p. 230.
[d] Forke (12), p. 232.   [e] Forke (12), p. 233.
[f] *Nan Chhi Shu*, ch. 54, p. 11b (Ku Huan's biography).

[g] As Maspero (12), p. 66; (13), p. 198, has underlined, the original welcome for Buddhism in China had depended on the interest of the Taoists in what they thought might be new techniques.

[h] Forke (12), pp. 237 and 250. Cf. p. 24 above.

[1] 法雅   [2] 康法朗   [3] 格義   [4] 孫綽   [5] 張融
[6] 周顒   [7] 顧歡   [8] 夷夏論   [9] 蕭子顯
[10] 孔老治世爲本釋氏出世爲宗   [11] 仙
[12] 仙化以變形爲上泥洹以陶神爲先   [13] 孟景翼   [14] 劉晝

(d) THE REACTION OF CHINESE NATURALISM

One of the chief causes of strain between Buddhism and the indigenous philosophies was the fact that though the Buddhists might combat the conception of an *ātman* or soul, they were forced to admit the existence of *something* individual which persisted through successive reincarnations and bore its load of varying good or evil *karma*. Hence they collided with Confucian scepticism and Taoist selflessness, since both the Chinese systems were truly *ucchedavāda*. The story of this controversy has been graphically told by Forke[a] and Maspero.[b] It started with the small tractate of the monk Hui-Yuan[1] (+333 to +416) *Hsing Chin Shen Pu Mieh*[2] (The Destructibility of the (Bodily) Form and the Indestructibility of the Spirit). Towards the end of the +5th century it reached a climax with the celebrated essay of Fan Chen in +484, the *Shen Mieh Lun* (On the Destructibility of the Soul), of which we have already spoken (p. 387). 'Man's substance', he said, 'is substance which possesses consciousness' (*Jen chih chih yu chih yeh*[3]).[c] Fan Chen is to be considered one of the most acute of all Chinese thinkers,[d] and a worthy successor of Wang Chhung. His essay provoked a flood of replies; among them may be counted the *Shen Pu Mieh Lun*[4] (On the Indestructibility of the Soul) of Chêng Tao-Chao[5] (d. +516); and the *Kêng Sêng Lun*[6] (On Reincarnation) of Lo Chün-Chang[7] (+6th).[e] In the same century, Fu Yi (already mentioned, p. 387) was outstanding in his attacks on Buddhism; Wright (3) has made a special study of him. It was philosophic Taoism fused with Confucianism which he advocated as the alternative to Buddhism for the ideology of the newly unified empire. But the fusion had to await the Sung.

Another doctrine introduced by Buddhism which was basically antagonistic to the indigenous philosophies was that of the illusory nature of the visible world (*māyā*), and its corresponding theoretical form of subjective idealism. This took time to develop, but as soon as the Thang began it was in full flower. Lu Hui-Nêng[8] for example (+638 to +713),[f] the sixth and last Chhan Buddhist patriarch, crystallised it in a famous passage, addressed to two monks who were discussing whether a flag was moving by itself or whether it was being moved by the wind. 'Neither the flag nor the wind is moving', he said, 'there is only a movement within your minds (*Pu shih*

---

[a] (12), pp. 260 ff.   [b] (12), p. 77.

[c] *Liang Shu*, ch. 48, p. 7b. Cf. the remark of Locke that there was nothing contradictory or scandalous about the suggestion that God might have 'given to some systems of matter, fitly disposed, a power to perceive and think' (*Essay concerning Human Understanding* (1687), IV, iii, 6).

[d] An important paper on him by Hou Wai-Lu & Chi Hsüan-Ping (1) analyses his materialism and suggests that in some of his statements he anticipated the famous maxim of William of Ockham eight centuries later.

[e] One is reminded of the pamphlets and counter-pamphlets of the time of de la Mettrie, *Man a Machine, Man Not a Machine*, and so on, in France more than a thousand years later; cf. Needham (13), p. 177; (14). Cf. p. 386 above.

[f] Often known simply as Hui-Nêng (Forke (12), p. 360).

[1] 慧遠   [2] 形盡神不滅   [3] 人之質有知也   [4] 神不滅論

[5] 鄭道昭   [6] 更生論   [7] 羅君章   [8] 盧慧能

fêng tung, pu shih fan tung, jen chê hsin tung[1]).'[a] Later we shall see that subjective idealism found fertile soil in a Confucianism which would not follow Chu Hsi.

By the time that we come to the Sung it is easy to find in Neo-Confucian writings controversial statements against Buddhist thought, for the issues had clarified and sharpened; I shall reproduce one or two of such passages. The Thang and Sung Taoists said less, partly because they were extremely busy in adapting Buddhist liturgical practices wholesale to their own rather artificially organised religion; and partly perhaps because, apart from the second flowering of their philosophy under the Thang, which continued to work with Chinese concepts only mildly affected by Buddhist ideas, they were more engaged with alchemy and other practical arts which necessitated a realistic, if not materialist, attitude towards the external world.

Hu Yin,[2] for example (+1093 to +1151)[b] considered that ice and glowing coals would mix better than Confucianism and Buddhism. In the *Sung Yuan Hsüeh An* he is reported as saying:

Buddhism looks upon emptiness as the highest (*khung wei chih*[3]) and upon existence as an illusion (*yu wei huan*[4]). Those who wish to learn the true Tao must take good note of this. Daily we see the sun and moon revolving in the heavens, and the mountains and rivers rooted in the earth, while men and animals wander abroad in the world. If ten thousand Buddhas were to appear all at once, they would not be able to destroy the world, to arrest its movements, or to bring it to nothingness. The sun has made day and the moon night, the mountains have stood firm and the rivers have flowed, men and animals have been born, since the beginning of time—these things have never changed, and one should rejoice that this is so. If one thing decays, another arises. My body will die, but mankind will go on. So all is *not* emptiness![c]

And he continues:

The teaching of the sages, it is true, considered the mind as the root of things (*hsin wei pên*[5]), and so does Buddhism, but not in the same way. The sages taught men to order rightly their minds. That which every right-minded man has in common with others is called the pattern (of human-hearted behaviour) (*li*[6]), and righteousness (*i*[7]). When one is grounded in these principles then the substance and operation of one's mind is complete. But Buddhism teaches that the mind should concentrate on the doctrine (*fa*[8]), deny the existence of the world (*chhi mieh thien ti*[9]) and consider it as a dream and a delusion (*mêng huan*[10]).[d]

Chu Hsi (+1130 to +1200), the greatest of the Neo-Confucian philosophers, whom we shall study fully in later sections, fought continually against Buddhism.[e] There

[a] I owe thanks to my friend Dr Wang Ching-Hsi for this reference from the *Kao Sêng Chuan*, pt. 3, ch. 8. It is also given by Forke (12), p. 364.   [b] G 826; Forke (9), pp. 135 ff.
[c] Ch. 41, pp. 6*a*, 7*b*, tr. Forke (9), eng. auct.   [d] Tr. Forke (9), eng. auct.
[e] See the useful discussions of Bruce (2), pp. 245 ff.; Chhen Jung-Chieh (1); and especially Graf (2), vol. 1, pp. 216 ff., 229 ff.

[1] 不是風動不是旛動仁者心動   [2] 胡寅   [3] 空爲至   [4] 有爲幻
[5] 心爲本   [6] 理   [7] 義   [8] 法   [9] 起滅天地   [10] 夢幻

it will be suggested that the Neo-Confucians (the Hsing-Li School) arrived at what was essentially an organic view of the universe. Composed of matter-energy (*chhi*[1]) and ordered by the universal principle of organisation (*Li*[2]), it was a universe which, though neither created nor governed by any personal deity, was entirely real, and possessed the property of manifesting the highest human values (love, righteousness, sacrifice, etc.) when beings of an integrative level sufficiently high to allow of their appearance, had come into existence. This was a world-outlook consonant with science indeed, and could not but be deeply inimical to the world-denying metaphysics of Buddhist asceticism. Bruce (2) brings together a number of telling passages from Chu Hsi[a] which illustrate this, but here I will offer some others.

In the *Chu Tzu Chhüan Shu* we find the following:

Liao Tzu-Hui[3] wrote as follows to Chu Hsi: 'There is only one *Li*[2] of Heaven and Man.[b] The root and the fruit are identical.[c] When the Tao of man('s nature) is perfected, the Tao of heaven is also perfected.[d] But the realisation of the fruit does not mean separation (*li*[4]) from its root.[e] Even those whom we regard as sages spoke only of perfecting (the relationships of human life).

Now the Buddhists discard Man and discourse (only) on Heaven, thus separating the fruit from the root, as if (they were the two horns of a dilemma of which) you must choose one and reject the other.[f] The presence of the Four Terminals (*tuan*[5]) (feelings)[g] and the Five Permanent Things (*chhang*[6]) (cardinal virtues) they regard as masking *Li*[2]. The indispensable relationships between father and son, prince and minister, husband and wife, elder and younger, they regard as accidental. They even go so far as to speak of Heaven and Earth, the Yin and Yang, with men and other creatures, as *phantasmal transformations* (*huan hua*[7]).[h] They have never so much as enquired (into their reality), but simply assert the (mental) nature of the Great Emptiness (*thai hsü*[8]).

But there are not two *Li*[2] in the universe. How then can the Buddhists take Heaven and Man, the root and the fruit, summarily asserting the one and denying the other, and call this a Tao? When their perceptions are so small and incomplete and partial, what possibility is there of the familiar doctrine of a perfect union between the transcendental and the lowly? (*chhê shang chhê hsia*[9]).'[i]

Here, then, the all-embracingness of the organic cosmic pattern is affirmed. Man and his society grow up out of the environing patterns of lesser complexity. More-

---

[a] See also the discussion of Fêng Yu-Lan (*1*) in Bodde (3), pp. 45 ff.; Fêng (*1*), vol. 2, p. 566.

[b] Because the whole universe, including man in it, forms one Great Pattern or Organism.

[c] The fruit is what we should call 'phenomena of high integrative level', the root is what we should call 'low level', undifferentiated, cosmical phenomena.

[d] Because Heaven has then brought into existence the highest levels of organisation which we know of in the universe: human social life and its values.

[e] For, as we should say, human existence is grounded in the worlds of electrons, atoms, living cells and organs, in fact, the successive stages of complexity.

[f] This is a reference to their metaphysical idealism and to their denial of individual selfs.

[g] Cf. *Mêng Tzu*, II (1), vi, 5–7.   [h] *Māyā*.

[i] Ch. 46, p. 7a, tr. Bruce (1), p. 280, mod. The last sentence could have been written by Whitehead himself. Chu Hsi replied that he substantially agreed with his correspondent.

[1] 氣   [2] 理   [3] 廖子晦   [4] 離   [5] 端   [6] 常   [7] 幻化

[8] 太虛   [9] 徹上徹下

over, the external world is real, and not an illusion.[a] The emergent levels are referred to by Chu Hsi in an adjacent passage (he is writing to Chan Chien-Shan[1]):

You say, 'For the Buddhists, apart from the One Intelligence,[b] there are no distinctions. For them, phenomena have no (real) existence. But for us Confucians, of all phenomena there are none which are not due to the Heavenly *Li*.[2]' This statement is all right, because for us Confucians also, the distinctions are not apart from Intelligence. But within this Intelligence there are the differences of height and depth as of Heaven and Earth; giving an infinite variety of things, not even the smallest hairbreadth of which can be changed.[c]

The Neo-Confucian reaction to Buddhism is summarised so well in the *Pei-Chhi Tzu I*,[3] a kind of philosophical glossary of Hsing-Li technical terms, written by Chhen Shun[4] about the time of the death of his master Chu Hsi (+1200), that it demands quotation, though rather long.[d] We give it a running commentary.

(1) Taoism and Buddhism were formerly prevalent, but now are still more so. The teaching of these two schools is roughly similar, but Buddhism is much more obscure than Taoism.

(2) Lao Tzu's chief point was *wu wei* (no action contrary to Nature), but the Buddhists exalt Emptiness. Lao Tzu said that Nothingness was important because out of Nothingness came Things. He recommended ataraxy, and occupying oneself with unworldly things so as to refine one's body. He and his disciples were sick of the vulgar way of rushing about on all kinds of business, so they frequented the mountains and the forests, and undertook the alchemy of the spirit and nourishing the *chhi*, according to the theory of embryonic respiration[e]—in this way they could leave their material bodies just as snakes come forth from moulted skins. They also wanted to ride on the clouds, flying on cranes above the nine heavens. It was only because of the transmutation of their *chhi* that they were able to become so light as to do this. Thus the doctrines of Lao Tzu are not really deceitful to men.

> It is interesting to note that he gives a rather favourable account of Taoism. In one sense, Neo-Confucianism might be considered a joint Confucian-Taoist reaction to Buddhism.

(3) But Buddhism appeals even to women and girls in the remotest mountain valleys, leading people to ascetic practices and even to ways of destroying the body.[f] From these things they cannot be converted. Buddhism has two harmful approaches; with its life-death guilt-happiness theory it cheats the foolish people; and with its high-sounding talk about philosophical virtue, it cheats the scholars.... As for us, we should be very clear in the mind about *Li*[2] and *I*,[5] and thus so settled in opinion that we cannot be shaken.

---

[a] Other sharply-worded statements opposing the doctrine of *māyā* are to be found in the *Chin Ssu Lu*, ch. 13, pp. 57, 58 (tr. Graf (2), vol. 2, pp. 715, 718).

[b] *Tathāgata-garbha.*

[c] That is, altered from the properties which their positions within the Great Pattern confer upon them. Ch. 46, p. 19*a*, tr. Bruce (1), p. 302, mod.

[d] Ch. 2, p. 39*a*, tr. auct. Numbering of sections ours.

[e] Cf. p. 144.

[f] Mutilation of one's own body, and even suicide, though rather rare in early Buddhism, were quite a prominent feature of medieval Chinese Buddhism (and modern too, cf. Reichelt (1), p. 274).

[1] 詹兼善   [2] 理   [3] 北溪字義   [4] 陳淳   [5] 義

Ordinary people are deceived by the life-death guilt-happiness theory.[a] They are terrified of going to hell after death, and also they pray for a good rebirth later; they perform mortifications and even mutilate themselves in order to acquire merit (lit. good causal fruits); thus, they think, they will be able to avoid many punishments after death, and be reborn as worthy persons, with all their descendants enjoying wealth and honours, rather than as beggars or animals. These ideas are maliciously propagated, and foolish men and women all believe them.

(4) Moreover, as for the wheel of existence, assuredly there is no such thing. As Chhêng I[b] said, it would be impossible to take *chhi* (matter-energy) which has already returned to its disaggregated form (and to reconstitute the same individual with it). This is true indeed. The vast continuum of *chhi* in all Nature moves and flows, producing the myriad things. Former collocations pass away, and later ones succeed them. Former ones decay and later ones grow up. In endless motion do these changes proceed, but the original *chhi* (of an entity) is certainly not collected together again to form the basis of a new (entity). A Yang returning is not the same Yang which went away. The sages established the hexa-grams and explained them. Although they said that the Yang returns, this must be inter-preted as meaning that the outer *chhi* decays and a new inner *chhi* is formed (in due course replacing the former).

> The Neo-Confucians were at one with the Taoists in denying that impermanence implied unreality (cf. p. 405 above).

But the Buddhists say that the *chhi* returns, as if in a circle, producing men and things. Now this does not agree with the *Li*[1] of creativeness. If the idea of a revolving wheel were true, there would have to be a constant number of things and men, and the *chhi* would simply oscillate to and fro. On such a view Nature would lose its creativeness. Only by understanding the *Li*[1] of the creativity of Heaven and Earth can we appreciate how weak the Buddhist doctrine is.

> Here he defends the creativeness (*ta tsao hua*[2]) of the Great Pattern of Nature, and denies any limitation of it.

> Next Taoist ataraxy confronts Hīnayāna concern for personal salvation in *arhat-ship*.

(5) Man is born between Heaven and Earth. He has the *chhi* of heaven and earth for his body and partakes in the *Li*[1] of heaven and earth by virtue of his nature (*hsing*[3]). If we trace it back to its origin we can know how man arises; if we follow it to its end, we can know how he dies. The men of old said that if a man could attain the true rightness, he might willingly die; and that if in the morning one understood the Tao, in the evening one could die without regret. A man searching after the Tao and the Li, once he feels that he has understood all, then he has no sorrow, and as he dies he is content to let the two forces and the five elements just disperse and melt away. This is peaceful death and natural growth, following the same transformations as those of heaven and earth. This is to be a disciple of Nature. But a person with selfish desires and self-love, who has not been able to get rid of them, is in contradiction with Nature (and will not peacefully die).

---

[a] *Karma* and rebirth.
[b] Cf. pp. 457, 471 below. See *ECCS, Honan Chhêng shih I Shu*, ch. 1, p. 3*a*, *Wai Shu*, ch. 7, p. 2*b*, ch. 10, p. 4*b*.

[1] 理   [2] 大造化   [3] 性

Now he proceeds to a sceptical attack on the Buddhist hells and paradises, finding no obvious place for them in the cosmic structure (as the Neo-Confucians conceived of it). The argument is reminiscent of post-Renaissance Christian theologians being laughed out of the belief in a material hell and heaven by the growing acceptance of the scientific view of the world. But we are here still in the +12th century.

(6) As for the theory of *karma*, it is absolute nonsense. Abundant 'proofs' have been supplied, but all are false. As Ssuma Kuang said, in ancient times no one dreamed anything about the ten kings of the underworld.[a] He was quite right; it was only the introduction of Buddhism which spread these ideas in men's minds.

Between heaven and earth, the wind and the thunder are the only formless phenomena. Real things all have form and substance. Houses, for example, are built with wood from forests and bricks from kilns—all are visible and tangible material things. But as for the Buddhist paradises and hells, where could their materials be obtained from?

Moreover, heaven is only aggregated *chhi*. The higher up it is, the more rapidly it rotates, like a howling wind. I really cannot imagine where (in such a world) their paradise could be, nor what could support it. Similarly, the earth is suspended in empty space in the midst of the heavens, and below it there is nothing but water down to depths profound. I really do not know where the so-called hell could be situated under the earth.

Furthermore, what they call happiness can be obtained with 'underworld money', and guilt can be pardoned with it. If the spiritual beings were righteous, they would not thus be greedy for bribes.

The whole thing was originally a pure invention to induce people to do good, and to frighten them from doing evil. Rustics and ignorant people are so concerned about their own personal fate that they readily incline to such ideas. But what is extraordinary is that even emperors such as Thang Thai Tsung, with all their wisdom, could not avoid the temptation of Buddhism.

(7) Scholars who read books only want to get a smattering of history in order to write essays. They care nothing about the (truths) established by the elaborate observations of the sages and worthies. Hence their minds are not settled, and Buddhist doctrines are able to attract and convince them. Even Han Yü and Pai Chü-I, for example, though very clever men, occupied themselves mainly with literature, poetry, and the like, and thus were quite unable to demonstrate the weak points of Buddhism. Han Yü attacked it only for its denial of social relationships; this is, of course, important, but it is not the root of the disease.

(8) What the Buddhists call the Mysterious (*hsüan miao*[1]) is simply the same thing as Kao Tzu's[2] saying, 'The (human) nature is what man is born with.'

Kao Tzu[b] was referring to sensation and action. He considered that what makes the eye see, and the ear hear, and so on, was a quick and vital consciousness (*ling-huo chih chih-chio*[3]). This is always at work behind the activity of the sense organs (*chhang tsai mu chhien tso yung*[4]). This '*sensus communis*'[c] we call the *hsing*[5] (human nature). To understand this is to understand the Tao (of man's reactions).

He is attacking the main Buddhist citadel of metaphysical subjective idealism.

[a] Cf. Reichelt (1), p. 72.
[b] Contemporary of Mencius; cf. p. 17 above.
[c] In Aristotelian terminology; Skr. *manas* (see p. 405 above). Cf. also p. 196 above.

[1] 玄妙   [2] 告子   [3] 靈活之知覺   [4] 常在目前作用   [5] 性

Now the Buddhists on the other hand exalt all this, and broaden it out (to a universality of mind).[a] This is the most fundamental point from which all their misconceptions arise. The denial of social relationships, in comparison with this, is a trivial matter.

(In a word) the profoundest error of the Buddhists, and the source of the vagueness of their thought on human affairs, *is their substitution of hsing*[1] *for chhi*.[2]

> He thus declares for materialism, informed by the principle of organisation.

For example, they say that a dog has the *hsing*[1] of Buddha. It can certainly wag its tail when it is called, and hence they say it has *hsing*. This is reducing man and animals to a single level. And yet that is what they call the 'inextinguishable'....

> As we should say, the neurological organisation of the lower animals and man is not equivalent.

(9) Now the sages and worthies from ancient times have all said that *hsing*[1] is simply (part of) *Li*.[3] The function of seeing and hearing is a matter of *chhi*. But to see what should be seen, and to hear what should be heard, is *Li*. For example, that a hand can grasp something, is *chhi*. But whether the hand holds a book for reading, or gesticulates (to call somebody) (is *Li*[3]). How can we not make such a distinction? Actions (of men) have to be classified as right or not right. What is right corresponds with the original *hsing*[1] (proper to human nature). What is not right arises from the creaturely self-will (*ssu*[4]) arising from the (urge to preserve the bodily) form (*hsing*[5]) and the *chhi*.[2]

The Buddhist doctrine may at first sight look similar to our views, but actually there is a profound difference. Our scholars sharply distinguish between the *Li* (organising pattern) and the *hsing-chhi*,[5,2] (material formed organism). *Li* is of course subtle and difficult to imagine. But the Buddhists simply *confound Chhi with Hsing*[1] and so leave it.

> In other words, they project mentality or spirit to fill the whole world, considering matter to be unreal.

Recapitulating, Chhen Shun objects fundamentally to the metaphysical idealism of the Buddhists, which he feels to be incompatible with the scientific Neo-Confucian view of the world.[b] He ridicules the pictures of corporeal hells and paradises, finding no room for them in the Neo-Confucian cosmology. Calling upon Taoist ataraxy, he uses it to shame *hīnayāna* concentration on personal salvation, and though he does not say so, he doubtless included the *mahāyāna* preoccupations in the same condemnation, for though they might be deeply compassionate in motive, their compassion was misplaced. The aim of release from the wheel of existence was not the proper way for man to react to Nature. Then he attacks the reincarnation doctrine, denying that any individual collocation of *chhi*, once dispersed, could ever be reformed again, and using a very interesting argument that to believe in a cyclical recurrence of births

---

[a] *Cittamātra* (see p. 405 above).

[b] On this point, if on this alone, Neo-Confucians and Jesuits coincided. Trigault has preserved for us an interesting encounter which Matteo Ricci had long afterwards with a Buddhist metaphysician (Gallagher tr. p. 340). They parted in mutual incomprehension. But the Neo-Confucian doctrine of immanent Li and Tao was also a great stumbling-block for Ricci (pp. 95, 342).

[1] 性   [2] 氣   [3] 理   [4] 私   [5] 形

left no place for the infinite creativity and novelty of Nature.[a] Finally, he agrees with other Confucians in disliking the Buddhist denial of the age-hallowed Chinese social relationships of family love and official hierarchic loyalty, but quite properly he does not consider this as fundamental as their philosophical position. In sum, the Neo-Confucian opposition to Buddhism was essentially that of a scientific view of the world combating a world-denying ascetic faith.

Such were the reactions of realistic Chinese Confucian-Taoist thought to the Buddhist challenge. It is for us, however, to attempt some estimate of the influence which Buddhism exerted on Chinese science and scientific thought. There can be little doubt that on the whole its action was powerfully inhibitory. While in propitious circumstances the doctrine of inevitable ethical causation might conceivably have been extended to cover the whole field of natural causation, this certainly never took place. Perhaps any beneficial influence which it might have had was wholly overshadowed by the doctrine of *māyā*, for how could a mere phantasmagoria invite serious scientific study? How could the mentality which averted the eyes from it, and which sought salvation in eternal release from it, encourage the investigation of it? And the negative attitude of Buddhism was as marked in what it refused to discuss, as in its positive doctrines, for cosmogony was among the problems regarded as unknowable and impenetrable.[b] Alas, the 'World', for the Buddhists, was not only 'the world, the flesh, and the devil', but the world of Nature itself. According to early Buddhist rules,[c] the monk should keep the doors of his senses guarded, and if he should see anything, devote no attention to its characteristics and details. Buddhism was not interested in co-ordinating and interpreting experience, or finding reality in the fullest and most harmonious statement of the facts of experience,[d] but in seeking some kind of 'reality' behind the phenomenal world, and then brushing the latter away as a useless curtain.

There must, of course, have been exceptions to this, particularly after *mahāyāna* doctrine had arisen and was concentrating attention on the relief of the pain and suffering of all creatures. Undoubtedly it gave an impetus to the study of the sciences allied to medicine.[e] This may be seen, for example, in the biography of the Central Asian missionary monk Fo-Thu-Têng[1] (fl. +310),[f] translated and commented by A. F. Wright (2). On an earlier page, details were given of the books of Indian learning in subjects such as pharmaceutical botany which were made available in Chinese before the Thang (Sect. 6*f*). But all were lost during the medieval period,

---

[a] How far the Neo-Confucians did justice to the more ancient forms of Buddhist doctrine on immortality has been discussed by Bodde (11).

[b] One is reminded of the converse of this, in one of Sir John Hill's satirical remarks about a subject discussed by the Royal Society—'A subject very suitable indeed for philosophical discussion, because impossible to be determined!' (*c.* +1675).

[c] Thomas (1), p. 46.

[d] Which might stand as a good enough description of Neo-Confucianism.

[e] See also the article 'Ping'[2] in *Hobogirin*, vol. 3, p. 224.

[f] *Kao Sêng Chuan*, ch. 9, *Taishō* ed. pp. 383.2 ff.

[1] 佛圖澄　　[2] 病

and since the elements in developed Chinese science which can be traced to Indian sources are surprisingly few, it does not seem that Buddhism played an important part in moulding it, though assuredly some science was carried from India to China by the monks. Perhaps this was because the Buddhist communities of monks and their lay supporters in China tended generally to form a rather closed system, by no means so intermixed with the indigenous magma of social and intellectual life as were Taoism and Confucianism.[a]

In his famous lecture on 'Evolution and Ethics' T. H. Huxley gave a not unsympathetic account of Buddhist thought, equating the inherited character of individual men with the load of *karma*.[b] Although he made it clear that he considered Buddhism as a whole an indefensible escape from the world of reality, his observations were later seized upon by Lafcadio Hearn, who in a characteristically elegant but philosophically vague essay (1), sought to show that Buddhist thought had anticipated the recognition of the 'ethical significance of the inexplicable laws of heredity'.[c] Such an attempt to show that Buddhist thought is similar to, or at least not incompatible with, the world-outlook of modern science, was carried further a decade later in a more elaborate work by Dahlke (1), and is now renewed by my friend John Blofeld (1).[d] I regret that I find myself unable to regard these efforts as more than *tours-de-force* of religious apologetic. The question of compatibility within an individual personality is of course one thing (as we know from the celebrated case of Faraday), and that of historical effects and influences quite another. It was natural that in modern times Buddhists should seek to reconcile their faith with modern science, just as in Europe floods of literature have been devoted to the same problem as it affected Christianity.

In this connection I should like to cite the works of another friend, Wang Chi-Thung,[1] the venerable engineer, one of the few men who mastered the old learning under the Chhing examination system, and also the natural sciences of the new world. Himself a student of the famous Buddhist scholar[e] Yang Wên-Hui,[2] he has tried, in books such as *Yin Ming Ju Chêng Li Lun Mo Hsiang*[3] (Elucidations of the Buddhist Classics), to reconcile science and Buddhism.

These apologetics take as their starting-point the given situation, and do not go into the matter historically. Yet it remains strange that the law of *karma* was never extended so as to give rise to the concept of scientific law. 'The operation of *karma*', as Streeter said,[f] 'was conceived not juristically as the punishment of a continuing

---

[a] Forke (12), p. 186, is surely right in his judgment that Buddhist thought remained to the end a 'foreign body' in Chinese civilisation.

[b] He relied upon the *hīnayāna* presentations of Rhys Davids and Oldenberg. This identification failed to take account of the fact that all the changes and chances of this mortal life, as well as inherited character, were to be put down to the individual's *karma* load.

[c] Though for science they are neither inexplicable nor have ethical significance.

[d] Following contemporary Chinese Buddhists such as Ouyang Ching-Wu and the Abbot Thai-Hsü (cf. Chhen Jung-Chieh, 4).

[e] 1837 to 1911, Hummel (2), p. 703.

[f] (1), p. 282.

[1] 王季同   [2] 楊文會   [3] 因明入正理論摸象

ego, but naturalistically in terms of a law of cause and effect, which was thought of almost as mechanistically as in the physical sciences.' Without anticipating here what will be said at a later point (Sect. 18) about the differentiation of the concepts of juridical and scientific law in East Asia as contrasted with Europe, one cannot but call attention to the remarkable failure of Buddhist ideas of law to give rise to natural science. There were presumably two reasons for this. First there was no incentive to do any serious thinking about the non-human, non-moral universe, conceived as it was in terms of *māyā*, a kind of disagreeable cinema performance which one was compelled to watch, or going on in a hall from which one had the greatest difficulty in getting out.[a] Secondly, though the operation of the 'law' of cause and effect as such may seem to modern minds quite obviously neutral morally, the moral functions attributed to it were really the only part which interested the Buddhists at all. In a sense, impersonal cosmic inevitability was only a superficial dress with which they clothed their profound religious belief in divine justice. It was therefore useless as a catalyst of causal science.[b]

(e) INFLUENCES OF BUDDHISM ON CHINESE SCIENCE AND SCIENTIFIC THOUGHT

There are certain specific theories associated with Buddhism, it must be admitted, which probably had a broadening effect upon Chinese minds, and might perhaps have predisposed them for modern science. One was the conviction of the infinity of space and time, of a plurality of worlds, and of almost endless lapses of time, reckoned in *kalpas* (*chieh*[1]).[c] Buddhist writings often speak of the enormous number of beings existing, for instance, in a single drop of water or a mote of dust.[d] This may be illustrated by a passage from a work at least as early as the +6th century, the *Lokasthiti Abhidharma Śāstra* (*Li Shih A-Pi-Than Lun*[2]),[e] which is mainly concerned with the motions of the sun and moon. It might almost be talking of the 'light-years' of modern astronomy.[f]

---

[a] European parallels are not entirely absent. For other-worldly Christians in the early Middle Ages the world was only an examination-hall, and the Order of Nature its furniture' (Raven (1), p. 49).

[b] I find that others such as J. Bissett Pratt (1) concur in this analysis.

[c] Cf. Fêng Yu-Lan (1), vol. 2, pp. 354, 372. These ideas, moreover, exerted some influence on Chinese mathematical technique, especially with regard to the expression of very large numbers (see Sect. 19g).

[d] Cf. McGovern (2), p. 48. Against all these imaginative vistas there were naturally protests of Neo-Confucian common sense; cf. *Chin Ssu Lu*, ch. 13, p. 58 (tr. Graf (2), vol. 2, p. 719).

[e] N 1297, TW 1644. This work was translated into Chinese by Paramārtha in +558. The passage quoted is found in *TSCC, Chhien hsiang tien*, ch. 14, *Thien pu wai pien*, p. 11a.

[f] Cf. the study which Mus (1) has made of the idea of the reversibility of time in Buddhist mythology. The 'open' quality of this thought-world can hardly be appreciated without an acquaintance with the rigidly limited Aristotelian world-outlook which Galileo and Kepler had so much difficulty in breaking through. Cf. Pagel (8).

[1] 劫　　　[2] 立世阿毗曇論

Monks enquired of Buddha the Illustrious how distant Jambūdvīpa was from the Brahma World. Buddha replied, 'It is very far. If, for instance, on the fifteenth day of the ninth month at full moon a man in the Brahma World should throw down a square stone a thousand feet long and broad, it would do no harm for a long time, for only in the following year at the same date would the stone reach Jambūdvīpa.' [a]

Another notion allows us to go further, and to say that it was probably responsible for the recognition of the true nature of fossils in China long before they were understood in Europe; [b] this was the theory of recurrent world-catastrophes or conflagrations, in which sea and land were turned upside down, and all things returned to a state of chaos before redifferentiating into a normal world again. [c] Four phases of these cycles were recognised, differentiation (*chhêng*[1]), stagnation (*chu*[2]), destruction (*huai*[3]) and chaos or emptiness (*khung*[4]). Later (pp. 485, 487) we shall see how this theory was taken over by the Neo-Confucians. And in Section 23 on geology its heuristic value for palaeontology will clearly appear. But it may well be held that both these theories were broadly Indian rather than specifically Buddhist, so that Buddhism conveyed them to China rather than itself inventing them. [d] In any case we cannot afford to dismiss them with a superior smile, for in our own time some of the most eminent astronomers (such as de Sitter) have found reasons for thinking that our universe may have undergone successive cycles of expansion and contraction.

Another point at which Buddhism made contact with Chinese scientific thought— it would be going rather too far to say either that it stimulated it in this direction, or that it added much to it—was in all that concerned the processes of biological change. This naturally involved both phylogeny and ontogeny. The doctrine of reincarnation or metempsychosis naturally aroused interest once more in those remarkable transformations in which the Chinese had always believed, generalising their correct observations of metamorphosing insects to imaginary metamorphoses of frogs and birds. [e] If birds could turn into mussels (see on, in Section 39 on zoology), it was less surprising that men might do so too (if their load of bad *karma* was sufficiently heavy) or into *pretas* (hungry ghosts) if it was worse. Such were the ends of life-cycles, but the beginnings of life-cycles were equally interesting, and a certain tendency therefore manifested itself for a re-examination of embryology. Although the inhibitory factors operating on Chinese science prevented much of importance being done, we may

---

[a] Tr. Forke (6), p. 141.

[b] This recognition occurred at least as early as the Thang dynasty (+8th century).

[c] A good statement of it is found in the *Lang Huan Chi*[5] (On the Cyclical Recurrence of World Catastrophes) by I Shih-Chen[6] of the Liao dynasty (+10th century); the passage is quoted in *TSCC*, *Chhien hsiang tien*, ch. 7, *Tsa lu*, p. 9*b*. This is the book which compares man to a tapeworm, which could not know that there were other men beside its host; so also there may be many universes other than ours. Cf. McGovern (2), pp. 45 ff. On Chinese medieval ideas concerning parasites see Hoeppli & Chhiang I-Hung (2).

[d] There are traces of the conflagration cycle doctrine in Greek thought, e.g. the *ecpyrosis* (ἐκπύρωσις) of Heraclitus (Diogenes Laertius IX, 7–9; Lovejoy & Boas (1), pp. 79, 83) and the Stoics.

[e] Cf. Reichelt (1), p. 75.

[1] 成   [2] 住   [3] 壞   [4] 空   [5] 瑯環記   [6] 伊世珍

yet trace a distinct parallel between the influence of Buddhist ideas in this connection in China, and the strong influence exerted upon European 17th- and 18th-century embryology by Christian theological theories (entry of the soul into the embryo, transmission of original sin, etc.).[a] Let us give an example.

In the *Mêng Chai Pi Than*[1] (Essays from the Mêng Hall) of Chêng Ching-Wang,[2] we have an early +12th-century (Sung) elaboration of the famous passage of *Chuang Tzu* already quoted (p. 78) on biological transformations.[b] The author follows Chuang Tzu's thought, tries to analyse the changes, links them with the conception of the 'ladder of souls' (cf. p. 23), arrives at the idea of innate tendencies, and ends by a distinctively Buddhist interpretation. Chêng Ching-Wang wrote:

Chuang Chou said, 'All things arise from germs (*chi*[3]) and go back to germs.' This is also recorded in the *Lieh Tzu* book, which has a more complete statement. When I lived in the mountains and quietly observed the transformations of things, I saw many examples of it. The outstanding ones are that earthworms turn into lilies,[c] and that wheat, when it has rotted, turns into moths.[d] From the ordinary principles of things (*wu li*[4]) we cannot analyse these phenomena. (One would suppose that) whenever such a transformation occurs, there must be some perception (*chih*[5]) which brings about an inclination (*hsiang*[6]) for it.

Now the change from the earthworm into the lily is a change from a thing which possesses perception (*chih*) into a thing which has none. But the change from wheat grains into moths is just the opposite.[e] When the earthworm winds itself in the earth into a ball during the stage when it is intending to change, the shape of the lily (bulb) is already formed. Wheat (grains) are changed into moths in one night; they appear like flying dust.

According to the Buddhists, these changes are brought about by extremely real and pure intentions. From these general and specific causes (*yin yuan*[7]) such phenomena arise. Take the everyday fact of the hen hatching the egg, for example; we know that the egg comes from the hen itself, but how can you explain the fact that a hen can hatch a duck's egg, and even as Chuang Tzu records,[f] that a hen can hatch a swan's egg?

[a] Cf. Needham (2), p. 182.

[b] Ch. 18 (Legge (5), vol. 2, pp. 9, 10); cf. also Hu Shih (2), p. 135. Mem. also *Lieh Tzu*, ch. 1, p. 6*b* (Wieger (7), p. 73; R. Wilhelm (4), pp. 4, 115).

[c] The plant is *pai ho*[8] (R682; *Lilium tigrinum*). One wonders whether the basis for this mistaken idea could not have been the very interesting *hsia tshao tung chhung*,[9] an insect larva which is parasitised by a fungus, and out of which, therefore, in the dried specimens (some of which I have), a stalk is seen growing. This double (plant-animal) drug is mentioned, not in the *Pên Tshao Kang Mu*, but in the *Pên Tshao Kang Mu Shih I*[10] (ch. 5, p. 27*b*), an amplification of it written by Chao Hsüeh-Min[11] in +1769; and it is therefore not discussed by Read. It is interesting to find in his memoirs that General Stilwell was told of this phenomenon by his Chinese military colleagues, but refused to believe it. We shall return to it in the Pharmacological Section (45). If Chêng Ching-Wang made some of his observations in Szechuan, he might well have met with it.

[d] It should be remembered that the demonstration of the absence of spontaneous generation in insects was not given till the experiments of Francisco Redi in 17th-century Italy (cf. Singer (1), p. 433).

[e] Because whereas the earthworm-lily transformation is a step downwards in the scale of creation, the wheat-moth transformation is a step upwards. In the first case the sensitive soul is dropped, and the new being has only a vegetative soul; in the second case the new beings will require new sensitive souls.

[f] *Chuang Tzu*, ch. 23 (Legge (5), vol. 2, p. 78).

[1] 蒙齋筆談　　[2] 鄭景望　　[3] 幾　　[4] 物理　　[5] 知　　[6] 向
[7] 因緣　　[8] 百合　　[9] 夏草冬蟲　　[10] 本草綱目拾遺　　[11] 趙學敏

As for the change from the wheat (grains) into the moths, they are actually produced from the 'seeds' (*chung*[1]) of the moths, and the wheat is first altered (*hua*[2]);[a] if not, it could not change itself into moths.[b]

From the above argument, whenever intentions (*nien*[3]) grow up, whether good or bad, there must be some result. Hou Chi[c] was born from a footprint; Chhi[d] from a stone; these things are undoubtedly true. The *Chin Kuang Ming Ching*[4] records[e] that with constant flowing, water is changed into fish, which all have life from Heaven; this is beyond doubt. Unfortunately, I fear that many people do not believe it.[f]

Here, then, we have, in the early +12th century, a real effort to observe, and to understand, the nature of biological transformations, and it is obviously connected with Buddhist ideas concerning metempsychosis.

The case is similar for embryology. Hübotter (2) has translated and annotated a Buddhist sūtra on pregnancy and foetal development, which we shall later examine. Verging on the same subject is the *Yuan Jen Lun*[5] (Discourse on the Origin of Man)[g] by the monk Ho Tsung-Mi[6] (+779 to +841),[h] which has been translated by Haas (1). More on the Taoist side, but doubtless influenced by the same current, are the discussions in the *Sêng Shen Ching*[7] (Canon on the Generation of the Spirits in Man),[i] which must be earlier than +500, and on which we have a valuable paper by Gauchet (1). Similar material is contained in the +11th-century *Lo Shan Lu*[8] of Li Chhang-Ling.[9] We shall have more to say about all these in Section 43 on anatomy and embryology.

A biological element can be found in Buddhist iconography. Fig. 47 shows a statue in the Temple of the Sleeping Buddha at Suchow (Chiu-chhüan) in Kansu province; the monk is undergoing a spiritual metamorphosis, moulting off the 'old man', and he is pulling apart the former skin with his hands.

---

[a] The writer seems to have some idea of preliminary decay in mind.

[b] Perhaps his idea is that in 'downward' transformations, the superior being has an 'intention' to go down. He must be thinking therefore rather of the mahayanist sacrifice of a *bodhisattva* voluntarily entering a plane of creation lower than that to which his enlightenment would entitle him, than of the forced descent of a *karma*-laden soul, since he speaks of 'extremely real and pure' intentions. On the other hand, in 'upward' transformations such as the wheat into the moths, the wheat cannot change itself, but the seeds of the moth can change it, i.e. use it. Here Chêng Ching-Wang had no inkling of insect eggs, as might seem at first sight; he was thinking of 'seeds of mothness' which had been in the wheat all the time, and the development of which was now set on foot by some 'perfuming' influence coming from outside (cf. p. 408 above, and Fêng Yu-Lan (1), vol. 2, p. 305). Perhaps there is here an echo of another mahayanist doctrine, that for upward transformations in the spiritual world the help of already enlightened *bodhisattvas* is necessary. On eggs and seeds see further, pp. 481, 487 below.

[c] G 664; legendary agriculture-hero.

[d] Mayers (1), p. 387; legendary ruler.

[e] This is the *Suvarṇa-prabhāsa Sūtra* already mentioned (p. 405).

[f] Ch. 2, p. 2a, tr. auct. In the last sentences he is arguing that all these things had the seeds of the beings which were evoked from them by 'perfuming'.

[g] N 1594, TW 1886.

[h] Forke (12), p. 366.

[i] *TT* 162 and 315; commentaries *TT* 393–395.

[1] 種   [2] 化   [3] 念   [4] 金光明經   [5] 原人論   [6] 何宗密
[7] 生神經   [8] 樂善錄   [9] 李昌齡

PLATE XIX

Fig. 47. Metamorphosis in Buddhist iconography. One of the statues in the temple of the Sleeping Buddha at Suchow (Chiu-chhüan), Kansu, suggests the 'putting off of the old man' as a snake or insect sheds its skin (orig. photo., 1943).

But behind all these points of tentative contact between Buddhist ideas and the developing interest in the sciences of Nature there lay the fact that Buddhism had introduced to China a wealth of highly sophisticated discussions concerning logic and epistemology. Indeed, these schools of philosophy constitute a veritable labyrinth.[a] The Buddhist (and other Indian) theoretical systems were often at least as subtle as the great philosophies of Europe, and very few have yet received the attention of modern logicians armed with all the aids which mathematical symbolism now provides. Buddhist logic (*pramāṇa*) was not Aristotelian, and its theories of knowledge were not the same as the epistemology of Kant or Locke. But in the present context the most important point is that there were Indian schools not only of formal, but also of dialectical logic, and that China received some at least of the teaching of both of these. The latter tendency thus strongly reinforced the indigenous current of dialectical thought which we have studied in the Taoist[b] and the Mohist[c] schools.

The earliest school of formal logic in India was the Nyāya-Vaiśeshika. With its five-membered inductive-deductive syllogism and its special features such as 'contraposition', it reached the Chinese through the translations of Paramārtha (Chen-Ti[1]), who was working in Nanking during the first half of the +6th century. His *Ju Shih Lun*[2] was the *Tarka-śāstra* of Vasubandhu (+5th), and into this treatise[d] other logical works of that school were absorbed. Further books, by the great Dignāga (late +5th), were translated by Hsüan-Chuang, whose disciple Khuei-Chi[3] became one of the most outstanding logicians in all Chinese history,[e] but Dignāga's best work, the *Pramāṇa-samuccaya*, was never put into Chinese.[f] Nor did the Chinese have access to the much later developments of Indian logic from the +13th century onwards (the Navya-Nyāya schools), which Ingalls (1) has recently interpreted using modern methods, and which seem to have anticipated Western symbolic logic by several centuries.

The dialectical logic was the work of the school of Nāgārjuna (early +2nd century), called the Mādhyamikas,[g] already mentioned (p. 404) in connection with the *māyā* concept. They sought to show that every possible syllogism (or affirmation) is a fallacy because it evokes its contrary (entailed inference or counter-syllogism,

---

[a] The reader is referred for further information to the works of Stcherbatsky (3), de la Vallée Poussin (8), and Walleser (1).

[b] See Sect. 10 above.   [c] See Sect. 11 above.

[d] N 1252, TW 1633.

[e] His commentaries are still regarded as the principal Chinese medieval work on logic. See Stcherbatsky (1), vol. 1, pp. 52 ff.

[f] Apart from the great work of Stcherbatsky (1), which considers Buddhist logic in all its aspects, there is the monograph of Tucci (1), who translated several pre-Dignāga Chinese logical texts. There is a book by Sugiura (1) on Hindu logic as preserved in China and Japan, but as it was produced by a Japanese who knew little English, and edited by friends who lacked all sinological knowledge, the names and references are garbled and useless throughout. The labour of identifying them and checking the text would be very great, but might some day be worth while.

[g] This was the San Lun[4] or Khung Tsung[5] school of China and Japan. For its later fortunes see Takakusu (2), pp. 96 ff.

[1] 眞諦   [2] 如實論   [3] 窺基   [4] 三論   [5] 空宗

prāsangika).[a] Hegel expressly referred[b] to Indian predecessors of his logic of contradiction, and it was the considered opinion of Stcherbatsky[c] that his guess was entirely justified by what we now know of the Mādhyamika systems. The *Mādhyamika-śāstra* was first translated into Chinese by Kumārajīva in +409 as the *Chung Lun*[1] (Discourse on the Middle Way).[d] It reached the peak of its influence just about two centuries later, with the famous commentary[e] written by Chi-Tsang[2] (the *Chung Kuan Lun Su*[3]). One of the first Chinese to expound this dialectical logic was the monk Sêng-Chao[4] (+384 to +414), a pupil of Kumārajīva's, and a thinker who had deeply studied the writings of the Taoist fathers. His *Chao Lun*[5] (Discourses of Brother Chao)[f] exemplified the Middle Way (Chung Tao[6]) as understood in his time by a series of antitheses and syntheses. All things exist (*yu*[7]) in one sense, but in another they do not (*fei*[8]); neither assertion is ultimately justified, and really there is neither being nor non-being (*pu chen khung*[9]). Again, things seem to move (*tung*[10]) or to remain at rest (*ching*[11]), but really there is neither movement nor quiescence in the ordinary sense (*pu chhien*[12]).

The logic of contradiction was greatly elaborated and systematised by Chi-Tsang (+549 to +623) with his *Erh Ti Chang*[13] (Essay on the Theory of the Double Truth).[g] His teacher, Fa-Lang,[14] had distinguished between mundane truth (*shih*[15]) and absolute truth (*chen*[16]). The highest level of truth was to be reached through a succession of negations of negations until nothing remained to be either affirmed or denied. The Double Truth was set forth at three levels:

MUNDANE	ABSOLUTE
(1) Affirmation of being (*yu*[17])	Affirmation of non-being (*khung*[18])
(2) Affirmation of either being or non-being (*i yu i khung*[19])	Denial of both being and non-being (*fei yu fei khung*[20])
(3) Either affirmation or denial of both being and non-being	Neither affirmation nor denial of both being and non-being

Chi-Tsang spoke of the gradual renouncing of all ordinary beliefs 'like a framework that leads upwards from the ground'. One can see that much of this dialectic was a kind of systematisation in a new form of what had for centuries been the implicit content of Taoist thought. But having reached this point, Chinese Buddhism fell

---

[a] See Conze (3).
[b] (1) (Lasson ed.), vol. 1, p. 68; (4), vol. 1, pp. 141 ff.   [c] (1), vol. 1, p. 425.
[d] Eng. tr. in Stcherbatsky (2), Germ. tr. in Walleser (2). It is N1179, TW1564.
[e] TW 1824.
[f] TW 1858. Now available in Eng. tr. by Liebenthal (1). See also Fêng Yu-Lan (1), vol. 2, pp. 258 ff.; Bodde (14), p. 59.
[g] TW 1854. Cf. Fêng Yu-Lan (1), vol. 2, pp. 293 ff.; Bodde (14), p. 58; Chhen Jung-Chieh (4), p. 102.

[1] 中論   [2] 吉藏   [3] 中觀論疏   [4] 僧肇   [5] 肇論   [6] 中道
[7] 有   [8] 非   [9] 不眞空   [10] 動   [11] 靜   [12] 不遷
[13] 二諦章   [14] 法朗   [15] 世   [16] 眞   [17] 有   [18] 空
[19] 亦有亦空   [20] 非有非空

rapidly under the domination of the Mere Ideation school so powerfully propagated by Hsüan-Chuang, and instead of pursuing the possibilities of the dialectic, returned[a] to the formal logic of Khuei-Chi and such successors as he had.

There seems to be no evidence that these highly developed modes of thought had any effect upon scientific speculation or study in Thang China. Yet there they were. If hard things were said by the founding fathers of modern science[b] about Aristotelian logic, it was not that it was intrinsically erroneous but that it was so universally used to 'demonstrate' theories about Nature which were based on premises fundamentally false. In a world where the dynamics of natural change were appreciated and could (at least to some extent) be dealt with, the logic of Hegel and his successors arose. The logic of Chi-Tsang was a precious instrument, but it lay in the wilderness, with no one to pick it up and use it.[c]

Sometimes the very absence of an assured body of scientific knowledge worked in favour of Buddhism. Thus Yen Chih-Thui,[1] who lived from +531 to +606, and was a high official under the Sui and the preceding minor dynasties, wrote a kind of Economica for his family, the *Yen shih Chia Hsün*[2] (Yen's Advice to his Family). In this he asks many (then unanswerable) questions about astronomical, meteorological and other scientific subjects, rather in the manner of the *Thien Wên* (Questions about Heaven) of Chhü Yuan; and ends by saying that since we know so little, the mythological stories in the sūtras about the Buddhas and *bodhisattvas* may well be true.

(f) TANTRISM AND ITS RELATION WITH TAOISM

This is by no means all that can be said, however, on the relations of Buddhism to science. Since every Yang must have its Yin, there was an obverse to Buddhism, startlingly different from the ascetic practice and idealist philosophy we have so far been discussing. This 'Taoist department' of Buddhism was Tantrism.[d]

The Tantras[e] (*ta chiao*[3] or *shen pien*[4]) were late sacred texts on the borderline between Hinduism and Buddhism, produced in India not earlier than the +6th century. The practices accompanying them were sometimes open (*dakṣiṇacaryā*[5]) and sometimes esoteric (*vāmacaryā*[6]); and at first sight odd indeed. Worship (*bhakti*) of personal gods was prominent, but more characteristic was the strongly magical element, including 'words of power' (*mantras*[7] or *dhāraṇīs*[8]), talismans (*yantras*), amulets (*kavacas*), hand-gestures (*mudrās*[9]) and other charms. These *i kuei*[10] overlap tantric texts; the *Saddharma-puṇḍarīka Sūtra*, for example, has a whole chapter of

---

[a] Cf. Stcherbatsky (5).　　　　　　　　　　[b] See pp. 200 ff. above.

[c] The success of dialectical materialist thought in China in our own time will probably lead to a revaluation of medieval dialectics. It is noteworthy that at least one European Buddhist scholar was led to the study of the Mādhyamikas from the dialectical philosophy of Marx (see Conze, 1, 2).

[d] For a recent general account, see the book of Dasgupta (1).

[e] The word *tantra* also means a textile web with its warp and weft.

[1] 顏之推　　　[2] 顏氏家訓　　　[3] 大教　　　[4] 神變　　　[5] 達嚫拏遮唎耶
[6] 縛摩遮唎耶　　　[7] 曼怛羅　　　[8] 陀隣尼　　　[9] 慕捺囉　　　[10] 儀軌

dhāraṇīs. Tantrism adopted as its symbolic forms what one might call 'electrical' imagery; it was known as the 'way of the thunderbolt' (*vajrayāna; chin kang chhêng*[1]). One can see at once that one is in the presence of a system of thought closely akin to the shamanist and magical side of ancient Taoism, and hence, on the principle that magic and science were originally united in a single undifferentiated complex of manual operations, here, if anywhere, Buddhism may have produced some contribution to science.

It is then of great interest to find that just as ancient and early medieval Taoism was deeply interested in the phenomena of sex, so also this was central to Tantrism. The *vajra* (thunderbolt or lightning flash) was identified with the male external generative organ, the *liṅgam* (*sêng chih*[2]), while the lotus, *padma* (*lien*[3])—so characteristic of Buddhist iconography—was identified with that of the female, the *yoni* (*nü kên*[4]). Essentially the theological doctrine was that the mystical or divine energy of a god (or of a Buddha) resided in his female counterpart, from whom he received it in an eternal embrace. There had to be one of these *śakti*,[5] therefore, for each god or Buddha.[a] The logical conclusion followed that the earthly *yogi* seeking for perfection must also embrace his *yoginī*, in a sexual union (*maithuna*) prepared for and conducted with special rites and ceremonies (*cakra*). There followed also the worship of women (*strīpūjā*)[b] as a preliminary to *maithuna*. The whole forms a remarkable parallel to the practices of early medieval Taoism (cf. pp. 149, 151), though Buddhism seems to have come a long way from its origin when we find the phrase *Buddhatvaṃ yoṣidyonisamāśritam*—Buddheity is in the female generative organs.[c] Naturally Victorian scholars spoke of Tantrism with bated breath, but we may well question whether these ideas, which after all we cannot judge by the canons of a civilisation which has had two thousand years of Pauline anti-sexuality, were not quite reasonably associated with the magical-scientific view of the world. I would remind the reader only of the great, though sometimes unsuspected, part which sexual symbolism has played in the language of the alchemists.[d] May it not have been that the very conception of chemical reaction arose by analogy from the congress of the human sexes?

One of the most important Buddhist Tantric texts is the *Guhyasamāja-tantra* or *Tathāgataguhyaka* (ed. B. Bhattacharya, 1), which is certainly not earlier than the +7th century.[e] We are not surprised to meet with a good deal about the control of respiration (*prāṇāyāma*) in it. The *śakti* element is also very strong, and indeed here the union of the sexes is said to be of the essence of Tantrism. The theory was that

---

[a] Hence the statues of gods and *śaktis* in sexual union, so common in Tibetan art. The most important Buddhist goddess here was Tārā[6] (de Blonay) or Kuan Shih Yin Mu.[7] She was the *śakti* of Avalokiteśvara,[8] so perhaps her femininity was transferred to him.

[b] Cf. the description by H. H. Wilson (1), pp. 160 ff.

[c] De la Vallée Poussin (5).

[d] The alchemical reference comes out clearly in de la Vallée Poussin's discussion of Tantrism, (4), p. 131. Cf. Jung (1).   [e] Winternitz (1). TW 885.

[1] 金剛乘   [2] 生支   [3] 蓮   [4] 女根   [5] 舍支   [6] 陀羅
[7] 觀世音母   [8] 阿縛盧枳低濕伐邏

emptiness (*śūnyatā*) was of male quality, while compassion (*karuṇā*) was female; in order, therefore, to achieve unity (*advaya*) a sexual act was required. This looks almost like a symbolisation of the two basic trends in Buddhism, the nihilistic philosophy on the one hand, and the warm-hearted love for all beings on the other. In that sense one may feel that the best in Buddhism was derived from its Yin, or *śakti*, side. It is interesting that Tantrism, like Taoism, encouraged woman adepts, and we find names such as those of Lakṣmīnkara (fl. +729) and Sahayayoginī (fl. +765) in the lists of its leaders.

Indian Buddhist Tantrism appears to have come to China in the +8th century. As Chou I-Liang (1) points out, in an interesting paper, it was not that magic formulae (*dhāraṇīs*) had failed to arrive much earlier. Sūtras including these had been translated as early as +230 by Chu Lü-Yen,[1] and in +313 by Chu Fa-Hu[2] (Dharmarakṣa), both Indian monks, as well as by many others; the spells included methods of rain-making, getting water from rocks, finding springs and sources, stopping storms, etc. Just as in Taoism, one real discovery or sound observation probably accompanied a hundred imaginary wish-fulfilments. The field has been so uncultivated that much research will be required to assess the place of these practices in the history of science. In the Thang the traffic greatly increased, largely owing to the labours of three Indian monks, Subhākarasiṃha (Shan Wu-Wei[3]) (+636 to +735), who came to China in +716; Vajrabodhi[4] (Chin-Kang-Chih[5]) (d. +732); and Amoghavajra[6] (A-Mou-Ka[7]) or Pu-Khung[8] (d. +774). But the Chinese were also active; monks such as Chih-Thung[9] wrote much on Tantrism, and the great traveller I-Ching[10] translated a Tantric Sūtra,[a] the *Ta Khung Chhüeh Chou Wang Ching*.[11] But the most important Tantrist was the monk I-Hsing[12] (+672 to +717), the greatest Chinese astronomer and mathematician of his time, and this fact alone should give us pause, since it offers a clue to the possible significance of this form of Buddhism for all kinds of observational and experimental sciences. It would be surprising if there were no alchemical connections, but the subject is difficult to investigate, because, for obvious reasons, Tantrists did not advertise their ways. Thus, for example, Shan Wu-Wei approved of the statues showing sexual union, but warned that they were not to be placed in the public halls of temples. So also in India, Tantrists employed a 'twilight-language' with allusions not intelligible to the uninitiated (*saṃdhyābhāṣā*), a tantric 'slang'.[b]

At first sight, then, Tantrism seems to have been an Indian importation to China. But closer inspection of the dates leads to a consideration, at least, of the possibility that the whole thing was really Taoist.[c] In Section 10*i* (pp. 150 ff. above), we saw that Taoist sexual theories and practices were flourishing between the +2nd and the

[a] The *Mahāmāyurī-vidyārājñī*; Great Peacock Queen of Spells (TW 985 ff.).
[b] Cf. Shadidullah (1).
[c] Cf. Bagchi (1, 4); Lévi (4).

[1] 竺律炎   [2] 竺法護   [3] 善無畏   [4] 跋日羅菩提   [5] 金剛智
[6] 阿目佉跋折羅   [7] 阿牟伽   [8] 不空   [9] 智通   [10] 義淨
[11] 大孔雀咒王經   [12] 一行

+6th centuries in China, definitely before the rise of the cult in India, and its re-importation (if it was a re-importation) by the Buddhists. Bhattacharya (2) significantly tells us here that the principal localities associated with Buddhist Tantrism were in Assam.[a] This reminds us that one of Pelliot's most remarkable memoirs (8) concerned a Sanskrit translation of the *Tao Tê Ching*.[b] It was made for Bhāskara Kumāra, king of Kāmarūpa (Assam), who had asked Wang Hsüan-Tshê for it in +644. A very living account of the work being done, with all the difficulties which the translation involved, exists in the *Chi Ku Chin Fo Tao Lun Hêng*[1] (Critical Collection of Discourses on Buddhist Doctrine in various Ages)[c] under date +647. Pelliot translated this. In Tantric literature, moreover, China (Mahācīna) occupies a very important place as being the seat of a cult Cīnācārya which worshipped a goddess called Mahācīnatārā[2] (Bagchi, 1). Sages such as Vasiṣṭha were said to have travelled there to gain initiation into this cult, in which women played a very prominent part. Possibly, therefore, Tantrism was another instance of foreigners amiably instructing Chinese in matters with which the Chinese were already quite familiar. However, the sexual element in Indian religion had from ancient times been so marked that Buddhist Tantrism may equally well be considered a kind of hybrid of Buddhism and Hinduism. The *śakti* idea is certainly ancient (cf. Das, 1).

In any case, it is possible to find detailed parallels of much precision between Taoism and Tantrism. It will be remembered that in Section 10 (p. 149), mention was made of the Taoist practice of *huan ching*,[3] 'making the *ching*, or seminal essence, return'. In this method, pressure was exerted on the urethra at the moment of ejaculation in such a way as to force the seminal discharge into the bladder, whence it was afterwards voided with the urine; the Taoists imagined, however, that it made its way up into the brain, which it nourished in some marvellous way. Now in Bose's book on the post-Caitanya Vaiṣṇavite (Hindu) Sahajiyā[d] cult of Bengal, still existing, we find that an exactly similar method is used. In this sect, where the rites of *maithuna*[e] are a kind of elaborately stylised and ritualised physical love, whether of couples married (*svakīya*) or otherwise (*parakīya*), 'the semen is made to go upwards to the region of Paramātma'. Though the physiological technique is not clearly described, the correspondence is too close to be accidental. There is, moreover, an epithet, *ūrdhvaretas* (lit. meaning 'upward semen'), which occurs commonly in the *Mahābhārata* and the

---

[a] Dr J. Filliozat tells me that there are Tamil legends about Tantric adepts voyaging between India and China; one from the *Sattakāṇḍam* is given in Mariadassou (1), p. 15.

[b] Kumārajīva is said to have made a commentary on the *Tao Tê Ching* (Bagchi, 1). Foreign envoys often requested copies of the Taoist classics, e.g. the Japanese in +735 (*Tshê Fu Yuan Kuei*, ch. 999, p. 18b).   [c] +661 to +664; N 1471, TW 2104, ch. 2, § 10.

[d] Bagchi (1), discussing the meaning of this word, likens it to Tao interpreted as Nature or the Order of Nature. The Sahajayāna cult of Buddhism flourished mainly between the +7th and the +12th centuries, and almost certainly goes back to the transmission of Taoism to Assam already mentioned. Its oldest text, the *Hevajra-tantra* (TW 892), is thought to be of the late +7th century. The Sahajiyā cult in Hinduism seems to have started in the +11th. Cf. Dasgupta (2), p. 221; Sastri (1), vol. 2, pp. 303 ff.   [e] See Renou (1), p. 183.

[1] 集古今佛道論衡   [2] 摩訶至那多羅   [3] 還精

Rāmāyaṇa epics, and which has often been translated 'chaste' or 'continent', but which may well have reference to this technique.[a]

In literary sources there are further indications of it. Walter (1), who translated the *Haṭhayoga Pradīpikā* of Swatmeram Swami, and de la Vallée Poussin,[b] commenting on this and other texts, have described how the *prāṇa* (= *chhi*?) must be made to mount upwards towards the heart by means of a certain vessel (*suṣumnā*).[c] Here perineal (*kanda*) pressure is part of the respiratory exercises, but these were closely connected, in both civilisations, with the sexual techniques. Tucci, in his monumental book on Tibetan painted scrolls,[d] says that the sexual act was controlled by respiratory mechanisms (*prāṇāyāma*) 'in such a manner that the semen goes its way backwards, not descending but ascending, till it reaches the "thousand-petalled lotus" at the top of the head'. This is also described in another Tantric book, the *Śubhāṣitasaṃgraha* (ed. Bendall).[e] Besides, quite apart from these detailed technical resemblances, Tantric literature is full of paradoxes similar to those so characteristic of early Taoism.[f]

Chinese Tantrism and its relations with early Taoism are certainly not easy matters to investigate. In the famous Ming novel *Chin Phing Mei*[1] (tr. Egerton) there seems to be no reference to it. But the continuation, the *Hsü Chin Phing Mei*,[2] contains, in chapters 35, 36 and 37, a remarkable description of what are obviously sexual temple rites of early Taoist character, though ascribed to Lamaism, and associated especially with the Chin Tartars (+ 1115 to + 1234). This book is again of uncertain date, but must be placed some time very early in the Chhing dynasty, i.e. about + 1660, and whoever was its author took the highly Taoist pseudonym Tzu Yang Tao Jen.[3] According to the description, the directors of the ceremonies were women, the chief priestess being known as Pai Hua Ku.[4] The Tantric element is demonstrated by the two keywords *khung*,[5] emptiness, and *sê*,[6] love. The name of the doctrine appears as Ta Hsi Lo Chhan Ting Chiao.[7] There is mention of bronze statues of gods in union with their *śakti* consorts, of aphrodisiac drugs (*hsieh yao*[8]), hypnotic dances, public hierogamy, and promiscuous unions on the part of the assembly—just as in the ancient Taoist accounts. One wonders what the source was from which the Purple Yang Taoist got his material in the 17th century.[g]

Summing up, then, one may say that Tantrism presents an aspect of Buddhism which has so far been quite insufficiently investigated. Scholars were formerly

---

[a] Personal communication from Mr D. R. Shackleton Bailey.   [b] (4), p. 143.

[c] Woodroffe (1, 2); Renou (1), p. 185. This has an obvious connection with the 'ascent of Kuṇḍalinī'.

[d] (3), vol. 1, p. 242.   [e] See also Bagchi (3).

[f] See, for example, the passages of the *Kulārṇava-tantra* translated by Renou, p. 179.

[g] Perhaps it was Chêng Ssu-Hsiao[9] who in his *Hsin Shih*[10] of about + 1295 describes Lamaist rites of this kind as taking place in the Fo-Mu-Tien[11] temple (cf. van Gulik (3), p. 96). Chêng, however, is known to have been extremely anti-Mongol, so he may have been setting down what he considered the scandals of the Yuan (see Kuwabara, 1). One version of the *Hsü Chin Phing Mei* has been translated by Kuhn (1).

[1] 金瓶梅   [2] 續金瓶梅   [3] 紫陽道人   [4] 百花姑   [5] 空

[6] 色   [7] 大喜樂禪定教   [8] 邪藥   [9] 鄭思肖   [10] 心史

[11] 佛母殿

deterred by an attitude to sex so drastically different from that of occidental culture; they will now be repelled by the vast mass of apparently unprofitable and nonsensical charms and magic. But it must be remembered that out of the morass of magic grew up the flowers of true knowledge of Nature—as in magnetism, pharmacy, chemistry and medicine itself. I would therefore venture to say that Tantrism represents one of the fields of research in which interesting discoveries concerning the early history of science in Asia are most likely to be made.

(g) CONCLUSIONS

All in all, however, the problem of analysing fully the antagonistic effects of Buddhism on East Asian science, remains. Perhaps it sprang from deeper causes than any which have so far in this Section been put into words. In the last resort, Buddhism was a profound rejection of the world, a world which, each in their different ways, both Confucianism and Taoism accepted. The Buddhists had what was essentially a 'sour-grapes' philosophy; from the transience of all earthly joys and pleasures they deduced their unreality and worthlessness, but it was a *non sequitur*. Inhabitants as we are of a world so much more liberated from pain and fear by true knowledge of Nature, and its application in machines, it is extremely difficult to place ourselves in the position of the ancient and medieval Buddhists. The insecurity of life was then so great, disease and death were everywhere, life was cheap; and the little nuclei of human happiness, the lovers or the young parents of children, could be exploded in a moment by drought, by flood, or by the activities of warring armies, without the hope of finding one another again except by merest chance. In such circumstances it was understandable that men and women should centre their hopes, if not on another world, at least on a creed and a way of life which did not depend upon the security of this; it was understandable that the *meditatio mortis* should flourish; and it was natural that they should unite in calling the visible world ugly because they could not make it happy.[a] The only surprising thing is that throughout these centuries Buddhism did not meet with even greater success, and that so large a number of Confucians continued to make the fundamental and significant affirmation that life *was* worth living in the well-ordered society, and that however bleak the immediate prospect might be, men would always arise who would practise what Confucius had taught as to *how* society could be well ordered.[b] Similarly, the Taoists, walking, as ever, outside society, refused to give up their naturalistic and realistic world-picture. The external world was, for them, real and no illusion; the sage, by following after its phenomena, would learn how to control them. A sexual element was at the heart of all things,

[a] Cf. the striking juxtaposition of texts made by Havelock Ellis (1), p. 208.

[b] As a single example, one might take Su Chhiung,[1] who served the Northern Chhi and died in +581. He was renowned for the humanity of his administration, the reduction of taxes, acquittal of innocent persons, encouragement of trade, and so on (*Pei Shih*, ch. 86).

[1] 蘇瓊

and asceticism, in so far as it was valuable at all, was simply a means to an end, the attainment of material immortality, so that the enjoyment of Nature and her beauty might have no end.

Here is the keynote. One of the pre-conditions absolutely necessary for the development of science is an acceptance of Nature, not a turning away from her. If the scientist passes the beauty by, it is only because he is entranced by the mechanism. But other-worldly rejection of this world seems to be formally and psychologically incompatible with the development of science.

On modern Buddhism in China there is of course a mass of literature, ranging from the almost century-old but still valuable book of Edkins (4) to the brief but vivid description of Blofeld (1) in 1948.[a] Intermediate in date are the semi-popular book of Johnston (2)[b] and the interesting, though tantalising, work of Reichelt (1), whose first-hand experience allowed him to do for Chinese Buddhism what Shryock (1) did for Confucianism. One has only to read Reichelt's descriptions of the liturgical beauties of the worship in Buddhist abbeys to realise that this religion supplied a certain factor in Chinese life which no other system did. The present writer, though never himself stirred by Buddhism as by the whole Taoist-Confucian complex, acknowledges freely the great and solemn loveliness of Buddhist religious buildings, and the hospitality which he, in common with millions of other men, has received at the hands of Buddhist monks. The judgment which he has often been tempted to make, that Buddhism developed all the vices of Christianity with few of its virtues, would be, on maturer consideration, too cruel, and one must rest in the conviction that at any rate during the + 1st millennium Buddhism was a great civilising force in Asia. For Central Asia this term is surely appropriate, but for China, which already had a civilisation of high order, matters were a little different; Buddhism there introduced that element of universal compassion which neither Taoism nor Confucianism, rooted as they were in family-ridden Chinese society, could produce. Despairing its philosophy might be, and for the scientist perverse, but its later practice was often plainly and recognisably that of universal love.[c]

[a] Cf. also Hamilton (2). Compendia of mythology will be found in Maspero (11) and Doré (1), pt. II, vols. 6, 7, 8; pt. III, vol. 15.

[b] The Japanese scene is out of our province, but their point of view on Buddhism may be approached by means of the little book of B. L. Suzuki (1).

[c] Cf. the sūtras quoted by D. T. Suzuki (1), pp. 366 ff.

16. CHIN AND THANG TAOISTS, AND SUNG NEO-CONFUCIANS

So GREAT was the heterodoxy of Taoism that even today there is apparently no good history of it after the period of the philosophers. In Western languages there is nothing corresponding to Shryock's study of the development of the State cult of Confucius, nor to the large amount of material which has been written on the Neo-Confucians of the Sung. Even in Chinese such books as those of Hsü Ti-Shan (1) and Chia Fêng-Chen (1) say practically nothing about later developments. It is hardly to be believed that not a single one of the 1464 books in the *Tao Tsang* has been translated.[a]

(a) TAOIST THOUGHT IN THE WEI AND CHIN PERIODS

(1) WANG PI AND THE REVISIONISTS

It seems that the proto-scientific teachings of the Taoist philosophers were the first to die out for want of appreciation. Han Fei, the early −3rd-century Legalist philosopher, devoted two of the chapters[b] of his *Han Fei Tzu*[1] to commenting on a number of chapters of the *Tao Tê Ching*, interpreting them as near to Legalist doctrines as possible. The practice of making the ambiguities of Lao Tzu serve one's own purposes may be said, therefore, to date at least from him. The first commentator, properly so called, of the *Tao Tê Ching* was a person of uncertain date, who used the sobriquet Ho Shang Kung[2] (the Old Gentleman of the Riverside); the *Sui Shu* bibliography placed him *c.* −160, but it is considered that he must have been of the Later Han, probably +1st or +2nd century. A translation of his commentary by Erkes (4) has been appearing in recent years. It may be said that a good deal of Lao Tzu's anti-feudalism was appreciated by Ho Shang Kung, who knew, for example, that *hun*[3] meant a united community, and said that the 'learned' (Confucians) should be opposed because their thoughts do not agree with the science of the Tao. 'He who wishes to be honoured must look for his basis in lowliness', and 'When the people have not enough, and the prince too much, that is robbery'. But of the proto-scientific aspect there is no understanding, and correspondingly there are many traces of increasing attention to the practices of the cultivation of the body.

---

[a] Except in so far as the *Tao Tsang* reprints versions of the texts of the great philosophers. There are perhaps a very few exceptions to this statement, as we shall see in the Section on alchemy.
[b] Chs. 20 and 21 (tr. Liao (1), pp. 169 ff.).

[1] 韓非子   [2] 河上公   [3] 混

In order to see what happened to Taoist thought in the Wei and Chin dynasties one cannot do better than follow the recent book of Fêng Yu-Lan (2).[a] Inevitably the Confucians encouraged what might be called a thorough revisionism in favour of religious-mystical interpretations—anything else would have threatened their own supremacy and the stability of the bureaucratic social order. Both scientific observations and democratic collectivism were shocking to them, so a far-reaching distortion of the ideas of the Taoist philosophers took place.[b] In the +3rd and +4th centuries there developed a Mystical School (*Hsüan Hsüeh*[1]) led by the new commentators of the Taoist books, Hsiang Hsiu,[2] Kuo Hsiang[3] and Wang Pi.[4] All were essentially Confucians, in that they considered Confucius superior to the Taoist sages, but it seems doubtful whether they were quite so mystical as Fêng Yu-Lan represents them, since there is a tradition that Hsiang Hsiu[c] at least practised alchemy. The Tao was now interpreted as 'non-being',[d] and Confucius praised because he did not try to speak about the unspeakable. The commentary of Hsiang Hsiu and Kuo Hsiang on *Chuang Tzu* states[e] that though his words were both true and sublime, they were useless for action in human society. This was doubtless so for the society to which Hsiang Hsiu and Kuo Hsiang[f] were obliged to conform. Wang Pi certainly did not appreciate the meaning of the old doctrine of ataraxy, of being liberated from emotions by comprehending the inevitable processes of Nature. The commentators returned to the praise of Yao and Shun (the typical Confucian sage-kings),[g] and urged that if his interior life was mystically sound, a man might occupy the highest offices at court.[h] Fêng Yu-Lan described[i] their work as 'an effort to turn the early Taoists' original theories of the solitary and contemplative life into a philosophy of the world fit for ordinary beings in it, combining what is outside the world with what is inside it'. He thought they succeeded. But in fact the whole Taoist system was emasculated for continued existence in, and adaptation to, a milieu in which the Confucian conventions were dominant.[j]

The implicit conservatism of Wang Pi[k] and his colleagues has recently been brought out in a remarkable monograph by Petrov (1).[l] Nevertheless, Wang Pi, whose life

[a] Tr. E. R. Hughes. See esp. pp. 130 ff. Cf. Fêng Yu-Lan (1), vol. 2, pp. 168 ff.
[b] I first realised the importance of this in a conversation with my friend Dr Chêng Tê-Khun.
[c] G693. He was one of the 'Seven Sages of the Bamboo Grove' (*chu lin chhi hsien*[5]), on whom see Balazs (2). Another of this group of Taoist poets and wits, Hsi Khang, besides studying alchemy, worked at a forge, which was considered very ungentlemanly.
[d] Fêng Yu-Lan (1), vol. 2, p. 208.
[e] In the Introduction (cf. Fêng Yu-Lan (1), vol. 2, p. 171).
[f] d. +312 (G 1062).
[g] Ch. 1 (*Pu Chêng* ed. ch. 1A, p. 12a), tr. Bodde in Fêng Yu-Lan (1), vol. 2, p. 234.
[h] There was no doubt an element of Confucian humanitarianism here, inspired by Mahāyāna Buddhism. Just as the *Bodhisattvas* returned from the threshold of perfection to plunge again as saviours into the mortal world, so the Confucian sages were supposed to have understood all that Lao Tzu and Chuang Tzu understood, and yet to have continued their efforts to help human society. The only thing wrong with this argument was its premise. [i] (2), p. 146.
[j] This is almost admitted by Fêng Yu-Lan (1), vol. 2, p. 175. [k] G 2210.
[l] English résumé by A. F. Wright (1).

¹ 玄學   ² 向秀   ³ 郭象   ⁴ 王弼   ⁵ 竹林七賢

lasted only twenty-three years (+226 to +249), was an extraordinary man.[a] It seems that he was partly responsible for the process whereby the Book of Changes (*I Ching*) became a repository of abstract conceptions. In his *Chou I Lüeh Li*[1] (Outline of the System used in the *I Ching*) he said: 'If something has the meaning of "firmness" (the definition of the first *kua*, *Chhien*[2]), there is no need to introduce the symbol of the horse to explain it. If something is in the category of "compliance" (the definition of the second *kua*, *Khun*[3]) there is no need to symbolise this by a cow.' Section 13*g* gave in full these abstract conceptions as they finally developed, and described the unfortunate effect which they had on Chinese scientific thought.[b] In the writings of Wang Pi we find many technical terms which later on had a great career, for example, the distinction between substance[c] (*thi*[4]) and operation[d] (*yung*[5]); the use of the word *li* which later became so important in Neo-Confucianism (*so i jan chih li*,[6] 'the reason why things are so'); and hardness (*kang*[7]) and softness (*jou*[8]), which were later central in the philosophy of Shao Yung (see on, p. 455).

These mixed Confucian-Taoist schools became known as the 'Philosophic Wit' or 'Pure Conversation' (*chhing than*[9]) groups.[e] Epigrams were prized and mundane matters avoided. Although the Confucianisation of Taoism attracted the best intellects of the age, some of Wang Pi's contemporaries reacted by emphasising the social radicalism of the Taoist complex. Hsi Khang[10] the poet (who scandalised the Confucians by his skill as a metal-worker), with his colleagues among the 'Seven Sages', was one such figure, but the refusal to accept the bonds of conventional morality and social institutions led far beyond in the direction of a Taoist antinomianism. Wang Têng,[11] Juan Chi[12] and Huwu Yen-Kuo[13] were representatives of this romantic (*fêng liu*[14]) movement. One must remember that most of the information available about them comes from their enemies, and that they were probably not so much hedonists as believers in Taoist methods of 'nourishing the vitality'[f] who shocked conventional people above all by their individualist scale of values.[g]

(2) PAO CHING-YEN AND THE RADICALS

During this period, indeed, the conservative revisionists did not have it all their own way. We have seen that the Han commentator, Ho Shang Kung, still appreciated a good deal of the political position of the ancient Taoists. Then in the late +3rd or early +4th century we come upon the singular figure of Pao Ching-Yen,[15] the most

---

[a] See the paper of Thang Yung-Thung (*1*).   [b] See above, pp. 311 ff., 322 ff.

[c] Or 'content'. Lit. 'basic body'.

[d] Or 'function' or even 'form', or (scholastic) 'accident' (cf. Graf (*2*), vol. 1, p. 254).

[e] The translations are conventional ones; 'Discussions of abstract and unworldly matters' would be better. Material is available on this subject by Fan Shou-Khang (*1*, *2*); Yü Hsüan (*1*) and Chhen Yin-Kho (*1*).   [f] Cf. pp. 67, 141 ff. above.   [g] Cf. Balazs (*2*).

---

¹ 周易略例   ² 乾   ³ 坤   ⁴ 體   ⁵ 用   ⁶ 所以然之理

⁷ 剛   ⁸ 柔   ⁹ 清談   ¹⁰ 嵇康   ¹¹ 王澄   ¹² 阮籍

¹³ 胡毋彦國   ¹⁴ 風流   ¹⁵ 鮑敬言

radical thinker of all the medieval Chinese centuries.[a] Nothing is known of his life, for we only meet with him in the 48th chapter of Ko Hung's *Pao Phu Tzu* (*Wai Phien*,[1] the so-called 'outer chapters'[b] of *Pao Phu Tzu*, which deal with social and political matters, in contrast with the alchemical content of the 'inner chapters'). The 48th chapter is an unusually long one, and is entirely taken up with a dialogue between Pao Ching-Yen and Ko Hung.[c] The former is therefore generally assumed to have been a contemporary of the latter, but he may have belonged to an earlier generation of the first half of the +3rd century, while the possibility is not altogether excluded that he may have been a character invented by Ko Hung, into whose mouth he could put political doctrines for which he did not dare to accept responsibility himself.

In Pao Ching-Yen the old Taoist aversion to feudalism, now slightly modified to face the feudal bureaucratism which had grown up, seems to have lost none of its force.

The Confucians say that Heaven created the people, and planted lords over them. But why should illustrious Heaven be brought into the matter, and why should it have given such precise instructions? The strong overcame the weak and brought them into subjection, the clever outwitted the simple, and made them serve them—this was the origin of lords and officials, and the beginning of mastery over the simple people. Corvée labour was imposed by the strong upon the weak and by the clever on the simple. Heaven had nothing whatever to do with it....(p. 1*a*).

In ancient times there were no lords and officials (*wu chün wu chhen*[2]). Men (spontaneously) dug wells for water, and ploughed fields for food. Man in the morning went forth to his labour (without being ordered to do so) and rested in the evening. People were free and uninhibited and at peace; they did not compete with one another, and knew neither shame nor honours. There were no paths on the mountains, and no bridges over waters nor boats upon them, nor were the rivers made navigable. Thus invasions and annexations were not possible,[d] nor did soldiers gather together in large companies in order to attack one another in organised war (p. 1*b*).

Power and profit were not the mainsprings of human activity, there were no insurrections or other misfortunes, no weapons, and no fortified places with moats. The myriad beings participated in a mysterious equality (*hsüan thung*[3]) and forgot themselves in the Tao. Contagious diseases did not spread, and long life was followed by natural death. The hearts of men were pure and innocent, ruses and deceits were not born. Having enough to eat, the people were contented, patted themselves on the belly, and wandered about for pleasure. How could the wrenching of the people's goods from them have been possible? How could mutilating punishments even have been imagined?

He goes on to describe the decrease of simplicity and honesty and the growth of artificiality, in traditional Taoist style.

---

[a] Forke (12), p. 224; Balazs (2).   [b] *TT* 1173.
[c] It deserves a full annotated translation.

[d] Note how clearly Pao Ching-Yen confirms the interpretation given in the Section on Taoism of the famous ideal condition of old when people in one village would hear the barking of dogs in other villages, yet never bother to go there (p. 100 above).

[1] 抱朴子外篇   [2] 無君無臣   [3] 玄同

This is why I say: Who can make sceptres without destroying the Uncarved Block, the Natural Jade? Why so much fuss about altruism and justice (*jen i*[1]) if the Tao and its Virtue had not already been ruined? How is it that tyrants like Chieh and Chou could burn men alive, assassinate censors, cut high officials in pieces, tear out hearts and break bones, exhaust all the possibilities of evil in their cruel tortures? If these tyrants were simply men of the commons, how could they give rein to their natures even if naturally cruel? The reason why they can put their cruelty into practice, give full scope to their perversity, and cut up the Empire like butchers, is because of this status of 'prince' which authorises them to follow their good pleasure. Once the relation of prince and subject is established—the ill-will of the masses grows day by day. Then arise the revolts of slaves, tumults in the mud and dust; then do the rulers tremble on high in their ancestral temples, and the people are harassed and distressed. The people are to be shut up in rites and ordinances, corrected by laws and punishments. You might as well try to protect yourself from the howling storm with a handful of earth or set up the palm of your hand as a dyke against a tidal wave (pp. 2*a*, *b*, 3*a*).[a]

It might be Gerrard Winstanley speaking of 'kingly power and Norman tyranny', or in an earlier generation, John Ball. With the Taoist ideal State, Pao Ching-Yen contrasts the actual. The desires of the lords are insatiable; they monopolise women in their inner apartments and throw away upon useless extravagances money derived from the bitter labour of the people. The people are hungry and cold, while in the stores of the lords and officials there is abundant clothing and food. So long as these social inequalities persist, all laws and ordinances, however just, will be valueless. It is evident that Pao Ching-Yen, whoever he was, had a clear insight into the origin of social institutions and war, and fully sounded again the political note of the early Taoists.[b] But after his time we do not hear it ring out for many centuries.[c]

Of course, there are anecdotes to be found which hint fairly clearly at views like those of Pao Ching-Yen. I append a contemporary one from the *Shih Shuo Hsin Yü* about a Chin official named Wang Hsiu.[2]

A monk named I[3] asked Wang Hsiu, 'Would you consider that the sage can have private preferences (or prejudices, *chhing*[4])?' Wang Hsiu replied, 'No.' 'Then', said the monk, 'can he be compared to a dumb wooden post?' Wang answered, 'He is like the counting-rods. Like them, he has no prejudice or design, but those who make use of him have.' Then the monk said, 'But who can make use of sages?' Wang turned and went away without giving any answer.[d]

---

[a] Tr. Forke (12), Balazs (2); eng. auct.

[b] 'He was the first Chinese thinker who dared to issue forth from the nebulous utopianism of the early Taoists, to place himself squarely on political ground, and to formulate in concrete fashion the struggle against despotic absolutism' (bureaucratic feudalism), writes Balazs (2).

[c] The literature has of course not been well explored with these ideas in mind. Towards the end of the +9th century there was the Taoist who wrote the *Wu Nêng Tzu*[5] book; he seems to have been fully in Pao Ching-Yen's tradition. Excerpts have been translated by Hsiao Kung-Chhüan (1). Cf. Forke (12), p. 326, and Soymié (1), p. 345. In this connection Thang Chen[6] is also to be mentioned (+1630 to +1704); Forke (9), p. 494.

[d] Ch. 4, p. 26*a*., tr. auct.

[1] 仁義　　　[2] 王修　　　[3] 意　　　[4] 情　　　[5] 无能子　　　[6] 唐甄

(3) KO HUNG AND SCIENTIFIC THOUGHT

Nevertheless, the experimental traditions of ancient Taoism not only continued, but flourished more and more in contrast to the declining fate of its political and philosophical doctrines. Throughout the period from the end of the Three Kingdoms (c. +270) to the beginning of the Sui dynasty (+580) alchemy and its related arts were cultivated to a hitherto undreamed-of extent. This was the period of Ko Hung[a] (fl. +325), the greatest alchemist in Chinese history, from whom we shall quote a remarkable passage. Unfortunately, these three centuries are among the least known of all, presumably partly because after the invention of paper at the beginning of the +2nd century a great many books were written on this non-durable material and have failed to survive. Nevertheless, it is probable that a considerable number of the works in the *Tao Tsang* date from these centuries, and it is urgently necessary that serious study should be given them, to establish their dates, and to elucidate their content. Juan Yuan's *Chhou Jen Chuan* (Biographies of Mathematicians and Scientists) gives particulars of no less than forty-four men between the end of the Han and the beginning of the Sui, and further understanding of their contribution to the development of scientific thought is much to be desired.

While we must not here anticipate what will have to be said about Ko Hung[1] and the alchemy of his age in the appropriate place (Section 33), the earlier chapters of his *Pao Phu Tzu*[2][b] contain some scientific thinking at what appears to be a high level. The mind is groping its way towards a mastery of the complexity of Nature, and it is more impressed by the diversity of phenomena than by any generalisations which would unify them; in other words, the prevailing atmosphere is empirical. There is an argument going on about the possibility of lengthening life or achieving material immortality by artificial means.

Someone said to Ko Hung: 'Even (Lu) Pan[c] and (Mo) Ti could not make sharp needles out of shards and stones. Even Ou Yeh[d] could not weld a fine blade (lit. a Kan Chiang sword) out of lead or tin. The very gods and spirits cannot make possible what is really impossible; Heaven and Earth themselves cannot do what cannot be done. How is it possible for us human beings to find a method which will give constant youth to those who must grow old, or to revive those who must die? And yet you say that (by the power of alchemy) you can cause a cicada to live for a year, and an ephemeral mushroom to survive for many months.[e] Don't you think you are wrong?...'[f]

---

[a] G 978. Wieger (4) devotes a whole chapter (the 52nd) of his survey to Ko Hung, but though easily accessible, it must be taken, like all his other expositions, *cum grano salis*.

[b] Lit. 'Book of the Master who is able to Preserve or Cherish Solidarity'.

[c] G 1424; semi-legendary mechanic of the State of Lu.

[d] Semi-legendary metallurgist.

[e] This (and certain stories also) suggests that the +4th-century alchemists were as familiar with the use of 'experimental animals' as modern workers on drugs and vitamins.

[f] Ch. 2, p. 2a of the *Tao Tsang* edition of *Pao Phu Tzu* (*Nei Phien*).

[1] 葛洪   [2] 抱樸子

Pao Phu Tzu answered: '[The rumbling thunder is inaudible to the deaf, and the three luminaries are invisible to the blind. Is it right then to say that the thunder is quiet and the sun pale? Yet the deaf say there is no roar and the blind say there is nothing there. Still less can they appreciate the harmonies of music, or the designs of the mountains and dragons on the imperial robes.... A mind which has become a prey to imbecility will reject even the Duke of Chou or Confucius, to say nothing of the teaching of the holy *hsien*.[1] The opposition of life to death, and of beginning to end, is indeed a feature of all natural phenomena (*ta thi*[2]), but when scrutinised in detail, however, they sometimes reveal that no such sharp antitheses exist. There are dissimilarities and uniformities, differences in length and shortness, now thus, now otherwise, changes and transformations into a thousand forms occur, things curious and strange show infinite variety, things may look equivalent but their details are different.][a] Things which in the end appear different may have the same root and origin. Things cannot all be spoken of in one way (*wei kho i yeh*[3]). Things which have a beginning generally also have an end, but this is not a universally applicable principle (*fei thung li*[4]). Thus it may be said that everything grows in summer, but shepherd's purse (*chhi*[5])[b] and wheat fade then. It may be said that everything withers in winter, but bamboos and the arbor-vitae bush (*pai*[6])[c] flourish then.[d] It may be said that everything comes to an end, as it begins, but heaven and earth have no end.[e] It is generally said that life is followed by death, but tortoises and cranes live almost for ever. In summer the weather is supposed to be hot, but we often have cool days then; in winter the weather is supposed to be cold, but mild days occur. A hundred rivers flow to the east, but one large river flows north.[f] The earth by nature is quiet, but sometimes it trembles and crumbles. Water by nature is cool, but there are hot springs in Wên Ku.[g] Fire by nature is hot, but there is a cool flame upon Hsiao Chhiu mountain.[h] Heavy things ought to sink in water, but in the South Seas there are floating hills of stone.[i] Light things ought to float, but in Tsang Kho there is a stream in which even a feather sinks.[j] No single generalisation can cover such multitudes of things,

[a] The portion of the text between square brackets is omitted in some editions. To what follows there is an interesting parallel passage in the *Lun Hêng*, ch. 10 (tr. Leslie (1), superseding Forke (4), vol. 2, pp. 1ff.).   [b] *Capsella bursapastoris*; R 478.   [c] *Thuja orientalis*; R 791.

[d] The flourishing of the chrysanthemum in autumn became a common literary allusion, and its petals were eaten to promote longevity. A letter of Tshao Phei about +220 (*CSHK*, San Kuo sect. ch. 7, p. 4a) says that this was done by the −3rd-century poet Chhü Yuan.

[e] Note the absence of influence of Buddhist ideas concerning periodical world catastrophes or conflagrations.

[f] Presumably the Yellow River above and below Ninghsia. But he could have meant the Gan-chiang south of the Poyang Lake, or the Hsiang-chiang south of the Tung-thing Lake. Ko Hung would hardly have known the Yangtze in wild North Yunnan.

[g] These were mentioned already in the *Mu Thien Tzu Chuan*, ch. 1.

[h] This must surely refer to natural gas. Elsewhere in the *Pao Phu Tzu* there is an amplification of this account, which the *Shuo Fu* thought it worth while to reproduce (ch. 8, p. 49b). It said the natives made charcoal with this *tzu-jan huo*.[7] The subject was always found interesting, and in the late 17th century Fang I-Chih treated of it (*Thung Ya*, ch. 48, p. 18a). Virey, writing of natural gas in 1821, spoke of 'fire which does not consume, and from which no warmth is felt'. A wavering pale blue flame under low pressure can indeed give this impression, though really no cooler than ordinary flame. True cool flames can be produced only under laboratory conditions. Cf. Anon. (31).

[i] Perhaps a reference to floating islands. Cf. a +10th-century Arabic text in Ferrand (1), vol. 1, p. 149.

[j] This links on with the many references to 'weak water' in ancient Chinese writings; I am inclined to think that they all referred to natural petroleum seepages, which can include some fractions of quite low boiling-point (see Sect. 23b).

[1] 仙   [2] 大體   [3] 未可一也   [4] 非通理   [5] 薺   [6] 柏

[7] 自然火

as these examples show (*Wan shu chih lei pu kho i i kai tuan chih, chêng ju tzhu yeh*[1])....Thus it is not to be wondered at that the *hsien* does not die like other human beings.'[a]

Someone else said: 'It may be admitted that the *hsien* differs very much from ordinary men, but just as the pine tree compared with other plants is endowed with an extremely long life, so may not the longevity of the *hsien* as exemplified in Lao Tzu and Phêng Tsu be after all a (special) endowment from Nature? One cannot believe that anyone could *learn* to acquire longevity such as theirs.'

Ko Hung replied: 'Of course the pine belongs to a kind different from other trees. But Lao Tzu and Phêng Tsu were human beings like ourselves. Since they could live so long, we can also.'[b]

Being still unsatisfied, someone protested: 'If the medicine which you employ were of the same substance as our own body, it might be efficacious. But I shall never be convinced of the efficacy of a medicine of different origin such as the pine or cypress.'

Ko Hung replied:...'If you drink a boiled extract of hair or skin it will not cure baldness. (So a medicine of the same nature as the body may be ineffective.) But on the other hand we live on the five grains. (So a medicine of a quite different nature to the body may be effective.)'[c]

Admittedly there is much in the *Pao Phu Tzu* which is wild, fanciful and superstitious, but here[d] we have a discussion scientifically as sound as anything in Aristotle, and very much superior to anything which the contemporary occident could produce.[e]

Another passage brings out well the mixture of strange beliefs with true facts which was characteristic of Ko Hung and other Taoist alchemists. Confucian conventionalists, like European scholastic rationalists, denied the beliefs and ignored the facts; Ko Hung, a true Paracelsus one thousand years before the rhapsodical experimentalist of Einsiedeln, fascinated by the facts was prepared to believe in the possibility of almost anything. We find him saying:

As for the art of Change, there is nothing it cannot accomplish. The body of man can naturally be seen, but there are means to make it invisible; ghosts and spirits are naturally invisible yet there are means whereby they can be caused to appear. These things have been repeatedly done.

Water and fire, which are in the heavens, may be obtained by the burning-mirror and the

[a] Ch. 2, pp. 3*b*, 4*a*.
[b] Ch. 3, p. 1*a*.
[c] Ch. 3, pp. 9*b*, 10*a*. Tr. Chikashige (1); Feifel (1); slightly mod.
[d] The passage was appreciated and gave rise to derivative parallels, e.g. in *Chin Lou Tzu*, ch. 5 p. 13*a* (*c.* +550).
[e] Cf. Sarton (1), vol. 1, p. 344. It is significant that Ko Hung, who attacked almost all the philosophical paragons of Chinese thought, reserved his praise for Wang Chhung (Forke (12), p. 112), though this did not prevent him from criticising severely Wang Chhung's ideas on astronomy and cosmology (*Chin Shu*, ch. 11); cf. Sect. 20 below. One of the elements in Ko Hung's greatness was his beautiful Chinese prose style, which commended him to scholars whose Confucianism would not otherwise have predisposed them in his favour. We trust that some traces of this may be found in the translations given. I have been glad to find that my estimate of Ko Hung corresponds with that of Forke (p. 207) who speaks of his 'grosser dialektischer Schärfe'.

[1] 萬 殊 之 類 不 可 以 一 概 斷 之 正 如 此 也

dew-mirror.[a] Lead, which is white, can be turned into a red substance.[b] This red substance can again be whitened to lead. Clouds, rain, frost, and snow, which are all the *chhi* of heaven and earth, can be duplicated exactly and without any difference, by chemical substances.[c]

Creatures which fly and run, and creatures which crawl, all derive a fixed form from the Foundation of Change. Yet suddenly they may change the old body and become totally different things.[d] Of these changes there are so many thousands and tens of thousands that one could never come to the end of describing them.

Man is the noblest of all creatures, yet men or women may be transformed into cranes, stones, tigers, monkeys, sand or turtles.[e] Similarly, the transformation of high mountains into abysses, and the making of peaks out of deep valleys, are examples of change in huge things.[f] Change is inherent in the nature of Heaven and Earth. Why then should we think that gold and silver cannot be made from other things?

Now, for example, the fire from the burning-mirror and the water from the dew-mirror are in no way different from ordinary fire and water. Dragons originating from snakes,[g] and fat (*kao*[1]) from *mao-san*[2],[h] are no different from ordinary dragons and fat.[i]

The basis of all these changes originates from effective influences (*kan chih*[3]) of one thing on another, of a purely natural kind. Unless one knows exhaustively the natural principles (*li*[4]) (of things) and their properties, one cannot know where they are going (*chih kuei*[5]) (i.e. what their intrinsic tendencies are). Unless one understands the beginning and observes the end (i.e. studies causes and effects) one can never get behind the appearances of phenomena.

Narrow-minded and ignorant people take the profound as if it were uncouth, and relegate the marvellous to the realm of fiction. To these people anything that was not spoken of by Chou Kung and Confucius, and not mentioned in the classics, is untrue. What narrow-mindedness and ignorance![j]

In a valuable article, Ware (1) has collected what the *Wei Shu* and the *Sui Shu* (dynastic histories) had to say about Taoism, and some of this material forms a useful background to the time of Ko Hung. 'As for transforming gold (*hua chin*[6])', the *Wei Shu* says,[k] 'melting jade (*hsiao yü*[7]), using talismans (*hsing fu*[8]) and preparing

---

[a] For these mirrors see Section 26g on physics. It was thought that just as the burning-mirror would set things on fire, the fire coming in some way from the heavens, so the dew which collected on a mirror left out at night was 'moon-water', hence the essence of Yin, and naturally of powerful virtue.

[b] Litharge, though the usual term for cinnabar is used. The word *tan* can mean many chemical substances.

[c] He refers to vapours, flames, sublimation, distillation, etc.

[d] Insect metamorphosis.

[e] At first sight this statement refers to the numerous beliefs (to be mentioned in Section 39 on zoology) about transformations of one animal or plant species into another. But under cover of these popular beliefs, Ko Hung may have meant that the material substance of the human body is not destroyed but finds its way into a thousand other natural objects.

[f] This is a very early statement of geological cataclysms, and may imply Buddhist influence.

[g] I think he was referring to the metamorphosis of newts and salamanders from the legless stage.

[h] This plant was not known to Li Shih-Chen, nor is it listed in Bretschneider (1).

[i] Ko Hung must be speaking of fat of vegetable origin. He probably made a potassium soap from it, and saw no difference between this and fat or soap of animal origin.

[j] Ch. 16, p. 42a, tr. auct., adjuv. Wu & Davis (2).   [k] Ch. 114, pp. 32b ff.

[1] 膏   [2] 茅檨   [3] 感致   [4] 理   [5] 指歸   [6] 化金
[7] 銷玉   [8] 行符

talisman water (*chhih shui*[1]), efficacious recipes and marvellous formulae existed by thousands and tens of thousands.' In +400 the Northern Wei emperor 'established a professorship of Taoism and alchemy (*hsien jen po shih*[2]) at the capital Phing-chhêng[3] in Shansi, and a Taoist workshop for the concoction of medicinal preparations. The Western Mountains were allocated to supply firewood (for the furnaces). It was ordered that those guilty of capital offences test (the preparations),[a] but since it was not their original intention (to obtain immortality) many died without proving the efficacy of the drugs.' There was a conspiracy of Chou Tan,[4] the imperial physician, to stop the activities of this laboratory, but he did not succeed against the Taoist Chang Yao,[5] and 'the preparation of drugs was carried on without respite'. This was only a few years before the démarche of Khou Chhien-Chih[6] (between +423 and 428), whereby, as we have already seen (p. 158), he obtained the title of Taoist 'Pope'. Under Wei Wên-Hsiu[7] experimental work was still going on in +448.

But to sketch further this background to the scientific thought of the time would anticipate too much the Section on chemistry.

During these centuries, just as the Confucians reinterpreted the books of Taoist philosophy, so the Taoists appropriated the Book of Changes (*I Ching*), as was natural enough in view of its age-old use for divination (cf. p. 349), and elaborated it into an attempt at a general scientific theory. This movement started in the Later Han, if we may accept the date of +142 as valid for the appearance of the *Tshan Thung Chhi*[8] (Book of the Kinship of the Three) by Wei Po-Yang,[9] which is regarded as the first book on alchemy in Chinese (and, indeed, in any other) history.[b] Significantly, its full title was *Chou I Tshan Thung Chhi*[10] (Book of the Kinship of the Three; or, the Accordance of the *Kua* of the Book of Changes with the Phenomena of Composite Things). In this work we find an elaborate correlation between the eight trigrams (*kua*) and the denary cycle of 'stems' (*kan*), serving to symbolise the various stages of the movements of the sun and moon, and hence the supposed fluctuations, waxings and wanings of the Yang and Yin influences in the world. We have given elsewhere (pp. 262, 313, 332) the details of this system. The alchemists considered it of great importance in connection with the choice of the precise times at which to conduct their experimental operations. Another early presentation of it is contained in the commentary on the *I Ching* of Yü Fan[11] (+164 to +233) preserved in Li Ting-Tso's[12] *Chou I Chi Chieh*[13] (Collected Commentaries on the Book of Changes), put together some time in the Thang.[c]

[a] Another kind of experimental animal, said to have been not unknown to Herophilus and Erasistratus at Alexandria.
[b] Tr. Wu & Davis (1).
[c] Cf. Fêng Yu-Lan (1) in Bodde (4), p. 116; and Fêng (1), vol. 2, p. 426.

[1] 勅水   [2] 仙人博士   [3] 平城   [4] 周澹   [5] 張曜
[6] 寇謙之   [7] 韋文秀   [8] 參同契   [9] 魏伯陽   [10] 周易參同契
[11] 虞翻   [12] 李鼎祚   [13] 周易集解

(b) TAOIST THOUGHT IN THE THANG AND SUNG PERIODS; CHHEN THUAN AND THAN CHHIAO

This kind of thinking may perhaps be said to have reached its climax in Chhen Thuan,[1] a now somewhat shadowy figure of great prominence during the series of short-lived dynasties between the Thang and the Sung.[a] He took his degree under the Later Thang in +932, and was first called to court under the Later Chou in +954. The second Sung emperor treated him with great honour from +976 to +984, but though undoubtedly an alchemist and a 'mutationist' (as we have been calling these philosophers who speculated about the trigrams and hexagrams) he pleaded ignorance, and retired as much as he could to a solitary mode of life.[b] He died in +989. There is much about him in the *I Thu Ming Pien*[2], a book written by Hu Wei[3] in +1706 which showed that Chhen Thuan was responsible for many later mutationist interpretations, and for the forms of the Ho Thu[4] and Lo Shu[5] (ancient magic squares; cf. Section 19d on mathematics) as we have them today. This book was of great importance, since it placed the history of these forms of what might be called para-scientific philosophy on a sound critical basis, and destroyed the traditional view that they went back to remote antiquity.[c] The *I Thu Ming Pien* (Clarification of the Diagrams in the Book of Changes) is essential for the study of the development of Chinese thought.

Hu Wei quotes[d] Chu Hsi as saying:

The Book of Changes is simply a matter of Yin and Yang. Chuang Tzu held the same opinion,[e] not without deep thought. Even those who have spoken of medical techniques and the Taoist methods of 'nourishing the vital spirits' have never been able to dispense with the Yin and Yang. Now Wei Po-Yang in his *Tshan Thung Chhi* seems to be the origin of Hsi-I's (i.e. Chhen Thuan's) learning.[f]...The 'prior to Heaven' arrangement of the trigrams started with Chhen Thuan, but he in his turn must have had some teacher. In fact, the magical techniques of *hsiu*[6] (meditation) and *lien*[7] (transformation) were already contained in the *Tshan Thung Chhi*. Now *tshan* means mixing, *thung* means penetrating, and *chhi* means uniting or coinciding. Thus it communicates the principles of the Book of Changes, and its meaning coincides with them.

Wei Po-Yang's book borrows the terms of 'prince' and 'minister' to mean inside and outside. The *kua Li*[8] and *Khan*[9] refer respectively to mercury and to lead. The *kua Chhien*[10] and *Khun*[11] refer respectively to the measurements of quantities and the cauldrons and vessels employed in the operations. He uses the terms 'father' and 'mother' to mean the beginning and the end respectively. He uses the expression 'embrace of husband and wife' to mean the marriage and intercourse (of substances), and with 'man' and 'woman' he reveals change and new production (i.e. chemical processes). Analysing the Yin-Yang system, he deduces the

---

[a] G 257; Forke (12), pp. 336 ff.   [b] Cf. *TH*, p. 1568.
[c] Hummel (2), p. 336.
[d] *I Thu Ming Pien*, ch. 3, p. 3a; tr. auct.
[e] Ch. 33 (Thien Hsia); cf. Legge (5), vol. 2, p. 216.   [f] In *CTYL*.

[1] 陳摶   [2] 易圖明辨   [3] 胡渭   [4] 河圖   [5] 洛書   [6] 修
[7] 鍊   [8] 離   [9] 坎   [10] 乾   [11] 坤

theory of *fan-fu*,[1] actions and reactions. By *Hui* (the last day of the month) and *Shuo* (the first day of the month) he means upwards and downwards. When he speaks of the *kua* and their *hsiao* (i.e. the individual lines of each *kua*), he means change and transformation. Following the handle of the Northern Dipper (the Great Bear) he selects the circling stars. He separates the morning and evening by noting the hours on the clepsydra (water-clock, *kho-lou*[2]). In all this there is absolutely nothing which does not depend on the symbols of the Book of Changes.[a] Therefore Wei Po-Yang's book is called the *Chou I Tshan Thung Chhi*.[b]

Here the explanation of the alchemical symbolism is interesting, and of course it will be again referred to in Section 33 on chemistry. There are obvious analogies in European alchemy. But what is relevant to the present discussion is the way in which the *I Ching* hexagrams were conceived of as so deeply 'embedded' in Nature. This was characteristic of Chinese proto-scientific or para-scientific thought, and must be considered one of the most important factors inhibitory to the development of truly scientific interpretations of Nature (cf. pp. 329 ff. above).

Yet all this prepared the way for the Neo-Confucian synthesis of the Sung. It did so not merely because the Taoist liking for diagrams stimulated the Sung thinkers to make their own, for this was after all not a very important matter, but rather by emphasising the through-and-through naturalness of Nature. For these later medieval Taoists there was no force in Nature which man could not, if he knew the right techniques, control. If there was 'supernaturalism', it was not that of the over-whelmingly 'other', before which man can only bow down, but that of inferior spirits within the natural order, which man could make to serve him.[c]

During the Thang dynasty there was a second flowering of true Taoist philosophy. Between the +6th and the +10th centuries we can find a number of books which, with the new knowledge acquired by Taoist experimental researches as a background, revived and expanded many of the old doctrines. These are worth taking a good look at, and deserve much more study than they have received. One of these is the *Kuan Yin Tzu*,[3] also known as the *Wên Shih Chen Ching*[4] (True Classic of the Word),[d] written by an unknown Taoist (perhaps Thien Thung-Hsiu[5] of the +8th century) late in the Thang, or perhaps under one of the dynasties of short duration immediately succeeding the Thang, and edited with commentary by Chhen Hsien-Wei,[6] who called

---

[a] The word here translated 'depend on' is *chha*,[7] which means to boast, or to be amazed at, so perhaps it would have been better to say 'nothing which cannot amazingly be found in'.

[b] *Chou I Tshan Thung Chhi Khao I*, ch. 1, pp. 1a, 2a.

[c] The late remains of this mentality have proved a source of much irritation to some. In a characteristically disgruntled passage, Wieger (4), p. 421, wrote: 'Miracles prove nothing to the Taoists. For them, there are none. Everything is possible if you have the right formula....No proofs, no doubts, no surprise. One simply notes that the magician must have had some very powerful technique.' We may of course sense a certain persistence here of intellectual primitivity; 'For the primitive mind', wrote Lévy-Bruhl, 'everything is miraculous, or rather nothing is, and therefore everything is credible and there is nothing either impossible or absurd' (1), p. 377. Cf. p. 284 above.

[d] The chapter numberings used here are from the *Wên Shih Chen Ching* version.

[1] 反復　　[2] 刻漏　　[3] 關尹子　　[4] 文始眞經　　[5] 田同秀
[6] 陳顯微　　[7] 詑

himself Pao I Tzu,[1] in +1254.[a] The *Thien Yin Tzu*[2] (Book of the Heaven-Concealed Master),[b] written by Ssuma Chhêng-Chên[3] about +700, is also interesting, but more obscure.[c] Then there is the *Hua Shu*[4] (Book of Transformations), attributed to Than Chhiao[5] of the +10th century[d] (perhaps incorrectly, but not likely to be later), and a work of much importance.

Of Nature, the *Kuan Yin Tzu* says:[e]

Nature may be compared to a vast ocean. Thousands and millions of changes are taking place in it. Crocodiles and fish are essentially of the same substance as the water in which they live. Man is (lit. I am) crowded together with the myriad other things in the Great Changingness, and his (lit. my) nature is one with that of all other natural things. Knowing that I am of the same nature as all other natural things, I know that there is really no (separate) self, no (separate) personality, no (absolute) death and no (absolute) life.[f]

Next we find that the old 'pre-Socratic' doctrine of the processes of alternating aggregation and dispersion as one of the main causes of the coming-into-being and passing-away of material and living things, is prominent in the Thang Taoists. Undoubtedly this was the channel through which the Neo-Confucians of the Sung, who made much use of it, obtained it. Note, however, the magical, if Baconian, tenor of the following passage:[g]

Man's might can conquer the changes of Nature, make thunder in winter and ice in summer, make the dead walk and the dry wood blossom, confine a spirit in a bean,[h] and catch a (big) fish in a cupful of water, open doors in paintings and make images speak. It is pure *chhi* which changes the myriad things; where it agglomerates (*ho*[6]) it causes life; where it disperses (*san*[7]) it causes death. What has never been aggregated or dispersed has never been alive or dead. The guests (living things, or phenomena) come and go, but their material basis remains unchanged (*yu chhang tzu jo*[8]).[i]

This is an extremely scientific statement. And what follows is strikingly like the classical definition of atomic materialism that all changes are apparent only, being due to the combinations and recombinations of fundamental particles which themselves do not change.

[a] Wieger (4) is, strangely enough, one of the few who have appreciated it; he termed it a 'masterly development of the Fathers' and gave a paraphrase of some of it in his ch. 65. He dates it as of +742. Forke (12), p. 349, agrees with this high valuation. But there has long been a suspicion that the book may really derive from the Sung, and Sun Ting[9] has been named as possibly its author.

[b] *TT* 1014.   [c] G 1748.   [d] G 1869.

[e] Ch. 7, p. 5*a*, tr. auct. To avoid repetition, it should be mentioned that this attribution applies to all translations from the *Kuan Yin Tzu* and the *Hua Shu* in this Section.

[f] I suspect that this has reference to the discovery of the persistence of reflex action and response to stimulation in isolated parts of animals, also mentioned in the same book (cf. Sect. 43).

[g] Ch. 7, p. 2*b*.

[h] This refers to a legend about Kuo Pho (cf. Forke (12), p. 360).

[i] *Yu*, the old term for a post-station, would mean the unchanging caravanserai. We have not come across another instance of this curious metaphor.

[1] 抱一子   [2] 天隱子   [3] 司馬承貞   [4] 化書   [5] 譚峭
[6] 合   [7] 散   [8] 郵常自若   [9] 孫定

The changes occurring in the myriad things are all due to the *chhi*, but whether they are hidden or whether they can be seen, the *chhi* remains a unity. The sage knows that the *chhi* itself is one, and never changes.[a]

One is almost tempted to dub such a statement a premonition of the first law of thermodynamics.

Throughout these Thang writers there is an appreciation of the subtle reactions of Man and Nature, each affecting the other. This still sometimes took an ancient phenomenalistic form (cf. p. 378 above), but often reached deeper levels. Perhaps relevant here is a passage from a much later writer, Liu Chhi,[1] who said (+ 1235), in his *Kuei Chhien Chih*[2] (On Returning to a Life of Obscurity):

There's an old saying in the *Tso Chuan*: 'The efforts of Man can conquer Nature, but Nature can also conquer Man.' I used to doubt this, but now I think I see what was meant. For example, in the bitter winter cold, a person waiting alone in a large hall feels very gloomy and after a certain time will have to leave, but if someone else comes and chats with him, they both will feel quite warm. This is because the *chhi* of Man can overcome the cold of Nature. Moreover, human beings know how to control their surroundings. In winter people make fires and wear padded clothes against the cold; in summer they seek cool breezes at the tops of towers or sit surrounded with ice and fans. Generally the richer and more influential people are, the more independent of Nature they can make themselves. But going further back, the reason of their becoming influential was due to natural causes (i.e. achievements of their forbears, or fortunate fate).[b]

In other words, some men, at least, can control their natural environment, but they themselves were moulded by Nature acting at the human level.

The following statement on change is worthy of note:

During the expiration and inspiration of a single breath of a man, the sun travels 400,000 *li*. This movement may be considered really rapid. But it takes a sage to discard the appearance of Nature's unchangingness.[c]

Pao I Tzu (the commentator) glosses:...The sun, the moon, and the five planets move away from or towards each other, following each other or separating. The sage can measure these movements so as to construct his calendar, and everyone sees them, yet no one can understand them. This the *Yin Fu Ching*[d] says also. Nothing is so rapid as the changes of Nature....Nature does not stay still for a single moment, and the myriad things are undergoing continuous change. Mountains and rivers change every day, yet the foolish people think they are quite stable. Time is fresh every day, yet the foolish people think that everything is as it was. What is lost flies back into obscurity. Myself in the present is not the same as myself in the past, and vice versa. How can the events of the present be pinned down and preserved? Why will people not realise that in the course of a single breath Nature has already travelled 400,000 *li*?

[a] Ch. 7, p. 4a. A similar excellent statement by Chêng Ssu-Hsiao occurs in *Sung I Min Lu*,[3] ch. 13, p. 3a.   [b] Ch. 12, p. 11a, tr. auct.

[c] *Kuan Yin Tzu*, ch. 7, p. 3b.   [d] See immediately below, p. 447.

[1] 劉祁   [2] 歸潛志   [3] 宋遺民錄

Then there is a passage on potentiality and actuality. Kuan Yin Tzu said:

The *chhi* involves a time factor. That which is not *chhi* knows no day nor night. Form has a spatial factor. That which has no form has no south nor north. What is 'not-*chhi*'? It is that which produces *chhi*. For instance, if a fan is agitated, a wind is produced and the *chhi* becomes palpable as wind. What is 'not-form'? It is that which produces form. For instance, if wood is bored (to make fire), fire is produced, and form becomes visible as fire.[a]

This is interesting because it shows the Thang Taoists groping after something more fundamental in the universe than *chhi* (matter) or *hsing* (form); their forefathers would have been content with the term Tao, but now something more precise was needed, and in the *Li* of the Neo-Confucians (see on, p. 473) it was found. As we shall see, this was thought of as a kind of four-dimensional pattern in the universe, according to which things were brought from potentiality into actuality. That such actualisation, coming-into-being, or growth, is often much more prolonged and apparently difficult than decay and passing-away, was recognised in another passage. It reminds one very much of William Harvey's embryological meditation: 'For more, and abler, operations, are required to the Fabrick and erection of Living creatures, than to their dissolution, and the plucking of them down; for those things that easily and nimbly perish, are slow and difficult in their rise and complement.'[b] So Master Kuan Yin: 'The construction of things is difficult; the destruction of things by the Tao is easy. Of all things under Heaven there is none that does not reach its completion with difficulty, and none that is not easily destroyed.'[c]

Some appreciation of the methods of induction and deduction may be found in these books. Thus in the *Kuan Yin Tzu* we have:

Ordinary people are bewildered by names; they see the things but not the Tao of each thing. The (Confucian) worthy (*hsien*[1]) analyses principles; he sees the Tao, but not the individual things. But the true (Taoist) sage (*shêng jen*[2]) unites himself with Heaven; he sees neither the Tao's nor the things, for one Tao includes all the separate Tao's. If you do not apply it to individual things you reach the Tao of all things—if you apply it to the individual Tao's then you understand the things.[d]

Is this not trying to say that ordinary people are not interested in generalisations at all, while the Confucians are interested in *a priori* generalisations? On the other hand, the Taoist, seeing neither the generalisations nor the individual phenomena, sees in fact both, looking sometimes at the particular and sometimes at the general. Again:

Those who understand the Tao (to which I refer) will know the Way of Heaven, understand the divine power of Nature, comprehend the destinies of things, and penetrate Nature's mysteries; all from observation of phenomena. Having this Tao you can by analysis attain

---

[a] Ch. 3, p. 12*a*.
[c] Ch. 1, p. 6*b*.

[b] *De Gen. Anim.* (Eng. tr.), 1653, p. 206.
[d] Ch. 8, p. 8*b*.

[1] 賢   [2] 聖人

the same result (*hsi thung shih*[1]), though dealing with things of many different names (*hsün i ming*[2]). Having this Tao you can also unite all the different results (*chhi thung shih*[3]), and forget all the different names (*wang i ming*[4]) (of the diverse phenomena).[a]

Compare with this what the *Hua Shu* says:

Amber cannot attract rotten mustard fragments. Cinnabar cannot enter (react with?) bad (unsuitable?) metal. The lodestone cannot attract 'exhausted' iron (perhaps some other metal is meant?). The primal *chhi* will not catch fire from a pottery kiln. Thus the great man is skilful at using the best of the Five Elements. Utilising the divine powers of the myriad things, he can get the highest rewards from Nature and from man, and can ride on the glory of the horses of the winds. His principle is that he neglects the (individual) shapes, and seeks the essence (which is common to many of them).[b]

Nevertheless, the old mistrust of logic and ratiocination is still dogging the Taoists' footsteps. In the *Kuan Yin Tzu* we read:

The wisest man of all knows that human knowledge cannot grasp the things of Nature, therefore he looks as if he were foolish. The best dialectician of all knows that argument will not succeed in describing the things of Nature, therefore he seems to stammer. The bravest man of all knows that courage cannot overcome the things of Nature, therefore he seems to be afraid.[c]

And now, in the Thang, the old theme of the necessity of being without partiality or preconceptions is taken up again. The same book says:

The sages learnt social order from bees, textiles and nets from spiders, ceremonies from praying rats, and war from fighting ants. Thus the sages were taught by the myriad things, and in their turn taught the worthies, who taught the people. But only the sages could understand the things (in the first place); they could unify themselves with natural principles, because they had no prejudices and preconceived opinions (*wu wo*[5]).
(The Tao is like) Chaos (*hun*[6]), or like an ocean, or like wandering in the Great Beginning (*thai chhu*[7]). Sometimes it (is to be studied in) metals, sometimes in jade, sometimes in manure, sometimes in earth or mud, sometimes in flying birds, sometimes in running animals, sometimes in the mountains and sometimes in the abysses. (The sage studies) every point, and evaluates (every change). So he seems (to the ignorant) to be like a madman or a fool.[d]

And the following paragraph from the +8th-century *Yin Fu Ching*[8] (Harmony of the Seen and the Unseen), probably by Li Chhüan,[9] shows that the general scientific

[a] Ch. 1, p. 2*a*.
[c] Ch. 9, p. 9*b*.
[b] P. 13*a*.
[d] Ch. 3, pp. 15*b*, 18*a*.

[1] 析同實 [2] 狥異名 [3] 契同實 [4] 忘異名 [5] 無我
[6] 渾 [7] 太初 [8] 陰符經 [9] 李筌

picture of the universe had been by no means entirely lost, though very little progress had been made in basic theory:

The spontaneous Tao (operates in) stillness, and so it was that heaven, earth, and the myriad things, were produced. The Tao of heaven and earth (operates like the process of) steeping (*chhin*[1]) (i.e. when chemical changes are brought about gently, gradually and insensibly as in dyeing or retting).[a]

(Thus it is that) the Yin and the Yang alternately conquer each other, and displace each other (*hsiang thui*[2]),[b] and change and transformation proceed accordingly.

Therefore the sages, knowing that the spontaneous Tao cannot be resisted, follow after it (observing it), and use its regularities. Statutes and calendrical tables drawn up by men cannot embody (the fullness of) the insensibly-acting Tao, yet there is a wonderful machinery by means of which all the heavenly bodies are produced, the eight trigram symbols,[c] and the sexagenary cycle. This is a spiritual machinery indeed, a ghostly treasury. All these things, together with the arts of the Yin and the Yang in their mutual conquests (to him who understands the Tao) come forward into bright visibility.[d]

Not only is there much proto-scientific material in these books, but also certain hints that the ancient cooperative political doctrines had somehow not been entirely forgotten.[e] Thus the *Kuan Yin Tzu* says:

In a shooting-match when two archers compete, you can see who is skilful and who is poor. When two people play chess you can see that one wins and the other loses. But if two persons meet in the Tao nothing is shown or expressed, there is no skill or lack of it, no winning and no losing.[f]

And the *Hua Shu* likens primitive society to the ant community:

Ants have a prince, and all share in common a palace as big as a fist. All of them meet on one platform, and store their grains of food in common. They all share the meat from a single insect. One who makes a fault is killed by all. For these reasons they attain a state when one mind (heart) interpenetrates them all, and therefore also one spirit, and therefore also one *chhi*, and therefore also one form. Therefore if one is ill all are ill, if one feels pain all feel pain. Under such conditions how could there be complaint? How could there be rebellion? This was the unity, too, of ancient (human) civilisation.[g]

The Ta Thung[3] principle, however, receives an alchemical interpretation.[h]

---

[a] The insensibility of chemical change in aqueous medium had attracted the attention of much earlier philosophers. The *Huai Nan Tzu* has a long paragraph on dyeing as an example of natural transformations (ch. 2, tr. E. Morgan (1), p. 41); and the *Mo Tzu* devotes a whole chapter to it as a pattern for the process of influence among human beings (ch. 3, tr. Mei (1), p. 9). Cf. also pp. 57, 383 above.

[b] This has been explained in Section 13 *d* on fundamental scientific theories (see above, p. 288).

[c] This also, in Section 13 *g*.

[d] Tr. Legge (5), vol. 2, p. 264, much mod.

[e] Cf. *Kuan Yin Tzu*, ch. 2, p. 15 *b*: 'We do not pay special respect to gentlemen, nor do we despise ordinary people.'   [f] Ch. 1, p. 7 *b*.

[g] P. 21 *a*.   [h] *Hua Shu*, p. 12 *b*.

[1] 浸   [2] 相推   [3] 大同

The two following passages are interesting as showing a mixture of magic, experimentation, bodily culture, and the invulnerability complex; and both have a sting in the tail, suggesting that techniques should be used rather for the understanding of Nature than for benefiting human society. Both are from *Kuan Yin Tzu*:

There are in the world many magical arts; some prefer those which are mysterious, some those which are understandable, some the powerful ones and some the weak ones. If you grasp (apply) them, you may be able to manage affairs; but you must let them go in order to attain the Tao.[a]

The Tao originates in Non-Being....Things originate in Being, but the Tao controls their hundred actions. If you attain the height of the Tao you can benefit humanity; if you attain the loneliness of the Tao you can establish your personality. If you realise that the Tao is not in time, you can take one day as a hundred years and conversely. If you know that it is not in space, you can reckon one *li* as a hundred *li*, and conversely. If you know that the Tao, which has no *chhi*, controls the things which have *chhi*, you can summon the wind and the rain. If you know that the Tao, which is formless, can change the things that have form, you can change the bodies of birds and animals. If you can attain the purity of the Tao, you can never be implicated in things; your body will feel light, and you will be able to ride on the phoenix and the crane. If you can attain to the homogeneity (*hun*[1]) of the Tao, nothing will be able to attack you; your body will be dark, and you will be able to caress crocodiles and whales.

Being is Non-Being and Non-Being is Being; if you know this, you can control the ghosts and demons. The Real is Empty and the Empty is Real; if you know this, you can enter metal and stone. Above is Below, and Below is Above; if you know this, you can watch the stars and the dawn. The Old is New and the New is Old; if you know this you will have no need of the tortoise-shell and the divining-sticks. Others are I and I am Others; if you know this you can see through their minds and bodies. Things are I and I am Things; if you know this, you can succeed in placing the Dragon and Tiger (Yang and Yin, Male and Female) within your breast....

If you know that the *chhi* emanates from the mind[b] you will be able to attain spiritual respiration,[c] and will succeed with the alchemical transmutations of the stove....

If you unify yourself with all things, you can go unharmed through water and fire.

Only those who have the Tao can perform these actions—and, better still, not perform them, though able to perform them![d]

Lastly, a pendant which shows that the Taoists were still at odds with the Confucians on the subject of the place of sex in the universe (cf. p. 151 above). The *Kuan Yin Tzu* says:[e]

It is the natural principle (*li*[2])[f] of the world that men lead and women follow, that female animals run and male animals chase after them, that males sing and females respond. The

[a] Ch. 1, p. 4*b*.
[b] Here is a trace of Buddhist influence and a premonition of the metaphysical idealism later to be described (p. 507).
[c] Cf. the *Thai Hsi Ching* (Manual of Embryonic Respiration) already mentioned, p. 144.
[d] Ch. 7, p. 1*a*.   [e] Ch. 3, p. 19*b*.
[f] Note the use of this word, so important for the Neo-Confucians.

[1] 混   [2] 理

(Taoist) sage speaks and acts according to these natural principles, but the (Confucian) 'worthies' (invented the Rites) as a bondage.[a]

This also suggests that the Taoists were maintaining doctrines which were quite unacceptable to the orthodox Confucian bureaucracy.

Kuan Yin Tzu said: If you know that something is false (*wei*[1])[b] you are not bound to expose it. It is like a clay ox or a wooden horse; if you yourself are no longer deceived by it, very well. Just let it alone![c]

Of all these books, the most original from the point of view of the philosophy of science is probably the *Hua Shu*. Than Chhiao (if he was really its author) developed a special kind of subjective realism, in which he emphasised that though the external world was real, our knowledge of it was so deeply affected by subjective factors that its full reality could not be said to have been seized (this, of course, is an attempt to express his point of view in modern terms).[d] First he considers an infinite regress of images of an object in oppositely placed plane mirrors.[e] The form and colour of the object (*hsing*[2]) is perfectly retained in each of the successive images (*ying*[3]). Since it can exist without them, it is not alone and in itself complete (*shih*[4]), but since they perfectly reproduce its form and colour, they are not in themselves empty (*hsü*[5]); or, as might be said in modern terms, it is not fully real, but they are not fully unreal. Now that which is neither real nor not-real, concludes Than Chhiao, is akin to the Tao.[f]

He then takes a biological example. For the owl, he says, the night is bright and the day dark; for the hen the converse is true, as for ourselves.[g] Which of the two, he asks, in good Taoist style, is to be considered 'normal' and which 'abnormal'? In fact, one cannot assume that daytime is bright and fit for sense-perceptions, while the night is not—it depends on the nature of the sense-organs. The inference is that the colours which we see and the sounds we hear are not really present, but are constructs

---

[a] Cf. Blake's 'Priests in black gowns were walking their rounds, And binding with briers my joys and desires'.

[b] Note the use of the same word as that for the 'Great Lie' (*Tao Tê Ching*, ch. 18), above, p. 109.

[c] Ch. 8, p. 8*b*. Again we meet with the tendency for the 'enlightened' to leave popular superstition or convention alone, cf. below, p. 491.

[d] Cf. Forke (12), p. 338.

[e] P. 2*b*.

[f] This piece in the *Hua Shu* betrays a +10th-century interest in optics as well as in epistemology. Cf. the metaphor of Indra's Net, referred to elsewhere (p. 499). One Thang Buddhist, Fa-Tsang, actually set up a system of ten plane mirrors with a statue at the centre in order to demonstrate his meaning to his disciples; cf. Fêng Yu-Lan (1), vol. 2, p. 353. The infinite regress of images in plane mirrors may well have stimulated the typically Buddhist idea of the power possessed by *Bodhisattvas* of 'multiplying' themselves in emanation form. I saw many examples of these multiplication emanations in the frescoes at the Tunhuang cave-temples; and also numerous pictures of monks meditating on mats in front of objects looking like electric heaters on stands, which I suspect were mirrors. We may return to this subject in Section 26*g* on optics; here I will only refer to the interesting paper of Demiéville (1) on the mirror in Buddhist thought.

[g] P. 3*a*. This early appreciation of the principle of 'oecological niches' is interesting also for the historian of biology.

---

[1] 偽　　[2] 形　　[3] 影　　[4] 實　　[5] 虛

of our own organs of sense. This is not without interest as an anticipation of Locke's distinction between primary and secondary qualities,[a] some eight centuries before him. Than Chhiao next refers[b] to the phenomena of optical illusions and of attention. A man may shoot at a striped stone, he says, under the impression that it is a tiger, or at a ripple on the water, under the impression that it is a crocodile. Moreover, even if these animals are really there, his attention may be so concentrated on them that he will simply not see the stones or the water beside them. The inference is that none of these things are really real, in the sense that they insist on being perceived, but all things are imaginary, in the sense that we pick out certain elements of the environing world with which to form our world-picture. This may be extended to life and death themselves. Only the Tao (the substratum of all sense-impressions of all beings) is truly real. This was indeed an attempt at an epistemology.

Another way in which Than Chhiao expresses it is by saying that our senses are like four (transparent) lenses.[c] One is shaped like a jade sceptre ($kuei^1$) and what is seen through it appears small; the second is shaped like a pearl (chu^2), and what is seen in it appears large; the third is shaped like a whetstone ($chih^3$), and the image given appears upright; the fourth is shaped like a bowl ($yü^4$), and what is seen therein appears inverted. But if one makes a right use of these instruments, and tests their information by other means, then one finds that there is no such thing as large or small, short or long, beautiful or ugly, desirable or hateful. Everything is relative. Our sense-organs by themselves give us no absolute picture of the external world.[d]

As to causation in Nature, Than Chhiao was clear that determining factors had to be looked for. Taking examples from human technology, he showed that they might be quite inconspicuous.

The control of a ship carrying ten thousand bushels of cargo is assured by means of a piece of wood no more than eight feet in length (the rudder).[e] The action of a crossbow-catapult of one thousand _chün_ (a pull of some 20 tons weight) depends upon an apparatus (the trigger) which is no more than an inch in length. With one eye one can see the vast expanse of the heavens; and millions of people can be governed by one emperor. The Great Emptiness (of the heavens) though (seemingly) boundless, yet has limits. The Greatest of Heights, though (apparently) vast and infinitely far, yet has its home boundaries.[f] If you can realise the connectedness ($kang^5$)[g] of heaven and earth; if you can understand the 'field of force'

---

[a] Whewell (2), p. 278.   [b] P. 3 _b_.

[c] P. 3 _a_. See below, Sect. 26_g_, where the physical significance of the passage will be elucidated.

[d] From this position the step was of course not very far to subjective idealism, either in its Buddhist or Confucian form (see above, p. 410; below, p. 507). The _Kuan Yin Tzu_ says, 'How can we know whether Heaven and Earth are not simply our own thoughts?' (ch. 1, p. 10_b_). But most of these Thang-Sung Taoist speculations do not venture beyond Chuang Tzu's queries concerning dreams, and who is dreaming about whom.

[e] This reference (+950) to what can only have been the stern-post rudder, will be referred to again in that connection (Section 29_g_ below).

[f] Presumably a reference to zenith, equator or ecliptic.

[g] An important technical term; cf. below, pp. 554 ff.

¹ 圭   ² 珠   ³ 砥   ⁴ 盂   ⁵ 綱

(*fang*[1]) of the Yin and the Yang; if you can know the hidden storehouse (*tsang*[2]) of seminal essence and spirit (*ching shen*[3]); then you can overcome (i.e. change) the numbers (*shu*[4]) (written in the book) of destiny, you can prolong your life, and you can turn all things upside down (*fan fu*[5]) (i.e. control Nature).[a]

Here once again was the old programme of Tsou Yen and Liu An.

(*c*) LI AO AND THE ORIGINS OF NEO-CONFUCIANISM

Under the Thang dynasty (+7th to +9th centuries) Taoism was indeed prosperous. But it must not be supposed that the Taoists had it all their own way—they had to compete with the Buddhists, and there were also a number of important Confucian scholars whose role in the preparation of Neo-Confucianism has often been overlooked (as Fêng Yu-Lan has shown).[b] One such man was Wang Thung[6] (+584 to +617),[c] whose life was almost exactly contemporary with the Sui dynasty, just preceding the Thang. To him is attributed the *Yuan Ching*[7] (Treatise on Origins), a chronicle history written in imitation of the Spring and Autumn Annals, and extending from +290 to the beginning of the Sui in +589.[d] He is said to have had a large number of pupils and to have exerted much influence. Then in the following century there was Han Yü,[8] the famous official and prose writer,[e] whose expostulations with the emperor against worship of Buddhist relics have already been referred to (p. 387). More important philosophically, however, seems to have been Li Ao[9] (d. +844),[f] who, in his *Fu Hsing Shu*[10] (Essay on Returning to the Nature), used a number of technical terms which afterwards became rather important in Neo-Confucianism. Such, for example, could be considered motion (*tung*[11]) and rest (*ching*[12]); as basic concepts these go back to the Book of Changes, but Li Ao used them in a psychological context.

There is one remark in the *Fu Hsing Shu* which is rather revealing about the origins of Neo-Confucianism. 'Although writings dealing with the Nature and with Destiny are still preserved', said Li Ao, 'none of the scholars understands them, and therefore they all plunge into Taoism or Buddhism. Ignorant people say that the followers of the Master (Confucius) are incapable of investigating the teachings on the Nature and Destiny, and everybody believes them.' This strongly suggests that during the Thang time the Confucians began to feel acutely the lack of a cosmology to offset that of the Taoists, and a metaphysics to compete with that of the Buddhists. Sung Neo-Confucianism was, in a word, the elaboration of this syncretism, and it was only

---

[a] P. 9*b*, tr. auct.

[b] In Bodde (4), and Fêng (1), vol. 2, pp. 407ff.

[c] Wên Chung Tzu; G2239; Forke (12), p. 274.

[d] This book, which should have interesting information concerning the period just discussed is, however, considered to be of later date (+11th century) and to have really been written by Juan I.[13]

[e] +768 to +824; G632.   [f] Forke (12), p. 297.

[1] 房   [2] 藏   [3] 精神   [4] 數   [5] 反覆   [6] 王通   [7] 元經
[8] 韓愈   [9] 李翱   [10] 復性書   [11] 動   [12] 靜   [13] 阮逸

possible by the borrowing of various elements from the thought of the other two schools. As Bruce has said, 'Neo-Confucianism rescued the ethical teaching of the classics from threatened oblivion, by bringing it into close relationship with a reasoned theory of the universe.'[a] How sane this was, and how remarkably it foreshadowed our modern conception of levels of organisation, each with their appropriate phenomena in rising complexity, I shall shortly attempt to show.

On the one hand there was Taoist naturalism. Its defect was that it was not very much interested in human society. Recognising clearly that ethical considerations were irrelevant to scientific observations and scientific thought (cf. p. 49), it had offered no explanation of how the highest human values manifested in society could be related to the non-human world. Hsün Chhing had said, 'They see Nature, and fail to see Man' (p. 28). On the other hand, there was Buddhist metaphysical idealism. This was one stage worse, for it was interested neither in human society nor in Nature. Both were elements in the vast conjuring trick from which all beings should escape, and should be helped to escape. An illusory phantasmagoria does not invite scientific study or encourage public justice. But it was no help to return to antique Confucianism, for its total lack of cosmology and philosophy no longer satisfied the demands of a maturer age. From all this there was only one way out, the way that was taken by the Neo-Confucians culminating in Chu Hsi, namely, to set, by a prodigious effort of philosophical insight and imagination, the highest ethical values of man in their proper place against the background of non-human Nature, or rather within the vast framework (or, to speak like Chu Hsi himself, the vast *pattern*) of Nature as a whole.[b] On such a view, the nature of the universe is, in a sense, moral, not because there exists, somewhere outside space and time, a moral personal deity directing it all, but because the universe has the property of bringing to birth moral values and moral behaviour when that level of organisation has been reached at which it is possible that they should manifest themselves. Although modern evolutionary philosophers tend to think of this process in terms of one long development, while the Neo-Confucians thought that there were many successive developments, occurring after each successive world conflagration, the basic conception was, I believe, the same.

Hence the perplexities which have assailed Western students of Neo-Confucianism. Jesuits have been outraged because the Neo-Confucians denied, in so many words, a personal God.[c] Protestant theologians have essayed to detect a kind of pantheism in Neo-Confucian thought.[d] One of them pointed out that Chu Hsi's materialism

---

[a] (1), p. 25.

[b] As Chhêng Hao said, 'The human-hearted man is one with heaven and earth and all the myriad things (*Jen chê i thien ti wan wu i thi yeh*[1])'; *ECCS, Honan Chhêng shih I Shu*, ch. 2A, p. 2a, ch. 4, p. 5a.

[c] But the Benedictine, Olaf Graf, has known how to appreciate the Neo-Confucian world-view; he makes a detailed comparison of Chu Hsi with Thomas Aquinas and Spinoza.

[d] Even Chinese scholars have expressed some surprise and regret that the Neo-Confucians 'mixed up' logical, ethical and scientific concepts so much (e.g. Fêng Yu-Lan (1) in Bodde (3), p. 50; Fêng (1), vol. 2, p. 571). Yet the most scientific of philosophies must account for the emergence of ethics in a natural world. Almost alone among modern sinologists, Graf understands the Neo-Confucian achievement of an ethics grounded in Nature, (2), vol. 1, pp. 33, 81, 90, etc.

[1] 仁者以天地萬物一體也

differed from occidental materialism, saying that in the latter, material substance obeys its own laws, which are unethical, while for the Neo-Confucians the material is subject to the ethical. Truly, Neo-Confucian materialism was not mechanical. Billiard-balls in fortuitous concourses have never been an element of Chinese thought. But what the Neo-Confucians did was to recognise the moral as fundamentally planted in Nature, and arising out of Nature by an emergent evolution, when, as we should say, conditions are present in which the moral can appear.[a] I shall suggest, therefore, that though knowing nothing of Hegelian dialectics, the Neo-Confucians approximated quite closely to the world-concepts of dialectical or evolutionary materialism and of the Whiteheadian philosophy of organism which is so cognate to it.[b]

On this view it would seem that Neo-Confucianism borrowed much less from Buddhism than from Taoism. As a synthesis it really joined hands with the latter; the effect of the former was mainly to exert the stimulus that convinced the Chinese that 'something had to be done'. Nevertheless, Buddhist influence showed itself in a certain interest of the later Confucians in meditation techniques. Some have said that Li Ao and others of his time wanted to lead men towards a Confucian type of 'Buddhahood', but if this is going a little far, there is no doubt that many Confucians wrote about the desirability of concentration (chih[1]), contemplation (kuan[2]),[c] mystically acquired understanding (chih[3]) and absorption (ting[4]). What is particularly interesting is that just as the Taoists had earlier appropriated the 'Confucian' Book of Changes, so the *Chung Yung* (Doctrine of the Mean) became a kind of common ground between Confucians and Buddhists. The monk Chih-Yuan,[5] for example, who died in +1002, took the name of Chung Yung Tzu[6] (Master Doctrine-of-the-Mean) and wrote a commentary on it. So also the monk Chhi-Sung[7] (d. +1072) wrote a book in explanation of the *Chung Yung*.

---

[a] In doing this they were certainly not unmindful of the crude ideas of the Han phenomenalists (see above, pp. 247, 378), for whom also there had been a kind of unity of the cosmic and ethical orders (Maspero (12), pp. 108, 109). But of course it was a far cry from saying that the environing universe reacted to human ethical choices, to saying that human ethical choices were natural events which occurred when the universe had developed beings capable of making them. There is perhaps more of a parallel between the Neo-Confucian world-view and that of the Greek pre-Socratic thinkers, as outlined in a famous passage of Plato's (*Laws*, 889), which he wrote with the object of discrediting them for ever. Nature is blind and purposeless until the advent of Man. Before his coming, all things are under pure chance and necessity; with his coming come design, purpose and technique. This also had been an evolutionary naturalism. But the Chinese never recognised blind and purposeless chance at any time or any level; before Man and beneath him there were simply other organisms, each following its Tao, and orchestrated (by no composer) into the Tao of all things. I speak, of course, of the main current of Chinese philosophy, not of exceptional men such as Wang Chhung and Liu Khun.

[b] On this relationship cf. Needham (9).

[c] Note this inward-looking use of the ancient Taoist term for the outward-looking observation of omina in Nature (p. 56 above).

---

[1] 止   [2] 觀   [3] 智   [4] 定   [5] 智圓   [6] 中庸子
[7] 契嵩

(d) THE NEO-CONFUCIANS

Between the Thang Taoists and the Sung Neo-Confucians there were a number of transitional figures worth some attention. It was not that they were chronologically earlier than the Neo-Confucians, for Shao Yung[1] (+1011 to +1077) was almost exactly a contemporary of Chou Tun-I (see below); and Chhêng Pên,[2] whose dates we lack, is to be placed in the middle of the +11th century—but they were in the Taoist tradition rather than of the new synthetic school. Nevertheless, Shao Yung was a friend of its chief leaders.

(1) CHU HSI AND HIS PREDECESSORS

Shao Yung, who in characteristic Taoist fashion refused all official responsibility, was an original but fanciful thinker. His *Huang Chi Ching Shih Shu*[3] (Book of the Sublime Principle which governs all Things within the World)[a] was partly composed of elaborate diagrams in which cosmological and ethical ideas mingled,[b] but which were found to be so hard to understand that they were replaced by a descriptive account from the hands of his son and another philosopher.[c] Shao Yung also wrote an interesting philosophical dialogue, *Yü Chhiao Wên Tui*[4] (The Conversation of the Fisherman and the Woodcutter), in which he spoke of the uniformity of natural phenomena, the finiteness of 'forms' (*hsing*[5]), and the infinity of 'matter' (*chhi*[6]); adding his views on the hexagrams. It would repay study and translation. Shao Yung retained the Tao as the name for the universal principle of Nature, but gave a very high place to 'number' conceived in a Pythagorean way; the Tao first makes numbers, then forms, and then fills these with matter. This was associated in him with a strong tendency to metaphysical idealism, in which connection we shall have to speak of him again, and in this sense he was not on the main line of Neo-Confucian development, which could rather be described as dualistic realism. Moreover, though much of his systematisation used terms similar to theirs, they were differently arranged. According to him, the two primary manifestations (*liang i*[7]) of the Tao were motion (*tung*[8]) and rest (*ching*[9]); the former generated the Yin and Yang of the heavens, the latter generated two entities newly systematised by Shao Yung, namely, softness (*jou*[10]) and hardness (*kang*[11]),[d] i.e. the Yin and Yang in their earthly manifestations. Thus Earth was a mixture of these while Heaven was a mixture of Yin and Yang.

[a] In *Hsing-Li Ta Chhüan*, chs. 7–13.

[b] Compare the diagrams of another Sung thinker, Li Kuo-Chi[12] (born *c.*+1170), which appear in the *Pai Chhuan Hsüeh Hai*[13] collection under the title of *Shêng Mên Shih Yeh Thu*[14] (Diagrams of Matters discussed in the Schools of the Sages). Cf. Martin (7).

[c] Forke (9), p. 21.

[d] This conception must have been a Taoist one, for later we shall see that Chu Hsi spoke of 'what the Taoists call the "hard wind"'. Actually *kang* and *jou* were ancient terms, in medical texts at any rate

[1] 邵雍　　[2] 程本　　[3] 皇極經世書　　[4] 漁樵問對　　[5] 形
[6] 氣　　[7] 兩儀　　[8] 動　　[9] 靜　　[10] 柔　　[11] 剛　　[12] 李國紀
[13] 百川學海　　[14] 聖門事業圖

The four generated entities were called the four secondary manifestations (*ssu hsiang*[1]), and each of them could exist in two qualities, strong or weak. With the eight components thus available Shao Yung derived, in a more or less arbitrary and fanciful way similar to the ancient type of thinking described in Section 13, all kinds of phenomena.[a] One can at any rate remember that this was in the +11th century, and what is worth noting is the extremely concrete and physical nature of his world-picture. Motion and rest, and to a lesser extent softness and hardness, were taken over as basic conceptions by the Neo-Confucians. Shao Yung fully believed in the Indian idea, introduced with Buddhism, that there were periodical world-catastrophes in which the world was dissolved into chaos and then remade anew (see below, p. 485). One medieval notion, very clear in him, is that of the microcosm and the macrocosm,[b] and there seems no reason to attribute it to any but indigenous sources.

Perhaps the most interesting idea of Shao Yung's from the point of view of the history of science was embodied in his expression *fan kuan*,[2] or 'objective observation'. In science (*wu li chih hsüeh*[3]),[c] he said, 'there are often things which one cannot understand. One must not attempt to "force" them (into a scheme), for "forcing" them brings in one's self (and one's prejudices) (*chhiang thung, tsê yu wo*[4]), and thus one loses the (objective) principles, and falls into artificial constructions (*tsê shih li erh ju yü shu*[5]).'[d] It is only unfortunate that Shao Yung did not bear this wisdom more in mind in the construction of his own theoretical schemes. Another useful emphasis of his was on the general community of observers; we are not restricted, he said, to personal observations, but we can use the eyes of all men as our eyes, and their ears as our ears, and thus we can form a connected body of understanding of Nature.

Chhêng Pên, author of the *Tzu Hua Tzu*,[6] concealed his true identity under the name of a philosopher of the Chou period, whose book, if he ever wrote one, had already been lost when the *Chhien Han Shu* bibliography was compiled—and fathered his text, which must have been written early in the +11th century, on the ancient figure. While the text is short and not very clear, it gives some insight into the Taoist discussions which were then going on; it speaks of empty space (*hsü*[7] and *khung tung*[8]), in which there are no barriers (*ai*[9] and *wu*[10]) to the movement of bodies; and of equilibrium (*phing*[11]), in which there is no tendency for bodies to move in any direction. 'Chhêng Pên' has frequent references to three fundamental forces, rhythms,

---

[a] Details in Forke (9), p. 27.   [b] Forke (9), p. 34. Cf. above, pp. 294 ff.
[c] Note the appearance of the Neo-Confucian technical term, *Li*.
[d] *Hsing-Li Ta Chhüan*, ch. 12, p. 4*b*. He also wrote (p. 3*a*): 'Look at things from the point of view of things, and you will see (their true) nature; look at things from your own point of view, and you will see (only your own) feelings; for nature is neutral and clear, while feelings are prejudiced and dark (*I wu kuan wu hsing yeh, i wo kuan wu chhing yeh; hsing kung erh ming, chhing phien erh an*[12]).' This was in line with the call of the ancient Taoists to 'unprejudiced' observation (see pp. 48, 60, 89). The avoidance of personal preconceptions in the investigation of Nature was one of their tenets. Hocking considered it as very characteristic of the Neo-Confucians, and we may recall the remark by him quoted already on p. 59.

[1] 四象   [2] 反觀   [3] 物理之學   [4] 強通則有我
[5] 則失理而入於術   [6] 子華子   [7] 虛   [8] 空洞   [9] 閡
[10] 忤   [11] 平   [12] 以物觀物性也以我觀物情也性公而明情偏而暗

or impulses,[a] though they are not clearly explained. One of his most interesting peculiarities is his attribution of geometrical forms to the five elements; thus water is straight, fire pointed, earth round, wood curved, and metal square. There is nothing in the text to suggest that the writer is speaking of shapes of *particles*, but if not, it is hard to imagine what he had in mind. The system of course includes identifications of the elements with the organs of the body, but has other more interesting pharmacological and physiological hints. An annotated translation of *Tzu Hua Tzu* would surely be worth while, as it would throw light on a period of Taoist scientific thought of which relatively little is known.

We must now consider the main school of Sung Neo-Confucian philosophers. Its five leading personalities were living at various overlapping periods occupying almost exactly the +11th and +12th centuries. In order to think of them in some historical perspective, therefore, one should remember that the first four were contemporaries of al-Bīrūnī[b] and Ibn Sīnā (Avicenna)[c] and 'Umar al-Khayyāmī (Omar Khayyam);[d] while the fifth and greatest was living at the same time as William of Conches,[e] Abū Marwān ibn Zuhr (Avenzoar),[f] Gerard of Cremona,[g] Ibn Rushd (Averroes),[h] and Maimonides.[i] The Neo-Confucians were thus working at about the same time as the height of the translation movement in Europe which gave access to Greek thought, and they accomplished their great synthesis of Confucian, Taoist and Buddhist elements just before the greatest synthesiser of European scholastic Christian-Aristotelian thinking entered upon his career.[j] If the contemporaneity of these two synthetic enterprises is but a coincidence, it is a rather remarkable one.[k]

Of Shao Yung, who may be considered as the precursor of the Sung school, something has just been said. The main body of five men started with Chou Tun-I[1] (+1017 to +1073),[l] a scholar who preferred philosophical study to high official rank.[m] One of his friends, Chhêng Hsiang, had two sons, the elder Chhêng Hao[2] (+1032 to +1085),[n] and the younger Chhêng I[3] (+1033 to +1107),[o] who both attained great

---

[a] Perhaps this fact is connected with the 'trinitarian' doctrines of the Taoist religion stabilised about the same time (cf. p. 160).

[b] Philosopher and geographer, one of the greatest of Islamic scientists; Sarton (1), vol. 1, p. 707.

[c] The greatest of all Islamic physicians; Sarton (1), vol. 1, p. 709.

[d] Persian mathematician and poet; Sarton (1), vol. 1, p. 759.

[e] French scholastic philosopher, astronomer and meteorologist; Sarton (1), vol. 2, p. 197.

[f] Greatest physician of Muslim Spain; Sarton (1), vol. 2, p. 231.

[g] Greatest of translators from Arabic into Latin; Sarton (1), vol. 2, p. 338.

[h] Most illustrious of western Muslim philosophers; Sarton (1), vol. 2, p. 355.

[i] Greatest of medieval Jewish philosophers; Sarton (1), vol. 2, p. 369.

[j] Chu Hsi died in +1200; Thomas Aquinas was born in +1225.

[k] I cannot help feeling that the Neo-Confucians deserve a much greater place in the history of science than that accorded them by Sarton (cf. (1), vol. 2, p. 295).

[l] G 425; Chou I-Chhing (1); Bruce (2), p. 18; Forke (9), p. 45; biography, Eichhorn (2).

[m] Although the general practice in this book is to neglect literary and other secondary names, the Sung philosophers are so often referred to by them that it will be desirable to mention them here. Thus Chou Tun-I is perhaps even more commonly known as Chou Lien-Hsi, or just Lien-Hsi.[4]

[n] G 278; Bruce (2), p. 41; Forke (9), p. 69; Graham (1). Literary name Chhêng Ming-Tao.[5]

[o] G 280; Bruce (2), p. 45; Forke (9), p. 85; Graham (1). Literary name Chhêng I-Chhuan.[6]

---

[1] 周敦頤   [2] 程顥   [3] 程頤   [4] 濂溪   [5] 程明道   [6] 程伊川

renown as philosophers. The fourth member of the group was the uncle of these two brothers, Chang Tsai[1] (+1020 to +1076);[a] he was probably the one most responsible for the introduction of acceptable elements from Taoism and Buddhism into Neo-Confucian thought.[b] Lastly came Chu Hsi[2] (+1131 to +1200),[c] the supreme synthetic mind in all Chinese history. Exactly how far he went in the study and practice of Taoism and Buddhism we do not know,[d] but it is certain that he was deeply learned in both systems, and often refers to elements of their doctrines, some of which became incorporated in the philosophical synthesis which he made. His official career was a very chequered one, periods of imperial favour alternating with resignations, retirement and deprivation of honours. Enormous personal literary output, great ability at organising the research and writing of others, unusual lucidity of expression,[e] and an unswerving fidelity to a clear and definite world-picture, placed him without question as one of the greatest men in the whole development of Chinese thought. Forke[f] lists the comparisons which have been made of Chu Hsi with occidental figures; these include Aristotle, Thomas Aquinas, Spinoza, Leibniz, and Herbert Spencer, and it is a tribute to Chu Hsi that such suggestions have no absurdity about them. To my mind St Thomas and Herbert Spencer are Chu Hsi's nearest equivalents, the one because he was, after all, a man of the medieval age, occupied with the systematisation rather than the radical transformation or supersession of beliefs which had a long history behind them; and the other because he uncompromisingly affirmed a thoroughly naturalistic view of the universe, all the more astonishing in that it lacked that vast background of assured experimental and observational knowledge of Nature which made Spencer's dogmatism understandable if not acceptable. Indeed, I shall suggest that Chu Hsi's philosophy was fundamentally a philosophy of organism, and that the Sung Neo-Confucians thus attained, primarily by insight, a position analogous to that of Whitehead, without having passed through the stages corresponding to Newton and Galileo. They thus present a parallel with the Mohist and Taoist thinkers of the Warring States period, who may be said to have attained gleams of dialectical logic, thereby anticipating Hegel, without ever having passed through the logic of Aristotle and the scholastics.[g] That such comparisons can arise in the mind is in itself the best appreciation of the Chinese achievement, but perhaps it was not by these flights of genius that modern natural science as we know it could come into existence.

[a] G 117; Bruce (2), p. 50; Forke (9), p. 56. He is more commonly referred to as Chang Hêng-Chhü.[3]

[b] For Englishmen it is striking to note that all this philosophical activity, so remarkably modern and scientific in tone, as we shall see, was going on about the time of the Norman Conquest.

[c] G 446; Bruce (2), p. 56; Forke (9), p. 164. Literary name Chu Yuan-Hui.[4]

[d] The question is discussed by le Gall (1), p. 9, and by Bruce (2), p. 63. Cf. Graf (2), vol. 1, p. 214.

[e] 'His sentences', Warren (1) has well said, 'are veritable crystals.' And the 'rationalism' of the Neo-Confucians is apparent even in their style of writing. Their statements constantly begin with the words 'chih shih'.[5] 'It is simply a matter of...' so and so.

[f] (9), pp. 199ff. The parallels with Aquinas and Spinoza are worked out in great detail by Graf (2) vol. 1, pp. 246ff., 256ff., 262ff., 279ff., who compares jen with the amor dei intellectualis (pp. 285, 286)

[g] Others have appreciated this, too, though they do not all like it, cf. Bernard-Maître (2), p. 37.

[1] 張載　　　[2] 朱熹　　　[3] 張橫渠　　　[4] 朱元晦　　　[5] 只是

The writings of the Neo-Confucian school were collected together by the Ming emperor Yung-Lo in a compendium known as the *Hsing-Li Ta Chhüan*[1] (Collected Works of the Hsing-Li School), of +1415. In the 18th century, under the emperor Khang-Hsi, this was condensed into smaller compass as the *Hsing-Li Ching I*[2] (Essential Ideas of the Hsing-Li School).[a] Details of the lives and sayings of the philosophers are found in the *Sung Yuan Hsüeh An*[3] (Schools of Philosophers of the Sung and Yuan dynasties), written in the Chhing time. Another selection of the writings of the Neo-Confucians is the *Sung Ssu Tzu Chhao Shih*[4] (Selections from the Writings of the Four Sung Philosophers, i.e. excluding Chu Hsi), edited by Lü Jan[5] in +1536.

Among modern writings the most accessible treatment of Sung philosophy is that of Fêng Yu-Lan (1, vol. 2), which has been translated by Bodde.[b] A much shorter account, in Chinese, is that of Chia Fêng-Chen (1).

We shall refer to the writings of the lesser figures of the Neo-Confucian school as we come to them, but before going further a word must be said about the principal works stemming from Chu Hsi himself. During his own lifetime he produced a large number of books in the period ranging from +1159 to +1188.[c] Some were commentaries and explanations of the works of his immediate predecessors, such as the *Chin Ssu Lu*[6] (Summary of Systematic Thought)[d] of +1176. Three years before he had written a famous, though extremely short, essay on the 'Diagram of the Supreme Pole' of Chou Tun-I (*Thai Chi Thu Chieh I*[7]). In view of the traces of chemical thought found in this essay, which appears in translation on pp. 462, 463 below, it is of interest (as we saw) that Chu Hsi wrote a commentary on the *Tshan Thung Chhi*[8] of Wei Po-Yang;[9] it was entitled the *Tshan Thung Chhi Khao I*[10] and appeared in +1197. Wei Po-Yang's treatise dates from the +2nd century and is of cardinal importance for the history of alchemy (see on, Sect. 33). During the half-century after the great philosopher's death in +1200, a number of collections of verbatim records of his conversations, and of his writings, especially letters to numerous disciples and opponents, were made by various scholars.[e] All these were combined and edited in +1270 by Li Ching-Tê[11] under the title *Chu Tzu Yü Lei*[12] (Classified Conversations

---

[a] Wieger (2), pp. 198 ff., has some fifty pages of translations from these two compendia, but as usual without any exact attributions or references to the positions of the texts or the names of their authors. Nevertheless a glance through them will show those who cannot procure or read the original, something of how the main ideas described in the rest of the present Section were developed by later or minor members of the Neo-Confucian school.

[b] The 13th chapter (on Chu Hsi) appeared separately, in Bodde (3), as also did the 10th (on Neo-Confucian indebtedness to Buddhism and Taoism) in Bodde (4).

[c] Full details in Forke (9), pp. 169, 170. Some of the works which bear Chu Hsi's name were of course large enterprises in which he must have acted as organising editor, e.g. the historical condensation *Thung Chien Kang Mu* already so often mentioned (Sects. 5a, 6h).

[d] This has been translated, with an analysis both copious and enlightening, by Graf (2). Lü Tsu-Chhien[13] was joint author with Chu Hsi. It is considered the *Summa* of Neo-Confucianism.

[e] Details in Wylie (1), p. 68.

[1] 性理大全	[2] 性理精義	[3] 宋元學案	[4] 宋四子抄釋
[5] 呂柟	[6] 近思錄	[7] 太極圖解義	[8] 參同契 [9] 魏伯陽
[10] 參同契考異	[11] 黎靖德	[12] 朱子語類	[13] 呂祖謙

of Chu Hsi),[a] and later in a collection called *Chu Tzu Wên Chi*[1] (Selected Writings of Chu Hsi). In +1713 the emperor ordered a selection and condensation of the philosophical opinions to be made, and this appeared as the *Chu Tzu Chhüan Shu*[2] (Collected Works of Chu Hsi).[b]

(2) THE 'SUPREME POLE'

Chou Tun-I was more of a teacher than a writer, and left little behind. His fame depends upon a very short exposition of a cosmical diagram, the *Thai Chi Thu Shuo* (Explanations of the Diagram of the Supreme Pole), upon which Chu Hsi, taking it as fundamental for his thought, wrote several commentaries with similar titles.[c] Fig. 48 shows the diagram of Chou Tun-I. His exposition is as follows:[d]

(1) That which has no Pole! And yet (itself) the Supreme Pole! (*Wu chi erh thai chi*[3]).

(2) The Supreme Pole moves and produces the Yang. When the movement has reached its limit, rest (ensues). Resting, the Supreme Pole produces the Yin. When the rest has reached its limit, there is a return to motion.[e] Motion and rest alternate, each being the root

---

[a] Referred to herein as *CTYL*.   [b] Referred to herein as *CTCS*.

[c] The bibliography of these writings is liable to cause great confusion, and therefore requires a few words of explanation. We have (*a*) the diagram itself, *Thai Chi Thu*,[4] and (*b*) Chou Tun-I's own philosophical exposition of it, the *Thai Chi Thu Shuo*.[5] A descriptive exposition of (*a*) was written by Chu Hsi; this is (*c*) the *Thai Chi Thu Chieh I*,[6] preceded by (*d*) a preface. But he also wrote (*e*) a philosophical commentary on (*b*), the *Thai Chi Thu Shuo Chieh*[7] or *Chu*,[8] and an additional essay (*f*) the *Thai Chi Shuo*;[9] together with (*g*) a postface to Chou Tun-I's works, the *Thai Chi Thu Thung Shu Hou Hsü*.[10] Here we are concerned with (*a*), (*b*), (*c*) and (*e*). Besides these writings many verbatim records (*h*) of Chu Hsi's philosophical discourses on the diagram remain in works such as the *Chu Tzu Chhüan Shu*, ch. 49, pp. 8*b* ff. (tr. le Gall (1), pp. 99 ff.), and the *Sung Yuan Hsüeh An*, ch. 12, pp. 1*b* ff., which reproduces the famous discussion between Chu Hsi and Lu Hsiang-Shan (cf. p. 508 below).

The texts may conveniently be found (apart from the great compilation *Hsing-Li Ta Chhüan*, ch. 1) in the following sources. The *Hsing-Li Ching I* (imperial edition of +1717, ed. Li Kuang-Ti[11]) prints the material in the following order: (*d*) ch. 1, p. 1*a*, (*a*) and (*c*) pp. 2*a* ff., (*b*) pp. 4*a* ff., inserting between the paragraphs as commentary both (*e*) and a *chi shuo*[12] collected from (*h*); then bringing up the rear with (*f*) pp. 15*a* ff. The *Sung Ssu Tzu Chhao Shih* heads its Chou Tun-I section, ch. 1, with (*a*), (*c*), and (*b*) in that order, adding (*g*) at pp. 14*a* ff. The *Sung Yuan Hsüeh An* gives (*a*) and (*b*) only, ch. 12, p. 1*b*. The *Hui-An hsien-sêng Chu Wên Kung Chi* gives (*f*) in ch. 67 (*SPTK* ed., p. 1142.2) and (*g*) in ch. 75 (p. 1389.1). The first chapter of the *Chin Ssu Lu* opens with (*b*) and (*e*).

Translators have often failed to indicate the precise nature of their texts. Already in 1876 v. d. Gabelentz (2), using and reproducing a Manchu version as well as the Chinese, put into German (*d*), pp. 11 ff., (*a*), (*b*) and (*e*), pp. 30 ff., and (*f*), pp. 82 ff.; his version remains of interest today. Forke (9), p. 48, made another German translation of (*b*) only. The well-known English one of Bruce (2) gives (*a*) and (*b*) at pp. 128 ff. and (*c*) at pp. 132 ff. Most recently Chou I-Chhing (1) has given (*b*) and (*e*) together in French, with parallel Chinese, pp. 154, 210 ff. Here we give (*b*) in full, and most of (*c*).

[d] We retain Forke's numbering of the paragraphs. The translation of both texts is based on those just mentioned but not identical with any of them.

[e] Bruce (2) preferred 'energy' and 'inertia' to motion and rest. This raises the whole problem of how 20th-century minds are to take 12th-century concepts. I feel that Bruce's choice was too Spencerian, and that he read into Chou Tun-I's words ideas which are too precise.

[1] 朱子文集   [2] 朱子全書   [3] 無極而太極   [4] 太極圖
[5] 太極圖說   [6] 太極圖解義   [7] 太極圖說解   [8] 太極圖說註
[9] 太極說   [10] 太極圖通書後序   [11] 李光地   [12] 集說

of the other. The Yin and Yang take up their appointed functions (*fên*[1]),[a] and so the Two Forces are established (*liang i li*[2]).[b]

(3) The Yang is transformed (*pien*[3]) (by) reacting (*ho*[4]) with the Yin,[c] and so water, fire, wood, metal and earth are produced.[d] Then the Five *Chhi* diffuse harmoniously, and the Four Seasons proceed on their course.

(4) The Five Elements (if combined, would form), Yin and Yang. Yin and Yang (if combined, would form) the Supreme Pole.[e] The Supreme Pole is essentially (identical with) that which has no Pole. As soon as the Five Elements are formed, they have each their specific nature.

(5) The true (principle) of that which has no Pole, and the essences of the Two (Forces) and the Five (Elements), unite (react) with one another in marvellous ways, and consolidations (*ning*[5]) ensue. The Tao of the heavens perfects maleness and the Tao of the earth perfects femaleness. The Two *Chhi* (of maleness and femaleness), reacting with and influencing each other (*chiao kan*[6]),[f] change and bring the myriad things into being. Generation follows generation, and there is no end to their changes and transformations.

(6) It is man alone, however, who receives the finest (substance) and is the most spiritual of beings. After his (bodily) form has been produced, his spirit develops

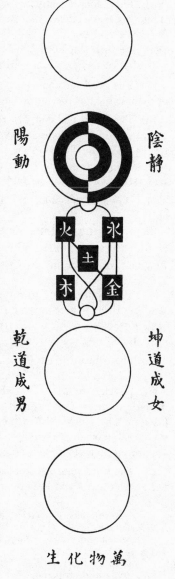

---

[a] Note the use of the ancient word for apportionment of duties and benefits in pre-feudal and feudal society; and hence the close parallel with Gk *moira* (Cornford (1), p. 15). Cf. p. 107 above.

[b] Previous translators prefer 'Modes', but we have never been able to understand what the word means. I think 'Forces' is not too strong.

[c] We believe that it is not forcing Chou's words too much to speak of 'reacting', for union involving transformation.

[d] Note that this is the 'Cosmogonic' Enumeration Order (p. 257 above).

[e] This was a great point of difference between the Neo-Confucians and the Buddhists. The *wu chi* or *thai chi* is *not* something 'empty' (*khung*,[7] Skr. *śūnya*), since it possesses, and can unfold, all the patterns of the Yin and Yang and the Five Elements (cf. *CTCS*, ch. 49, p. 14*a* (tr. le Gall (1), p. 109); Fêng Yu-Lan (1) in Bodde (3), p. 14; Fêng (1), vol. 2, p. 538).

[f] Here the metaphor is undoubtedly chemical (cf. the sexual symbolism of the alchemists).

[1] 分   [2] 兩儀立   [3] 變   [4] 合
[5] 凝   [6] 交感   [7] 空

Fig. 48. The 'Diagram of the Supreme Pole' (*Thai Chi Thu*) of Chou Tun-I (+1017 to +1073). For description and explanation see text. The second circle from the top is marked, on the left 'Yang, motion', on the right 'Yin, quiescence'. Below are the five elements. The second circle from the bottom is marked, on the left 'The Tao of *Chhien*, perfecting maleness', on the right, 'The Tao of *Khun*, perfecting femaleness'. Below the lowest circle is written 'The myriad things undergoing transformation and generation'.

consciousness; (when) his five agents are stimulated and move, (there develops the) distinction between good and evil, and the myriad phenomena of conduct appear.[a]

(7) The sages ordered their lives by the Mean, by the Correct, by Love and Righteousness. They adopted ataraxy as their dominant attitude, and set up the highest possible standards for mankind. Thus it was that 'the virtue of the sages was in harmony with that of heaven and earth,[b] their brightness was one with that of the sun and moon, their actions were one with the Four Seasons, and their control over fortune and misfortune was one with that of the gods and spirits.'[c]

(8) The good fortune of the noble man lies in cultivating these virtues; the bad fortune of the ignoble man lies in proceeding contrary to them.

(9) Therefore it is said, 'In representing the Tao of Heaven one uses the terms Yin and Yang, and in representing the Tao of Earth one uses the terms Soft and Hard;[d] while in representing the Tao of Man, one uses the terms Love and Righteousness.' And it is also said, 'If one traces things back to their beginnings, and follows them to their ends, one will understand all that can be said about life and death.'

(10) Great is the (Book of) Changes! (Of all descriptions) it is the most perfect.

Before saying anything on the significance of this truly credal statement, Chu Hsi's commentary of +1173 must be given. This *Thai Chi Thu Chieh I* refers to the diagram, of which no direct mention is made by Chou Tun-I.

(a) The uppermost figure represents that of which it is said, 'That which has no Pole! And yet (itself) the Supreme Pole!' It is the original substance (*pên thi*[1])[e] of that motion which generates the Yang (force), and of that rest which generates the Yin (force).
It should be regarded neither as separate from, nor as identical with,[f] the Two Forces.

(b) The concentric circles in the second figure symbolise motion giving rise to Yang and rest giving rise to Yin. The complete circle in the centre symbolises the substance which does this (equivalent to the circle of the first figure). The semicircles on the left indicate the motion which produces Yang; this is the operation (*yung*[2]) of the Supreme Pole when moving (*hsing*[3]). The semicircles on the right indicate the rest which produces Yin; this is the substance (*thi*[4]) when at rest (*li*[5]). Those on the right are the root from which those on the left are produced, and vice versa (i.e. Yang generating Yin, and Yin generating Yang).

(c) The third figure symbolises the transformations of the Yang and Yin forces in union with each other,[g] and thus the generation of the Five Elements. The diagonal line from left

---

[a] Good evolutionary doctrine, and sound embryology.

[b] A naturalistic account of the highest human values.

[c] Quotation from the *I Ching* (Wên Yen, *Chhien* sect.), tr. Wilhelm (2), Baynes tr. vol. 2, p. 15; Legge (9), App. IV.

[d] Technical terms from Shao Yung's system.

[e] A technical term perhaps borrowed from Buddhism. It was used to translate Skr. *ātmakhatva*, *dharmatā*, substance. Cf. p. 481 below.

[f] Understandably enough, Bruce (2) felt the credal character of this statement so strongly that he wrote 'not to be confounded with' in the first instance, and 'procession' in the second, but we must of course eject this intrusion of occidental theology.

[g] This seems to us again a clear statement of the idea of the production of novelty by the interaction of two reactive agents—the thought is chemical.

[1] 本體   [2] 用   [3] 行   [4] 體   [5] 立

to right symbolises the transformations of the Yang, and that from right to left symbolises the unions of the Yin.

Water is predominantly Yin (*Yin shêng*[1]) and its place is therefore on the right. Fire is predominantly Yang (*Yang shêng*[2]) and its place is therefore on the left. Wood and Metal are modifications (lit. tender shoots, *chih*[3]) of the Yang and the Yin respectively, and therefore they are placed to the left and right under Fire and Water.[a] Earth is of a mixed nature (*chhung chhi*[4]), therefore it is placed centrally. The crossing of the lines above the positions of Fire and Water indicates that the Yin generates Yang and *vice versa*. (The order of their generation is indicated by the intersecting lines connecting the Five Elements), Water being followed by Wood, Wood by Fire, Fire by Earth, Earth by Metal, and Metal again by Water, in an endless unceasing round,[b] so that the five *Chhi* spread abroad and the four seasons revolve.

(*d*) The Five Elements all come from the Yin and Yang (Forces). The five different things (fit in to) the two realities without the slightest excess or deficiency[c] (*wu shu erh shih wu yü chhien*[5]). And the Yin and the Yang (go back to) the Supreme Pole (perfectly), neither one of them being more or less elaborate than the other,[d] nor more or less fundamental than the other (*ching tshu pên mo wu pi tzhu*[6]).

The Supreme Pole is essentially the same as that which has no Pole. Noiseless, odourless, it exists everywhere in the universe. As soon as the Five Elements are generated, they have each their specific natures. Since these *chhi* are different, the tangible matter (*chih*[7]) (which manifests them) is also different. Each sort has its completeness, and this there is no gainsaying.

The small circle below, connected by the four lines with the Five Elements above, indicates that which has no Pole, in which all are mysteriously unified, as indeed again cannot be denied.

(*e*) The fourth figure represents (the operations of the *chhi* of Yin and Yang exhibited in) the principles of (heavenly) maleness (*chhien*[8]) and of (earthly) femaleness (*khun*[9]) (which

---

[a] This idea of the secondary nature of Wood and Metal as opposed to Water and Fire seems to go back to the Thang Taoist who wrote the *Kuan Yin Tzu*, ch. 1, pp. 9*a*, 12*a*, *b*.

[b] Note that this is the 'Mutual Production' Enumeration Order (cf. p. 257 above).

[c] This is an extremely interesting passage. Excess and defect played an important part in Aristotle's biological thinking, as was brought out in a brilliant essay by d'Arcy Thompson (1). Here the thought probably is that if the Five Elements could be combined (as perhaps they are at the end of each cosmic cycle, see below, p. 486) there would be absolutely nothing lacking to make up the totality of the Yin and Yang, nor would there be anything left over. This is likely because Chu Hsi is following out, evidently, the inverse argument of Chou Tun-I in his para. 4. But the words may have the undertone of meaning that in each of the Five Elements there is absolutely nothing but Yin and Yang (in different proportions, of course, according to the preceding para. *c*), because, since the sentence has no verb, 'are composed of' might be inserted instead of 'fit in to'. It would have been tempting to translate it: 'The five (things), although different, each have the two realities without the slightest excess or deficiency', which would have implied that their different properties were due to a kind of varying 'stereoisomeric' arrangement—but we were told just above that some elements were 'more full of Yin' than others. On the general principle see p. 566 below.

[d] Or 'neither one of them being finer or coarser'.

---

[1] 陰盛　　[2] 陽盛　　[3] 稺　　[4] 沖氣　　[5] 五殊二實無餘欠
[6] 精粗本末無彼此　　[7] 實　　[8] 乾　　[9] 坤

pervade the universe), each having their own natures, but (both going back to) the one Supreme Pole, (as indicated by the reproduction of the original circle).

(*f*) The fifth figure represents the birth and transformation of the myriad things in their sensible forms, each of which has its own nature. But, (as indicated again by the reproduction of the original circle), all the myriad things go back to the one Supreme Pole.

> There follows a further page, giving the applications of the diagram to ethics and human affairs, which is here omitted.

The most arresting statement in Chou Tun-I's credo is that of its first paragraph. There has been difficulty in understanding the import of the five epigrammatic characters ever since the days when the Neo-Confucians themselves were trying to expound their thought. It is now agreed that the particle is a copula expressing not temporal succession but paradoxical identity. It is also clear that this opening statement is essentially that of a synthetic philosophy uniting in itself the streams of Taoist and Confucian thought, for *wu chi*[1] comes from the *Tao Tê Ching*[a] and *thai chi*[2] is a phrase in the Book of Changes.[b] Chu Hsi himself reaffirms the identity, saying that the *wu chi* is not something outside or beyond the *thai chi*.[c] Nor is the *thai chi* something outside or beyond the world; it constitutes and resides in the myriad things.[d] V. d. Gabelentz faithfully rendered the form of the statement with his 'Ohne Prinzip, dabei Urprinzip', but the full content was not expressed thereby. Zenker recalled[e] Jacob Boehme's 'Ungrund und doch Urgrund', but that was mystical theism and says too much. Other translators have favoured the sense of 'limit' for *chi*, concluding in such terms as 'The Boundless! And also the Supreme Ultimate!' But although 'infinite' is a possible (and usual) meaning for *wu chi*,[f] it lets slip the essential significance of *chi*[g] as not merely any boundary, but a polar or focal point on a boundary. Conscious of this, Chou I-Chhing writes 'Sans-Faîte—et Faîte Suprême', but even 'summit' or 'ridge-pole'[h] fails to allow for the fact that *chi* was from of old the technical term for the astronomical pole.[i] Around the Pole Star all man's universe revolved.

---

[a] Ch. 28 (cf. Waley (4), p. 178).

[b] Hsi Tzhu Chuan (Ta Chuan), pt. 1, ch. 11; tr. Wilhelm (2), Baynes tr. vol. 1, p. 342; Legge (9), App. III. But in *Chuang Tzu* also, ch. 6 (Legge (5), vol. 1, p. 243).

[c] *Thai Chi Thu Shuo Chieh*, para. 1 (v. d. Gabelentz (2), p. 33; Chou I-Chhing (1), p. 155).

[d] *CTCS*, ch. 49, p. 9*b* (le Gall (1), p. 101).

[e] (1), vol. 2, p. 216; cf. Forke (9), p. 50.

[f] It was employed in their translations by both Bruce (2), p. 128, and le Gall (1), p. 112. But Chu Hsi explains in a remarkable letter to Lu Hsiang-Shan (*CTCS*, ch. 52, pp. 48*b*, 50*a*) that Chou Tun-I was *not* implying 'boundless' in the same sense as Lao Tzu and Chuang Tzu; by *wu chi* he meant rather 'nowhere in particular but invisibly everywhere'.

[g] The etymological origin of *chi* (K910) is unfortunately obscure.

[h] The reference to the ridge-pole of a house was made by Chu Hsi himself, *CTCS*, ch. 49, p. 13*a* (le Gall (1), p. 107).

[i] See below, Sect. 20*e*.

[1] 無極   [2] 太極

It is possible to follow this thought further in the commentary on the *Thai Chi Thu Shuo* of one of the later Sung Neo-Contucians, Jao Lu.[1] He wrote:[a]

The term *thai chi* expresses the majesty of the universal Pattern (*Thien Li chih tsun*[2]).[b] The word *chi* means axis or pivot (*shu*[3]), knot or node (*niu*[4]), root (*kên*[5]) or basis (*ti*[6]);[c] as we say in common speech *shu chi*,[7] or *kên chi*.[8]... The word *thai* means so great that nothing can be added (to it),[d] and expresses the fact that it is the Great Pivot and the Great Basis of the universe. All things, however, which bear this name, such as the North (celestial) Pole, the South (celestial) Pole, the ridge of a house, the 'Capital of Shang',[e] or the four compass-point directions, have visible forms and locations to which we can point, but this *chi* alone is without form, and has no relation to space.[f] Master Chou therefore added the term *wu* (*wu chi*), expressing the fact that it is not (confined to) any *form* such as that of a nodal pivot or a basic root,[g] yet none the less is really the Great Nodal Pivot and the Great Basic Root of the Universe.[h]

And Chu Hsi's mind was dwelling perhaps on the polar axis when in conversation he compared *thai chi* to the central longitudinal axis of a candlestick on his table.[i]

What then, in terms comprehensible to us, were these Sung philosophers affirming? Surely the conception of the entire universe as a single organism. We must think of the *chi* as a kind of organisation centre. After all, in ancient times, natural philosophers could not be sure that the astronomical pole was a mere geometrical point; for them it was the nearest they could approach to the bearings of the very world-axle itself, and the position of the emperor on earth was established in its image. The Taoist *wu chi* was an affirmation that the true and entire universe depended on no such cardinal point, for every part of it took the leadership in turn, as we saw from Chuang Tzu's parable of the parts of the body (p. 52 above), and from many Taoist references to involuntary (autonomic) physiological processes. The Confucian *thai chi*, on the other hand, was a recognition of immanent power informing the wholeness of the universe, and present everywhere within it.[j] In some such terms, perhaps, we think

[a] *Hsing-Li Ching I*, ch. 1, p. 5a.
[b] So also said Chu Hsi, *CTCS*, ch. 49, p. 8b (le Gall (1), p. 99).
[c] These four words had all been used by Chu Hsi, *CTCS*, ch. 49, p. 11a (le Gall (1), p. 103), and *Thai Chi Thu Shuo Chieh*, para. 1 (Chou I-Chhing, p. 155).
[d] Cf. *CTCS*, ch. 49, p. 18b (le Gall (1), p. 118).
[e] One of the poems in the *Shih Ching* (IV, iii, 5; Karlgren (14), p. 266, no. 305) refers to the Shang capital as the centre (*chi*) of the whole world.
[f] Emphasised often by Chu Hsi, e.g. *CTCS*, ch. 49, p. 11a (le Gall (1), p. 104), ch. 52, p. 48b.
[g] Its incorporeality was often stated by Chu Hsi, e.g. *CTCS*, ch. 49, p. 11a; *Thai Chi Thu Shuo Chieh*, para. 1.
[h] Tr. auct., adjuv. Bruce (2), p. 134. Here the thought approaches the idea of a Great Motor, and reminds us of the Aristotelian conception of an Unmoved Mover (cf. also p. 322 above).
[i] *CTCS*, ch. 49, p. 13a (le Gall (1), p. 108).
[j] The *thai chi*, wrote Chu Hsi, is at the centre of all things, but not their centre (*CTCS*, ch. 49, p. 18b; le Gall (1), p. 118). Yet it is also as vast as space and as eternal as time (*CTCS*, ch. 49, p. 13b; le Gall (1), p. 108). Cf. Pascal: 'une sphère dont le centre est partout, la circonférence nulle part' (*Pensées*, vol. 1, p. 134).

[1] 饒魯　　　[2] 天理之尊　　　[3] 樞　　　[4] 紐　　　[5] 根　　　[6] 柢
[7] 樞極　　　[8] 根極

today of a universal process such as the increase of entropy. Thus we arrive at the idea of the world as indeed a single organism, no particular part of which can be identified as permanently 'in control'.

Modern minds have become accustomed to thinking (or consciously not thinking) in these terms; the world has long been full of poles, foci and centres—vortices, magnetic fields, Descartes' pineal gland, then cells and their nuclei, organisation and induction centres of embryos, centres of social control in operations of war or peace. But they are secondary to the organisms of which they form part, not superior to them. Chu Hsi wrote:[a] 'If one peers into its mystery, the *thai chi* seems a chaotic and disorderly wilderness lacking all sign of an arranger (*chhung mo wu chen*[1]),[b] yet the Li (fundamental pattern) of motion and rest, and of Yin and Yang, is fully contained within it.'

Innumerable smaller organisms were also contained within it, and indeed composed it. Some of these were more highly organised than others. In fact the world was no more undifferentiated for the Neo-Confucians than for modern organic philosophy; it manifested a series of integrative levels of organisation, wholes at one level being parts on the next. A clear statement of this conception appears in the ninth paragraph of the *Thai Chi Thu Shuo*, which indicates the inapplicability of categories outside the level to which they belong. Though entirely natural, the highest human values are relevant only at the human level.

In Neo-Confucian writings it is even possible to find a technical term for 'level of organisation'. In the *Chin Ssu Lu* we read:[c]

In the great pattern of Heaven and Earth there is nothing isolated (*tu*[2]), and every thing must have its opposite (*tui*[3]); and this is spontaneously so, not the result of any (purposive) arrangement. Pondering often on this at night, I could not help experiencing deep joy.

[Comm.] Someone asked how the *Thai Chi* could have an opposite. (The philosopher) answered 'The Yin-and-Yang is the opposite of the *Thai Chi*, for the latter is the invisible Tao (within all forms), while the former is the visible instrument (composing all forms). Thus there is clearly (what we might call) a "horizontal opposition" (*hêng tui*[4]).'[d]

Here the thought evidently concerns the relation of two of the different levels of the *Thai Chi Thu* diagram. But the opening sentence strangely prefigures both Whitehead's 'prehension'[e] and Hegel's antitheses and negations.

In sum, the identity of *thai chi* and *wu chi* was (as we might say) a recognition of two things, first, the existence of a universal pattern or field determining all states and transformations of matter-energy, and secondly, the omnipresence of this pattern. The motive power could not be localised at any particular point in space and time. The organisation centre was identical with the organism itself.

When one takes a further look at this pithily expressed system of Nature, one cannot

---

[a] *Thai Chi Thu Shuo Chieh*, para. 2 (v. d. Gabelentz (2), p. 40; Chou I-Chhing (1), p. 156).
[b] Here 'arranger' is represented by the royal plural.
[c] Ch. 1, p. 15 (para. 25).
[d] Tr. auct., adjuv. Graf (2), vol. 2, p. 71. Cf. *ECCS; Honan Chhêng shih I Shu*, ch. 11, p. 3*b*.
[e] See, e.g., Whitehead (2), pp. 197, 226, 232.

[1] 沖漠無朕   [2] 獨   [3] 對   [4] 橫對

but admit that the Sung philosophers were working with concepts not unlike some of those which modern science uses. No doubt the idea of the two fundamental forces in the universe was an ancient hypostatisation of the two sexes of the human and other species, possibly affected by the parallel (though very different) dualism of Iran; no doubt the Sung thinkers did little more than tabulate its logical consequences. But the more one reads them the more one comes to feel that they attained an insight (albeit admittedly as through a glass darkly) into those two profoundly rooted aspects of matter which appeared out of the experiments of Gilbert and Volta as negative and positive electricity, and which in our own age have proved to constitute, in such forms as protons and electrons, the components of all material particles. It was something which of course they could not express, but it was nevertheless a true insight. Here again the Chinese shot an arrow close to the spot where Bohr and Rutherford were later to stand, without ever attaining to the position of Newton.

Another deep conviction which clearly emerges from the words of the Neo-Confucians is that Nature worked in a wave-like manner. Each of the two Forces rose to its maximum in turn and then fell away, leaving the field to its opposite; moreover, they generated each other, in a way reminiscent of that 'interpenetration of opposites' expounded by dialectical philosophers in modern times. The constant references to motion and rest, occurring in alternate periods, the motion rising to a maximum degree and then returning to a null point,[a] express a legitimate scientific abstraction. Later on, in Section 26b on physics, we shall look more closely at China as a place of origin of wave-conceptions.

Thirdly, we have seen in the foregoing quotations a rather clear notion of the production of new things by reactions which one hardly hesitates to call chemical. In one place, as noted, an alchemical symbolism is certainly used. Martin (5, 6), in papers written now nearly a century ago, but which are still well worth reading, was not so far off the mark in speaking of Neo-Confucian philosophy as Cartesianism four hundred years before Descartes. It is frequently said that before the Neo-Confucians China had no metaphysics in the strict sense, but one might claim that if they did introduce metaphysics, it was of a kind very congruent with physics.

For the *Thai Chi Thu* a Taoist origin is probable. Some have thought that it originated with Chhen Thuan[1] (d. +989),[b] the famous Wu Tai expositor of the *I Ching*, and that it reached Chou Tun-I through Chhung Fang[2] and Mu Hsiu.[3] This may well be possible.[c] But even further back, a diagram very similar to that of Chou Tun-I is

---

[a] Later on, the technical term *chi*,[4] 'continuation-point', was adopted by Chu Hsi to indicate these points of maximal and minimal motion (see *CTCS*, ch. 49, p. 2a; le Gall (1), p. 85).

[b] G257. Cf. p. 442 above. See A. C. Graham (1).

[c] Chou Tun-I is recorded as having written for a Taoist temple the following epigram to be engraved on stone, in which Chhen Thuan is mentioned. 'Since reading the "Instructions concerning the Medicine (of Immortality) of the Adept Ying", I agree with Hsi-I (Chhen Thuan), having comprehended the creative mechanism of the Yin and Yang. (So) the child, when he has come forth from his mother, can attain the mastery. If the seminal essence (*ching*[5]) and the spirit (*shen*[6]) are at one with each other, then even the most minute (mysteries) can be known.' Cf. Chou I-Chhing (1), pp. 53, 190.

[1] 陳摶　　[2] 种放　　[3] 穆修　　[4] 繼　　[5] 精　　[6] 神

to be found in the early +8th-century Taoist book *Shang Fang Ta Tung-Chen Yuan Miao Ching Thu*[1] (Diagrams of the Mysterious Cosmogonic Classic of the Tung-Chen Scriptures).[a] Buddhist influence would have reached him through his teacher, Shou-Yai.[2]

Chou Tun-I's other extant work, the *I Thung Shu*[3] (Fundamental Treatise on the Book of Changes)[b] is at first sight concerned wholly with ethical matters remote from natural science—the sage and his role in society, his wisdom, the rites, music, and so on. The argument of the book revolves around the technical term *chhêng*,[4] the normal meaning of which is 'sincerity'. But here it clearly denotes more than the usual ethical connotations of this word at the human social level, and indeed it has been elevated to the rank of a cosmological principle.[c] One might find an earlier parallel; the cosmological use of *li*[5] in the *Hsün Tzu* book.[d] This needs further explanation.

The *Chung Yung*, punning, says,[e] 'He who is sincere (*chhêng*[4]), perfects (*chhêng*[6]) himself'. This gives the clue that Chhêng is a quality essentially capable of inhering in an individual and not arising only from the relations between individuals. It is therefore a question rather of what might be called 'integrity', of being sincere with oneself, of not deluding oneself nor acting contrary to one's true nature.[f] The *Chung Yung* also says,[g] 'Sincerity is the Tao of heaven; to apply oneself to sincerity is the Tao of man', indicating that it transcends the human sphere. Heaven has Chhêng because it faithfully follows its true nature and does nothing against its Tao; it is perfectly itself. In this way we come to the realisation that Chhêng is achieved when every organism fulfils with absolute precision whatever its function may be in the higher organism of which it forms a part.[h] Only by following its inner law and light can it

---

[a] *TT* 434.

[b] Tr. Grube (5); Eichhorn (1); Chou I-Chhing (1).

[c] I follow, as far as it goes, the interesting special study which has been made of this by Chou I-Chhing (1), esp. pp. 93, 101, 102. Realisation that the significance of the word could not be confined to the ethical level has led sinologists to translate it in various ways, e.g. 'perfection' (Wieger, 4), 'truth' (Bruce, 2; Zenker, 1), 'realness' (Hughes, 2), 'Wahrhaftigkeit', 'veracity' (Grube, 5; Eichhorn, 1). If the present interpretation is acceptable, these may all be laid aside. 'Sincerity' commended itself to Couvreur and Legge, but the word is so untranslatable and at the same time so important that it probably ought to be retained in mere transliteration, like Tao and Li. This we do.

[d] Cf. pp. 27, 283, 287 ff. above.

[e] Ch. 25 (Legge (2), p. 282).

[f] This has often been taken as fundamental in aesthetics. We dislike to see a wooden object imitated slavishly in reinforced concrete, for example. The material is not being true to itself. Cf. Collingwood (1), p. 65. Nevertheless there has always been a tendency in technology for shapes of objects in a new material to imitate the shapes which they had in an older one. Such objects are called by archaeologists and ethnologists 'skeuomorphs' (see R. U. Sayce (1), pp. 80ff.). In bronze at first they imitated stone, in pottery they imitated wickerwork baskets or buffalo horns, and so on. One would like to know whether any aesthetic judgments of this kind formed part of the background of the Chinese philosophical thinkers.

[g] Ch. 20 (Legge (2), p. 277).

[h] This was certainly applied by Chinese thinkers to such natural movements as those of the heavenly bodies. The sun did not 'put the shades to flight' and chase the moon out of the sky; the moon knew how to withdraw discreetly when the proper time came. Still more impressively, so did the sun. Cf. p. 283.

[1] 上方大洞眞元妙經圖   [2] 壽涯   [3] 易通書   [4] 誠
[5] 禮   [6] 成

so act.[a] This state of affairs is quite familiar to modern philosophies of organism, which do not seem however to have adopted a special term for it.

The conception of Chhêng as a cosmic principle seems inescapably present already in the *Chung Yung*, a work parts of which may well be as old as the −5th century.[b] Here, for example, is the opening passage of chapter 26:

Thus perfect integrity (*chih chhêng*[1]) never ceases (for a moment).[c] Now if that be so, then it must be extended in time (*chiu*[2]); if extended in time, then capable of demonstration (*chêng*[3]); if capable of demonstration, then extended in space-length (*yu yuan*[4]); if extended in length, then extended in depth (*po hou*[5]); if extended in depth, then extended in height-visibility (*kao ming*[6]). And this quality of extension in depth is what makes material things supportable (from below); this quality of extension in height-visibility is what makes things coverable (from above); while the extension in time is what makes them capable of coming to completion. So depth pairs with (*phei*[7]) earth, height-visibility with heaven, and space plus time makes limitlessness (*wu chiang*[8]). This being the nature of Chhêng, it is not visible and yet clearly to be seen, does not move and yet (brings about) change, takes no action (*wu wei*[9]) and yet brings (all things) to their completion.[d]

Of course Legge,[e] following late commentators, took Chhêng in its purely human-psychological sense of 'sincerity', applying all the expressions to the qualities of the sage and his virtues. But its status as an all-pervading organic principle seems too clear to be overlooked.

The cosmic significance of Chhêng was developed fifteen centuries later in many places of the *I Thung Shu*. Just as the first of the *kua* (*Chhien*) symbolises the beginning of all things, so it is the origin of Chhêng,[f] which comes into being with them.[g] It is something pure and perfect (*shun sui*[10]),[h] and in acting it exerts no force (*wu wei*[9]).[i] It is, or generates, the germ of all good and evil (*chi shan o*[11]);[j] and like the Tao it has a virtue (*tê*[12]), which transmutes private loves to universal human-heartedness, rightness into righteousness, and natural human patterns (*li*[13]) into social order (*li*[14]).[k] This is why it is the foundation of sageliness.[l] Diffusing outwards it evokes beginnings and developments; ebbing, it leaves permanent gains.[m] When at rest, it is as if it did not exist; when in action, its existence is manifest.[n] Like the individual patterns which

---

[a] Chou I-Chhing suggests, p. 97, that this conception might explain that long-debated phrase of Mencius (II (I), ii, II) about his 'hao jan chih chhi',[15] the *chhi* of the vast universe which he felt moving in him. The context of the passage supports such a view. Perhaps what in Christendom men called 'obedience to the will of God' was approximately represented in Chinese thought by 'love and righteousness as part of Nature's plan'.

[b] The oldest parts are attributed to Khung Chi[16] (Tzu-Ssu[17]), the grandson of Confucius.

[c] It has just previously been said that Chhêng is the end as well as the beginning, and ground, of things.　　　　　[d] Tr. Hughes (2), p. 132, mod.

[e] (2), p. 283.　　　　　[f] Ch. 1 (Chou I-Chhing tr. p. 163).　　　　　[g] Ch. 1 (tr. p. 164).

[h] Note the use of the ancient Taoist technical term, cf. p. 106 above.

[i] Ch. 3 (tr. p. 166).　　　　　[j] Ch. 3 (tr. p. 166).

[k] Ch. 3 (tr. p. 166).　　　　　[l] Chs. 1 and 2. Cf. p. 507 below.

[m] Ch. 1 (tr. p. 164).　　　　　[n] Ch. 2 (tr. p. 165).

[1] 至誠	[2] 久	[3] 徵	[4] 悠遠	[5] 博厚	[6] 高明
[7] 配	[8] 無疆	[9] 無爲	[10] 純粹	[11] 幾善惡	[12] 德
[13] 理	[14] 禮	[15] 浩然之氣	[16] 孔伋	[17] 子思	

orient themselves to its influence, it belongs to the category of unseen (spiritual, *shen* [1]) things in the universe. Of these Chou Tun-I gives a very dialectical definition:[a]

> Entities of which it can be said that when they move they are not at rest and when at rest they do not move, are material things (*wu* [2]). Entities of which it can be said that they move and yet at the same time that they do not move, or that they are at rest and yet not at rest, are spiritual principles (*shen* [1]). Such entities are neither unmoving nor unresting. Material things are those which do not everywhere pervade. Spiritual principles are the mystery of the world and all that is therein.[b]

We should say 'organising relations'. Chou further wrote: 'That which germinates in the minutest invisible particles and (at the same time) fills the vast realm of boundless space, is the spiritual (*Fa wei pu kho chien chhung chou pu kho chhiung chih wei shen* [3]).'[c] And:

> Radiant is Chhêng, because a pure spirit;
> Marvellous are the Shen, because responsive;
> Abscondite are the Chi,[4] because exceedingly minute.[d]

If these conceptions were really those of the Neo-Confucians, it should be possible to confirm them in the words of Chu Hsi. And indeed, in his *Thai Chi Thu Shuo Chieh*, he wrote:[e] 'By the activity of the Supreme Pole (*thai chi*), the Chhêng goes forth into all things (*chhi tung yeh chhêng chih thung yeh* [5])'—and all things work together for good. The parallel has seemed to some[f] so striking as to suggest that Master Chou would willingly have accepted the phrase '*Wu Chhêng erh Thai Chhêng*'. For while the universe was all spontaneity and uncreatedness yet at the same time it was all order (*ta hua;* [6] *ta shun* [7]),[g] that sublime order produced by the harmony of individual wills, the intuitive faithfulness of organisms to their own natures.

It will be seen, therefore, that this was a philosophic naturalism with great relevance to the natural sciences. And once again it fused together, in what was an implicitly evolutionary scheme, the natural world of the Taoists and the moral world of the Confucians.[h]

A few words may be added here about the other members of the Neo-Confucian school. While sceptical of Buddhist 'immortality' theories, Chhêng Hao showed (alone

---

[a] Not at all surprising in view of what we have seen of Buddhist logic, p. 424 above.
[b] Ch. 16 (Chou tr. p. 174), here tr. auct.
[c] Ch. 3 (tr. auct.).   [d] Ch. 4 (tr. auct.).
[e] Para. 1 (v. d. Gabelentz (2), p. 36; Chou I-Chhing (1), p. 155).
[f] Chou I-Chhing (1), p. 102.
[g] *I Thung Shu*, ch. 11 (Chou tr. p. 171). Chou Tun-I here uses the expression *ta hua* almost as we should say progress, or social evolution, saying that men cannot see its foot-prints or know how it comes about, for it is a phenomenon of *shen* [1] (organising relations).
[h] Chou I-Chhing, too, has recognised this, (1), p. 142.

[1] 神   [2] 物   [3] 發微不可見充周不可窮之謂神   [4] 幾
[5] 其動也誠之通也   [6] 大化   [7] 大順

of the main group of Sung thinkers) a tendency to metaphysical idealism[a] (cf. on, p. 507). In the *Erh Chhêng Chhüan Shu*[1] (Collected Works of the Two Chhêng brothers) he says, 'I contain the myriad things all within myself.'[b] His brother, Chhêng I, has several scientific points worthy of note, but is not so important for those aspects of Sung philosophy as a whole in which we are here interested.[c]

Chang Tsai paid particular attention to one concept which we shall constantly meet with in this Section, namely, the formation of all things and living creatures by processes of condensation or aggregation of the Chhi, and their destruction by processes of dispersion and disaggregation. The same technical terms are used as in Wang Chhung, a thousand years earlier, indeed without much development of thought.[d] Chang Tsai shows his naturalism by the statement in his *Hsi Ming*[2] (The Inscription on the Western Wall),[e] 'My body is of the same substance as that of heaven and earth; my nature is of the same organising (lit. leading) principle which controls heaven and earth (*Ku thien ti chih sai, wu chhi thi; thien ti chih shuai, wu chhi hsing*[3]).' Foreshadowing the use of the term *Li*[4] by Chu Hsi as the principle of cosmic organisation, Chang Tsai uses the expression *thai ho*,[5] the 'Great Harmony', and in a very materialist sense.[f] For him, as for the other Neo-Confucians, the world contained nothing supernatural.[g]

The conception of the formation of things by the aggregation of the universal 'matter-energy', Chhi, was fully taken over by Chu Hsi himself. In the *Chu Tzu Chhüan Shu*[h] he says that the Chhi 'condenses to form solid matter' (*Chhi chi wei chih*[6]). And again, the Chhi 'is able by condensing to form material objects' (*Chhi tsê nêng ning chieh tsao tso*[7]). What was new in Chu Hsi's thought was the association of the condensation process with Yin and the dispersion process with Yang. 'When the Yin *chhi* flows and streams forth (*liu hsing*[8]), that is Yang; when the Yang *chhi* condenses and congeals (*ning chü*[9]), that is Yin.'[i] The Yang (male or positive) principle thus became associated with expansion; the Yin (female or negative) principle with contraction. More questionable in its results was another identification of Chu Hsi's, namely, that of expansion and contraction, or dispersion and condensation, with the terms for the two 'souls' of man, and for the two kinds of 'ghosts and demons' which had been transmitted from remote antiquity. This is a much more important matter than would appear at first sight, but its consideration will be postponed until a later place in this Section (p. 491).

---

[a] Forke (9), pp. 76, 77, 78.
[b] *Honan Chhêng shih I Shu*, ch. 11, p. 3*a*; cf. ch. 2A, p. 16*a*, and 2B, p. 6*b*. See Graham (1).
[c] See elsewhere in this book, p. 568, and also in Sects. 19*k*, 38 and 49.
[d] He uses the same parallel of the forming and melting of ice (*Hsi Ming*, ch. 2, p. 6*b*).
[e] Tr. Eichhorn (3).
[f] Cf. Forke (9), p. 60.
[g] Cf. Forke (9), p. 68.
[h] Ch. 49, pp. 1*a*, 2*b*.   [i] *CTCS*, ch. 49, p. 34*a*.

[1] 二程全書   [2] 四銘   [3] 故天地之塞吾其體天地之帥吾其性   [4] 理
[5] 太和   [6] 氣積爲質   [7] 氣則能凝結造作   [8] 流行   [9] 凝聚

After the Sung, these doctrines of expansion and condensation became part of the universal background of Chinese thought. In the Ming dynasty, Kao Phan-Lung[1] (+1562 to +1626)[a] criticised them, and writers of Taoist flavour such as Chuang Yuan-Chhen[2] (Shu Chü Tzu;[3] the 'Hempseed Master')[b] expounded them as a matter of course. Even in the Chhing, long after the coming of the Jesuits, they were being taken as something of profound importance in the pre-Socratic manner, as by Sun Chhi-Fêng[4] (+1584 to +1675)[c] and Lu Lung-Chhi[5] (+1630 to +1692).[d]

(3) THE STUDY OF UNIVERSAL PATTERN; THE CONCEPTS *CHHI* (MATTER-ENERGY) AND *LI* (ORGANISATION)

We can now proceed to a more systematic examination of the naturalist philosophy of Chu Hsi. The first question which presents itself is the exact interpretation of the two fundamental concepts with which he worked, Chhi[6] and Li.[7] There is no doubt that in general these two terms represent the material and non-material elements respectively in a basically naturalistic universe. We have often had occasion to observe the use of the word *chhi*[6] in Chinese thought, for it can be found in the writings of almost every author who was even indirectly concerned with Nature since the beginning of the ancient philosophical schools.[e] Though in many ways analogous to the Greek *pneuma* ($\pi\nu\epsilon\hat{\upsilon}\mu\alpha$), I have preferred to leave it untranslated, since the significance which it had for Chinese thinkers cannot be conveyed by any single English word. It could be a gas or vapour, but also an influence as subtle as those which 'aethereal waves' or 'radioactive emanations' have implied for modern minds. There has been general agreement among sinologists that the best translation for the new sense in which Chu Hsi uses it is simply 'matter',[f] but it must be remembered that he also has another term *chih*,[8] which means matter in its solid, hard and tangible state. Though *chih* is a form of *chhi*, *chhi* is not always *chih*, for matter can exist in tenuous non-perceptible forms.

On the other hand, there has been much disagreement concerning the interpretation of *li*.[7] An early tendency was to translate it by 'form', as le Gall[g] and Zenker did; but Forke[h] was undoubtedly right in saying that this reads into the thought of the Sung Neo-Confucians an Aristotelianism which was not there. It will be convenient to postpone for a few moments the detailed criticism of this view. Equally bad was the interpretation of Bruce (1, 2), Hackmann (1), Henke (1), Warren (1), and Bodde (3, 4),[i] who all translated it as natural (scientific) 'law'; this prejudges the whole question as to whether the Chinese at any time developed the conception of laws of nature.

---

a Forke (9), p. 429. Cf. p. 506 below.   b Forke (9), p. 449.
c Forke (9), p. 468.   d Forke (9), p. 491.   e Cf. pp. 22, 41, 76, 150, 238, 275, 369.
f Forke (9); le Gall (1); Bruce (2), p. 102. The translation of *chhi* as 'vital force' by Chhen Jung-Chieh (1) will not do.
g See, for example, p. 31.   h (9), p. 171, and especially (11).
i In his recent translation of Fêng Yu-Lan (1), vol. 2, Bodde has abandoned this and writes simply 'Principle'. We still prefer *Li* untranslated.

¹ 高攀龍   ² 莊元臣   ³ 叔苴子   ⁴ 孫奇逢   ⁵ 陸隴其
⁶ 氣   ⁷ 理   ⁸ 質

We must not anticipate here a problem the discussion of which will occupy us at a later stage (Sect. 18). It is, however, necessary to point out that the word *li*[1] (K 978) in its most ancient meaning signified the 'pattern' in things, the markings in jade[a] or the fibrous texture of muscle, and only later acquired its standard dictionary meaning of 'principle'. This is confirmed by Chu Hsi himself,[b] who instanced the strands in a thread, or the grain in bamboo, or the bamboo strips themselves woven into basketwork.

Now since this pattern was to be understood as the universal cosmic pattern, containing in itself all smaller and more limited patterns (*Gestalten*), it must include, as it did in Neo-Confucian philosophy, all the phenomena of social life, and of mind, and of the highest manifestations of human virtue and intellect. But these only make their appearance at what we would now call the higher levels of organisation. Hence sinologists have tended to ascribe almost divine attributes to *Li*,[1] since whatever this principle of the universe was, it must, they thought, be something 'higher' than the highest manifestations of virtue and intellect, in fact the highest values known to man.[c] Forke wrote unhesitatingly *Vernunft*, Reason; '*Li*', he said, 'is the rational as opposed to the material principle, in fact Reason, which creates and masters Matter.'[d] On this view, the Neo-Confucian position was analogous to that of Giordano Bruno, who spoke less of 'law' than of 'reason' (*ratio, raggione*) as forming the inherent natures of all things, which make them behave as they do.[e] But reason is also unacceptable as a translation of Li.[f] It implies consciousness, even personality, and while it was quite natural as a term for the organising forces of the universe in a civilisation saturated with theistic conceptions, it cannot be extended to the concept of Li without severe distortion of meaning. Bruce went further, and fared worse, by taking Li as divine 'law' and by reading into both Chhi and Li all kinds of 'spiritual' attributes, so much so indeed, as to end by representing the Neo-Confucians as frank theists.[g]

[a] As a verb, it meant to cut jade according to its natural markings. Such, at any rate, are the usual views, but some scholars (e.g. Demiéville, 3a) believe that the earliest meaning of the word was the pattern in which the fields were laid out for cultivation according to the lie of the land.

[b] *CTCS*, ch. 46, p. 12b; Bruce (1), p. 290.

[c] The same perplexity is shown by Fêng Yu-Lan (1), vol. 2, p. 571, and in Bodde (3), p. 50, where the complaint is made that Chu Hsi mixes up the ethical and the logical-scientific. Of course every Chinese philosopher has done this, but the Neo-Confucians did what every philosophy must ultimately do, namely, bring these two into some relation. This can best be done on an evolutionary basis, in which it can be seen that the good and other high human values come into existence when a certain high level of organisation is reached. Cf. the discussion in Waddington *et al.* (1).

[d] (9), p. 171: '*Li* ist das rationale Prinzip im Gegensatz zum materiellen; die Vernunft, welche den Stoff schafft und beherrscht.' So also Chhen Jung-Chieh (1), Forke (11). Cf. Yasuda (1).

[e] Personal communication from Dr Dorothea Singer.

[f] I was glad to find that Graf (2), vol. 1, p. [4], agrees with us in rejecting all these translations. Of course *Li* is that in natural things which makes them *intelligible* to the rational mind (cf. his pp. 256ff.). It has been very common to accept 'Law' and 'Reason' for *Li*; e.g. Feifel (1), p. 135, translating *Pao Phu Tzu*, ch. 2; Laufer (17), translating the entry on amber in the *Pên Tshao Kang Mu*; and Chhen Jung-Chieh (4, 5). Lin Yü-Thang (5), p. 247, accepts both 'Reason' and 'laws of Nature' on the same page for Li, but neither will do.

[g] On Bruce, see Graf (2), vol. 1, pp. 242 ff.

[1] 理

All previous interpretations of Neo-Confucianism, however, lacked the background (in the perspective of which it is now possible to set it) of modern organicist philosophy, of which one may cite Whitehead as the outstanding western representative.[a] On the organic view of the world, the universe is one which simply has the property of producing the highest human values when the integrative level appropriate to them has arisen in the evolutionary process. Whether it is necessary to endow the universe, or some creativity 'behind' the phenomenal universe, with 'spiritual' qualities as high as, or higher than (if that is imaginable), those which we know at the highest levels of organisation, is a question which is perhaps outside the field of philosophy, and certainly outside that of natural science. Here it is not possible to outline the philosophy of organism in a few sentences, and the reader must be referred to the works of the philosophers just mentioned.[b] From the point of view of the scientist, at any rate, the levels of organisation can be described as a temporal succession of spatial envelopes; thus there were certainly atoms before there were any living cells, and living cells themselves contain and are built up of atoms.[c] It would, of course, be absurd to suggest that Chu Hsi and his Neo-Confucian colleagues talked like this, or even to interpret what they said as implying any of these detailed conceptions, still less to translate their words accordingly. But I am prepared to suggest, in view of the fact that the term Li always contained the notion of pattern, and that Chu Hsi himself consciously applied it so as to include the most living and vital patterns known to man, that something of the idea of 'organism' was what was really at the back of the minds of the Neo-Confucians,[d] and that Chu Hsi was therefore further advanced in insight into the nature of the universe than any of his interpreters and translators, whether Chinese or European, have yet given him credit for.[e]

Although the discussion of Bruce (2) on the interpretation of Chu Hsi's philosophy in terms of natural science is now quite outdated, he did well, in my opinion, to include energy with matter in the interpretation of Chhi. Today we know (too surely for our peace of mind) that matter and energy are interconvertible. And, as I have

a (1–4). Though there are many others, e.g. Engels, Lloyd-Morgan, Smuts, Sellars, and so on.

b Elsewhere I have tried to expound the philosophy of organism as it appears to a present-day scientist (Needham (9, 10), reprinted in (3), esp. pp. 178 ff., 185, 233 ff.).

c This may be the place to refer to the dangers of le Gall's (1) discussion and translation. In many sentences (pp. 31, 34, 74, 80, 102) he speaks unhesitatingly of 'atoms', though so far as I can see there is absolutely no authority from any text of Chu Hsi for doing so. We also have molecules (p. 37) with even less justification. Typically Thomistic medieval terms such as 'noble' (p. 83) and 'dignity' (pp. 84, 88), and Aristotelian words such as 'form' (p. 84), appear in the translation, and would certainly mislead a reader not himself possessing any knowledge of Chinese. Needless to say, Wieger (2), e.g. p. 188, continues all this. On le Gall see Graf (2), vol. 1, pp. 240, [20].

d Graf (2) well says that if one were able to explain to Chu Hsi the parallelism which modern science has established between the solar systems and the orbits of particles within each individual atom, he would indeed find it wonderful, but he would not be particularly surprised, so greatly does it exemplify his own conceptions of the unity and universality of Li, of the natural principle of Order inhering in matter at all levels (vol. 1, p. 76).

e My interpretation, to judge from certain expressions in the study of Chu Hsi made by an eminent modern philosopher, W. E. Hocking, and from conversation with him in the autumn of 1942, would not have met with his disapproval.

elsewhere pointed out,[a] the two fundamentals in the modern view of the universe, as the natural scientist and the organic philosopher sees it, are Matter-Energy on the one hand, and Organisation, the principle of Organisation, on the other. If, therefore, it were indispensable to translate the Li[1] of Chu Hsi into English, 'Organisation' or 'Principle of Organisation' would be the choice which I should make. But we shall adhere to our practice of leaving untranslated those fundamental Chinese words which it is almost impossible to translate, and speak therefore of Li, as of Chhi, or Yin and Yang. All that need be added is that at least the attempt should be made to re-appraise Chu Hsi's philosophy of Li in the light of the philosophy of organism. If, as I believe, he was feeling his way towards such a philosophy, it was a very remarkable achievement for the age in which he thought and wrote, i.e. the +12th century; and he might thus be said to have accomplished much more than Thomas Aquinas from the point of view of the history of science.

This is the point at which we can return to examine more closely the suggestion that Li[1] and Chhi[2] can be equated with the Form and Matter of Platonic-Aristotelian philosophy. This suggestion has recently been revived,[b] but I believe it to be entirely unacceptable. It is true that form was the factor of individuation, that which gave rise to the unity of any organism and its purposes; so was Li. But there the resemblance ceases. The form of the body was the soul; but the great tradition of Chinese philosophy had no place for souls, and in Neo-Confucianism, as we shall see,[c] the spiritual *pneumata* of man were thought to lose themselves after death among the circum-ambient vapours. The distinctive importance of Li is precisely that it was not intrinsically soul-like or animate. Again, Aristotelian form actually conferred substantiality on things, but although atomic particles were just as unpopular in Fukien as in Macedonia, the Chhi was not brought into being by Li, and Li had only a logical priority. Chhi did not depend upon Li in any way.[d] Form was the 'essence' and 'primary substance' of things, but Li was not itself substantial or any form of Chhi or *chih*.[3] Li was not more real than Chhi, and neither was illusory or subjective (*huan*[4] or *khung*[5]), nor was Chhi potentially Li as matter was potentially form. In spite of common interpretations of the famous phrase *hsing erh shang*,[6] I believe that Li was not in any strict sense metaphysical, as were Platonic ideas and Aristotelian forms, but rather the invisible organising fields or forces existing at all levels within the natural world. Pure form and pure actuality was God, but in the world of Li and Chhi there was no *Chu-Tsai*[7] whatsoever.

---

[a] (11), reprinted in (6), p. 199; (28).
[b] By Fêng Yu-Lan (1), vol. 2, pp. 482, 507, 542; Graf (2), vol. 1, pp. 66, 77, 255, e.g.
[c] P. 490 below.
[d] Doubtless the Aristotelian material and efficient causes would have been acceptable to the Neo-Confucians as concerned with Chhi, while the formal cause would have pertained to the realm of Li. The final cause might well have puzzled them, yet, as we saw above (p. 289), Chinese thought had long contained the idea of operation backwards in time.

[1] 理   [2] 氣   [3] 質   [4] 幻   [5] 空   [6] 形而上
[7] 主宰

This non-theistic quality prohibits us again from comparing Neo-Confucianism too closely with the pantheism of the Stoics. The Stoic logos often tends, we find, to present itself to the thought of Western scholars when they meet with the philosophy of Chu Hsi and his associates,[a] and the comparison has been explicitly made.[b] 'Hoc ubi supponimus', wrote Brucker in 1744, after giving quite a good account of Neo-Confucianism, 'ovum ovo non erit similius, quam Stoica sunt Sinensibus.'[c] This conviction probably accounts for the acceptance by so many translators, as we have just seen, of 'Reason or *Vernunft* for *Li*.[1] The Generative Reason (*logos spermatikos*, λόγος σπερματικός), individuated in all forms and shapes, all life and all intelligence, is certainly a conception somewhat parallel to the principle of order, Li, in every particular collocation of Chhi. Chu Hsi would hardly have rejected the idea of this as a seed of organisation in each thing, and would have appreciated the Johannine 'light which lighteth every man that cometh into the world'. But the seminal logos was God himself as the organic principle of the cosmic process, which he directed to a rational and moral end,[d] and for the later Stoics God was identical with universal matter as well as with the creative force fashioning it. Their philosophy, in fact, could escape neither from the theistic preconception, nor from the bifurcation of organisms into bodies and souls; it flourished in a European medium where such assumptions were part of the unnoticed intellectual background. But in the Chinese philosophic tradition the need for a Supreme Being had never been felt,[e] and the distinction between Li and Chhi was not at all the same as that between soul and body, for souls were composed of subtler kinds of Chhi, and the very arrangement of the parts of bodies in space and time, with all their interactions, was Li in manifestation and effect.

Long after I had written the above paragraphs I learnt that Professor Demiéville in Paris had reached similar conclusions,[f] and had adopted the word *Ordonnancement* as a translation of Li. His views are available in a short report (3a) of great interest, in which he emphasises the tendency of the Buddhists to transcendentalise and supernaturalise the originally naturalistic organicism of Han and pre-Han times. Hence the metaphysical undertones which the word had acquired by the time it was utilised by Chu Hsi, and from which he himself was perhaps never quite able to liberate it. Elsewhere[g] we have seen not a few instances of the way in which the word

---

[a] In October 1954, in one week, the point was raised by Dr A. C. Bouquet of Cambridge and Mrs Martha Kneale of Oxford, quite independently.

[b] By Garvie (1).   [c] Vol. 5, p. 897.

[d] The words are those of Inge (2). See p. 534 below.

[e] There was therefore no psychological urge to retain God as a name for the universe of organised matter long after it had ceased to be philosophically possible to think of him as a distinct personal being.

[f] Cf. p. 290 above. Long ago (1888) Martin (6) used the expression 'organising principle', but no one paid any attention to this insight. Even earlier (1815) Morrison (1) in his dictionary gave 'a principle of organisation' as one of the chief meanings of the word. Hughes (2), p. 50, also implicitly affirms the organicistic character of Chu Hsi's thought. Now Graf (2) also fully accepts the view of Li as *Ordnungsprinzip* and *Bauplan* (vol. 1, pp. 44, 76, 248 ff., e.g.). Chou I-Chhing (1), p. 71, accepts 'pattern'.

[g] Pp. 51, 73, 272, 276, 322, 328, 408, 411 ff., 438 ff., 449.

[1] 理

was used by thinkers of earlier centuries.[a] The organismic note is apparent already in a sorites in chapter 55, one of the oldest parts of *Kuan Tzu* (cf. Haloun, 5). 'Names (*ming*[1]) derive from reality (*shih*[2]); reality derives from pattern (*li*[3]) (Haloun translates "structure"). Pattern derives from properties (*tê*[4]) (of things); properties derive from harmony (*ho*[5]); and harmony derives (ultimately) from congruity (*tang*[6]) (i.e. the "fit" of all natural things).' The Han school of Ritualists (as in the *Li Chi*) saw in *li*[3] simply a principle of order, of the right and proper disposition and distribution of things, whether on the cosmic or the social plane. The *I Ching* appendices indicate that man should know how to conform himself to the natural order, how to understand his place in nature and society, fulfilling the duties (*i*[7]) associated with his lot (*fên*[8]).[b] In the −4th century Mencius uses *li*[3] to describe the harmonious cooperation of an orchestra.[c] For Hsün Chhing and Han Fei the *li*[3] of any particular thing is its configuration, its specific form, and all the data about it which permit one to handle it successfully; all these individual *li*[3] being subsumed in the great Tao[9] which itself has no 'fixed specificity' (*ting li*[10]), and so runs through all particular things. The Taoists did not disagree. Nor was there any essential change from the time of Liu An to that of Wang Pi and Ho Yen.

With Hsi Khang in the +3rd century, however, a different note is heard, for the 'mysterious Li' (*miao Li*[11]) is 'something cut off from ordinary discourse', and to be apprehended only by mystical experience. This was in line with the mystical interpretations of ancient Taoism by Hsiang Hsiu and Kuo Hsiang, already described,[d] which Demiéville is perhaps justified in terming 'supernaturalism'. *Li*[3] was becoming something like a metaphysical absolute, and this was what Phei Wei[e] combated as 'nihilistic'. It only remained for the Buddhists to appropriate the word and detach it as thoroughly as they could from naturalist thought. Thus the monk Chih-Tun[12] (+314 to +366) identified Li with *prajñā*, the ineffable, unchangeable, supramundane absolute. As Demiéville well says, the original Chinese conception was one of a universal order *in* the world, an explanation *of* the world and not its negation. The Buddhists 'denaturalised' it, placing it above or behind the world, which to them was an illusion only. Chu Tao-Sêng[13] (d. +434) identified *Li*[3] with *buddhatā*, and Sêng-Chao[14] (d. +414) equated it with *ārya-satyāni*, the four Buddhist dogmas; in general, it was used to assist the translation of Sanskrit terms for the absolute.

[a] It would take a special monograph to trace the use of this word before the Sung. It had always meant the general principles of things in the world, as in books such as the Han *Thai Hsüan Ching* of Yang Hsiung (cf. p. 329) or the *Chou I Lüeh Li* of Wang Pi. Phei Wei[15] (+267 to +300), who opposed the Taoism of his time with his *Chhung Yu Lun*[16] (Discourse on the Primacy of Being), used the word in a sense prefiguring that of the Neo-Confucians very closely. But no one had ever tried to define it so precisely as the Neo-Confucians did.

[b] Cf. p. 107 above, and p. 550 below. In another report, Demiéville (3*b*) draws a parallel between *fên*[8] and Indian *svadharma* or *svakarman*, and Stoic *kathekon* or *officium* (Arnold (1), pp. 301 ff.).

[c] *Mêng Tzu*, v (2), i, 6.

[d] P. 433 above.   [e] Cf. p. 386.

¹ 名   ² 實   ³ 理   ⁴ 德   ⁵ 和   ⁶ 當   ⁷ 義
⁸ 分   ⁹ 道   ¹⁰ 定理   ¹¹ 妙理   ¹² 支遁   ¹³ 竺道生
¹⁴ 僧肇   ¹⁵ 裴頠   ¹⁶ 崇有論

The work of Chu Hsi, therefore, was to remove Li^1 from most of its Buddhist contexts, and to restore its ancient naturalist significance, immanent rather than transcendent. The precise degree to which he was able to do this remains a matter for minute future research; certainly his critics of later centuries often believed that he did not entirely succeed in divesting the concept of its religious-metaphysical undertones.

With regard to Chu Hsi's organicism, it is well also to bear in mind the thesis of a modern Chinese philosopher, Chang Tung-Sun (1), that while European philosophy tended to find reality in *substance*, Chinese philosophy tended to find it in *relation*.[a] This might throw light on many characteristic features of the thought of both civilisations. Hughes[b] and Chang Tung-Sun have both linked it with the personalisation of the deity in Europe and the impersonality of 'Heaven' in China—we shall later glance at the great consequences which this difference may have brought in its train.[c] Behind the metaphysical idea of 'substance', Hughes points out, lies the logical idea of 'identity', and Western philosophers laid down as a basic principle of thought that a thing cannot both be and not be at the same time. Chinese philosophers, on the other hand, laid down that a thing is always 'becoming' or 'de-becoming'; all the time on its way to being something else. Already in the sections on the Taoists (10e), Logicians (11) and Buddhists (15e) we have seen abundant examples of this tendency to skip the stage of formal logic and go straight to the stage of Hegelian logic.[d] So also an emphasis on 'relation' most appropriately describes Chu Hsi's appreciation of the organising principle according to which parts combine in wholes. It is Hughes' further contention that by giving so important a place in his scheme to Chhi, Chu Hsi cleared the way for an emphasis on substance in China, just as later Leibniz was perhaps the first in Europe to clear the way for an emphasis on relation.[e] Chinese and European thought would thus have reached a synthesis in the 17th century, unacknowledged by Western historians.

[a] This he traces to linguistic differences between Chinese and the Indo-European languages (cf. Sect. 2 above and Sect. 49 below). A guide to the literature of current discussions about these differences will be found in H. Franke (5), p. 43. On relation and substance cf. p. 199 above.

[b] (2), pp. 52, 169.

[c] Pp. 580 ff. below.

[d] If my intuition is not at fault, the instinctive modes of expression of many of my Chinese colleagues still mediate this thought-form characteristic of their great civilisation. Where a westerner would say 'yes' or 'no', they are likely to answer 'well, not exactly'. When Mark Twain said that the reports of his death had been greatly exaggerated, he was being more Chinese than he knew. The possibility arises that the harshness of Aristotelian logic was what ultimately stimulated Europeans to the invention of so many technical terms. In China every event or phenomenon could have an *ad hoc* description of its own.

[e] This remark may seem somewhat surprising, for Leibniz has often been considered (as by Russell) so wedded to subject-predicate logic that he could make nothing of relations. But we are justified by a paradox. It is true that Leibniz did insist that all the characters of a substance belong to it in itself, and as it were in its own right. On the other hand, for him substance has no character except its representation of other substances, its nature as a 'living mirror'. Thus what began as a denial of relation ended by presenting a system of relations without any terms. Thanks are due to Mrs Martha Kneale for pointing out the necessity of this clarification.

¹ 理

But it is high time that we listened to Chu Hsi himself. The forty-ninth chapter of his collected works opens with the following statement: 'Throughout the universe there is no Chhi without Li, nor is there any Li without Chhi.'[a] With due regard to what was said above, this affirmation does remind us of the Aristotelian doctrine of form and matter.[b] For indeed Aristotle held that there could be form without matter, though no matter without form. But according to him, the only entities which possessed form without matter were the divine prime mover, the fifty-five intelligent demiurges which moved the spheres, and perhaps the rational soul of man. Some of these are factors in which experimental science has never been very much interested. On the other hand, he maintained that there could be no matter without form, for however pure the matter was (even the chaotic primal menstrual matter which was the raw material of the embryo), it was always composed of the elements, that is to say, it was always hot, cold, dry or wet, and hence had a minimum of form. Apart from the fundamental differences between the conceptions of Aristotle and Chu Hsi, their thoughts were here running along parallel lines. In its medieval way, the affirmation of the universal interpenetration of Li and Chhi mirrors the standpoint of modern science. Form is not the perquisite of the morphologist. It exists as the essential characteristic of the whole realm of organic chemistry, and cannot be excluded either from 'inorganic' chemistry or nuclear physics. But at that level it blends without distinction into Order as such. Similarly, Matter is no longer as simple as philosophers thought, and is interconvertible with energy. We must therefore finally give up all the old arguments about form and matter, and speak only of Energy and Organisation. And since Chu Hsi admitted none of the Aristotelian exceptions in which Organisation ('pure form') could be conceived of as existing without matter-energy, he was closely in accord with the organic world-view of modern natural science. For is not the natural world wholly composed of energy and order?

The chapter continues:

Someone asked about the relations of Li and Chhi. The philosopher answered,[c] 'Master I-Chhuan (Chhêng I) spoke well when he said that Li is one, but its functions (fên[1])[d] are manifold. Consider heaven and earth and the myriad things—they have but one unitary Li. As for men, each of them possesses (individualised in himself) the one unitary Li.'[e]

Throughout heaven and earth there is Li and there is Chhi. Li is the Tao[2] (organising) all forms from above (hsing erh shang[3]), and the root from which all things are produced. Chhi

---

[a] In all the following translations from *CTCS* the versions of le Gall (1), Bruce (1) or Forke (9) have been used with whatever modifications were judged desirable, and Englished from the French or German for the present book; this is mentioned to avoid constant repetition of the usual attributions.

[b] The subsequent sentences of this paragraph are partly taken from another place (Needham (12), p. xvi). See also Peck (1), p. xii.

[c] Chu Hsi is frequently referred to in this way.

[d] Note the use of the word anciently signifying the part or lot of man in primitive collectivist or feudal society, cf. pp. 107. [e] *CTCS*, ch. 49, p. 1*b*.

[1] 分   [2] 道   [3] 形而上

is the instrument (*chhi*[1]) (composing) all forms from below (*hsing erh hsia*[2]), and the tools and raw material (*chü*[3])[a] with which all things are made.[b] Thus men and all other things must receive this Li at the moment of their coming into being, and thus get their specific nature (*hsing*[4]); so also they must receive this Chhi, and thus get their form (*hsing*[2]).[c]

So far the text clearly justifies the interpretation of Chhi as matter-energy and Li as cosmic principle of organisation. The next question was whether there was any precedence or priority as between them, and the texts show that Chu Hsi's doctrine on this was a little hesitant.

First there was Li and later there was Chhi. This is what the *I Ching* means when it says, 'One Yin and one Yang go to make the Tao.' The 'nature' (resulting) naturally possesses love and righteousness (because these are qualities appropriate to it).

First there is the Li of Heaven, then there is the Chhi. The Chhi agglomerates to form *chih* (*chhi chi wei chih*[5]) and that is the preparatory raw material for the 'nature'.

Someone asked whether Li or Chhi came first. The philosopher answered, 'Li is never separated from Chhi. But Li is above all form (non-material) while Chhi is below all form (material). If one has to speak of above and below in this way, there could hardly but be a before and after. Li has no form, but Chhi is gross and contains (impure) sediments (*cha tzu*[6]).'[d]

Yet one cannot really speak of any priority or posteriority of time as between Li and Chhi; it is only if one insists on considering their origins that one has to say that Li came first. Li is not some kind of separate thing, it has (necessarily) to inhere in Chhi. If there were no Chhi, Li would have no way of manifesting itself and no dwelling-place. Chhi can produce the Five Elements, but Li can produce (also) Love and Righteousness, Good Customs and Wisdom.[e]

Someone asked again about what the philosopher had said concerning whether Li or Chhi came first, and he replied, 'It is useless to try to express the matter in that way, seeking now whether Li came first and Chhi afterwards, or whether it was the other way round. It is a thing which we cannot investigate. If, however, I may express a conjecture, it is that the activity of Chhi depends absolutely on that of Li, and that wherever Chhi agglomerates (*chü*[7]) Li is present. Chhi, condensing (*ning*[8]),[f] can form beings; Li is without will or intention (*wu chhing i*[9]), it makes no plans (*wu chi tu*[10]), it forms no beings (*wu tsao tso*[11]), but wherever Chhi is accumulated and gathered together, there is Li in its midst. Now of all beings between heaven and earth, men, plants, trees, birds and beasts, there are none

---

[a] Note the use of a word for 'preparatory raw material' which we have already met with in the 'Water Chapter' of the *Kuan Tzu* book, cf. p. 42.

[b] Fêng Yu-Lan (1), vol. 2, pp. 508, 535, and Bodde (3) interpret these two sentences as referring to the metaphysical and to the physical respectively. But the philosophy of organism, to which I believe Chu Hsi's ideas were related, does not involve an ontological decision. Li might be a non-material principle, but it was part of the natural physical universe and certainly not subjective.

[c] *CTCS*, ch. 49, p. 5*b*.

[d] Note the use of the same terms as for the heavy silt in the centrifugal cosmology, see p. 373 above. This famous passage again is often interpreted by Western scholars in such a way as to make Li 'metaphysical'; here I purposely avoid this in order to emphasise the 'naturalness' of organising relations.

[e] *CTCS*, ch. 49, p. 1*a, b*.   [f] Wang Chhung's favourite term, cf. pp. 369 ff.

[1] 器   [2] 形而下   [3] 具   [4] 性   [5] 氣積爲質   [6] 渣滓
[7] 聚   [8] 凝   [9] 無情意   [10] 無計度   [11] 無造作

that do not come from seeds;[a] but if in white soil there should come forth some creatures (by spontaneous generation) that is an effect of Chhi. As for Li, it is a world pure, empty, vast and limitless, having no forms which could be perceived; obviously it could bring no creatures into being. But Chhi produces everything by fermentation (*yün niang*[1]) and aggregation (*ning chü*[2]).'[b]

Someone else objected, saying, 'You speak of Li as first and Chhi as second, but it seems that one cannot apportion to either of them priority or posteriority.' The philosopher replied, 'I do wish to retain a sense in which Li is first (and Chhi second). But you can never say that here and now is Li while tomorrow there will be Chhi. And yet there is (in some sense) a before and an after.'

It was asked whether Li existed before heaven and earth, at the ultimate beginning. The philosopher replied, 'Certainly it did. There was nothing else. Heaven and earth came into existence because of it; and without it they could not have come into existence, nor men, nor other beings—everything would have lacked support and foundation. And as soon as Li existed, Chhi existed also, and the Chhi moved, flowed, blossomed forth, and nourished everything.' It was asked whether it was not rather Li which engendered and nourished everything. But the philosopher answered that though these were all functions of Chhi, it could not carry them out if Li did not exist. Yet Li has no form or substance (*thi*[3]).

'Is not "substance" a rather forced and improper word?'[c]

'Yes indeed.'

'Are Li and Chhi both limitless?'

'How could one assign limits to either of them?'

Someone asked yet again whether Li was prior and Chhi posterior, and the philosopher replied, 'Fundamentally one cannot say that there is any difference between them in time, but if one goes back in thought to the beginning of all things, one cannot help imagining that Li was first and Chhi came after.'[d]

It is to be hoped that the intrinsic interest of the above excerpts (which, it must be remembered, are, like most of the other remains, the verbatim reports of students rather than the connected writing of the master) will atone for their length. There seems to have been some ambiguity in the minds of the participants in the discussion, since the cosmogonic issue was so easily confused with the metaphysical one; 'before' and 'after' could also be interpreted as 'reality' and 'appearance'. On the latter point Chu Hsi was determined not to fall into idealism, but he did not want to be a (mechanical) materialist either, and was therefore evidently anxious not to be pushed into saying either that matter-energy arose from organisation, or vice versa. He nevertheless inclined to the former view, as we have seen, presumably because it was so difficult to think of organisation as a category perfectly independent of mind,[e]

---

[a] He uses the word *chung*.[4] That he meant seeds in the biological sense and not the Buddhist-Stoic sense (cf. pp. 408, 422 above) is indicated by his immediately following remark about spontaneous generation.   [b] *CTCS*, ch. 49, pp. 2*b*, 3*a*.

[c] It was derived from a passage in the *I Ching*, Great Appendix, ch. 4 (R. Wilhelm (2), Baynes tr. vol. 1, p. 319), but it had become involved in Buddhist technical terminology (see p. 462 above)

[d] *CTCS*, ch. 49, p. 3*a, b*.

[e] Though here the ancient Taoists could have given him moral support. Cf. pp. 51 ff., 54, 302.

¹ 醞釀   ² 凝聚   ³ 體   ⁴ 種

and to get rid of the idea that a plan implies a planner who must be prior in time and superior in status to that which is planned. Hence he would have laid himself somewhat open to theistic interpretations[a] if he had not expressly disclaimed them, at least so far as a personal deity was concerned.[b] At bottom, Chu Hsi remained a dualist, in the sense that matter-energy and organisation were coeval and of equal importance in the universe, 'neither afore nor after other', though the residue of belief in some slight 'superiority' on the part of the latter was extremely difficult to discard. I take it that the reason for this was unconsciously social; since in all forms of society of which the Neo-Confucians could conceive, the planning, organising, arranging, adjusting administrator, was socially superior to the farmer and the artisan occupied with, and hence the representatives of, Chhi. If Chu Hsi could have liberated himself fully from this prejudice he would have anticipated by eight hundred years the standpoint of organic materialism with its dialectical and integrative levels.[c]

All this was bound to be reflected in Chu Hsi's epistemology. Without allowing ourselves to stray too far into the realms of pure philosophy, it is well to take note of an epigrammatic formulation in one of his discourses. 'Cognition (or apprehension) is the essential pattern of the mind's existence, but that there is (something in the world) which can do this, is (what we may call) the spirituality inherent in matter (*So chio chê, hsin chih li yeh; nêng chio chê, chhi chih ling yeh*[1]).'[d] In other words, the mind's function is perfectly natural, something which matter has the potentiality of producing when it has formed itself into collocations with a sufficiently high degree of pattern or organisation.

As the principle of Organisation, it is Li which prevents the processes of Nature from falling into confusion.

Someone said, 'With regard to Li being inherent in the Chhi, by what effects can we see that it dwells there?' The philosopher replied, 'Take, for example, the Yin and Yang and the Five Elements; the reason why they do not make mistakes in their counting, and do not lose the threads of their weaving (i.e. do not fall into irremediable disorder), is because of Li. And if Chhi did not agglomerate at specific times, Li would have nothing to permeate and through which to manifest itself.'[e]

Li is also said to be identical with Thai Chi,[2] the 'Supreme Pole'[f] of which so much has already been said in relation to the Neo-Confucians before Chu Hsi. Every thing or being has its share of the Thai Chi, manifesting itself in a myriad ways; without this,

---

[a] Such as that of Bruce (2).   [b] See on, p. 492.

[c] In this connection there is an interesting parallel between Chu Hsi and Hegel. Just as Hegel ended by extolling the Prussian State of his time as the crown of dialectical evolution, so Chu Hsi's philosophy became the orthodoxy of the mandarinate and was felt to be strongly reactionary by Chinese scholars of the later 17th- and 18th-century schools, who violently attacked it. Cf. p. 514 below.

[d] *CTYL*, ch. 1, p. 40*b*, tr. auct. One could almost write 'emergent from'. On the Neo-Confucian theories of knowledge much could be said; cf. *ECCS, Honan Chhêng shih I Shu*, ch. 25, p. 2*a*.

[e] *CTCS*, ch. 49, p. 2*b*.

[f] *CTCS*, ch. 49, p. 8*b*.

[1] 所覺者心之理也能覺者氣之靈也   [2] 太極

individual things or beings could not have come into existence.[a] The Thai Chi is described as Li endowed with the properties of motion or rest (energy or inertia),[b] corresponding to the active Yang and the passive Yin.

This appreciation of the motion of the universe was often uppermost in Chu Hsi's mind.

When Chhi moves, Li moves also; the two are perpetually in mutual dependence, and never separated from one another. In the beginning, before any being existed, there was only Li, then when it moved it generated the Yang and when it rested it generated the Yin. Upon reaching the extremest point of rest it began to move once more, and at the extremest point of motion it began to return to rest once more (*ching chi fu tung; tung chi fu ching*[1]).[c] Following a cyclical process, it flows on, ever turning and returning (*hsün huan liu chuan*[2]).[c] Li being truly limitless, Chhi participates in its infinity. After the heavens and the earth were formed, it was this active principle (Li) which imparted to them their gyratory movement. Each day has its diurnal revolution, and each month and year their revolutions, (of the heavenly bodies). And it is the same active principle (Li) which rolls the world around (*kun chiang chhü*[3]).[d]

We have already studied Chu Hsi's statement of the 'centrifugal cosmogony'[e] (p. 373). He goes on to say:

The heavens revolve without resting. Dawn and night revolve as if on well-polished bearings (*kun chuan*[4]). The earth is like a bridge (*chio*[5]) in the middle. If the heavens stopped for a single instant, the earth would fall to destruction....

It was asked whether the heavens consist of tangible matter (*hsing chih*[6]). The philosopher replied, 'It is like a wind blowing spirally,[f] tenuous below but getting hard towards the top. The Taoists call it the "hard wind" (*kang fêng*[7]).[g] People commonly say that the heavens have nine layers (spheres), each one of which has a different name.[h] This is not right; it is more like a spiral with nine turns. Below, the Chhi is gross and dull, above, it is pure and brilliant.[i]

Shao Yung used to say that heaven was associated with Hsing (form) and earth attached to Chhi. He constantly emphasised this because he was afraid that some people might seek for

---

[a] It is likened to the moon, which, though one, mirrors itself in an infinite number of reflections on the waters of earth below (*CTCS*, ch. 49, p. 10*b*). This is the Buddhist metaphor of Indra's Net (cf. Fêng Yu-Lan (1), vol. 2, pp. 353, 541, and in Bodde (3), p. 18), each node of which reflected all the others. Cf. Whitehead (2), p. 202; there is 'a focal region where the thing is, but its influence streams away from it throughout the utmost recesses of space and time'

[b] *CTCS*, ch. 49, p. 12*a*.

[c] These passages are classical statements of the approximation of the alternations of Yin and Yang to a wave-theory; cf. below in Section 26*b* on physics.

[d] *CTCS*, ch. 49, pp. 9*b*, 10*a*.   [e] *CTCS*, ch. 49, p. 19*a*.

[f] This idea probably originated from the observation of waterspouts or dust-devils (cf. p. 81).

[g] This probably originated from observation of the properties of jets of air issuing from metallurgical or kitchen bellows. More on p. 25*b* of ch. 49. Cf. Sects. 27*b, j*.

[h] The nine-sphere theory goes back to Chhü Yuan[8] (−332 to −295; G 503) and his *Thien Wên*[9] (Questions about the Heavens). See Sect. 20*d*.

[i] *CTCS*, ch. 49, p. 19*a, b*.

¹ 靜極復動動極復靜   ² 循環流轉   ³ 滾將去   ⁴ 輥轉   ⁵ 橋
⁶ 形質   ⁷ 剛風   ⁸ 屈原   ⁹ 天問

some place outside heaven and earth. But there is no such thing as 'outside heaven and earth'. For their form has a boundary (*yai*[1]) but their Chhi has no boundary.

It is because the Chhi (in the form of aerial matter) is (capable of being) extremely condensed and hard that it is able to support the earth. If that were not so, the earth would fall. At the exterior of the (aerial) Chhi there must be some kind of hard shell, very thick, which retains and fortifies the Chhi.[a]

Heaven with its Chhi depends on the form of the earth, and earth with its form hangs in the midst of heaven's Chhi. Earth is surrounded by the heavens, and it is the one thing in the midst of the heavens.[b]

Thus it is difficult to estimate how nearly Chu Hsi's world-picture came to our own. The heavens are spoken of at one moment as having a hard outer shell, and at another of being limitless. The importance of rotational forces was, however, quite clearly appreciated. It is doubtful whether Chu Hsi envisaged other bodies like the earth also situated in the midst of the heavens.

To conclude this discussion of Chu Hsi's conception of Li, it is interesting to note what he said about the world of quantity.

Someone asked about the relation of Li to number. The philosopher said, 'Just as the existence of Chhi follows from the existence of Li, so the existence of numbers follows from the existence of Chhi. Numbers, in fact, are simply the distinction of objects by delimitation.[c]

There was here the germ of something which could have revolutionised Chinese science—the missing mathematisation of hypotheses concerning Nature. But it is only a momentary flash, we hear no more about it, and it is to be feared that the 'numbers' referred to were rather the sterile Pythagorean numerological symbols earlier discussed (pp. 268 ff.) than any mathematics helpful to natural science.

Interesting also is the relation between Li and Tao in Neo-Confucian philosophy. The discussion on this is at the beginning of chapter 46 of the *Chu Tzu Chhüan Shu*. Chu Hsi goes back to etymology and reminds his students that the original meaning of *tao* was 'way', while that of *li* was the graining or pattern of markings (*Gestalt*) in any natural object.[d] 'The term *tao*', he says, 'refers to the vast and great, the term *li* includes the innumerable vein-like patterns included in the Tao.' Thus Tao was to be used only for the pattern of the whole cosmic organism, while Li could mean also the minute patterns of small individual organisms.[e] But in accordance with the Confucian tradition which the Neo-Confucians could not desert, the term *tao* was used more frequently for the Tao of man in human society than for the Tao of non-human Nature. To find the Tao of man, one has to look within one's self.[f] Bruce,[g] carried away by his theistic tendencies, here misinterpreted a passage[h] in

---

[a] *CTCS*, ch. 49, p. 21*b*. Cf. p. 415. [b] *CTCS*, ch. 49, p. 25*a*.
[c] *CTCS*, ch. 49, p. 5*a*. [d] P. 1*b*.
[e] Bruce (1), pp. 269, 270.
[f] Ch. 42, p. 13*a* (Bruce (1), p. 32); ch. 46, p. 5*a* (Bruce (1), p. 276).
[g] (2), pp. 163, 171. [h] Ch. 42, p. 13*b*.

[1] 涯

which Chu Hsi says that in all the world there are no men, and indeed no creatures, which do not know the principles of Love, Righteousness, Good Customs and Wisdom. It is clear, however, from other passages[a] that what Chu Hsi had in mind was that Nature was capable of bringing these high-level qualities into manifestation when the appropriate integrative level was reached, with organisms capable of manifesting them. It cannot be said that Chu Hsi lacked all conception of the evolutionary view of the world, for as we shall shortly see there were in his world-picture periods when more and more complex organisms were gradually coming into existence—the only way in which his outlook differed from our own (if indeed this is a real difference) was that he believed in alternating cycles of creation and destruction.[b] After each destruction, the centrifugal cosmogonic process began all over again, and new evolutionary successions arose. When, therefore, Bruce described Chu Hsi as saying that 'the universe is pervaded by Moral Law', he was using a theological idiom which expresses Chu Hsi's thought much less well than that of emergent morality.

The relations of Tao and Li were discussed further by one of Chu Hsi's pupils, Chhen Shun[1] (+1153 to +1217), who restored a rather more Taoist flavour to the word Tao by laying emphasis on its cosmic all-pervadingness.[c] In general it is clear, however, that Chu Hsi's doctrine of Li and Chhi had reconciled the divergent uses of the term Tao by the ancient Taoists and Confucians (cf. pp. 36 ff. and 8 ff.). The Tao of human society was now seen to be that part of the Tao of the cosmos which makes itself manifest at the organic level of human society, not before, and not elsewhere.[d] In this way the two greatest indigenous schools of Chinese thought attained a synthesis.

(4) EVOLUTIONARY NATURALISM IN A CYCLICAL SETTING

The idea that the universe passed through alternating cycles of construction and dissolution was common ground for most of the Neo-Confucians.[e] It seems to have been systematised first by Shao Yung[2] (+1011 to +1077)[f] who started to apply the duodenary cycle of hour and compass-point characters to its various phases.[g] There

[a] Especially ch. 42, p. 29a, which I shall quote later (p. 568) in another connection.

[b] Fairly certainly, I think, an Indian idea introduced by Buddhism; Eliade (1, 2). It passed also, with Persian modifications, to the civilisation of the Mediterranean region (cf. Cumont, 2). And thus it became known to the founders of modern geology (see Lyell (1), vol. 1, p. 23).

[c] Forke (9), p. 215. It may be mentioned in passing that Chhen Shun was the originator of the name by which Neo-Confucian philosophy was always subsequently known in Chinese, hsing-li,[3] i.e. '(human) nature, and Nature'.

[d] It is interesting to note that in Chu Hsi's writings there are polemics against the Taoist conceptions of the word, esp. ch. 46, p. 3a (Bruce (1), p. 273), which rested on complete misunderstandings of Lao Tzu. In the +12th century the political sarcasm of Tao Tê Ching, ch. 18, was not appreciated by the medieval philosophers, who took it quite au pied de la lettre, and were shocked by it. Cf. Bruce (2), p. 167.

[e] Possibly it started from the idea, common in many ancient peoples, that the precession of the equinoxes was an oscillation or nutation rather than a continuous change. There was also Plato's suggestion that sometimes a god stops turning the world, whereupon it runs backwards until he starts again (Eisler (1), p. 121).

[f] Forke (9), pp. 26 ff. Cf. p. 455 above.　　　　　　[g] Bruce (2), p. 159.

[1] 陳淳　　　[2] 邵雍　　　[3] 性理

are many subsequent statements of it, and le Gall (1) reproduces two, one from Hsü Lu-Chai[1] (+1209 to +1281)[a] a thinker of the generations immediately following Chu Hsi, and one from Wu Lin-Chhuan[2] (+1249 to +1333), most of whose activity fell within the Yuan dynasty. I think it is worth while to give here the latter's statement,[b] with annotations.

The cosmic period (yuan[3]) is one of 129,600 years, divided into 12 hui[4] of 10,800 years each.[c] When heaven and earth, in their revolutions, attain the eleventh hui (hsü[5]), all things are closed down, and all men and beings between heaven and earth come to nothingness. After 5400 years the position hsü is past, and when the middle of the twelfth hui (hai[6]) is reached, that heavy and gross matter which, in solidifying, had formed the earth, becomes dissipated and rarefied, joining with the tenuous matter which had formed the heavens, and uniting in one single mass; this is called Chaos (hun-tun[7]).[d] This mass then acquires an accelerating rotational movement, and when the position hai is coming to its end, the material reaches its darkest and most dense condition.

At the point[e] chêng,[8] the Great Period begins again and a new era opens; it is the beginning of the first hui, tzu.[9] Undifferentiated chaos persists, hence it is called the Great Beginning (Thai Shih[10]) and also the Great Oneness (Thai I[11]). Thenceforward, light gradually increases. After another 5400 years, in the middle of the position tzu,[9] the lightest part of the mass separates and rises, forming sun and moon, planets and fixed stars. These are the signs of heaven. 5400 years more and tzu comes to an end. Thus it is said 'Heaven is opened (constituted) in tzu.' However, the heavier portions of the Chhi, though remaining at the centre, have still not condensed to form the earth, so as yet it does not exist.

When the middle of the second hui, chhou,[12] is reached, the heaviest Chhi condenses forming earth and rocks, and its liquid part becomes water, which flows and does not solidify, while its caloric part becomes fire, burning and never going out. Water, fire, earth and rocks each have their special forms and constitute the earth. Thus it is said 'Earth is opened (constituted) in chhou.' 5400 years more and chhou comes to an end.

Another 5400 years, and the middle of the third hui, yin,[13] is reached, and now human beings begin to be born between heaven and earth. Thus it is said 'Man is born in yin.'[f]

---

[a] Forke (9), p. 286.

[b] Hsing-Li Ta Chhüan, ch. 26 (not in the Hui Thung edition).

[c] The yuan which Wu Lin-Chhuan uses here is not the same as that of the astronomers. The San Thung calendar of −7 made it 4617 tropical years, and the Ssu Fên calendar of +85 made it 4560. Both these were rationally based on recurrences of lunations, eclipse periods, etc. (cf. Chatley, 16). But the yuan here referred to is one of the smaller Indian kalpas. Although it equals 36 Babylonian saros periods, its origin was probably arbitrary, in that the hui is the same as the 'Great Year' which Aetius ascribes to Heracleitus (Burnet (1), vol. 4, p. 156; Freeman (1), p. 116). This was arrived at by taking 30 as the shortest time in which a man could become a grandfather, i.e. one generation, and multiplying that by 360. Cf. Chatley (15), p. 48.

[d] Note the persistence of the ancient Taoist term.

[e] This is one of four cosmic cyclical points in a system favoured by the Sung schools, yuan, hêng, li and chêng.[14] The first corresponded to the beginning of spring in the year cycle, and the others to the beginning of summer, autumn and winter respectively.

[f] Tr. le Gall (1), pp. 27, 127; eng. auct. Cf. p. 372 above.

¹ 許魯齋   ² 吳臨川   ³ 元   ⁴ 會   ⁵ 戌   ⁶ 亥
⁷ 混沌   ⁸ 貞   ⁹ 子   ¹⁰ 太始   ¹¹ 太乙   ¹² 丑
¹³ 寅   ¹⁴ 元亨利貞

The account of Hsü Lu-Chai[a] is very similar, save that he applies the names of two *kua* of the *I Ching* to the two phases. The period of differentiation, reconstruction, and development comes under the eleventh *kua*, *Thai*;[1] the period of dedifferentiation, destruction and retrogression comes under the twelfth *kua*, *Phi*[2] (cf. p. 315). One cannot overlook the relation between these conceptions and the predilection of the Chinese mind for wave-forms, as in the alternating inversely proportional dominance of Yin and Yang. On the whole these cosmological descriptions followed a weary round of unsupported speculation, but it would be going too far to say that they contributed nothing to Chinese science, for apart from their implied naturalism, they helped it (as we shall later see, Sect. 23) to arrive at advanced notions in geology, and indeed to the recognition of the true nature of fossils, much earlier than in Europe. Chu Hsi himself distinctly stated this, and echoes of discussions about the exact lengths of time involved in the cosmic periods are to be found in the *Chu Tzu Chhüan Shu*.[b]

Indeed, as Forke (9) has noted, 'plutonic' and 'neptunian' ideas of world structure arose clearly in the minds of Chu Hsi and his school as they meditated on the recurrent world-catastrophes in which they believed. The 'centrifugal cosmogony' which we studied in Section 14*d* was of course supposed to occur at the opening phases of each new period. The separation of water and fire involved convulsions of the earth, as light and movement conquered darkness and stillness.[c]

What of Chu Hsi's opinions on the origin of life? He believed that spontaneous generation had once played a great part in producing life, and still took place to a certain extent:

Someone asked how the first men were produced. The philosopher answered that they were formed from Chhi by transformation of the subtlest parts of the Yin and Yang and the Five Elements, uniting and producing (human bodily) shapes. This is what the Buddhists call spontaneous generation (*hua sêng*[3]). And still there are many creatures which are thus engendered, for example, lice.[d]

At the beginning of the generation of beings, the most subtle parts of the Yin and the Yang condensed to form two (components), like the spontaneous appearance of lice, which burst forth (under the influence of warmth). But when two individuals, one male and one female, had been brought into being, their succeeding generations came from seeds, and this is the most universal process.[e]

---

[a] *Hsing-Li Ta Chhüan (Hui Thung)*, ch. 26, p. 18b; le Gall (1), pp. 31, 128.
[b] E.g. *CTCS*, ch. 49, p. 20b; cf. Forke (9), p. 182.
[c] Cf. le Gall (1), p. 34.
[d] *CTCS*, ch. 49, p. 20a. Cf. p. 422 above.
[e] *CTCS*, ch. 49, p. 26a. Here again he used the word *chung*[4] and there can be little doubt that he really had fertilised eggs in mind. For a brief account of the doctrine of spontaneous generation, especially of parasites, in medieval China and Europe, see Hoeppli & Chhiang I-Hung (1). What modern biochemistry is able to say concerning the origin of life may be learnt from the stimulating articles of Pirie (1). On the whole subject see further in Section 39.

[1] 泰   [2] 否   [3] 化生   [4] 種

As to the nature of the lower animals, he clearly recognised that the categories and values applicable to human society were not applicable to them. The behaviour of the social Hymenoptera, he says, speaking in a quite modern way, shows a gleam (*i tien*[1]) of righteousness (*i*[2]); that of mammals in care for their offspring a gleam (*i tien*) of love (*jen*[3]).[a] Animals have a material constitution opaque and gross (*chhi hun cho*[4])—we should say a low level of neurological organisation—through which the full possibilities of nature (*hsing*[5]) cannot manifest themselves, just as the light of the sun or moon is partially obscured by the walls of a mat-shed (*phou wu*[6]).[b] All animals behave as they do, not by consciousness nor choice, but because of the specific *Tao*[7] or *Li*[8] which they have to follow.[c] Thus when consciousness appears at the human level, it is not something quite unconnected with man's material composition.

Someone asked whether consciousness (*chih-chio*[9]) is an inward stirring of something spiritual (*hsin chih ling*[10]) or due to the activity of Chhi (*chhi chih wei*[11])? The philosopher answered, 'It is not entirely a question of Chhi (matter-energy), because the Li of consciousness exists beforehand. Li alone is not conscious (*Li wei chih-chio*[12]) but when Li is combined with Chhi, then consciousness arises. Take for example the flame of this candle, it is because it receives so much good wax that we receive so much light.'[d]

Modern science could find little to quarrel with in these views, which, it must be remembered, date from the middle of the +12th century. And the highest human virtues are profoundly natural, not supernatural, the highest manifestations, as we should say, of the evolutionary process. Aligning himself with that numinous train of thought which runs through the Orphics and the pre-Socratics in Europe,[e] and which we have referred to in connection with Hsün Chhing's book,[f] Chu Hsi could on occasion speak of love (the principle of aggregation in the universe) as the motive force of all things. 'The mind of heaven and earth (*thien ti chih hsin*[13]),' he says,[g] 'which gives birth to all things, is love (*jen*[14]). Man, in being endowed with Chhi, receives this mind of heaven and earth, and thereby his life. Hence tender-heartedness and love are part of the very essence of his life (*sêng Tao yeh*[15]).' This explains Chu Hsi's insistence on the oneness of the Li of Heaven and Man[h] (*thien jen wu erh Li*[16]).[i]

---

[a] *CTCS*, ch. 42, p. 26a; Bruce (1), p. 59; cf. Bruce (1), pp. 211 ff. Cf. Chhêng Hao in *Sung Yuan Hsüeh An*, ch. 13, p. 20a; Forke (9), p. 81. See also pp. 568 ff. below.

[b] *CTCS*, ch. 42, p. 27a; Bruce (1), p. 61; cf. p. 570 below. See also Graf (2), vol. 1, pp. 77 ff.

[c] *CTCS*, ch. 46, p. 9a; Bruce (1), p. 283; cf. Bruce (2), p. 164. Chu Hsi may well have had in mind what we now call instinctive behaviour.

[d] *CTCS*, ch. 44, p. 2a.

[e] Cf. pp. 39, 151 above.   [f] Cf. p. 27 above.

[g] *CTCS*, ch. 44, p. 13b; Forke (9), p. 187; Bruce (1), p. 182.

[h] Echoing Wang Chhung's emphatic statement on the same theme, cf. p. 368.

[i] *CTCS*, ch. 46, p. 7a; Bruce (1), p. 280.

---

¹ 一點   ² 義   ³ 仁   ⁴ 氣昏濁   ⁵ 性   ⁶ 蔀屋
⁷ 道   ⁸ 理   ⁹ 知覺   ¹⁰ 心之靈   ¹¹ 氣之爲   ¹² 理未知覺
¹³ 天地之心   ¹⁴ 仁   ¹⁵ 生道也   ¹⁶ 天人無二理

Just as the transition from the lower to the higher animals, with its corresponding increase in the 'gleams' of higher values, depended on the relative purity of their Chhi, so also differences between goodness and badness in human beings were explained by the composition or *krasis* (κρᾶσις), to use a Hippocratic word, of their respective constitutions. This is what Chu Hsi calls the 'inequality of their endowment of Chhi (*chhi ping pu thung*[1])'.[a] He does not interpret this in a fatalistic way, as Wang Chhung did, but urges that by using the Li in himself a man may achieve greater virtue than would be manifested by the simple action of his Chhi. This doctrine of unequal material endowment was also known in medieval Europe, as the visions of Chu Hsi's contemporary, St Hildegard of Bingen, bear witness;[b] and foreshadowed modern genetics. Moreover, the preparation of the infinitely various constitutions is not the result of 'forethought' and design on the part of the universe, but of chance:

Someone also asked, 'When Heaven brings into being saints and sages, is it only the effect of chance (*ou jan*[2]) and not a matter of design?' The philosopher replied, 'How could Heaven and Earth say: "We will now proceed to produce saints and sages"? It simply comes about that the required quantities (of Chhi) meet together in perfect mutual concordance (*chhia hsiang tshou chu*[3]—a mechanical rather than a chemical metaphor), and thus a saint or a sage is born. And when this happens it looks as if Heaven had done it by design.'[c]

The 'organismic' quality of Chu Hsi's thought comes out very well when he is discussing, in opposition to the Buddhists, the nature of human social organisations:

Under heaven, only the principles of Tao and Li exist, and we cannot but follow them unto the end. The Buddhists and the Taoists, for example, even though they would destroy the social relationships (i.e. by becoming monks and cutting themselves off from the world) are nevertheless quite unable to escape from them. Thus, lacking (the relationship of) father and son, they nevertheless pay respect to their own preceptors (as if they were fathers) on the one hand, while they treat their novices as their sons on the other. The elder among them become elder-brother preceptors, while the younger become younger-brother preceptors. And yet (in so doing) they are clinging to something false, whereas it is the (Confucian) sages and worthies who have preserved the reality.[d]

Here he indicates that the quality of social organisations and human relationships is such that however one may wish to get away from them it is impossible to do so. By setting up a monastic community instead of the family one merely institutes a new and different form of community, in fact, a different type of social organism.

[a] *CTCS*, ch. 43, p. 4*b*; Bruce (1), p. 85.
[b] See p. 19 for references.
[c] *CTCS*, ch. 43, p. 30*b*.
[d] *CTYL*, ch. 126, p. 8. Tr. Bodde in Fêng Yu-Lan (1), vol. 2, p. 568; also Bodde (3), p. 48.

[1] 氣稟不同   [2] 偶然   [3] 恰相湊著

(5) The Denial of Immortality and Deity

As for death and survival after death, Chu Hsi was quite clear that individual human spirits did not survive.

Someone asked whether, at the time of death, a man's consciousness is dissipated and scattered (*san*[1]). The philosopher answered that it was not merely dissipated, but completely finished. The Chhi (of his body) comes to an end, and so does his consciousness.[a]

The opinion of the Buddhists, he said, that human spirits may survive as ghosts, and be reincarnated in later human beings, is absolutely wrong.[b] 'That which dies disappears and does not return. Of changeless in the universe there is nothing but Li. No creatures are eternal, all are subject to change and mortality.'[c]

The Neo-Confucians adopted in this connection a remarkable rationalisation of the ancient terms used in Confucian times for the spirits and demons, retaining them but giving them technical meanings. The system may be represented in tabular form:[d]

Table 21. *Rationalisation of Confucian terms by the Neo-Confucians*

Associated with Yang 陽	Associated with Yin 陰
chhi 氣 used in its old sense as the 'breath of life'	*ching* 精 the seminal essence
hun 魂 the 'warm' part of the spirit or soul, which at death ascends to mingle with the Chhi of the heavens	*pho* 魄 the 'cold' part of the spirit or soul, which at death descends to mingle with the Chhi of earth
shen 神 the ancient term for a god, now used to express the concepts of:	*kuei* 鬼 the ancient term for a demon, now used to express the concepts of:
shen 伸 expansion, disaggregation and *san* 散 dissipation, dispersion	*chhü* 屈 contraction, aggregation and *chü* 聚 collection, condensation

[a] *CTCS*, cn. 51, p. 30*b*.
[b] *CTCS*, ch. 51, p. 19*b*. Cf. the interesting discussion by Bodde (11).
[c] *CTCS*, ch. 51, p. 34*a*. See further on this subject, Forke (9), p. 188, and le Gall (1), p. 89.
[d] It is based on *CTCS*, ch. 51, pp. 5*b*, 19*a*, 21*b*, 22*b*, etc.; cf. le Gall (1), pp. 72–8; Forke (9), p. 190; Bruce (2), p. 243.

[1] 散

There was nothing new in the idea that the human spirit was composed of two parts, one which ascended and one which descended at death; this theory is found already in the *Li Chi*.[a] The innovation of the Neo-Confucians was to utilise these terms to express rather clear physical conceptions, and to apply them in describing natural phenomena. 'When wind, rain, thunder and lightning occur', says Chu Hsi,[b] 'this is the operation of *shen*[1] (gods—or expansive forces). When the wind goes down, the rain ceases, the thunder is ended, and the lightning flashes no more, this is the operation of *kuei*[2] (demons—or, alternatively, contractive forces).' Le Gall rightly points out[c] that the whole of this system of identifications was, if not unnatural, very unfortunate, since it had the result of allowing the mass of the people to continue employing the idiom of superstitious folk religion,[d] while the scholar and the official could, without altering his terminology, explain the world of phenomena on a purely naturalistic basis.[e] This whole situation cannot be properly evaluated without remembering the background of Chinese bureaucratic society, and we are reminded of several incidents already referred to, such as Hsün Chhing's remark that the enlightened are not taken in by conventional ceremonies such as praying for rain or reliance on divination (p. 365), and Liu Khun's firm conviction that what the common people interpreted as his miraculous powers were in truth only the results of chance (p. 367). It may be that when, at the end of our book, we are able to look back upon the course of Chinese thinking in its social context, we shall feel that this serious failure to elaborate new terminology instead of merely rationalising ancient words with all their religious undertones, was one of the most unfortunate aspects of the social milieu in which Chinese science struggled for birth.[f] And obviously it paralleled that European tendency which is seen, for example, in Cicero's *De Natura Deorum*, and in many 18th-century statements, according to which religion is all very well for the mass of the people, indeed, even a socially valuable fraud, but quite unnecessary for the cultured patrician.[g]

We come lastly to the question of theism. What was the position of the greatest

[a] Ch. 21; Legge (7), vol. 2, p. 220.

[b] *CTCS*, ch. 51, p. 2*b*.

[c] P. 74.

[d] Even Chu Hsi himself (*CTCS*, ch. 51, p. 3*b*) admitted the existence of 'dishonest and depraved *kuei-shen*', which whistle on the roofs or hit people in the dark, or to whom it is customary to offer exorcistic sacrifices. Though they were, for the Neo-Confucians, simply manifestations of natural forces they might well be somewhat alarming. Here we catch a glimpse of a very unidealised + 12th century; and see something of what it must have cost the Neo-Confucians to maintain their rationalism amidst the encircling gloom. Typical passages condemning superstitious ideas about ghosts and devils are to be found in the *Chin Ssu Lu*, ch. 3, pp. 57, 60 (tr. Graf (2), vol. 2, pp. 249, 262).

[e] A parallel to this might be found in the terminology of the Sung algebraists (see Sect. 19*i*), where age-old words such as *thien*, *yuan*, *jen* and *wu* were used to denote the unknowns. This rhetorical-positional system delayed the invention of symbolic notation.

[f] It is remarkable that, as technical terms, *kuei* and *shen* have found continued use in the works of Chinese philosophers still living (see Chhen Jung-Chieh (4), pp. 37, 247, 248, 258).

[g] Cf. Farrington (3, 5).

[1] 神　　[2] 鬼

synthetic philosopher whom China ever produced, concerning the nature of God? Let the 49th chapter of the *Chu Tzu Chhüan Shu* speak for itself:

Question. It is said (in the classics), 'The Ruler Above infuses a spirit of virtue into the people.' And also, 'Heaven will give important charges to those who are meritorious.' And again, 'Heaven helps the people, giving them (good) princes.' And again, 'Heaven produces all creatures, and treats them according to their capacities; the good receive the hundred felicities; the bad receive the hundred calamities.' And again, 'When Heaven is about to send some extraordinary calamity, it first sends an extraordinary man who foresees it.' I ask whether these and similar passages mean that there exists above the blue sky a real master and governor (*chu-tsai*[1]);[a] or whether. Heaven having no mind (consciousness, *hsin*[2]), it is *Li*[3] that is responsible?

Answer (of the philosopher). These passages have all the same meaning—it is Li alone which acts thus. Chhi, in its eternal revolutions, has always had successive periods of growth and decay, of decay and growth, following each other in an endless round. There was never a decay which was not followed by a growth.[b]

Question. Regarding the mind (*hsin*[2]) of Heaven and Earth; is it to be considered active or inert (*wu wei*[4])?

Answer. One cannot say that it is not active, but it does not think and will after the manner of human beings.

Question. Regarding further the mind (*hsin*) and Li of Heaven and Earth; does Li here mean the universal principle of organisation (*Tao Li*[5]), and does *hsin* mean master and governor (*chu-tsai*[1])?

Answer. Hsin certainly implies master and governor, but this is nothing other than Li, for Li is never separated from *hsin* nor *hsin* from Li.

Question. Can *hsin* here be considered as meaning ruler?

Answer. Just as man (*jen*[6]) resembles heaven (*thien*[7])—(a pun on the two characters, implying that man is a microcosm)—so *hsin*[2] corresponds to *ti*,[8] ruler....

Ti (ruler) is nothing else than Li considered as ordering all things (*Ti shih Li wei chu*[9]).

The blue sky is called heaven; it revolves continuously and spreads out in all directions. It is now sometimes said that there is up there a person who judges all evil actions; this assuredly is wrong. But to say that there is no ordering (principle) would be equally wrong.[c]

It is therefore quite clear that Chu Hsi did not approve of the conception of a personal God.[d] His standpoint fixed Confucian orthodoxy. Later on[e] we shall have occasion to inquire how far, with all its similarity to modern scientific naturalism, it really contributed to the development of the scientific world-outlook in China.

---

[a] Note the use of the same term as in *Chuang Tzu*, cf. p. 52.
[b] *CTCS*, ch. 49, p. 4*a*.
[c] *CTCS*, ch. 49, pp. 22*b*, 25*a*.
[d] This was appreciated much better by le Gall, the Jesuit, who disliked Chu Hsi's philosophy, than by Bruce, who read into it his Protestant theology. Cf. also ch. 43, pp. 34*b*, 35*a*; Bruce (2), p. 298; Forke (9), p. 179.
[e] Sect. 18.

[1] 主宰   [2] 心   [3] 理   [4] 無爲   [5] 道理   [6] 人   [7] 天
[8] 帝   [9] 帝是理爲主

But let no one suppose that the Neo-Confucian conception of *Thien* (Heaven) was a coldly rational one. Chu Hsi's world outlook possessed a markedly numinous quality.[a] Of this many illustrations could be given, but perhaps the following[b] may suffice:

(Fu) Shun-Kung asked about the Five Sacrifices, saying that he supposed they were simply a duty, a manifestation of great respect; it was not necessary (to believe that) any spirit was present. (The philosopher) answered: '(No spirit, say you?) Speak of the mysterious perfection of the ten thousand things and you have spoken of the Spirit (*Shen yeh chê, miao wan wu erh yen chê yeh*[1]).[c] Heaven and earth and all that is therein—all is Spirit! (*Ying thien ti chih chien chieh shen*[2]).'[d]

(e) NEO-CONFUCIANISM AND THE GOLDEN PERIOD OF NATURAL SCIENCE IN THE SUNG

In the foregoing pages I have ventured to interpret the philosophy of Neo-Confucianism as an attempt at a philosophy of organism, and by no means an unsuccessful one. Before making up his own mind as to the validity of this interpretation (in so far as that can be done on the basis alone of the material which we adduce in this book), the reader should turn to Section 18*f*, on the history of the concepts of juridical and (scientific) natural law, where we have placed further important passages from the writings of Chu Hsi and other Neo-Confucians. But whether or not he will feel inclined to accept my interpretation, there can at least be no doubt that the Neo-Confucian view of the world was one extremely congruent with that of the natural sciences.

It is therefore worth while to emphasise here once again that this period, that of the Sung dynasty, was precisely that which saw the greatest flowering of indigenous Chinese science. At an earlier stage, it was argued (p. 161) that if the interpretation of ancient Taoism adopted in this book were correct, that philosophy should have shown some connections with practical science; and in fact it did indeed, since many aspects of Chinese science, such as alchemy, pharmaceutical botany, zoology, and the physics of magnetism, are patently Taoist in inspiration. So also, if we are right about the tendency of Neo-Confucianism, one might expect that it would be accompanied by a great development of scientific work. And, in fact, the instances of this which can be adduced are embarrassingly numerous.

---

[a] Graf, who fully agrees with this, (2), vol. 1, pp. 288 ff., finds a comparison in the German poet Hölderlin. Englishmen might think of Blake. There is no doubt that Neo-Confucianism was an inspiration to generations of scholar-officials.

[b] Deservedly noted by Chhen Jung-Chieh (4), p. 255.

[c] Chu Hsi was quoting from the Shuo Kua appendix of the *I Ching*, ch. 6, though the original saying differed by one character and did not embody the idea of immanence so clearly. R. Wilhelm (2), Baynes tr., vol. 1, p. 291, misunderstood it; Couvreur (2) sub *miao* has it right.

[d] *CTCS*, ch. 39, p. 21*a*, tr. auct.

[1] 神也者妙萬物而言者也　　　[2] 盈天地之間皆神

In considering the ensuing brief survey of scientific achievements in the Sung we need only remember that roughly speaking the whole of the period from +1000 to +1100 was occupied by the lives of the founders of Neo-Confucianism, while the succeeding century closely corresponded with the life of Chu Hsi, and the impetus of the movement continued strongly until the fall of the Sung dynasty about +1275. Furthermore, it should be remembered that we have seen evidence of the preparation of the ground for Neo-Confucianism during the Thang and early Sung, in the +9th and +10th centuries.

When in the course of the researches from which this book derives, I came to consider Li Ao[1] (+775 to +844) in connection with the beginnings of Neo-Confucian philosophy, I found his name familiar, and upon looking into the appropriate card index I found indeed that he was there, but as a pharmaceutical botanist, who had written a tractate on *Polygonum multiflorum* (the *Ho Shou Wu Chuan*[2]); as well as his philosophical work, the *Fu Hsing Shu*. This incident might be considered symbolical. Taoism and Confucianism were now joining forces, in the face of the challenge of Buddhism, to evolve a unitary world-picture. In this there would be as much room for experimental and observational science as for humanistic philosophy.

If we run over the great scientific names of this period, the man whom we meet at once is Shen Kua[3] (+1030 to +1093), in whose book occurs the first definitely dated mention of the magnetic compass, the first account of the construction of relief maps, and numerous descriptions of fossils, with recognition of their nature, besides many other valuable scientific contributions. In mathematics there were many names, Liu I[4] (fl. +1075), Li Yeh[5] (+1178 to +1265), Chhin Chiu-Shao[6] (fl. +1244 to +1258) and Yang Hui[7] (fl. +1261 to +1275), to name but a few. These were the men who worked out Sung algebra and constituted the most advanced mathematical school anywhere in the world at that time. In astronomy there was Su Sung[8] (+1020 to +1101), whose elaborately illustrated book on the armillary sphere we still possess. It was in +1247 that the famous Suchow planisphere was inscribed on stone, and it is interesting to note that in its text it makes use of Neo-Confucian technical terms, such as *thai chi*,[9] the Supreme Pole. In geography and cartography the period was preluded by Chia Tan[10] (+730 to +805) and closed with Chu Ssu-Pên[11] (+1273 to +1320), both among the greatest geographers of any country and any age. Between the lives of these two men, in +1137, two famous maps were inscribed on stone for the College of Fênghsiang in Shensi; they will be found reproduced in Section 22d on geography.

The absence of outstanding individual names in chemistry is compensated for by the fact that a large proportion of the alchemical and chemical books in the *Tao Tsang* were written during the Sung (see on, Section 33). It is from this time also that we derive our earliest remaining illustrations of Chinese chemical apparatus. Moreover, it must be remembered that Chu Hsi himself, as we have seen, wrote on the oldest

---

[1] 李翱　　　[2] 何首烏傳　　　[3] 沈括　　　[4] 劉益　　　[5] 李冶　　　[6] 秦九韶
[7] 楊輝　　　[8] 蘇頌　　　[9] 太極　　　[10] 賈耽　　　[11] 朱思本

alchemical book, the *Tshan Thung Chhi* (+2nd century). Meanwhile, in botany and zoology the output was extraordinary. In the Wu Tai and Sung periods no less than nine out of the total number of great books of the *Pên Tshao* class were issued, including some by very notable authors such as Khou Tsung-Shih[1] (fl. +1080 to 1125) and Thang Shen-Wei[2] (fl. +1040 to +1095). Moreover, this was the time of maximum production of the separate monographs on highly specialised subjects such as that of Li Ao referred to above. Here the type-specimen is the *Chü Lu*[3] (Orange Record) of Han Yen-Chih[4] (fl. +1178). Later on, Section 38 on Botany will mention many such specialised works of Sung date. Nor were the agriculturalists idle. In +1149 there was a valuable *Nung Shu*[5] by Chhen Fu,[6] which led up to the splendid book of the same title by Wang Chen[7] in +1313.

This last-named work describes printing with movable type, as also had the book by Shen Kua in the +11th century, which reminds us that the period opened with the general use of printing at the end of the +9th. Similarly, it was in the Sung dynasty that the system first grew up of printing many small books in one large collection. The first of these, the *Pai Chhuan Hsüeh Hai*, dates from the end of the +12th century, and about a quarter of the hundred books collected in it are of some scientific interest.

In medicine the period was also fruitful,[a] as the names of Chhen Yen[8] (fl. +1180), Chhien I[9] (fl. +1068 to +1078), Liu Wan-Su[10] (fl. +1200) and Li Kao[11] (fl. +1220 to +1250) bear witness. Their achievements will be touched on in Section 44 on medical science. Here one must not forget the name of Sung Tzhu[12] (fl. +1247), the founder of forensic medicine, not only in China, but in the whole world.

I will only add the fact that in two other fields, architecture and military technology, the basic books were produced at this time. China's greatest work on architecture, the *Ying Tsao Fa Shih*, was compiled by Li Chieh,[13] who died in +1110. And the great encyclopaedia of warlike arts, including much information on the uses of explosives, incendiary techniques, poisonous smokes, etc., the *Wu Ching Tsung Yao*, was also a product of a Sung writer, Tsêng Kung-Liang.[14]

The conclusion is therefore not a far-fetched one that Neo-Confucian philosophy, essentially scientific in quality, was accompanied by a hitherto unparalleled flowering of all kinds of activities in the pure and applied sciences themselves.

[a] See especially the recent reviews of Li Thao (1, 2).

[1] 寇宗奭   [2] 唐慎微   [3] 橘錄   [4] 韓彥直   [5] 農書

[6] 陳旉   [7] 王禎   [8] 陳言   [9] 錢乙   [10] 劉完素   [11] 李杲

[12] 宋慈   [13] 李誡   [14] 曾公亮

(f) CHU HSI, LEIBNIZ, AND THE PHILOSOPHY OF ORGANISM

Nevertheless all these achievements did not bring Chinese science to the level of Galileo, Harvey and Newton. After a certain stagnation in the Yuan and Ming dynasties it becomes quite evident that apart from some unforeseen train of events beyond historical probability, Chinese civilisation was not going to produce 'modern' theoretical science. The last act of the drama of indigenous Chinese thought was played out in a rather sterile metaphysical controversy between the idealists and the materialists.[a] At the end of the +16th century the first emissaries of post-Renaissance occidental civilisation reached the Chinese capital, and Chinese scholars were invited to join with their European colleagues in assisting the 'new, or experimental, philosophy' in its transformation of the world. The rest of the story belongs to the history of modern science in East Asia, and lies outside the scope of this book.

On the bringing of European mathematics, science and technology to the Chinese by the Jesuits there exists a large literature.[b] Much information on this movement, and the part played by that great man Fr. Matteo Ricci in it, will be found in the book of Bernard-Maître (1). So dazzling has this epic been made[c] that many must have been tempted to suppose that European thought derived little or no stimulus from that vast edifice of Chinese philosophy which the Jesuit fathers rightly sought to understand. Nevertheless, I believe that the contribution of Chinese thought, summed up in Neo-Confucianism, to European thinking was greater than has yet been fully realised, and may in the end turn out to be no less than the debt which the Chinese owe to those who brought them the science and techniques of 17th- and 18th-century Europe. Some of the best minds of Europe gave themselves in due course to the study of Chinese philosophy by means of the Jesuit despatches, as may be read in the interesting works of Bernard-Maître,[d] Chu Chhien-Chih (1) and Hughes.[e] But while the manifold influence of Chinese culture on European culture has been much discussed, notably in the books of Pinot (1), Creel (4), Maverick (1), Reichwein (1), etc., the full significance of the philosophical contribution has not yet, I think, been appreciated.[f]

At the conclusion of the part of the present volume devoted to the Yin-Yang and Five-Element theories, and the system of 'correlative thinking' which they formed, it

---

[a] It will be the subject of the following Section. Its history was appropriately written by Sun Chhi-Fêng,[1] in his *Li Hsüeh Tsung Chuan*[2] (General Chronicles of Philosophy) of about +1650.

[b] Lach (3) is a good guide.

[c] It has perhaps been made too dazzling. Though the Jesuits transmitted a knowledge of Galileo's telescope, they did not transmit Copernican heliocentric theory, and thus retarded, rather than advanced, Chinese astronomy; see Pasquale d'Elia (1); Duyvendak (6); Szczesniak (1, 2). Cf. Sect. 20j below.

[d] (2), pp. 153 ff.   [e] (2), pp. 5, 22, 167 ff.

[f] We have already mentioned (p. 374) the suggestion of Martin (6) that the centrifugal cosmogony systematised by the Neo-Confucians may have influenced Descartes in his theory of vortices in the physical aether.

¹ 孫奇逢   ² 理學宗傳

was suggested (p. 291) that after the systematisation of the Chinese world-picture by Chu Hsi and the Neo-Confucians, its *organic* quality was transferred into the stream of occidental philosophical thought through the intermediation of Gottfried Wilhelm Leibniz (1646 to 1716). If this is true its importance can hardly be overestimated. Since in the present section[a] much evidence has been presented which suggests that Neo-Confucianism was really and basically a philosophy of organism, we are now in a position to take up and develop further the suggestions made on p. 303.

Among the great thinkers of the European 17th century Leibniz was the one who was most interested in Chinese thought. His interest in China has given rise to a considerable literature (e.g. Merkel (1), O. Franke (7), Lach), and here we need only briefly recapitulate the principal facts. When he was barely twenty he read such books as G. Spizel's *De Re Litteraria Sinensium Commentarius*, and later Fr. Athanasius Kircher's *China Monumentis Illustrata*. The former is a very small book dealing with the characters (though it gives but few), which Spizel recognised as ideographic like the ancient Egyptian; there is mention of Yin and Yang, the *I Ching*, the five elements, the abacus and the study of alchemy.[b] Kircher's book deals more with architecture, roads, bridges and the like. In +1666 Leibniz published his *De Arte Combinatoria* (cf. Couturat (1), C. I. Lewis) which made him the father of symbolic or mathematical logic, the stimulus for its ideas coming admittedly from the ideographic nature of Chinese characters. Later on (Sect. 49) we shall recur to this a little more fully. In +1687 Leibniz read the *Confucius Sinarum Philosophus*, as we know from extant letters which he wrote to the Landgraf v. Hessen-Rheinfels about it. Two years later, on a visit to Rome, he met Fr. Grimaldi, a Jesuit on leave from China, and afterwards sent him lists of questions to which he hoped to receive answers. Indeed, throughout his life he was in constant touch with the Jesuits, receiving and sending much manuscript material; some of the Jesuit descriptions were edited and published by Leibniz himself in the *Novissima Sinica, Historiam Nostri Temporis Illustratura* of +1697.[c] In +1700 Fr. Bouvet sent him a detailed analysis of the *I Ching*, an event from which flowed one of the most remarkable examples of Chinese-

[a] And in Section 18 below on the development of ideas of Natural Law, especially pp. 558, 565.

[b] Throughout his life there was no branch of Chinese science which did not interest Leibniz. In +1669 he considered that Chinese medicine was at least as good as that of Europe (and for that time he was not far wrong). Leibniz was a great founder and propagator of Academies (he founded that of Berlin), and one of his main objects in this was to exchange scientific information with China. After proposing in that same year a 'Societät in Teutschland zu Aufnehmen d. Künste u. Wissenschaft', he suggested in +1670 a 'Société Philadelphique' which would be an international order of scientists ('exemplo Jesuitarum') and would have scientific liaison offices (!) in the Far East. Among other things these would participate in a world magnetic survey; cf. Harnack (1a) vol. 1, pt. 1, p. 30n., (1b) vol. 4, pt. 1, p. 552; Couturat (1).

[c] This was the year in which Leibniz wrote the often-quoted words (apropos of the Chinese edict of tolerance for Christianity): 'If this continues, I think the Chinese will soon surpass us in sciences and arts; this I do not say in envy of their glory for I rejoice with them, but to induce us to learn from them their courtesy and that admirable art of government which no other nation in the world possesses as they do. For we live so disorderly that it seems to me that just as we send missionaries to them to teach them true theology, they should be asked to send to us sages who would teach us their art of government, and that *natural theology* which they have taken to such a high pitch of perfection' (Pinot (1), p. 335). I italicise two words the significance of which will shortly appear.

European intellectual contact, as we saw at the conclusion of Section 13 *g*. Until the very end of his life, sixteen years later, Leibniz took a prominent part in defending the standpoint of the Jesuits in the Rites Controversy, and this was to some extent synonymous (for him) with defending Neo-Confucian thought.

Now it might be said that the part played by Leibniz in the history of philosophy was that of a bridge-builder. The antagonistic viewpoints of theological idealism on the one hand and of atomic materialism on the other had been an antinomy which European thought had never succeeded in solving. The development of Leibniz himself was an example of this split-personality of Europe. He first grew up in Aristotelian-Thomist theological scholastic vitalism, but then went over (as he tells us himself in his autobiographical fragments)[a] to 'atoms and the void', i.e. to Lucretian-Cartesian mechanical materialism, a system of thought which had always tended, however disguised, to atheism.[b] Essentially this was the same situation which formed the background of the attempts at more satisfactory syntheses which continued during the next two centuries. In the Newtonian age, mechanical materialism (even if ornamentally presented, as in deism) would still do, but as the 19th century began, the progress of science itself began to break its bounds. Hence the Hegelian dialectic and all that followed. The world of Darwin, Freud and Einstein was almost as different from that of the 17th century as it in turn had been from what had gone before. Hence the flowing tide which manifested itself as philosophies of *organism*, whether of Marx and Engels with their integrative levels, or of Lloyd Morgan and Smuts with their emergent evolution, or of the biologists for whom classical mechanism and vitalism are no longer a live issue, or of Whitehead himself and the full organic view of the world. If these ideas are traced backwards in the thought of Europe they lead to Leibniz, and then they seem to disappear. What we have now to discuss is whether that is precisely because his own, the first, great attempt at a synthesis which should surmount the dichotomy of *either* theological vitalist idealism *or* mechanical materialism, was strongly stimulated by, if not indeed derived from, the organic world-outlook which we have found to be characteristically Chinese.

This is a great theme, which deserves better justice than can be done it within the framework of such a book as this. It would be difficult without an *ad hoc* investigation to attempt even an estimate of how much stimulus Leibniz received from Chinese philosophy, and such an investigation would not be easy because he was so unsystematic a writer, and so much of his writing remains only in the form of correspondence and fragments, some apparently still only in manuscript. But something can be said.

Carr[c] tells us that in Leibniz's own account he wanted a realism, but not a mechanical one. Against the Cartesian view of the world as a vast machine, Leibniz proposed the alternative view of it as a vast living organism, every part of which was also an

---

[a] E.g. his letter to de Remond (10 Jan. 1714), *Philos. Schriften*, ed. Gerhardt, vol. 3, p. 606.
[b] Cf. Wiener's account (1) of the historical context of Leibniz's logic.
[c] (1), p. 146, and (2).

organism.[a] This was the picture finally presented (in 1714, at the very end of his life) in the short but brilliant treatise posthumously published, the *Monadology*. These monads of which he considered the world to be composed were indissoluble organisms participating as parts of higher organisms.[b] There were different levels of monads. It might almost be said that the monads were the first appearance of organisms upon the stage of occidental philosophy.[c] The hierarchy of monads and their 'pre-established harmony' resembled the innumerable individual manifestations of the Neo-Confucian Li in every pattern and organism. Each monad mirrored the universe like the nodes in Indra's Net (cf. p. 450).[d] By the aid of this hierarchical universe Leibniz hoped to overcome the antinomy between theological vitalism on the one hand and mechanical materialism on the other. If he was the first of a long line of thinkers to feel deep dissatisfaction with this alleged 'either-or', was it perhaps the Neo-Confucian synthesis which hinted to him a more excellent way?

One has no difficulty in finding echoes of Chinese thought in his philosophy. When he says 'Every portion of matter may be conceived of as a garden full of plants or a pond full of fish; but every stem of a plant, every limb of an animal, and every drop of sap or blood is also such a garden or pond',[e] we feel that here is Buddhist speculation seen through a Neo-Confucian glass, yet meeting (*mirabile dictu*) with the experimental verifications seen through the microscope by Leeuwenhoek and Swammerdam, verifications of which Leibniz knew well and to which he admiringly refers.[f]

When Leibniz speaks of the difference between machines and organisms as lying in the fact that every constituent monad of the organism is somehow alive and cooperating in a harmony of wills,[g] we are irresistibly reminded of that 'harmony of wills' which we noted (p. 283) as characteristic of the Chinese system of 'correlative thinking' in which the whole universe in all its parts spontaneously cooperates without direction or mechanical impulsion. One monad, as Latta puts it,[h] influences another only ideally (i.e. in a manner of speaking), not *ab extra*, but through inner pre-established conformity or harmony. Such words would most perfectly apply to the type of relations between things and events conceived in the Chinese system of correlative thought, where everything happens according to plan, yet nothing is the mechanical cause of anything else. Leibniz's pre-established harmony, a doctrine

[a] Carr (1), pp. 178, 204.

[b] Carr (1), p. 18.

[c] B. Russell (2), pp. 604 ff. It is at first sight disturbing to find that monads are defined as without parts, but Leibniz used the word 'parts' in a rather special way. He refers to sand grains in a heap of sand as 'parts', thus defining a part as an unorganised member of a non-organismic aggregate, and says 'a thing which has parts is not a unity' (in Kortholt (1), vol. 2, p. 445).

[d] Whitehead (1), p. 95. As for the ancient metaphor, cf. the study of Pettazzoni (1) on the bodies of gods covered with eyes.

[e] *Monadology*, sect. 67 (ed. Carr, p. 116).

[f] As an indication of the influence of Leibniz's thought it may be mentioned that after the actual discovery of living cells by Schleiden and Schwann in 1839, they were referred to by the great physiologist Johannes Müller as 'organic monads'. See E. S. Russell (1), pp. 170ff. For further parallels between Leibniz and Buddhist philosophy, see Stcherbatsky (1), vol. 1, pp. 114, 199.

[g] Carr (1), p. 112; and also the monograph of H. L. Koch (1).

[h] (1), p. 42.

which was really an effort to solve the body-mind problem stated in the imperfect terms of the 17th century, has not lasted as such, but one can understand its place in an organicism of that date, and its congruency with traditional Chinese thought is too striking to be overlooked. One might well refer here to the vital passage from Tung Chung-Shu (— 2nd century) quoted above (p. 281), in which the comparison is made between causation as conceived in the universe of correlative thought, and the resonance of two musical instruments at some distance from one another. Eighteen centuries later Leibniz has recourse to a somewhat similar analogy when he emphasises an experiment made by his friend Huygens, who attached two or more pendulums to the same piece of wood and found that if originally out of step they would come before long to swing in time with one another.[a] The transmission of vibrations through the wood was not as unknown to Leibniz as the transmission of the sound-waves had been to Tung Chung-Shu, but it is rather striking that both should have employed a somewhat similar analogy to indicate their ideas of the method of working of an organic universe.

Yet another echo of Chinese thought may be sensed in the passage where Leibniz says:[b] 'There is neither absolute birth nor complete death in the precise meaning of the separation of soul and body. What we call births are developments and unfoldings, and what we call deaths are foldings and shrinkages.' How many times have we not heard the Taoists talking of dissipation and condensation, and saying that there is no real creation or destruction, only densification and rarefaction (cf. pp. 40, 76, 107, 369). In Leibniz the processes are reversed, so that the old thought, which was really a naturalistic explanation of corporealisation and decorporealisation, joins hands with the new microscopical discoveries of Malpighi and Swammerdam on early embryonic development, and hence with the great debate about preformation versus epigenesis.[c] In this connection we shall not forget the term *chi*[1] which Chuang Tzu and others (pp. 43, 78, 469, 470) used to mean the 'germs' of things.

It is indeed fortunate that we possess the considered opinions of Leibniz himself on Chinese philosophy. In + 1701 two books became available, written by a 'dissenting' Jesuit, i.e. a Jesuit who did not share the attitude of most of his colleagues towards Chinese thought and ceremonies; and a Franciscan. The question was very complex. Fr. Ricci and most of his followers, going by the sense of the classical texts themselves, had concluded that the ancient Chinese expression *Shang Ti*[2] (the Ruler Above) could be used as a translation for the God of the Christians, that the *kuei shen*[3] or *thien shen*[4] could be used for angels, and that *ling hun*[5] could be used for soul. But, of course, as we have seen, the first of these conceptions had long lost its original anthropomorphic character, and the Neo-Confucians had made it metaphorical for *Li*.[6] Similarly, the Neo-Confucians interpreted the *kuei shen* as natural causes,

---

[a] Latta (1), pp. 45, 332.
[b] *Monadology*, sect. 73 (ed. Carr, p. 123).
[c] Cf. Needham (2); A. W. Meyer (1).

[1] 幾   [2] 上帝   [3] 鬼神   [4] 天神   [5] 靈魂   [6] 理

and considered the *hun* soul perishable. The texts thus said one thing and the Neo-Confucian commentaries said something quite different. Fr. Ricci and (at first) the majority held to the texts, but Fr. Nicholas Longobardi (the Jesuit) and Fr. Antoine de Ste Marie (the Franciscan) thought it better to accept the commentaries. In the first case Chinese thought needed only a minimum of revealed religion to assume the status of Catholic Christianity; in the second China was a land of atheists and agnostics. We can see now that to a large extent Ricci was right, and if the Jesuits had persisted in interpreting the ancient texts along these lines ultimate historical researches would have justified them. But Longobardi was equally right in his estimate of Neo-Confucianism.[a] The book of Longobardi was entitled *Traité sur Quelques Points de la Religion des Chinois*, and that of de Ste Marie *Traité sur Quelques Points Importans de la Mission de la Chine*. The first was concerned rather with doctrine, the second with ceremonial and usages.[b] Both give a very vivid picture of the anxious discussions which went on among missionaries and scholars in China. But the remarkable thing is that we have Leibniz's marginal annotations on both, printed in an edition of his miscellaneous papers by Kortholt in 1735. This material is followed by a long letter from Leibniz to de Remond, who was then Counsellor to the Regent (the Duke of Orleans) and Chief of Protocol, written about a year before his death in 1716, in which he takes up many aspects of Chinese thought.[c] While he defends, in general, the 'Jesuit' view, i.e. that of Ricci—and his marginal notes on Longobardi and de Ste Marie are all critical, sometimes amusingly and trenchantly so—it can easily be seen that he has long been much stimulated by Chinese thought, and that he has derived a great deal more from it than simply a conviction of its congruency with Christian philosophy. (See Fig. 49.)

Longobardi makes great complaint that the Chinese recognised no 'spiritual substances' as distinct from matter, i.e. no God, no angels, no reasonable soul,[d] but Leibniz, seeking for a naturalism whence an immanent God would not be excluded, finds this universal association of the material component with the spiritual (organisational) component perfectly justifiable.[e] Longobardi objects to the way in which the Chinese make the 'physical principle' of the universe in some way the same as the 'moral principle' of human virtue and other 'spiritual' things (i.e. deriving thus the highest human and social values from roots in the non-human, even non-living, world), but Leibniz is much attracted by it.[f] Longobardi, having (perhaps wrongly) inter-

[a] These problems put the European theologians in a dreadful fix. If Ricci was right, the natural religion of the Chinese was not in need of revelation and grace. If Longobardi was right, the argument from universal consent was shattered. And, worse, the interdependence of morality and religion fell to the ground, since this people without religion had the reputation of being the best moralists in the world. Cf. Pascal, *Pensées*, vol. 2, p. 70.

[b] Of course there were all kinds of political and other intrigues behind this and related controversies, cf. Pinot (1), p. 312 and elsewhere.

[c] Bernard-Maître (10), Merkel (1) and Lach (1) have referred to this paper and quoted from it, but without elucidating its importance in the history of philosophy. Cf. Brucker (1), vol. 5, p. 877.

[d] Kortholt ed. pp. 170, 212.

[e] P. 415.

[f] Pp. 420, 424.

preted the expression *thai hsü*[1] as referring to space, Leibniz says,[a] 'One must conceive of space, not as a substance with parts,[b] but as the *order* of things, in that they are considered as existing together (in a pattern), and as proceeding from the immensity of God, in that all things at every moment depend on it.' Later, Leibniz says, regarding Neo-Confucian naturalism, 'Thus the Chinese, far from being blameworthy in the matter, merit praise for believing that things come into being because of natural predispositions, and by a *pre-established order*. Chance has nothing to do with it, and to speak of chance seems to be introducing something which is not in the Chinese texts.'[c] Here Leibniz put his finger on a very fundamental point. Longobardi repeatedly says that on the Chinese world-view, the universe has come into being by chance.[d] He says this because he is unable to imagine any kind of materialism or naturalism other than the Lucretian-Cartesian mechanical materialism which, with its chance clash of atoms, was one of the two polar opposites of European thought. But Leibniz is beginning to see that there could be a naturalism which is not mechanical, but (as later men would say) organic or dialectical.

Leibniz was better informed than some later European sinologists. Thus he says,[e] 'The Li[2] is called the natural rule of Heaven, because it is by its operation that all things are governed by weight and measure conformably with their estates. This rule of Heaven is called Tien Tao.'[3] And for a closing passage, one may instance the prophetic statement where he hints that the discoveries of modern science were more congruent with Neo-Confucian organic naturalism than with the spiritualism of Europe:

Thus we may applaud the modern Chinese[f] interpreters when they reduce the government of Heaven to natural causes, and when they differ from the ignorant populace, which is always on the look out for supernatural (or rather supra-corporeal) miracles, and Spirits like *Deus ex machina*. And we shall be able to enlighten them further on these matters by informing them of the new discoveries of Europe, which have furnished almost mathematical reasons for many of the great marvels of Nature, and have made known the true systems of the macrocosm and the microcosm.[g]

It is not, of course, to be suggested that the stimulus of Chinese organicism was the only one which led Leibniz to his new philosophy. For example, he himself found

---

[a] P. 421.   [b] Remembering Leibniz's peculiar use of the word 'parts'.   [c] P. 434.

[d] E.g. p. 198, where he says, 'They imagine that from the primal matter *Li*[2] (a misunderstanding of course), air (*chhi*[4]) came forth naturally and by chance...'; to which Leibniz's marginal note is 'Why? It might have come forth by reason.' The misunderstanding about Li and Chhi is put right by Leibniz, when he says (p. 434), 'The subject of all generations and corruptions (alternately) assuming and divesting itself of diverse qualities or accidental forms...is not the *Li*,[2] but rather the protogenous Air (*chhi*[4]), in which the Li produces the primitive Entelechies, or substantial operative virtues which are the constitutive principle of spirits.'

[e] P. 447.

[f] Notice that he expressly says the 'modern' Chinese, indicating that he has the Neo-Confucian philosophical commentators in view, and not the writers of the ancient texts.

[g] P. 466.

[1] 太虛   [2] 理   [3] 天道   [4] 氣

PLATE XX

✿ ✿ ✿   413

LETTRE XVIII.
DE MONS. DE LEIBNIZ
SVR LA
PHILOSOPHIE CHINOISE
A
MONS. DE REMOND,
Conseiller du Duc Regent, et Introducteur des
Ambaſſadeurs.

SECTION PREMIERE
DV SENTIMENT DES CHINOIS
DE DIEV.

Fig. 49. Title-page of Leibniz's *Letter on Chinese Philosophy* (from Kortholt).

points of contact between his own position and that of the Cambridge Platonists,[a] a school of theologians and philosophers[b] who had taught and written in the middle decades of the 17th century.[c] Such men as Benjamin Whichcote, Henry More, and Ralph Cudworth, whose inspiration had been drawn from Plotinus no less than the *Theologica Platonica* of Marsilio Ficino and the Florentine Academy, stood in sharpest opposition to the rising influence of mathematical mechanism and materialism in the age of Bacon, Descartes, and Hobbes. It was the 'inorganic' world which gave to modern natural science its first triumphs, but the Cambridge Platonists, and their biological friends such as John Ray and Nehemiah Grew, could not, as it were, agree to forget for a while the problems of organic form in living things. Thus Cudworth believed that if Nature was to be coherent and intelligible it could be explained neither by the random movements of matter in space, nor by successions of arbitrary and incalculable acts of God. Together with his Cambridge colleagues, therefore, he developed a philosophy of science which came very close to being organic in the modern sense. Nature as a whole was 'plastic', 'spermatical' or 'vital', not mechanical. Each individual thing had an indwelling formative organising 'plastic nature', the unconscious deputy of God within it. As Cassirer has well said;[d] for Cudworth, all events in the universe depended not on forces operating from without, but on formative principles acting from within[e] (how strangely Chinese the doctrine sounds). Cudworth himself wrote:

Wherefore since neither all things are produced fortuitously or by the unguided Mechanism of Matter, nor God himself may reasonably be thought to do all things Immediately and Miraculously, it may well be concluded, that there is a *Plastick Nature* under him, which as an Inferior and Subordinate Instrument doth drudgingly execute that Part of his Providence which consists in the Regular and Orderly Motion of Matter; yet so as there is also besides this a Higher Providence to be acknowledged, which presiding over it doth often supply the Defects of it, and sometimes overrule it; forasmuch as this *Plastick Nature* cannot act Electively or with Discretion.[f]

Modern biologists find the organicism of the Cambridge Platonists attractive.[g] An experimentalist acquainted with the strange limitations as well as the wonderful capacities of specific morphogenetic processes can indeed appreciate their realisation of the fact that 'the plastic nature cannot act electively or with discretion' beyond a certain point. The Cambridge philosophers speculated that plastic natures would

[a] As he acknowledged in (for instance) his *Consideration sur les Principes de Vie et sur les Natures Plastiques* (*Philos. Schriften*, ed. Gerhardt, vol. 6, p. 544).
[b] For historical and bibliographical details see Tulloch (1) and Powicke (1).
[c] Their philosophy has recently been evaluated anew in books by Cassirer (2) and Raven (1).
[d] (2), p. 140.
[e] Cudworth wrote (1), vol. 1, p. 283, of 'that plastic principle in particular animals, forming them as so many little worlds'. One must bear in mind that his *True Intellectual System of the Universe* was written about +1671, the year before the publication of Malpighi's great work on the microscopic anatomy of the developing chick embryo. Precursors like Henry Power had long been in print, and the new world revealed by the microscope of van Leeuwenhoek was just becoming known. The instrument is admiringly referred to by Cudworth (1), vol. 1, p. 218.
[f] (1), vol. 1, pp. 223ff.
[g] Cf. Arber (2), pp. 202ff. and Raven (1). I share in some degree their admiration.

not only account for the laws of motion which the physicists and mechanists were establishing, but would 'extend further to the regular disposal of matter in the formation of plants and animals and other things in order to that apt coherent frame and harmony of the whole universe.'[a]

The Cambridge thinkers wanted understanding and contemplation of Nature, not control over it; they sought synthesis, not analysis. But their relevance to the present argument depends on how far they ever freed themselves from the animism of the Neo-Platonists. Although the plastic nature was 'such a thing as doth not know, but only do...',[b] it was also often spoken of as an 'inward and living soul' in the material object.[c] With all their biological insight, sadly lacking otherwise in the early Newtonian period, the Cambridge divines and naturalists remained fundamentally vitalist, and the substitution of archaei for souls did not really help.[d] Spiritualism had been ingrained for centuries in Europe, as we saw from the fortunes of the macrocosm-microcosm analogy there.[e] Confronted with a mathematised physical universe, it had either to retire into the fastnesses of ecclesiastical authority, or (more nobly) to send the rational theologians into a counter-attack in which vitalism was opposed to mathematics. Yet in the 17th century the only path truly leading beyond Descartes (and his apparently irretrievable bifurcation of Nature) did not turn away from mathematics, it passed directly through its midst;[f] this was the path which Leibniz took. It could only have been taken in the light of an organicism from which every animistic residue, every component other than the pure organising relations themselves, had disappeared. Perhaps Neo-Confucian Li showed the way for the purification of Neo-Platonic plastic nature.

In a word, therefore, I propose for further examination[g] the view that Europe owes to Chinese organic naturalism, based originally on a system of 'correlative thinking',

---

[a] Cudworth (1), vol. 1, p. 226.

[b] (1), vol. 1, p. 240. Cudworth is quoting directly from Plotinus, *Enneads*, II, 3, xvii. And he goes on to claim William Harvey as on his side (*De Gen. Anim.* ex. 49), not entirely justifiably.

[c] (1), vol. 1, p. 236. On p. 232 we read that the plastic nature is no other than the Aristotelian vegetative soul. On p. 272 the general conclusion is reached that the plastic nature 'is either a lower faculty of some conscious soul, or else an inferior kind of life or soul by itself...'. For (p. 255), it is neither matter, forms nor accidents, but incorporeal.

[d] The *archaeus*, a new name used by the 'Chymists and Paracelsians', but not much differing from the plastic nature, is mentioned in vol. 1, p. 232.

[e] Pp. 294 ff. above.　　　　　　　　　　[f] The phrase is Cassirer's (2), p. 133.

[g] The assessment of the extent to which Neo-Confucian philosophy directly influenced Leibniz will involve detailed biographical researches. If it should be considered, as it often now is, that all the essentials of his system were worked out in the *Discourse on Metaphysics* (written in the winter of 1685–6), the terminology of monads being alone missing; then this was accomplished in the year before he read the *Confucius Sinarum Philosophus*. It was not till 1689 that during his six months' stay in Rome, Leibniz established those close relations with the Jesuits of the China Mission which so long afterwards continued. But his interest in China then dated back more than twenty years, to the early days at Nuremberg when he had read Spizel and Kircher, and worked on the 'project of a universal character'. Before accepting the view (suggested, for instance, by Mrs Martha Kneale in private correspondence) that Leibniz's organic philosophy was developed largely under the influence of Spinoza, and that the Chinese ideas were taken up by him only as an unexpected and extraordinary confirmation of his own thought, we should know more about his contacts during the four years he spent in Paris before his call to the Librarianship at Hanover came to him in 1676. Could he not have been personally acquainted there or elsewhere with Jesuit translators? Couplet returned from China in 1682.

brought already to brilliant statement in the Taoist philosophers of the −3rd century, and systematised in the Neo-Confucian thinkers of the +12th, a deeply important stimulus, if it was no more, in the synthetic efforts which began in the 17th century to overcome the European antinomy between theological vitalism and mechanical materialism.[a] The great triumphs of early 'modern' natural science were possible on the assumption of a mechanical universe—perhaps this was indispensable for them— but the time was to come when the growth of knowledge necessitated the adoption of a more organic philosophy no less naturalistic than atomic materialism. That was the time of Darwin, Frazer, Pasteur, Freud, Spemann, Planck and Einstein. When it came, a line of philosophical thinkers was found to have prepared the way—from Whitehead back to Engels and Hegel, from Hegel to Leibniz—and then perhaps the inspiration was not European at all.[b] Perhaps the theoretical foundations of the most modern 'European' natural science owe more to men such as Chuang Chou, Chou Tun-I and Chu Hsi than the world has yet realised.

[a] Graf (1, 2) has drawn attention to similarities between Chu Hsi and Spinoza. Kant's categorical imperative has a remarkably Mencian ring. In their combination of rationalism, humanism and mysticism, Rousseau, Blake, Hölderlin and Shelley were often profoundly Chinese without being aware of it.

[b] It is interesting that Conger (2) has seen a distinct congruity between the organic naturalism of modern science and the *philosophia perennis* of Asia, especially China.

17. SUNG AND MING IDEALISTS, AND THE LAST GREAT FIGURES OF INDIGENOUS NATURALISM

AFTER THE death of Chu Hsi there was little development of Neo-Confucianism. Some followers tried to apply his principles in specific fields, and thus we shall have to come back to Chen Tê-Hsiu,[1] for instance (+1178 to +1235), in connection with the history of biology (in Sect. 39). Others elaborated special theories, e.g. Hsü Lu-Chai (+1209 to +1281) and Wu Lin-Chhuan (Wu Chhêng[2]) (+1249 to +1333)[a] who have already been mentioned (p. 486) in connection with the cyclical theory of universal catastrophes. Others again busied themselves with collecting and publishing Chu Hsi's remains.

(a) THE SEARCH FOR A MONISTIC PHILOSOPHY

On the whole the predominant effort of the thinkers of the +14th to the +16th centuries seems to have been devoted to attaining some kind of monism, in other words, to achieving a greater degree of unity, in some cases almost pantheistic, by asserting the ultimate identity of Li and Chhi. This was done, for example, by Wu Chhêng, who considered that the distinction between them was purely subjective.[b] In the Ming dynasty Lo Chhin-Shun[3] (+1465 to +1547) held the same position in his *Khun Chih Chi*[4] (Convictions Reached after Hard Study) of +1531.[c] So also, in the next generation, Yang Tung-Ming[5] (+1548 to +1624), who wrote the *Hsing-Li Pien I*[6] (Doubts and Discussions concerning the Hsing-Li Philosophy); and his contemporary Kao Phan-Lung[7] (+1562 to +1626).[d] 'One can say', wrote Yang Tung-Ming, 'that the nature of social values and Li arise out of energy and matter; but not that energy and matter arise out of social values and Li (*Chin wei i li chih hsing chhu yü chhi chih, tsê kho; wei chhi chih chih hsing chhu yü i li, tsê pu kho*[8]).' These men were all consciously in opposition to the tradition of metaphysical idealism which had come to a climax with Wang Yang-Ming about +1500.

[a] Nagasawa (1), pp. 231, 265, 271; Forke (9), pp. 286 and 290.
[b] Forke (9), p. 292.
[c] Forke (9), pp. 332, 340.
[d] Already mentioned, p. 472.

[1] 眞德秀　　　[2] 吳澄　　　[3] 羅欽順　　　[4] 困知記　　　[5] 楊東明
[6] 性理辨疑　　　[7] 高攀龍　　　[8] 今謂義理之性出于氣質則可謂氣質之性出于義理則不可

(b) THE IDEALISTS; LU HSIANG-SHAN AND
WANG YANG-MING

This is the tradition at which we must now look. I do not propose to devote much space to it, since subjective and metaphysical idealism was no more helpful to the natural sciences in China than in any other civilisation. It seems to me that its popularity simply added another weight to the scales against Chinese science.

The responsibility seems to lie at the door of Buddhism. In ancient Chinese thought there is no evidence of the existence of metaphysical idealism—the famous passage in *Chuang Tzu* about the dream of the butterfly[a] was surely intended as sceptical poetry rather than philosophy; and the exhortations to 'sincerity' (*chhêng*[1]) in the *Chung Yung*,[b] where it is said that the perfectly sincere man forms a 'trinity' with Heaven and Earth, seem to have been seized upon by medieval idealist philosophers as a support for their views in default of anything more convincing.

The Thang dynasty saw the first real growth of idealism in the work of Buddhist teachers such as Lu Hui-Nêng (+638 to +713), whose views we have already mentioned (p. 410), and Ho Tsung-Mi (+779 to +841), whose *Yuan Jen Lun* (Discourse on the Origin of Man) has also been referred to (p. 422). Metaphysical idealism in China may thus be said to have been mostly a development of the Indian philosophy of *māyā*, the unreality of the external world, though before long it was, as we shall see, taken over by some of the Confucian schools. The Taoists were little concerned, though some of their speculations about the subjectivity of sense-perceptions, optical illusions and the like, as in the case of the *Hua Shu* of Than Chhiao (see pp. 450 ff.), may perhaps be considered to show Buddhist influence, and certainly added their quota to the feeling of uneasiness which typical Chinese realism or materialism was by this time developing. Another Thang Taoist book, the *Kuan Yin Tzu* (cf. pp. 443 ff.), says in one place:[c] 'How can we know that Heaven and Earth do not possess consciousness? (*An chih chin chih thien ti fei yu ssu chê hu?*[2]).'

It is curious that the thinker who is generally regarded as the earliest of the Neo-Confucians had strong tendencies to idealism, namely, Shao Yung (+1011 to +1077).[d] Thus, in his *Yü Chhiao Wên Tui* (Conversation of the Fisherman and the Woodcutter), he says: 'The myriad things are all in myself (*wan wu i wo yeh*[3])'.[e] Elsewhere: 'All natural changes and all human affairs arise in the mind (*wan hua wan shih sêng hu hsin yeh*[4])'.[f] Shao Yung's idealist views were elaborated by his son[g] Shao Po-Wên.[5] But the only member of the main Neo-Confucian group who was influenced by this

[a] Ch. 2, tr. Fêng Yu-Lan (5), p. 64.
[b] Ch. 22, tr. Legge (2), p. 279. Cf. p. 469 above.
[c] Ch. 2, p. 10*b*, tr. Forke (14), p. 147; referring to the dream of Chuang Tzu.
[d] See pp. 455 ff.   [e] P. 3*a*.
[f] *Hsing-Li Ta Chhüan*, ch. 12, p. 11*b*.   [g] Forke (9), p. 41; (14), p. 150.

[1] 誠   [2] 安知今之天地非有思者乎   [3] 萬物亦我也
[4] 萬化萬事生乎心也   [5] 邵伯溫

was Chhêng Hao (+1032 to +1085). His sayings as reported in the *Erh Chhêng Sui Yen*[1] (Essential Words of the Two Chhêng Brothers) contain many statements such as 'there is nothing in the universe that is not in myself. He who knows that everything is in his (mind), will be able to bring everything to completion (*mo fei wo yeh. Chih chhi chieh wo, ho so pu chin*[2]).'[a] Chhêng Hao's doctrines were continued by a number of pupils, ranging from the famous Yang Shih[3] (+1053 to +1135)[b] to men of lesser note, such as Hsieh Liang-Tso[4] (+1060 to +1125),[c] Lü Ta-Lin[5] (c.+1044 to c.+1090)[d] and Wang Phin[6] (+1080 to 1150).[e]

But the greatest idealist of the Sung was undoubtedly Lu Chiu-Yuan[7] (+1138 to +1191),[f] the great contemporary and opponent of Chu Hsi. Several special monographs have been devoted to him.[g] He stated the philosophical doctrine with much greater emphasis and precision than any of his predecessors. Among his writings, collected after his death, in the *Hsiang-Shan Chhüan Chi*,[8] passages such as the following are found: 'Space and time are (in) my mind, and it is my mind which (generates) space and time (*Yü-chou pien shih wu hsin, wu hsin chi shih yü-chou*[9]).'[h] Elsewhere we find: 'The myriad things are condensed into a space, as it were, of a cubic inch, filling the mind. Yet, emanating from it, they fill the whole of time and space (*Wan wu sên jan yü fang-tshun chih chien, man hsin; erh fa, chhung sai yü-chou*[10]).'[i] Such expressions as these led Forke (14) to the statement that Lu Hsiang-Shan anticipated Kant's affirmation of the subjectivity of space and time by six centuries, and this seems to be justified. Its negative value to the development of the natural sciences is another matter. Lu Hsiang-Shan thus placed the Neo-Confucian principle of organisation (*Li*[11]) wholly within the experiencing mind. After many years of controversy with Chu Hsi, the two men had to agree to differ; their systems were indeed irreconcilable. On account of his philosophical views it was natural that Lu Hsiang-Shan should be accused of partiality to Buddhism, but although he, like other late Confucians, adopted from the Buddhists various techniques of meditation, both he and they always continued to affirm the duties of man in the world of affairs, and to deny the Buddhist doctrine of salvation as escape from the world. According to a statement frequently found,[j] Buddhist meditation leads towards extinction, Confucian meditation towards action.

---

[a] Ch. 1, p. 12*b*. See further in A. C. Graham (1).
[b] Forke (9), pp. 104 ff. Literary name Yang Kuei-Shan.[12]
[c] Forke (9), pp. 111 ff.
[d] Forke (9), pp. 117 ff.
[e] Forke (9), p. 154.
[f] Literary name Lu Hsiang-Shan;[13] Forke (9), pp. 232 ff.
[g] E.g. that of Huang Hsiu-Chi (1).   [h] Ch. 22, p. 8*b*.
[i] Ch. 34, p. 38*b*.   [j] Quoted also by Wieger (2), p. 225.

[1] 二程粹言   [2] 莫非我也知其皆我何所不盡   [3] 楊時   [4] 謝良佐
[5] 呂大臨   [6] 王蘋   [7] 陸九淵   [8] 象山全集
[9] 宇宙便是吾心吾心卽是宇宙   [10] 萬物森然於方寸之間滿心而發充塞宇宙
[11] 理   [12] 楊龜山   [13] 陸象山

Lu Hsiang-Shan had a succession of pupils who perpetuated his doctrines, for example, Yang Chien[1] (+1140 to +1225)[a] and Wei Liao-Ong[2] (+1178 to +1237).[b] His influence extended down into the Ming dynasty, with Chhen Hsien-Chang[3] (+1428 to +1500)[c] and his 'Heaven and Earth are set up by my (mind), the myriad changes come forth from my (mind), and space and time are of my (mind) (*Thien ti wo li, wan hua wo chhu, erh yü-chou tsai wo i*[4])'.[d] This Chhen Hsien-Chang was an elder contemporary of the philosopher who is generally considered the chief representative of late Chinese idealism, namely, Wang Shou-Jen[5] (+1472 to +1528).[e]

Apart from the careful study of the thought of Wang Yang-Ming by Forke (9) and numerous monographs in Chinese, there are studies by Wang Chhang-Chih (1) and Henke (1), while some of the writings have been translated (Henke, 2, and a couple of pages in Wieger, 2, 4). Wang Yang-Ming did not express his idealism in the same terms as his predecessors, though he constantly described himself as a follower of Lu Hsiang-Shan. In the selected works, *Yang-Ming hsien-sêng Chi Yao*,[6] one may read: 'The master of the body is the Mind; what the Mind develops are Thoughts; the substance of Thought is Knowledge; and those places where the thoughts rest are Things (*i chih so tsai pien shih wu*[7]).'[f] For Wang Yang-Ming the external world was not of lesser reality than the world of imagination, but all material objects were unquestionably the product of the thought of the world-spirit,[g] with which the thoughts of all individual men were, in some way or other, identical. Hence the great emphasis which he placed upon inborn intuition, *liang chih*,[8] apart from which there could be, in his view, no knowledge. This was often conceived of in a very ethical way, as moral intuition, for which Wang Yang-Ming felt that he had Mencian authority. If, therefore, he anticipated the idealism of Berkeley by some two hundred years (and many of his arguments are remarkably like those of later European idealists), he may also be said to have anticipated the categorical imperative of Kant by a still longer period. Wang Yang-Ming was a considerable poet, and some of his poetical writings, long commonplaces in China, may now almost be said to have become a part of world literature, for instance:

> Everyone has a Confucius in his heart
> Sometimes visible but sometimes hidden
> Without many words one may point to what it is
> The innate knowledge of goodness which admits of no doubts.[h]

---

[a] Forke (9), p. 250.
[b] Forke (9), p. 256.   [c] Forke (9), p. 355.
[d] *Ming Ju Hsüeh An*, ch. 5, p. 6b.
[e] Forke (9), p. 380. Literary name Wang Yang-Ming,[9] by which he is much better known.
[f] Ch. 1, p. 8b.
[g] *Thien ti ti hsin*,[10] or *thien ti chien ling ming*.[11]
[h] Wieger (2), p. 260.

[1] 楊簡   [2] 魏了翁   [3] 陳獻章   [4] 天地我立萬化我出而宇宙在我矣
[5] 王守仁   [6] 陽明先生集要   [7] 意之所在便是物   [8] 良知
[9] 王陽明   [10] 天地的心   [11] 天地間靈明

Unfortunately, all this, however sublime, was most inimical to the development of natural science. In a passage which has been perhaps too often quoted, but which is indispensable here, Wang Yang-Ming discussed the interpretation of the famous phrase ko wu,[1] the 'investigation of things',[a] which Chu Hsi had made much use of, though even in his interpretation of it the phrase had meant mainly the study of human affairs, with observation of Nature taking a secondary place. The passage runs:

In former years I discussed this with my friend Chhien, saying, 'If to be a sage or a man of virtue one must investigate everything under Heaven, how can anyone at present acquire such tremendous strength?' Pointing to some bamboos in front of the pavilion, I asked him to investigate them. So both day and night Chhien (sat and) investigated the principles of the bamboos. After three days he had exhausted his mind and thought, so that his mental energy was fatigued and he became ill. At first I said that this was because his energy and strength were insufficient, so I myself undertook to carry on the investigation. But day and night I was unable to understand the principles of the bamboos, until after seven days I also became ill because of having been wearied and burdened by thoughts. Thus we both sighed and concluded that we could not be either sages or men of virtue, lacking the great strength required for carrying on the investigation of things. And moreover, during the three years which I spent amongst the tribesfolk[b] I found that no one could possibly investigate everything in the world. And I came to the conclusion that research could only concentrate introspectively on one's self. This leads to a wisdom within the reach of every man.[c]

It is not unjustified to say that this famous passage demonstrates the incapacity of some Ming scholars to grasp the most elementary conceptions of scientific method; they could have learnt better from the men of the Han, such as Wang Chhung and Chang Hêng, or from the Sung, such as Shen Kua. One cannot but trace the influence of Buddhism here, with its emphasis on introspective meditation.

The tradition continued with Wang Chi[2] (+1498 to +1583)[d] and Thang Chen[3] (+1630 to +1704),[e] but so far as science was concerned, its effect had already been fully exerted.

The lives of these two men occupied the whole of the +16th and +17th centuries. But the idealist doctrine was by then no longer in the ascendant; a great reverse movement was under way, which, though accepting the philosophy of Chu Hsi as the highest orthodoxy, tended to criticise him for not having been materialistic enough. In this movement there took part the men whom I have called the last great figures of indigenous naturalism.

---

[a] Cf. Vol. 1, p. 48.

[b] Wang Yang-Ming was in banishment in Kweichow province for some years after +1506, i.e. among the Miao, Lolo and other 'barbarian' tribes.

[c] *Yang-Ming hsien-sêng Chi Yao*, ch. 2, p. 20b; *Wang Wên Chhêng Kung Chhüan Shu* (Collected Works), pt. 3, ch. 3, pp. 28b, 29a; tr. Henke (2), p. 177, mod.

[d] Forke (9), p. 415.

[e] Forke (9), p. 493. Yet see p. 436 above.

[1] 格物   [2] 王畿   [3] 唐甄

(c) THE REAFFIRMATION OF MATERIALISM; WANG CHHUAN-SHAN

Pre-eminent among these was one of the earliest of them, Wang Fu-Chih,[1] better known by his literary name as Wang Chhuan-Shan[2] (+1619 to +1692).[a] This excellent scholar served the Ming dynasty as long as any of its organisation remained, and then, refusing to take any public office under the Manchus, retired to a mountain near Hêngyang where he spent the rest of his life in study and writing. It seems that at one point he met one or other of the Jesuits, but little perceptible Western influence can be traced in his thought.[b]

Philosophically he was a materialist and sceptic, strongly combating the idealist tradition of Lu Hsiang-Shan and Wang Yang-Ming on the one hand, and the various forms of superstition in Chinese thought on the other. Thus he wrote against astrology and phenomenalism, and almost the only classical author to whom he gave praise rather than criticism was Wang Chhung. Although adhering to Neo-Confucianism in general, he was disinclined to accept its theories of cosmological cycles (which, as we have seen, were doubtless Buddhist in origin), and indeed dismissed all cosmogonic speculation as being outside the realm of that which could be observed or usefully discussed.[c] In a sense, therefore, he returned to the more ancient Confucian position, though now, so to speak, on a more sophisticated level. His naturalist thought is contained mostly in his commentaries on the *I Ching*, e.g. the *Chou I Wai Chuan*,[3] and in a few smaller books such as the *Ssu Wên Lu*[4] (Record of Thoughts and Questionings), and the *Ssu Chieh*[5] (Wait and Analyse).

For Wang Chhuan-Shan, reality consisted of matter in continuous motion, and he emphasised the materialist interpretation of Chu Hsi's philosophy, *Li*[6] (the principle of organisation in the universe) having no more important a position than *Chhi*,[7] matter-energy, as we should say. 'Apart from phenomena', he wrote, 'there is no Tao (*Hsiang wai wu Tao*[8]).'[d] Wang Chhuan-Shan's most interesting contribution to Chinese scientific thought (though it might be considered implicit in some of Chuang Chou's statements)[e] was perhaps his emphasis on what would today be called the principle of dynamic equilibrium. Forms (*hsing*[9]), he said, for certain periods of time remain recognisably the same, but their material composition (*chih*[10]) is in process of continual

[a] Forke (9), p. 484; Hummel (2), p. 817. There is a small book about him by Hou Wai-Lu (2).

[b] His collected works, *Chhuan-Shan I Shu*,[11] have been issued twice, but few of his writings were published in his own lifetime.

[c] E.g. *Ssu Wên Lu* (*Wai Phien*), p. 25a.

[d] *Chou I Wai Chuan*, ch. 6, p. 5a; *Chêng Mêng Chu*[12] (Commentary on the *Chêng Mêng* of Chang Tsai), ch. 6, p. 2b. Cf. *ECCS, Honan Chhêng shih I Shu*, ch. 4, p. 4b; *Sui Yen*, ch. 1, p. 1a.

[e] Wang Chhuan-Shan wrote brilliant commentaries on the Taoist classics, especially his *Lao Tzu Yen*[13] (Generalisations on Lao Tzu), and his *Chuang Tzu Chieh*[14] (Analysis of Chuang Tzu).

[1] 王夫之   [2] 王船山   [3] 周易外傳   [4] 思問錄   [5] 俟解
[6] 理   [7] 氣   [8] 象外無道   [9] 形   [10] 質   [11] 船山遺書
[12] 正蒙注   [13] 老子衍   [14] 莊子解

change, as, for instance, in a flame or a fountain. Since he unhesitatingly applied this to all life-forms, he may be said to have clearly, though intuitively, appreciated the existence of metabolism. The five elements were simply the basis (*tshai*[1]) of all the different kinds of substance, and the myriad forms had no 'unchanging material substratum' (*ting chih*[2]); on the contrary, it was in constant change as long as they persisted. As for coming into being and passing away, he believed that 'things disperse and return to the Great Undifferentiatedness (*thai hsü*[3]), that is to say, to the origin of the Generative Force (*yin yün*[4]) of Nature. They are not absolutely extinguished.'[a] Or again: 'Life is not creation from nothing, and death is not complete dispersion and destruction (*Sêng fei chhuang yu, erh ssu fei hsiao mieh*[5]).' And: 'The *I Ching* speaks of "coming" and "going"; not of "birth" and "destruction" (*I yüeh wang lai pu yüeh sêng mieh*[6]).' These general ideas of 'assembly' of parts, as it were, and of the 'return of parts to store', though no doubt deriving from the ancient conceptions of aggregation and dispersion (which, as we saw in the section on philosophical Taoism (pp. 40 ff., 371 ff.), go back to the −4th century), acquire in 17th-century thinkers such as Wang Chhuan-Shan a quality of precision and conviction which raises them to the level of an intuitive appreciation of the law of the conservation of matter. Had he ever known of the exact occidental statements of this principle, he would undoubtedly have recognised it as his own thought. His natural philosophy as a whole is frequently known as the 'Theory of the Generative Power of Nature' (*Yin Yün Sêng Hua Lun*[7]).

Most of Wang Chhuan-Shan's study and writing, however, was occupied with historical questions. Here his materialism naturally showed itself, though allied with the burning patriotism of a man who had supported the cause of the Ming until the very end, and who wished his epitaph to be only 'The last of the servants of the Ming'. In his *Tu Thung Chien Lun*[8] (Conclusions on Reading the Mirror of History, of Ssuma Kuang), and his *Sung Lun*[9] (Discourse on the Sung Dynasty) and other works, he clearly distinguished between ancient feudalism and feudal bureaucratism, elaborated a theory of social evolution,[b] exalted national heroes and denounced traitors in all ages, and dissected the failings of bureaucratic society. This last question he took further in a number of shorter books, especially the *Huang Shu*[10] (Yellow Book), the *O Mêng*[11] (The Nightmare), and the *Sao Shou Wên*[12] (Questions of a Head-Scratcher), in which he strongly attacked the corruption inherent in the official bureaucracy. At the same time he saw the potential importance of the merchant class, and maintained that bureaucratism had held back its development, which, however, would have been for the good of the country. We shall have to return to these very modern viewpoints of Wang Chhuan-Shan's in Section 48 on the economic

---

[a] *Chêng Mêng Chu*, ch. 1, pp. 3b ff.; ch. 3, pp. 1b ff.
[b] These were part of his 'Theory of Historical Changes' (*Ku Chin Yin Pien*[13]).

[1] 材   [2] 定質   [3] 太虛   [4] 絪縕   [5] 生非創有而死非消滅
[6] 易曰往來不曰生滅   [7] 絪縕生化論   [8] 讀通鑑論   [9] 宋論
[10] 黃書   [11] 噩夢   [12] 搔首問   [13] 古今因變

and social background. In view of all that has been said it is not very surprising that he should be regarded by contemporary Chinese marxists and adherents of dialectical philosophy as an indigenous forerunner of Marx and Engels. This interpretation, which does not lack plausibility, has been put forward, for instance, by Yang Thien-Hsi (1) and Fêng Yu-Lan (6).

Wang Chhuan-Shan's materialist approach was paralleled by men such as Lu Lung-Chhi[1] (+1630 to +1692),[a] and continued by others of whom more must be said.

(d) THE REDISCOVERY OF HAN THOUGHT; YEN YUAN, LI KUNG AND TAI CHEN

Two contemporaries of Wang Chhuan-Shan were particularly important in the materialist movement, though their emphases were quite different—Yen Yuan[2] (+1635 to +1704)[b] and Li Kung[3] (+1659 to +1733).[c] The group which they founded became known as the 'Yen-Li School' or the 'Han Hsüeh Phai',[4] the 'Back-to-the-Han Movement'. They attacked Sung Neo-Confucianism on a number of grounds and tried to get back to the ideas of the Han scholars. They thus prepared the way, as many studies have shown, for the philosophy of Wang Chhuan-Shan's great 18th-century successor, Tai Chen[5] (+1724 to +1777).[d] Tai Chen was early interested in scientific questions and when only twenty wrote a short book on the use of the calculating rods,[e] then prepared an important commentary on the technological part of the *Chou Li*, the *Khao Kung Chi Thu Chu*[6] (Commentary on the Artificers' Record). Later in life he was very active in the recovery of old works on mathematics. A prominent scholar, he was one of those who served as compilers of the Imperial Manuscript Library, the *Ssu Khu Chhüan Shu*.[f] But he was also the greatest of the few philosophical thinkers which the Chhing dynasty produced. Those of his books here relevant are the *Yuan Shan*[7] (Original Goodness) of +1776, and the *Mêng Tzu Tzu I Su Chêng*[8] (Explanation of the Meaning of Mencian Terms) of +1772.

It had been the work of Yen Yuan and Li Kung to bring about a realisation of the extent to which Sung Neo-Confucianism had been impregnated with Taoism and Buddhism. Tai Chen now set about the construction of a materialistic monism, such

---

[a] Forke (9), p. 489; Hummel (2), p. 547.

[b] Literary name Yen Hsi-Chai.[9] Forke (9), p. 526; Hummel (2), p. 912.

[c] Literary name Li Shu-Ku.[10] Forke (9), p. 539; Hummel (2), p. 475.

[d] Cf. M. Freeman (1); literary name Tai Tung-Yuan.[11] There is a special study of his thought by Hu Shih (10). Forke (9), p. 552; Hummel (2), p. 695; Demiéville (3d).

[e] The *Tshê Suan*.[12]

[f] Tai Chen was the centre of a famous controversy which went on until our own time involving charges of plagiarism, but it seems now to have been settled by Hu Shih (5), in Hummel (2), entirely in favour of Tai's scholarly integrity.

---

[1] 陸隴其　　　[2] 顏元　　　[3] 李塨　　　[4] 漢學派　　　[5] 戴震

[6] 考工記圖注　　　[7] 原善　　　[8] 孟子字義疏證　　　[9] 顏習齋

[10] 李恕谷　　　[11] 戴東原　　　[12] 策算

as might remain if these foreign elements were cast out. As Fang Chao-Ying says,[a] he 'boldly thrust aside the concept of Li^1 as a Heaven-sent entity, lodged in the mind, and took the outright materialist position that Chhi alone is able to account for all phenomena—not only the basic instincts and oft-condemned emotions of man, but all the highest manifestations of man's nature'. Tai Chen thus returned to the older conception of the Tao (transmitted indeed by Yang Tung-Ming and Wang Chhuan-Shan) which understood by it the Order of Nature, as exhibited in the phenomena explained by the Yin and Yang and the five elements. He also laid great emphasis on the re-discovery of the meaning of 'pattern' in the term Li, a meaning which had been almost lost by the accretion of much poetical and moral paraphrasing. The influence of Buddhist (and perhaps also of Christian) theology, with its supernaturalism, had led to a kind of 'transcendentalisation' of Li, but now Tai Chen firmly restored it to the position of immanence in Chhi (matter-energy) which Chu Hsi and his school had sought, with partial success, to give it. Moreover, Neo-Confucianism (like Hegelianism later on, and perhaps this is no accident) had been only too susceptible of being made the tool of political privilege,[b] when its universal pattern-principle was misunderstood as universal law, and confused with the ground of positive law to inculcate law-abidingness at all costs. There had thus developed a tendency to justify the activities of the government of the day by viewing them as natural corollaries of the universal 'laws' of Nature. From such vulgarisations Tai Chen broke completely away, severing any assumed connection with legal law and restoring the original meaning of Li as Pattern, Organisation, or Structure, in Nature.

'These principles of things', Fang Chao-Ying continues, 'cannot (in Tai Chen's view) be adequately revealed by (Buddhist) introspection or meditation,[c] nor will they come to man in a flash of "sudden enlightenment", as the Sung philosophers had maintained. They can be known only by "wide learning, careful investigation, exact thinking, clear reasoning, and sincere conduct."...Reason is not something super-imposed by Heaven on man's physical nature; it is exemplified in every manifestation of his being, even in the so-called baser emotions.' Here again Tai Chen was extremely modern in his viewpoint. Vulgar Neo-Confucianism had been Buddhicised almost to the extent of maintaining that man's natural desires were essentially evil, and should be minimised or suppressed. For Tai Chen, on the other hand, the ideal society would be one in which these desires and feelings could be freely expressed without injury to others. He insisted that even the great qualities of fellow-feeling, righteousness, decorum and wisdom are simply extensions of the fundamental instincts of nutrition and sex, or the natural urge to preserve life and postpone death, and that they are not to be sought apart from these urges. Virtue will therefore not be the

---

[a] In the entry in Hummel (2).

[b] Cf. Hughes (2), p. 51. The same process occurred in Japan, hence the attacks of radical thinkers upon Neo-Confucianism there in the early +18th century. For example, Andō Shōeki[2] (see Norman (1), vol. 1, pp. 134 ff.). See also p. 482 above.

[c] Cf. the passage quoted from Wang Yang-Ming above, p. 510.

[1] 理   [2] 安藤昌益

absence or suppression of desires, but their orderly expression and fulfilment.[a] In these respects, Tai Chen, though a contemporary of Rousseau and almost of Blake, would have found himself at home in a post-Freudian world at least as much as they.

Tai Chen held that the social consequences of regarding Li[1] as a heaven-sent principle of illumination in the individual nature ($hsing$[2]) had worked great harm in Chinese society. While recognising that the thought of Li was present in the humblest man, enhancing his dignity and giving him in effect a higher law to which he could appeal when dispassionate analysis failed to win for him freedom from injustice and oppression, Tai Chen found its effects much less satisfactory when appealed to as the justification for the subjective judgments of the magistrate. No man's private opinion, he urged, should be called Li. And as Fang Chao-Ying points out, we have here an understanding that scientific proof is public, not private.

Among the dissatisfactions which the Yen-Li School had felt with Neo-Confucianism was its predominantly bookish character. Yen Yuan, rediscovering the ancients, found abundant reason for thinking that their educational methods had been much more practical. Accordingly, when, after himself studying and practising medicine, he was asked in +1694 to take charge of a new kind of school, he brought about what might have been a revolution in Chinese education by introducing technological and practical subjects. The Chang Nan Shu Yuan,[3] as it was called, had not only a gymnasium, but also halls filled with machines of war for demonstration and practice, special rooms for mathematics and geography, an astronomical observatory, and facilities for learning hydraulic engineering, architecture, agriculture, applied chemistry and pyrotechnics.[b] Unfortunately, the growing school was entirely destroyed by a severe flood after a few years, and there was no time to reorganise and rebuild before Yen Yuan's death in +1704. Although it may well be that Yen Yuan's enterprise owed some stimulus to the Jesuits, it would have been remarkably advanced in character even for Europe in the last decade of the 17th century, and it would be very desirable to collect, translate and publish all possible textual material relating to this College. It seems likely that we shall have to term Yen Yuan the Comenius of China.

This was, of course, by no means the first time in Chinese history that an attempt had been made to orientate education towards practical affairs rather than book-learning. As we have already seen (Vol. 1, p. 139), Wang An-Shih, in the Sung dynasty, introduced papers on hydraulic engineering, medicine, botany and geography into the imperial examination system, but this was one of the reforms which did not outlast him.

[a] 'Duty (pi-jan[4]) is not the contrary of Nature (tzu-jan[5]) but its fulfilment'; Demiéville (3d). Cf. Aquinas' 'Gratia non tollit Naturam sed perficit et supplicit defectum Naturae'.
[b] Forke (9), p. 529.

[1] 理　　　[2] 性　　　[3] 漳南書院　　　[4] 必然　　　[5] 自然

(e) THE 'NEW, OR EXPERIMENTAL, PHILOSOPHY'—
HUANG LÜ-CHUANG

It must be remembered that we are now in the period when the full weight of the introduction of modern post-Renaissance science by the Jesuits was making itself felt. In view of the theological character of the channel, it is a fact of particular interest that the indigenous Chinese tradition of naturalism was still so strong as to raise up thinkers such as Tai Chen and Hung Liang-Chi,[1][a] whose world-outlook was really more in accord with that of modern science than was the world-outlook of the contemporary Jesuits.

We know that during the first millennium of our era the flow of techniques and inventions had been mainly from East to West. During the 17th and 18th centuries the reverse process was taking place. Tai Chen's interest in mathematical and scientific matters has already been noted, but it may be added that according to his follower Ling Thing-Khan[2] (+1757 to +1809)[b] who wrote the *Tai Tung-Yuan Shih Chuang*[3] about him, it was he who recommended widely the Archimedean screw as a water-raising device. In classical Chinese technology the principle of the screw had been unknown (as we shall see in due detail in Section 27 on engineering), but now it appeared in the 'occidental dragon-tail water-raising machine' (*hsi-jen lung wei chhê fa*[4]), and a short work was composed by Tai Chen on the subject, the *Lo Tsu Chhê Chi*[5] (Record of the Class of Helical Machines).

Although this period transgresses the limits of the present plan I cannot forbear from giving a glimpse of its scientific and technological atmosphere. Chang Yin-Lin (2) has resurrected an obscure book, the *Chhi Chhi Mu Lüeh*[6] (Enumeration of Strange Machines) written by Tai Jung[7] in +1683. Most of this work is concerned with the remarkable machines and instruments constructed by his friend Huang Lü-Chuang.[8] Huang made (and/or described) barometers and thermometers, a humidity meter with dial-pointers turning left and right, mirrors, siphons, microscopes and magnifying glasses, various automata, some kind of bioscope, a crank or pedal cart or bicycle, perhaps partly worked by springs, which could go eighty *li* in one day, together with an 'automatic' fan, improvements to water-raising machinery, water-piping, etc. Tai Jung relates that

At Kuang-ling in Chiangsu, Huang Lü-Chuang and I lived some time together. We learn 'occidental' geometry, trigonometry and mechanics; and his ingenuity was greatly improved thereby.

---

[a] Literary name Hung Chih-Tshun[9] (+1746 to +1809); Forke (9), p. 562; Hummel (2), p. 373. We shall meet with him again on account of his theories of population, which have earned for him the name of the Chinese Malthus.

[b] Hummel (2), p. 514.

[1] 洪亮吉   [2] 凌廷堪   [3] 戴東原事狀   [4] 西人龍尾車法
[5] 蠡族車記   [6] 奇器目略   [7] 戴榕   [8] 黃履莊   [9] 洪稚存

Huang Lü-Chuang made many very ingenious machines and was never exhausted. Some people were astonished at such strange things, and thought that he must have some magical books or teachers. But I lived all the time with him, and used to joke familiarly with him, and I never saw any such books, and I know that he had no such teacher. He used to say, 'What is so strange about these things? Heaven and Earth and all creatures are strange things. Moving like the sky, stationary like the earth, intelligent like men, leaving traces behind as all things do—how could any natural thing not be considered strange? But none of these things are strange on their own account; there must be some source which is the governor and master—extremely strange (in our eyes), yet not strange to itself—just as paintings have to have a painter and buildings an architect. This may be called strangest of all.' So I was astonished at the grandeur of his words.[a]

Are not Aristotle and Boyle speaking here through the mouth of a 17th-century Chinese? Yet Chuang Tzu and Shen Kua seem to be no less present.

Perhaps this may be allowed to symbolise the conclusion of the story of the development of scientific thought in China, from its earliest beginnings among the hundred schools of philosophers until it merges in the 17th century with the world-wide unity of modern science.

[a] Tr. auct.

18. HUMAN LAW AND THE LAWS OF NATURE IN CHINA AND THE WEST

(a) INTRODUCTION

AMONG THE ELEMENTS of the Chinese intellectual climate which merit a close examination in connection with the background of Chinese scientific thought, it is essential to include the conception of law. In Western civilisation the ideas of natural law (in the juristic sense) and of the laws of Nature (in the sense of the natural sciences) go back to a common root.[a] What development, one may ask, paralleled this in the thought of the Chinese? Was it more difficult for them to reach the conception of laws of Nature obeyed by every created thing? For without doubt one of the oldest notions of Western civilisation was that just as earthly imperial lawgivers enacted codes of positive law, to be obeyed by men, so also the celestial and supreme rational creator deity had laid down a series of laws which must be obeyed by minerals, crystals, plants, animals and the stars in their courses. Unfortunately, if one turns to the best books and monographs on the history of science, asking the simple question, when in European or Islamic history was the first use of the term 'laws of Nature' in the scientific sense, it is extremely hard to find an answer.[b] By the + 18th century it was of course current coin—most Europeans are acquainted with these Newtonian words of + 1796:

> Praise the Lord, for he hath spoken,
> Worlds his mighty voice obeyed;
> Laws, which never shall be broken,
> For their guidance he hath made.[c]

But this could not, in fact, have been written by a Chinese scholar of the autochthonous tradition. Why?

The necessary discussion falls into four parts: first, an introductory description of the basic concepts; secondly, a brief account of the development of Chinese law and jurisprudence; thirdly, a summarised history of the differentiation in Europe of the ideas of natural law and the laws of Nature; and lastly, a comparison of the unfolding of thought regarding these matters in China and the West. One aim must be to see whether there is anything here which could properly be classed among the factors in Chinese civilisation which inhibited the indigenous rise of modern science and technology.

---

[a] I first realised the importance of this subject when reading a paper by Ginsberg (1), which, however, deals with the matter from the philosophical rather than the historical point of view. In the following argument, in order to avoid confusion, I propose to reserve the expression 'natural law' for juridical natural law, i.e. that law which it is natural for all men to obey even in the absence of positive statutes; and 'law or laws of Nature' for law in the sense in which the term is used in the natural sciences.

[b] I doubt not that there exists some monograph which specifically addresses itself to this question, but I have been unable to find it. The best review is that of Zilsel (1).

[c] Foundling Hospital Collection (*English Hymnal*, no. 535; cf. no. 466).

(b) THE COMMON ROOT OF THE NATURAL LAW OF THE JURISTS, AND THE LAWS OF NATURE OF SCIENCE

Scholars unversed in the history of jurisprudence turn naturally to the well-known book of Maine (1).[a] He first explains that the earliest law was the case-law of unwritten custom in primitive societies. Their usages were not commands and there was little sanction save the moral disapproval of the society if they were transgressed, but gradually a body of judgments grew up after the differentiation of society into classes; the 'dooms' of Teutonic, or the *themistes* (θέμιστες) of Homeric, chieftains. With the growth of State power these judgments could more and more afford to overstep the bounds of the precepts which the society had formerly followed, and continued to follow, as being, for it, demonstrably based on universally acceptable ethical principles. And thus the will of the lawgiver could embody in codes of enacted statutes, not only laws which had as their basis the immemorial customs of the folk, but also laws which seemed good to him for the greater welfare of the State (or the greater power of the governing class) and which might have no basis in *mores* or ethics. This 'positive' law partook of the nature of the commands of an earthly ruler, obedience was an obligation, and precisely specified sanctions followed transgression. This is undoubtedly represented in Chinese thought by the term *fa*,[1] just as the customs of society based on ethics (e.g. that men do not normally, and also should not, murder their parents), or on ancient tabus (e.g. incest), are represented by *li*,[2] a term which, however, includes in addition all kinds of ceremonial and sacrificial observances. Pollock remarks in passing that 'no one has heard of a nation which, having acquired a body of legislation, reverted from it to pure customary law',[b] but the defeat of the school of Legalists in ancient China, which we have followed in Sections 6 and 12, would surely almost be a case in point. For, as we there saw, positive law was reduced to a minimum from the Han dynasty onwards, and custom returned to its former position of dominance. In a sense, perhaps, this reversion might be considered analogous to that universal process which seeks ways of modifying the administration of justice in accordance with advancing cultural level, but instead of such devices as legal fictions,[c] equity, or amending legislation, Confucian jurists exalted ancient custom, arbitration and compromise, confining positive law to purely penal (criminal) purposes.

We learn further from Maine that in Roman law two parts were recognised, on the one hand the civil coded law of a specific people or State (positive law), *lex legale*, in the later phrase; and on the other hand the law of nations (*jus gentium*), more or

---

[a] Sir Henry Maine was a pioneer in the study of the history of law, but his book *Ancient Law*, first published in 1861, has needed much revision since. The 1916 edition has elaborate commentaries by Pollock (1), and more recently there has been an interesting revaluation of Maine's views by Robson (1). See also Stone (1) in this connection.

[b] (1), p. 22. For the juridical aspects of folk custom in Europe cf. Maunier (1).

[c] Though these were known in later Chinese law; cf. Escarra (1), p. 65.

[1] 法   [2] 禮

less equivalent to natural law (*jus naturale*). The *jus gentium* was presumed to follow the *jus naturale* if the contrary did not appear. Their identity was assumed in Roman law, though not very safely, for (*a*) some customs would certainly not be self-evident to natural reason, and (*b*) there were rules which deserved to be recognised by all mankind but in fact were not (e.g. the undesirability of slavery). The traditional origin of this 'natural law' was the increasing residence at Rome of merchants and other foreigners, who were not citizens and therefore not subject to Roman law, and who wished to be judged by their own laws. The best that the Roman jurisconsults could do was to take a kind of lowest common denominator of the usages of all known peoples, and thus attempt to codify what would seem nearest to justice to the greatest number of people. Thus it was that the conception of natural law originated. There are other slightly different versions of the process (cf. Buckland,[a] Nettleship (1), Jolowicz[b]), but for our purpose this will do. Natural law was thus the mean of what all men everywhere felt to be naturally right, and 'there came a time', as Maine says, 'when from an ignoble appendage of the *jus civile*, the *jus gentium* came to be considered a great, though as yet imperfectly developed, model to which all law ought as far as possible to confirm'.[c]

The distinction is found in Aristotle, who speaks of positive law as *dikaion nomikon* (δίκαιον νομικόν) and of natural law as *dikaion physikon* (δίκαιον φυσικόν).[d] He says:

Political justice is of two kinds, one natural (*physikon*) and the other conventional (*nomikon*). A rule of justice is natural when it has the same validity everywhere, and does not depend on our accepting it or not. A rule of justice is conventional when in the first instance it may be settled in one way or the other indifferently—though having once been settled it is not indifferent; for example that the ransom for a prisoner shall be a *mina*, or that a sacrifice shall consist of a goat and not of two sheep, etc....Some people think that all rules of justice are merely conventional, because whereas (a law of)[e] nature is immutable and has the same validity everywhere, as fire burns both here and in Persia, rules of justice are seen to vary. That rules of justice vary is not absolutely true, but only with qualifications....But nevertheless there is such a thing as natural justice as well as justice not ordained by nature, and it is easy to see which rules of justice, though not absolute, are natural, and which are not natural but legal and conventional, both sorts alike being variable.[f]

The passage is very interesting, for it refers to the fact that quantitative and ethically indifferent matters can only be settled by positive legislation (cf. p. 210),[g] and also

---

[a] (1), pp. 52 ff.   [b] (1), pp. 100 ff.

[c] (1), p. 55.

[d] There is a valuable discussion of the ideas of φύσις and νόμος in Lovejoy & Boas (1), pp. 103 ff., 185 ff. The −5th-century sophists set up a strong antithesis between *nomos* (human law or custom) and *physis* (i.e. Nature conceived as a force older and in some sense more valid than any conventional human law). This has been fully studied by Heinimann (1). For a fuller treatment of Aristotle's thought in relation to natural law, see J. N. Frank (1), pp. 358 ff.; (2), pp. 94 ff., 119 ff.

[e] The word law here is not in the text.

[f] *Nicomach. Eth.* v, vii, tr. Rackham (1), p. 295.

[g] Cf. Pollock (1), p. 74, 'Natural justice may tell me not to drive recklessly, but cannot tell me which is the right side of the road.' 'Rules involving number or measure cannot be fixed by natural justice alone.'

trembles on the verge of speaking of laws of Nature in the scientific sense. Now in the Chinese context there could hardly be a *jus gentium*, for owing to the 'isolation' of Chinese civilisation there were no other *gentes* from whose practices an actual universal law of nations could be deduced, but there was certainly a natural law, namely, that body of customs which the sage-kings and the people had always accepted, i.e. what the Confucians called *li*.[1]

(c) NATURAL LAW AND POSITIVE LAW IN CHINESE JURISPRUDENCE; THE RESISTANCE TO CODIFICATION

These preliminaries having been completed, we may now proceed to an account of the main features of the history of law in Chinese civilisation.

The peoples of Western civilisation [says Escarra][a] have lived throughout under the Graeco-Roman conception of law. The mediterranean spirit, while central to the patrimony of the Latin peoples, has also inspired large parts of the law of Islam, as also of the Anglo-Saxon, Germanic, and even Slavonic, nations. In the West the law has always been revered as something more or less sacrosanct, the queen of gods and men, imposing itself on everyone like a categorical imperative, defining and regulating, in an abstract way, the effects and conditions of all forms of social activity. In the West there have been tribunals the role of which has been not only to apply the law, but often to interpret it in the light of debates where all the contradictory interests are represented and defended. In the West the jurisconsults have built, over the centuries, a structure of analysis and synthesis, a corpus of 'doctrine' ceaselessly tending to perfect and purify the technical elements of the systems of positive law. But as one passes to the East, this picture fades away. At the other end of Asia, China has felt able to give to law and jurisprudence but an inferior place in that powerful body of spiritual and moral values which she created and for so long diffused over so many neighbouring cultures, such as those of Korea, Japan, Annam, Siam, and Burma. Though not without juridical institutions, she has been willing to recognise only the natural order, and to exalt only the rules of morality. Essentially purely penal[b] (and very severe), sanctions have been primarily means of intimidation. The State and its delegate the judge have always seen their power restricted in face of the omnipotence of the heads of clans and guilds, the fathers of families, and the general administrators, who laid down the duties of each individual in his respective domain, and settled all conflicts according to equity, usage, and local custom. Few indeed have been the commentators and theoreticians of law produced by the Chinese nation, though a nation of scholars.

The Lü Hsing[2] chapter in the *Shu Ching* (Historical Classic) may be regarded as the oldest notice of a legal code in China, but though of the Chou period its date is quite uncertain.[c] The oldest datable codification of Chinese law known to us, therefore, is

---

[a] (1), p. 3.
[b] Escarra certainly exaggerates here. Chinese law was above all administrative law.
[c] Ch. 47, tr. Medhurst (1), p. 312; Legge (1), p. 254; Karlgren (12), p. 74. See also Escarra (1), p. 87.

[1] 禮   [2] 呂刑

that related in the *Tso Chuan* for −535. Here, at the beginning of the story, appears that uncompromising objection to codification which characterised Confucian thought throughout Chinese history. In the text we read:

In the third month the people of the State of Chêng made (metal cauldrons on which were inscribed the laws relating to) the punishment (of crimes). Shu Hsiang[1] wrote[a] to Tzu-Chhan[2] (i.e. Kungsun Chhiao,[3] prime minister[b] of Chêng), saying:
'Formerly, Sir, I took you as my model. Now I can no longer do so. The ancient kings, who weighed matters very carefully before establishing ordinances, did not (write down) their system of punishments, fearing to awaken a litigious spirit among the people. But since all crimes cannot be prevented, they set up the barrier of righteousness (*i*[4]), bound the people by administrative ordinances (*chêng*[5]), treated them according to just usage (*li*[6]), guarded them with good faith (*hsin*[7]), and surrounded them with benevolence (*jen*[8])....But when the people know that there are laws regulating punishments, they have no respectful fear of authority. A litigious spirit awakes, invoking the letter of the law, and trusting that evil actions will not fall under its provisions. Government becomes impossible....Sir, I have heard it said that a State has most laws when it is about to perish.'[c]

The situation was repeated later in the same century. For the year −513 the *Tso Chuan* says:

In the winter Ju-Pin was fortified. The inhabitants of the State of Chin were forced[d] to contribute 480 catties of iron to make cauldrons on which the penal laws were inscribed. They were those of Fan Hsüan-Tzu.[9][e] Confucius said, 'I fear that Chin is going to destruction. If its government would observe the laws which its founder prince received from his brother, it would direct the people rightly....Now the people will study the laws on the cauldrons and be content with that; they will have no respect for men of high rank.'[f]

Thus from the beginning the supple and personal relations of *li* were felt to be preferable to the rigidity of *fa*.[g] As one of the inserted chapters of the *Shu Ching* says, 'Virtue has no invariable rule but fixes on that which is good as its law. And goodness itself has no constant resting-place, but accords only with perfect sincerity.'[h] Through the centuries there descended the idea enshrined in proverbial wisdom— *Li i fa, sêng i pi*;[10] 'for each new law a new way of circumventing it will arise'.[i] And

---

[a] He was the brother of Duke Hsüan of Lu.
[b] G 1029.
[c] *Tso Chuan*, Duke Chao, 6th year (tr. Couvreur (1), vol. 3, p. 116, eng. auct.); cf. Granet (5), p. 461.
[d] By the minister Chao Yang.[11]   [e] Prime minister.
[f] *Tso Chuan*, Duke Chao, 29th year (tr. Couvreur (1), vol. 3, p. 456, eng. auct.).
[g] It is interesting that these first Chinese codifications antedate the first Roman 'Twelve Tables' by nearly a hundred years, as Balazs (6) has pointed out.
[h] Ch. 15, Hsien Yu I Tê, probably Han or Chin in date, tr. Medhurst (1), p. 153; Legge (1), p. 100.
[i] I am indebted to Dr A. W. Hummel for a reminder of this proverb.

[1] 叔向   [2] 子產   [3] 公孫僑   [4] 義   [5] 政   [6] 禮   [7] 信
[8] 仁   [9] 范宣子   [10] 立一法生一弊   [11] 趙鞅

this was wisdom. Too often we forget that the control of fraud and abuses was simply not possible until modern science brought it within the reach of organised societies. Later, as in the case of the measurement of specific gravity (Sect. 26c) or in that of the discovery of individual differences of finger-prints (Sect. 43), we shall see that the Chinese themselves took important steps in this direction. But perhaps the proverbial wisdom is wisdom still.

There is, of course, another source from which information about the legal practice of the Chou dynasty can be obtained, namely, the inscriptions on bronzes, some of which are elaborate accounts of disputes at law. Only a beginning has, however, as yet been made in their study. H. Maspero (10) has shown that from the −7th century onwards a distinction was made between civil (*sung*[1]) disputes (regarding property— *i huo tshai hsiang kao*[2]) and criminal (*yü*[3]) cases (*i tsui hsiang kao*[4]). There was much use of oaths, either taken solemnly before the gods and spirits (*mêng*[5]) or without the sacrifice of victims (*shih*[6]). Similarly, Granet[a] has collected a good deal of information on the laws of inheritance.

Nothing whatever is known of the provisions of the early codes, and the system which proved to be really the ancestor of all the later ones was that of Li Khuei (see above, p. 210). Li Khuei[b] was a minister of the State of Wei about −400; his code was called the *Fa Ching*[7] (Juristic Classic). Although it has long been lost, the headings of its contents have been preserved;[c] they concerned: (1) robbery (*tao fa*[8]); (2) brigandage (*tsei fa*[9]); (3) imprisonment (*chhiu fa*[10]); (4) arresting (*pu fa*[11]); (5) miscellaneous matters (*tsa fa*[12]); and (6) definitions (*chü fa*[13]). These divisions are found in all subsequent codes. In the first emperor's time, Hsiao Ho[14] (d. −193)[d] added three *shih lü*[15] sections as follows: (7) regulations concerning the census, the family and marriage (*hu lü*[16]); (8) regulations concerning corvée labour (*hsing lü*[17]); and (9) regulations concerning military service (lit. the imperial stables) (*chiu lü*[18]). The laws connected with the imperial household (such as the *Yüeh Kung Lü*[19] (Royal Palace Regulations) drawn up by Chang Thang,[20] and the *Chhao Lü*[21] (Court Regulations) drawn up by Chao Yü[22]) were added, and the whole built up into the great Code of the Han dynasty by Shusun Thung[23] and other jurists. But it was completely lost long before the Sui. Nevertheless, a good idea of Han practice may be gained by reading the Hsing Fa Chih[24] (Record of Law and Punishments) of the *Chhien Han Shu* (ch. 23).[e] Each one of the later dynastic histories contains such a chapter.

---

[a] (3), p. 377.
[b] Giles (G 1164) places him a century too late.
[c] See Escarra (1), p. 91; and on the whole story, Pelliot (13).
[d] G 702.
[e] Tr. Andreozzi (1); Vogel (1); superseded by Hulsewé (1).

[1] 訟	[2] 以貨財相告	[3] 獄	[4] 以罪相告	[5] 盟	[6] 誓
[7] 法經	[8] 盜法	[9] 賊法	[10] 囚法	[11] 捕法	[12] 雜法
[13] 具法	[14] 蕭何	[15] 事律	[16] 戶律	[17] 興律	[18] 廏律
[19] 越宮律	[20] 張湯	[21] 朝律	[22] 趙禹	[23] 叔孫通	
[24] 刑法志					

The Chin, Liu Sung, Chhi, Liang and other dynasties had codes, but little is known of them and all were lost before the Sui.[a] From the indications which we have about them, however, it is certain that they followed the model of the *Fa Ching* and the Han Code in being almost entirely concerned with criminal matters and government ordinances regarding taxation. Civil law remained extremely undeveloped. Over the centuries there are numerous records of imperial edicts softening the rigours of the earliest codes; for example, the abrogation of the practice of exterminating the whole families of criminals in −178[b] and under the Northern Wei in +474,[c] or the mildening of mutilative punishments in −144.[d] Such was also the spirit of the code which Sui Wên Ti ordered Su Wei[1] and Niu Hung[2] to prepare in +583. In +624, with little change, it became the great Code of the Thang, the *Thang Lü Su I*,[3] issued under the editorship of Chhangsun Wu-Chi,[4] and now the oldest extant Chinese code.[e] One Sung code, *Sung Lü Wên*[5] of +1029, is lost, though mnemonic verses based on it survive;[f] another, the *Hsing Thung*,[6] still exists. The Mongols introduced numerous alterations without touching anything fundamental;[g] their system, described in chapter 102 of the *Yuan Shih* (History of the Yuan Dynasty) has been translated by Ratchnevsky (1). Then after the *Ta Ming Lü*[7] of +1374 came the *Ta Chhing Lü Li*[8] of +1646, and it is upon this that most of the studies of Chinese law by Europeans have been made. Much of it was translated by Staunton (1) and more by Boulais (1). Alabaster (1) commented very favourably upon its practical application, and Plath (1) and Werner (2) reproduced selections of its provisions.[h]

The other main class of Chinese juristic writings was the series of books recording decisions (*phan*[9]) in *causes célèbres*. Typical of these are the *I Yü Chhien Chi*[10] (First Collection of Doubtful Law Cases) by Ho Ning,[11] and the *I Yü Hou Chi*[12] (Second Collection) by his son Ho Mêng;[13] both written between +907 and +960. Another Sung work of the same kind was the *Chê Yü Kuei Chien*[14] (Tortoise Mirror of Case Decisions) compiled by Chêng Kho.[15] 'But there lacked in China', says Escarra, 'that tradition of jurisconsults succeeding one another through the centuries, whose opinions, independent of the positive law, and whatever its practical applications might be, built up, on account of their methodical, doctrinal, and scientific character,

---

[a] The juristic chapter of the *Sui Shu* (ch. 25) has been translated by Balazs (8) with excellent notes, and he is working on that of the *Chin Shu* (ch. 30).

[b] By Han Wên Ti (*TH*, p. 328).   [c] By Thopa Hung (*TH*, p. 1147).

[d] By Han Ching Ti (*TH*, p. 373). We shall note a particular case of this kind in Section 43.

[e] Translation and discussion by Bünger (1). It was given an official commentary in +737 (Niida & Makino).

[f] These are available in the contemporary collection *Chen Pi Lou Tshung Shu*,[16] edited by Shen Chia-Pên.[17]

[g] Cf. Riazanovsky (1).

[h] It was, furthermore, the basis of Kohler's studies (1, 2) in comparative jurisprudence.

[1] 蘇威	[2] 牛弘	[3] 唐律疏議	[4] 長孫無忌	[5] 宋律文
[6] 刑統	[7] 大明律	[8] 大清律例	[9] 判	[10] 疑獄前集
[11] 和凝	[12] 疑獄後集	[13] 和㠓	[14] 折獄龜鑑	[15] 鄭克
[16] 枕碧樓叢書	[17] 沈家本			

the "theory" or speculative part of law. China had no "Institutes", manuals, or treatises.[a] A jurisconsult such as Tung Chung-Shu, liturgiologists like the elder and younger Tai, codifiers like Chhangsun Wu-Chi...did not accomplish works parallel to those of a Gaius, a Cujas, a Pothier or a Gierke.'[b]

It may well be that in this, and in the passage previously cited, Escarra has somewhat overstated his case. Current research on the history of Chinese law is indicating that the actual practice of jurisprudence and legislation gave rise to a more abundant literature, written largely by Confucian scholars, than has been thought. A work on the Legal Customs of the Han (Han I[1]) was presented to the throne by Ying Shao[2] in the year of his death (+195); it had 250 chapters. In the Chin Shu[c] we read that by the end of the +3rd century there were ten schools of glossators, deriving from the teaching of various famous scholars of the Han[d] such as Chêng Hsüan,[3] Ma Jung,[4] and Chhen Chhung.[5] The corpus of their writings had more than 26,000 paragraphs and over seven million characters. And it is stated that these interpretations were commonly used in practice.

At the same time it must also be recognised that in certain respects Chinese legal mentality was sometimes ahead of the European. In our account of anatomy and medicine (Sects. 43, 44) we shall take notice of the remarkably early appearance of forensic medicine in China, with the books of the type of the Hsi Yuan Lu[6] (The Washing Away of Wrongs) from the Sung (+1247 onwards).[e] Given the pattern of Chinese thought as we have already sketched it, this was not an unnatural development, for the dictates of li[7] would assure (at any rate theoretically) that the utmost possible effort should be made to prevent the fixation of guilt on the innocent,[f] and

---

[a] On the School of Beirut for example, see Collinet's account.

[b] (1), p. 359.   [c] Ch. 30, p. 5b.

[d] Biographies in Hou Han Shu, ch. 76.

[e] Credit for this was freely given to China by Europeans such as Harland (2) a century ago.

[f] In this connection may be mentioned the very striking testimony of the first Europeans to enter China in the pre-Jesuit period, i.e. throughout the +16th century. Pires was there in +1518, Pinto (perhaps) some time before +1558 and the anonymous Portuguese whose account was printed by Alvares about the same time, Pereira 1549 to 1553, and Mendoza before +1585. Most of these men saw the inside of Chinese prisons for considerable periods, and all agree in extolling the conscientiousness with which justice was administered in that country. Hudson wrote, (1), p. 244, 'There is no endeavour in these accounts to minimise the cruelties of the Chinese law—the horrors of the prisons, the use of torture and the prevalence of punishment by flogging. But such things were the commonplace of Europe also until the nineteenth century. What seems to have struck the sixteenth-century observers as most remarkable in the operation of Chinese law was the system of reviewing all cases in which a death sentence had been pronounced. As the Anonymus says: "They take all possible pains to avoid condemning any to death." This is hardly in accord with the commonly held belief that human life has always been cheaper in China than in Europe.' Moreover, there is earlier confirmation from non-European sources. In +1420 an embassy had gone to the Ming court from Shāh Rukh, the son of Tīmūr, and a narrative of it was written by Ghiyāth al-Dīn-ī Naqqāsh. Although he thought some of the punishments terrible, he wrote: 'The people of Cathay in all that regards the treatment of criminals proceed with extreme caution. There are twelve courts of justice attached to the Emperor's administration; if an accused person has been found guilty before eleven of these, and the twelfth has not yet concurred in the condemnation, he may still have hopes of acquittal. If a case requires a reference involving a six months' journey or even more, still, as long as the matter is not perfectly clear,

---

[1] 漢儀   [2] 應劭   [3] 鄭玄   [4] 馬融   [5] 陳寵   [6] 洗寃錄   [7] 禮

the overwhelmingly empirical character of the Chinese study of Nature would result in the compilation of all possible methods and tests which might enable the magistrate to decide a criminal case, though the standard of scientific criticism available was quite insufficient to sort out the sound tests from those which rested on a merely superstitious basis. But in comparison with the primitive ideas still persisting in Europe up to the +18th century, the Chinese practice was much more civilised.[a]

(1) LAW AND PHENOMENALISM; THE UNITY OF THE ETHICAL AND COSMIC ORDER

This rapid survey of the history of Chinese law[b] has confirmed the conclusion already suggested at the end of the Section on the ancient school of Legalists (p. 214) and at the beginning of the present Section. The struggle between systematic law and law administered by men paternalistically judging every new case on its own merits and in accordance with *li*[1]—as Wu Ching-Hsiung put it in the title of a valuable paper (2)—was settled decisively in favour of the latter. But one would not appreciate the full force of the word *li* if one failed to recognise that the customs, usages and ceremonials which it summed up were not simply those which had empirically been found to agree with the instinctive feelings of rightness experienced by the Chinese people 'everywhere under Heaven'; they were those which, it was believed, accorded with the 'will' of Heaven, indeed with the structure of the universe. Hence the basic disquiet aroused in the Chinese mind by crimes, or even disputes, because they were felt to be disturbances in the Order of Nature. Already the Hung Fan[2] chapter of the *Shu Ching* (Historical Classic), which if not written in the early Chou time is quite old enough for our purpose, indicates that excessive rain is a sign of the emperor's injustice, prolonged drought indicates that he is making serious mistakes, intense heat accuses him of negligence, extreme cold of lack of consideration, and strong winds (curiously enough) show that he is being apathetic.[c]   In the *Chou Li* (Record of the Rites of

---

the criminal is not put to death but only kept in custody.' Allowing for some exaggerations and mis-apprehensions, the testimony of the Persian is exactly the same as that of the Portuguese. Other aspects of Ming justice which greatly impressed Pereira were the care with which the magistrates themselves took notes of cases, the fully public nature of the hearings, and the effective measures taken to prevent false witness. See Yule (2), vol. 1, p. 281; and Boxer (1), pp. xxix, 17, 19, 158ff., 166, 175ff.

[a] On the history of forensic medicine cf. Balthazar & Dérobert (1). A cognate point is the recent demonstration of Bünger (2), that Chinese legal practice concerning lunatics, invalid persons and negligents, was much more humane than 19th-century sinologists, misled by wrong translations, realised.

[b] The writings of Escarra (1, 2) are the only ones available in a Western language which discuss the general history of law in China. There are two great compendia of Chinese texts, however, the *Li-Tai Hsing Fa Khao*[3] of Shen Chia-Pên,[4] and the *Chiu Chhao Lü Khao*[5] of Chhêng Shu-Tê.[6] The *Chung-Kuo Li-Tai Fa Chia Chu Shu Khao* by Sun Tsu-Chi[7] is a bibliography of 574 works on Chinese juris-prudence. The best monographs in Chinese on its history are those of Yang Hung-Lieh:[8] *Chung-Kuo Fa Lü Fa Ta Shih* and *Chung-Kuo Fa Lü Ssu-Hsiang Shih*.

[c] Ch. 24, Legge (1), p. 148; Medhurst (1), p. 206; Karlgren (12), p. 29. Probably −5th to −3rd centuries.

[1] 禮   [2] 洪範   [3] 歷代刑法考   [4] 沈家本   [5] 九朝律考
[6] 程樹德   [7] 孫祖基   [8] 楊鴻烈

Chou) and in many other ancient texts there is upheld the idea that punishments can only be carried out in autumn, when all things are dying; to execute criminals in the spring would have a deleterious effect on the growing crops. In trying to visualise what this 'phenomenalism' meant to the ancient Chinese, Eberhard (6) has made the interesting suggestion that they thought of Heaven and Earth as if the sequences of their phenomena proceeded along two parallel strands in time, as if in two parallel wires, and that perturbations in one sequence affected the other as if by a kind of inductance.

Many examples of this 'phenomenalist' world-picture were seen in the Section on Wang Chhung and the Sceptics (p. 378), and in that on the fundamental ideas of Chinese science (p. 247). Wang Chhung had to maintain that excessive seasonal heat and cold did *not* depend on the ruler's joy and anger, that plagues of tigers and grain-eating insects were *not* due to the wickedness of secretaries and minor officials. In the *Tso Chuan*[a] it is said that the abundance of animals depends upon the proper performance of their duties by the State officials. Similar ideas are frequently found in *Huai Nan Tzu*,[b] and numerous statements of them, collected from many ancient books, are given in the works of Granet. In a word, the emperor embodied in himself (and, by extension, in his bureaucracy) that system of semi-magical relationships between man and the cosmos which it had been the function, in very primitive times, of the folk-festivals and folk-ceremonies to maintain in good order.

I do not wish to suggest that this world-outlook was exclusively Chinese. It has an obvious connection with the theories of the microcosm and the macrocosm which were discussed above (pp. 294 ff.). In his great work *From Religion to Philosophy*, Cornford noted a similar, though less elaborated, conviction among the Greeks (as in Hesiod, Aeschines, Sophocles, Herodotus, etc.), and, indeed, this was almost the starting-point from which he set out on his exploration of the Greek ideas of Destiny and Nature. It has been defined as the conviction of the unity of the ethical and the cosmic order. Roughly speaking, Cornford came to the conclusion that this hypostatisation of the moral order had been a kind of projection of the internal relationships of primitive tribal collectivism on to external Nature,[c] with which human society was felt as continuous. He drew attention to a striking remark of Iamblichus: 'Themis in the realm of Zeus, and Dike in the world below, hold the same place and rank as Nomos in the cities of men; so that he who does not justly perform his appointed duty may appear as a violator of the whole order of the universe.'[d] This covered sufficiently the heavens, the earth and human society upon it.[e] But the idea of justice and law was even extended to the relations between the parts of the human body itself. Who began this presentation I do not know, but as Temkin (1) has pointed out, it is strikingly clear in Galen. In his time (+2nd century) the old notion of *isonomia* (ἰσονομία), equality, was interpreted as meaning, not that every person in the State,

[a] Duke Chao, 29th year (Couvreur (1), vol. 3, p. 452).
[b] E.g. Morgan tr. pp. 55, 82, 84.   [c] Cf. esp. his p. 55.
[d] *Vit. Pythag.* IX, 46.
[e] See below, p. 571, on *dike* and *ṛta*.

or every part of the body, had equal claims, but that each ought to share according to its rank.[a] Galen uses the concept of justice (*dikaia*, δικαία; *dikaiosune*, δικαιοσύνη) again and again to explain the anatomy of the body.[b] Parts differ in size; this is only just, as Nature has apportioned their size according to their usefulness.[c] Some parts have few nerves; this, too, is just, as they do not need much sensitiveness.[d] Every organ of the organism receives its just due or share[e] from Nature; 'Is not Nature most just in everything?'[f]

We are thus in presence of a thorough parallelism at the three levels of cosmos, human society and individual body. But it seems that the Western conception was deeply different from the Chinese. The former saw justice and law at all levels, closely associated with personalised beings, enacting laws or administering them. The latter saw only that righteousness embodied in good custom represented the harmony necessary for the existence and function of the social organism. It recognised also a harmony in the function of the heavens, and, if pressed, would have admitted one in the functions of the individual body also, but these harmonies were spontaneous, not decreed. Discord in one was echoed by dysharmony in the others. Now doubtless in European thought the ideas expressed by Iamblichus and Galen had been important elements of the stream which formed the *koinos nomos*, the Universal Law, of the Stoics. On the contrary, in China, the phenomenalist conviction of cosmic-ethical unity gave no stimulation whatever to the idea of laws of Nature. Indeed, those who, like Wang Chhung in the +1st century, most strongly advocated the world-outlook of scientific naturalism, were totally opposed to the basic belief, as well as to the extravagances, of phenomenalism, attacking it (and them) on the essentially Copernican ground that the implied anthropocentrism was nonsense. In Europe the rejection of geocentrism, and hence anthropocentrism, came much later, and by then it was possible to reject it while retaining and intensifying the notion of universal law in Nature.

If, then, all crimes and disputes were looked upon in ancient China, not primarily as infractions of a purely human, though imperial, legal code, but rather as ominous disturbances in the complex network of causal filaments by which mankind was connected on all sides with surrounding Nature, it was perhaps the very subtlety of these which made positive law seem so unsatisfactory. The preface of the +7th-century Thang code suggests that it is dangerous and ominous to 'leave *li* and engage in legally fixed punishments (*chhu li ju hsing*[1])'.

---

[a] Nothing else could have been expected from Hellenistic class-stratified society.

[b] *De Usu Partium*, v, 9; 1, 17, 22; 11, 16.

[c] III, 10.

[d] v, 9.

[e] An evident connection with the old idea of *moira*; see above, p. 107.

[f] This Galenic physiological *justitia* is perpetuated in the +16th and +17th centuries by men such as Fludd, van Helmont and Marcus Marci (Pagel (1), pp. 284 ff.). Paracelsus applies it to the 'ladder of souls' (cf. p. 22) and the right of man to make use of the animal creation (*De Pestilitate*, 1603, p. 327) —information for which I thank Dr Pagel.

[1] 出禮入刑

In this conception [Escarra well writes],[a] there is no place for law in the Latin sense of the term. Not even rights of individuals are guaranteed by law. There are only duties and mutual compromises governed by the ideas of order, responsibility, hierarchy and harmony. The prince, assisted by the sages, ensures the dominance of these throughout the realm. The supreme ideal of the *chün tzu*[1][b] is to demonstrate in all circumstances a just measure, a ritual moderation; as is shown in the Chinese taste for arbitration and reciprocal concessions. To take advantage of one's position, to invoke one's 'rights', has always been looked at askance in China. The great art is to give way (*jang*[2])[c] on certain points, and thus accumulate an invisible fund of merit whereby one can later obtain advantages in other directions.

Hence the lack of 'positivisation' (Gernet) of primitive customary law in China, and the failure of the Legalists. As Granet says, 'The Sophists did not succeed in persuading the Chinese that there could exist necessarily contradictory terms. Nor did the Legalists succeed in getting them to accept the idea of unvarying regulations and sovereign Law.'[d]

The truth of what has just been said will certainly be appreciated by everyone who has lived in China. To this day 'the Chinese method is, in practice, to fix responsibility in terms, not of "who has done something" but of "what has happened". When something has once happened, responsibility must be assigned; and hence there is always an underlying tendency to try to prevent decisive things from happening, and to diffuse responsibility.'[e] Escarra gives[f] a revealing verbatim account of a member of a merchant-guild[g] council replying to the questions of a foreign assessor in a treaty-port mixed court in 1926. Against all suggestions he stuck to his point that the guild members could accept decisions of the Supreme Court in Peiping only if they seemed to them in accordance with *li*,[3] and it had therefore to be admitted that the Supreme Court was not a sovereign force in the occidental sense.[h]

It is interesting to note what 19th-century English juristic historians thought of this kind of thing. They became involved because the development of Indian law had followed a course somewhat parallel to that of China, and Maine himself was for

[a] (1), p. 17.
[b] See p. 6 above.　　　　　　　　　　　[c] See pp. 61 ff. above.
[d] (5), p. 471.　　　　　　　　　　　　　[e] Lattimore (6), p. 80.
[f] (1), p. 81.　　　　　　　　　　　　　　[g] Cf. Sect. 48 below.
[h] Padoux (1) also emphasised this dialogue. It may go without saying, from all the foregoing, that the institution of counsel and advocates never grew up in China until our own time. The rich could, it is true, be represented before the magistrate by deputies (*kho*,[4] lit. guests or clients) who might argue their case, as we know from Wang Fu's[5] +2nd-century *Chhien Fu Lun*;[6] Balazs (1). In a sense, all Chinese law was administrative. There were neither feudal lords nor merchant princes whose disputes required settlement by due process of law with its accompanying pleadings and advocacies. The imperially appointed bureaucrat could not be sued, and individuals took very good care that no dispute should come before him unless it was absolutely unavoidable. A remarkably good popular account of the history of Chinese law and how it differed from other systems will be found in Wigmore (1), vol. 1, pp. 141 ff.; cf. also Hughes (6). Anyone interested in recent Chinese law, which has been greatly 'modernised', may consult Chêng Thien-Hsi (2); Schlegelberger (1), vol. 1, pp. 328 ff.; Escarra (1, 3); and Meijer (1). No study of law in China since the establishment of the People's Republic has yet become available.

[1] 君子　　　[2] 讓　　　[3] 禮　　　[4] 客　　　[5] 王符　　　[6] 潛夫論

a number of years legal adviser to the Indian government. Holland,[a] in discussing the views of Maine (2, 3), said that 'he asks in what sense it is true that the village customs of the Punjab were enforced by Ranjit Singh[b]....He denies that Oriental Empires, whose main function is the levying of taxes,[c] busy themselves with making or enforcing legal rules....He would almost restrict to the Roman Empire, and the States which grew out of its ruins, the full applicability of the Austinian conception of positive law.' Holland goes on to say that disobedience to village or provincial custom must either be forcibly repressed by the local authority, in which case it has effectively the force of law, or else acquiesced in, in which case the empire is, strictly speaking, lawless, and constitutes 'an arbitrary force acting upon a subject-mass imperfectly bound together by a network of religious and moral scruples'. Finally, he admits the difficulty of judging Asian systems of society by European criteria, saying, 'It is convenient to recognise as laws only such rules as can reckon on the support of a sovereign authority, though there are states of society in which it is difficult to ascertain as a fact what rules answer to this description.' These observers probably hardly made enough allowance for what arbitration, compromise and face-saving devices were capable of in the hands of Chinese magistrates. Waley[d] has strikingly written, 'No Chinese magistrate, after passing what he knew to be an unfair sentence, would have pointed out with a glow of pride (as sometimes happens outside China) that he had faithfully administered the law of the land.'[e]

(2) SOCIAL ASPECTS OF LAW, CHINESE AND GREEK

Yet the full significance of the distinction between li[1] and fa[2] cannot be understood unless their relationships to social classes are appreciated. In feudal times it was natural enough that the feudal lords should not consider themselves subject to the positive laws which they themselves gave forth; li, therefore, was the 'code of honour' of the ruling groups, and fa the ordinances (e.g. concerning corvée duty) to which the common people were subject. This is enshrined in the famous passage in

[a] (1), p. 52.
[b] The great Sikh leader; see V. A. Smith (1), pp. 614, 692.
[c] Echo of François Bernier here.
[d] (12), p. 141.
[e] Padoux, in his brilliant introduction to Liang Chhi-Chhao (2), well recognised these qualities of Chinese law, saying that such Western tags as 'dura lex sed lex' or 'summum jus, summa injuria', or 'fiat justitia, ruat coelum', had no meaning for the Chinese. Nor could they have been expected to appreciate that characteristic of European law which made Sir Walter Raleigh write, 'Sir Thomas More said (whether more pleasantly or truely I know not) that a trick of Law had no lesse power than the wheele of fortune, to lift men up, or cast them downe.' Yet Padoux and others (after the first World War) did not hesitate to call for a revival of the School of Legalists (cf. pp. 204 ff. above) so that 'Chinese mentality' might be made to approximate more closely to that of the West. It is ironical that Hegel's description of what he supposed to be Chinese law and morality fitted the Legalists alone. His chapter on China in the *Philosophy of History* was, alas, almost entirely composed of errors and misapprehensions.

[1] 禮　　　[2] 法

the *Li Chi* (Record of Rites):[a] '*Li*[1] does not reach down to the people; *hsing*[2] (punishments, or penal statutes) do not reach up to the great officers (*Li pu hsia shu jen; hsing pu shang ta fu*[3]).' This throws further light on the opposition to codification in the −6th century; Shu Hsiang and Confucius were opposing codification not only as a prelude to 'litigiousness' or 'obstructionism' on the part of the commoners, but also as embodying the danger of encroachment of fixed laws upon the whole class of the feudal nobility. Such an extension, we have seen (p. 212), was ultimately carried out by the Legalists in paving the way for the triumph of bureaucratism. And since it was the Confucians who in later ages operated the bureaucratic machine, they too, as Balazs (6) has well shown, became jurists of positive law. Yet the fluidity of *li*[1] retained for centuries so much of its original social prestige, and was so much more in accord with the general trend of Chinese philosophy than the rigidity of *fa*,[4] that even after bureaucratism had long been solidly established, the former dominated over the latter. This reveals another meaning of the phrase from the code of the Thang quoted a page or two above—'he who leaves *li*[1] will fall into *hsing*',[2] i.e. if one does not follow the *mores* felt to be ethically right, one will find oneself caught in the net[b] of criminal law. Chhen Chhung[5] said in +94 that *li* and *hsing* were like the outer surface and the lining of the same garment.[c] In the end, says Balazs (6), 'the elasticity and nuanced flexibility of *li* invariably worked out in favour of the privileged bureaucratic governing class, and late Confucianism often strengthened, instead of relieving, the arbitrary character of the laws to the detriment of the people'. Gradations in punishments according to rank in the official hierarchy persisted into the code of the Chhing.

No doubt we may look upon *li* and *fa* in more than one sociological context. Centralising and centrifugal tendencies were very delicately balanced in ancient and medieval Chinese society. *Fa* suited the bureaucratic irrigation administrators; *Tao* suited the self-contained rural communities; perhaps *li* was the ultimate compromise between the centre and the periphery of the social organism.[d]

As has been pointed out by Frank,[e] ancient Greek, as opposed to ancient Roman law, shared to a considerable extent the characteristic Indian and Chinese preference for equity and arbitration as opposed to abstract formulae. This he calls the 'individualisation of cases'. In an earlier work[f] he had made the interesting suggestion that one may see in the Roman 'quest for a practically unrealisable legal certainty' a certain masculine element, while in the milder Asian dominance of equity and the flexible determination of all cases on their individual merits a certain feminine element manifested itself. It is certainly notable that the Roman legal system arose in a society in which the power of the father (*patria potestas*) was carried to the extreme. Certainly in most cultures the father stands for the strict rules which the child is supposed to

[a] Ch. 1, p. 35a (Legge (7), vol. 1, p. 90). [b] Cf. p. 556 below.
[c] *Hou Han Shu*, ch. 76, p. 9b. Cf. Boodberg (3).
[d] This suggestion is due to Mr S. Adler. [e] (1), p. 378. [f] (3), p. 263.

[1] 禮   [2] 刑   [3] 禮不下庶人刑不上大夫   [4] 法   [5] 陳寵

obey, while the mother stands for lenience and the principle that 'circumstances alter cases'. On the one hand there is the ideal of the closed, static and consistent system of law, on the other, the 'feminine' attitudes of flexibility, tact, understanding and intuition. How striking is this suggestion in the light of what we have seen in Section 10d, i on the Taoists, the whole of whose philosophy and symbolism was permeated by an emphasis on the feminine. We saw, too,[a] that in the realm of nature-philosophy, Han Confucianism adopted a great deal of the Taoist thought of the Warring States period, just as, later on, Neo-Confucianism was deeply affected by the Taoism of the Thang. Could we not go so far as to say that when Han Confucianism triumphed over the excessive maleness of the Chhin Legalists, it did so partly by accepting from Taoism an attitude to law which rejected the search for a 'code fixed beforehand', and granted to magistrates the widest freedom to follow principles of equity, arbitration and 'natural law'?

The question of the relative roles of equity, arbitration or 'individualisation' as against rigid positive law is still far from being a dead issue. As we may see from the discussion of Frank (4) on 'legal pragmatism', the argument continues in our own time in the form of the weight to be placed on the findings of trial courts or courts of first instance on the one hand, and upper appellate courts on the other. The former are able to take into account many things which the latter cannot—the psychology of jurors and defendants, the difficulties of fact-finding, the 'wordless language' of witnesses (what it is about some of them which carries conviction to their hearers)—things which give concrete meaning to the trial judge's 'sovereignty'. Courts of appeal can only work on rules and their interpretations; they cannot re-hear the case in its original freshness.

Summing up, we may remember that the distinction between natural law and positive law has left many traces in European legal terminology itself. The physic justice of Aristotle, so closely connected with universal morality, comes down to us as *jus*, A.S. *riht*, *droit*, *diritto*, *recht* and *pravo*—it is the *donné* of Gény,[b] and China's *i*[1] and *li*.[2][c] The nomic justice of Aristotle, laid down by specific legislative authority, comes down to us as *lex*, law, *gesetz*, etc.—it is the *construit* of Gény, and China's *fa*.[3] And in China, *li*[2] was, for the greatest part of history, enormously more important than *fa*.[3]

---

[a] Above, p. 247.   [b] Cf. Wortley (1).

[c] It must of course be understood that 'right' in the sense of the 'Rights of Man' was not a concept characteristic of Chinese thought, which emphasised duties, compromise and unselfishness in the interests of harmony. But the Chinese had very clear ideas of what constituted action morally 'right'. It must also be remembered that the Mencian justification of the people's right to overthrow tyrants was quoted again and again through Chinese history; doubtless because the scholars could not usually speak openly and had to quote a 'sacred text'. Moreover, there was always 'right' in the sense of privilege, since punishments were graded. Throughout Chinese history, age gave varying degrees of protection against the law's rigour, and similar immunities covered imperial relatives and high officials. These could be cancelled in particular cases by higher authority. Such privileges had something in common with the 'benefit of clerks' in medieval Europe, but the modification of 'equality under law' by the demands of filial piety was characteristically Chinese.

[1] 義   [2] 禮   [3] 法

(d) STAGES IN THE MESOPOTAMIAN-EUROPEAN DIFFERENTIATION OF NATURAL LAW AND LAWS OF NATURE

We turn now to the third part of the argument, the stages ot development within Western civilisation of the ideas of natural law and the laws of Nature.[a]

There can be little doubt that the conception of a celestial lawgiver 'legislating' for non-human natural phenomena has its first origin among the Babylonians. Jastrow gives[b] the translation of Tablet no. 7 of the Later Babylonian Creation Poem, in which the sun-god Marduk (raised to a position of central importance about the same time as the unification and centralisation under Hammurabi, *c*. −2000) is pictured as the giver of law to the stars. He it is 'who prescribes the laws for (the star-gods) Anu, Enlil (and Ea), and who fixes their bounds'. He it is who 'maintains the stars in their paths' by giving 'commands' and 'decrees'.[c] The same idea occurs also very early in India.[d]

The pre-Socratic philosophers of Greece speak much of necessity (*ananke*, ἀνάγκη), though not of law (*nomos*, νόμος) in Nature. But 'the Sun' Heracleitus says[e] (*c*. −500 'will not transgress his measures; otherwise the Erinyes, the bailiffs of Dike (the goddess of justice) will find him out'. Here the regularity is accepted as an obvious empirical fact, but the idea of law is present, since sanctions are mentioned (Guérin, 1). Anaximander, too (*c*. −560),[f] speaks of the forces of Nature 'paying fines and penalties to each other'. Heracleitus refers[g] to a 'divine law' (*theios nomos*) by which all human laws are 'nourished'. This may have covered non-human Nature as well as human society, since it is 'common to all things', all-powerful and all-sufficing. But the conception of Zeus Nomothetes in the older Greek poets pictures him as giving laws to gods and men, not to the processes of Nature, for he himself was not truly a Creator.[h] Demosthenes, however (−384 to −322, living thus between the generation of Mo Ti and that of Mêng Kho), uses the word law in its most general sense when he says:[i] 'Since also the whole world, and things divine, and what we call the seasons, appear, if we may trust what we see, to be regulated by Law and Order.'

[a] By far the best account of this subject known to me is that of the late Edgar Zilsel (1). I was made aware of it (by my friend Dr Jean Pelseneer) only after the first draft of the following pages had been prepared; but there was nothing to alter, and only a few points to add; our conclusions were the same. Pelseneer himself (1) has a brief but valuable discussion of the subject.

[b] (2), pp. 441 ff.

[c] Cf. also Eisler (1), p. 233. Later on, in Section 20*e*, we shall find unexpected clarity and precision regarding these 'bounds' and 'paths'.

[d] *Ṛg Veda*, x, 121.　　　　　　　　　　　　[e] Diels-Freeman (1), p. 31; Freeman (1), p. 112.

[f] Diels-Freeman (1), p. 19; Freeman (1), p. 63.

[g] Diels-Freeman (1), p. 32. Cf. Pohlenz (1), who shows a semantic relationship between *nomos* and *moira* (*fên*); see p. 107 above.

[h] Cornford (1), p. 27; Guérin (1).

[i] *Adv. Aristog*. B, p. 808 (quoted by Holland, 1). Cf. Pindar, fr. 152.

Nevertheless, Aristotle never used the law-metaphor,[a] though, as we have already noted (p. 520), he occasionally comes within an inch of doing so. Plato uses it only once,[b] in the *Timaeus*,[c] where he says that when a person is sick, the blood picks up the components of food 'contrary to the laws of nature' (παρὰ τοὺς τῆς φύσεως νόμους). But the conception of the governance of the whole world by law seems to be peculiarly Stoic. Most of the thinkers of this school maintained that Zeus (immanent in the world) was nothing else but *koinos nomos* (κοινὸς νόμος), Universal Law[d] (e.g. Zeno, fl. –320; Cleanthes, fl. –240; Chrysippus, d. –206; Diogenes, d. –150). To some extent this idea may have been implicit in the word 'cosmos' which Platonists, Pythagoreans and Peripatetics had all used (Dodds, 1). But strong support for the new and more definite conception was probably derived from Babylonian influences, since we know that astrologers and star-clerks from Mesopotamia began about –300 to spread through the Mediterranean world. Among these one of the most famous was Berossos, a Chaldean who settled in the Greek island of Cos in –280.[e] Zilsel, alert for concomitant social phenomena, notes that just as the original Babylonian conceptions of laws of Nature had arisen in a highly centralised oriental monarchy, so in the time of the Stoics, a period of rising monarchies, it would have been natural to view the universe as a great empire, ruled by a divine Logos.[f]

Since, as is known, the Stoic influence at Rome was great, it was inevitable that these very broad conceptions should have their effects in the development of the idea of a natural law common to all men whatever might be their cultures and local customs. Cicero (–106 to –43), of course, reflects this, saying: 'Naturalem legem divinam esse censet (Zeno), eamque vim obtinere recta imperantem prohibentemque contraria';[g] and elsewhere: 'The universe obeys God, seas and land obey the universe, and human life is subject to the decrees of the Supreme Law.'[h] Curiously, it is in Ovid (–43 to +17) that we find the clearest statements of the existence of laws in the

---

[a] Checked by Bonitz's index. Zilsel (1) draws attention to the very interesting fact that in the only place where the word does occur in Aristotle (*Physics*, 193 a 15) the sense is just contrary to *nomos* as laws of Nature. Aristotle points out that if a wooden bed is buried in the ground, and sends up a shoot, what is produced is wood and not a bed. He then contrasts the perishable and artificial shape of the bed with its permanent and natural material by calling the *former* 'a mere arrangement according to law'. This agrees, of course, with what we saw a few pages above on his distinction between 'physic justice' and 'nomic justice'.

[b] Checked by Ast's index.

[c] 83 E.

[d] Zeller (1), pp. 143, 161; E. V. Arnold (1), pp. 220, 272, 385, 402, 407; Vinogradov (1), vol. 2, pp. 40 ff. There are many analogies between the Stoics and the Mohists.

[e] Cf. the monograph on him by Schnabel (1); and Eisler (1), p. 77. Cf. Dodds (1), p. 245.

[f] This seems sound, but if 'oriental' monarchies of Mesopotamian type could so readily generate the idea of celestial laws of nature, why should this not have happened also in China, where even in the feudal period there was some degree of centralisation, and far more after the unification of Chhin Shih Huang Ti?

[g] *De Natura Deorum*, I, 14 (tr. Brooks, p. 30), 'Zeno considers natural law to be divine, commanding men to do right with the same force as it forbids them to do the contrary.'

[h] *De Legibus* (tr. Keyes, p. 461). Note the parallel here with ch. 25 of the *Tao Tê Ching*, discussed on p. 50.

non-human world. He does not hesitate to use the word *lex* for astronomical motions. Speaking of the teaching of Pythagoras,[a] he says·

> in medium discenda dabat, coetusque silentum
> dictaque mirantum magni primordia mundi
> et rerum causas, et quid natura docebat,
> quid deus, unde nives, quae fulminis esset origo,
> Juppiter an venti discussa nube tonarent,
> quid quateret terras, *qua sidera lege mearent*,
> et quodcumque latet....

Most translators have failed to do justice to this remarkable statement; Dryden turned it thus:

> What shook the stedfast earth, and whence begun
> The Dance of Planets round the radiant Sun....

while King simply left the phrase out altogether. Elsewhere Ovid, complaining of the faithlessness of a friend, says that it is monstrous enough to make the sun go backward, rivers flow uphill, and 'all things proceed reversing nature's laws' (*naturae praepostera legibus ibunt*).[b]

In this connection the origin and fate of the word 'astronomy' is interesting. As Zilsel points out, this compound term could not have been coined or used if there had not been a tacit recognition of quasi-juridical laws controlling the movements of the celestial bodies. Recently a special study has been devoted to the history of the word by Laroche (1). Astronomy and astrology were at first synonymous, and the former was familiar to Aristophanes as early as the −5th century.[c] Later usage seemed to follow the chance preferences of individual authors; Plato wanted to settle on the term astrology, but it was already acquiring the significance of 'astro-mancy'. In the +5th century, Latin encyclopaedias for monks explain 'astronomy', literally translating the term, as the science dealing with the 'law of the stars' (*lex astrorum*);[d] but the significance might be rather that of the laws which the stars gave to every man in fixing his fate, than that of the laws which they themselves had to obey in their motions.

E. V. Arnold[e] has ventured upon the speculation that the Cynic school of philosophers of Hellenistic times may have been influenced by Buddhism, since it is established that King Asoka sent missionaries 'with healing herbs and yet more healing doctrine' from India to Ptolemy II of Egypt, Antiochus of Syria, and other rulers, just before −250.[f] It is tempting to consider, therefore, the possibility that

---

[a] In the *Metamorphoses*, xv, 66 ff. I owe the reference to this passage to Dr Charles Singer and Mr Henry Deas.

[b] *Tristia*, 1, 8, 5. There are other passages in Ovid very close to these statements (cf. the index of Deferrari, Barry & McGuire).

[c] *Clouds*, 194, 201.

[d] Cassiodorus, *Inst.* 2, 7; Isidorus, *Diff.* 2, 152.

[e] (1), pp. 14, 17.   [f] V. A. Smith (2), p. 174.

the *koinos nomos* of the Stoics may have had something to do with the Buddhist universal 'law' of *karma*, but we have already seen that this was never applied to non-moral, non-human phenomena,[a] and it will be more convenient to return to it later.

Far more certain as another contributory line of thought was that which emanated from (or was transmitted from the Babylonians by) the Hebrews. The idea of a body of laws laid down by a transcendent God and covering the actions both of man and the rest of Nature is frequently met with, as Singer (5) and many others have pointed out.[b] Indeed, the divine lawgiver was one of the most central themes of Israel. It would be difficult to overestimate the effect of these Hebrew ideas on all occidental thinking of the Christian era—'The Lord gave his decree to the sea, that the waters should not pass his commandment' (Psalm 104)—'He hath made them fast for ever and ever; he hath given them a law which shall not be broken' (Psalm 148). Furthermore, the Jews developed a kind of natural law applying to all men as such, somewhat analogous to the *jus gentium* of Roman law, in the 'Seven Commandments for the Descendants of Noah' (Isaacs, 1). This was liable to conflict with Talmudic law (Teicher, 1).

We have spoken of the Stoics and Cynics, but have not mentioned the most important scientific school of all, the Epicureans. It is remarkable, indeed, that Democritus and Lucretius, who so powerfully advocated natural and causal explanations, never spoke of laws of Nature. There is only one place in the *De Rerum Natura* where Lucretius uses the term in its later sense.[c] Denying the existence of chimaeras; he says that members of a body can combine only if they are adapted to each other, all animals are 'bound by these laws' (*teneri legibus hisce*). Zilsel acutely points out that on Epicurean theology laws of Nature would be impossible in the strictest sense, since the gods had not created the world and took no interest in it; perhaps this was why the Epicureans spoke of principles but not of laws. Another point where the history of biology enters in is the work of Galen (+129 to +201), whose *De Usu Partium*, in which he seeks to demonstrate the teleological significance of every part of the human body, has been held to involve an approximation to the idea of laws of Nature.[d]

Christian theologians and philosophers naturally continued the Hebrew conceptions of a divine lawgiver. In the early centuries of Christianity statements in which laws of non-human Nature are implicit are not difficult to find. For example, the oratorical apologist Arnobius (*c.* +300), arguing that Christianity is nothing monstrous, says that since its introduction there have been no changes in 'the laws initially established'.[e] The (Aristotelian) elements have not changed their properties. The structure

[a] Above, p. 419.
[b] Cf. Isaiah 40. 12 and 22; 45. 5 and 7; Jeremiah 5. 22; Proverbs 8. 9; Psalms 104. 9; Job, 56. 10; 28. 26; 38. 10 and 11, 31–3. For a discussion of the Hebrew words used for 'law', 'boundary', etc. see Zilsel (1).
[c] II, 719. Checked by Paulson's index.
[d] Singer & Singer (1), Singer (6). See above, p. 528.
[e] *Adv. Gentiles*, 1, 2.

of the machine of the universe (presumably the astronomical system) has not dissolved. The rotation of the firmament, the rising and setting of stars, has not altered. The sun has not cooled. The changes of the moon, the turn of the seasons, the succession of long and short days, have neither been stopped nor disturbed. It still rains, seeds still germinate, trees still put forth leaves and shed them in the autumn, and so on.

We are still in the stage, however, before a sharp separation between (human) natural law and (non-human) laws of Nature has come about. In the early centuries of the Christian era there are two statements of particular interest which show the ideas in their more or less undifferentiated state. In the *Constitution* of Theodosius, Arcadius and Honorius of +395 there is a passage forbidding anyone to practise augury on pain of punishment for high treason: 'Sufficit ad criminis molem naturae ipsius leges velle rescindere, inlicita perscrutari, occulta recludere, interdicta temptare'[a]—it is impious to tamper with the principles which keep the secret laws of Nature from men's eyes. This is strikingly similar to the prohibition of the Chhan-Wei[1] books of augury in China (cf. pp. 380–2), but here its interest is that it suggests the existence of laws of Nature, connected indeed with the course of human affairs, but not concerned with morality.

The second statement is a famous one of Ulpian, the eminent Roman jurist (d. +228)[b] whose work occupies so large a part of the Justinian *Corpus Juris Civilis*,[c] of +534. 'Jus naturale', he says in the first paragraph of the *Digest*, 'est quod natura omnia animalia docuit. . . .'

Natural law is that which all animals have been taught by Nature; this law is not peculiar to the human species, it is common to all animals which are produced on land or sea, and to fowls of the air as well. From it comes the union of man and woman called by us matrimony, and therewith the procreation and rearing of children; we find in fact that animals in general, the very wild beasts, are marked by acquaintance with this law.[d]

Historians of jurisprudence are at pains to explain that this never had any influence on subsequent legal thinking. That may well be the case,[e] but it was accepted by medieval writers and commentators, and clearly expresses the idea of animals as quasi-juristic individuals obeying a code of laws laid down by God. At this point we are very close to the idea of the laws of Nature as the divine legislation which matter (including animal life) obeys.

As the Christian centuries went on it was inevitable that natural law should come to be identified with Christian morality. St Paul had clearly recognised it.[f] St Chrysostom

[a] *Cod. Theod.* XVI, Tit. x, 12 (cited by Bryce (1), vol. 2, pp. 112 ff.); cf. Bréhier (1).

[b] Ledlie (1) has given a good account of him.

[c] The *Corpus* comprises the *Digest* (legal literature), about one-third of which was written by Ulpian, the *Institutes* (students' books), and the *Codex* (enacted laws).

[d] Tr. Monro (1), vol. 1, p. 3.

[e] Yet one cannot help feeling that it may have had some connection with the medieval trials of animals in courts of law, to which I shall shortly refer (p. 574).   [f] Ep. Rom. 2. 14.

[1] 讖緯

(early +5th century) had seen in the ten Hebrew commandments a codification of natural law, and with the *Decretum* of Franciscus Gratianus (+1148) the identification, never afterwards departed from by orthodox canonists, was complete.[a] It was, moreover, as Pollock (2) says, the universal medieval belief that commands of princes contrary to natural law were not binding on their subjects, and could therefore lawfully be resisted. This doctrine, summarised in the phrase 'Positiva lex est infra principantem sicut lex naturalis est supra', bore much fruit at the time of the rise of Protestantism, and the 'right of rebellion against un-Christian princes' had no small part to play in the beginnings of modern European democracy (Gooch, 1). It is interesting to note how precisely it corresponds with the Confucian doctrine, expressed in Mencius,[b] that subjects have a right to dethrone the ruler who ceases to act according to *li*; and the similarity was certainly not lost upon European social thinkers who read the Latin translations by the Jesuits of the Chinese classics after +1600.

The systematisation of all this is of course in Thomas Aquinas.[c]

There is a certain Eternal Law, to wit, Reason, existing in the mind of God, and governing the whole universe....For law is nothing else than the dictate of the practical reason ('dictamen practicae rationis') in the ruler who governs a perfect community. Now it is manifest that if, as we have already seen, the world is ruled by divine providence, the whole universe is a community governed by the divine reason. And so this Reason, thus ruling all things, and existing in God the governor of the universe, has the nature of Law.[d]

Just as the reason of the divine wisdom, inasmuch as by it all things were created, has the nature of a type or idea; so also, inasmuch as by such reason all things are directed to their proper ends, it may be said to have the nature of an eternal law....And accordingly the eternal law is nothing else than the reason of the divine wisdom regarded as directive of all actions and motions. ('Lex aeterna nihil aliud est quam summa ratio divinae sapientiae, secundum quod est directiva omnium actuum et motionum.')[e]

Every law framed by man bears the nature of a law only in the extent to which it is derived from the Law of Nature. But if on any point it is in conflict with the Law of Nature, it at once ceases to be a law; it is a mere corruption of law.[f]

St Thomas thus pictured four systems of law, the *lex aeterna*, governing all things always;[g] the *lex naturalis*, governing all men; and the *lex positiva*, laid down by human

---

[a] Cf. the article by Jacob (1).

[b] E.g. *Mêng Tzu*, v, (2), ix.

[c] +1225 to +1274. Cf. Salmond (1); Carlyle & Carlyle (1), vol. 1; vol. 5, pp. 37 ff.

[d] *Summa*, 1, (2), Q. 91, art. 1. With Graf (2), vol. 1, p. 274, it would be seductive to take Tao as the Chinese equivalent of *lex aeterna*, but really there is between them a great gulf fixed.

[e] *Summa*, 1, (2), Q. 93, art. 1.   [f] *Summa*, 1, (2), Q. 95, art. 2.

[g] The expression 'laws of Nature' is found in Maimonides (+1135 to +1204) as Singer & Singer (1) have pointed out. He attributed it (erroneously) to Aristotle, and accepted the rule of such laws for the sublunary sphere though not for other parts of the universe; cf. L. Roth (1), p. 61; and A. Cohen (1). There is a translation of the *Guide for the Perplexed* by M. Friedlander (1); see esp. pt. 2, chs. 19–24. On Islamic thought concerning 'laws of Nature' I have been able to find out very little. Dr Zaki Validi Togan, however, informs me that there is some discussion of the subject in the *Tukhfat al-Faqīr* (The Gift of the Poor Man) written by Shams al-Ijī in Persia about +1397 (unique MS in the Jami Library at Istanbul, no. 231). The whole question would repay attention by Arabists and Iranists. Cf. Togan (1).

lawgivers (*divina* if canon law inspired by the Holy Spirit working through the Church; *humana* if common law enacted by princes and legislatures).[a] In the last of the three quotations we have a very close parallel to what the Confucians (in other terms) urged against the Legalists. If *fa* were contrary to *li* it must be false *fa*.

When the scholastic synthesis was dynamited by the Reformation natural law began to undergo its greatest development, and a basis of universal human reason was substituted for the former basis of divine will. The secularisation of natural law, with the rise of nationalism from +1500 onwards, has been described by Gierke (1) and others. It lived on in many forms; in England it was equated with Chancellor's equity, and it became particularly important in cosmopolitan, mercantile and international relations. Just as it is thought to have had its origin among the merchants at Rome, so in the 17th century it returned to a commercial milieu, as a book such as Malynes' *Lex Mercatoria* shows. Foreign merchants came, it was said, within the king's jurisdiction, but this was exercisable 'secundum legem naturae que est appelle par ascuns Ley Marchant, que est ley universal par tout le monde' (Pollock, 2). After this the extension of the principle to international law by its founder Grotius came about quite naturally (Figgis, 1).

(e) THE ACCEPTANCE OF THE LEGISLATIVE METAPHOR IN RENAISSANCE NATURAL SCIENCE

But what of the scientists and their laws of Nature? We are now in the 17th century, and with Boyle and Newton the concept of laws of Nature, 'obeyed' by chemical substances and planets alike, is fully developed. Very little investigation has been made, however, of the exact points at which it differentiated from the synthesis of the schoolmen. The lexicographers say that the first use of the expression in its scientific sense occurs in the first volume of the *Philosophical Transactions* of the Royal Society (+1665). Thirty years later Dryden inserts it gratuitously in his translation of the 'Felix qui potuit rerum cognoscere causas' of Virgil's *Georgics* (II, l. 490)— it has become a commonplace. Robson, in his excellent book *Civilisation and the Growth of Law*, regarded it as a specifically 17th-century idea, present in the philosophies of Spinoza and Descartes as well as in the 'new, or experimental, philosophy' of the natural scientists. It is the merit of Zilsel to have disentagled clearly the stages through which the idea at last came into its own. We find also in jurists such as Huntington Cairns a recognition of the parallel development in the 17th century of secularised natural law based on human reason, and the mathematical expression empirical laws of Nature.

[a] Zilsel (1), p. 257, gives a slightly different interpretation of the thought of St Thomas, based, in part, on other citations from the *Summa Theologica*. The thought of St Thomas combines, as he says, so much seeming logical exactness with so much empirical vagueness, that a special analysis would be required to attain a definitive view of the Thomistic position on these subjects.

Perhaps the first thinker who drew forth parallel laws of the non-human world from the scholastic *lex aeterna* was Giordano Bruno (1548 to 1600). His use of the term 'laws of Nature' is rare, and he generally speaks of *ratio* or *raggione*.[a] Rudolf Eisler cites two passages, however. In the first he is still scholastic, speaking of 'lex in mente divina, quae est ipsa rerum omnium dispositio'.[b] But elsewhere he says that God is to be sought for 'in inviolabili intemerabilique naturae lege'.[c] Bruno's world-conception approached almost more closely than that of any other European thinker to the 'organic causality' which we have seen (pp. 288 ff., 304) was characteristic of classical Chinese thought. Bruno ascribes all motion, and indeed all change of state, to the inevitable reaction of a body to its environment. He does not conceive the action of the environment as taking place mechanically, but rather regards the onset of change in a given body as a function of the nature of that body itself, a nature so constituted as to necessitate that particular reaction to that particular set of environmental circumstances. He thus visualised the phenomena of the universe of Nature as a synthesis of freely developing innate forces impelling to eternal growth and change. Bruno spoke of the heavenly bodies as *animalia* pursuing their courses through space, believing that inorganic as well as organic entities were in some sense animated. The *anima* constitutes the *raggione* or inherent law which, in contradistinction to any outward force or constraint, is responsible for all phenomena and above all for all motion.[d] The thought is extremely Chinese, even if vitiated by the characteristic animism of Europe.

There is no doubt that the turning-point occurs between Copernicus (1473 to 1543)[e] and Kepler (1571 to 1630).[f] The former speaks of symmetries, harmonies, motions,[g] but never in any place of laws. Gilbert, in his *De Magnete* (1600), does not speak of laws either, though he enunciates certain generalisations about magnetism for which the term would have been most suitable.[h] Francis Bacon's position is complex; in the *Advancement of Learning* (1605) he speaks of the 'Summary law of Nature' as the highest possible knowledge, but doubts whether it can be attained by man;[i] while in the *Novum Organon* (1620) he uses the term law as synonymous with Aristotelian substantial form.[j] He had thus really advanced no further than the scholastics. Galileo, like Copernicus, never uses the expression Laws of Nature, whether in his *jugendarbeit* on mechanics of 1598 or in his *Discourses and Mathematical Demonstrations on Two New Sciences* (1638), which was the beginning of modern mechanics and mathematical physics. What would later have been called laws appear as 'proportions', 'ratios', 'principles', etc.[k] The same remarks apply both to Simon

[a] Dorothea Singer (1), and personal communication, Oct. 1949.
[b] *Acrotismus* (+1588), *Opera*, pp. 1880 ff.
[c] *De Immenso*, VIII, 10.   [d] *De Immenso*, I, 1; *Opera*, p. 204.
[e] Cf. Dampier-Whetham (1), p. 119; Pledge (1), p. 36; Armitage (1).
[f] Cf. Dampier-Whetham (1), p. 139; Pledge (1), p. 39.
[g] And of anomalies, a word which contained the idea, but perhaps unconsciously.
[h] II, 32, p. 99.   [i] *Works*, Ellis & Spedding ed., p. 44.
[j] II, 17; *Works*, Ellis & Spedding ed., p. 321.
[k] Zilsel (1) believes that this was because Galileo still clung to the traditional deductive mathematical form of exposition as used by Archimedes and Euclid.

Stevin (whose works are of 1585 and 1608), and to Pascal (1663); the law metaphor was not used by them.

By a remarkable paradox, Kepler, who discovered the three empirical laws of the planetary orbits, one of the first occasions on which the laws of Nature were expressed in mathematical terms, never himself spoke of them as laws, though he used the phrase in other connections. Kepler's first and second 'laws', given in the *Astronomia Nova* of 1609, are paraphrased in long expositions;[a] the third, published in *Harmonices Mundi* (1619), is called a 'theorem'.[b] Yet he speaks of 'law' in connection with the principles of the lever,[c] and in general uses the word as if it were synonymous with measure or proportion.[d]

Since laws of Nature played so large a part in the astronomical sciences, it has been natural to search mostly among the Renaissance astronomers for the first mentions of them. It does not seem to have been pointed out hitherto that a very early reference occurs in connection with quite another group of sciences, geology, metallurgy and chemistry. In his *De Ortu et Causis Subterraneorum* of 1546, Georgius Agricola, discussing the Aristotelian theory of the participation of the element water in the composition of metals, wrote:

But what proportion of 'earth' is in each liquid from which a metal is made, no mortal can ever ascertain, or still less explain, but the one God has known it, Who has given sure and fixed laws to Nature for mixing and blending things together.[e]

It seems worthy of note that this conception should have come to the front at least as early in chemistry as in astronomy.

Meanwhile, an important step in the clarification of the concept had been made by the Spanish theologian Suarez, who in his *Tractatus de Legibus* (1612) made a sharp distinction between the world of morality and the world of non-human Nature, maintaining that the idea of law applied only to the former. He opposed the Thomistic synthesis because it disregarded this distinction. 'Things lacking reason', he says,[f] 'are, properly speaking, capable neither of law nor of obedience. In this the efficacy of divine power and natural necessity...are called law by a *metaphor*.' This was clear thinking, and reminds us of the difficulty which the Chinese had in extending the concepts of *li*[1] and *fa*[2] to the non-human world.

---

[a] III, 59f.

[b] v, 3 (*Opera*, ed. Frisch), vol. 5, p. 280.

[c] *Opera*, vol. 3, p. 391.

[d] For much fuller details on Kepler, see Zilsel (1), p. 265; Kepler, like Bruno, conceived of planets as partly animate, and raised the question of 'whether the laws are such, that they can probably be known to the planet'.

[e] 'Sed quota terrae portio in quoque humore, ex quo efficitur metallum, insit, nemo mortalium unquam mente cernere potest, nedum explicare: sed novit deus unus qui naturae certas et definitas quasdam leges dedit res inter se miscendi et temperandi.' Tr. Hoover & Hoover (1), p. 51.

[f] I, 1, sects. 1, 2.; II, 2, sects. 4, 10, 12, 13.

[1] 禮   [2] 法

In Descartes the idea of laws of Nature is as well developed as later in Boyle and Newton. The *Discours de la Méthode* (1637) speaks of the 'laws which God has put into Nature'.[a] The *Principia Philosophiae* (1644) concludes by saying that it has discussed 'what must follow from the mutual impact of bodies according to mechanical laws, confirmed by certain and everyday experiments'. So also in Spinoza. The *Tractatus Theologico-Politicus* (1670) distinguishes the laws 'depending on the necessity of Nature' from laws resulting from human decrees. Moreover, Spinoza agrees with Suarez that the application of the term 'law' to physical things is based on a metaphor—though for different reasons, since Spinoza was a pantheist who could not have believed in the naïve picture of a celestial lawgiver.

Zilsel sees one essential component in the development of 17th-century laws of Nature in the empirical technologies of the 16th century. He points out that the higher craftsmen of that time, the artists and military engineers (of whom Leonardo da Vinci was the supreme example) were accustomed not only to experimentation, but also to expressing their results in empirical rules and quantitative terms. He instances the small book *Quesiti ed Inventioni* of Tartaglia (1546)[b] in which quite exact quantitative rules were given for the elevation of guns in relation to ballistics.[c] 'These quantitative rules of the artisans of early capitalism are, though they are never called so, the forerunners of modern physical laws.' They rose to science in Galileo.

Here the most fundamental problem is why, after so many centuries of existence as a theological commonplace in European civilisation, the idea of laws of Nature attained a position of such importance in the 16th and 17th centuries. It is, of course, only a part of the whole problem of the rise of modern science at that time. How was it, asks Zilsel, that in the modern period, the idea of God's reign over the world shifted from the exceptions in Nature (the comets and monsters which had disturbed medieval equanimity) to the unvarying rules? His answer, which must surely be in principle the right one, is that since the idea of a reign over the world had originated from a hypostatisation into the divine realm of men's conceptions of earthly rulers and their reigns, we should look at concomitant social developments to reach an understanding of the change which now took place. Evidently with the decline and disappearance of feudalism and the rise of the capitalist State there occurred a disintegration of the power of the lords and a great increase in the power of centralised royal authority. We are familiar indeed with this process in Tudor England and 18th-century France; and while Descartes was writing, the English Commonwealth was taking the process even further, towards an authority which was centralised but no longer royal. If, then, we may relate the rise of the Stoic doctrine of Universal Law to the period of the rise of the great monarchies after

---

[a] See also his letter to Mersenne of 15 Aug. 1630 (reproduced by Lefebvre (1), p. 200) in which the royal analogy is explicitly made. 'Ne craignez point, je vous prie de publier que c'est Dieu qui a établi ces lois en la nature, ainsi qu'un roi établit des lois en son royaume.'

[b] Eighteen years before the birth of Galileo.

[c] Cf. the article by E. J. Walter (1). On the essential features in the rise of modern science, see further Section 19*k* below.

Alexander the Great, we may find it equally reasonable to relate the rise of the concept of laws of Nature at the Renaissance to the appearance of royal absolutism at the end of feudalism and the beginning of capitalism. 'It is not a mere chance', says Zilsel, 'that the Cartesian idea of God as the legislator of the universe, developed only forty years after Jean Bodin's theory of sovereignty.'[a] Thus the idea, which had originated in a milieu of 'oriental despotism', was preserved in rudimentary form through two thousand years, to awake to new life in early capitalist absolutism. Yet this brings us face to face with the paradox that in China, where 'imperial absolutism' covered an even longer period, we hardly meet with the idea at all. How this could be affords the subject of the rest of this Section.

For the present purpose, then, it suffices to say that between the time of Galen, Ulpian and the Theodosian Constitution on the one hand, and that of Kepler and Boyle on the other, the conceptions of a natural law common to all men, and of a body of laws of Nature common to all non-human things, had become completely differentiated. With this established we are in a position to see in what way the development of Chinese thought on natural law and the laws of Nature differed from that of Europe.

(f) CHINESE THOUGHT AND THE LAWS OF NATURE

We examined the fundamental ideas of Chinese scientific thought in the series of Sections dealing with the ancient and medieval philosophical schools, and in Section 13, where the long-enduring theories of the Yin and Yang and the Five Elements were described. It will be remembered that the Taoist thinkers, profound and inspired though they were, failed, perhaps because of their intense mistrust of the powers of reason and logic, to develop anything resembling the idea of laws of Nature. With their appreciation of relativism and the subtlety and immensity of the universe, they were groping after an Einsteinian world-picture, without having laid the foundations for a Newtonian one. By that path science could not develop. It was not that the Tao, the cosmic order in all things, did not work according to system and rule; but the tendency of the Taoists was to regard it as inscrutable for the theoretical intellect. It would not perhaps be going too far to say that this was one reason why, when to them was consigned the care of Chinese science through the centuries, this science had to remain on a purely empirical level. Moreover, it is not irrelevant that their social ideals had less use than those of any other school for positive law. Seeking to go back to primitive tribal collectivism, where nothing was formulated and written down, but everything worked well in communal cooperativeness, they could not have been interested in the abstract law of any lawgivers.

The Mohists, on the other hand, or rather the late Mohists, together with the Logicians, strove mightily to perfect logical processes, and took the first steps in applying them to zoological classification and to the elements of mechanics and

[a] *De la République* (1577).

optics. We do not know why this scientific movement failed; perhaps it was because the Mohists' interest in Nature was too strongly bound up with their practical aims in military technology, at any rate, as we saw (pp. 165, 202), these schools had few survivors after the upheavals of the first unification of the empire. It will be remembered that the proper translation of their technical term *fa*[1] (identical with 'law' as used by the Legalists) gave us some pause during the discussion of the logic of the *Mo Ching* (cf. pp. 173–5), but so far as can be seen the conclusion there reached, that the term was used by the Mohists to mean causative factors somewhat resembling the Aristotelian causes, holds good. They seem to have approached no nearer than the Taoists to the idea of laws of Nature.

(i) THE WORDS *FA* (POSITIVE LAW), *LI* (GOOD CUSTOMS, *MORES*) AND *I* (JUSTICE)

With the Legalists (*Fa Chia*) and Confucians we are in the realm of pure sociological interest, for neither of these schools had any curiosity about Nature outside and surrounding man. As we have seen, the Legalists laid all their emphasis on positive law (*fa*[1]), which was to be the pure will of the lawgiver, irrespective of what the generally accepted *mores* or morality might be, and capable of running quite contrary to it if the welfare of the State should so require. The law of the Legalists was at any rate precisely and abstractly formulated. As against this the Confucians (Ju Chia) adhered to the body of ancient custom, usage and ceremonial, which included all those practices, such as filial piety, which unnumbered generations of the Chinese people had instinctively felt to be right—this was *li*,[2] and we may equate it with natural law.[a] In other words, the *li* was the sum of the folkways whose ethical sanctions had risen into consciousness.[b] Moreover, it was necessary that this 'right' behaviour be taught, rather than enforced, by paternalistic magistrates. Moral suasion was better than legal compulsion.[c] Confucius had said[d] that if the people were given laws and levelled by punishments, they would try to avoid the punishments but would have no sense of shame; while if they were 'led by virtue' they would spontaneously avoid disputes and crimes. The *Li Chi* (Record of Rites)[e] speaks, in symbolism appropriately taken from hydraulic engineering, of good customs as dykes or embankments, saying that while it is easy to know what has already happened, it is difficult to know what is going to happen. Good customs, therefore, more flexible

---

[a] I note that this identification is expressly approved by modern Chinese jurists such as Hsiao Ching-Fang (1); cf. his p. 66. Cf. also Hummel (3); Bodde (7); Creel (4), p. 175. Hu Shih (9) has not been available to us.

[b] Cf. Sumner's *Folkways*, and Kroeber (1), p. 266.

[c] Bodde (7), p. 25.

[d] *Lun Yü*, II, iii.

[e] Ch. 30, tr. Legge (7), vol. 2, p. 284.

[1] 法   [2] 禮

than formulated laws, prevent disturbances before they arise, while laws can only operate after they have arisen. Hence one can understand the point of view which after the victory of the Confucians over the Legalists came to dominate Chinese thinking, that since correct behaviour in accordance with *li*[1] always depended on the circumstances, such as the status of the acting parties in social relationships, to publish laws beforehand which could take insufficient account of the complexity of concrete circumstances, was an absurdity.[a] Hence the severe restriction, which we have already noted, of codified law to purely criminal provisions.[b]

While it is convenient in this discussion to draw the contrast between *fa*[2] and *li*,[1] it is sure that the earliest form of this distinction was between *fa* and *i*,[3] a term which is generally translated as 'justice', and which certainly originally meant that which seemed just to the natural man. Innumerable passages could be quoted to show this. Perhaps the *locus classicus* is the final section of the *Ta Hsüeh* (Great Learning), where it is said: 'In a State, financial gain is not (real) gain—justice is gain (*Tzhu wei kuo, pu i li wei li, i i wei li*[4]).'[c] Another important place is *Hsün Tzu*, chapter 16, where there is a long discussion of *i* in contrast to *fa*.[d] The *Wên Tzu* book links the two together well by saying:[e] 'Laws (should) arise out of justice, and justice arises out of the common people and must correspond with what they have at heart (*Fa sêng yü i, i sêng yü chung, shih ho hu jen hsin*[5]).' This is the Confucian view, that law cannot exist without demonstrable ethical sanction. The Legalists held just the opposite.

A typical crux was whether or not it was right for children to delate parents.[f] When Confucius was on his travels, he met a feudal lord of Chhu State, the Duke of Shê (Shê Kung), who was sufficiently sympathetic to make discussion possible. The Duke supported what was later on to be the Legalist view of the matter, while Confucius of course maintained that 'the father should conceal the misconduct of the son' and vice versa.[g] Mencius naturally followed him in this,[h] and the mutual protection of close relations is stated implicitly in the *Hsiao Ching* (Filial Piety Classic), which is probably of Chhin and Han date.[i]

[a] Cf. Creel (4), pp. 151, 161.

[b] See the excellent chapter 'Custom and Law in the Universal Empire' (of China) in H. Wilhelm (3), pp. 65 ff. The law was not intended to protect property or persons, but only good customs between persons. Bodde (7), too, makes a significant point when he says that Western law would have seemed 'cold and mechanical' to Chinese jurists. Exactly, for the spirit of Chinese jurisprudence was akin to the spirit of Chinese philosophy, not mechanical but organic.

[c] Hughes (2), pp. 102, 163; Legge (2), p. 244.

[d] Ch. 16, pp. 14 a ff.; tr. Dubs (8), p. 171.

[e] Ch. 21, p. 31 a; cf. Forke (13), p. 352. On *i* see also Boodberg (3).

[f] This has been referred to already (Sect. 7 b) and the parallel with the dialogue *Euthyphro* of Plato noted.

[g] *Lun Yü*, XIII, xviii. Cf. Balazs (8), pp. 193 ff., for a particularly interesting case of a somewhat similar kind in the Sui.

[h] *Mêng Tzu*, VII (1), xxxv. [i] Ch. 9 (Legge (1), p. 476).

[1] 禮   [2] 法   [3] 義   [4] 此謂國不以利爲利以義爲利
[5] 法生于義義生于衆適合乎人心

The distinction between i^1 and fa^2 was remembered throughout Chinese history. One might say that i^1 was something that stood behind $li,^3$ as its justification, its inward and spiritual grace. In the Thang, for instance, cases were judged (a) according to the code ($lü^4$); (b) according to $li,^3$ i.e. by reference to Confucian classical texts[a] dealing with ethically and customarily right behaviour; and (c) according to $i.^1$ An example of the latter occurs in one of the writings of the poet Pai Chü-I.[b] A's wife was married to him for three years without bearing a child. A's parents wanted to have her divorced, and were justified according to the *Li Chi*, but the wife pleaded that she had no home to go to. Judgment was that although li^3 permitted such a divorce, i^1 made it impossible on overriding grounds of humanity. This illustrates the clash which could occur between, one might almost say, a lower and a higher conception of natural law. In Thang times, however, the main clash was between $lü^4$ and $li,^3$ particularly in vendetta cases, such actions being forbidden by $lü^4$ and enjoined by $li.^3$ In Sung times the main clash was between $lü^4$ and imperial edicts ($chao^5$), since the latter often authorised heavier penalties than the code permitted.[c] But for our present purpose the important point is that i^1 was even more heavily tied to human-heartedness than $li,^3$ and neither invited extension into the non-human world.

Now it is the argument of the present section that the term fa^2 was never applied in the sense of the laws of Nature until quite recent times—or at least that cases of its use in this regard are astonishingly rare. Matthews' dictionary gives[d] a translation of the expression *pan fa*[6] as 'laws of planetary motion', but this is probably due to a mistranslation of certain passages in the *Kuan Tzu* book (chapters 7, 66, 67) where the laws of man are said to be modelled after the regularities of the heavens.[e] We shall return to this point presently. The only example known to us, in all ancient and medieval Chinese literature, of the use of the word fa^2 for the processes of Nature occurs in *Chuang Tzu*, chapter 22, in a passage which has already been quoted in another connection.[f] In three eight-syllable phrases Chuang Chou praises the silence of the all-effecting universe:

Thien ti yu ta mei erh pu yen,[7]	Heaven and earth have the greatest beauty, but they are silent,
Ssu shih yu ming fa erh pu i,[8]	The four seasons have manifest *laws*, but they do not discuss them,
Wan wu yu chhêng li erh pu shuo.[9]	The ten thousand things have perfect intrinsic principles of order, but they do not talk about them.[g]

[a] In the Thang, reference could be made to Taoist classical texts also, such as the *Tao Tê Ching*, which during that dynasty ranked on a level with their Confucian counterparts.

[b] *Pai Hsiang Shan Chi*, ch. 50, pp. 6. 7 (judgement no. 22).

[c] I am much indebted to Dr Arthur Waley for the information contained in this paragraph.

[d] M 4886.

[e] The *Tzhu Yuan* and other Chinese dictionaries know nothing of such a meaning of *pan fa*,[6] explaining the phrase as the selection of important rules or laws, and the carving of them on boards.

[f] P. 70 above.   [g] Tr. auct. adjuv. Lin Yü-Thang (1), p. 68.

¹ 義   ² 法   ³ 禮   ⁴ 律   ⁵ 詔   ⁶ 版法
⁷ 天地有大美而不言   ⁸ 四時有明法而不議   ⁹ 萬物有成理而不說

But does this unequivocally mean law? One of the difficulties of the subject is that from the beginning, or at any rate from a very long way back, the word *fa*[1] also meant 'method' and 'model', and this might be a better translation here. An intensive search for other passages where the word might seem to be employed in the sense of laws of Nature should certainly be undertaken.

One of the texts which will have to be investigated in this connection is the *Ho Kuan Tzu*[2] (Book of the Pheasant-Cap Master).[a] This work is extremely difficult to date because it is highly composite; much of it must be about −4th century, and most is not later than the Later Han (*c.* +2nd century), but about a seventh of it is an incorporated commentary of the +4th or +5th. By the +7th century the text was more or less as we now have it.[b] Until it has been critically established, interpretations are premature, yet there seem to be strangely interesting passages. For example: 'Unity is the *fa* for all (*I wei chih fa*[3]).' 'The unitary *fa* having been established, all the myriad things conform to it (*I chih fa li, erh wan wu chieh lai shu*[4]).'[c] '*Fa* seals (moulds) all things, yet does not boast about it; such is the Tao of Heaven (*Fa chang wu erh pu tzu hsü chê; Thien chih Tao yeh*[5]).'[d] In any investigation, the senses of 'mould' and 'law' will have to be carefully distinguished.

(2) THE PHRASE *THIEN FA* (NATURAL LAW) AND THE WORD *MING* (DECREE)

It is not contended that *fa* in the phrase *Thien fa*,[6] 'the laws of Heaven', did not have the meaning of juristic natural law, something like *li*.[7] An early example of this occurs in the *Tso Chuan*[e] under date −515, where a feudal leader says, 'If you, my kinsmen by birth and marriage, will rally round me according to the Law of Heaven (*shun Thien fa*[8])....' But this is not a law of Nature in the scientific sense; it concerns human affairs and human society. A close Greek parallel would be the passage in Plato's *Gorgias*[f] in which the phrase *nomos tes physeos* (νόμος τῆς φύσεως) is put into the mouth of Callicles, a character who defends the old antithesis of *nomos* and *physis*. Here the words say 'law of Nature' but what they refer to is the 'natural right of the stronger'.[g] The contrast is instructive as well as the similarity.

In this realm of ideas it is possible to find many expressions in which Heaven is said to give commands—the phrase *Thien ming*[9] is almost a commonplace, and particularly frequent in certain writers like Tung Chung-Shu, who inclined to greater

[a] We have met with this before: Sect. 1 (Vol. 1, p. 10). Cf. Wieger (2), p. 330.
[b] I am indebted to the late Professor G. Haloun for details on the history of this book.
[c] Ch. 5, p. 12*b*.　　　[d] Ch. 4, p. 8*b*.　　　[e] Duke Chao, 26th year (Couvreur (1), vol. 3, p. 415).
[f] 483 E. I thank Professor E. R. Dodds for calling my attention to this.
[g] Cf. phrases which seem to come easily to the pens of modern European writers: 'But the irrevocable law of nature must have its way; the better race must gradually supplant the inferior one....' (Gill (1), p. 113.)

[1] 法　　　[2] 鶡冠子　　　[3] 一爲之法　　　[4] 一之法立而萬物皆來屬
[5] 法章物而不自許者天之道也　　　[6] 天法　　　[7] 禮　　　[8] 順天法
[9] 天命

personalisation of Heaven than the majority of scholars. *Ming*, decree, is nothing but an ancient graph of a mouth, a tent and a person kneeling (K 762, 823). 'Heaven, when it constituted man's nature', says Tung Chung-Shu,[a] 'commanded him to practise love and righteousness (*Thien chih wei jen hsing ming, shih hsing jen i*[1]).' But what we are trying to catch a glimpse of is Heaven commanding non-human things to behave as they do, commanding the stars, for example, to rotate nightly in the sky.

Apparently it never does. 'The king respectfully carries forward the purpose (ideas) of Heaven above,' says Tung Chung-Shu,[b] 'thus conforming to its Decree (*Ku wang chê, shang chin yü chhêng Thien i, i shun ming yeh*[2]).' He and his people might well do so, but the stars did not, and we find ourselves facing the same paradox again that the conception of *Thien fa*,[3] like *li*,[4] did not apply outside human society.[c]

K 762, 823

It is true, of course, that there were occasional and exceptional presentations in which the principle of *li*[4] was extended to cover the behaviour of all things in the universe without exception. The chief instance of this poetical kind of philosophy is found in the *Hsün Tzu* (−3rd cent.), and we have already described it (p. 27 above). What was much more common than his conception of *li*[4] as functioning in realms beyond the sublunary and human world was the conviction that it had in some sense come to man from there. Essentially this was equivalent to giving a 'divine' authority to human ethical concepts, and later on, in the evolutionary world of the Neo-Confucians, the idea rose to its highest status as that of a universe which had the property of producing moral behaviour when the sufficient degree of organisation had arisen at which it could manifest itself. Doubtless the most typical statement of *li*[4] as 'heavenly' occurs in the *Li Chi*,[d] which, in a passage of the −1st or −2nd century, says:

> From all this it follows that *li*[4] has its origin in the Great Unity (*Thai I*[5]).
> This, differentiating, became Heaven and Earth.
> Revolving, it became the Yin and the Yang.
> Changing, it manifests itself in the Four Seasons,
> Dispersing, it appears in the form of the gods and spirits.
> Its revelations are called Destiny (*chhi chiang yüeh ming*[6]).
> Its authority is in Heaven (*chhi kuan yü Thien yeh*[7]).[e]

And the writer adds that while *li*[4] is rooted in Heaven, its movement reaches to the Earth. All this amounted to saying that in some way or other human moral order

---

[a] *Chhun Chhiu Fan Lu*, ch. 3; cit. Fêng Yu-Lan (1), vol. 2, pp. 38, 48, tr. Bodde.
[b] *Chhien Han Shu*, ch. 56, p. 16a, cit. Fêng Yu-Lan (1), tr. Bodde; vol. 2, p. 49, cf. p. 6a.
[c] I have to thank Professor Derk Bodde for an interesting discussion which led to this paragraph, the translations in which are his. He draws attention to a curious story in the Chhan-Wei apocrypha (*YHSF*, ch. 56, pp. 3a, 50b, 51a) about miraculous writing from Heaven on a city-gate, and Confucius 'preparing laws for the Han'. See immediately below, p. 550.   [d] Ch. 9 (Li Yün), p. 66a.
[e] Tr. auct. adjuv. Legge (7), vol. 1, p. 388; R. Wilhelm (6), p. 40.

[1] 天之爲人性命使行仁義   [2] 故王者上謹於承天意以順命也   [3] 天法
[4] 禮   [5] 太一   [6] 其降曰命   [7] 其官於天也

had superhuman (not necessarily supernatural) authority. Such a conviction did not raise the question of the intrinsic control of non-human Nature. And certainly the use of the word 'law' in Legge's version of the last sentence was unjustifiable and should not be retained.

Sometimes the word *fa*[1] seems to be applied to mathematical or natural regularities when a closer look shows that it only refers to the fixing of quantitative metrological standards by decree of positive law.[a] An example of this is in the *Yin Wên Tzu*, which describes[b] four types of law.

Of law there are four kinds. The first is called the immutable law (for example, that which governs the relations of) prince and minister, superior and inferior. The second is called the law which adjusts the customs of the people (for example, that which governs the relations of) the capable and the rustic, likeness and unlikeness. The third is called the law which governs the masses (for example, that which bestows) honours and rewards, punishments and fines. The fourth is called the law of correct balance (for example, that which has to do with) calendrical science, acoustics, the degrees of the circle, balances and weights.[c]

Here the first kind of law is certainly (juristic) natural law, and the second is akin to it, analogous to that discussed in a previous paragraph, and connected with the natural processes whereby differently gifted people find their own level in society. The third covers both natural and positive law. In the fourth, the borderline with true laws of Nature is approached fairly closely, since the sizes which bells or pitch-pipes had to be if properly tuned, or the measured movements of the planets, had nothing to do with earthly law, whether natural or positive. But it is most probable that what the writer had here in mind was the action of the ruler in promulgating those sizes and measurements which his proto-scientific advisers recommended to him as nearest to the ideal, and in deciding, quite arbitrarily, upon the standards of weights and lengths.[d]

Pending further investigation, then, we may take it that the term *fa*,[1] in a sense analogous to the positive law of human societies, was rarely or never used for the laws of Nature by Chinese thinkers. Yet, as O. Franke (6) has so well emphasised, they had a profound conviction of the great unity of heaven and earth. It is therefore somewhat strange that while, for the Chinese, law could not be said to be *in* non-human Nature, there are a number of statements that the laws of human society were, or should be, modelled *on* non-human Nature. We have just come upon such a statement in *Kuan Tzu*,[e] but perhaps the most important passage is in the *Chung Yung*

---

[a] It is interesting in this connection that in ancient Chinese mathematics (see Section 19 below) the denominator of a fraction was called *fa*, and we shall there suggest that this was because it represented the ruler or scale by which the value of the fraction was determined.

[b] P. 1*b*.   [c] Tr. Escarra & Germain, p. 21; eng. auct. mod.

[d] Cf. pp. 209 ff. above. Even the calendar had constantly to be readjusted by imperial order.

[e] P. 546 above.

[1] 法

(Doctrine of the Mean),[a] where it is said of Confucius that he handed on the traditions of the ancient sage-kings. 'From above they took (as a model for the) laws, the seasonal (motions of the) Heavens. Below they followed the waters and the earth (*Shang lü Thien shih, hsia hsi shui thu*[1]).'[b] One presumes that they did so on account of the regularity of the heavens, the persistence of the waters, and the firmness of the earth. The *Kuan Tzu* passage[c] compares the instruments of peace and war with the warmth and cold of the seasons. The passage from *Chuang Tzu* just quoted[d] goes on to say that the sages modelled themselves on heaven and earth.[e] Tung Chung-Shu repeatedly says that kings should do so.[f] The only obvious conclusion is that we have here a poetical or metaphorical derivation of human laws, the qualities of which were thought of as mirroring certain desirable qualities seen in non-human Nature. But the paradox remains that it should never have occurred to anyone as odd that law could be derived from where no law existed. Clearly an intuitive conception of the emergence of novelty at the human level was extremely strong in classical Chinese thought.

(3) THE WORD *LÜ* (REGULATIONS, AND STANDARD PITCH-PIPES)

Throughout this section we have to insist continually upon the distinction between *li*[2] and *fa*[3]. Neither of these words was easily applicable to non-human Nature. But we have just come upon one ancient Chinese word which does seem to link the spheres of non-human phenomena and human law. This word is *lü*.[4] We have often noticed it in the paragraphs on the development of Chinese legal codes (pp. 523 ff. above), where, with its usual dictionary meaning, it stands for 'statutes' and 'regulations'. This sense is undoubtedly quite old, as the phrase in *Kuan Tzu* may witness: 'the laws serve to distinguish each person's portion and place, and to put a stop to quarrels (*lü chê so i ting fên chih chêng yeh*[5]).'[g] Here the idea is very close to that of *moira* and

[a] Ch. 30 (tr. Legge (2), p. 291; Hughes (2), p. 139); neither seems quite to do justice to the text.

[b] Tr. Hughes (2), Legge (2), mod.

[c] Ch. 7 (Pan Fa), and ch. 66 (Pan Fa Chieh), opening sentences.

[d] Ch. 22 (Legge (5), vol. 2, p. 61).

[e] Another passage is in one of the essays of the Han scholar Chia I (−2nd century), who says that the ancient kings maintained the principle of rewarding the good and punishing the evil as solidly as iron and stone, and as regularly as the four seasons (*Chhien Han Shu*, ch. 48, p. 21a, *Ta Tai Li Chi*, ch. 46; tr. R. Wilhelm (6), p. 175). Cf. the memorial of Chang Min[6] about +80 in *Hou Han Shu*, ch. 74, pp. 7b, 8a.

[f] *Chhun Chhiu Fan Lu*, chs. 44, 45, cit. and discussed by Fêng Yu-Lan (1), vol. 2, pp. 47ff. Again, in one of the Han apocrypha, the *Chhun Chhiu Wei Han Han Tzu*,[7] Confucius is made to say 'I have examined the historical records, drawn upon ancient charts, and investigated and collected the mutations of Heaven, in order to institute laws for the emperors of the Han dynasty' (*Ku Wei Shu*, ch. 12, p. 1b). If Bodde (in Fêng Yu-Lan (1), vol. 2, p. 128) is right in taking this reference to Heaven to mean portents and anomalies, then the thought concerned occasional 'celestial reprimands' rather than astronomical regularity. Obviously, man's obedience to intermittent portents is one step further removed from any law given to the stars, than his copying regular stellar motion.

[g] Ch. 52. Another passage is in the *Tso Chuan*, Duke Hsüan, 12th year (Couvreur (1), vol. 1, p. 617). On *fên* see Hu Yen-Mêng (1).

¹ 上律天時下襲水土   ² 禮   ³ 法   ⁴ 律
⁵ 律者所以定分止爭也   ⁶ 張敏   ⁷ 春秋緯漢含孳

the other Greek entities discussed by Cornford (1). But the word had also a quite different meaning, namely, the series of standard bamboo pitch-pipes used in ancient music and acoustics, and the twelve semitones which these pipes represented. What connection could there have been between the laws of sound and the laws of human lawgivers?

The word *lü*[1] (K 502) has as its right-hand phonetic a sign which was certainly in the most ancient times a hand holding a writing implement, and for its radical the word *chhih*[2] which meant a step with the left foot (paralleling *chhu*,[3] a step with the right foot).[a] This suggests an original connection with the notation of a ritual dance.[b] Later on, since the twelve semitones were made to correspond with the months of the year, the word became linked with the calendar, and thus is found associated with the word *li*[4] in titles of chapters on calendrical science, such as the 'Lü Li Chih'[5] of the *Chhien Han Shu*. Since details about the standard pitch-pipes will be given later in the Section on physics (acoustics, Sect. 26*h*), they will not be dealt with here.[c] The question at issue is how the conception of laws, statutes or regulations can have been derived from, or even associated with, the word for the standard musical tones.

Perhaps the etymological considerations just mentioned hold one clue. It would not be so far a step from the directions for music and ritual dancing laid down by a diviner or priest-magician (indeed a shamanist *wu*[6]) to the directions for conduct of other behaviour, especially organised military behaviour, laid down by a temporal ruler. There was a logical analogy between what dancing would do against the spirits and what drilling and weapon-practising would do against human enemies.[d] Some kinds of dances certainly involved the carrying and brandishing of weapons.[e] It is thought that originally there were five stations around the dancing-floor which in time gave their names to a certain quality of sound, according to the instrument stationed in each place, and later to a difference in pitch.[f]

This, however, is not the only connection between the musical tones and military affairs. Many references exist in Chou books (the *Tso Chuan*, the *Shang Chün Shu*, etc.) to the use of drums in battle as the signal for advance, and of the beating of

---

[a] Khang-Hsi Dictionary, followed by Couvreur (2) and others.

[b] Another view is that the radical of *lü* is half of the character *hsing*, 'to go', a diagram of a cross-roads (see pp. 222, 229 above). In this case the primary meaning of *lü* would simply be 'public announcement of government ordinances'. But in order to gain the connection with standardised musical tones, the ritual dance prescribed by authority is still required.

[c] Cf. Levis (1), p. 63; Chavannes (1), vol. 3 (ii), pp. 630 ff.; Soulié de Morant (1), p. 12; K. Robinson (1).

[d] We have many traces of this in our own culture, for example, the varied ceremonial dances in which swords are used, either for imitating combat (the 'pyrrhic' type) or first held hilt-and-point by a row or ring of dancers and finally locked round the head of a victim (the 'sacrificial' type). Compare the Greek dance of the Kuretes, and the Roman Salii. For the traditions which have persisted in England until our own time, see Kennedy (1) embodying the conclusions of Needham (8).

[e] Granet (1), pp. 171 ff.; cf. pp. 132, 134 above.

[f] K. Robinson (personal communication). The different notes were certainly associated with different dances, as we know from the description of the department of the Master of Music (*Ta Ssu Yo*[7]) in the *Chou Li* (ch. 6, pp. 2*a* ff. (ch. 22); Biot (1), vol. 2, pp. 29 ff.).

---

[1] 律   [2] 彳   [3] 亍   [4] 歷   [5] 律歷志   [6] 巫   [7] 大司樂

suspended slabs of metal (predecessors of gongs) as the signal for retreat. But besides this, it seems that the pitch-pipes themselves were taken into battle, or at least to the field headquarters of the commander. Granet has drawn attention[a] to passages in the *Chou Li* (Record of the Rites of Chou) which deal with the duties of the Grand Annalist (*Thai Shih*[1]) and the Grand Instructor (*Ta Shih*[2]). It is said of the latter:[b] 'When the army is assembled (and marches forth), he takes the standard pitch-pipe tubes in order to determine the "note" of the army, and thus to announce its good or evil fortune.' And of the former:[c] 'When the army is assembled (and marches forth), he takes with him the Times of Heaven (the commentator says that this means that he takes care of the *shih*[3] or diviner's board, in order to ascertain the times of heaven—*Thai Shih pao shih i chih thien shih*[4]).[d] And he rides in the same chariot as the Grand Instructor (*Ta Shih*[2]).' The standard tubes and the diviner's board were thus important instruments which travelled in the same chariot under the care of two high officials. Were it not for the fact that the pipes must have been very difficult to blow, and that their flute-like notes could have carried only a short distance, it would be possible to believe that they formed a more elaborate code of signals than the drum and the gong. But they must rather have been used for divination, since the commentator quotes, after the first of the above passages, some sentences from a lost military work, the *Ping Shu*,[5] describing the blowing of the pitch-pipes as a method of divination at headquarters in order to learn what success the combat units were having and what heart they were in.[e]

A general connection is nevertheless obvious between musical notes on the one hand, and regulations for ritual dancing and military activity on the other. There was also the fancied connection between the pipe lengths and certain numbers which were involved in calendrical calculations. Alternative links between the two senses of *lü* may be sought in the relation of metrology to positive law (cf. Sect. 12) and in the use of bamboo tubes for making the handles of writing-brushes. In any case, there is nothing here which suggests that the Chinese ever thought of the semitone intervals of the standard pitch-pipes as originating from, or constituting, any kind of law in the non-human phenomenal world. The fact that what we now regard as a branch of physics stood at the origin of a word which took on the sense of human legal ordinances, has thus several probable explanations, and does not, in short, mean that ancient Chinese thinking here contained the elements of the conception of laws of Nature.

[a] (5), p. 209.
[b] Ch. 6, p. 14*a* (ch. 23); Biot (1), vol. 2, p. 51; tr. auct.
[c] Ch. 6, p. 42*b* (ch. 26); Biot (1), vol. 2, p. 108, tr. auct.
[d] This is an interesting mention of the ancestor of the magnetic compass (cf. Sect. 26*i*).
[e] Biot (1), vol. 2, p. 51. Chavannes (1), vol. 3, pp. 293 ff., considered that the first seven pages of Ssuma Chhien's treatise on the musical tubes in the *Shih Chi*, ch. 25, constituted a part of this lost military work. Cf. *I Ching*, 7th hexagram (K. Robinson, 2). For Indian parallels in the use of musical instruments for military divination see E. W. Hopkins (1), p. 199.

¹ 大史   ² 大師   ³ 式   ⁴ 大史抱式以知天時   ⁵ 兵書

(4) THE WORD *TU* (MEASURED DEGREES OF CELESTIAL MOTION)

If, at this stage, a reader should happen to glance at the accepted translation of the astronomical chapter of the *Shih Chi* (Historical Records), written about −95, he might well come upon the following passage:

As for me [Ssuma Chhien refers to himself], I have studied the memoirs of the annalists, and have examined the movements (of the heavenly bodies). During the past hundred years it has never happened that the five planets have made their appearances without (from time to time) moving backwards, and when they move backwards they are at the full and change their colours. And moreover there are definite times when the sun and moon are veiled or eclipsed, and when they move to the north or the south. These are *general laws*.[a]

In the light of the whole discussion of this Section, he will then turn to the Chinese text fairly certain that whatever Ssuma Chhien actually said, he did not speak of general laws in the sense of the scientific laws of Nature. Now the actual expression he used is *tu*[1] (*tzhu chhi ta tu yeh*[2]), and this word therefore demands notice.[b]

The primary meaning of *tu*[1] is 'degrees of measurement', and that this is overwhelmingly its commonest use appears not only from the lexicographers but also from the indexes or concordances which have been made for many of the most important ancient Chinese books. Its etymology, such as might be deduced from oracle-bone forms (K 801), does not throw any light on how it came to mean this. Nevertheless, its implication may be that of 'law', especially when it is found in combinations such as *chih tu*[3] or *fa tu*,[4] 'systematic rules and laws'. Couvreur (2) gives examples of these uses from the *I Ching* (Book of Changes) where the former combination occurs, and from the *Shu Ching* (Historical Classic) where *tu*[1] occurs alone in the sense that certain people had 'gone beyond the bounds' or 'transgressed'.[c] There is of course a close semantic connection between 'law' and 'measure', for every law has a certain quantitative aspect; 'how far' we say 'is it true that such-and-such an action comes under the scope of such-and-such a provision of the law', or 'measures must be taken, by means of bye-laws, to curb such-and-such a practice which is growing up'. But this quantitative aspect tends to remain metaphorical until legislators set out to make positive law, independent of morality, as, for instance (cf. p. 210), when a Chhin Shih Huang Ti begins to regulate the gauge of chariot-wheels. Still, there are to be found,

[a] Ch. 27, p. 43 a, tr. Chavannes (1), vol. 3, p. 409; eng. auct. Italics ours.

[b] Chavannes' translation was followed by Veith (1), p. 135, in her version of the *Huang Ti Su Wên Nei Ching*, ch. 8, p. 36b, where one of the characters in the dialogue speaks of *Thien tu*.[5] Moreover, the Han apocryphal treatise *Chhun Chhiu Wei Shuo Thi Tzhu*[6] (*Ku Wei Shu*, ch. 11, p. 5b) speaks of *hsing chhen chih tu*[7] which Bodde (in Fêng Yu-Lan (1), vol. 2, p. 124) translates as 'rules governing the stars and planets'.

[c] Ch. 34 (To Shih), one of the genuinely Chou chapters; see Legge (1), p. 198; Medhurst (1), p. 258; Karlgren (12), pp. 54, 56.

[1] 度   [2] 此其大度也   [3] 制度   [4] 法度   [5] 天度
[6] 春秋緯說題辭   [7] 星辰之度

among the writings of the philosophers of the Warring States and Han periods, numerous analogies between law in human societies, and the carpenter's square, the compasses and the plumb-line.[a]

More important is the fact, pointed out by Couvreur (2), that *tu*[1] may be considered a definite technical term for the movements of the heavenly bodies. The word was used throughout Chinese history for each of the $365\frac{1}{4}$ degrees into which the celestial sphere was divided, and for many other scales of divisions, such as the hundred parts of a day or night as shown by the clepsydra (water-clock). Revealing is the phrase used by Tung Chung-Shu in his *Chhun Chhiu Fan Lu*[b] of about the same time as Ssuma Chhien, where he says '*Thien Tao yu tu*':[2] the Tao of Heaven has its regular measured movements.[c] The general conclusion to which we must come is that on the strictest standards of the philosophy of science Chavannes was not justified in translating the word *tu*,[1] standing alone, as 'general laws'. It would have been preferable to say: 'These phenomena all have their regular measured (or measurable) recurrent movements.'

One wishes that it were possible to ask of Ssuma Chhien the question, 'In using the word *tu*,[1] measured degrees, did you mean it to have the undertone of "law"? If so, whose law?' I believe it is exceedingly unlikely that he would reply, 'The laws of Shang Ti' (the Ruler Above); and almost certain that he would say it was '*tzu-jan tu*',[3] natural measured movement, or '*Thien Tao tu*',[4] the movements of the (impersonal) Tao of Heaven. He might even complain, indeed, that we were taking him too seriously, for the phrase *ta-tu* in his last sentence could also mean 'This is, broadly speaking, the long and the short of it'.[d]

(5) THE EXPRESSION *CHI-KANG* (NET, OR NEXUS, OF NATURAL CAUSATION)

Still keeping within the realm of ancient Chinese astronomical thought, there is to be found, in an obscure fragment of early date, a discussion very much to our purpose. This is the so-called *Chi Ni Tzu*,[5] contained in the famous collection of fragments made by Ma Kuo-Han.[e] We do not even know whether Chi Ni Tzu was a real person, or simply a character invented by whoever it was who wrote the *Chi Jan*[6] chapters or book attributed to Fan Li.[7] Fan Li himself was a historical person, a statesman of the southern State of Yüeh in the −5th century,[f] but from the internal evidence,

---

[a] See *Shih Chi*, ch. 23, p. 1a; Chavannes (1), vol. 3, p. 202. Cf. pp. 108, 209, 211 above, and Vol. 1, p. 164.
[b] Ch. 45, opening sentence. Cf. ch. 12, cit. Fêng Yu-Lan (1), vol. 2, p. 521. So also *Ho Kuan Tzu*, ch. 12, p. 1a.
[c] Cf. Waley (12), p. 21. The word *tu* has very generally an astronomical context, e.g. in the Chhan-Wei books (cf. *YHSF*, ch. 53, p. 47a) or *Chin Shu*, ch. 11, p. 9b.
[d] Cf. his use of the phrase in *Shih Chi*, ch. 8, p. 2a.   [e] *YHSF*, ch. 69, pp. 19a ff.
[f] We shall meet with him in other connections, cf. Sects. 41, 42.

[1] 度   [2] 天道有度   [3] 自然度   [4] 天道度   [5] 計倪子
[6] 計然   [7] 范蠡

the discussions which Chi Ni Tzu, or Chi Yen,[1] carried on with Kou Chien,[2] King of Yüeh,[3] can hardly have been written before the time of Tsou Yen (late – 4th century).[a] There seems indeed no reason why part at least of these chapters should not be a Han fabrication, but it has to be admitted that they contain rather archaic material, such as the names of the gods of the five elements, and in view of their origin they may perhaps be placed in the late – 4th or early – 3rd centuries, and considered to embody a southern tradition of naturalism. In any case their exact date and provenance does not affect the present argument.

In the Nei Ching[4] chapter we find the following:

The King of Yüeh said, 'Since you discuss human affairs so brilliantly, perhaps you can tell me whether natural phenomena (wu[5]) have diabolical or auspicious meanings (in relation to man)?'

Chi Ni answered, 'There are the Yin and the Yang. *All things have their chi-kang*[6] (i.e. their fixed positions and motions with regard to other things in the web of Nature's relationships). The sun, moon, and stars signify punishment or virtue, and their changes indicate fortune and misfortune. Metal, wood, water, fire and earth conquer each other successively; the moon waxes and wanes alternately. *Yet these normal (changes) have no ruler or governor (mo chu chhi chhang*[7]). If you follow it (Heaven's Way) virtue will be attained; if you violate it there will be misfortune....[b] All affairs must be managed following the course of Heaven and Earth and the Four Seasons with reference to the Yin and Yang. If these principles are not carefully used, State affairs will get into trouble. Man when born does not know the day of his death. If you want to change the normality of Heaven and Earth you will simply unleash mischief, fall into poverty and shorten your life. Thus the sage rejects bribes and obtains a (good) response, but the mass of foolish men strive after wealth and honour (at all costs), not knowing what direction they should take.'

The King said, 'Excellent'.[c]

Here it would have been easy for the unwary to translate *chi-kang*[6] as laws of Nature.[d] Forke (13) used the words 'bestimmte Wandlungen an feste Regeln gebunden', fixed changes, governed by definite rules. And the lexicographers admit the meaning of (human) laws for this expression,[e] while its later use in the specific juristic sense of natural law is not uncommon.

It is obvious that we have to deal here with an analogy from textiles; both words have the silk radical (Rad. 120). *Chi*[8] combines 'silk' with 'self', it comes from an uncertain bone graph (K 953 *i*) and means 'to disentangle silk threads one from the other, to put in order, to regulate, rule, law, norm, regular series, cycle of years,

---

[a] Ssuma Chhien knows nothing of Chi Ni Tzu, but the *Wu Yüeh Chhun Chhiu* does (Forke (13), p. 500). Tradition would make him the teacher of Fan Li. But what he is made to say here about the Yin and Yang and the five elements can hardly be earlier than Tsou Yen's time (see pp. 232, 238). The chapters of the *Shih Chi* which speak of Kou Chien and Fan Li (31 and 41) have been translated by Pfizmaier (13, 19).

[b] One sentence omitted as corrupt and incomprehensible.   [c] Ch. 1, p. 4*b*, tr. auct.

[d] Cf. the use of similar words in *Chuang Tzu*, ch. 14, p. 1*a* (Legge (5), vol. 1, p. 345): *kang-wei*.[9]

[e] Couvreur (2) cites *Shih Ching* and *Shih Chi* for this.

[1] 計研   [2] 句踐   [3] 越   [4] 內經   [5] 物   [6] 紀綱

[7] 莫主其常   [8] 紀   [9] 綱維

conjunction of the sun and moon, inscribed annals'. We know that the cycle of years in question is the Jupiter cycle, and significantly Chi Ni Tzu speaks about this, giving it as twelve years, elsewhere in the fragment. *Kang*[1] combines 'silk' with 'net', and the ancient graph shows for the phonetic a net and a man (K697*a, c, e*; cf. also K744*b*). From its original meaning, the cord forming the selvedge of a net, it came to mean 'rule, regulate, dispose, put in order, direct', especially when used[a] with *chi*.[2] The analogous word *wang*[3] (K742 *l, a'*), though restricted more closely to the meaning of 'net', came to imply punishments, and hence law, perhaps because of its analogical use in chapter 73 of the *Tao Tê Ching*.[b] Then *ching*,[4] the warp, is used occasionally for the consistent principles on which Heaven gave life to Man.[c] On the basis of these undertones the translation of the expression *chi-kang*[5] in the above quotation is adopted.[d]

It is striking that a number of the interpretations of the words in question imply an active verb, to disentangle, to set in order, to rule, to make (?) laws. But Chi Ni Tzu very kindly relieves us from the anxiety as to whether the idea of a 'disentangler' or lawgiver was at the back of his mind, by saying in an immediately following sentence that these normal motions in the universe have *no* Master or Governor. This express denial seems to exclude the idea of a Setter in Motion. It is the first time that we have met with it in this Section, but it will not be the last.

Moreover, the conception of a net is close to that of a vast pattern. There is a web of relationships throughout the universe, the nodes of which are things and events. Nobody wove it, but if you interfere with its texture, you do so at your peril. In the following pages we shall be able to trace the later developments of this Web woven by no weaver, this Universal Pattern, until we reach, with the Chinese, something approaching a developed philosophy of organism.[e]

(6) THE WORD *HSIEN* (CONSTITUTION)

Another place where the expression *chi-kang* occurs is the short astronomical tractate *Ling Hsien*[6] of Chang Hêng[7] (+78 to +139).[f] This would be four or five centuries later than the probable date of Chi Ni Tzu, and now the phrase seems to imply the

---

[a] Couvreur (2) cites *Shih Ching* (Ta Ya).

[b] 'Heaven's net is wide, it lets nothing slip through.' In later Chinese writings, the 'net' becomes a common metaphor for the law, cf. *Kung Chhi Shih Hua*, ch. 3, p. 1 *a*.

[c] *Chhun Chhiu Fan Lu*, ch. 10, cit. and discussed by Fêng Yu-Lan (1), vol. 2, p. 517.

[d] Perhaps it had some connection with the use of the *quipu* (cf. pp. 100, 327), as my friend Professor Chhen Shih-Hsiang suggests.

[e] It is quite natural to find the expression *chi-kang* used by the Neo-Confucian philosophers in the Sung; by then it seems to have acquired a relevance mainly to social patterns (cf. *Chu Tzu Hsüeh Ti*[8] (The Aims of Chu Hsi's Teaching), compiled by Chhiu Chün[9] about +1475, ch. 2, pp. 37*b* ff.).

[f] I call it 'short', but there is no telling what its original length was, since we only have a five-page fragment of it preserved in the collections of Ma Kuo-Han (*YHSF*, ch. 76) and Yen Kho-Chün (*CSHK*, Hou Han section, ch. 55). A partial translation is given in Sect. 20*d* below.

[1] 綱   [2] 紀   [3] 網   [4] 經   [5] 紀綱   [6] 靈憲   [7] 張衡
[8] 朱子學的   [9] 邱濬

network of celestial (equatorial) declination-circles and hour-circles. It may also possibly refer, however, to the beginnings of the grid system in quantitative cartography, which, as we shall see in Section 22d on geography (in Vol. 3), seem to go back to Chang Hêng. His biography in the *Hou Han Shu* says[a] that he cast a network (of coordinates) over all Heaven and Earth and made calculations with it (*wang lo thien ti erh suan chih*[1]). This other book of his, long lost, bore the title *Suan Wang Lun*[2] (Book of the Mathematics of the (Coordinate) Network).

But even more interesting is his use of the word *hsien* in the title of his *Ling Hsien*. This word in modern usage means 'constitution' (political and legal, of a State), and is derived from an oracle-bone graph (K 250) of uncertain significance. Its ancient meaning was 'law' or 'model'. If, then, we translate the title of the book as 'The Spiritual (or Mysterious) Constitution (of the Universe)', we may well wonder to what extent the idea of laws of Nature lay behind Chang Hêng's use of it. Who had laid down this constitution? No answer to this problem is contained in the text of the fragment itself, which opens by speaking of tracing (*pu thien lu*[3]) the mysterious tracks (*ling kuei*[4]) of the heavenly bodies in their normal motions (*thien chhang*[5]) around the polar axis (*shu*[6]). Instruments (*i*[7]) graduated in degrees (*li tu*[8]) will measure them. Chang Hêng then describes a cosmogony in several stages,[b] gives some celestial measurements, and says that both the motions of the heavens as well as the unusual occurrences, such as eclipses and comets, portend good or evil fortune (to the State).

We can only conclude that what Chang Hêng had in mind was an 'organisation' of the heavens corresponding to the organisation of the imperial government, with its ranks of officials and its administrative rules on earth; and perhaps the word *hsien* was here used in the sense of model. It has already been pointed out (pp. 546, 550 above) that there are other statements in which it is said that the rules which the prince lays down on earth should be, in some sense, modelled on the regularities of the motions of the heavens. In any event, Chang Hêng's title constitutes another of those cases where ancient Chinese conceptions hovered on the brink of the idea of laws of Nature, without ever clearly formulating it.

(7) THE WORDS *LI* (PATTERN), AND *TSÊ* (RULES APPLICABLE TO PARTS OF WHOLES)

So far, then, we have not found in Chinese thought any clear evidence of the idea of law in the strict sense of the natural sciences. Still keeping to the schools which considered themselves Confucian, we must turn next to the Neo-Confucians of the Sung dynasty, already described in Section 16d. There we saw that Chu Hsi and the

[a] Ch. 89, p. 2a, commentary.
[b] Quoting the *Tao Tê Ching*, and touching on the centrifugal theory with mention of *chhing*[9] and *cho*[10] (light and heavy) elements.

[1] 網絡天地而算之　　　[2] 算罔論　　　[3] 步天路　　　[4] 靈軌　　　[5] 天常
[6] 樞　　　[7] 儀　　　[8] 立度　　　[9] 清　　　[10] 濁

other thinkers of his group made a great effort to bring all Nature and Man into one philosophical system, and we noted that the principal concepts with which they worked were *Li*[1] and *Chhi*.[2] The second corresponded approximately to matter, or rather to matter and energy, and the first was not far removed from the Taoist conception of the *Tao*[3] as the Order of Nature (cf. Section 10*b*), though the Neo-Confucians also used the term *tao*[3] in a slightly different and technical sense (cf. p. 484). Li could best be described as the ordering and organising principle in the cosmos. It has been equated, as we saw, with 'Reason' and with Aristotelian 'Form', while many have adopted the translation 'Law', but (in my judgment) such renderings are based on deep misconceptions, and in view of the great confusion which they are liable to cause, they should be abandoned.

The word Li (K 978), in its most ancient meaning, signified the pattern in things, the markings in jade or the fibres in muscle; as a verb it meant to cut things according to their natural grain or divisions. Thence it acquired the common dictionary meaning, 'principle'. It undoubtedly always conserved the undertone of 'pattern', and Chu Hsi himself confirms this, saying:

Li is like a piece of thread with its strands, or like this bamboo basket. Pointing to its rows of bamboo strips, the philosopher said, One strip goes this way; and pointing to another strip; Another strip goes that way. It is also like the grain in the bamboo—on the straight it is of one kind, and on the transverse it is of another kind. So also the mind possesses numerous principles (*li*).[a]

Li, then, is rather the order and pattern in Nature, not formulated law. But it is not pattern thought of as something dead, like a mosaic; it is dynamic pattern as embodied in all living things, and in human relationships and in the highest human values. Such dynamic pattern can only be expressed by the term 'organism', and as has already been suggested in Section 16*f*, Neo-Confucian philosophy was in fact a scheme of thought striving to be a philosophy of organism.

We must nevertheless carefully examine the grounds on which Bruce identified Li with Universal Law. It will introduce us to another word which we have not yet met in this connection, neither *fa*[4] nor *li*,[5] neither *Li*[1] nor *lü*.[6] In the opening paragraph of the 42nd chapter of the *Chu Tzu Chhüan Shu*, there is the following dialogue:

Question. In distinguishing between the four terms Heaven (*thien*[7]), Fate (*ming*[8]), the Nature (*hsing*[9]) and *Li*;[1] would it be correct to speak as follows? In the term Heaven, the reference is to spontaneous naturalness (*tzu-jan*[10]). In the term Fate, the reference is to its flowing through and pervading the universe, and being present in all things. In the term Nature, the reference is to that complete provision which any specific thing must have before

[a] *CTCS*, ch. 46, p. 12*b*, tr. Bruce (1), p. 290.

[1] 理   [2] 氣   [3] 道   [4] 法   [5] 禮   [6] 律   [7] 天
[8] 命   [9] 性   [10] 自然

it can come into being. In the term Li, the reference is to the fact that *every event and thing has each its own rule of existence* (*shih shih wu wu ko yu chhi tsê*[1]). And taking them all together, may it not be said that Heaven (i.e. the natural universe as a whole) is Li, that Fate is in fact the Nature (i.e. the constitution of a thing or a man), and that the Nature is in fact also Li? Is this not correct?

Answer. You are right. But people say today that Heaven has no reference to the material heavens, whereas in my view this cannot be left out of account.

The philosopher continued: Li is Heaven's 'substance' (*thien chih thi*[2]), Fate is Li in operation (*Li chih yung*[3]), the Nature is what man receives, and sensitivity (*chhing*[4]) is the Nature in operation.[a]

The operative word here is evidently *tsê*,[5] which has been translated 'rules of existence'. There can be little doubt that Chu Hsi's interlocutor had in mind a famous passage in the *Shih Ching* (Book of Odes),[b] thus translated by Legge:[c]

> Heaven, in giving birth to the multitudes of the people,
> To every faculty and relationship annexed its law (*yu wu yu tsê*[6])
> The people possess this normal nature (*ping i*[7])
> And (consequently) love its normal virtue (*hao shih i tê*[8]).

This verse was quoted by Mencius,[d] and is referred to again, by Chu Hsi himself, at a later place in the same chapter 42,[e] where he gives his opinion, speaking of human desires, that likes and dislikes themselves are the 'things', i.e. Legge's 'faculties and relationships', while to like that which is good and to dislike that which is evil are the 'rules of existence' (Legge's 'laws', *tsê*[5]). In other words, though the psychological context introduces unnecessary complexity, we have to deal with neutral natural phenomena or properties on the one hand, and their regular tendency to behave in a certain specific manner on the other.[f]

K 906

There is no doubt that we are here once again in the no-man's-land between scientific law ('laws of Nature') and juridical law, indeed, natural law in the legal sense; or rather we are back again in those shadowy regions where the concepts are in a highly undifferentiated state. A discovery of much interest is reserved for us, therefore, when we take a look at the etymology of the word *tsê*[5] (K 906), for we find that the ancient writing of the character on bones and bronzes shows a cauldron and a knife—in other words, the very act of incising codes of laws on ritual cauldrons, as described in the two passages quoted for the −6th century at the beginning of this section.[g] The character's radical should have continued to be that for cauldron

[a] P. 1*a*, *b*, tr. auct. adjuv. Bruce (1), p. 3.
[b] III, iii, 6 (Chêng Min).
[c] (8), p. 541. Cf. Karlgren (14), p. 228; Waley (1), p. 141, who prefer freer interpretations.
[d] *Mêng Tzu*, VII (1), vi. 8 (Legge (3), p. 279).
[e] Ch. 42, p. 24*b*.
[f] Cf. Fêng Yu-Lan (1), vol. 2, pp. 466 and 501 on Shao Yung, and p. 503 on the Chhêng brothers. Every individual thing has its own pattern, i.e. its Li.
[g] See p. 522.

[1] 事事物物各有其則 [2] 天之體 [3] 理之用 [4] 情 [5] 則
[6] 有物有則 [7] 秉彝 [8] 好是懿德

(Rad. 206), but it was corrupted into that for cowry-shell (Rad. 154). It therefore becomes of importance to follow the fortunes of this word throughout the development of Chinese thinking. Everyone who reads Chinese at all is familiar with it in its common meaning as a consequential particle, 'so', 'then', 'in that case', but it has conserved to this day a variety of secondary usages connected with laws and regulations, e.g. the expressions *chhang tsê*,[1] unvarying laws; or *shui tsê*,[2] a Customs tariff. The use of it in ancient writings seems much commoner in legal-administrative connections than in any 'scientific' sense. Thus in the *Chung Yung* we have 'his words will be a rule for the empire (*yen erh shih wei thien hsia tsê*[3])'; in the *Li Chi* there are *nei tsê*,[4] i.e. domestic rules (*oeconomica*); the *Chou Li* has 'he governs the cantons and districts by means of the eight regulations (*i pa tsê chih tu pi*[5])'.[a] Nearer to what we are looking for is another passage in the *Shih Ching*, which speaks of 'obeying the laws of the Ruler (Above) (*shun Ti chih tsê*[6])', but this is again of human behaviour. The most significant text is that of the *I Ching*, which under the *Wên Yen*[7] explanation of the first kua *Chhien*,[8] has the words: 'When *chhien* and *yuan* appear in all nine, one can see the laws of Heaven (*nai chien Thien tsê*[9]).'[b] R. Wilhelm duly wrote[c] 'so erblickt man das Gesetz des Himmels', and passed on apparently unaware of the interest of the passage.

If this were all that could be adduced in support of the interpretation of Bruce (and it is a good deal more than the evidence he himself gave) the identification of Chu Hsi's *Li*[10] with the *tsê*[11] of the Odes would not be very convincing. But even if we must ultimately part company with Bruce and Henke, the great interest of the subject calls for investigation of any text which could throw light on the scientific use of the word *tsê*,[11] and there is indeed more to be found. In the works of the scholar-poet Chhü Yuan[12] (−332 to −295)[d] there is an astronomical poem, the *Thien Wên*[13] (Questions about Heaven) in which occur the words *huan tsê chiu chhung, shu ying tu chih*,[14] which *may* be translated, 'As to the circular rule of the Nine Storeys (or layers) of Heaven, who made the plan and measurement of it?' Unfortunately, it is not quite clear whether the word *tsê*[11] is here used in its consequential sense or as a noun.[e] The latter interpretation seems to have been adopted by the Thang poet Liu Tsung-Yuan[15] (+773 to +819),[f] who wrote an essay *Thien Tui*[16] (Answers about Heaven),[g] which was intended to answer one by one the somewhat rhetorical questions

---

[a] I cite these usages on the authority of Couvreur (2).
[b] *Ching*, pt. 1, p. 7*b*.
[c] (2), vol. 2, p. 12; Baynes tr., vol. 2, p. 16.   [d] *Chhu Tzhu Pu Chu*,[17] ch. 3, p. 2*a*.
[e] Forke (6, p. 136); Conrady & Erkes (1), Erkes (8) and Edkins (1) considered it as a consequential word, for they rendered the sentence, 'The vault of Heaven has the shape of Nine Storeys....'
[f] G 1361.
[g] Contained in *TSCC, Chhien hsiang tien*, ch. 11, *i wên* 2, p. 2.

---

¹ 常則   ² 稅則   ³ 言而世爲天下則   ⁴ 內則
⁵ 以八則治都鄙   ⁶ 順帝之則   ⁷ 文言   ⁸ 乾   ⁹ 乃見天則
¹⁰ 理   ¹¹ 則   ¹² 屈原   ¹³ 天問   ¹⁴ 圜則九重孰營度之
¹⁵ 柳宗元   ¹⁶ 天對   ¹⁷ 楚辭補註

of Chhü Yuan. So also a late commentator on the *Thien Tui* glossed *tsê fa yeh*:[1] *tsê* here means law or method. Perhaps a clearer case is that of a phrase in the biography of the great later Han astronomer Chang Hêng[2] (+78 to +139):[a] *Thien pu yu chhang tsê*,[3] 'The steps of heaven (i.e. the number of degrees passed through by planets and constellations in a given time, their risings and settings, etc.) follow unvarying rules.'[b] This is undoubtedly the kind of thought which led his contemporary Wang Chhung to give support to individual astrology (cf. pp. 356, 384), and Chang Hêng's remark (for it is in a recorded speech of his) occurs in an astrological context. On the other hand, it is extremely interesting that the understanding of *tsê*[4] in natural phenomena is sometimes, in other texts, despaired of. Thus the *Huai Nan Thu* book says:

The Tao of Heaven operates mysteriously and secretly (*Thien Tao hsüan mo*[5]); it has no form or shape (*wu yung*[6]); it is beyond all particular definite rules (*wu tsê*[7]); it is so great that you can never come to the end of it, it is so deep that you can never fathom it.[c]

And again eight centuries later by Liu Tsung-Yuan, in the passage just referred to, where he says: 'Heaven has no colour of any kind, no centre and no sides—how can you hope to find (lit. see) its *tsê*?'[4]

I have noted two other instances of the denial or doubt that Nature works according to *tsê*.[4] The first is from the memorial ode on Chhü Yuan written by Chia I,[8] and therefore dates from about −170.[d]

Heaven and Earth are like a smelting-furnace, the forces of natural change are the workmen (*tsao hua wei kung*[9]), the Yin and the Yang are the fuel, and the myriad things are the metal. Now it runs together (*ho*[10]), now it disperses (*san*[11]), sometimes moving and sometimes resting. But *there is no fixed law* (*an yu chhang tsê*[12]), and to the thousand changes and the myriad transformations there is no end (*wei shih yu chi*[13])[e]....

The second is from the commentary of Wang Pi[14] on the *I Ching* (Book of Changes), and must therefore date from the close neighbourhood of +240. Explaining the 20th kua, *Kuan*,[15] meaning view or vision, he says:[f]

The general meaning of the Tao of 'Kuan' is that one should not govern by means of punishments and legal pressure, but by looking forth one should exert one's influence (by example) so as to change all things. Spiritual rule is without form and invisible (*Shen tsê wu hsing chê yeh*[16]). *We do not see Heaven command the four seasons*, and yet they never swerve

[a] G 55.   [b] *Hou Han Shu*, ch. 89, p. 5*a*.
[c] Ch. 9, p. 1*b*, tr. auct.   [d] Repeated in *Shih Chi*, ch. 84, pp. 12*b*, 13*a*.
[e] Tr. Forke (12), eng. auct. mod. Cf. Edkins (1), p. 225. The last phrase is a quotation from *Chuang Tzu*, ch. 6, cf. Legge (5), vol. 1, p. 243; Fêng Yu-Lan (5), p. 116.
[f] *Shih-san Ching Chu Su* ed., ch. 4, p. 20*b*.

[1] 則法也   [2] 張衡   [3] 天步有常則   則   [5] 天道玄默
[6] 無容   [7] 無則   [8] 賈誼   [9] 造化爲工   [10] 合   [11] 散
[12] 安有常則   [13] 未始有極   [14] 王弼   [15] 觀   [16] 神則无形者也

from their course (*Pu chien Thien chih shih ssu shih, erh ssu shih pu thê*[1]). So also we do not see the sage ordering the people about, and yet they obey and spontaneously serve him.[a]

This is perhaps the most illuminating passage of all. We have a flat denial of the conception of orders issued to the four seasons (and hence the courses of the stars and planets) by some celestial lawgiver. The thought is extremely Chinese. Universal harmony comes about not by the celestial fiat of some King of Kings, but by the spontaneous cooperation of all beings in the universe brought about by their following the *internal* necessities of their own natures. Tsê is the internal rule of existence embodied in each individual thing, whereby it conforms to its position within the whole of which it is a part. One begins to see how deeply rooted in ancient Chinese ideas was the Neo-Confucian philosophy of organism. In Whitehead's idiom the 'atoms do not blindly run' as mechanical materialism supposed, nor are all entities specifically directed on their paths by divine intervention, as spiritualistic philosophies have supposed; but rather all entities at all levels behave in accordance with their position in the greater patterns (organisms) of which they are parts. The conception of internal necessity was stated in so many words by Chang Tsai[2] in his *Chêng Mêng*.[3] 'All rotating things', he said,[b] with reference to the heavens, 'have a spontaneous force (*chi*[4]) and thus their motion is not imposed upon them from outside (*tung fei tzu wai yeh*[5]).'[c] One can now realise how mistaken would be the view that *tsê*[6] meant anything like the laws of Nature in the Newtonian sense, and how dangerous it would be to assume that such an interpretation could properly explain the thought of the Neo-Confucians about *Li*.[7]

(8) NON-ACTION AND LAWS OF NATURE

The affirmation that Heaven does not *command* the processes of Nature to follow their regular courses is indeed linked with that root idea of Chinese thought, *wu wei*,[8] non-action, or unforced action. The legislation of a celestial lawgiver would be *wei*, a forcing of things to obedience, a firm imposition of sanctions. Nature shows a ceaselessness and regularity, yes, but it is not a *commanded* ceaselessness and regularity. The Tao of Heaven is a *chhang Tao*,[9] the cosmic Order of Nature is an unvarying order, as Hsün Tzu says,[d] but that is not the same thing as affirming that anyone ordered it to be so.[e]

[a] Tr. auct. Cf. Thang Yung-Thung (1).
[b] Tshan Liang chapter (ch. 4 in *Sung Ssu Tzu Chhao Shih*, p. 6a).
[c] Of course it is not impossible to find European formulations of a similar kind. In +1571 Peter Severinus wrote of an 'innata lex', in good mystical Paracelsian style. For this reference I am indebted to my friend Dr W. Pagel. Cf. the ideas of Giordano Bruno above, p. 540.
[d] *Hsün Tzu*, ch. 17, p. 1a (tr. Dubs (8), p. 173; Forke (13), p. 223). Cf. *TTC*, ch. 1.
[e] Reasons for leaving Tao untranslated and for understanding it as the Order of Nature have already been given (pp. 6, 36). The suggestion of Hu Shih (4), p. 64, that Tao could be translated as laws of Nature is absolutely inadmissible (cf. Forke (13), p. 271). O. Franke (6), in his otherwise so brilliant essay on the cosmic conceptions of the ancient Chinese, mixes all kinds of ideas together, for example, when he says: 'Wie die ratio (*logos*, λόγος) im Gesetz (*lex*) und im Recht (*jus*) ihre Form findet, so äussert das

¹ 不見天之使四時而四時不忒   ² 張載   ³ 正蒙   ⁴ 機
⁵ 動非自外也   ⁶ 則   ⁷ 理   ⁸ 無爲   ⁹ 常道   ¹⁰ 德

Thus in the *Li Chi* (Record of Rites)[a] there is an apocryphal conversation between Confucius and Duke Ai of Lu. The Duke asked what was the most valuable thing to note about the ways of Heaven:

The Master replied, 'The most important thing about it is its ceaselessness. The sun and moon follow each other round from east to west without ceasing; such is the Tao of Heaven. Time goes on without interruption; such is the Tao of Heaven. *Without any action being taken*, all things come to their completion; such is the Tao of Heaven (*Wu wei erh wu chhêng, shih Thien Tao yeh*[1]).[b]

Here again, then, is a denial, if an implicit one, of any heavenly creation or legislation. It should be noted, in passing, that although the concept of *wu wei*[2] was emphasised particularly by the Taoists, it was part of the common ground of all ancient Chinese systems of thought, including the Confucians.

(9) THE CHINESE DENIAL OF A CELESTIAL LAWGIVER AN AFFIRMATION OF NATURE'S SPONTANEITY AND FREEDOM

It may be worth while following out this digression a little further. It is not at all difficult to find passages which confirm the conception of Heaven acting according to *wu wei*;[2] it runs throughout the *Tao Tê Ching* (e.g. chapter 37), where we find the significant statement (in chapter 34) that though the Tao produces, feeds and clothes the myriad things, it does not lord it over them (*erh pu wei chu*[3]), and asks nothing of them. The idea is, in fact, a Taoist commonplace, and appears in such books as the *Wên Tzu*[4][c] and many later writings. The *Lü Shih Chhun Chhiu* affords us a little further insight into the working methods of the Tao of Heaven.

In Chapter 94 we read:

The operations of Heaven are profoundly mysterious (*Thien chih yung mi*[5]). It has water-levels for levelling, but it does not use them; it has plumb-lines for setting things upright, but it does not employ them (*Yu chun pu i phing; yu shêng pu i chêng*[6]).[d] It works in deep stillness....

Thus it is said, Heaven has no form and yet the myriad things are brought to perfection. It is like the most impalpable of featureless essences, and yet the myriad changes are all brought about by it. (So also the sage is busied about nothing, and yet the thousand executives of State are effective in the highest degree.)[e]

Tao im *tê*[10] oder im *li* [which *li* is not clear] seine universalistische Wirkung, und die "gesamte Welt" umfasst auch in China den himmlischen Staat so gut wie den irdischen.' Cf. Hegel (4), vol. 1, p. 141.

[a] Ch. 24 (Legge (7), vol. 2, p. 269).

[b] Tr. Forke (13), p. 173; eng. auct. Italics ours. Cf. pp. 68 ff. above.

[c] The *Wên Tzu* (Book of Master Wên) is considered a work of Han date or even later, but may contain a considerable amount of pre-Chhin material. The passage referred to above appears in ch. 1, p. 11*b* (cf. Forke (13), p. 338).

[d] Cf. *Pao Phu Tzu*, ch. 1 (Feifel (1), p. 118).

[e] The brackets here indicate, not that the sentence is not in the text, but that it interrupts the sequence of thought which we are trying to follow.

¹ 無爲而物成是天道也　　²無爲　　³而不爲主　　⁴文子
⁵ 天之用密　　⁶有準不以平有繩不以正

This may be called the *untaught teaching*, and the *wordless edict* (*Tzhu nai wei pu chiao chih chiao, wu yen chih chao*[1]).[a]

And this is echoed by the words of Chhêng Hao: 'The laws of Heaven are wordless but they keep faith; divine law has majesty untinged with wrath (*Thien tsê pu yen erh hsin, shen tsê pu nu erh wei*[2]).'[b]

Such a conception is undeniably sublime.[c] But how profoundly incompatible it is with the conception of a celestial lawgiver. The movements of the celestial bodies proceeded, in the one case, according to teachings which no one had ever taught, and according to edicts which no one had ever issued, or even put into words. But the laws of Nature which Kepler, Descartes, Boyle and Newton believed that they were revealing to the human mind (the very word 'revealing' is symptomatic of the spontaneous background of occidental thought), were edicts which *had* been issued by a supra-personal supra-rational being. The fact that this was later generally recognised to be a metaphor does not mean that it may not have had great heuristic value at the beginning of modern science in Europe.

Not only were there no divine edicts in the great tradition of Chinese thought, but no divine creator who could have issued them. There is a striking passage in the +3rd-century commentary of Hsiang Hsiu and Kuo Hsiang on the *Chuang Tzu*, dealing with the famous conversation between Penumbra and Shadow[d] (given on p. 51 above).

Some people say that the penumbra is dependent upon the shadow, the shadow upon the bodily form, and the bodily form upon an Originator of things (*tsao wu chê*[3]).[e] But we venture to ask whether this Originator is or is not? If he is not, how can he have originated things (which are)? If he is, then (being one of these things), he could not have originated the universe of bodily forms. Hence only after we have realised that all the bodily forms are things of themselves (*tzu wu*[4]), can we begin to talk about the origination of things. Within the realm of things, there is nothing within the Mystery, even the penumbra, which is not 'self-transformed' (*tu hua*[5]).[f] Hence the origination of things has no lord (*wu chu*[6]); all things originate themselves (*wu ko tzu tsao*[7]). Everything produces itself, and does not depend on anything else. This is the very normality of the universe.[g]

[a] Ch. 94 (vol. 2, p. 43), cf. Forke (13), p. 541.
[b] *Sung Ssu Tzu Chhao Shih, Erh Chhêng* sect., ch. 3, p. 1*a*.
[c] Cf. *Lun Yü*, XVII, xix: 'Who ever heard of Heaven speaking? The four seasons come round, innumerable beings are born and grow; does Heaven ever say anything?'
[d] Ch. 2 (Legge (5), vol. 1, p. 197).
[e] The words mean literally 'founder of things'. Bodde boldly translates 'Creator', but we hesitate to do so, because it is highly doubtful whether in ancient China there existed the full conception of creation *ex nihilo*. The writers evidently did not consider the idea of a transcendent personal creator; if they had, they would certainly have found many reasons for rejecting it. Graf (2), vol. 1, p. 86, agrees with us on this.
[f] Perhaps the stimulus for the formation of this technical term came from Buddhism, for the *pratyeka-buddhas* of Hīnayāna who sought only for their own salvation were called *tu-chio*.[8] Theirs was an autogenous, if not automatic, enlightenment.
[g] Ch. 2 (*Pu Chêng* ed., ch. 1B, p. 36*b*), tr. Bodde, in Fêng Yu-Lan (1), vol. 2, p. 210; mod.

[1] 此乃謂不教之教無言之詔   [2] 天則不言而信神則不怒而威   [3] 造物者
[4] 自物   [5] 獨化   [6] 無主   [7] 物各自造   [8] 獨覺

(10) *Li* AND *Tsê* IN NEO-CONFUCIANISM; THE PHILOSOPHY OF ORGANIC LEVELS

It would be of much value for the history of Chinese scientific thought to concentrate investigation on other occurrences of the expression *Thien tsê*.[1] But so far as our observations have gone, it is not a common one.[a] The word *tsê* seems to represent a borderline conception. Legal it certainly always was,[b] and human too, but though occasionally applied in a scientific or proto-scientific sense, such a use did not seem to 'catch on'. Here the best support is Chang Hêng.[c] The *I Ching* passage [d] is not very decisive, since in a book on divination, where the formation and transmutation of the hexagrams was supposed to mirror the changing processes of the real world, a strong poetical and symbolic element would be natural, and the implicit relevance to human affairs is brought out by Chu Hsi's commentary, which says: 'that the hard should be able to act yieldingly, is Heaven's law (*kang erh nêng jou, Thien chih fa yeh*[2])'. The human relevance is also obvious in the *Shih Ching* verse.[e] Chu Hsi must have pondered much on these classical texts. The extent to which his idea of *Li*[3] involved the conception of laws of Nature can hardly be assessed until more is known as to the consensus of emphasis of the passages which he is likely to have had in mind. Nevertheless, there is one feature of the crucial dialogue given at the beginning of this discussion, which suggests to me that laws of Nature in the sense of scientific generalisations were not meant, namely, the sentence in italics—'every event and thing has each *its own* rule of existence' (p. 559). It is not said that every event and thing obeys general laws or rules valid for many other similar events and things. The thought is therefore much more applicable to individual events and things as organisms.[f] There is no absolute contradiction, but a difference of emphasis, and this agrees with the very distinct statement of Wang Pi (p. 561).

A further insight into what the Neo-Confucians meant by Li and *tsê*[4] may be obtained by looking into the *Pei-Chhi Tzu I*[5] (Philosophical Glossary of Neo-Confucian Technical Terms), written by Chhen Shun,[6] an immediate pupil of Chu Hsi, about the time of the latter's death (+1200). He analyses, in a beautifully clear passage,[g] the meaning of Li:

(1) *Tao*[7] and *Li*[3] are roughly the same, but two words are used, and a distinction between them can be made. The difference is that Tao is what prevails (at the) human (level). In

---

[a] There seems to be nothing like it in places where one would expect to find it, such as the *Lü Shih Chhun Chhiu*, the *Lun Hêng*, or the ancient scientific fragments collected by Ma Kuo-Han. We have also failed to find it in the astronomical chapters of the dynastic histories.

[b] Cf. the numerous instances of its employment thus in the *Tso Chuan*.

[c] P. 561 above. [d] P. 560 above. [e] P. 559 above.

[f] Cf. *Chung Yung*, XIII, 2, where the *tsê* of an axe-handle has generally been translated 'pattern' (Legge (2), p. 257; Hughes (2), p. 111).

[g] Ch. 2, p. 5*b*. The numbering of the stages of the argument is ours.

[1] 天則　[2] 剛而能柔天之法也　[3] 理　[4] 則　[5] 北溪字義
[6] 陳淳　[7] 道

comparison with Li, Tao is broader, and Li more profound. Li has the definite (*chhio jan*[1]) meaning of unchangeableness, so although the Tao has run through all the centuries (as a principle of variable human organisation), the Li in all this time has never changed.

(2) Li is formless; how could it be seen? Li (Pattern or Organisation) *is a natural and unescapable law* (*i ko tang jan chih tsê*[2]) of affairs and things.

It is a Patterning Law (*li tsê*[3]).

It is a Standardising Law (*chun tsê*[4]).

It is a Modelling Law (*fa tsê*[5]).

It conveys the idea of certainty and fixity (*chhio ting*[6]) and unchangeableness (*pu i*[7]).

The meaning of 'natural and unescapable' (*tang jan*[2]) is that (human) affairs, and (natural) things, *are made just exactly to fit into place* (*chêng tang ho tso chhu*[8]).

The meaning of 'law' (*tsê*[5]) is that the fitting into place (*chhia hao*[9]) *occurs without the slightest excess* (*wu kuo*[10]) *or deficiency* (*wu pu chi*[11]).[a]

(3) For instance, stopping[b] at benevolence is the natural unescapableness of the ruler. Stopping at respect is the natural unescapableness of the minister. Stopping at paternal love is the natural unescapableness of a father. Stopping at filial piety is the natural unescapableness of a son.

Or in the case of the foot supporting the weight of the body, this supporting is the natural unescapableness of the foot. Or in the case of the hand, its ability to make polite motions of greeting is its natural unescapableness.

Again, it is like the Impersonator of the Dead[c] who simply sits in the middle of the ceremony; this is the natural unescapableness of one who sits. And on the other hand it is like the Sacrificer, who stands during the ceremony; this is the natural unescapableness of one who stands.

(4) The men of old, investigating things to the utmost, and searching out Li, wanted to elucidate the natural unescapableness of (human) affairs and (natural) things; and this simply means that what they were looking for was all the exact places where things precisely fit together. Just that.

(5) If we compare *Li*[12] with *Hsing*[13] (human nature), Li is the Li which permeates (non-human) things, while Hsing is the Li which permeates human selfs.

The Li which permeates (non-human) things is that (universal) Tao Li which is common to heaven and earth and to all human beings and to all things.

But that which permeates human selfs is that which has already the quality of specific individuation.

(6) If we compare *Li*[12] with *I*[14] (righteousness), Li is what (organises) the substance (*thi*[15]), while I is the same thing in function (or operation) (*yung*[16]).

---

[a] We have already met with this important conception in several earlier Sections; cf. pp. 270, 286, 463, 489. See also *Sung Yuan Hsüeh An*, ch. 90, pp. 2b, 3a; *Hsing Li Ching I*, ch. 9, p. 29b.

[b] The reference is to the famous ancient phrase in the *Ta Hsüeh*, I, I, 'stopping at the highest excellence (*chih yü chih shan*[17])'—and not going on beyond it by sophistical arguments.

[c] In the ancient ceremonies for the dead, a living person thus acted. Cf. *Shih Ching*, Legge (1), pp. 300, 365ff.; *Li Chi*, Legge (7), vol. 1, pp. 62, 69, vol. 2, pp. 152, 240ff.; *Mêng Tzu*, VI (1), v, 4.

[1] 確然	[2] 一箇當然之則	[3] 理則	[4] 準則	[5] 法則
[6] 確定	[7] 不易	[8] 正當合做處	[9] 恰好	[10] 無過
[11] 無不及	[12] 理	[13] 性	[14] 義	[15] 體	[16] 用	[17] 止於至善

Li permeating things is the natural unescapableness of them; I is how to handle this Li (or direct or administer it).

Thus Chhêng Tzu said, 'In things it is Li; in handling things it is I.'[a]

It would hardly be possible to have more striking confirmation of the interpretation of Li adopted earlier (in the Section on Neo-Confucianism, p. 475) as the 'principle of organisation' in the universe. There is 'law' implicit in it, but this law is the law to which parts of wholes have to conform by virtue of their very existence as parts of wholes. And this is true whether they are material parts of material wholes, or non-material parts of non-material wholes. The most important thing about parts is that they have to fit precisely into place with the other parts in the whole organism which they compose, without, as Chhen Shun says, the slightest excess or deficiency. There is nothing here about the fiat of any Controller. Such laws as these were not the statutes of a celestial lawgiver analogous to an earthly prince, but arose, in the thought of the Neo-Confucians, directly out of the nature of the universe. Nor is there anything here which could remind us of fortuitous concourses of atoms, obeying only the statistical laws of their own chaos, and in no way affected by the patterns which they generate in the chance succession of Nature's kaleidoscopic figures.

In the first paragraph the Tao seems to be considered as something which, though possessing an inner consistency, has allowed, through the ages, a certain amount of 'play'. But the cosmic organisation has been whole and unchanging. It is, says the second paragraph, in effect, a Great Pattern in which all lesser patterns are included, and the 'laws' which are involved in it are *intrinsic* to these patterns, whatever their degree of complexity, not extrinsic to them, and dominating them, as the laws of human society constrain individual men. The laws of the Neo-Confucian organic philosophy would thus be internal to the individual organisms at all levels, just as in later occidental philosophy it was felt that the laws of an ideal State should be written, not on tables, but in the hearts of its citizens. And thus leading from within (a profoundly Taoist contribution to this thought) the *tsê*[1] would generate the pattern, standardise its manifestation, and model its form.

In the third paragraph a number of instances of natural unescapableness are given. In the case of the ruler, for instance, the meaning is that he *has* to act like that by the necessity of things, in order to succeed in being himself. If he did not, he would turn into something else, and inevitably come to grief—we are reminded of one of the early Christian definitions of the Devil, ὁ ἀντιτάττων τοῖς κοσμικοῖς (Hippolytus)—'he who resists the cosmic process'.[b] And so for the other instances given. Among them it is important to remark one relating to a purely biological organism, namely, the foot as part of the body.

In the fourth paragraph it is suggested that all those who have throughout the ages sought for meaning in the universe have really been looking for the Great Pattern.

[a] Tr. auct. Cf. *ECCS, I-Chhuan I Chuan*, ch. 4, p. 20*b*.　　　　[b] Cf. p. 283.

[1] 則

Finally, the Principle of Organisation is traced in its specific individuation at human level in human nature, and in its active manifestation there in the world of human relations and the administration of things.

We conclude, therefore, that 'law' was understood in a Whiteheadian organismic sense by the Neo-Confucian School. One could almost say that 'law' in the Newtonian sense was completely absent from the minds of Chu Hsi and the Neo-Confucians in their definition of *Li*;[1] in any case it played a very minor part, for the main component was 'pattern', including pattern living and dynamic to the highest extent, and therefore 'organism'. In this philosophy of organism all things in the universe were included; Heaven, Earth and Man have the same Li.[a]

What exactly this phrase means is important for our argument. The following passage[b] expounds it:

Someone asked the question: In the relation of parents and offspring in tigers and wolves, of sovereign and minister in bees and ants, in the gratitude to their creation of jackals and otters,[c] and in the faculty of discrimination of water-fowl and doves;[d] though the ethical principle (*i li*[2]) is present in one way only, yet if we thoroughly investigate the phenomena, we find that these creatures unerringly possess these ethical principles. On the other hand, all men possess humaneness (the Decree of Heaven) (*Thien ming*[3]) in its entirety, but it is so obscured by creaturely desire, and by the material endowment, that sometimes they are not as well able as these animals to attain to their complete and perfect development. How do you explain this?

(The philosopher) answered: It is only in these specific directions that these animals are intelligent, and there it is concentrated. But man's intelligence is comprehensive, embracing everything in some degree, but diffused, and therefore more easily obscured.

Someone else asked the question: Can dried and withered things [as we should say— inorganic things] also possess a natural endowment? (*hsing*[4]).

(The philosopher) answered: They also possess Li from the first moment of their existence; therefore it is said, 'There is nothing under Heaven which does not have its own natural endowment.'[e]

Walking up some steps, the philosopher said: The bricks of these steps have the Li of bricks. Sitting down, he said: A bamboo chair has the Li of a bamboo chair. You may say (he went on) that dried and withered things (*khu kao chih wu*[5]) are without the vital impulse

[a] *Chu Tzu Chhüan Shu*, ch. 46, p. 7a (Bruce (1), p. 280); cf. the passage from Wang Chhung cited on p. 368.

[b] *Chu Tzu Chhüan Shu*, ch. 42, pp. 29a ff. For an occidental parallel to its opening theme cf. the fragment of Plutarch, *De Amore Prolis* (Lovejoy & Boas (1), p. 404).

[c] These animals were supposed to spread out their prey as if sacrificing to the gods before eating it. In fact, otters are accustomed to consume only a part of their prey, and leave the rest on the river-bank; the false interpretation of this had begun as early as the time when the *Li Chi* was written (see Legge (7), vol. 1, pp. 221, 251). It is mentioned in other Sung philosophers, notably Chhêng I (see p. 457, and Forke (9), p. 97). See also p. 488 above on the theory of the 'gleam'.

[d] These animals were noted to be of monogamous habit. In general, the part played in the history of thought by observations of unusual animal behaviour, has hardly received the attention which it deserves. Recently Gudger (1–4) and Burton (1) have discussed numerous so-called legends about the behaviour of wild animals, especially birds and mammals. Cf. Marshall (1), J. B. S. Haldane (1) and Friedmann & Weber.

[e] Close parallels to this, worth noting, are in *CTYL*, ch. 1, p. 30a.

[1] 理   [2] 義理   [3] 天命   [4] 性   [5] 枯槁之物

(*sêng i*[1]), but not that they are without the *Li*[2] of specific existence. For example, rotten wood is useless for anything except for putting in the cooking-stove. It is without the vital impulse. And yet each kind of wood as it burns has its own fragrance, each differing from the other. It is (its) Li which originally constituted it so.

It was further asked: Is there Li, then, in dried and withered things?

(The philosopher) answered: As soon as the object exists, (a) Li is inherent in it. Even in the case of a pen—though not produced by Heaven (directly), but by Man, who takes the long soft hairs of the hare and makes them into brush-pens—as soon as it exists Li is inherent in it.

It was further asked: How can a pen possess Love and Righteousness?

(The philosopher) answered: In small things like these there is no need for such distinctions as that between Love and Righteousness.[a]

Someone said: Birds and beasts, as well as men, all have perception and vitality (*chih-chio*[3]), though with different degrees of penetration. Is there perception and vitality also in the vegetable kingdom?

(The philosopher) answered: There is. Take the case of a plant; when watered, its flowers shed forth glory; when pinched, it withers and droops. Can it be said to be without perception and vitality? Chou Tun-I refrained from clearing away the grasses from in front of his window, because, he said, 'their vital impulse is just like my own'.[b] In this he attributed perception and vitality to plants. But the vitality of the animals is not on the same plane as man's vitality, nor is that of plants on the same level as that of animals. Take, for example, the case of the drug rhubarb (*ta huang*[4]),[c] when swallowed, it acts as a purgative; while aconite (*fu tzu*[5])[d] has heating properties—their vitality (specific natural endowment) can (each) follow only one road.

It was asked whether decayed (vegetable) material (*fu pai chih wu*[6]) also had (such a specific natural endowment)?

(The philosopher) answered: Yes, indeed it has. If it is burnt to ashes by fire and then heated with water, the liquid will be bitter and caustic. Then he smiled and said: Only today I met the gentlemen of Hsinchow, who maintained that vegetable things have no natural endowment, and now tonight you are suggesting that vegetable things have no *hsin*[7] (lit. mind, i.e. specific nature).[e]

This passage is interesting in many respects.[f] We see Chu Hsi, just as Chuang Chou, 1400 years before him,[g] maintaining that the Tao (here the Li) runs through all things in the universe, that the universe is orderly, and, in a sense, rational. But not therefore intelligible in the scientific as opposed to the philosophical sense, and not necessarily following rules capable of being formulated in a precise and abstract way by man. Chu Hsi expresses, however, what amounts to a conception of ethics in terms of levels of organisation. 'Inorganic' objects have their place, relatively low,

---

[a] Tr. Bruce (1), p. 64; mod.

[b] *ECCS*; *Honan Chhêng shih I Shu*, ch. 3, p. 2a. Cf. *Sung Yuan Hsüeh An*, ch. 14, p. 5b.

[c] *Rheum officinale* (R 582).   [d] *Aconitum autumnale* (R 532a).

[e] *CTCS*, ch. 42, pp. 31b ff., tr. Bruce (1), p. 68; mod.

[f] Cf. Forke (9), pp. 172, 193.   [g] Cf. pp. 38, 47, 50, 66, 76 above.

[1] 生意   [2] 理   [3] 知覺   [4] 大黃   [5] 附子   [6] 腐敗之物

[7] 心

in the overall pattern, and Chu Hsi is clearly feeling his way towards a classification of chemical properties—in his examples of potash and of alkaloidal drugs he steps into the opening of that long avenue which led to the inorganic and organic chemistry of our own time.[a] But ethical and moral phenomena, properly so-called, only begin to appear when a sufficiently high level of organisation is reached, first incompletely and one-sidedly in animals, and then fully in man. Chu Hsi says in so many words that moral concepts are not applicable to 'inorganic' objects. Yet he himself, in a fashion somewhat parallel to the ancient Taoists,[b] cannot find terms other than 'natural endowment' and even 'mind' when he wants to describe the properties of chemical substances. 'Specificity' was doubtless what he sought.[c]

We seem to be in presence, then, in the latter part of the +12th century, of a point of view rather similar to that which Ulpian had expressed in Europe nearly a millennium before, and which had been incorporated into the Justinian Digest.[d] But the profound difference is that while Ulpian had spoken quite uncompromisingly of *law*, Chu Hsi relies chiefly on a technical term the primary meaning of which is *pattern*. For Ulpian (as for the Stoics) all things were 'citizens' subject to a universal law; for Chu Hsi all things were 'dancers' in a universal pattern.[e] On the whole, it does not appear possible to find more than traces of the concept of laws of Nature in the greatest of Chinese philosophical schools, the Neo-Confucians of the Sung.

(g) LAW IN BUDDHIST THOUGHT

What of Buddhism? It had been against Buddhist philosophy that the Neo-Confucians, in reacting, had produced their great synthesis. As we have seen in Section 15, Buddhist philosophy, while denying the existence of a soul or spirit which could persist after the dissolution of the *skandhas* (*yün*[1]), i.e. the material and mental components of the individual, nevertheless retained the Hindu or Brahminical theory of transmigration. Hence the doctrine of *karma* (*yin yuan*[2]) which stated that 'as soon as a sentient being (man, animal, or god) dies, a new being is produced in a more or less painful and material state of existence, according to the *karma*, the desert or merit, of the being who has died.... The *karma* of the previous set of *skandhas*, or sentient being, determines the locality, nature and future of the new set of *skandhas*, of the

---

[a] Chu Hsi would have appreciated greatly the fact that the formulae of organic chemistry are patterns. Cf. p. 474 above.

[b] See p. 43 above, where the failure of the Taoists to elaborate mineralogical technical terms was noted.

[c] He may have been encouraged to identify as precisely as he could the degree of organisation present in non-living things, by the Thang Buddhist doctrine that 'even inanimate things possess the Buddha-nature (*wu chhing yu hsing*[3])'. This had been said, for instance, by Chan-Jan[4] in the +8th century (cf. Fêng Yu-Lan (1), vol. 2, pp. 385, 551). But it was too pantheistic to be more than a reminder for Chu Hsi's resolute naturalism.

[d] Cf. p. 537 above.

[e] Cf. p. 287 above. See also pp. 191, 196, 270, 281, 368, 453 and 488.

¹ 蘊   ² 因緣   ³ 無情有性   ⁴ 湛然

new sentient being' (Rhys Davids (1), p. 101). In this way, morality, as the Buddhists conceived of it, was set right at the heart of the universal scheme of things, and this was the Law (*fa*[1]) of which one hears so much in connection with Buddhism.[a]

There is no doubt that with its conception of *karma* Buddhism emphasised cause and effect in Nature very strongly, even though with a purely moral reference. Here we are still in the primitive undifferentiated stage of law. In the ancient Indian conception, the virtues and vices of an individual had their unescapable results in the endowment with which another linked individual began his or her existence. In the ancient Chinese conception, the virtues and vices of human leaders had their unescapable results in natural calamities or the behaviour of the weather. Human morality was in fact still inextricably bound up with the phenomena of non-human Nature. 'The operation of *karma*', Streeter rightly said,[b] 'was conceived not juristically as the punishment of a continuing ego, but naturalistically in terms of a law of cause and effect, which was thought of as mechanistically as in the physical sciences.'

One might therefore be tempted to suppose that the law of the Buddhists could have led rather easily to the development of the idea of laws of Nature dissociated from the moral-ethical element. In an interesting paper,[c] Rhys Davids (3) maintained that besides animism, another fundamental belief should be distinguished in ancient religious thought—he suggested the name 'normalism'. By this he meant all those types of belief which concerned, not souls, spirits, gods or demons, but certain regularities of cause and effect, certain unchanging patterns of action in the universe. In this second category he included the Tao of the Taoists. Other scholars also have pointed out that this is characteristic of most of the ancient Asian thought-systems, for it is possible to liken Tao as the Order of Nature to the *Ṛta* of the Indian Vedas (*c.* −11th century), to *Arta* (Old Persian) and to *Asha* (Avestan Persian), all of which contain the meanings: motion, rhythmic motion (of the heavens), order, cosmic order, moral order, the right, etc. (cf. Cornford,[d] Filliozat[e]). We spoke above of the Chinese feeling that crimes or disputes were ominous infractions of the Order of Nature; similar ideas were present in ancient India and Persia. Vedic *druh* and Persian (Avestan) *drug* were terms for anything which militated against the established cosmic order; heresy, impiety, sin, pathological influences, all often personified as demons.[f] To this the corresponding Chinese word is perhaps *ni*,[2] as we shall see in Section 44 on medicine, and elsewhere (Section 46). Cornford (1) has added the interesting point that Greek *Dike* (cf. above, pp. 283, 527, 533) originally meant 'the Way', just as Tao did.

But these ancient forms of 'normalism' or cosmic organicism, as we might call it, developed in different directions. In India they were soon overlaid with a multiplicity

---

[a] Cf. the practice, of which we have already seen examples, of calling eminent Buddhist philosophers 'Masters of the Law'. See also Rhys Davids (2).   [b] (1), p. 282.
[c] Entitled 'Cosmic Law in Ancient Thought'.   [d] (1), pp. 172 ff.
[e] (1), pp. 42, 52, 76, 79.   [f] Cf. p. 567 above.

[1] 法   [2] 逆

of personifications—Vayu (the wind)[a] or Varuna became the 'masters' of Ṛta (Ṛtaspati)—and Ṛta was hidden behind the pullulations of the Hindu pantheon. China took the contrary way, and the aversion from personification enabled the great schools of philosophical Taoism to flourish, emphasising that the realm of human morality was only a part, even a very small part, of the operations of the Tao in all Nature. Buddhism took a third way, retaining the impersonality of Ṛta, but applying it exclusively to the moral sphere in the law of *karma*. None of this takes us as far as laws of Nature in the true scientific sense, for which conception order, pattern, causality and regularity do not quite suffice. Scholars such as Berriedale Keith[b] categorically deny that Buddhist philosophy ever thought of extending to the non-moral sphere a strict applicability of its belief in causality. We have already suggested that one basic reason why the law of *karma* could not lead to a scientific conception of laws of Nature was because of the parallel doctrine (not indeed, strictly speaking, Buddhist only, though China derived it from India through Buddhist channels) that the visible world was all illusion, *māyā* (*mi*[1] or *huan wang*[2]).[c] It was precisely from the pains and miseries inherent in existence in this visible world of Nature that the Buddhists desired to set men free. The last thing, therefore, to which their philosophy could invite, was a dispassionate study of the phenomena of non-human Nature—that was the very Wheel of Illusory Existence from which they offered a way of escape. Hence it is not surprising that neither in India nor in China did the idea of laws of Nature arise from this source.[d]

(h) ORDER WHICH EXCLUDES LAW

At the end of our investigation, therefore, we have to conclude that none of the words in ancient and medieval Chinese texts which have tempted translation as 'laws of Nature', give us any right so to translate them.[e] Granet (5) was correct in his conclusion that the Chinese world-outlook was running along quite different lines, and that the Chinese notion of Order positively excluded the notion of Law.

[a] There is here, as Filliozat shows, an important connection with 'pneumatic' theories in medicine and natural science.

[b] (1), pp. 96, 112, 178.

[c] Keith (1), p. 261.

[d] A point which may be of great sociological importance arises here. Evidence is slowly accumulating which may show that Buddhism as a religion was particularly associated, at least in several dynasties, with the merchant class. To anyone familiar with the Tunhuang cave-temples, located as they are at a site where contributions from merchant caravans must have come in over many centuries, this suggestion has much plausibility. If it should be substantiated, it would constitute yet another factor opposing the spontaneous development of modern science and technology in China; not only were the merchants unable to achieve a position of power in the society and state, but even if they had been able to do so, they would have been handicapped by a Nature-denying religion. I owe this point to my friend Dr E. Balazs. For a recent introductory account of the Tunhuang site see Vincent (1).

[e] After I had come to this conclusion I found that it had the strong support of Forke (9), who wrote (p. 384): 'Der Begriff des Naturgesetzes ist der chinesischen Denkart fremd.' Graham (1), p. 76, agrees.

[1] 迷   [2] 幻妄

So unconscious has the idea of laws of Nature been among Europeans, however, that not a few sinologists have unsuspectingly read into texts the word law when in fact there was no word there to justify it. For example, Gale, in his translation of the *Yen Thieh Lun* (Discourses on Salt and Iron), wrote,[a] 'The Tao hung its laws in the heavens and spread its products on the earth, etc.' All that the text[b] says is *Tao hsüan yü thien...*,[1] i.e. 'The Tao (the Order of Nature) is hung up (manifest) in the heavens....' Couvreur[c] and Forke[d] make *Thien tao*[2] into 'Heaven's laws'. Similarly, Hansford,[e] translating a passage of the *Thien Kung Khai Wu*, makes Sung Ying-Hsing say, 'I do not understand by means of what natural law this is effected.' But the Chinese simply has *wu li*.[3] Another instance in which the word *li* tempted a translator to use the expression laws of Nature occurs in the interesting *Yen Lien Chu*[4] (The String of Pearls Enlarged) by the great writer Lu Chi[5] (*c.* +290).[f] The passage: 'I have heard that what is possible according to the Laws of Nature can be performed by natural forces; what is not possible according to the Laws of Nature, natural forces will not perform. For example, strong fire will melt metal, but it will not burn a shadow; and intense cold can freeze the sea, but not the wind'[g] renders the words: *Chhen wên, li chih so khai, li so chhang ta; shu chih so sai, wei yu pi chhiung*.[6] Here the idea of Laws is clearly not present, and the following alternative may be offered: 'I have heard that when in the Great Pattern of things there is a way open, (natural) forces will always penetrate through; but when according to the Numbers (of the processes of the universe) the way is blocked, then even forces royal in might find the bounds which they cannot pass.'[h]

Sinologists are wont to translate almost any word by 'laws'. The *Kuo Yü*[i] says that at night and especially at the autumn equinox, the emperor or a high official must watch the heavens for prognosticatory purposes (*chiu chhien Thien hsing*[7]). Couvreur[j] makes this: 'He carefully observes and respects astronomical laws', but though the commentator says[k] that *hsing* here means *fa*,[8] the obvious translation would be, 'He reverently collects the admonishings of the Heavens', which is something quite different. Again, Dubs[l] gives the following version of *Hsün Tzu*:[m] 'Two nobles

---

[a] (1), p. 109.

[b] Parallel passage in one of the Han apocryphal treatises, *Shang Shu Wei Hsüan Chi Chhien*[9] (*Ku Wei Shu*, ch. 4, p. 5 a). Again, Bodde (in Fêng Yu-Lan (1), vol. 2, p. 124) could not resist 'rules and regulations' for *chieh tu*[10] instead of 'fixed times and regular motions'.

[c] (1), vol. 3, p. 181 (*Tso Chuan*, Duke Chao, 11th year, −530), and again, p. 673. Cf. p. 547 above.

[d] (4), vol. 2, p. 392, translating *Lun Hêng*, ch. 69. So also Vacca (10), translating *Chuang Tzu*, systematically; and Chou I-Chhing (1), p. 171. Cf. Brucker (1), vol. 5, p. 869.

[e] (1), pp. 62, 63.

[f] Not to be confused with his contemporary the naturalist, whose name differs only by having the jade instead of the wood radical.

[g] *Wên Hsüan*, ch. 55, the 49th aphorism of the work cited; tr. E. von Zach (1), eng. auct.

[h] Tr. auct.   [i] *Lu Yü*, ch. 2, p. 15 a.

[j] (2), p. 805.   [k] Wei Chao of the +3rd century.

[l] (8), p. 124.   [m] Ch. 9, p. 3 b.

[1] 道懸於天   [2] 天道   [3] 物理   [4] 演連珠   [5] 陸機

[6] 臣聞理之所開力所常達數之所塞威有必窮   [7] 糾虔天刑   [8] 法

[9] 尚書緯璇璣鈐   [10] 節度

cannot serve each other; two commoners cannot employ each other—this is a law of nature.' Here the relevant words in the text are *Thien shu yeh*,[1] which would be more faithfully rendered 'Such are the Numbers of Heaven', i.e. the unalterable numerical data of fate. Veith, translating the *Huang Ti Su Wên Nei Ching*, inserts the expression 'laws of Nature' several times when there are no words in the text to which it could correspond.[a] Of course, free translations are always more attractive than literal ones, but they are liable to suffer from the unconscious intellectual background of the free translator, and there are occasions on which this may matter a great deal.[b] The time has come for a rigorous effort to follow Chinese modes of thought.

(i) Judicial Trials of Animals; contrasting European and Chinese Attitudes to Biological Abnormalities

Before concluding, we may glance at a striking illustration of the difference in outlook between China and Europe in the matter of laws of Nature. It is generally known that during the European Middle Ages there were a considerable number of trials and criminal prosecutions of animals in courts of law, followed frequently by capital punishment in due form. Evans (1) and Hyde (1) have gone to the trouble to collect a large amount of information on these cases, building on the earlier work of Berriat St Prix (1), Ménebréa (1), and von Amira (1). Their frequency follows a curve with a well-marked peak at the 16th century, rising from three instances in the 9th to about sixty in the 16th, and falling to nine in the 19th century; and it seems doubtful whether this is due, as Evans suggests, to lack of adequate records for the earlier periods. The peak corresponds to the witch-mania (Withington, 1). The trials fall into three types: (*a*) the trial and execution of domestic animals for attacking human beings (e.g. the execution of pigs for devouring infants); (*b*) the excommunication, or rather anathematisation, of plagues or pests of birds or insects; and (*c*) the condemnation of *lusus naturae*, e.g. the laying of eggs by cocks. It is the last two which are most interesting for the present theme. In 1474 a cock was sentenced to be burnt alive for the 'heinous and unnatural crime' of laying an egg, at Basel; and there was another Swiss prosecution of the same kind as late as 1730. One of the reasons for the alarm involved was perhaps that *œuf coquatri* was thought to be an ingredient in witches' ointments, and that the basilisk or cockatrice, a particularly venomous animal, hatched from it.[c]

---

[a] Ch. 9, p. 39*b*, (1), p. 137; ch. 9, p. 41*a*, (1), p. 138.

[b] Graf (2), vol. 1, p. 287, well writes, 'Noch immer ist der abendländische Mensch der im Grunde doch ein wenig naiv-anmassenden Ansicht, die Geschichte seines europäischen Denkens sei die Philosophiegeschichte der Menschheit schlechthin'.

[c] Needham (2), p. 85; Robin (1), p. 86. The legend appears first in Alexander Neckham (Sarton (1), p. 385) late in the +12th century. In 1710 Lapeyronie suggested that what were taken for cock's eggs were small almost yolkless eggs laid by hens suffering from diseases obstructing their oviducts. But as L. J. Cole (1) points out, sex-reversals may so completely approximate the plumage of a hen to that of a cock that it would have been assumed to be a cock in days before the understanding of the anatomy of the sexual organs.

[1] 天數也

The interest of the story lies in the fact that such trials would have been absolutely impossible in China. The Chinese were not so presumptuous as to suppose that they knew the laws laid down by God for non-human things so well that they could proceed to indict an animal at law for transgressing them. On the contrary, the Chinese reaction would undoubtedly have been to treat these rare and frightening phenomena as *chhien kao*[1] (reprimands from Heaven),[a] and it was the emperor or the provincial governor whose position would have been endangered, not the cock. Let us quote chapter and verse. In the long Wu Hsing Chih[2] (Record of (Derangements of) the Five Elements) in the *Chhien Han Shu* (History of the Former Han Dynasty) there are several references to sex-reversals in poultry[b] and in man.[c] These were classified under the heading of 'green misfortunes' (*chhing hsiang*[3]) and thought of as connected with the activities of the element Wood.[d] They foreboded serious harm to the rulers in whose dominions they occurred.

We are considering what might be called the dominant attitudes of the respective civilisations. In the less marked, or recessive, attitudes, behaviour characteristic of the other can be found. Thus it is not impossible to find in late Chinese folklore examples of animals being brought before magistrate's courts, as in certain stories of the *Chhih Pei Ou Than*[4] (Chance Stories told North of the Lake) by Wang Shih-Chên[5] of the late Ming and early Chhing periods.[e] But these generally concern the repentance of tigers for having killed men; they are patently Buddhist in inspiration, and in any case belong to the first type of prosecution mentioned above, in which at least a tort or criminal action had been committed against man. The important cases for the present argument are those of the third type, where no harm had been done to him. Conversely, as regards the second of the three types, it is interesting that the European medieval attitude wavered. Sometimes the fieldmice or locusts were considered to be breaking God's laws, and therefore subject to human prosecution, conviction, and punishment, while at other times the view prevailed, doubtless urged by preaching friars and bishops, that these animals had been sent to admonish men to repentance and amendment. This might be called a 'Chinese' reaction.

The extent to which such an attitude involved resignation to the heavenly visitation on the one hand, as opposed to active measures to combat it on the other, varied a good deal in China. A famous minister, Yao Chhung,[6] of the Thang (+650 to +721), urged in a memorial to the emperor concerning the locust plagues of +716,

---

[a] Cf. the Section on Wang Chhung and the Sceptics, pp. 378 ff.

[b] Ch. 27BA, pp. 20a, b ff.

[c] Ch. 27CA, p. 18b. Also *Lun Hêng*, ch. 7: 'Men occasionally turn into women and vice versa' (mis-tr. in Forke (4), vol. 1, p. 327).

[d] Cf. Eberhard (6), pp. 22, 32, 36. So also in the *Hsin Thang Shu* (New History of the Thang Dynasty), chs. 34–6 (tr. Pfizmaier, 67, pp. 30, 31) under dates +687, +689 and +854. Examples could doubtless be adduced from every dynastic history.

[e] I owe this reference to the kindness of my friend Professor W. Eberhard.

---

[1] 譴告　　[2] 五行志　　[3] 青祥　　[4] 池北偶談　　[5] 王士禎

[6] 姚崈

that they were entirely 'natural' and not the result of a 'phenomenalist' reprisal on the part of Heaven.[a] This being accepted, he organised nation-wide counter-measures. It is recorded that the emperor Thai Tsung about a century previously (+628) publicly ate a dish of fried locusts in order to demonstrate that they were not something sacred sent from Heaven as a punishment.[b] But these were practical reactions, not prosecutions at law.

Somewhat related to this whole matter is the English law of 'deodands' or 'banes', under which inorganic, inanimate objects, or animals, which caused the death of human beings, were forfeited to the Church or the Crown.[c] 'Omnia quae movent ad mortem sunt Deo danda' (Bracton). This law, not abolished till 1846, perhaps originated from the same complex of ideas, namely, that non-living things could, like human beings, transgress the laws of God. There can have been no parallel to this in Chinese jurisprudence.

(2) DOMINANCE PSYCHOLOGY AND EXCESSIVE ABSTRACTION

Pondering over differing Eastern and Western conceptions of law in relation to the living world, it may occur to us that some difference of emphasis might arise according to whether man has to do chiefly with the animal or with the vegetal world. That contrasting attitudes originate, even in the abstract sphere which is the subject of this Section, from pastoral as opposed to agricultural life, has been suggested by André Haudricourt.[d] The shepherd and the cowherd beat their beasts, and take up an active attitude of command over their flocks and herds. God is imagined as a 'Good Shepherd' leading his flock into satisfying pastures. But the shepherd is not far from the legislator, and pastoral dominance over animals consorts well with legislation over things as well as men. Maritime usages strongly reinforce this command-psychology, for the safety of all in a ship doubtless required from the earliest times an unquestioning obedience of the many to the experienced one. Hence law in Nature would have been derived from the masteries of shepherds and sea-captains as well as kings. But when man has to do primarily with plants, as in pre-dominantly agricultural civilisations, the psychological conditions are quite different—often the less he interferes with the growth of his crops the better. Until the harvest he does not touch them. They follow their Tao, which leads to his benefit. Is not the conception of *wu wei* ('no action contrary to Nature')[e] deeply congruent with peasant life? In Mencius there is a famous story:

Let us not be like the man of Sung. There was once a man of Sung, who was grieved that his growing corn was not longer, so he pulled it up. Returning home, looking very stupid,

---

[a] *Chiu Thang Shu*, ch. 96, pp. 2b ff.; *Thang Yü Lin*, ch. 1, p. 22a. He met with strong opposition.
[b] *Chih Huang Chhüan Fa* (Complete Handbook of Locust Control), ch. 3, pp. 20b–22a.
[c] Robson (1), p. 85; Pollock & Maitland (1), vol. 2, p. 473.
[d] Personal communication, 2 Jan. 1951, and in the article of de Hetrelon (1).
[e] So often mentioned, cf. pp. 68 ff. above.

he said to his people, 'I am tired today. I have been helping the corn to grow long.' His son ran to look at it, and found the corn all withered.[a]

Agricultural civilisations would therefore not be expected to show the dominance-psychology and the notion of a divine legislator which is perhaps connected with it. If, indeed, this notion began in Babylonia,[b] it was no doubt because the ancient economy of the fertile crescent was a mixed one, and certainly much of its spread was due to that pre-eminently pastoral people, the Hebrews.

The point is perhaps worth emphasising by quoting from a famous essay of Liu Tsung-Yuan (+773 to +819), a Thang writer of naturalistic interests. It concerns a famous market-gardener, familiarly known as Camel-Back Kuo,[c] who was extremely successful in his methods.

One day a customer asked him how this was so, to which he replied: 'Old Camel-Back cannot make trees live or thrive. He can only let them follow their natural tendencies. In planting trees, be careful to set the root straight, to smooth the earth around them, to use good mould and ram it down well. Then, don't touch them, don't think about them, don't go and look at them, but leave them alone to take care of themselves, and Nature will do the rest. I only avoid trying to make my trees grow. I have no special method of cultivation, no special means for securing luxuriance of growth. I just don't spoil the fruit. I have no way of getting it either early or in abundance. Other gardeners set with bent root, and neglect the mould, heaping up either too much earth or too little. Or else they like their trees too much and become anxious about them, and are for ever running back and forth to see how they are growing; sometimes scratching them to make sure they are still alive, or shaking them to see if they are sufficiently firm in the ground; thus constantly interfering with the natural bias of the tree, and turning their care and affection into a bane and a curse. I just don't do those things. That's all.'

'Can these principles of yours be applied to government?' asked his listener. 'Ah', replied Camel-Back, 'I only understand market-gardening; government is not my trade. Still, in the village where I live, the officials are constantly issuing all kinds of orders, apparently out of compassion for the people, but really to their injury. Morning and night the underlings come round and say, "His Honour bids us urge on your ploughing, hasten your planting, supervise your harvest. Do not delay with spinning and weaving. Take care of your children. Rear poultry and pigs. Come together when the drum beats. Be ready when the rattle goes." Thus we poor people are badgered from morning till night. We haven't a moment to ourselves. How could anyone develop naturally under such conditions? It was this that brought about my deformity. And so it is with those who carry on the gardening business.'

'Thank you', said the listener. 'I simply asked about the management of trees, but I have learnt about the management of men. I will make this known, as a warning to government officials.'[d]

[a] *Mêng Tzu*, II (1), ii, 16; tr. Legge (3), p. 66.
[b] See p. 533 above.
[c] Actually Kuo was a writer himself, on horticulture and forestry (cf. Section 41 below), and a famous expert on grafting.
[d] Tr. H. A. Giles (12), in Chiang Fêng-Wei (1).

Thus in the Thang, just as in the Chou,[a] the Taoist artisan gives sound advice to rulers, with the background thought that all things spontaneously work together for good, without the necessity of intervention of divine or other legislators.

Another point of contrast between Chinese and European conceptions of law involves not biology, but mathematics. In Section 19 on mathematics we shall see that in contrast with the Greek gift for geometry, Chinese mathematics was algebraic and algorismic. Now there is something suspiciously similar between the abstractness of Euclidean geometry and the abstractness of Roman law.[b] In Roman law a *vinculum* or contract between two persons, that which they agree upon between themselves, was considered to have no possible bearing on any third person. But for Chinese law such an abstractness was inconceivable; an agreement could not be considered in isolation from the attendant concrete circumstances, the position and obligations of the persons in society, and the effects which it might have on other persons. Just as Greek geometry dealt with pure and abstract figures, the size of which was quite immaterial once the axioms and postulates had been accepted, so Roman law dealt with codified abstractions. But the Chinese preferred to think only of concrete numbers (though, as in algebra, they might not be any particular numbers), and of concrete social circumstances.

(3) THE COMPARATIVE PHILOSOPHY OF LAW IN CHINA AND EUROPE

The only general examination of the comparative philosophy of law in China and Europe which we have seen is a recent and stimulating paper by Dorsey (1).[c] Unfortunately, it is not based, in our opinion, on a sufficiently sure foundation from the sinological point of view. He is fully justified in his firm conviction of the importance of *li*[1] as against *fa*[2] in Chinese thought and practice, and he notes correctly the attitudes of the Legalists and Confucians. He is right, I am sure, in emphasising the direct verifiability of the good customs to which the Confucians appealed, as opposed to the necessity of taking on trust what the codifiers, whether European or Chinese, built up into their positive law. One can contrast the non-codified customary, demonstrable ethical *droit* of the Confucian paternalistic sage-kings with the codified, enacted, non-demonstrable, non-ethical law of the *Fa Chia*. He notes that a word such as 'person' has a different significance in Confucian *li*[1] with its ever-variable flexibility from that which it has in the deductively formulated abstract phraseology of Roman law. On the other hand, his contention that Chinese law was based on non-human Nature is less convincing; it rests on a misapprehension of what the 'study of things' (*ko wu*[3])[d] meant in Chinese tradition, and on some perhaps rather

---

[a] Cf. above, p. 122.

[b] I owe this point to a conversation with my friend Dr Meredith Jackson.

[c] Cf. the symposium by J. A. Wilson *et al.*

[d] Cf. Vol. 1, p. 48. Although this phrase has been taken up to mean natural science in modern times, it generally implied, until the present century, the study of human affairs. See a note by C. Mao (1).

[1] 禮   [2] 法   [3] 格物

exaggerated remarks of Granet's[a] about the imitation of animals in the primitive mating festivals. But more fundamental is his main conclusion, that Chinese law differed from European law because the Chinese apprehended Nature in a different way from Europeans. This rests on the general view of Northrop that while the Greeks developed the way of knowing Nature by postulation and scientific hypothesis, the Chinese approached Nature throughout their history only by direct inspection and aesthetic intuition.

Such a view is, we fear, contradicted by almost all the facts brought together in the present work. There is no good reason for denying to the theories of the Yin and Yang, or the Five Elements, the same status of proto-scientific hypotheses as can be claimed by the systems of the pre-Socratic and other Greek schools. What went wrong with Chinese science was its ultimate failure to develop out of these theories forms more adequate to the growth of practical knowledge, and in particular its failure to apply mathematics to the formulation of regularities in natural phenomena.[b] This is equivalent to saying that no Renaissance awoke it from its 'empirical slumbers'. But for that situation the specific nature of the social and economic system must be held largely responsible, and differences in the apprehension of Nature as such cannot, as we see it, explain the differences between Chinese and European conceptions of law.

In Europe natural law may be said to have helped the growth of natural science because of its *universality*. But in China, since natural law was never thought of as law, and took a very social name, *li*,[1] it was hard to think of any law as applicable outside human society, though relatively *li*[1] was much more important in society than the natural law of Europe. When order and system and pattern were visualised as running through the whole of Nature, it was generally not as *li*[1] but as the *Tao*[2] of the Taoists or the *Li*[3] of the Neo-Confucians, philosophical principles neither of which had juristic content.[c]

Again, in Europe positive law may be said to have helped the growth of natural science because of its *precise formulation*. This was encouraging because of the idea that to the earthly lawgiver there corresponded in heaven a celestial one, whose writ ran wherever there were material things. In order to believe in the rational intelligibility of Nature, the Western mind had to presuppose (or found it convenient to presuppose) the existence of a Supreme Being who, himself rational, had put it there. The Chinese

[a] (2), pp. 93, 229. Cf. Fêng Yu-Lan (1), vol. 2, p. 85, discussing Tung Chung-Shu. Cf. also pp. 549 ff. above; kings 'modelling their laws on Nature'.

[b] This we shall study at the end of Sect. 19 in the next volume. An interesting little point here to which we have not seen attention called before is the fact that the Han legal code in nine sections was called the *Chiu Chang*.[4] It is probably only a coincidence that one of the most important mathematical books of the Han is the *Chiu Chang Suan Shu*[5] (Nine Chapters on the Mathematical Art). Could there have been any idea of a parallel between law and computation? It is very unlikely.

[c] Of course sinologists have sometimes unthinkingly translated Tao as 'law', e.g. Forke (4), vol. 2, p. 157, in the remark of Wang Chhung 'Thien jen thung Tao',[6] Heaven and Man have the same Tao. So also Dubs (19), p. 272. But I believe, as pointed out above, p. 573, that this is absolutely inadmissible.

[1] 禮　　[2] 道　　[3] 理　　[4] 九章　　[5] 九章算術　　[6] 天人同道

mind did not think in these terms at all. Imperial majesty corresponded, not to a legislating creator, but to a polar star, the focal point of universal ever-moving pattern and harmony not made with hands, even those of God. And the pattern was rationally intelligible because it was incarnate in Man.

This brings us back to the conclusions which we reached at the end of the Sections on the philosophical schools. The Taoists, though profoundly interested in Nature, distrusted reason and logic. The Mohists and the Logicians fully believed in reason and logic, but if they were interested in Nature it was only for practical purposes. The Legalists and Confucians were not interested in Nature at all. Now this gulf between empirical nature-observers and rationalist thinkers is not found to anything like the same extent in European history. Whitehead has suggested that this was perhaps because European thought was so dominated by the idea of a supreme creator being, whose own rationality guaranteed intelligibility in his creation. Whatever may be the needs of mankind now, such a supreme God had inevitably to be personal then. This we do not find in Chinese thought. Even the present-day Chinese term for laws of Nature, *tzu-jan fa*,[1] 'spontaneous law', is a phrase which so uncompromisingly retains the ancient Taoist denial of a personal God that it is almost a contradiction in terms.

(4) VARYING CONCEPTIONS OF DEITY

Here we cannot investigate the ancient Chinese conceptions of God. An immense literature exists on the subject, for Christian missionaries in the last few centuries engaged in much debate as to the correct translation of European terms.[a] Most of this is now not worth the paper on which it was written, since at that time sinological studies were in their infancy. We know that the most ancient terms for God in Chinese were *Thien*[2] (Heaven) or *Shang Ti*[3] (the Ruler Above), though other terms were used, e.g. *Tsai*[4] (Governor) by Chuang Chou.[b] *Thien* (K 361) is undoubtedly an anthropomorphic graph (presumably of a deity) in its most ancient form,[c] and *Ti* (K 877) has been thought anthropomorphic also (Hopkins, 11), though modern views regard it, with Wu Ta-Chhêng, as a loan-word from one depicting the stem and basis of a flower. *Tsai* (K 965) on the other hand shows, according to Kuo Mo-Jo, an elderly person (cf. *hsin*[5], K 382) under a roof. Much sinological work is being done on the extent to which there was in ancient China a personalistion of these conceptions, and it is hard to summarise such conclusions as have been reached.[d] Many theories are in the field: some think, for instance, that *Shang Ti* was a transcendentalisation of the function of the Emperor or bronze-age High King (Creel, 3); others consider that he was a personification of the calendrical order of the seasons (Granet, 4) or a vegetation 'Corn King' (Schindler, 1); a third view, represented by Fitzgerald, looks upon

---

[a] See Cordier (2), pt. I, sect. xi, and Suppl.   [b] Cf. p. 52.
[c] Though the Roman Catholic Church adopted *Thien Chu*,[6] 'the Master of Heaven'.
[d] See especially Schindler (2); Thien Chhih-Kang (1); Forke (13), pp. 30, 34; Grube (3).

[1] 自然法   [2] 天   [3] 上帝   [4] 宰   [5] 幸   [6] 天主

him, and upon *Thien*, as symbols of the Original Ancestor. Creel (1) expresses the now generally received opinion that *Shang Ti* is the older of the two, being associated with the Shang period, while *Thien* is rather a later Chou term. According to Fu Ssu-Nien (2), the expression *Shang Ti* occurs only once on the oracle-bones so far examined, and there it refers to Ti-Ku,[1] the mythical High Ancestor of the Shang people. What we know of their sacrificial customs indicates a high proportion of offerings to ancestors (Chhen Mêng-Chia, 2, 3). Tai Kuan-I (1) believes that the name *Shang Ti* was taken over by the Chinese from the Miao peoples. Creel

K 361   K 877   K 965   K 382

has suggested that the concept eventually implied the *company* or *multitude* of ancestors, which became as impersonal as any earthly corporation.[a] But in any case two things are clear: (*a*) that the de-personalisation of God in ancient Chinese thought took place so early and went so far that the conception of a divine celestial lawgiver imposing ordinances on non-human Nature never developed; and (*b*) that the highest spiritual being ever known and worshipped had not been a Creator in the sense of the Hebrews and the Greeks.[b]

It was not that there was no order in Nature for the Chinese, but rather that it was not an order ordained by a rational personal being, and hence there was no conviction that rational personal beings would be able to spell out in their lesser earthly languages the divine code of laws which he had decreed aforetime. The Taoists, indeed, would have scorned such an idea as being too naïve for the subtlety and complexity of the universe as they intuited it.[c] Human rational personal beings had another faith; the universal order was intelligible because they themselves had been produced by it. They were indeed its highest component patterns—'Heaven, Earth, and Man have the same Li',[d] 'The human-hearted man is with Heaven and Earth a unity',[e] 'With Heaven and Earth together the Sage forms a trinity'.[f]

It is extremely interesting that modern science, which since the time of Laplace has found it possible and even desirable to dispense completely with the hypothesis of a God as the basis of the laws of Nature, has returned, in a sense, to the Taoist outlook. This is what accounts for the strangely modern ring in so much of the writing of that

[a] Cf. letters which are couched in terms such as 'It is thought that...'.

[b] As Eitel (3) remarked, more than seventy years ago, 'The idea of creation out of nothing has ever remained entirely foreign to the Chinese mind, so much so that there is no word in the language to express the idea of creation *ex nihilo*.'

[c] Examples can be found of thinkers in China who maintained a belief in the personality of 'Heaven', e.g. Chang Shih[2] (+1133 to +1180), cf. Forke (9), p. 263; but they were exceptional. Chuang Tzu often speaks of the 'Author of Change' (*tsao hua chê*[3]) or 'of Things' (*tsao wu chê*[4]) but the references are poetical and even somewhat mocking. See Section 23 below.

[d] See p. 488 above.   [e] See p. 453 above.

[f] *Chung Yung*, ch. 22 (Legge (2), p. 280). Cf. especially p. 281 above, with its cross-references.

[1] 帝嚳   [2] 張栻   [3] 造化者   [4] 造物者

great school. But historically the question remains whether natural science could ever have reached its present stage of development without passing through a 'theological' stage.

In the outlook of modern science there is, of course, no residue of the notions of command and duty in the 'laws' of Nature. They are now generally thought of as statistical regularities, valid only in given times and places, descriptions not prescriptions, as Karl Pearson put it in a famous chapter. The exact degree of subjectivity in the formulations of scientific law has been hotly debated during the whole period from Mach to Eddington, and such questions cannot be entered into here. The problem is whether the recognition of such statistical regularities and their mathematical expression could have been reached by any other road than that which science actually travelled in the West. Was the state of mind in which an egg-laying cock could be prosecuted at law necessary in a culture which should later have the property of producing a Kepler?

(i) CONCLUSIONS

To sum up, therefore, we may say that the conception of laws of Nature did not develop from Chinese juristic theory and practice for the following reasons. First, the Chinese acquired a great distaste for precisely formulated abstract codified law from their bad experiences with the school of Legalists during the period of transition from feudalism to bureaucratism. Secondly, when the system of bureaucratism definitively set in, the old conceptions of *li*[1] proved more suitable than any others for Chinese society in its typical form, and thus the element of natural law became relatively more important in Chinese than in European society. But the fact that so little of it was expressed in formal legal terms, and that it was overwhelmingly social and ethical in content, made any extension of its sphere of influence to non-human Nature impossible. Thirdly, the autochthonous ideas of a supreme being, though certainly present from the earliest times, soon lost the qualities of personality and creativity. The development of the concept of precisely formulated abstract laws capable, because of the rationality of an Author of Nature, of being deciphered and re-stated, did not therefore occur.

The Chinese world-view depended upon a totally different line of thought. The harmonious cooperation of all beings arose, not from the orders of a superior authority external to themselves, but from the fact that they were all parts in a hierarchy of wholes forming a cosmic pattern, and what they obeyed were the internal dictates of their own natures. Modern science and the philosophy of organism, with its integrative levels, have come back to this wisdom, fortified by new understanding of cosmic, biological and social evolution. Yet who shall say that the Newtonian phase was not an essential one? And lastly there was always the environment of Chinese social and economic life, out of which arose the transition from feudalism to bureaucratism just

[1] 禮

mentioned, and which could not but condition at every step the science and philosophy of the Chinese people. Had these conditions been basically favourable to science, any inhibitory influences of the kind considered in this Section would no doubt have been overcome. All we can say of that science of Nature which then would have developed is that it would have been profoundly organic and non-mechanical.

What manner of disciplines the sciences of ancient and medieval China actually were, the next volume will begin to tell.

BIBLIOGRAPHIES

A Chinese books before +1800
B Chinese and Japanese books and journal articles since +1800
C Books and journal articles in Western Languages

In Bibliographies A and B there are two modifications of the Roman alphabetical sequence: transliterated *Chh-* comes after all other entries under *Ch-*, and transliterated *Hs-* comes after all other entries under *H-*. Thus *Chhen* comes after *Chung* and *Hsi* comes after *Huai*. This system applies only to the first words of the titles. Moreover, where *Chh-* and *Hs-* occur in words used in Bibliography C, i.e. in a western language context, the normal sequence of the Roman alphabet is observed.

When obsolete or unusual romanisations of Chinese words occur in entries in Bibliography C, they are followed, wherever possible, by the romanisations adopted as standard in the present work. If inserted in the title, these are enclosed in square brackets; if they follow it, in round brackets. When Chinese words or phrases occur romanised according to the Wade-Giles system or related systems, they are assimilated to the system here adopted without indication of any change. Additional notes are added in round brackets. The reference numbers do not necessarily begin with (1), nor are they necessarily consecutive, because only those references required for this volume of the series are given.

ABBREVIATIONS

AA   *Artibus Asiae*

A/AIHS   *Archives internationales d'Histoire des Sciences* (continuation of *Archeion*)

AAL/RSM   *Atti d. r. Accademia dei Lincei,* (Rendiconti, Sci. Mor.)

AAN   *American Anthropologist*

ABAW/PH   *Abhandlungen d. bayerischen Akademie d. Wissenschaften, München* (Phil.-hist. Klasse)

ACF   *Annuaire du Collège de France*

AEPHE/SSR   *Annuaire de l'Ecole pratique des Hautes Etudes* (Sect. des Sci. religieuses)

AGMN   *Archiv. f. d. Geschichte d. Medizin u. d. Naturwissenschaften* (Sudhoff's)

AHR   *American Historical Review*

AJP   *American Journal of Philology*

AM   *Asia Major*

AMG   *Annales du Musée Guimet*

AMLN   *Midland Naturalist*

AMM   *American Mathematical Monthly*

AMS   *American Scholar*

AMSC   *American Scientist*

AN   *Anthropos*

ANNB   *Année Biologique*

AO   *Acta Orientalia*

AP   *Aryan Path*

APAW   *Abhandlungen d. preussischen Akademie d. Wissenschaften zu Berlin*

APDSJ   *Archives de Philosophie du Droit et de la Sociologie juridique*

AP/HJ   *Historical Journal, National Peiping Academy*

ARLC/DO   *Annual Reports of the Librarian of Congress* (Division of Orientalia)

ARSI   *Annual Reports of the Smithsonian Institute*

AR   *Archiv f. Religionswissenschaft*

AS   *Année Sociologique*

AS/BIHP   *Bulletin of the Institute of History and Philology, Academia Sinica*

AS/CJA   *Chinese Journal of Archaeology* (Academia Sinica)

ASEA   *Asiatische Studien; Etudes Asiatiques*

ASRZB   *Annales de la Société royale zoologique de Belgique*

BA   *Baessler Archiv* (Beiträge z. Völkerkunde herausgeg. a. d. Mitteln d. Baessler Instituts, Berlin)

BAFAO   *Bulletin de l'Association Française des Amis de l'Orient*

BAISP   *Bulletin de l'Académie impériale de St Petersbourg*

BCS   *Bulletin of Chinese Studies* (Chhêngtu)

BCSH   *Pai Chhuan Hsüeh Hai* (Hundred Rivers Sea of Learning)

BE/AMG   *Bibliographie d'Études* (Musée Guimet)

BEFEO   *Bulletin de l'Ecole Française de l'Extrême Orient*

BEHE/PH   *Bibliothèque de l'Ecole des Hautes Etudes* (Philol. et Hist.)

BIHM   *Bulletin of the (Johns Hopkins) Institute of the History of Medicine*

BIOS   *Abhandlungen z. theoretischen Biologie u. ihrer Geschichte sowie z. Philosophie d. Organischen Naturwissenschaften*

BJRL   *Bulletin of the John Rylands Library* (Manchester)

BLSOAS   *Bulletin of the London School of Oriental and African Studies*

BMFEA   *Bulletin of the Museum of Far Eastern Antiquities* (Stockholm)

BNI   *Bijdragen tot de taal-, land-, en volkenkunde v. Nederlandsch Indië*

BOR   *Babylonian and Oriental Record*

BR   *Biological Reviews*

BSEIC   *Bulletin de la Société des Etudes Indochinoises*

BSRBAP   *Bulletin de la Société royale Belge d'Anthropologie et de Préhistoire*

BUA   *Bulletin de l'Université de l'Aurore* (Shanghai)

BVSAW/PH   *Berichte über d. Verhandlungen d. sächsischen Akademie d. Wissenschaften zu Leipzig* (Phil.-hist. Klasse)

CC   *Chün Chung*

CCS   *Collectanea Commissionis Synodalis in Sinis*

CHJ   *Chhing-Hua Hsüeh-Pao* (Chhing-Hua (Ts'ing-Hua University) Journal)

CHLR   *China Law Review*

CIB   *China Institute Bulletin* (New York)

CIBA/M   *Ciba Review* (Medical History)

CIMC/MR   *Chinese Imperial Maritime Customs* (Medical Report Series)

CJ	Chinese Journal of Science and Arts	HITC	Hsüeh I Tsa Chih
CKKSH	Chung-Kuo Kho-Hsüeh	HJAS	Harvard Journal of Asiatic Studies
CLPRO	Current Legal Problems	HMA	Hermathena
CMJ	China Medical Journal	HMSO	Her Majesty's Stationery Office (London)
CN	Centaurus		
CNRS	Centre Nationale de la Recherche Scientifique (Paris)	HOS	Harvard Oriental Series
		HTR	Harvard Theological Review
CR	China Review	HWTS	Han Wei Tshung-Shu
CRAIBL	Comptes Rendus de l'Académie des Inscriptions et Belles-Lettres		
		IHQ	Indian Historical Quarterly
CRR	Chinese Recorder	ILN	Illustrated London News
CSPSR	Chinese Social and Political Science Review	IPR	Institute of Pacific Relations
		ISIS	Isis
D	Discovery		
DIO	Diogenes	JA	Journal Asiatique
DVN	Dan Viet Nam	JAFL	Journal of American Folklore
DWAW/PH	Denkschriften d. k. Akademie d. Wissenschaften, Wien (Vienna) (Phil.-hist. Klasse)	JAOS	Journal of the American Oriental Society
		JBC	Journal of Biological Chemistry
		JBTS	Journal of the Buddhist Text Society
EE	Electrical Engineering	JEFDS	Journal of the English Folk-Dance and Song Society
EI	Encyclopaedia of Islam (ed. Houtsma et al.)		
		JEGP	Journal of English and Germanic Philology
ENB	Ethnologisches Notizblatt (Kgl. Mus. f. Völkerkunde, Berlin)	JEM	Journal of Experimental Medicine
ENG	Engineering	JFI	Journal of the Franklin Institute
EPJ	Edinburgh Philosophical Journal	JH	Journal of Heredity
ER	Erasmus	JHI	Journal of the History of Ideas
ERE	Encyclopaedia of Religion and Ethics (ed. Hastings)	JMH	Journal of Modern History
		JMLOL	Journal of Mammalology
ES	Encyclopaedia Sinica (ed. Couling)	JOSHK	Journal of Oriental Studies, Hongkong University
ESEJ	Etudes de Sociologie et d'Ethnologie Juridique (Institut de Droit Comparé)		
		JP	Journal of Philology
		JPOS	Journal of the Peking Oriental Society
ESS	Encyclopaedia of Social Sciences (ed. Alvin Johnson)		
		JPS	Journal of Psychology
ETC	Etcetera; a Review of General Semantics	JRAI	Journal of the Royal Anthropological Institute
		JRAS	Journal of the Royal Asiatic Society
ETH	Ethnos	JRAS/NCB	Journal of the North China Branch of the Royal Asiatic Society
FASIE	France-Asie; Revue Mensuelle de Culture et de Synthèse Franco-Asiatique		
		JRSA	Journal of the Royal Society of Arts
		JS	Journal des Savants
FEQ	Far Eastern Quarterly	JSCL	Journal of the Society of Comparative Legislation
FFC	Folklore Fellows Communications		
FJHC	Fu Jen (University) Hsüeh Chih	JSHB	Journal Suisse d'Horlogerie et Bijouterie
FLS	Folklore Studies (Peiping)		
FMNHP/AS	Field Museum of Natural History (Chicago) Publications; Anthropological Series	JSI	Journal of Scientific Instruments
		JTVI	Journal of the Transactions of the Victoria Institute
GBA	Gazette des Beaux-Arts	JWCBRS	Journal of the West China Border Research Society
GGM	Geographical Magazine		
GHA	Göteborgs Högskolas Årsskrift	JWCI	Journal of the Warburg and Courtauld Institutes
GR	Geographical Review		
GUJ	Gutenberg Jahrbuch	JWH	Journal of World History
GWI	Geschichte in Wissenschaft und Unterricht		
		KDVS/HFM	Kongelige Danske Videnskabernes Selskab (Hist.-filol. Meddelelser)
HH	Han Hiue (Han Hsüeh): Bulletin du Centre d'Etudes Sinologiques (Franco-Chinois) de Pékin	KHS	Kho-Hsüeh
		KS	Keleti Szemle
		KSP	Ku Shih Pien

L	*Leonardo*
LG	*Literary Guide*
LHP	*Lingnan Hsüeh-Pao* (Lingnan University Journal)
LQR	*Law Quarterly Review*
M	*Mind*
MAAA	*Memoirs of the American Anthropological Association*
MBH	*Medical Bookman and Historian*
MCB	*Mélanges Chinois et Bouddhiques*
MCHSAMUC	*Mémoires concernant l'Histoire, les Sciences, les Arts, les Mœurs et les Usages, des Chinois, par les Missionaires de Pékin, Paris, 1776–1814*
MCM	*Macmillan's Magazine*
MCMU	*Memoirs of the Carnegie Museum* (Pittsburgh)
MDGNVO	*Mitteilungen d. deutschen Gesellschaft f. Natur- u. Völkerkunde Ostasiens*
MHJ	*Middlesex Hospital Journal*
MLN	*Modern Language Notes*
MN	*Monumenta Nipponica*
MRASP	*Mémoires de l'Académie royale de Sciences* (Paris)
MRDTB	*Memoirs of the Research Dept. of Tōyō Bunko* (Tokyo)
MS	*Monumenta Serica*
MSAF	*Mémoires de la Société (Nat.) des Antiquaires de France*
MSOS	*Mitteilungen d. Seminar f. orientalischen Sprachen* (Berlin)
N	*Nature*
NCR	*New China Review*
NDL	*Notre Dame Lawyer*
NGM	*National Geographic Magazine*
NGWG/PH	*Nachrichten v. d. k. Gesellschaft d. Wissenschaften z. Göttingen* (Phil.-hist. Klasse)
NH	*Natural History*
NLIP	*Natural Law Institute Proceedings* (Notre Dame University)
NQCJ	*Notes and Queries on China and Japan*
O	*Observatory*
OAA	*Orientalia Antiqua*
OAZ	*Ostasiatische Zeitschrift*
OB	*Orientalistische Bibliographie*
OC	*Open Court*
OL	*Old Lore; Miscellany of Orkney, Shetland, Caithness and Sutherland*
OLL	*Ostasiatische Lloyd*
OLZ	*Orientalische Literatur-Zeitung*
OR	*Oriens*
OSIS	*Osiris*

P	*Politica*
PA	*Pacific Affairs*
PAAAS	*Proceedings of the American Academy of Arts and Sciences*
PAS	*Proceedings of the Aristotelian Society*
PBA	*Proceedings of the British Academy*
PC	*People's China*
PEW	*Philosophy East and West* (University of Hawaii)
PHR	*Philosophical Review*
PL	*Philologus; Zeitschrift f. d. klass. Altertums*
PM	*Presse Médicale*
PNHB	*Peking Natural History Bulletin*
PP	*Past and Present*
PR	*Princeton Review*
PRSM	*Proceedings of the Royal Society of Medicine*
QBCB/E	*Quarterly Bulletin of Chinese Bibliography* (English edition)
QRSIACE	*Quarterly Review of the Sun Yat-Sen Institute for the Advancement of Culture and Education*
RHR/AMG	*Revue de l'Histoire des Religions* (Annales du Musée Guimet)
RMM	*Revue de Métaphysique et de Morale*
RP	*Revue philosophique*
RPLHA	*Revue de Philologie, Littérature et d'Histoire anciennes*
RR	*Review of Religion*
S	*Sinologica*
SA	*Sinica*
SAM	*Scientific American*
SBE	*Sacred Books of the East* Series
SBGAEU	*Sitzungsberichte d. berliner Gesellschaft f. Anthropol., Ethnol. und Urgeschichte*
SCI	*Scientia*
SCM	Student Christian Movement
SG	*Shinagaku*
SHAW/PH	*Sitzungsberichte d. Heidelberger Akademie d. Wissenschaften* (Phil.-hist. Klasse)
SM	*Scientific Monthly* (formerly *Popular Science Monthly*)
SPAW/PH	*Sitzungsberichte d. preussischen Akademie d. Wissenschaften* (Phil.-hist. Klasse)
SPCK	Society for the Promotion of Christian Knowledge
SRIMR	*Scientific Reports, Rockefeller Institute of Medical Research* (New York)
SS	*Science and Society*
SSE	*Studia Serica* (West China Union University Library and Historical Journal)
SW	*Sociological World* (Yenching)

SWAW/PH	*Sitzungsberichte d. k. Akademie d. Wissenschaften, Wien* (Vienna) (Phil.-hist. Klasse)	*VS*	*Variétés Sinologiques*
SWJA	*Southwestern Journal of Anthropology* (U.S.A.)	*WCYK*	*Wên Chê Yüeh Khan* (Literary and Philosophical Monthly)
SZUQB	*Szechuan University Quarterly Bulletin*	*WHNP*	*Wên-Hsüeh Nien Pao* (Literary Annual)
		WR	*World Review*
TAPS	*Transactions of the American Philosophical Society*	*WUJAP*	*Wuhan University Journal of Arts and Philosophy*
TAS/J	*Transactions of the Asiatic Society of Japan*	*YAHS*	*Yenching Shih-Hsüeh Nien Pao* (*Yenching Annual of Historical Studies*) or *Yenching Historical Annual*
TFTC	*Tung Fang Tsa Chih*		
TH	*Thien Hsia* (Shanghai)		
TMIE	*Travaux et Mémoires de l'Institut d'Ethnologie* (Paris)	*YCHP*	*Yenching Hsüeh-Pao* (*Yenching Journal of Chinese Studies*)
TNS	*Transactions of the Newcomen Society*	*YJSS*	*Yenching Journal of Social Studies*
TP	*T'oung Pao*	*Z*	*Zalmoxis; Revue des Etudes religieuses*
TYG	*Tōyō Gakuho*		
		ZAW	*Zeitschrift f. d. alttestamentliche Wissenschaft*
UPLR/ALR	*University of Pennsylvania Law Review and American Law Register*	*ZDMG*	*Zeitschrift d. deutsch. morganländischen Gesellschaft*
		ZFE	*Zeitschrift f. Ethnologie*
VAG	*Vierteljahrsschrift d. astronomischen Gesellschaft*	*ZVRW*	*Zeitschrift f. d. vergleichende Rechtswissenschaft*
VBW	*Vorträge d. Bibliothek Warburg*		

A. CHINESE BOOKS BEFORE +1800

Each entry gives particulars in the following order:

(a) title, alphabetically arranged, with characters;

(b) alternative title, if any;

(c) translation of title;

(d) cross-reference to closely related book, if any;

(e) dynasty;

(f) date as accurate as possible;

(g) name of author or editor, with characters;

(h) title of other book, if the text of the work now exists only incorporated therein; or, in special cases, references to sinological studies of it;

(i) references to translations, if any, given by the name of the translator in Bibliography C;

(j) notice of any index or concordance to the book if such a work exists;

(k) reference to the number of the book in the *Tao Tsang* catalogue of Wieger (6), if applicable;

(l) reference to the number of the book in the *San Tsang* (Tripiṭaka) catalogues of Nanjio (1) and Takakusu & Watanabe, if applicable.

Words which assist in the translation of titles are added in round brackets.

Alternative titles or explanatory additions to the titles are added in square brackets.

It will be remembered (p. 585 above) that in Chinese indexes words beginning *Chh-* are all listed together after *Ch-*, and *Hs-* after *H-*, but that this applies to initial words of titles only.

Where there are any differences between the entries in these bibliographies and those in vol. 1, the information here given is to be taken as more correct.

References to the editions used in the present work, and to the *tshung-shu* collections in which books are available, will be given in the final volume.

ABBREVIATIONS

C/Han	Former Han.
H/Han	Later Han.
H/Shu	Later Shu (Wu Tai).
H/Thang	Later Thang (Wu Tai).
J/Chin	Jurchen Chin.
L/Sung	Liu Sung.
N/Chou	Northern Chou.
N/Chhi	Northern Chhi.
N/Sung	Northern Sung (before the removal of the capital to Hangchow).
N/Wei	Northern Wei.
S/Chhi	Southern Chhi.
S/Sung	Southern Sung (after the removal of the capital to Hangchow).

Ao Yü Tzu Hsü Hsi So Wei Lun 鏊隅子歔欷瑣微論.
　　Whispered Trifles by the Tree-stump Master.
　　Sung, c. +1040.
　　Huang Hsi 黃晞.

Chai Ching
　　See *Huang Ti Chai Ching.*

Chan Kuo Tshê 戰國策.
　　Records of the Warring States.
　　Chhin.
　　Writer unknown.

Chao Lun 肇論.
　　Discourses of Brother Chao
　　[dialectical philosophy interpreting the Mādhyamika Buddhist doctrines with the help of Taoist ideas].
　　Chin, c. +400.
　　Sêng-Chao 僧肇.
　　Tr. Liebenthal (1).
　　TW/1858.

Chê Yü Kuei Chien 折獄龜鑑.
　　Tortoise Mirror of Case Decisions.
　　Sung.
　　Chêng Kho 鄭克.

Chen Kao 眞誥.
　　True Reports.
　　Liang, early +6th century (but the earliest material contained in it is dated +365).
　　Thao Hung-Ching 陶弘景.

Chêng Lun 政論.
　　On Government.
　　H/Han, +155.
　　Tshui Shih 崔寔.

Chêng Mêng 正蒙.
　　Right Teaching for Youth.
　　Sung, c. +1076.
　　Chang Tsai 張載.

Chêng Mêng Chu 正蒙注.
　　Commentary on the *Right Teaching for Youth* (of Chang Tsai).
　　Chhing, c. +1650.
　　Wang Chhuan-Shan 王船山.

Chêng shih Hsing An 鄭氏星案.
　　Astrological opinions of Mr Chêng.
　　Yuan
　　Chêng Hsi-Chhêng 鄭希誠.

Chi Chung Chou Shu 汲冢周書.
　　The Books of (the) Chou (Dynasty) found in the Tomb at Chi.
　　See *I Chou Shu.*

Chi Jan.
　　See *Chi Ni Tzu.*

Chi Ku Chin Fo Tao Lun Hêng 集古今佛道論衡.
　　Critical Collections of Discourses on Buddhist Doctrine in various Ages.
　　Thang, +661 to +664.
　　See Pelliot (8).

Chi Ku Chin Fo Tao Lun Hêng (cont.)
Tao-Hsüan 道宣.
N/1471, TW/2104.

Chi Ku Lu 集古錄.
Collection of Ancient Inscriptions.
Sung, *c.* +1050.
Ouyang Hsiu 歐陽修.

Chi Ku Lu Pa Wei 集古錄跋尾.
Postscript to the *Collection of Ancient Inscriptions*.
Sung, *c.* +1060.
Ouyang Hsiu 歐陽修.

Chi Ni Tzu 計倪子.
[=*Fan Tzu Chi Jan* 范子計然.]
The Book of Master Chi Ni.
Chou, −4th century.
Attrib. Fan Li [Chi Jan] 范蠡 [計然].

Chi Shan Chi 霽山集.
Poetical Remains of the Old Gentleman of Chi Mountain.
Sung, end +13th century.
Lin Ching-Hsi 林景熙.

Chih Huang Chhüan Fa 治蝗全法.
Complete Handbook of Locust Control.
See Ku Yen (1) in Bibliography B.

Chin Kang Ching 金剛經.
Vajracchedikā Sūtra [Kumārajīva's Condensation of the *Prajñāpāramitā Sūtra*]; Diamond-cutter Sūtra.
Chin, +405.
Kumārajīva 鳩摩羅什婆.
N/10-15; TW/235 ff.

Chin Kuang Ming Tsui Shêng Wang Ching 金光明最勝王經.
Suvarṇa-prabhāsa Sūtra; The Gold-Gleaming.
India, tr. into Chinese, +415.
N/127, 130; TW/663 ff.

Chin Lou Tzu 金樓子.
Book of the Golden Hall Master.
Liang, *c.* +550.
Hsiao I 蕭繹
(Liang Yuan Ti) 梁元帝.

Chin Phing Mei 金瓶梅.
Golden Lotus [novel]. (Cf. *Hsü Chin Phing Mei*.)
Ming.
Writer unknown.
See Hightower (1), p. 95.
Tr. Egerton (1); Kuhn (2).

Chin Shu 晉書.
History of the Chin Dynasty, [+265 to +419].
Thang, +635.
Fang Hsüan-Ling 房玄齡 *et al.*
A few chs. tr. Pfizmaier (54–57).

Chin Ssu Lu 近思錄.
Summary of Systematic Thought.
Sung +1175.

Chu Hsi & Lü Tsu-Chhien 朱熹, 呂祖謙.
Tr. Graf (1).

Ching-Khang Hsiang Su Tsa Chi 靖康緗素雜記.
Miscellaneous Records relating to the Ching-Khang reign-period (last year of the N/Sung dyn. +1126).
Sung, early +12th century.
Huang Chao-Ying 黃朝英.

Chiu Chang Suan Shu 九章算術.
Nine Chapters on the Mathematical Art.
H/Han, +1st century (containing much material from C/Han and perhaps Chhin).
Writer unknown.

Chiu Thang Shu 舊唐書.
Old History of the Thang Dynasty, [+618 to +906].
Thang and Wu Tai, +945.
Liu Hsü 劉昫.

Chiu Ting Shen Tan Ching Chüeh.
See *Huang Ti Chiu Ting Shen Tan Ching Chüeh.*

Chou I.
See *I Ching.*

Chou I Chi Chieh 周易集解.
Collected Commentaries on the *Book of Changes*.
Thang, between +740 and +900.
Ed. Li Ting-Tso 李鼎祚.

Chou I Lüeh Li 周易略例.
Outline of the System used in the *Book of Changes*.
San Kuo, *c.* +240.
Wang Pi 王弼.

Chou I Pên I 周易本義.
The Basic Ideas of the *Book of Changes*.
Sung, +1177.
Chu Hsi 朱熹.

Chou I Tshan Thung Chhi 周易參同契.
See *Tshan Thung Chhi.*

Chou I Wai Chuan 周易外傳.
Commentary on the *Book of Changes*.
Chhing, *c.* +1670.
Wang Chhuan-Shan 王船山.

Chou Li 周禮.
Record of the Rites of (the) Chou (Dynasty) [descriptions of all government official posts and their duties].
C/Han, perhaps containing some material from late Chou.
Compilers unknown.
Tr. E. Biot (1).

Chü Lu 橘錄.
Orange Record [citrus horticulture].
Sung, +1178.
Han Yen-Chih 韓彥直.
Tr. Hagerty (1).

Chu Ping Yuan Hou Lun 諸病源候論.
　Discourses on the Origin of Diseases [systematic pathology].
　Sui, c. +607.
　Chhao Yuan-Fang 巢元方.
Chu Shu Chi Nien 竹書紀年.
　The Bamboo Books [annals].
　Chou, −295 and before, such parts as are genuine. (Found in the tomb of An Li Wang, a prince of the Wei State, r. −276 to −245; in +281.)
　Writers unknown.
　Tr. E. Biot (3).
Chu Tzu Chhüan Shu 朱子全書.
　Collected Works of Chu Hsi.
　Sung (ed. Ming; *editio princeps* +1713).
　Chu Hsi 朱熹.
　Ed. Li Kuang-Ti 李光地 (Chhing).
　Partial trs. Bruce (1); le Gall (1).
Chu Tzu Hsüeh Ti 朱子學的.
　What Chu Hsi was aiming at in his Philosophy.
　Ming, c. +1475.
　Ed. Chhiu Chün 邱濬.
Chu Tzu Pien 諸子辨.
　Discussions on the (Authenticity of) the Writings of the (Ancient) Philosophers.
　Yuan, +1358.
　Sung Lien 宋濂.
Chu Tzu Wên Chi 朱子文集.
　Selected Writings of Chu Hsi.
　Sung.
　Chu Hsi 朱熹.
　Ed. Chu Yü 朱玉 (Chhing).
Chu Tzu Yü Lei 朱子語類.
　Classified Conversations of Chu Hsi.
　Sung, c. +1270.
　Chu Hsi 朱熹.
　Ed. Li Ching-Tê 黎靖德 (Sung).
Chuang Tzu 莊子.
　[= *Nan Hua Chen Ching*.]
　The Book of Master Chuang.
　Chou, c. −290.
　Chuang Chou 莊周.
　Tr. Legge (5): Fêng Yu-Lan (5); Lin Yü-Thang (1); Wieger (7).
　Yin-Tê Index no. (Suppl.) 20.
Chuang Tzu Chieh 莊子解.
　An Interpretation of *Chuang Tzu*.
　Chhing, late +17th century.
　Wang Chhuan-Shan 王船山.
Chün-Chai Tu Shu Chih 郡齋讀書志.
　Memoir on the Authenticities of Ancient Books, by (Chhao Kung-Wu) Chün-Chai.
　Sung, c. +1175.
　Chhao Kung-Wu 晁公武.
Chün-Chai Tu Shu Fu Chih 郡齋讀書附志.
　Supplement to Chhao Kung-Wu's Memoir on the Authenticities of Ancient Books.

　Sung, c. +1200.
　Chao Hsi-Pien 趙希弁.
Chung Kuan Lun Su 中觀論疏.
　Commentary on the *Mādhyamika Śāstra* [contains information on the Buddhist philosophical schools of the Chin period].
　Sui, c. +615.
　Chi-Tsang 吉藏.
　TW/1824.
Chung Lun 中論.
　Discourse on the Middle Way [tr. of the *Mādhyamika Śāstra* of Nāgārjuna, on dialectical logic].
　India, c. +120.
　Tr. into Chinese by Kumārajīva, (Chin) +409.
　Tr. Stcherbatsky (2); Walleser (2).
　N/1179; TW/1564.
Chung Shu Kuo Tho-Tho Chuan 種樹郭橐駝傳.
　The Story of Camel-Back Kuo the Fruit-Grower.
　Thang, c. +800.
　Liu Tsung-Yuan 柳宗元.
Chung Yung 中庸.
　Doctrine of the Mean.
　Chou (enlarged in Chhin and Han), −4th century, with additions of −3rd.
　Trad. attrib. Khung Chi (Khung Tzu-Ssu) 孔伋 (孔子思).
　Tr. Legge (2); Lyall & Ching Chien-Chün (1); Hughes (2).
Chhao shih Ping Yuan.
　See *Chu Ping Yuan Hou Lun*.
Chhêng Wei Shih Lun 成唯識論.
　Vijñapti-mātratā-siddhi; Completion of the Doctrine of Mere Ideation [by Vasubandhu 天親, +5th century, and ten commentators].
　India, late +5th.
　Tr. into Chinese and conflated, Hsüan-Chuang 玄奘, Thang, c. +650.
　Tr. de la Vallée Poussin (3).
　TW/1585.
Chhi Chhi Mu Lüeh 奇器目略.
　Enumeration of Strange Machines.
　Chhing, +1683.
　Tai Jung 戴榕.
Chhi Wei 七緯.
　Seven (Chhan-)Wei (Apocryphal Treatises). (Cf. *Ku Wei Shu*.)
　Han, prob. −1st century.
　Ed. Chao Tsai-Han 趙在翰. (Chhing, 1804.)
Chhien Chin Fang 千金方.
　The Thousand Golden Remedies [medical].
　Thang, c. +670.
　Sun Ssu-Mo 孫思邈.
Chhien Fu Lun 潛夫論.
　Complaint of a Hermit Scholar.
　H/Han, +140.
　Wang Fu 王符.

Chhien Han Shu 前漢書.
History of the Former Han Dynasty [−206 to +24].
H/Han, *c.* +100.
Pan Ku 班固, and after his death in +92 his sister Pan Chao 班昭.
Partial trs. Dubs (2), Pfizmaier (32-4, 37-51), Wylie (2, 3, 10), Swann (1), etc.
Yin-Tê Index no. 36.

Chhien Hsü 潛虛.
The Hidden Emptiness [*I Ching* divination].
Sung, +11th century.
Ssuma Kuang 司馬光.

Chhih Pei Ou Than 池北偶談.
Chance Conversations North of Chhih(-chow)
Chhing, +1691.
Wang Shih-Chen 王士禎.

Chhin-Ting Hseih Chi Pien Fang Shu 欽定協紀辨方書.
Imperial Compendium of Astrology.
Chhing, +1739.
Ed. Wang Yün-Lu 王允祿.

Chhin-Ting Ku Chin Thu Shu Chi Chhêng 欽定古今圖書集成.
See *Thu Shu Chi Chhêng*.

Chhin-Ting Shu Ching Thu Shuo 欽定書經圖說.
The *Historical Classic* with Illustrations.
Chhing (edition by imperial order, 1905).
Ed. Sun Chia-Nai 孫家鼐 *et al.*

Chhin-Ting Ssu Khu Chhüan Shu Chien Ming Mu Lu 欽定四庫全書簡明錄.
Abridged Analytical Catalogue of the Books in the *Ssu Khu Chhüan Shu* Encyclopaedia, made by imperial order.
Chhing, +1782.
[There are two versions of this: (*a*) ed. Chi Yün 紀昀, which contains mention of nearly all the books in the *Thi Yao*; (*b*) ed. Yü Min-Chung 于敏中, which contains only the books which were copied.]

Chhin-Ting Ssu Khu Chhüan Shu Tsung Mu Thi Yao 欽定四庫全書總目提要.
Analytical Catalogue of the Books in the *Ssu Khu Chhüan Shu* Encyclopaedia, made by imperial order.
Chhing, +1782.
Ed. Chi Yün 紀昀
Indexes by Yang Chia-Lo; Yü & Gillis.
Yin-Tê Index no. 7.

Chhing Ching Ching 清靜經.
Canon of Pure Calm.
San Kuo (Wu), *c.* +250.
Ko Hsüan 葛玄.

Chhing Nang Ao Chih 青囊奧旨.
Mysterious Principles of the Blue Bag (i.e. the Universe) [geomancy].
Thang, *c.* +880.
Attrib. Yang Yün-Sung 楊筠松.

Chhou Jen Chuan 疇人傳.
Biographies of (Chinese) Mathematicians (and Scientists).
Chhing, +1799.
Juan Yuan 阮元.
With continuations by Lo Shih-Lin 羅士琳, Chu Kho-Pao 諸可寶 and Huang Chung-Chün 黃鐘駿, in *HCCC*, ch. 159.

Chhü I Shuo Tsuan 祛疑說纂.
Discussions on the Dispersal of Doubts.
Sung, *c.* +1230.
Chhu Yung 儲泳.

Chhu Tzhu 楚辭.
Elegies of Chhu (State).
Chou (with Han additions), *c.* −300.
Chhü Yuan 屈原
(& Chia I 賈誼, Yen Chi 嚴忌, Sung Yü 宋玉, Huainan Hsiao-Shan 淮南小山 *et al.*).
Partial tr. Waley (23).

Chhüan Chih 泉志.
Treatise on Coinage [numismatics].
Sung, +1149.
Hung Tsun 洪遵.

Chhuan-Shan I Shu 船山遺書.
Collected Writings of Wang Fu-Chih (Chhuan-Shan).
Chhing, 2nd half +17th century; not printed till 19th.
Wang Chhuan-Shan 王船山.

Chhun Chhiu 春秋.
Spring and Autumn Annals [i.e. Records of Springs and Autumns].
Chou; a chronicle of the State of Lu kept between −722 and −481.
Writers unknown.
See Wu Khang (1); Wu Shih-Chhang (1).
Tr. Couvreur (1), Legge (11).

Chhun Chhiu Fan Lu 春秋繁露.
String of Pearls on the *Spring and Autumn Annals*.
C/Han, *c.* −135.
Tung Chung-Shu 董仲舒.
See Wu Khang (1).
Partial trs. Wieger (2); Hughes (1); d'Hormon (ed.).
Chung-Fa Index no. 4.

Chhun Chhiu Wei Han Han Tzu 春秋緯漢含孳.
Apocryphal Treatise on the *Spring and Autumn Annals*; Cherished Beginnings of the Han Dynasty.
C/Han, −1st century.
Writer unknown.

Chhun Chhiu Wei Shuo Thi Tzhu 春秋緯說題辭.
Apocryphal Treatise on the *Spring and Autumn Annals*; Discussion of Phraseology.
C/Han, −1st century.
Writer unknown.

Chhung Hsü Chen Ching.
　　See *Lieh Tzu*.
Chhung Yu Lun 崇有論.
　　Discourse on the Primacy of Being.
　　Chin, *c.* +290.
　　Phei Wei 裴頠.

Erh Chhêng Chhüan Shu 二程全書.
　　Complete Works of the Two Chhêng
　　　　Brothers [Neo-Confucian philosophers].
　　Sung, *c.* +1110; collected +1323.
　　Chhêng I and Chhêng Hao 程頤, 程顥.
　　Coll. Ed. Than Shan-Hsin 譚善心 (Yuan).
Erh Chhêng Sui Yen 二程粹言.
　　Essential Words of the Two Chhêng
　　　　Brothers [Neo-Confucian philosophers].
　　Sung, *c.* +1110; collected +1166.
　　Chhêng I and Chhêng Hao 程頤, 程顥.
　　Ed. Hu Yin 胡寅 (Sung).
Erh Ti Chang 二諦章.
　　Essay on the Theory of the Double Truth
　　　　[dialectical logic].
　　Sui, *c.* +610.
　　Chi-Tsang 吉藏.
　　TW/1854.
Erh Ya 爾雅.
　　Literary Expositor [dictionary].
　　Chou material, stabilised in Chhin and C/Han.
　　Compiler unknown.
　　Enlarged and commented on *c.* +300 by
　　　　Kuo Pho 郭璞.
　　Yin-Tê Index no. (Suppl.) 18.

Fa Yen 法言.
　　Model Sayings.
　　Hsin, +5.
　　Yang Hsiung 楊雄.
　　Tr. von Zach (5).
Fan Tzu Chi Jan 范子計然.
　　See *Chi Ni Tzu*.
Fang Kuang Ta Chuang Yen Ching 方廣大莊
　　嚴經.
　　Lalitavistara Sūtra; Extended Account of
　　　　the Sports of the Boddhisattva.
　　India, +1st century; tr. into Chinese
　　　　+5th century.
　　N/159, 160; TW/186, 187.
Fêng Su Thung I 風俗通義.
　　Popular Traditions and Customs.
　　H/Han, +175.
　　Ying Shao 應劭.
　　Chung-Fa Index no. 3.
Fu Hsing Shu 復性書.
　　Essay on Returning to the Nature.
　　Thang, *c.* +820.
　　Li Ao 李翱.

Han Chhang-Li hsien-sêng Chhüan Chi 韓昌黎
　　先生全集.
　　Collected Works of Han Yü (with the critical
　　　　notes of 500 scholars).

Thang, +824 (this ed. +1761).
　　Han Yü 韓愈.
Han Fei Tzu 韓非子.
　　The Book of Master Han Fei.
　　Chou, early −3rd century.
　　Han Fei 韓非.
　　Partial tr. Liao Wên-Kuei (1).
Ho Kuan Tzu 鶡冠子.
　　Book of the Pheasant-Cap Master.
　　A very composite text, stabilised by +629,
　　　　as is shown by one of the MSS found at
　　　　Tunhuang. Much of it must be Chou
　　　　(−4th century) and most is not later than
　　　　Han (+2nd century), but there are later
　　　　interpolations including a +4th- or
　　　　+5th-century commentary, which has
　　　　become part of the text and accounts for
　　　　about a seventh of it (Haloun (5), p. 88).
　　　　It contains also a lost 'Book of the Art of
　　　　War'.
　　Attrib. Ho Kuan Tzu 鶡冠子.
　　TT/1161.
Ho Shou Wu Chuan 何首烏傳.
　　Tractate on the Ho-shou-wu Plant
　　　　(*Polygonum multiflorum*, R576).
　　Thang, *c.* +840.
　　Li Ao 李翱.
Honan Chhêng shih I Shu 河南程氏遺書.
　　Collected Sayings of the Chhêng brothers
　　　　of Honan [Neo-Confucian philosophers].
　　Sung, +1168.
　　Ed. Chu Hsi 朱熹.
Hou Han Shu 後漢書.
　　History of the later Han Dynasty [+25 to
　　　　+220].
　　L/Sung, +450.
　　Fan Yeh 范曄. The monograph chapters
　　　　by Ssuma Piao 司馬彪.
　　A few chs. tr. Chavannes (6, 16); Pfizmaier
　　　　(52, 53).
　　Yin-Tê Index no. 41.
Hua Hu Ching 化胡經.
　　Book of (Lao Tzu's Conversions of)
　　　　Foreigners.
　　Thang.
　　Writer unknown.
Hua Shu 化書.
　　Book of the Transformations (in Nature).
　　H/Thang, *c.* +940.
　　Attrib. Than Chhiao 譚峭.
　　TT/1032.
Hua Yen Ching 華嚴經.
　　Buddha-avataṃsaka Sūtra; The Adornment
　　　　of Buddha.
　　India, tr. into Chinese +6th century.
　　TW/278, 279.
Hua Ying Chin Chhen 花營錦陣.
　　Varied Positions of the Flowery Battle.
　　Ming, +1610.
　　Writer unknown.
　　Tr. van Gulik (3).

Huai Nan Tzu 淮南子.
  [=*Huai Nan Hung Lieh Chieh* 淮南鴻
  烈解.]
  The Book of (the Prince of) Huai Nan
    [compendium of natural philosophy].
  C/Han, c. −120.
  Written by the group of scholars gathered by
    Liu An (prince of Huai Nan) 劉安.
  Partial trs. Morgan (1); Erkes (1); Hughes (1);
    Chatley (1); Wieger (2).
  Chung-Fa Index no. 5.
  TT/1170.

Huang Chi Ching Shih Shu 皇極經世書.
  Book of the Sublime Principle which governs
    all Things within the World.
  Sung, c. +1060.
  Shao Yung 邵雍.
  TT/1028.

Huang Shu 黃書.
  The Yellow Book.
  Chhing, late +17th century.
  Wang Chhuan-Shan 王船山.

Huang Ti Chai Ching 黃帝宅經.
  The Yellow Emperor's House-Siting Manual.
  L/Sung, +5th century.
  Wang Wei 王微.

Huang Ti Chiu Ting Shen Tan Ching Chüeh 黃
  帝九鼎神丹經訣.
  Explanation of the Yellow Emperor's *Manual
    of the Nine-Vessel Magical Elixir*.
  Thang or Sung.
  Writer unknown.
  TT/878.

Huang Ti Su Wên Ling Shu Ching 黃帝素問
  靈樞經.
  Pure Questions of the Yellow Emperor;
    The Canon of the Spiritual Pivot [medical
    and physiological].
  Perhaps Thang, +8th century.
  Writer unknown.

Huang Ti Su Wên Nei Ching 黃帝素問內經.
  Pure Questions of the Yellow Emperor; The
    Canon of Internal Medicine.
  Chhin or Han.
  Writer unknown.
  Partial tr. Veith (1).

Hui-An hsien-sêng Chu Wên Kung Chi 晦菴先
  生朱文公集.
  Collected Writings of Chu Hsi.
  Sung, c. +1200.
  Chu Hsi 朱熹.

Hung Fan Wu Hsing Chuan 洪範五行傳.
  Discourse on the Hung Fan chapter of the
    Shu Ching in relation to the Five Elements.
  C/Han, c. −10.
  Liu Hsiang 劉向.

Hung Ming Chi 弘明集.
  Collected Essays on Buddhism.
    (Cf. *Kuang Hung Ming Chi*.)

S/Chhi, c. +500.
  Sêng-Yu 僧祐.

Hsi Ching Tsa Chi 西京雜記.
  Miscellaneous Records of the Western
    Capital.
  Liang or Chhen, mid +6th century.
  Attrib. to Liu Hsin 劉歆 (C/Han) and to
    Ko Hung 葛洪 (Chin), but prob. Wu
    Chün 吳均.

Hsi Ming 西銘.
  The Inscription on the Western Wall (of
    his lecture-theatre).
  Sung, c. +1066.
  Chang Tsai 張載.
  Tr. Eichhorn (3).

Hsi Yuan Lu 洗冤錄.
  The Washing Away of Wrongs [treatise on
    forensic medicine].
  Sung, +1247.
  Sung Tzhu 宋慈.
  Partial tr. H. A. Giles (7).

Hsiang-Shan Chhüan Chi 象山全集.
  Collected Writings of Lu Chiu-Yuan
    (Hsiang-Shan).
  Sung, c. +1200.
  Ed. Lu Hsiang-Shan 陸象山.

Hsiao Ching 孝經.
  Filial Piety Classic.
  Chhin and C/Han.
  Attrib. Tsêng Shen (pupil of Confucius)
    曾參.
  Tr. de Rosny (2); Legge (1).

Hsiao Tai Li Chi.
  See *Li Chi.*

Hsiao Tao Lun 笑道論.
  Taoism Ridiculed.
  N/Chou, +6th century.
  Chen Luan 甄鸞.

Hsieh Chi Pien Fang Shu.
  See *Chhin Ting Hsieh Chi Pien Fang Shu.*

Hsieh Lüeh 蟹略.
  Monograph on the Varieties of Crabs.
  Sung, c. +1185.
  Kao Ssu-Sun 高似孫.

Hsin Lun 新論.
  New Discussions.
  H/Han, c. +20.
  Huan Than 桓譚.

Hsin Shih 心史.
  History of Troublous Times.
  Yuan, but not discovered until +1638.
  Chêng Ssu-Hsiao [So-Nan] 鄭思肖
    [所南].

Hsin Thang Shu 新唐書.
  New History of the Thang Dynasty [+618
    to +906].
  Sung, +1061.
  Ouyang Hsiu 歐陽修 & Sung Chhi 宋祁.

Hsin Thang Shu (cont.)
 Partial trs. des Rotours (1, 2); Pfizmaier (66–
 74).
 Yin-Tê Index no. 16.
Hsing Chin Shen Pu Mieh 形靈神不滅.
 The Destructibility of the (Bodily) Form
 and the Indestructibility of the Spirit.
 Chin, c. +400.
 Hui-Yuan 慧遠.
Hsing Li Ching I 性理精義.
 Essential Ideas of the Hsing-Li (Neo-
 Confucian) School of Philosophers.
 Chhing, +1715.
 Li Kuang-Ti 李光地.
Hsing Li Pien I 性理辨疑.
 Doubts and Discussions concerning the
 Hsing-Li (Neo-Confucian) Philosophy.
 Ming, c. +1600.
 Yang Tung-Ming 楊東明.
Hsing Li Ta Chhüan [*Shu*] 性理大全 [書].
 Collected Works of (120) Philosophers of
 the Hsing-Li (Neo-Confucian) School
 [Hsing=Human Nature; Li=the
 Principle of Organisation in all Nature].
 Ming, +1415.
 Ed. Hu Kuang 胡廣 *et al.*
Hsing Ming Su Yuan 星命溯源.
 Astrology traced back to its Origins.
 Thang, +8th century.
 Chang Kuo 張果.
Hsing Ming Tsung Kua 星命總括.
 General Descriptions of Stars and their
 Portents.
 Liao, c. +1040.
 Yehlü Shun 耶律純.
Hsing Thung 刑統.
 Legal Code.
 Sung, +959, officially adopted + 963.
 Tou I 竇儀.
Hsing Thung Fu 刑統賦.
 Legal Code in Mnemonic Rhyme.
 Sung, c. +1180.
 Fu Lin 傅霖.
Hsing Tsung 星宗.
 The Company of the Stars.
 Thang, +732.
 Chang Kuo 張果.
Hsü Chin Phing Mei 續金瓶梅.
 Golden Lotus, continued [novel].
 (Cf. *Chin Phing Mei*.)
 Chhing, +17th century.
 Tzu Yang Tao-Jen 紫陽道人.
 Tr. Kuhn (1).
Hsü Hsi So Wei Lun.
 See *Ao Yü Tzu Hsü Hsi
 So Wei Lun*.
Hsü Po Wu Chih 續博物志.
 Supplement to the *Record of the Investigation
 of Things*. (Cf. *Po Wu Chih*.)
 Sung, mid +12th century.
 Li Shih 李石.

Hsü Shen Hsien Chuan 續神仙傳.
 Supplementary Lives of the Hsien.
 (Cf. *Shen Hsien Chuan*.)
 Thang.
 Shen Fên 沈汾.
Hsü Shih Shuo 續世說.
 Continuation of the *Discourses on the Talk of
 the Times* [see *Shih Shuo Hsin Yü*].
 Sung, c. +1157.
 Khung Phing-Chung 孔平仲.
Hsü Yu Kuai Lu 續幽怪錄.
 Supplementary Record of Things Dark
 and Strange.
 Thang, c. +850.
 Li Fu-Yen 李復言.
Hsüan-Ho Po Ku Thu Lu.
 See *Po Ku Thu Lu*.
Hsüan Nü Ching 玄女經.
 Canon of the Mysterious Girl.
 Han.
 Writer unknown.
 Only as fragment in *Shuang Mei Ching An
 Tshung Shu*.
 Partial tr. van Gulik (3).
Hsüan Tu Lü Wên 玄都律文.
 Code of the Mysterious Capital [organisation
 of the Taoist Church].
 Ascr. Chin.
 Writer unknown.
 TT/185.
Hsüeh Ku Pien 學古編.
 On our Knowledge of Ancient Objects [seal
 inscriptions].
 Yuan, +1307.
 Wuchhiu Yen 吾邱衍.
Hsün Tzu 荀子.
 The Book of Master Hsün.
 Chou, c. −240.
 Hsün Chhing 荀卿.
 Tr. Dubs (7).

I Ching 易經.
 The Classic of Changes [Book of Changes].
 Chou with C/Han additions.
 Compiler unknown.
 See Li Ching-Chhih (1, 2); Wu Shih-
 Chhang (1).
 Tr. R. Wilhelm (2), Legge (9), de Harlez (1).
 Yin-Tê Index no. (Suppl.) 10.
I Chou Shu 逸周書.
 [=*Chi Chung Chou Shu*.]
 Lost Books of Chou.
 Chou, c. −3rd century. (According to Sui
 tradition, found in the tomb of An Li
 Wang in +281, but this is questioned.)
 Writers unknown.
I Chuan 易傳.
 Record of Symbols in the (*Book of*) *Changes*
 [for divination].
 C/Han, c. −30.
 Ching Fang 京房.

I Chuan 易傳.
Explanations of the (*Book of*) *Changes*.
N/Wei, *c.* +490.
Kuan Lang 關朗.

I Hsia Lun 夷夏論.
Discourse on the Barbarians and the
Chinese.
L/Sung, *c.* +470.
Ku Huan 顧歡.

I Hsüeh Chhi Mêng 易學啓蒙.
Introduction to Knowledge of the (*Book of*)
Changes.
Sung, +1186.
Chu Hsi 朱熹.

I Lin 易林.
Forest of Symbols of the (*Book of*) *Changes*
[for divination].
C/Han, *c.* −40.
Chiao Kan 焦贛.

I Lin 意林.
Forest of Ideas [philosophical encyclo-
paedia].
Thang.
Ma Tsung 馬總.
TT/1244.

I Lung Thu 易龍圖.
The Dragon Diagrams of the (*Book of*)
Changes.
Wu Tai, *c.* +950.
Chhen Thuan 陳搏.

I Shu Kou Yin Thu 易數鈎隱圖.
The Hidden Number-Diagrams in the (*Book
of*) *Changes* Hooked Out.
Sung, early +10th century.
Liu Mu 劉牧.

I Thu Ming Pien 易圖明辨.
Clarification of the Diagrams in the (*Book
of*) *Changes* [historical analysis].
Chhing, +1706.
Hu Wei 胡渭.

I Thung Shu 易通書.
Fundamental Treatise on the (*Book of*)
Changes [Neo-Confucian philosophy].
Sung, *c.* +1055.
Chou Tun-I 周敦頤.
Tr. Chou I-Chhing (1); Eichhorn (1).

I Tung Lin 易洞林.
Grottoes and Forests of the (*Book of*)
Changes [divination].
Chin, *c.* +300.
Kuo Pho 郭璞.

I Wei Chhien Tso Tu 易緯乾鑿度.
Apocryphal Treatise on the (*Book of*)
Changes; a Penetration of the Regularities
of Chhien (the first Kua).
C/Han, −1st century.
Writer unknown.

I Wei Chi Lan Thu 易緯稽覽圖.
Apocryphal Treatise on the (*Book of*)
Changes; Consultation Charts.

C/Han, −1st century.
Writer unknown.

I Wei Thung Kua Yen 易緯通卦驗.
Apocryphal Treatise on the (*Book of*)
Changes; Verifications of the Powers of
the Kua.
C/Han, −1st century.
Writer unknown.

I Yü Chhien Chi 疑獄前集.
First Collection of Doubtful Law Cases.
Wu Tai, between +907 and +940.
Ho Ning 和凝.

I Yü Hou Chi 疑獄後集.
Second Collection of Doubtful Law Cases.
Wu Tai, between +940 and +960.
Ho Mêng 和㠓.

Ishinhō 醫心方.
The Heart of Medicine [partly a collection
of ancient Chinese and Japanese books].
Japan, +982 (not printed till +1854).
Tamba no Yasuyori 丹波康賴.

Jen Wu Chih 人物志.
The Study of Human Abilities.
San Kuo (Wei), *c.* +235.
Liu Shao 劉邵.
Tr. Shryock (2).

Jih Chih Lu 日知錄.
Daily Additions to Knowledge.
Chhing, +1673.
Ku Yen-Wu 顧炎武.

Ju Shih Lun 如實論.
Tarka-Śāstra [treatise on formal logic].
India, +5th century.
Vasubandhu (Thien-Chhin 天親).
Tr. into Chinese by Paramārtha (Chen-Ti
眞諦), early +6th century, Liang.
N/1252; TW/1633.

Kao Sêng Chuan 高僧傳.
Biographies of Famous (Buddhist) Monks.
Liang, between +519 and +554.
Hui-Chiao 慧皎.
TW/2059.

Kêng Sêng Lun 更生論.
On Reincarnation.
N/Wei, +6th century.
Lo Chün-Chang 羅君章.

Khan Yü Man Hsing 堪輿漫興.
Agreeable Geomantic Aphorisms.
Ming, *c.* +1370.
Liu Chi 劉基.

Khang Tshang Tzu 亢倉子.
The Book of Master Khang Tshang.
Thang, +745.
Wang Shih-Yüan 王士元.

Khao Ku Thu 考古圖.
Illustrations of Ancient Objects.
Sung, +1092.
Lü Ta-Lin 呂大臨.

Khao Kung Chi 考工記.
The Artificers' Record [a section of the *Chou Li*].
Chou and Han, perhaps originally an official document of Chhi State, incorporated *c.* −140.
Tr. E. Biot (1).

Khao Kung Chi Thu Chu 考工記圖注.
Illustrated Commentary on the *Artificers' Record* (of the *Chou Li*).
Chhing, +1746.
Tai Chen 戴震.
In *HCCC*, chs. 563, 564.

Khun Chih Chi 困知記.
Convictions Reached after Hard Study.
Ming, +1531.
Lo Chhin-Shun 羅欽順.

Khung Tshung Tzu 孔叢子.
The Book of Master Khung Tshung.
Prob. H/Han or later.
Attrib. Khung Fu 孔鮒.

Khung Tzu Chia Yü 孔子家語.
Table Talk of Confucius
H/Han or more probably San Kuo, early +3rd century (but compiled from earlier sources).
Ed. Wang Su 王肅.
Partial trs. Kramers (1); A. B. Hutchison (1); de Harlez (2).

Ku Chin Hsing Shih Shu Pien Chêng 古今姓氏書辨證.
Investigations of the Origins of Clan and Family Names, New and Old.
Sung, +1134.
Têng Ming-Shih 鄧名世.

Ku Chin Wei Shu Khao 古今僞書考.
Investigation into Forged Books, New and Old.
Chhing, *c.* +1675.
Yao Chi-Hêng 姚際恆.

Ku Wei Shu 古微書.
Old Mysterious Books [a collection of the apocryphal Chhan-Wei treatises]. (Cf. *Chhi Wei.*)
Date uncertain, in part C/Han.
Ed. Sun Chio 孫瑴 (Ming).

Ku Yü Thu 古玉圖.
Illustrated Description of Ancient Jade Objects.
Yuan, +1341.
Chu Tê-Jun 朱德潤.

Ku Yü Thu Phu 古玉圖譜.
Illustrated Record of Ancient Jades.
Alleged to be Sung, +1176, but really a forgery; first edition +1712.
Attrib. Lung Ta-Yuan 龍大淵.
See Pelliot (22).

Kuan Shih Ti Li Chih Mêng 管氏地理指蒙.
Mr Kuan's Geomantic Indicator.

Ascr. San Kuo, +3rd century; prob. Thang, +8th century.
Attrib. Kuan Lo 管輅.

Kuan Tzu 管子.
The Book of Master Kuan.
Chou and C/Han. Perhaps mainly compiled in the Chi-Hsia Academy (late −4th century) in part from older materials.
Attrib. Kuan Chung 管仲.
Partial trs. Haloun (2, 5); Than Po-Fu *et al.*

Kuan Wu Phien 觀物篇.
Treatise on the Observation of Things.
Sung, *c.* +1060.
Shao Yung 邵雍.

Kuan Yin Tzu 關尹子.
The Book of Master Kuan Yin.
Thang, +742 (may be later Thang or Wu Tai).
Prob. Thien Thung-Hsiu 田同秀.

Kuang Hung Ming Chi 廣弘明集.
Further Collection of Essays on Buddhism. (Cf. *Hung Ming Chi.*)
Thang, *c.* +660.
Tao-Hsüan 道宣.

Kuang Shih Ming 廣釋名.
The Enlarged *Explanation of Names* [dictionary]. (Cf. *Shih Ming.*)
See Chang Chin-Wu (1) in Bibliography B.

Kuei Chhien Chih 歸潛志.
On Returning to a Life of Obscurity.
J/Chin, +1235.
Liu Chhi 劉祁.

Kuei Ku Tzu 鬼谷子.
Book of the Devil Valley Master.
Chou, −4th century? (perhaps partly Han or later).
Writer unknown.

Kuliang Chuan 穀梁傳.
Master Kuliang's Commentary on the *Spring and Autumn Annals*.
Chou (with Chhin and Han additions), late −3rd and early −2nd centuries.
Attrib. Kuliang Chhih 穀梁赤.
See Wu Khang (1).

Kung Chhi Shih Hua 碧溪詩話.
River-Boulder Pool Essays [literary criticism].
Sung, +1168.
Huang Chhê 黃徹.

Kung Kuo Ko 功過格.
Examination of Merits and Demerits.
Thang, +8th century.
Attrib. Lü Tung-Pin 呂洞賓.

Kungsun Lung Tzu 公孫龍子.
The Book of Master Kungsun Lung. (Cf. *Shou Pai Lun.*)
Chou, −4th century.
Kungsun Lung 公孫龍.
Tr. Ku Pao-Ku (1); Perleberg (1); Mei Yi-Pao (3).

Kungyang Chuan 公羊傳.
    Master Kungyang's Commentary on the
        Spring and Autumn Annals.
    Chou (with Chhin and Han additions), late
        −3rd and early −2nd centuries.
    Attrib. Kungyang Kao 公羊高, but more
        probably Kungyang Shou 公羊壽.
    See Wu Khang (1).

Kuo Yü 國語.
    Discourses on the (ancient feudal) States.
    Late Chou, Chhin and C/Han, containing
        early material from ancient written
        records.
    Writers unknown.

Lang Huan Chi 瑯環記.
    On the Cyclical Recurrence of World
        Catastrophes.
    Liao, +10th century.
    I Shih-Chen 伊世珍.

Lao Tzu Yen 老子衍.
    Generalisations on Lao Tzu.
    Chhing, late +17th century.
    Wang Chhuan-Shan 王船山.

Lêng-Ka A-Po-To-Lo Pao Ching 楞伽阿跋多
    羅寶經.
    Laṅkāvatāra Sūtra; The Entrance of the
        Good Doctrine into Lanka.
    India, +3rd century; tr. into Chinese +430
        and +433.
    Tr. D. T. Suzuki (2).
    N/175–7; TW/670 off.

Li Chi 禮記.
    [=*Hsiao Tai Li Chi*.]
    Record of Rites [compiled by Tai the
        Younger].
    C/Han, c. −50. The earliest pieces may
        date from the time of the *Analects*
        (c. −465/−450).
    Ed. Tai Shêng 戴聖.
    See Wu Shih-Chhang (1).
    Tr. Legge (7); Couvreur (3); R. Wilhelm (6).
    Yin-Tê Index no. 27.

Li Hai Chi 蠡海集.
    The Beetle and the Sea [title taken from the
        proverb that the beetle's eye view cannot
        encompass the wide sea—a biological
        book].
    Ming, late +14th century.
    Wang Khuei 王逵.

Li Hsü-Chung Ming Shu 李虛中命書.
    Book of Fate (-Calculation) of Li Hsü-
        Chung.
    Thang, +8th century.
    Li Hsü-Chung 李虛中.

Li Hsüeh Tsung Chuan 理學宗傳.
    General Chronicles of Philosophy [history
        of the Neo-Confucian school].
    Chhing, c. +1655.
    Sun Chhi-Fêng 孫奇逢.

Li Huo 理惑.
    The Resolution of Doubts.
    H/Han, +192.
    Mou Tzu 牟子.
    Tr. Pelliot (14).

Li Shih 隸釋.
    Collection of Han Inscriptions.
    Sung, +1167 to +1181.
    Hung Kua 洪适.

Li Shih A-Pi-Than Lun 立世阿毗曇論.
    Lokasthiti Abhidharma Śāstra; Philosophical
        Treatise on the Preservation of the World
        [astronomical].
    India, tr. into Chinese +558.
    Writer unknown.
    N/1297; TW/1644.

Li Tai Shen Hsien Thung Chien 歷代神仙
    通鑑.
    Survey of the Lives of the Hsien in all Ages
        (Cf. *Shen Hsien Thung Chien*.)
    Chhing, +1712.
    Hsü Tao 徐道 & Chheng Yü-Chhi 程毓奇.

Li Wei Chi Ming Chêng 禮緯稽命徵.
    Apocryphal Treatise on the *Record of Rites*;
        Investigation of Omens.
    C/Han, −1st century.
    Writer unknown.

Liang Shu 梁書.
    History of the Liang Dynasty [+502 to
        +556].
    Thang, +629.
    Yao Chha 姚察 and his son Yao Ssu
        Lien 姚思廉.

Lieh Hsien Chuan 列仙傳.
    Lives of Famous Hsien. (Cf. *Shen Hsien
        Chuan*.)
    Chin, +3rd or +4th century.
    Attrib. Liu Hsiang 劉向.
    Tr. Kaltenmark (2).

Lieh Tzu 列子.
    [=*Chhung Hsü Chen Ching*.]
    The Book of Master Lieh.
    Chou and C/Han −5th to −1st century.
        (Ancient fragments of miscellaneous
        origin finally cemented together with
        much new material about +380.)
    Attrib. Lieh Yü-Khou 列禦寇.
    Tr. R. Wilhelm (4); L. Giles (4); Wieger (7).
    TT/663.

Ling Hsien 靈憲.
    The Spiritual Constitution (or Mysterious
        Organisation) of the Universe [cosmo-
        logical and astronomical].
    H/Han, c. +120.
    Chang Hêng 張衡.
    YHSF, ch. 76.

Ling Pao Ching 靈寶經.
    Divine Precious Classic.
    San Kuo (Wu), c. +250.
    Ko Hsüan 葛玄.

Ling Shu Ching.
See *Huang Ti Su Wên Ling Shu Ching*.

Liu Jen Lei Chi 六壬類集.
Classified Collections on (Divination by) the Six Cardinal Points [geomancer's divining-board].
Chhing, based on Thang material.
Writer unknown.

Liu Jen Li Chhêng Ta Chhüan Chhien 六壬立成大全鈐.
Complete Key Tables of (Divination by) the Six Cardinal Points [geomancer's divining-board].
Chhing, based on Thang material.
Writer unknown.

Liu Thao 六韜.
The Six Quivers [treatise on the art of war].
H/Han, +2nd century, incorporating material as early as the −3rd.
Writer unknown.
See Haloun (5); L. Giles (11).

Liu Tzu 劉子.
The Book of Master Liu.
N/Chhi, c. +550.
Prob. Liu Chou 劉晝.
TT/1018.

Lo Lu Tzu.
Book of the Bead-string Master.
See *San Ming Hsiao Hsi Fu*.

Lo Shan Lu 樂善錄.
How Happiness comes to the Good.
Sung, +11th century.
Li Chhang-Ling 李昌齡.

Lo Tsu Chhê Chi 贏族車記.
Record of the Class of Helical Machines.
Chhing, late +18th century.
Tai Chen 戴震.

Lü Chai Shih Erh Pien 履齋示兒編.
Instructions and Miscellaneous Information for the Use of Children of his own Family, (by the Scholar of the) Right Comportment Library.
Sung, +1205.
Sun I 孫奕.

Lü Shih Chhun Chhiu 呂氏春秋.
Master Lü's Spring and Autumn Annals [compendium of natural philosophy].
Chou (Chhin), −239.
Written by the group of scholars gathered by Lü Pu-Wei 呂不韋.
Tr. R. Wilhelm (3).
Chung-Fa Index no. 2.

Lun Hêng 論衡.
Discourses Weighed in the Balance.
H/Han, +82 or +83.
Wang Chhung 王充.
Tr. Forke (4).
Chung-Fa Index no. 1.

Lun Yü 論語.
Conversations and Discourses (of Confucius), [perhaps Discussed Sayings, Normative Sayings, or Selected Sayings]; Analects.
Chou (Lu), c. −465 to −450.
Compiled by disciples of Confucius (chs. 16, 17, 18 and 20 are later interpolations).
Tr. Legge (2); Lyall (2); Waley (5); Ku Hung-Ming (1).
Yin-Tê Index no. (Suppl.) 16.

Mêng Chai Pi Than 蒙齋筆談.
Essays from the Mêng Hall.
Sung, +12th century.
Chêng Ching-Wang 鄭景望.

Mêng Chan I Chih 夢占逸旨.
Easy Explanation of the Principles of Oneiromancy.
Ming, +1562.
Chhen Shih-Yuan 陳士元.

Mêng Chhi Pi Than 夢溪筆談.
Dream Pool Essays.
Sung, +1086; last supplement dated +1091.
Shen Kua 沈括.

Mêng Tzu 孟子.
The Book of Master Mêng (Mencius).
Chou, c. −290.
Mêng Kho 孟軻.
Tr. Legge (3); Lyall (1).
Yin-Tê Index no. (Suppl.) 17.

Mêng Tzu Tzu I Su Chêng 孟子字義疏證.
Explanation of the Meanings of Mencian Terms.
Chhing, +1772.
Tai Chen 戴震.

Miao Fa Lien Hua Ching 妙法蓮華經.
Saddharma-puṇḍarīka Sūtra; The Lotus of the Wonderful Law.
India, c. +200; tr. into Chinese +5th century.
Tr. Soothill (3).
N/134, 136−9; TW/262 ff.

Ming Ju Hsüeh An 明儒學案.
Schools of Philosophers of the Ming Dynasty.
Chhing, c. +1700.
Huang Tsung-Hsi & Wan Ssu-Thung 黃宗羲, 萬斯同.

Ming Shu.
See *Li Hsü-Chung Ming Shu*.

Ming Tao Tsa Chih 明道雜志.
Miscellany of the Bright Tao.
Sung, late +11th century.
Chang Lei 張耒.

Mo Ching 墨經.
See *Mo Tzu*.

Mo Tzu (incl. *Mo Ching*) 墨子.
The Book of Master Mo.
Chou, −4th century.

Mo Tzu (cont.)
Mo Ti 墨翟 and disciples.
Tr. Mei Yi-Pao (1); Forke (3).
Yin-Tê Index no. (Suppl.) 21.
TT/1162.
Mou Tzu Li Huo.
See *Li Huo.*
Mu Thien Tzu Chuan 穆天子傳.
Account of the Travels of the Emperor Mu.
Chou, before −245. (Found in the tomb of
An Li Wang, a prince of the Wei State,
r. −276 to −245; in +281.)
Writer unknown.
Tr. Eitel (1); Chêng Tê-Khun (2).

Nan Chhi Shu 南齊書.
History of the Southern Chhi Dynasty
[+479 to +501].
Liang, +520.
Hsiao Tzu-Hsien 蕭子顯.
Nan Hua Chen Ching.
See *Chuang Tzu.*
Nei Ching.
See *Huang Ti Su Wên Nei Ching.*
Nung Shu 農書.
Treatise on Agriculture.
Sung, +1149; printed +1154.
Chhen Fu (Taoist) 陳旉.
Nung Shu 農書.
Treatise on Agriculture.
Yuan, +1313.
Wang Chen 王禎.

O Mêng 噩夢.
The Nightmare.
Chhing, late +17th century.
Wang Chhuan-Shan 王船山.

Pai Chhuan Hsüeh Hai 百川學海.
The Hundred Rivers Sea of Learning
[a collection of separate books; the
first *tshung-shu*].
Sung, late +12th or early +13th century.
Compiled and edited by Tso Kuei 左圭.
Pai Hu Thung Tê Lun 白虎通德論.
Comprehensive Discussions at the White
Tiger Lodge.
H/Han, c. +80.
Pan Ku 班固.
Tr. Tsêng Chu-Sên (1).
Pan-Jo Po-Lo-Mi-To Ching 般若波羅蜜多經.
Prajñāpāramitā Sūtra; The Perfection of
Wisdom.
India, c. +3rd century; tr. into Chinese
+5th century.
Writer unknown.
Trs. Lamotte (1); Conze (4).
N/19, 20, 935; TW/220.
Pao Phu Tzu 抱樸 (or 朴) 子.
Book of the Preservation-of-Solidarity
Master.

Chin, early +4th century.
Ko Hung 葛洪.
Partial trs. Feifel (1, 2); Wu & Davis (2), etc.
TT/1171–1173.
Pao Sêng Hsin Chien 保生心鑑.
Mirror of Medical Gymnastics.
Ming, +1506.
Hu Wên-Huan 胡文煥.
Pei-Chhi Tzu I 北溪字義.
(Chhen) Pei-Chhi's Analytic Glossary of (Neo-
Confucian) Philosophical Terms.
Sung, c. +1200.
Chhen Shun 陳淳.
Pên Chhi Ching 本起經.
Book of Origins.
Thang, +7th century.
Writer unknown.
Pên Tshao Kang Mu 本草綱目.
The Great Pharmacopoeia.
Ming, +1596.
Li Shih-Chen 李時珍.
Paraphrased and abridged tr. Read &
collaborators (1–7) and Read & Pak (1)
with indexes.
Pên Tshao Kang Mu Shih I 本草綱目拾遺.
Supplementary Amplifications of the *Great
Pharmacopoeia* (of Li Shih-Chen).
Chhing, +1769.
Chao Hsüeh-Min 趙學敏.
Pên Tshao Shih I 本草拾遺.
Omissions from Previous Pharmacopoeias.
Thang, c. +725.
Chhen Tshang-Chhi 陳藏器.
Phu Yao Ching 普曜經.
Lalitavistara Sūtra; Extended Account of
the Sports of the Boddhisattva.
India, +1st century; tr. into Chinese +5th
century.
N/160; TW/187.
Pi Shu Lu Hua 避暑錄話.
Conversations while Avoiding the Heat of
Summer.
Sung, +1156.
Yeh Mêng-Tê 葉夢得.
Pien Chêng Lun 辨正論.
Discourse on Proper Distinctions.
Thang, c. +630.
Fa-Lin 法琳.
Pien Huo Pien 辯惑編.
Disputations on Doubtful Matters.
Yuan, +1348.
Hsieh Ying-Fang 謝應芳.
Pien I Chih 辯疑志.
Notes and Queries on Doubtful Matters.
Thang.
Lu Chhang-Yuan 陸長源.
Po Ku Thu Lu 圖錄博古 (sometimes has *Hsüan-
Ho* 宣和, reign-period, at beginning of title).
Illustrated Record of Ancient Objects
[catologue of the archaeological museum
of the emperor Hui Tsung].

Po Ku Thu Lu (*cont.*)
    Sung, +1111.
    Wang Fu 王黻 *et al.*
Po Wu Chih 博物志.
    Record of the Investigation of Things.
    (Cf. *Hsü Po Wu Chih*.)
    Chin, *c.* +290.
    Chang Hua 張華.
Pu Shih Chêng Tsung Chhüan Shu 卜筮正宗
    全書.
    Encyclopaedia of Divination by the Tortoise-
    shell and the Milfoil.
    Chhing, +1709.
    Wang Wei-Tê 王維德.

San Kuo Chih 三國志.
    History of the Three Kingdoms [+220 to
    +280].
    Chin, *c.* +290.
    Chhen Shou 陳壽.
    Yin-Tê Index no. 33.
San Ming Hsiao Hsi Fu 三命消息賦.
    Essay on the Communications concerning
    the Three Kinds of Fate.
    Sung, +10th century.
    Lo Lu Tzu 珞琭子.
    Commentary by Hsü Tzu-Phing
    徐子平 (Sung).
San Ming Thung Hui 三命通會.
    Compilation of Material concerning the
    Three Kinds of Fate.
    Ming.
    Wan Min-Ying 萬民英.
San Tzu Ching 三字經.
    Trimetrical Primer.
    Sung, *c.* +1270.
    Wang Ying-Lin 王應麟.
    Tr. H. A. Giles (4), S. Julien (9).
Sao Shou Wên 搔首問.
    Questions of a Head-Scratcher.
    Chhing, late +17th century.
    Wang Chhuan-Shan 王船山.
Sêng Shen Ching 生神經.
    Canon on the Generation of the Spirits in
    Man.
    Pre-Sui, before +500.
    Writer unknown.
    Tr. Gauchet (1).
    TT/162 and 315; comm. *TT*/393–395.
Shan Hai Ching 山海經.
    Classic of the Mountains and Rivers.
    Chou and C/Han.
    Writers unknown.
    Partial tr. de Rosny (1).
    Chung-Fa Index no. 9.
Shang Chhing Tung-Chen Chin Kung Tzu Fang
    Thu 上清洞眞九宮紫房圖.
    Description of the Purple Chambers of the
    Nine Palaces of the Tung-Chen Heaven
    [parts of the microcosmic body corre-
    sponding to stars in the macrocosm].
    Sung, probably +12th century.

Writer unknown.
    TT/153.
Shang Chhing Wo Chung Chüeh 上清握中訣.
    Explanation of the Highly Pure Method of
    Grasping the Central Ones.
    H/Han (?).
    Attrib. Fan Yu-Chhung 范幼沖.
    TT/137.
Shang Chün Shu 商君書.
    Book of the Lord Shang.
    Chou, −4th or −3rd century.
    Attrib. Kungsun Yang 公孫鞅.
    Tr. Duyvendak (3).
Shang Fang Ta Tung-Chen Yuan Miao Ching
    Thu 上方大洞眞元妙經圖.
    Diagrams of the Mysterious
    Cosmogonic Classic of the Tung-Chen
    Scriptures.
    Thang, before +740.
    Writer unknown.
    TT/434.
Shang Shu Ta Chuan 尚書大傳.
    Great Commentary on the Shang Shu
    chapters of the *Shu Ching* (Historical
    Classic).
    C/Han, −2nd century.
    Fu Shêng 伏勝.
Shang Shu Wei Hsüan Chi Chhien 尚書緯璇
    璣鈐.
    Apocryphal Treatise on the *Historical Classic*
    The Linchpin of the Polar Axis.
    C/Han, −1st century.
    Writer unknown.
Shen Chien 申鑒.
    Precepts Presented (to the Emperor).
    H/Han, *c.* +190.
    Hsün Yüeh 荀悅.
Shen Hsiang Chhüan Pien 神相全編.
    Complete Account of Physiognomical
    Prognostication.
    Ming, *c.* +1400.
    Yuan Kung & Yuan Chung-Chhê
    袁珙, 袁忠徹.
Shen Hsien Chuan 神仙傳.
    Lives of the Divine Hsien. (Cf. *Lieh Hsien
    Chuan* and *Hsü Shen Hsien Chuan*.)
    Chin, early +4th century.
    Attrib. Ko Hung 葛洪.
Shen Hsien Thung Chien 神仙通鑑.
    Survey of the Lives of the Hsien. (Cf. *Li
    Tai Shen Hsien Thung Chien*.)
    Ming, +1640.
    Hsüeh Ta-Hsün 薛大訓.
Shen I Ching 神異經 (or *Chi* 記).
    Book of the Spiritual and the Strange.
    Prob. +4th or +5th century.
    Attrib. Tungfang Shuo 東方朔.
Shen Mieh Lun 神滅論.
    On the Extinction of the Soul.
    Liang, +484.
    Fan Chen 范縝.

Shen Pu Mieh Lun 神不滅論.
On the Indestructibility of the Soul.
Liang, *c.* +500.
Chêng Tao-Chao 鄭道昭.

Shen Thung Yu Hsi Ching 神通遊戲經.
Lalitavistara Sūtra; Extended Account of
the Sports of the Boddhisattva.
India, +1st century; tr. into Chinese +5th
century.
N/159, 160; TW/186, 187.

Shen Tzu 慎子.
The Book of Master Shen.
Date unknown, probably between +2nd
and +8th centuries.
Attrib. Shen Tao (Chou philosopher)
慎到.

Shêng Mên Shih Yeh Thu 聖門事業圖
Diagrams of Matters discussed in the
Schools of the Sages.
Sung.
Li Kuo-Chi 李國紀.

Shih Chi 史記.
Historical Record (down to −99).
C/Han, *c.* −90.
Ssuma Chhien 司馬遷, and his father
Ssuma Than 司馬談.
Partial trs. Chavannes (1); Pfizmaier (13–
36); Hirth (2); Wu Khang (1); Swann
(1), etc.
Yin-Tê Index no. 40.

Shih Ching 詩經.
Book of Odes [ancient folksongs].
Chou, −9th to −5th centuries.
Writers and compilers unknown.
Tr. Legge (1, 8); Waley (1); Karlgren (14).

Shih-erh Chang Fa 十二杖法.
The Method of the Twelve Chang
[geomancy].
Thang, *c.* +880.
Attrib. Yang Yün-Sung 楊筠松.

Shih Ku Wên 石鼓文.
The Stone Drum Inscriptions.
Sung, *c.* +1150.
Chêng Chhiao 鄭樵.

Shih Ming 釋名.
Explanation of Names [dictionary].
H/Han, *c.* +100.
Liu Hsi 劉熙.

Shih-san Ching Chu Su 十三經注疏.
The Thirteen Classics with Collected
Commentaries.
Sung, first edited +12th century.
Ed. Huang Thang 黃唐.

Shih Shuo Hsin Yü 世說新語.
New Discourse on the Talk of the Times
[notes of minor incidents from Han to
Chin]. (Cf. *Hsü Shih Shuo*.)
L/Sung, +5th century.
Liu I-Chhing 劉義慶.
Commentary by Liu Hsün 劉峻 (Liang).

Shih Su 詩疏.
Studies on the *Book of Odes*.
Thang, *c.* +640.
Khung Ying-Ta 孔穎達.

Shih-Tshu Lai Chi 石徂徠集.
Shih (Chieh's) Encouraging Exhortations.
Sung, *c.* +1045.
Shih Chieh 石介.

Shih Tzu 尸子.
The Book of Master Shih.
Ascr. Chou, −4th century; probably
+3rd or +4th century.
Attrib. Shih Chiao 尸佼.

Shih Wu Chi Yuan 事物紀原.
Records of the Origins of Affairs and Things.
Sung, *c.* +1085.
Kao Chhêng 高承.

Shou Pai Lun 守白論.
A Treatise in Defence of (the Doctrine of)
Whiteness (and Hardness)
Alternative title for *Kungsun Lung Tzu*,
(*q.v.*).

Shu Ching 書經.
Historical Classic [Book of Documents].
The 29 'Chin Wên' chapters mainly Chou
(a few pieces possibly Shang); the 21
'Ku Wên' chapters a 'forgery' by Mei Tsê
梅賾, *c.* +320, using fragments of
genuine antiquity. Of the former, 13 are
considered to go back to the −10th
century, 10 to the −8th, and 6 not before
the −5th. Some scholars accept only 16
or 17 as pre-Confucian.
Writers unknown.
See Wu Shih-Chhang (1); Creel (4).
Tr. Medhurst (1); Legge (1, 10); Karlgren (12).

Shu Ching Thu Shuo.
See *Chhin Ting Shu Ching Thu Shuo*.

Shu Chü Tzu 叔苴子.
Book of the Hemp-seed Master.
Ming, +15th or +16th century.
Chuang Yuan-Chhen 莊元臣.

Shu Pho 鼠璞.
Rats and Jade.
Sung, *c.* +1260.
Tai Chih 戴埴.

Shuang Mei Ching An Tshung Shu 雙梅景閣
叢書.
Double Plum-Tree Collection [of ancient
and medieval books and fragments on
Taoist sexual techniques].
See Yeh Tê-Hui 葉德輝 (1) in Bibliography B.

Shui Ching Chu 水經注.
Commentary on the *Waterways Classic*
[geographical account of rivers and canals
greatly extended].
N/Wei, late +5th or early +6th century.
Li Tao-Yuan 酈道元.

Shuo Wên Chieh Tzu 說文解字.
Analytical Dictionary of Characters.
H/Han, +121.
Hsü Shen 許慎.

Shuo Yuan 說苑.
Garden of Discourses.
Han, c. −20.
Liu Hsiang 劉向.

Sou Shen Chi 搜神記.
Reports on Spiritual Manifestations.
Chin, c. +348.
Kan Pao 干寶.
Partial tr. Bodde (9).

Sou Shen Hou Chi 搜神後記.
Supplementary Reports on Spiritual Mani-
festations.
Chin, late +4th or early +5th century.
Thao Chhien 陶潛.

Ssu Chieh 俟解.
Wait and Analyse.
Chhing, c. +1660.
Wang Chhuan-Shan 王船山.

Ssu Khu Chhüan Shu, etc.
See *Chhin Ting Ssu Khu Chhüan Shu*, etc.

Ssu Wên Lu 思問錄.
Record of Thoughts and Questionings.
Chhing, c. +1670.
Wang Chhuan-Shan 王船山.

Su Nü Ching 素女經.
Canon of the Immaculate Girl.
Han.
Writer unknown.
Only as fragment in *Shuang Mei Ching An
Tshung Shu*.
Partial tr. van Gulik (3).

Su Nü Miao Lun 素女妙論.
Mysterious Discourses of the Immaculate
Girl.
Ming, c. +1500.
Writer unknown.
Partial tr. van Gulik (3).

Su Shu 素書.
Book of Pure Counsels.
Ascr. Chhin or C/Han.
Attrib. Huang Shih Kung 黃石公.

Su Wên Ling Shu Ching.
See *Huang Ti Su Wên Ling Shu Ching*.

Su Wên Nei Ching.
See *Huang Ti Su Wên Nei Ching*.

Sui Shu 隋書.
History of the Sui Dynasty [+581 to +617].
Thang, +636 (annals and biographies);
+656 (monographs and bibliography).
Wei Chêng 魏徵 et al.
Partial trs. Pfizmaier (61–65); Balazs (7, 8);
Ware (1).

Sun Chho Tzu 孫綽子.
The Book of Master Sun Chho.
Chin, c. +320.
Sun Chho 孫綽.

Sung I Min Lu 宋遺民錄.
Sung officials who refused to serve the
Yuan Dynasty.
Ming.
Chhêng Min-Chêng 程敏政.

Sung Lun 宋論.
Discourse on the Sung Dynasty.
Chhing, late +17th century.
Wang Chhuan-Shan 王船山.

Sung Shu 宋書.
History of the (Liu) Sung Dynasty [+420
to +478].
S/Chhi, +500.
Shen Yo 沈約.
A few chs. tr. Pfizmaier (58).

Sung Ssu-Hsing Tzu-Wei Shu 宋司星子韋書.
Book of the Astrologer (Shih) Tzu-Wei of the
State of Sung.
Chou (Sung), early −5th century.
Shih Tzu-Wei 史子韋.

Sung Ssu Tzu Chhao Shih 宋四子抄釋.
Selections from the Writings of the Four
Sung (Neo-Confucian) Philosophers
[excl. Chu Hsi].
Sung (ed. Ming, +1536).
Ed. Lü Jan 呂柟.

Sung Yuan Hsüeh An 宋元學案.
Schools of Philosophers in the Sung and
Yuan Dynasties.
Chhing, c. +1750.
Huang Tsung-Hsi & Chhüan Tsu-Wang
黃宗羲; 全祖望.

Ta Chhing Lü Li 大清律例.
Penal Code of the Chhing Dynasty.
Chhing, +1646.
Compilers unknown; published under the
name of the emperor.
Tr. Staunton (1); Boulais (1).

Ta Chih Tu Lun 大智度論.
Commentary on the *Prajñāpāramitā Sūtra*.
See *Pan-Jo Po-Lo-Mi-To Ching*.

Ta Hsüeh 大學.
The Great Learning [The Learning of
Greatness].
Chou, c. −260.
Trad. attrib. Tsêng Shen 曾參, but
probably written by Yochêng Kho
樂正克, a pupil of Mencius.
Tr. Legge (2); Hughes (2).

Ta Khung Chhüeh Chou Wang Ching 大孔雀
咒王經.
Mahāmāyurī-vidyārājñī Sūtra; Great
Peacock Queen of Spells.
India, tr. into Chinese in Thang.
Tr. I-Ching 義淨.
TW/985 ff.

Ta Ming Lü 大明律.
Penal Code of the Ming Dynasty.

Ta Ming Lü (*cont.*)
Ming, +1373.
Compilers unknown.

Ta Pan-Jo Po-Lo-Mi-To Ching.
See *Pan-Jo Po-Lo-Mi-To Ching.*

Ta Tai Li Chi 大戴禮記.
Record of Rites [compiled by Tai the
Elder].
C/Han, stabilised H/Han, between +80 and
+100.
Attrib. ed. Tai Tê 戴德; in fact prob-
ably ed. Tshao Pao 曹褒.
See Legge (7).
Tr. Douglas (1), R. Wilhelm (6).

Tai Tung-Yuan Shih Chuang 戴東原事狀.
Some Account of Tai Chen (Tung-Yuan).
Chhing.
Ling Thing-Khan 凌廷堪.

Tao Tê Ching 道德經.
Canon of the Tao and its Virtue.
Chou, before −300.
Attrib. Li Erh (Lao Tzu) 李耳 (老子).
Tr. Waley (4); Chhu Ta-Kao (2); Lin Yü-
Thang (1); Duyvendak (18); and very many
others.

Tao Tsang 道藏.
The Taoist Patrology [containing 1464 Taoist
works].
All periods, but first collected and printed
in the Sung (+1111 to +1117). Also
printed in J/Chin (+1186 to +1191), Yuan,
and Ming (+1445, +1598 and +1607).
Index by Wieger (6), on which see Pelliot's
review.
Yin-Tê Index no. 25.

Tao Yen Nei Wai Pi Shê Chhüan Shu 道言內
外祕設全書.
Complete Book of the Established Inner
and Outer Doctrines of the Tao
[a compilation].
Chhing, +1717.
Compiler unknown.
Partial tr. Pfizmaier (81).

Têng Chen Yin Chüeh 登眞隱訣.
Instructions for Ascending to the True
Concealed Ones.
Liang, late +5th century.
Thao Hung-Ching 陶弘景.
TT/418.

Têng Hsi Tzu 鄧析子.
The Book of Master Têng Hsi.
Chou, ascr. −6th to −3rd (possibly as late as
+5th) century.
Attrib. Têng Hsi 鄧析.
Tr. H. Wilhelm (2).

Thai-Chhing Shen Chien 太清神鑑.
The Mysterious Mirror of the Thai-Chhing
Realm [treatise on physiognomy].
Named for +548; attrib. N/Chou about
+955, but probably Sung.
Attrib. Wang Pho 王朴.

Thai-Chhing Tao Yin Yang Sêng Ching 太清
導引養生經.
Manual of Nourishing the Life by
Gymnastics; a Thai-Chhing scripture.
Date uncertain.
Writer unknown.
TT/811.

Thai Chi Shuo 太極說.
Essay on the Supreme Pole.
Sung, *c.* +1175.
Chu Hsi 朱熹.
Tr. v. d. Gabelentz (2).

Thai Chi Thu Chieh I 太極圖解義.
Descriptive Exposition of the *Diagram of the
Supreme Pole.*
Sung, *c.* +1175.
Chu Hsi 朱熹.
Tr. Bruce (2).

Thai Chi Thu Shuo 太極圖說.
Explanation of the Diagram of the Supreme
Pole.
Sung, *c.* +1060.
Chou Tun-I 周敦頤.
Trs. v. d. Gabelentz (2); Forke (9);
Bruce (2); Chou I-Chhing (1).

Thai Chi Thu Shuo Chieh 太極圖說解 (or
Chu 註).
Philosophical Commentary on the *Explanation
of the Diagram of the Supreme Pole.*
Sung, +1173.
Chu Hsi 朱熹.
Trs. v. d. Gabelentz (2); Chou I-Chhing (1).

Thai Hsi Ching 胎息經.
Manual of Embryonic Respiration.
Date unknown.
Writer unknown.
TT/127.

Thai Hsüan Ching 太玄經.
Canon of the Great Mystery.
C/Han, *c.* +10.
Yang Hsiung 揚雄.

Thai I Chin Hua Tsung Chih 太一金華宗旨.
The Secret of the Golden Flower of the
Great Unity [Taoist manual of meditation
particularly associated with the Chin Tan
Chiao sect].
Chhing, +17th century.
Writer unknown.
In *Tao Tsang Hsü Pien Chhu Chi.*
Tr. R. Wilhelm & Jung (1).

Thai-Phing Kuang Chi 太平廣記.
Miscellaneous Records collected in the
Thai-Phing reign-period.
Sung, +981.
Ed. Li Fang 李昉.

Thai-Phing Yü Lan 太平御覽.
Thai-Phing reign-period Imperial
Encyclopaedia.
Sung, +983.
Ed. Li Fang 李昉.

Thai-Phing Yü Lan (*cont.*)
Some chs. tr. Pfizmaier (84–106).
Yin-Tê Index no. 23.

Thai Shang Huang Thing Wai Ching Yü Ching
太上黃庭外景玉經.
Excellent Jade Classic of the Yellow
Court.
San Kuo or Chin, +3rd or +4th century.
Writer unknown.
TT/329.

Thai Shang Kan Ying Phien 太上感應篇.
Tractate of Actions and Retributions.
Sung, early +11th century.
Attrib. Li Chhang-Ling 李昌齡.
Tr. Legge (5).
TT/1153.

Thai Shang San Thien Chêng Fa Ching 太上
三天正法經.
Exalted Classic of the True Law of the
Three Heavens.
Prob. Chin, before +4th century.
Writer unknown.
TT/1188.

Thang Lü Su I 唐律疏議.
Commentary on the Penal Code of the
Thang Dynasty [imperially ordered].
Thang, +653.
Ed. Chhangsun Wu-Chi 長孫無忌.

Thang Yü Lin 唐語林.
Miscellanea of the Thang dynasty.
Sung, collected *c.* +1107.
Wang Tang 王讜.

Thien Kung Khai Wu 天工開物.
The Exploitation of the Works of
Nature.
Ming, +1637.
Sung Ying-Hsing 宋應星.

Thien Ti Yin Yang Ta Lo Fu 天地陰陽大
樂賦.
Poetical Essay on the Supreme Joy.
Thang, *c.* +800.
Pai Hsing-Chien 白行簡.

Thien Tui 天對.
Answers about Heaven.
Thang, *c.* +800.
Liu Tsung-Yuan 柳宗元.

Thien Wên 天問.
Questions about Heaven [ode].
Chou, *c.* −300.
Chhü Yuan 屈原.
Tr. Erkes (8).

Thien Yin Tzu 天隱子.
Book of the Heaven-Concealed Master.
Thang, *c.* +720.
Ssuma Chhêng-Chêng 司馬承貞.

Thu Shu Chi Chhêng 圖書集成.
Imperial Encyclopaedia.
Chhing, +1726.
Ed. Chhen Mêng-Lei 陳夢雷 *et al.*
Index by L. Giles (2).

Thung Chien Kang Mu 通鑑綱目.
(Short view of the) Comprehensive Mirror
(of History for Aid in Government) Clas-
sified into Headings and Subheadings [the
Tzu Chih Thung Chien condensed].
Sung, +1189, begun +1172
Chu Hsi 朱熹 and his school.
With later continuations.
Partial tr. Wieger (1).

Thung Chih 通志.
Historical Collections.
Sung, *c.* +1150.
Chêng Chhiao 鄭樵.

Thung Chih Lüeh 通志略.
Compendium of Information [part of
Thung Chih (*q.v.*)].

Thung Shu.
See *I Thung Shu*.

Thung Ya 通雅.
General Encyclopaedia.
Ming and Chhing, finished +1636, pr. +1666
Fang I-Chih 方以智.

Thung Yuan Chen Ching.
See *Wên Tzu*.

Tsang Shu 葬書.
Burial Book.
Ascr. Chin, +4th century.
Attrib. Kuo Pho 郭璞.

Tshan Thung Chhi 參同契.
The Kinship of the Three; or, The
Accordance (of the *Book of Changes*)
with the Phenomena of Composite Things.
H/Han, +142.
Wei Po-Yang 魏伯陽.
Tr. Wu & Davis (1).

Tshan Thung Chhi Khao I 參同契考異.
A Study of the *Kinship of the Three*.
Sung, +1197.
Chu Hsi 朱熹 (originally using pseudonym
Tsou Hsin 鄒訢).

Tshao Mu Tzu 草木子.
The Book of the Fading-like-Grass Master.
Ming, +1378.
Yeh Tzu-Chhi 葉子奇.

Tshê Suan 策算.
On the Use of the Calculating-Rods.
Chhing, +1744.
Tai Chen 戴震.

Tso Chuan 左傳.
Master Tsochhiu's Enlargement of the
Chhun Chhiu (*Spring and Autumn Annals*).
Chou, compiled between −430 and −250,
but with additions and changes by
Confucian scholars of the Chhin and
Han, especially Liu Hsin. Deals with the
period −722 to −453. Greatest of the
three commentaries on the *Chhun Chhiu*,
the others being the *Kungyang Chuan* and
the *Kuliang Chuan*, but unlike them,
probably itself originally an independent
book of history.
Attrib. Tsochhiu Ming 左邱明.

Tso Chuan (*cont.*)

See Karlgren (8); Maspero (1); Chhi
Ssu-Ho (1); Wu Khang (1); Wu Shih-
Chhang (1); Eberhard, Müller &
Henseling.

Tr. Couvreur (1); Legge (11); Pfizmaier (1–12).

Tsun Sêng Pa Chien 遵生八牋.

Eight chapters on Putting Oneself in
Accord with the Life Force.

Ming, +1591.

Kao Lien 高濂.

Abr. tr. Dudgeon (1).

Tu Jen Ching 度人經.

Canon on (the Guidance of) Man through
the Stages of Birth and Rebirth.

Chin, early +4th century.

Writer unknown.

Partial tr. Gauchet (4).

TT/1 and 78.

Tu shih Hsing An 杜氏星案.

Astrological Opinions of Mr Tu.

Ming, c. +1470.

Tu Chhüan 杜全.

Tu Shu Chi Shu Lüeh 讀書記數略.

Register of Numerical Categories.

Chhing, +1707.

Kung Mêng-Jen 宮夢仁.

Tu Thung Chien Lun 讀通鑑論.

Conclusions on Reading the *Mirror of
Universal History* (of Ssuma Kuang).

Chhing, late +17th century.

Wang Chhuan-Shan 王船山.

Tung Hsiao Thu Chih 洞霄圖志.

Illustrated Description of the Tung Hsiao
(Taoist Temple at Hangchow).

Yuan, +1305, completed after +1306.
Completed by Mêng Tsung-Pao 孟宗寶
cf. Fu Lo-Shu (1)

Têng Mu 鄧牧.

Tung Hsüan Tzu 洞玄子.

Book of the Mystery-Penetrating Master.

Pre-Thang, perhaps +5th century.

Writer unknown.

In *Shuang Mei Ching An Tshung Shu*.

Tr. van Gulik (3).

Tzu Hua Tzu 子華子.

Book of Master Tzu-Hua.

Sung, early +11th century.

Chhêng Pên (ps.) 程本.

Wang Wên Chhêng Kung Chhüan Shu 王文成
公全書.

Collected Works of Wang Yang-Ming.

Ming, c. +1550.

Wang Yang-Ming 王陽明.

Wei Shih Erh-shih Lun 唯識二十論.

Vijñapti-mātratā-siddhi Vimśatikā;
Treatise in Twenty Stanzas on Mere
Ideation.

India, +5th century.

Vasubandhu (Thien-Chhin 天親), with
commentary by Dharmapāla (Hu-Fa 護法).

Tr. into Chinese in the Thang by Hsüan-
Chuang 玄奘.

Tr. Hamilton (1).

TW/1588ff.

Wei Shu 魏書.

History of the (Northern) Wei Dynasty
[+386 to +550, including the Eastern
Wei successor state].

N/Chhi, +554, revised +572.

Wei Shou 魏收.

See Ware (3). One ch. tr. Ware (1, 4).

Wên Hsüan 文選.

General Anthology of Prose and Verse.

Liang, +530.

Ed. Hsiao Thung (prince of the Liang) 蕭統.

Wên Shih Chen Ching 文始眞經.

See *Kuan Yin Tzu*.

Wên Tzu 文子.

[= *Thung Yuan Chen Ching*.]

The Book of Master Wên.

Han and later, but must contain pre-Chhin
material; probably took its present form
about +380.

Attrib. Hsin Yen 辛研 or 鈃.

Wu Hsing Ta I 五行大義.

Main Principles of the Five Elements.

Sui, c. +600.

Hsiao Chi 蕭吉.

Wu Nêng Tzu 无能子.

The Book of the Incapability Master.

Thang, +887.

Unknown Taoist.

TT/1016.

Wu Tshan Tsa Pien Hsing Shu 五殘雜變星書.

Book of the various Changes undergone
by the Wu Tshan asterism (and seventeen
other dangerous asterisms).

Chou or C/Han.

Writer unknown.

Wu Yüeh Chhun Chhiu 吳越春秋.

Spring and Autumn Annals of the States
of Wu and Yüeh.

H/Han.

Chao Yeh 趙曄.

Yang-Ming hsien-sêng Chi Yao 陽明先生
集要.

Selected Works of (Wang) Yang-Ming.

Ming, c. +1600.

Wang Yang-Ming 王陽明.

Partial tr. Henke (2).

Yang Sêng Yen Ming Lu 養生延命錄.

On Delaying Destiny by Nourishing the
Life.

Pre-Thang or early Thang, +5th or +7th
century.

Writer unknown.

TT/831.

Yeh Hsing Chu 夜行燭.

Candle in the Night.

Ming, c. +1390.

Tshao Tuan 曹端.

Yen Lien Chu 演連珠.
The String of Pearls Enlarged.
Chin, *c.* +290.
Lu Chi 陸機.

Yen shih Chia Hsün 顏氏家訓.
Mr Yen's Advice to his Family.
Sui, *c.* +590.
Yen Chih-Thui 顏之推.

Yen Thieh Lun 鹽鐵論.
Discourses on Salt and Iron [record of the debate on State control of commerce and industry of −81].
C/Han, *c.* −80.
Huan Khuan 桓寬.
Partial tr. Gale (1); Gale, Boodberg & Lin.

Yin Fu Ching 陰符經.
Harmony of the Seen and the Unseen.
Thang, +8th century.
Li Chhüan 李筌.
Tr. Legge (5).

Yin Wên Tzu 尹文子.
The Book of Master Yin Wên.
Ascr. Chou. Probably Han, including Warring States material.
Attrib. Yin Wên 尹文.
Partial tr. Masson-Oursel & Chu Chia-Chien.

Ying Tsao Fa Shih 營造法式.
Treatise on Architectural Methods.
Sung, +1097; printed +1103; revised +1145.
Li Chieh 李誡.

Yü Chao Shen Ying Chen Ching 玉照神應眞經.
See *Yu Chao Ting Chen Ching*.

Yü Chao Ting Chen Ching 玉照定眞經.
True Manual of Determinations by the Jade Shining Ones [astrology].
Ascr. Chin; *c.* +300; more probably Sung.
Attrib. Kuo Pho 郭璞; more probably Chang Yung 張顒.

Yü Chhiao Wên Tui 漁樵問對.
Conversation of the Fisherman and the Woodcutter.
Sung, *c.* +1070.
Shao Yung 邵雍.

Yü Chien 寓簡.
Allegorical Essays.
Sung.
Shen Tso-Chê 沈作喆.

Yü Fang Chih Yao 玉房指要.
Important Matters of the Jade Chamber.
Pre-Sui, perhaps +4th century.
Writer unknown.
Preserved only in *Ishinhō* (*q.v.*).
Partial tr. van Gulik (3).

Yü Fang Pi Chüeh 玉房祕訣.
Secret Instructions concerning the Jade Chamber.
Pre-Sui, perhaps +4th century.
Writer unknown.
Only as fragment in *Shuang Mei Ching An Tshung Shu* (*q.v.*).
Partial tr. van Gulik (3).

Yu Li Tzu 郁離子.
The Book of Master Yu Li.
Yuan, *c.* +1360.
Liu Chi 劉基.

Yü Shu Ching 玉樞經.
Canon of the Jade Pivot.
Yuan, +13th century.
Writer unknown.

Yuan Chen Tzu 元眞子.
The Book of the Original-Truth Master.
Thang, *c.* +770.
Chang Chih-Ho 張志和.

Yuan Ching 元經.
Treatise on Origins [chronicle history +290 to +589 in the style of the *Spring and Autumn Annals*].
Sui, *c.* +600.
Attrib. Wang Thung 王通, but may be by Juan I 阮逸 (+11th century).

Yuan Hsing 原性.
Essay on the Origin of (Man's) Nature.
Thang, *c.* +800.
Han Yü 韓愈.
Tr. Legge (3).

Yuan Jen Lun 原人論.
Discourse on the Origin of Man.
Thang, *c.* +800.
Ho Tsung-Mi 何宗密.
Tr. Haas (1).
N/1594, TW/1886.

Yuan Shan 原善.
On Original Goodness.
Chhing, +1776.
Tai Chen 戴震.

Yuan Shih 元史.
History of the Yuan (Mongol) Dynasty [+1206 to +1367].
Ming, *c.* +1370.
Sung Lien 宋濂 *et al.*
Yin-Tê Index no. 35.

Yün Chi Chhi Chhien 雲笈七籤.
Seven Bamboo Tablets of the Cloudy Satchel [a great Taoist collection].
Sung, +1025.
Chang Chün-Fang 張君房.
TT/1020.

B. CHINESE AND JAPANESE BOOKS AND JOURNAL ARTICLES SINCE +1800

Chang Chin-Wu (1) 張金吾.
Kuang Shih Ming 廣釋名.
The Enlarged *Explanation of Names*
[dictionary].
1814 (printed 1816).

Chang Hung-Chao (2) 章鴻釗.
*Ta-erh-wên ti Thien Tsê Lü yü Chuang Tzu
ti Thien Chün Lü* 達爾文的天擇律與
莊子的天鈞律.
Natural Selection Theories of Charles
Darwin and Chuang Chou.
HITC, 1927, **6** (no. 2), 1.

Chang Ping-Lin (1) 章炳麟.
Kuo Ku Lun Hêng 國故論衡.
Critical Discourses on History and Archaeo-
logy.
Kuo Hsüeh Chiang Hsi Hui, 1910.

Chang Tai-Nien (1) 張岱年.
Chung-Kuo Chih Lun Ta Yao 中國知論
大要.
Outline of Chinese Theories of Epistemo-
logy.
CHJ, 1934, **9**, 385.
Engl. abstr. *CIB*, 1936, **1**, 11.

Chang Tung-Sun (1) 張東蓀.
*Tshung Yen-Yü Kou Tsao shang Khan
Chung Hsi Chê-Hsüeh ti Chha-I* 從言語
構造上看中西哲學的差異.
On Philosophical Differences between China
and the West from the Standpoint of
Language Structure.
TFTC, 1938, **3**, 1.

Chang Tung-Sun (2) 張東蓀.
Kungsun Lung ti Pien-Hsüeh 公孫龍的
辯學.
The Logical Philosophy of Kungsun Lung.
YCHP, 1949, **37**, 27.

Chang Tung-Sun (3) 張東蓀.
Chih Shih yü Wên-Hua 知識與文化.
Epistemology and Culture.
Com. Press, Shanghai, 1940.

Chang Tung-Sun (4) 張東蓀.
Ssu-Hsiang Yen-Lun yü Wên-Hua 思想言
論與文化.
Thought, Language and Culture.
SW, 1938, **10** (no. 1).
Tr. by Li An-Chê as Chang Tung-Sun (1).

Chang Yin-Lin (2) 張蔭麟.
*Chung-Kuo Li-Shih shang chih 'Chhi Chhi' chi
chhi Tso-Chê* 中國歷史上之奇器及其作者
Scientific Inventions and Inventors in
Chinese History.
YCHP, 1928, *1* (no. 3), 359.

Chia Fêng-Chen (1) 賈豐臻.

Chung-Kuo Li-Hsüeh Shih 中國理學史.
History of Chinese Philosophy.
Com. Press, Shanghai, 1936.

Chiang Hêng-Yüan (1) 江恆源.
Chung-Kuo Hsien Chê Jen Hsing Lun 中國
先哲人性論.
Discussion of the Theories about Human
Nature in Ancient Chinese Philosophy.
Com. Press, Shanghai, *c.* 1930.

Chu Chhien-Chih (1) 朱謙之.
*Chung-Kuo Ssu-Hsiang tui yü Ou-chou Wên-
Hua chih Ying Hsiang* 中國思想對於
歐洲文化之影響.
The Influence of Chinese Thought on
Western Civilisation.
Com. Press, Shanghai, 1940.

Chu Pao-Chhang (1) 朱寶昌.
Wei Shih Hsin Chieh 唯識新解.
A New Interpretation of the *Vijñapti-
mātratā-siddhi Sūtra* of Vasubandhu.
YCHP, 1938, **23**, 93.

Chhen Mêng-Chia (1) 陳夢家.
Wu Hsing chih Chhi-Yüan 五行之起源.
On the Origin of the (Theory of the) Five
Elements.
YCHP, 1938, **24**, 35.

Chhen Mêng-Chia (2) 陳夢家.
*Ku Wên Tzu chung chih Shang Chou Chi-
Ssu* 古文字中之商周祭祀.
Sacrifices in the Shang and Chou periods as
seen in ancient inscriptions.
YCHP, 1936, **19**, 91.

Chhen Mêng-Chia (3) 陳夢家.
Shang Tai ti Shen-Hua yü Wu-Shu 商代
的神話與巫術.
Myths and Witchcraft of the Shang period.
YCHP, 1936, **20**, 486.

Chhen Phan (1) 陳槃.
Ku Chhan-Wei Shu Lu Chieh Thi 古讖緯
書錄解題.
Remarks on some Works of the Occult
Science of Prognostication in Ancient
China (the Chhan-Wei or Weft Classics).
AS/BIHP, 1945, **10**, 371; 1947, **12**, 35.

Chhen Phan (2) 陳槃.
Ku Chhan-Wei Shu Lu Chieh Thi 古讖緯
書錄解題.
Further Remarks on Some Works of the
Occult Science of Prognostication in
Ancient China.
AS/BIHP, 1948, **17**, 59; 1950, **22**, 85.

Chhen Phan (3) 陳槃.
Chhan-Wei Shih Ming 讖緯釋名.

Chhen Phan (3) (cont.)
The Origin of the Name Chhan-Wei
(Weft Classics).
AS/BIHP, 1946, **11**, 297.

Chhen Phan (4) 陳槃.
Chhan-Wei Su Yuan 讖緯溯源.
The Origin of the (content of the) Chhan-
Wei (Weft Classics) [attempted recon-
struction of a text of Tsou Yen].
AS/BIHP, 1946, **11**, 317.
Ref. W. Eberhard, OR, 1949, **2**, 193.

Chhen Phan (5) 陳槃.
Ku Chhan-Wei Chhüan I Shu Tshun Mu
Chieh Thi 古讖緯全佚書存目解題.
Remarks on Some Lost Works of the
Occult Science of Prognostication in
Ancient China.
AS/BIHP, 1947, **12**, 53; 1948, **17**, 65.

Chhen Phan (6) 陳槃.
Chhan-Wei Ming Ming chi chhi Hsiang
Kuan chih Chu Wên-Thi 讖緯命名及
其相關之諸問題.
Divinatory Terms and Kindred Questions.
LHP, 1949, **10** (no. 1), 19.

Chhen Phan (7) 陳槃.
Chan-Kuo Chhin Han chien Fang-Shih
Khao Lun 戰國秦漢間方士考論.
Investigations on the Magicians of the
Warring States, Chhin and Han periods.
AS/BIHP, 1948, **17**, 7.

Chhen Yin-Kho (1) 陳寅恪.
Thao Yuan-Ming chih Ssu-Hsiang yü
Chhing-Than chih Kuan-Hsi 陶淵明之
思想與清談之關係.
The Thought of Thao Yuan-Ming in
relation to the 'Philosophic Wit'
Schools.
Harvard-Yenching Inst., Peiping, 1945.

Chhêng Shu-Tê (1) 程樹德.
Chiu Chhao Lü Khao 九朝律考.
Investigation on the Laws of the Nine
Dynasties.
Shanghai, 1927.

Chhi Ssu-Ho (1) 齊思和.
Huang Ti chih Chih Chhi Ku Shih 黃帝
之制器故事.
Stories of the Inventions of Huang Ti.
YAHS, 1934, **2** (no. 1), 21.

Chhi Ssu-Ho (2) 齊思和.
Fêng-Chien Chih-Tu yü Ju Chia Ssu-Hsiang
封建制度與儒家思想.
Feudalism and Confucian Thought.
YCHP, 1937, **22**, 175.

Chhien Wên-Hsüan (1) 錢文選.
Chhien shih so Tshang Khan-Yü Shu Thi
Yao 錢氏所藏堪輿書提要.
Descriptive Catalogue of the Geomantic
Books collected by Mr Chhien.
Peking.
Cit. Wang Chen-To (5), p. 121.

Chhü Tui-Chih (1) 瞿兌之.
Shih Wu 釋巫.
On Chinese Witchcraft.
YCHP, 1930, **7**, 1327.

Creel, H. G. [Ku Li-Ya] (1) 顧立雅.
Shih 'Thien' 釋天.
On the Meaning and Origin of the word
'Heaven'.
YCHP, 1935, **18**, 59.

Fan Shou-Khang (1) 范壽康.
Wei Chin ti Chhing-Than 魏晉的清談.
'Philosophic Wit' in the Wei and Chin
dynasties (+3rd and +4th centuries).
Com. Press, Shanghai, 1936.
Also in WUJAP, 1936, **5**, 237.
Eng. abstr. CIB, 1936, **1**, 19.

Fêng Yu-Lan (1) 馮友蘭.
Chung-Kuo Chê-Hsüeh Shih 中國哲學史.
History of Chinese Philosophy (2 vols.).
Shen-chou, Shanghai, 1931 (vol. 1 only);
Com. Press, Shanghai & Chhangsha,
1934, 2nd ed. 1941.
Tr. Bodde.
Supplementary volume Chung-Kuo Chê-
Hsüeh Shih Pu, Com. Press, Shanghai,
1936, containing fifteen collected papers;
abstract by D. Bodde in Fêng-Yu-Lan (1),
vol. 1, second edition.

Fêng Yu-Lan (2) 馮友蘭.
Khung Tzu tsai Chung-kuo Li-Shih chung
chih Ti-Wei 孔子在中國歷史中之
地位.
The Place of Confucius in Chinese History.
YCHP, 1927, **2**, 233.
KSP, 1930, **2**, 194.

Fêng Yu-Lan (3) 馮友蘭.
Yuan Ju Mo 原儒墨.
On the Origin of the Confucians and Mohists.
CHJ, 1935, **10**, 279.
Eng. abstr. CIB, 1936, **1**, 1.

Fêng Yu-Lan (4) 馮友蘭.
Yuan Ming Fa Yin-Yang Tao-Tê 原名法
陰陽道德.
On the Origins of the Logicians, the
Legalists, the Naturalists, and the Taoists
[refuting the 'Ministries' legend; cf. Hu
Shih (6)].
CHJ, 1936, **11**, 279.
Eng. abstr. CIB, 1936, **1**, 1.

Fu Ssu-Nien (1) 傅斯年.
Shui shih 'Chhi Wu Lun' chih Tso chê?
誰是「齊物論」之作者?
Who wrote the 'Essay on the Identity of
Contraries' (in Chuang Tzu)?
AS/BIHP, 1936, **6**, 557.
Eng. abstr. CIB, 1937, **1**, 46.

Fu Ssu-Nien (2) 傅斯年.
Hsing Ming Ku Hsün Pien-Chêng 性命古
訓辨正.

A Critical Study of the Traditional Theories
of Human Nature and Destiny.
2 vols. Com. Press, Shanghai, 1940.
Academia Sinica Inst. Hist. Philol. Monogr.
Ser. B, no. 5.

Hou Wai-Lu (1) 候外廬.
*Chung-Kuo Ku Tai Ssu-Hsiang Hsüeh Shuo
Shih* 中國古代思想學說史.
Historical Reflections on Ancient Chinese
Philosophical thought.
Wên-fêng, Chungking and Kweiyang, 1944.
Hou Wai-Lu (2) 候外廬.
Chhuan-Shan Hsüeh An 船山學案.
The Teachings of Wang Chhuan-Shan.
San-yu, Chungking, 1944.
Hou Wai-Lu & Chi Hsüan-Ping (1) 候外廬,
紀玄冰.
*Wu Shih-Chi Mo Wei Wu Lun Chê Fan
Chen Yen-Chiu* 五世紀末唯物論者
范縝研究.
On the Materialism of Fan Chen at the end
of the Fifth Century.
CKKSH, 1950, **1**, 255.
Hu Chih-Hsin (1) 胡芝蕲.
Chuang Tzu Khao Chêng 莊子考證.
On the Authenticity of Chuang Tzu.
WHNP, 1937, **3**, 129.
Eng. abstr. *CIB*, 1938, **2**, 142.
Hu Shih (2) 胡適.
Than Than Shih Ching 談談詩經.
On the *Book of Odes*.
KSP, 1931, **3**, 576.
Eng. abstr. *CIB*, 1938, **3**, 72.
Hu Shih (3) 胡適.
Hsien Chhin Chu Tzu Chin Hua Lun 先秦
諸子進化論.
Theories of Evolution in the Philosophers
before the Chhin Period.
KHS, 1917, **3**, 19.
Hu Shih (4) 胡適.
Chung-Kuo Chê-Hsüeh Shih Ta Kang 中國
哲學史大綱.
History of Chinese Philosophy (vol. 1).
Shanghai, 1919.
Hu Shih (5) 胡適.
*Chhing Tai Han-Hsüeh Chia ti Kho-Hsüeh
Fang-Fa* 清代漢學家的科學方法.
The Scientific Method of the Scholars of the
'Han Learning' School in the Chhing
Dynasty.
KHS, 1920, **5** (no. 2), 125; (no. 3), 221.
Hu Shih (6) 胡適.
Chu Tzu pu Chhü yü Wang Kuan Lun
諸子不出於王官論.
The Philosophers did not come from the
Imperial Ministries.
KSP, 1933, **4**, 1.
Abstr. *CIB*, 1938, **3**, 80.
Hu Shih (8) 胡適.
Shuo Ju 說儒.

On the Ju (Confucians).
AS/BIHP, 1934, **4**, 233.
Eng. abstr. *CIB*, 1936, **1**, 1.
Hu Shih (9) 胡適.
*Chhing Tai Hsüeh-Chê ti Chih-Hsüeh Fang-
Fa* 清代學者的治學方法.
The Scientific Method of the Scholars of
the Chhing Dynasty.
In Collected Works (*Wên Tshun*),
1st series, vol. 2, p. 539.
Oriental Book Co., Shanghai, 1921.
Hu Shih (10) 胡適.
Tai Tung-Yuan ti Chê-Hsüeh 戴東原的
哲學.
The Philosophy of Tai Tung-Yuan (Tai
Chen).
Com. Press, Shanghai, 1927.
Huang Fang-Kang (1) 黃方剛.
Shih Lao Tzu chih Tao 釋老子之道.
On the Tao of Lao Tzu.
WUJAP, 1941, **7**, 41.
Huang Fang-Kang (2) 黃方剛.
*Chuang Tzu 'Thien Hsia' phien chung
Hui Shih shih Shih Chieh* 莊子「天下」篇
中惠施十事解.
Analysis of the Ten Paradoxes of Hui Shih
in the 'Thien-Hsia' chapter of *Chuang
Tzu*.
SZUQB, 1934, 149.
Hung Yeh (1) 洪業.
Li Chi Yin-Tê Hsü 禮記引得序.
On the Dates of Compilation of the *Li Chi*
and *Ta Tai Li Chi*.
Introduction to *Li Chi* Index in Harvard-
Yenching series, no. 21.
Hsieh Fu-Ya (N. Z. Zia) (1) 謝扶雅.
Thien Phing ho Tsou Yen 田駢和騶衍.
Thien Phing and Tsou Yen [two Warring
States philosophers].
LHP, 1934, **3**, 87.
Hsü Chung-Shu (1) 徐中舒.
Tsai Lun Hsiao-tun yü Yang-shao 再論小
屯與仰韶.
Further Remarks on Hsiao-tun and the
Yang-shao people.
In *An-yang Fa Chüeh Pao-Kao* 安陽發
掘報告.
Reports of the Excavations at An-yang
(ed. Li Chi), 1929, vol. 1 (pt. 3), p. 523,
esp. p. 539.
Hsü Ping-Chhang (1) 徐炳昶.
Chung-Kuo Ku Shih ti Chhuan-Shuo Shih-Tai
中國古史的傳說時代.
The Legendary Period in Ancient Chinese
History.
Chungking, 1943.
Hsü Ti-Shan (1) 許地山.
Tao Chiao Shih 道教史.
History of Taoism.
Com. Press, Shanghai, 1934.

Hsü Ti-Shan (2)　許地山.
Tao Chia Ssu-Hsiang yü Tao Chiao 道家
思想與道教.
Taoist Philosophy and Taoist Religion.
YCHP, 1927, **2**, 249.

Kao Hêng (1)　高亨.
Lao Tzu Chêng Ku 老子正詁.
Establishment of the text of the *Tao Tê
Ching*.
Khaiming, Shanghai, 1933; repr. 1948.

Khang Yu-Wei (1)　康有爲.
Ta Thung Shu 大同書.
Book of the Great Togetherness [socialism].
Conceived c. +1884, first pr. Chhang-hsing,
Shanghai, 1913; repr. 1935.

Ku Chieh-Kang (1)　顧頡剛.
Han Tai Hsüeh Shu Shih-Lüeh 漢代學術
史略.
Outline History of Learning in the Han
Dynasty.
Tung-fang, Chungking, 1944.

Ku Chieh-Kang (2) (ed.)　顧頡剛.
Ku Shih Pien 古史辨.
Discussions on Ancient History and
Philosophy [a collective work].
Vols. 1 to 3, and 5, Phu-shê, Peiping,
1916-31, 1935.

Ku Chieh-Kang (3)　顧頡剛.
*Chhun Chhiu Shih ti Khung Tzu ho Han
Tai ti Khung Tzu* 春秋時的孔子和
漢代的孔子.
The Confucius of the *Spring and Autumn
Annals* Period and the Confucius of the
Han Dynasty.
KSP, 1930, **2**, 130.
Eng. abstr. in CIB, 1938, **3**, 85.

Ku Chieh-Kang (4)　顧頡剛.
*Shih Ching tsai Chhun Chhiu Chan-Kuo
chien ti Ti-Wei* 詩經在春秋戰國間
的地位.
The Place of the *Book of Odes* in the
Spring and Autumn and Warring States
Periods.
KSP, 1931, **3**, 309.
Eng. abstr. CIB, 1938, **3**, 75.

Ku Chieh-Kang (5)　顧頡剛.
*Yü Chhien Hsüan-Thung hsien-sêng Lun
Ku Shih Shu* 與錢玄同先生論古
史書.
On (the Legendary Element in) Ancient
(Chinese) History—two letters to Chhien
Hsüan-Thung.
KSP, 1926, **1**, 59.
Eng. abstr. CIB, 1938, **3**, 67.

Ku Chieh-Kang (6)　顧頡剛.
*Wu Tê Chung Shih Shuo hsia ti Chêng-
Chih ho Li-Shih* 五德終始說下的政
治和歷史.

The Theories of the Rise and Fall of the
Five Elements in relation to Government
and History.
CHJ, 1930, **6**, 71.
Summary, W. Eberhard, SA, 1931, **3**, 136.

Ku Chieh-Kang (7)　顧頡剛.
*Chhan Jang Chhuan-Shuo Chhi yü Mo Chia
Khao* 禪讓傳說起於墨家考.
The Origin of the Voluntary Abdication
Legends from the Mohist School.
AP/HJ, 1936, **1**, 163.
Eng. abstr. CIB, 1936, **1**, 2.

Ku Yen (1)　顧彥.
Chih Huang Chhüan Fa 治蝗全法.
Complete Handbook of Locust Control.
Huan-chhêng, 1857.

Kubo Noritada (1)　窪德忠.
Dōkyō to Nihon no Minkan Shinkō 道教と
日本の民間信仰.
Taoism and Japanese Folk Religion.
JJE, 1953, **18**, 33.

Kuo Mo-Jo (1)　郭沫若.
Shih Phi Phan Shu 十批判書.
Ten Critical Essays.
Chün-i, Chungking, 1945.

Kuo Mo-Jo (4)　郭沫若.
Chhing Thung Shih-Tai 青銅時代.
On the Bronze Age (in China).
Shanghai, 1946, repr. 1947, 1951.

Kuwabara, Takeo (1) (ed.)　桑原武夫.
Rusō Kenkyū ルソー研究.
Essays on Rousseau [collective work by
several scholars].
Iwanami Shoten, Tokyo, 1951.

Li Chêng-Kang (1)　李證剛.
I Hsüeh Thao Lun Chi 易學討論集.
Collection of Treatises on the *Book of
Changes* [contains the translation by Liu
Pai-Min 劉百閔 of the Bouvet-Leibniz
letters].
Com. Press, Shanghai, 1941.

Li Chi (1) (ed.)　李濟.
An-yang Fa Chüeh Pao-Kao 安陽發掘
報告.
Reports of the Excavations at An-yang [one
of the Shang capitals].
Academia Sinica, 4 vols. consecutively
paged, vols. 1 and 2, Peiping, 1929;
vol. 3, Peiping, 1931; vol. 4, Shanghai,
1931.

Li Ching-Chhih (1)　李鏡池.
Chou I Kua Ming Khao Shih 周易卦名
考釋.
A Study of the Names of the Sixty-four
Hexagrams in the *Book of Changes*.
LHP, 1948, **9** (no. 1), 197 and 303.

Li Ching-Chhih (2)　李鏡池.
Chou I Shih Tzhu Hsü Khao 周易筮辭
續考.

A Further Study of the Explicative Texts
in the *Book of Changes*.
LHP, 1947, **8** (no. 1), 1 and 169.

Li Mai-Mai (*1*) 李麥麥.
*Chung-Kuo Ku Tai Chêng-Chih Chê-Hsüeh
Phi Phing* 中國古代政治哲學批評.
Considerations on Ancient Chinese Govern-
ment and Philosophy.
Shanghai, 1933.
Eng. abstr. *CIB*, 1938, **2**, 89.

Liang Chhi-Chhao (*1*) 梁啓超.
Yin Ping Shih Wên Chi 飲冰室文集.
Collected Essays.
Shanghai, 1926.

Liang Chhi-Chhao (*2*) 梁啓超.
Tzu Mo Tzu Hsüeh Shuo 子墨子學說.
Treatise on the Philosophy of Mo Tzu.
Shanghai, 1922.

Liang Chhi-Chhao (*3*) 梁啓超.
*Chung-Kuo Chin San-pai Nien Hsüeh Shu
Shih* 中國近三百年學術史.
History of Chinese Historical Scholarship
during the past Three Centuries.
Chungking, 1943.

Liang Chhi-Chhao (*4*) 梁啓超.
Yin-Yang Wu Hsing Shuo chih Lai Li
陰陽五行說之來歷.
On the Earliest Philosophical Use of the
Terms Yin and Yang and the Five Elements.
TFTC, 1923, **20** (no. 10); repr. *KSP*, 1935,
5, 343.

Liang Chhi-Chhao (*5*) 梁啓超.
Hsien Chhin Chêng-Chih Ssu-Hsiang Shih
先秦政治思想史.
History of Political Theory before the
Chhin Dynasty.
Peiping, 1924.

Liu Hsien (*1*) 劉咸.
Hainan Li Jen Wên Shen chih Yen-Chiu
海南黎人文身之研究.
On Tattooing among the Li People of
Hainan Island.
ETHS, 1936, **1**, 197.

Liu Ming-Shu (*1*) 劉銘恕.
*Han Wu Liang Tzhu Hua Hsiang chung
Huang-Ti Chhih-Yu Ku Chan Thu Khao*
漢武梁祠畫象中黃帝蚩尤古戰圖考.
A Study of the Fighting between Huang Ti
and the Chhih-Yu depicted in the rear
stone chamber of the Wu Liang tomb
Shrine of the Han Dynasty.
BCS, 1942, **2**, 341.

Liu Wên-Tien (*1*) 劉文典.
Chuang Tzu Pu Chêng 莊子補正.
Emended Text of *Chuang Tzu*, with com-
mentaries.
Com. Press, Shanghai, 1947.

Liu Wên-Tien (*2*) 劉文典.
Huai Nan Hung Lieh Chi Chieh 淮南鴻
烈集解.

Collected Commentaries on the *Huai Nan
Tzu* book.
Com. Press, Shanghai, 1923, 1926.

Lo Kên-Tsê (*1*) 羅根澤.
*Chan-Kuo Chhien Wu Ssu Chia Chu Tso
Shuo* 戰國前無私家著作說.
Absence of Books by Individual Writers
before the Warring States Period.
KSP, 1933, **4**, 8.
Eng. abstr. *CIB*, 1938, **3**, 82.

Lo Kên-Tsê (*2*) 羅根澤.
Chuang Tzu Wai Tsa Phien Than Yuan
莊子外雜篇探源.
Investigation of the Authorship of the
'Outer' and 'Miscellaneous' Chapters
of *Chuang Tzu*.
YCHP, 1936, **19**, 39.
Eng. abstr. *CIB*, 1937, **1**, 45.

Lo Kên-Tsê (*3*) (ed.) 羅根澤.
Ku Shih Pien 古史辨.
Discussions on Ancient History and
Philosophy [a collective work], vol. 4.
Phu-shê, Peiping, 1933.

Mei Ssu-Phing (*1*) 梅思平.
*Chhun Chhiu Shih Tai ti Chêng-Chih ho
Khung-Tzu ti Chêng-Chih Ssu-Hsiang*
春秋時代的政治和孔子的政治思想
Politics in the Spring and Autumn Period
and in the Thought of Confucius.
KSP, 1930, **2**, 161.
Eng. abstr. *CIB*, 1938, **3**, 83.

Niida, Naboru & Makino, Tatsumi (*1*) 仁井田陞,
牧野巽.
Ko-Tōritsu Sogi Seisaku Nendai-kō 故唐
律疏議製作年代考.
On the Date of Completion of the Official
Commentary on the Criminal Code of
the Thang dynasty.
TYG, 1931, **1**, 70; **2**, 50.

Phan Wei (*1*) 潘霨.
Wei Shêng Yao Shu 衛生要術.
Essential Techniques for the Preservation
of Health [based on earlier material on
breathing exercises, physical culture and
massage etc. collected by Hsü Ming-
Fêng 徐鳴峰].
1848, repr. 1857.

Shen Chia-Pên (*1*) (ed.) 沈家本.
Chen Pi Lou Tshung Shu 枕碧樓叢書.
Jade Pillow Tower Collection (of Sung
dynasty Juristic Books).
Shanghai, 1913.

Shen Chia-Pên (*2*) 沈家本.
Li Tai Hsing Fa Khao 歷代刑法考.
Investigation (and Collection of Documents)
on the History of (Chinese) Law.
Shanghai, 1900.

Shou Su (ps.) 守素.
　Mo Ching ti Lo-Chi Ssu-Hsiang 墨經的
　邏輯思想.
　The Logic of the Mohist Canon.
　CC, 1949 (n.s.), 3 (no. 6), 30.
Sun I-Jang (2) 孫詒讓.
　Mo Tzu Hsien Ku 墨子閒詁.
　Exposition of the Text of Mo Tzu.
　Shanghai, 1894.
Sun Tsu-Chi (1) 孫祖基.
　Chung-Kuo Li Tai Fa Chia Chu Shu Khao
　中國歷代法家著述考.
　Investigation of the History of Jurisprudence
　in China.
　Shanghai, 1934.

Takakusu, Junjiro & Watanabe, Kaigyoku (1) (ed.)
　高楠順次郎, 渡邊海旭.
　Ta Chêng Hsin Hsiu Ta Tsang Ching
　(Taishō Shinshū Daizōkyō) 大正新修大
　藏經.
　The Chinese Buddhist Tripiṭaka.
　Tokyo, 1924/1929. 55 vols.
　Catalogue by the same, Fascicule Annexe
　to Hōbōgirin, Maison Franco-Japonaise,
　Tokyo, 1931.
Than Chieh-Fu (1) 譚戒甫.
　Mo Ching I Chieh 墨經易解.
　Analysis of the Mohist Canon.
　Com. Press (for Wuhan University),
　Shanghai, 1935.
Thang Chün-I (1) 唐君毅.
　Hei-Ko-Erh ti Pien-Hua Hsing-erh-Shang
　Hsüeh, yü Chuang Tzu ti Pien-Hua Hsing-
　erh-Shang Hsüeh Pi Chiao
　黑格耳的變化形而上學與莊子的
　變化形而上學比較.
　A Comparison between the Hegelian Meta-
　physics of Change and Chuang Tzu's
　Metaphysics of Change.
　QRSIACE, 1936, 3, 1301.
　Eng. abstr. CIB, 1937, 1, 275.
Thang Chün-I (2) 唐君毅.
　Lun Chung Hsi Chê-Hsüeh chung Pen-Thi
　Kuan Nien chih I Chung Pien Chhien
　論中西哲學中本體觀念之一種變遷.
　Ontological Ideas (The One and the Many;
　Change and Permanence, etc.) in Chinese
　and Western Philosophy.
　WCYK, 1936, 1, 13.
　Eng. abstr. in CIB, 1937, 1, 36.
Thang Yung-Thung (1) 湯用彤.
　Chung-Kuo Fo Shih Ling Phien 中國佛
　史零篇.
　Notes on the History of Chinese Buddhist
　Thought.
　YCHP, 1937, 22, 1.
Tokiwa, Daijo (1) 常盤大定.
　Dōkyō Gaisetsu 道教概說
　Outline of Taoism.
　TYG, 1920, 10 (no. 3), 305.

Tokiwa, Daijo (2) 常盤大定.
　Dōkyō Hattatsu-shi Gaisetsu 道教.
　發達史概說.
　General Sketch of the Development of
　Taoism.
　TYG, 1921, 11 (no. 2), 243.
Tsuda, Sōkichi (1) 津田左右吉.
　Jukyō no Raigaku-setsu 儒教の禮樂說.
　The Doctrine of the Literati on the Rites
　and Music.
　TYG, 1932, 19, 1, 212, 354, 529; 1933, 20,
　61, 250, 351.
Tsumaki, Naoyoshi (1) 妻木直良.
　Dōkyō no Kenkyū 道教之研究.
　Studies in Taoism.
　TYG, 1911, 1 (no. 1), 1; (no. 2), 20; 1912, 2
　(no. 1), 58.

Ui, Hakuju (1) 宇井伯壽.
　Zenshūshi no Kenkyū 禪宗史の研究.
　Studies on the History of Zen Buddhism.
　3 vols. Tokyo, 1939-43.

Wang Chen-To (5) 王振鐸.
　Ssu-Nan Chih-Nan Chen yü Lo Ching
　Phan 司南指南針與羅經盤(下).
　Discovery and Application of Magnetic
　Phenomena in China, III (Origin and
　Development of the Chinese Compass
　Dial).
　AS/CJA, 1951, 5 (n.s., 1), 101.
Wang Chi-Thung 王季同.
　Yin Ming Ju Chêng Li Lun Mo Hsiang
　因明入正理論摸象.
　Elucidations of the Buddhist Classics.
　Com. Press, Chhangsha, 1940.
Wang Chin (1) 王璡.
　Chung-Kuo chih Kho-Hsüeh Ssu-Hsiang
　中國之科學思想.
　On (the History of) Scientific thought in
　China.
　Art. in Kho-Hsüeh Thung Lun 科學通論,
　Sci. Soc. of China, Shanghai, 1934.
Wang Tsu-Yuan (1) 王祖源.
　Nei Kung Thu Shuo 內功圖說.
　Illustrations and Explanations of Medical
　Gymnastics.
　1881.
Wei Chü-Hsien (1) 衛聚賢
　Chung-Kuo Khao Ku Hsüeh Shih
　中國考古學史
　History of Archaeology in China.
　Com. Pr. Shanghai, 1937.

Yang Chia-Lo (1) (ed.) 楊家駱.
　Ssu Khu Chhüan Shu Hsüeh Tien 四庫全
　書學典.
　Bibliographical Index of the Ssu Khu
　Chhüan Shu Encyclopaedia.
　World Book Co., Shanghai, 1946.

Yang Hung-Lieh (1) 楊鴻烈.
　　Chung Kuo Fa Lü Fa Ta Shih 中國法律
　　發達史.
　　History of Chinese Law.
　　Shanghai, 1930.
Yang Hung-Lieh (2) 楊鴻烈.
　　Chung-Kuo Fa Lü Ssu-Hsiang Shih 中國法
　　律思想史.
　　History of Chinese Jurisprudence.
　　Shanghai, 1936.
Yang Jung-Kuo (1) 楊榮國.
　　Khung Mo ti Ssu-Hsiang 孔墨的思想.
　　The Ideas of Confucius and Mo Tzu.
　　Sêng-huo, Peking, 1951.
Yang Shou-Ching (1) 楊守敬.
　　Li Tai Yü Ti Thu 歷代輿地圖.
　　Historical Atlas of China.
　　Canton, 1911.
Yang Thien-Hsi (1) 陽天錫.
　　Wang Chhuan-Shan ti Chê-Hsüeh 王船山
　　的哲學.
　　On the philosophy of Wang Chhuan-Shan.
　　CC, 1943, **7**, 395 and 424.
Yasuda, Jirō (1) 安田二郎.

Shushi no Sonzairon ni okeru 'Ri' no Seishitsu
　　ni tsuite 朱子の存在論に於ける
　　「理」の性質について.
　　The Concept of Li in Chu Hsi's Ontology
　　and Philosophy of Nature.
　　SG, 1939, **9**, 629.
Yeh Tê-Hui (1) (ed.) 葉德輝.
　　Shuang Mei Ching An Tshung Shu 雙梅景
　　闇叢書.
　　Double Plum-Tree Collection [of ancient
　　and medieval books and fragments on
　　Taoist sexual techniques].
　　Contains *Su Nü Ching, Hsüan Nü Ching,
　　Tung Hsüan Tzu, Yü Fang Pi Chüeh,
　　Thien Ti Yin Yang Ta Lo Fu*, etc.
　　(*qq.v.*).
　　Chhangsha, 1903 and 1914.
Yü Hsüan (1) 余遜.
　　*Tsao Chhi Tao Chiao chih Chêng-Chih
　　Hsin-Nien* 早期道教之政治信念.
　　Political Thoughts of the Early [Chin to
　　Thang] Taoists.
　　FJHC, 1942, 87.
　　Also sep., Peiping, 1942.

C. BOOKS AND JOURNAL ARTICLES IN WESTERN LANGUAGES

ABEGG, E., JENNY, J. J. & BING, M. (1). 'Yoga.' *CIBA/M*, 1949, **7** (no. 74), 2578.

ADAMS, F. (1) (tr.). *The Genuine Works of Hippocrates*, etc. 2 vols. Sydenham Society, London, 1849.

AIKEN, H. H. & HOPPER, G. M. (1). 'Automatic sequence-controlled calculator.' *EE*, 1946 (Sept., Oct., Nov.).

ALABASTER, SIR CHALONER (1). *Notes and Commentaries on Chinese Criminal Law.* Luzac, London, 1899.

ALBRIGHT, W. F. (1). 'Primitivism in Ancient Western Asia.' Supplementary essay no. 1, in Lovejoy & Boas (1), *q.v.*

ALLEY, REWI (2) (tr.). *The People Speak Out; translations of poems and songs of the People of China.* Pr. pub. Peking, 1954.

AMIOT, J. J. M. (3). On Lao Tzu. *MCHSAMUC*, 1787, **15**, 208.

VON AMIRA, K. (1). *Thierstrafen u. Thierprocesse.* Innsbrück, 1891.

ANDREOZZI, A. (1) (tr.). *Le Leggi Penali degli antichi Cinesi trad. del 'hin' fa-ce'* [Hsing Fa Chih], *o sunto storico degli leggi penali che fa parte della Storia della Dinastia dei Han.* Civelli, Florence, 1878. (Alternative title-page = *Le Leggi Penali degli antichi Cinesi, Discorso Proemiale sul Diritto e sui Limiti del Punire, e Traduzioni originali dal Cinese.*)

ANESAKI, M. (1). 'Early Christian Parallels in Buddhist Scriptures.' See Edmunds & Anesaki.

ANON. (2). 'Account of China by a Portuguese who was prisoner there for six years.' Printed as appendix to F. Alvarez, *Historia de las Cosas de Ethiopia*, 1561, quoted by G. T. Staunton in his edition of Mendoza, pp. xxxixff. See de Mendoza, Juan Gonzales.

ANON. (31). 'Le Feu Perpetuel de Bakou, par un voyageur Russe.' *JA*, 1833 (2e sér.), **11**, 358.

AQUINAS, THOMAS (1). *Summa Theologica.* Translation by English Dominicans, 27 vols. Burns, Oates & Washbourne, London, 1911.

ARBER, AGNES (1). 'Analogy in the History of Science.' In *Studies and Essays in the History of Science and Learning.* Sarton Presentation Volume, Schuman, New York, 1944.

ARBER, AGNES (2). *The Natural Philosophy of Plant Form.* Cambridge, 1950.

ARBERRY, A. J. (1) (tr.). *The Doctrine of the Sufis, a translation of the 'Kitāb al-Ta'arruf li-madhhab ahl al-tasawwuf' of Abū Bakr al-Kalabadhī.* Cambridge, 1935.

ARLINGTON, L. C. (1). 'Chinese and Western Cheiromancy.' *CJ*, 1927, **7**, 170 and 228.

ARMITAGE, A. (1). (*a*) *Sun, Stand thou Still; the Life and Work of Copernicus the Astronomer.* Schuman, New York, 1947. (*b*) *Copernicus, the Founder of Modern Astronomy.* Allen & Unwin, London, 1938.

ARNOLD, E. V. (1). *Roman Stoicism.* Cambridge, 1911.

AST, F. (1). *Lexicon Platonicum.* Leipzig, 1835.

AUNG, S. Z. & RHYS DAVIDS, C. A. F. (1) (tr.). *Points of Controversy; being a translation of the 'Katha-vatthu'.* Pali Text Soc. Series, no. 5. Oxford, 1915.

AUROUSSEAU, L. (1). Critique of N. Tsumaki's work on the history of Taoism. *BEFEO*, 1912, **12** (no. 9), 108.

AVALON, A. (ps.). See Woodroffe, Sir J.

AYSCOUGH, F. (1). 'The Cult of the Chhêng Huang Lao Yeh (Spiritual Magistrate of the City Walls and Moats).' *JRAS/NCB*, 1924, **55**, 131.

BACON, FRANCIS (1). *Philosophical Works*, ed. R. L. Ellis & J. Spedding. Routledge, London, 1905.

BACON, JOHN S. D. (1). *The Chemistry of Life.* Watts, London, 1944. (Thinker's Library, no. 103.)

BAGCHI, P. C. (1). *India and China; a thousand years of Sino-Indian Cultural Relations.* Hind Kitab, Bombay, 1944. 2nd ed. 1950.

BAGCHI, P. C. (2). (*a*) *Le Canon Bouddhique en Chine.* Geuthner, Paris, 1927. (Calcutta University Sino-Indica Series, no. 1.) (*b*) *Les Traducteurs et les Traductions.* Geuthner, Paris, 1938. (Sino-Indica Series, no. 2.)

BAGCHI, P. C. (3). *Studies in the Tantras.* Calcutta, 1939.

BAGCHI, P. C. (4). 'On Foreign Elements in the Tantra.' *IHQ*, 1931, **7**, 1. (rev. P. Pelliot, *TP*, 1932, 148.)

BAILEY, K. C. (1). *The Elder Pliny's Chapters on Chemical Subjects.* 2 vols. Arnold, London, 1929 and 1932.

BALAZS, E. (=S.) (1). 'La Crise Sociale et la Philosophie Politique à la Fin des Han.' *TP*, 1949, **39**, 83.
BALAZS, E. (=S.) (2). 'Entre Révolte Nihiliste et Evasion Mystique' (the Seven Sages of the Bamboo Grove, and Pao Ching-Yen). *ASEA*, 1948, **1**, 27 (sequel to Balazs, 1).
BALAZS, E. (=S.) (3). 'Buddhistische Studien; der Philosoph Fan Dschen [Fan Chen] und sein Traktat gegen den Buddhismus' (*Shen Mieh Lun*). *SA*, 1932, **7**, 220.
BALAZS, E. (=S.) (6). 'L'Esprit des lois au Moyen-Âge [en Chine].' Communication to the International Younger Sinologists' Group, Paris, July 1951 (typescript).
BALAZS, E. (=S.) (7) (tr.). 'Le Traité Économique du *Souei-Chou* [*Sui Shu*]' (Études sur la Société et l'Économie de la Chine Médiévale). *TP*, 1953, **42**, 113. Also sep. issued, Brill, Leiden, 1953.
BALAZS, E. (=S.) (8). 'Le Traité Juridique du *Souei-Chou* [*Sui Shu*].' *TP*, 1954. Sep. pub. as *Etudes sur la Société et l'Economie de la Chine Médiévale*, no. 2. Brill, Leiden, 1954 (Bibliothèque de l'Inst. des Hautes Etudes Chinoises, no. 9).
BALD, R. C. (1). 'Sir William Chambers and the Chinese Garden.' *JHI*, 1950, **11**, 287.
BALDWIN, ERNEST H. F. (1). *Dynamic Aspects of Biochemistry*. Cambridge, 1952 (2nd ed.).
BALFOUR, F. H. (1) (tr.). *Taoist Texts, ethical, political, and speculative* (incl. *Tao Tê Ching, Yin Fu Ching, Thai Hsi Ching, Hsin Yin Ching, Ta Thung Ching, Chih Wên Tung, Chhing Ching Ching, Huai Nan Tzu* ch. 1, *Su Shu* and *Kan Ying Phien*). Kelly & Walsh, Shanghai, n.d. but probably 1884.
BALSS, H. (2). *Albertus Magnus als Zoologe*. Münchner Drucke, München, 1928. (*Münchener Beiträge z. Gesch. u. Lit. d. Naturwiss. u. Med.* no. 11/12.)
BALTHAZAR, P. & DÉROBERT, L. (1). 'Médecine Légale.' In Laignel-Lavastine (1), *Histoire Générale de la Médecine, q.v.* (vol. 3, p. 451).
BARDE, R. (1). 'Recherches sur les Origines Arithmétiques du *Yi-King* [*I Ching*].' *A/AIHS*, 1952, **5**, 234.
BARDE, R. (2). 'La Divination par le *Yi-King* (*I Ching*).' MS.
BARNETT, L. D. (2) (tr.). *Śāntideva's 'Bodhicaryāvatāra'*. London, 1909.
BASU, B. N. (1) (tr.). *The 'Kāmasūtra' of Vātsyāyana* (prob. +4th century); revised S. L. Ghosh. Pref. by P. C. Bagchi. Med. Book Co., Calcutta, 1951 (10th ed.).
BECK, T. (1). *Beiträge z. Geschichte d. Maschinenbaues*. Springer, Berlin, 1900.
BEGRICH, J. (1). 'Literary and documentary analysis of the Jewish Doctrine of the Fall of Man.' *ZAW*, 1932, **50**, 93.
BEHANAN, KOVOOR T. (1). *Yoga; a Scientific Evaluation*. Secker & Warburg, London, 1937.
BENDALL, C. (1) (ed.). *Subhāṣitasaṃgraha*. Museon Ser. nos. 4 and 5. Istas, Louvain, 1905.
BENEDICT, R. (1). *Patterns of Culture* (ch. 6, 'Potlatch'). Mifflin, New York, 1934.
BENTHAM, JEREMY (1). *Theory of Fictions*, ed. C. K. Ogden. Kegan Paul, London, 1932.
BERGAIGNE, A. (1). *La Religion Védique d'après les Hymnes du Ṛg Veda*, 4 vols. *BEHE/PH*, 1878–97, nos. 36, 53, 54, 117.
BERGMANN, E. (1). *Erkenntnisgeist und Muttergeist*. Hirt, Breslau, 1933 (2nd ed.).
BERKELEY, E. C. (1). *Giant Brains; or Machines that Think*. Wiley, New York, 1949; Chapman & Hall, London, 1949.
BERNARD-MAÎTRE, H. (1). *L'Apport Scientifique du Père Matthieu Ricci à la Chine*. Mission Press, Hsienhsien, 1935. Eng. tr. *Matteo Ricci's Scientific Contribution to China*, tr. by E. T. C. Werner. Vetch, Peiping, 1935. Crit. Chang Yu-Chê, *TH*, 1936, **3**, 538.
BERNARD-MAÎTRE, H. (2). *Sagesse Chinoise et Philosophie Chrétienne; Essai sur leurs Relations Historiques*. Mission Press, Tientsin, 1935. Repr. Belles Lettres, Paris, 1953.
BERNARD-MAÎTRE, H. (6). 'Comment Leibniz découvrit le Livre des Mutations.' *BUA*, 1944 (3e sér.), **5**, 432.
BERNARD-MAÎTRE, H. (10). 'Chu Hsi's Philosophy and its Interpretation by Leibniz.' *TH*, 1937, **5**, 9.
BERNIER, FRANÇOIS (1). *Bernier's Voyage to the East Indies; containing The History of the Late Revolution of the Empire of the Great Mogul; together with the most considerable passages for five years following in that Empire; to which is added A Letter to the Lord Colbert, touching the extent of Hindustan, the Circulation of the Gold and Silver of the world, to discharge itself there, as also the Riches Forces and Justice of the Same, and the principal Cause of the Decay of the States of Asia—with an Exact Description of Delhi and Agra; together with* (1) *Some Particulars making known the Court and Genius of the Moguls and Indians; as also the Doctrine and Extravagant Superstitions and Customs of the Heathens of Hindustan*, (2) *The Emperor of Mogul's Voyage to the Kingdom of Kashmere, in 1664, called the Paradise of the Indies....* Dass (for SPCK), Calcutta, 1909. [Substantially the same title-page as the editions of 1671 and 1672.]
BERRIAT ST PRIX, J. (1). 'Rapport et Recherches sur les Procés et Jugements relatifs aux Animaux.' *MSAF*, 1829, **8**, 403.
BERRIEDALE KEITH, A. See Keith, A. Berriedale.

VON BERTALANFFY, L. (1). 'Das Weltbild d. Biologie.' In Moser's *Weltbild und Menschenbild.* Tyrode, Vienna, 1942.

VON BERTALANFFY, L. (2). *Problems of Life; an Evaluation of Modern Biological Thought.* Watts, London, 1952.

BERTHELOT, M. (1). *Les Origines de l'Alchimie.* Lib. Sci. et Arts, Paris, 1938 (repr.). 1st ed. 1884.

BERTHELOT, RENÉ (1). *La Pensée de l'Asie et l'Astrobiologie.* Payot, Paris, 1949.

BERTHOLD, O. (1). *Die Unverwundbarkeit in Sage u. Aberglauben der Griechen.* Toepelmann, Giessen, 1911. (Religionsgeschichtliche Versuche u. Vorarbeiten.)

BEVERIDGE, W. I. B. (1). *The Art of Scientific Investigation.* Heinemann, London, 1950.

BEZOLD, C. (1). 'Sze-ma Ts'ien [Ssuma Chhien] und die babylonische Astrologie.' *OAZ*, 1919, **8**, 42.

BEZOLD, C. (2). 'Astronomie, Himmelschau und Astrallehre bei den Babyloniern.' *SHAW/PH*, 1911, **2**, no. 18.

BEZOLD, C. & BOLL, F. (1). 'Reflexe astrologische Keilinschriften bei griechischer Schriftstellern.' *SHAW/PH*, 1911, **2**, no. 23.

BEZOLD, C., KOPFF, A. & BOLL, F. (1). 'Zenit- und Aequatorialgestirne am babylonischen Fixstern-himmel.' *SHAW/PH*, 1913, **4**, no. 11.

BHATTACHARYA, B. (1) (ed.). *Guhyasamājatantra, or Tathāgataguhyaka.* Gaekwad Orient. Ser. no. 53, Orient. Instit., Baroda, 1931.

BHATTACHARYA, B. (2). *Introduction to Buddhist Esoterism.* Oxford, 1932.

BIALLAS, F. X. (1). *Konfuzius und sein Kult.* Pekinger Verlag, Peking and Leipzig, 1928.

BIELENSTEIN, H. (2). 'The Restoration of the Han Dynasty.' *BMFEA*, 1954, **26**, 1–209, and sep. Göteborg, 1953.

BIOT, E. (1) (tr.). *Le Tcheou-Li ou Rites des Tcheou.* 3 vols. Imp. Nat., Paris, 1851. (Photographically reproduced, Wêntienko, Peking, 1939.)

BIOT, E. (3) (tr.). *Chu Shu Chi Nien* (Bamboo Books). *JA*, 1841 (3ᵉ sér.), **12**, 537; 1842, **13**, 381

BIRKBECK, W. J., DEARMER, P. *et al.* (ed.) (1). *The English Hymnal.* Oxford, 1906.

BISHOP, C. W. (9). 'The Ritual Bullfight.' *CJ*, 1925, **3**, 630. Reprinted *ARSI*, 1927, 447.

BLOFELD, J. (1). *The Jewel in the Lotus; an outline of present-day Buddhism in China.* Sidgwick & Jackson, London, 1948.

BLOFELD, J. (2) (tr.). *The Path to Sudden Attainment* (a Chhan Sūtra). Sidgwick & Jackson, London, 1948.

DE BLONAY, G. (1). 'Matériaux pour servir à l'Histoire de la Déesse Bouddhique Tārā.' *BEHE/PH*, Paris, 1895, no. 107.

BOAS, G. (1). *Essays on Primitivism and Related Ideas in the Middle Ages.* Johns Hopkins Univ. Press, Baltimore, 1948.

BODDE, D. (2). (*a*) 'The New Identification of Lao Tzu proposed by Prof. Dubs.' *JAOS*, 1942, **62**, 8. (*b*) 'Further Remarks on the Identification of Lao Tzu; a Last Reply to Prof. Dubs.' *JAOS*, 1944, **64**, 24.

BODDE, D. (3) (tr.). 'The Philosophy of Chu Hsi' (the 13th chapter of vol. 2 of Fêng Yu-Lan's *Chung-Kuo Chê-Hsüeh Shih*). *HJAS*, 1942, **7**, 1.

BODDE, D. (4) (tr.). 'The Rise of Neo-Confucianism and its Borrowings from Buddhism and Taoism' (the 10th chapter of vol. 2 of Fêng Yu-Lan's *Chung-Kuo Chê-Hsüeh Shih*). *HJAS*, 1942, **7**, 89.

BODDE, D. (5). 'Types of Chinese Categorical Thinking.' *JAOS*, 1939, **59**, 200.

BODDE, D. (6). 'The Attitude towards Science and Scientific Method in Ancient China.' *TH*, 1936, **2**, 139, 160.

BODDE, D. (7). (*a*) 'Dominant Ideas' (of Chinese Thought). In *China*, ed. H. F. McNair. Univ. of Calif. Press, Berkeley, 1946, p. 18. (*b*) 'Dominant Ideas in the Formation of Chinese Culture.' *JAOS*, 1942, **62**, 293.

BODDE, D. (8) (tr.). 'A General Discussion of the Period of Classical Learning' (the 1st chapter of vol. 2 of Fêng Yu-Lan's *Chung-Kuo Chê-Hsüeh Shih*). *HJAS*, 1947, **9**, 195.

BODDE, D. (9). 'Some Chinese Tales of the Supernatural; Kan Pao and his *Sou Shen Chi*.' *HJAS*, 1942, **6**, 338.

BODDE, D. (10). 'Again Some Chinese Tales of the Supernatural; Further Remarks on Kan Pao and his *Sou Shen Chi*.' *JAOS*, 1942, **62**, 305.

BODDE, D. (11). 'The Chinese View of Immortality; its Expression by Chu Hsi and its Relationship to Buddhist Thought.' *RR*, 1942, 369.

BODDE, D. (14). 'Harmony and Conflict in Chinese Philosophy.' In *Studies in Chinese Thought*, ed. A. F. Wright. *AAN*, 1953, **55** (no. 5), 19 (Amer. Anthropol. Assoc. Memoirs, no. 75).

BOHR, NIELS (1). 'Newton's Principles and Modern Atomic Mechanics.' In *The Royal Society Newton Tercentenary Celebrations.* Royal Society, London, 1947.

BOLL, F. (1). *Sphaera.* Teubner, Leipzig, 1904.

BOLL, F. & BEZOLD, C. (1). 'Antike Beobachtung färbiger Sterne.' *ABAW/PH*, 1918, **89** (30), no. 1.

BOLL, F., BEZOLD, C. & GUNDEL, W. (1). (a) Sternglaube, Sternreligion und Sternorakel. Teubner, Leipzig, 1923. (b) Sternglaube und Sterndeutung; die Gesch. u. d. Wesen d. Astrologie. Teubner, Leipzig, 1926.

BONITZ, H. (1). Index Aristotelicus. Berlin, 1870.

BOODBERG, P. A. (3). 'The Semasiology of some primary Confucian Concepts.' PEW, 1953, 2, 317.

BOSE, M. M. (1). The Post-Caitanya Sahajiyā Cult of Bengal. Univ. Press, Calcutta, 1930.

BOUCHÉ-LECLERCQ, A. (1). L'Astrologie Grecque. Leroux, Paris, 1899.

BOUCHÉ-LECLERCQ, A. (2). Histoire de Divination dans l'Antiquité. 4 vols. Leroux, Paris, 1879–82.

BOULAIS, G. (1) (tr.). Manuel du Code Chinois (transl. of Ta Chhing Lü Li). 2 vols. Shanghai, 1923 and 1924 (VS, no. 55).

BOUSSET, W. (1). 'Hauptprobleme der Gnosis' (Forsch. z. Relig. Lit. d. alt.- u. neuen Testaments, 110. 10). Vandenhoek & Ruprecht, Göttingen, 1907.

BOWDEN, B. V. (1) (ed.). Faster than Thought; a Symposium on Digital Calculating Machines. Pitman, London, 1953.

BOXER, C. R. (1) (ed.). South China in the Sixteenth Century; being the Narratives of Galeote Pereira, Fr. Gaspar da Cruz, O.P., and Fr. Martin de Rada, O.E.S.A. (1550–1575). Hakluyt Society, London, 1953 (Hakluyt Society Pubs. 2nd series, no. 106).

BOYLE, ROBERT (1). The Sceptical Chymist. Cadwell & Crooke, London, 1661.

BRACE, A. J. (1). 'Some Secret Societies in Szechuan.' JWCBRS, 1936, 8, 177.

BRAITHWAITE, R. B. (1). Scientific Explanation. Cambridge, 1953.

BRÉHIER, L. (1). La Philosophie de Plotin. Paris, 1928.

BRETSCHNEIDER, E. (1). Botanicon Sinicum; Notes on Chinese Botany from Native and Western Sources. 3 vols. Trübner, London, 1882 (printed in Japan). (Reprinted from JRAS/NCB, 1881, 16.)

BRITISH MUSEUM (1). Cuneiform Texts from Babylonian Tablets, etc. in the British Museum. Ed. E. A. Wallis Budge & C. J. Gadd, 41 vols. 1896–1931.

BROECK, RINAKER TEN J. R. & YÜ TUNG (1). 'A Taoist Inscription of the Yuan Dynasty' (+1329). TP,

BRONOWSKI, J. (1). A Man without a Mask. (William Blake.) Secker & Warburg, London, 1943.

BROOKS, F. (1) (tr.). Cicero's 'De Natura Deorum'. Methuen, London, 1896.

BRUCE, J. P. (1) (tr.). The Philosophy of Human Nature, translated from the Chinese, with notes. (Chs. 47, 48 and 49 of the Chu Tzu Chhüan Shu.) Probsthain, London, 1922.

BRUCE, J. P. (2). Chu Hsi and his Masters; an introduction to Chu Hsi and the Sung School of Chinese philosophy. Probsthain, London, 1923.

BRUCE, J. P. (3). 'The I Wei, a Problem in Criticism.' JRAS/NCB, 1930, 61, 100.

BRUCKER, J. C. (1). Historia Criticae Philosophiae. 6 vols. Breitkopf, Leipzig, 1742–1767 (Chinese material in vol. 5, pp. 846–906, and vol. 6, pp. 978–999). Eng. tr. (much abridged) W. Enfield, 2 vols. Johnson, London, 1791.

BRUNET, P. & MIELI, A. (1). L'Histoire des Sciences (Antiquité). Payot, Paris, 1935.

BRUNO, GIORDANO (1). De Imaginum Signorum et Idearum Compositione. 1597.

BRYCE, J. (1). 'The Law of Nature.' In Studies in History and Jurisprudence (vol. 2, pp. 112ff.). Oxford, 1901.

BUCKLAND, W. W. (1). A Textbook of Roman Law from Augustus to Justinian. Cambridge, 1932.

BÜNGER, K. (1). Quellen z. Rechtsgeschichte d. Thang-Zeit. Fu-jen Press, Peking, 1946. [Monumenta Serica Monographs, no. 9.]

BÜNGER, K. (2). 'The Punishment of Lunatics and Negligents According to Classical Chinese Law.' SSE, 1950, 9, 1.

BURKITT, F. C. (1). The Religion of the Manichees. Cambridge, 1925.

BURKITT, F. C. (2). Church and Gnosis. Cambridge, 1932.

BURNET, J. (1). Early Greek Philosophy. Black, London, 1908.

BURNETT, JAMES (LORD MONBODDO) (1). Of the Origin and Progress of Language. 3 vols. Edinburgh, 1773–1776.

BURTON, MAURICE (1). 'Animal Legends.' ILN, 1952, 221. 1, Foxes and Fleas, 228; 2, Hedgehogs and Apples, 264; 3, 'Funeral Processions' of Stoats, 300; 4, Turkey 'Ceremonials', 340; 5, Two Rats and an Egg, 462; 6, 'Play' of Stoats and Martens, 508; 7, Anting of Birds and Squirrels, 554; 8, Anting of Pangolins, 592; 9, Shrews, 676; 10, Bats and their Radar, 736; 11, Summing-up, 816, 820. Also Animal Legends, Muller, London, 1954.

BUSCH, H. (1). 'Hsün Yüeh, ein Denker am Hofe des letzten Han Kaisers.' MS, 1945, 10, 58.

BUSH, V. & CALDWELL, S. H. (1). 'Differential analyser.' JFI, 1945, 240, 255.

BUSHELL, S. W. (1). 'Ancient Roman Coins from Shansi.' JPOS, 1886, 1, 17.

BUSHELL, S. W. (2). Chinese Art. 2 vols. For Victoria and Albert Museum. HMSO, London, 1909. 2nd ed. 1914.

CAIRNS, HUNTINGTON (1). Law and the Social Sciences. Univ. of N. Carolina Press, Chapel Hill, 1935; Kegan Paul, London, 1935.

CAJORI, F. (1). (a) 'The Purpose of Zeno's Arguments on Motion.' *ISIS*, 1920, **3**, 7. (b) 'The History of Zeno's Arguments on Motion.' *AMM*, 1915, **22**.

CAMMANN, S. (1). 'Tibetan Monster Masks.' *JWCBRS*, 1940, A **12**, 9.

CARLYLE, R. W. & CARLYLE, A. J. (1). *A History of Medieval Political Theory in the West.* 6 vols. Blackwood, Edinburgh, 1927.

CARR, H. WILDON (1). *The Monadology of Leibniz, with an Introduction, Complementary and Supplementary Essays.* Favil Press, London, 1930.

CARR, H. WILDON (2). *Leibniz.* Benn, London, 1929.

CARUS, P. (1) (tr.). *The Canon of Reason and Virtue* [*Tao Tê-Ching*]. Open Court, Chicago, 1903.

CARUS, P. (2). *Chinese Thought.* Open Court, Chicago, 1907.

CASSIRER, E. (1). 'Galileo's Platonism.' In *Sarton Presentation Volume*, ed. M. F. Ashley-Montagu, p. 279. Schuman, New York, 1944.

CASSIRER, E. (2). *The Platonic Renaissance in England.* Eng. tr. from the German edition of 1932 by J. P. Pettegrove. Nelson, London, 1953.

CHALFANT, F. H. (1). 'Early Chinese Writing.' *MCMU*, 1906, **4**, 1.

CHAN WING-TSIT. See Chhen Jung-Chieh.

CHANG TUNG-SUN (1). 'A Chinese Philosopher's Theory of Knowledge.' *YJSS*, 1939, **1**, 155. Reprinted *ETC*, 1952, **9**, 203.

CHANG, Y. Z. (1). 'A Note on Sharawadgi.' *MLN*, 1930, **45**, 221.

CHAO WEI-PANG (1). 'The Chinese Science of Fate-Calculation.' *FLS*, 1946, **5**, 279.

CHAO WEI-PANG (2). 'The Origin and Growth of the *fu-chi*; the Chinese planchette.' *FLS*, 1942, **1**, 9.

CHASE, STUART (1). *Men and Machines.* London, 1929.

DE CHATEAUBRIAND, F. R. (1). *Les Natchez; Roman Indien.* Paris, 1827. Ed. with notes, by G. Chinard: Johns Hopkins University Press, Baltimore, 1932.

CHATLEY, H. (1). MS. translation of the astronomical chapter (ch. 3, Thien Wên) of *Huai Nan Tzu.* Unpublished. (Cf. note in *O*, 1952, **72**, 84.)

CHATLEY, H. (2). 'The Development of Mechanisms in Ancient China.' *TNS*, 1942, **22**, 117. (Long abstr. without illustr. *ENG*, 1942, **153**, 175.)

CHATLEY, H. (3). 'Science in Old China.' *JRAS/NCB*, 1923, **54**, 65.

CHATLEY, H. (5). 'Chinese Natural Philosophy and Magic.' *JRSA*, 1911, **59**, 557.

CHATLEY, H. (6). 'Magical Practice in China.' *JRAS/NCB*, 1917, **48**, 16.

CHATLEY, H. (7). 'Fêng-Shui.' In *ES*, p. 175.

CHAVANNES, E. (1). *Les Mémoires Historiques de Se-Ma Ts'ien* (Ssuma Chhien). 5 vols. Leroux, Paris, 1895–1905. (Photographically reproduced in China, n.d.)

    1895   vol. 1 tr. *Shih Chi*, chs. 1, 2, 3, 4.
    1897   vol. 2 tr. *Shih Chi*, chs. 5, 6, 7, 8, 9, 10, 11, 12.
    1898   vol. 3 (i) tr. *Shih Chi*, chs. 13, 14, 15, 16, 17, 18, 19, 20, 21, 22.
              vol. 3 (ii) tr. *Shih Chi*, chs. 23, 24, 25, 26, 27, 28, 29, 30.
    1901   vol. 4 tr. *Shih Chi*, chs. 31, 32, 33, 34, 35, 36, 37, 38, 39, 40, 41, 42.
    1905   vol. 5 tr. *Shih Chi*, chs. 43, 44, 45, 46, 47.

CHAVANNES, E. (6) (tr.). 'Les Pays d'Occident d'après le Heou Han Chou.' *TP*, 1907, **8**, 149. (Ch. 118, on the Western Countries, from *Hou Han Shu*.)

CHAVANNES, E. (7). 'Le Cycle Turc des Douze Animaux.' *TP*, 1906, **7**, 51.

CHAVANNES, E. (16). 'Trois Généraux Chinois de la Dynastie des Han Orientaux.' *TP*, 1906, **7**, 210. (Tr. ch. 77 of the *Hou Han Shu* on Pan Chhao, Pan Yung and Liang Chhin.)

CHÊNG CHIH-I (ANDREW) (1). *Hsüntzu's Theory of Human Nature and its Influence on Chinese Thought.* Inaug. Diss. Columbia Univ. New York. Printed Peking, 1928.

CHÊNG TÊ-KHUN (2) (tr.). 'Travels of the Emperor Mu.' *JRAS/NCB*, 1933, **64**, 142; 1934, **65**, 128.

CHÊNG THIEN-HSI (1). *China Moulded by Confucius; the Chinese way in western light.* Stevens, London, 1946.

CHÊNG THIEN-HSI (2). 'The Development and Reform of Chinese Law.' *CLPRO*, 1948, **1**, 170.

CHHEN HSIANG-CHHUN (1). 'Examples of Charms against Epidemics, with short Explanations.' *FLS*, 1942, **1**, 37.

CHHEN JUNG-CHIEH (CHAN WING-TSIT) (1). 'Neo-Confucianism.' In *China*, ed. H. F. McNair. Univ. of California Press, 1946, p. 254.

CHHEN JUNG-CHIEH (2). 'Trends in Contemporary [Chinese] Philosophy.' In *China*, ed. H. F. McNair. Univ. of California Press, 1946, p. 312.

CHHEN JUNG-CHIEH (3). 'An Outline and a Bibliography of Chinese Philosophy.' (Mimeographed notes), Dartmouth College, Hanover, New Hampshire, 1953; also *PEW*, 1954, **3**, 241, 337; rev. W. E. Hocking & R. Hocking, *PEW*, 1954, **4**, 175.

CHHEN JUNG-CHIEH (4). *Religious Trends in Modern China*. Columbia Univ. Press, New York, 1953 (Haskell Lectures, Chicago, 1950).

CHHEN JUNG-CHIEH (5). Contributions to *A Dictionary of Philosophy*, ed. D. D. Runes. Philos. Lib. New York, 1942. Notably *Chhi* (pneuma, matter-energy), p. 50; *Jen* (human-heartedness), p. 153; and *Li* (Neo-Confucian organic pattern), p. 168 [our definitions, not his]. Also *PEW*, 1952, **2**, 166.

CHHEN MÊNG-CHIA (1). 'The Greatness of Chou [Dynasty], *c.* −1027 to *c.* −221.' In *China*, ed. H. F. McNair (p. 54). Univ. of California Press, 1946.

CHHEN SHIH-HSIANG (1). 'In Search of the Beginnings of Chinese Literary Criticism.' *SOS*, 1951, **11**, 45.

CHHEN SHOU-YI (2). 'The Chinese Garden in 18th-century England.' *TH*, 1936, **2**, 321.

CHHEN TAI-O (P. J. ZEN) (1) (tr.). 'Le Chapitre 33 du Tchoang-Tse [*Chuang Tzu*].' *BUA*, 1949 (3ᵉ sér.), **10**, 104.

CHHIEN CHUNG-SHU (1). 'China in the English Literature of the Seventeenth Century.' *QBCB/E*, 1940 (n.s.), **1**, 351.

CHHIEN CHUNG-SHU (2). 'China in the English Literature of the Eighteenth Century.' *QBCB/E*, 1941 (n.s.), **2**, 7.

CHHU TA-KAO (2) (tr.). *Tao Tê Ching, a new translation*. Buddhist Lodge, London, 1937.

CHI HSIEN-LIN (1). 'Indian Physiognomical Characteristics in the Official Annals of the Three Kingdoms, Chin, and Southern and Northern Dynasties.' *SSE*, 1949, **8**, 96.

CHIANG FÊNG-WEI (1) (ed.). *Gems of Chinese Literature*. Progress Press, Chungking, 1942.

CHIKASHIGE, M. (1). *Alchemy and other Chemical Achievements of the Ancient Orient; the Civilisation of Japan and China in Early Times as seen from the Chemical* [and metallurgical] *Point of View*. Rokakuho Uchida, Tokyo, 1936 (in Engl.).

CHILDE, V. GORDON (14). 'Science in Preliterate Societies and the Ancient Oriental Civilisations.' *CN*, 1953, **3**, 12.

CHOU I-CHHING (1). *La Philosophie Morale dans le Neo-Confucianisme (Tcheou Touen-Yi)* [*Chou Tun-I*]. Presses Univ. de France, Paris, 1954. (Includes tr. of *Thai Chi Thu Shuo, Thai Chi Thu Shuo Chieh* and of *I Thung Shu*.)

CHOU I-LIANG (1). 'Tantrism in China.' *HJAS*, 1945, **8**, 241.

CHOW YIH-CHING. See Chou I-Chhing.

CIBOT, P. M. (3). 'Notice sur le Cong-Fou [Kung Fu], exercice superstitieux des *tao-che* [Tao Shih] pour guérir le corps de ses infirmités et obtenir pour l'âme une certaine immortalité.' *MCHS AMUC*, 1779, **4**, 441.

CLARKE, J. & GEIKIE, Sir A. (1). *Physical Science in the Time of Nero*. A transl. of Seneca's *Quaestiones Naturales*. Macmillan, London, 1910.

COEDÈS, G. (1) (tr.). *Textes d'auteurs grecs et latins relatifs à l'Extrême Orient depuis le 4ème siècle avant J. C. jusqu'au 14ème siècle après J. C.* Leroux, Paris, 1910.

COHEN, A. (1). *The Teachings of Maimonides*. Routledge, London, 1927.

COHEN, J. (1). 'On the Project of a Universal Character.' *M*, 1954, **63**, 49.

COKER, F. W. (1). *Organismic Theories of the State*. Columbia University Studies in History, Economics and Public Law. New York, 1910, no. 38.

COLE, L. J. (1). 'The Lay of the Rooster.' *JH*, 1927, **18**, 97.

COLLINET, P. (1). 'Histoire de l'Ecole de Droit de Beyrouth.' In *Etudes Historiques sur le Droit de Justinian*, vol. 2. Recueil Sirey, Paris, 1925.

COLLINGWOOD, R. G. (1). *Outlines of a Philosophy of Art*. Oxford, 1925.

COMENIUS (KOMENSKY), JAN AMOS (1). *A Reformation of Schooles, designed in two excellent Treatises; the first whereof summarily sheweth, the great necessity of a generall reformation of common learning; what grounds of hope there are for such a reformation, and how it may be brought to pass; followed by a Dilucidation answering certaine Objections made against the Endeavours and Means of Reformation in Common Learning, expressed in the foregoing discourse.* (Tr. Sam. Hartlib.) London, 1642.

COMRIE, L. J. (1). 'Recent Progress in Scientific Computing.' *JSI*, 1944, **21**, 129.

CONGER, G. PERRIGO (1). *Theories of Macrocosm and Microcosm in the History of Philosophy*. Inaug. Diss. Columbia Univ. New York, 1922.

CONGER, G. PERRIGO (2). 'A Naturalistic Garland for Radhakrishnan.' Art. in *Radhakrishnan; Comparative Studies in Philosophy presented in honour of his 60th Birthday*, ed. W. R. Inge *et al*. Allen & Unwin, London, 1951.

CONRADY, A. (1). 'Indischer Einfluss in China in 4-jahrh. v. Chr.' *ZDMG*, 1906, **60**, 335.

CONRADY, A. (2). 'Alte Westöstliche Kulturwörter.' *BVSAW/PH*, 1925, **77**, no. 3, 1.

CONRADY, A. (3). 'Zu *Lao-tzu*, cap. 6.' (The Valley Spirit.) *AM*, 1932, **7**, 150.

CONRADY, A. (4). '*Yih King* [*I Ching*] Studien.' *AM*, 1931, **7**, 409.

CONRADY, A. & ERKES, E. (1). *Das älteste Dokument zur chinesische Kunstgeschichte, Tien-Wên, die 'Himmelsfragen' d. K'üh Yüan* [Chhü Yuan], *abgeschl. u. herausgeg. v. E. Erkes.* Leipzig, 1931. Critiques: B. Karlgren, *OLZ*, 1931, **34**, 815; H. Maspero, *JA*, 1933, **222** (Suppl.), 59; Hsü Tao-Lin, *SA*, 1932, **7**, 204.

CONZE, E. (1). *Buddhism, its Essence and Development.* Cassirer, Oxford, 1953.

CONZE, E. (2). *The Scientific Method of Thinking; an Introduction to Dialectical Materialism.* Chapman & Hall, London, 1935.

CONZE, E. (3). 'The Ontology of the *Prajñāpāramitā*.' *PEW*, 1953, **3**, 117.

CONZE, E. (4) (tr.). *Selected Sayings from the 'Perfection of Wisdom'; Prajñāpāramitā.* Buddhist Soc. London, 1955. (The complete translation is distributed in typescript by the Society.)

CONZE, E., HORNER, L. B., SNELLGROVE, D. & WALEY, A. (ed.). *Buddhist Texts through the Ages* (anthology). Faber & Faber, London, 1954.

COOK, A. B. (1). *Zeus.* 3 vols. Cambridge, 1914, 1925, 1940.

CORDIER, H. (2). *Bibliotheca Sinica; Dictionnaire bibliographique des Ouvrages relatifs à l'Empire Chinois.* 3 vols. Ec. des Langues Orientales Vivantes, Paris, 1878–95. 2nd ed. 5 vols. pr. Vienna, 1904–24.

CORNFORD, F. M. (1). *From Religion to Philosophy; a Study in the Origins of Western Speculation.* Arnold, London, 1912.

CORNFORD, F. M. (2). *The Laws of Motion in Ancient Thought.* (Inaugural Lecture.) Cambridge, 1931.

CORNFORD, F. M. (4). 'Greek Natural Philosophy and Modern Science.' Art. in Needham & Pagel (1) (*q.v.*), repr. in Cornford (6), p. 81.

CORNFORD, F. M. (5). 'The Marxist View of Ancient Philosophy.' Lecture to the Classical Association, Cambridge, 1942; first printed, 1950, in Cornford (6). Crit. B. Farrington, *SS*, 1953, **17**, 289.

CORNFORD, F. M. (6). *The Unwritten Philosophy, and other Essays,* ed. W. K. C. Guthrie. Cambridge, 1950.

COUTURAT, L. (1). *La Logique de Leibniz.* Alcan, Paris, 1901.

COUVREUR, F. S. (1) (tr.). '*Tch'ouen Ts'iou' et 'Tso Tchouan'; Texte Chinois avec Traduction Française.* (*Chhun Chhiu* and *Tso Chuan*.) 3 vols. Mission Press, Hochienfu, 1914. Repr. Belles Lettres, Paris, 1951.

COUVREUR, F. S. (2). *Dictionnaire Classique de la Langue Chinoise.* Mission Press, Hsienhsien, 1890; photographically reproduced, Vetch, Peiping, 1947.

COUVREUR, F. S. (3) (tr.). '*Li Ki', ou Mémoires sur les Bienséances et les Cérémonies.* (*Li Chi*.) 2 vols. Hochienfu, 1913. Repr. Belles Lettres, Paris, 1950.

CRAMER, F. H. (1). 'Bookburning and Censorship in Ancient Rome; a chapter from the History of Freedom of Speech.' *JHI*, 1945, **6**, 157.

CREEL, H. G. (1). *Studies in Early Chinese Culture* (1st series). Waverly, Baltimore, 1937.

CREEL, H. G. (2). *The Birth of China.* Fr. tr. by M. C. Salles, Payot, Paris, 1937. (References are to page numbers of the French ed.)

CREEL, H. G. (3). *Sinism; A Study of the Evolution of the Chinese World-View.* Open Court, Chicago, 1929.
(Rectifications of this by the author will be found in (4), p. 86; he acknowledges that the chief mistake herein was the view that the cosmism, naturalism, and phenomenalism of the Han was of very early origin, rather than due to Tsou Yen and his school in the −4th century.)

CREEL, H. G. (4). *Confucius; the Man and the Myth.* Day, New York, 1949; Kegan Paul, London, 1951. Crit. D. Bodde, *JAOS*, 1950, **70**, 199.

CREEL, H. G. (5). 'Was Confucius Agnostic?' *TP*, 1935, **29**, 55.

CREEL, H. G. (6). *Chinese Thought from Confucius to Mao Tsê-Tung.* Univ. of Chicago Press, Chicago, 1953. Crit. J. Needham, *SS*, 1954, **18**, 373; Chhen Jung-Chieh (Chan Wing-Tsit), *PEW*, 1954, **4**, 181.

CROWTHER, J. G. (1). *The Social Relations of Science.* Macmillan, London, 1941.

CUDWORTH, RALPH (1). *The True Intellectual System of the Universe; wherein all the Reason and Philosophy of Atheism is confuted and its Impossibility demonstrated, with a Treatise concerning Eternal and Immutable Morality.* First pub. 1678; ed. and tr. J. Harrison from the Latin of J. L. Mosheim (1733), with the latter's notes and dissertations. Tegg, London, 1845.

CUMONT, F. (1). *Astrology and Religion among the Greeks and Romans.* Putnam, New York, 1912.

CUMONT, F. (2). 'La Fin du Monde selon les Mages Occidentaux.' *RHR/AMG*, 1931, **103**, 29.

CUNNINGHAM, R. (1). 'Nangsal Obum'. in *JWCBRS*, 1940, A, **12**, 35.

CURTIS, J. G. (1). *Harvey's Views on the Use of the Circulation of the Blood.* New York, 1915.

DAHLKE, P. (1). *Buddhism and Science.* Macmillan, London, 1913.

DALCQ, A. M. (1). 'Le Problème de l'Evolution [Phylogénétique], est-il près d'être résolu?' *ASRZB*, 1951, **82**, 117.

DAMPIER-WHETHAM, W. C. D. (1). *A History of Science, and its Relations with Philosophy and Religion.* Cambridge, 1929.

DANIELLI, J. F. & BROWN, R. (1) (ed.). *Physiological Mechanisms in Animal Behaviour.* Cambridge, 1950 (Symposia of the Society of Experimental Biology, no. 4). Contributions by E. D. Adrian, J. Konorski, K. S. Lashley, C. F. A. Pantin, P. Weiss and others.

DAS, S. K. (1). *Śakti or Divine Power.* Univ. of Calcutta, Calcutta, 1934.

DASGUPTA, S. B. (1). *An Introduction to Tantric Buddhism.* Univ. of Calcutta Press, Calcutta, 1950.

DASGUPTA, S. B. (2). *Obscure Religious Cults as Background of Bengali Literature.* Univ. of Calcutta Press, Calcutta, 1946.

DAVIDS, T. W. RHYS. See Rhys-Davids.

DEFERRARI, R. J., BARRY, M. & McGUIRE, M. R. P. (1). *A Concordance of Ovid.* Washington, 1939.

DEHERGNE, J. (1). 'Un Envoyé de l'Empereur Kang-Hi à Louis XIV, le Père Joachim Bouvet.' *BUA*, 1943 (3ᵉ sér.), 651.

DELHERM, M. & LAQUERRIÈRE, M. (1). 'Histoire de la Physiothérapie.' In Laignel-Lavastine (1), *Histoire Générale de la Médecine* (vol. 3, p. 593).

DEMIÉVILLE, P. (1). 'Le Miroir Spirituel.' *S*, 1947, **1**, 112.

DEMIÉVILLE, P. (3). 'Résumé des Cours de l'Année Scolaire; Chaire de Langue et Littérature Chinoises.' (a) *ACF*, 1947, **47**, 151 (on the formation of the vocabulary of Chinese philosophy, especially the word *li*); (b) *ACF*, 1948, **48**, 158 (on the word *fên*, and on *Chuang Tzu*, ch. 2); (c) *ACF*, 1949, **49**, 177 (on *subitisme* and *gradualisme*); (d) *ACF*, 1950, **50**, 188 (on some thinkers of the Chhing period, e.g. Ku Yen-Wu and Tai Chen).

DENNYS, N. B. (1). *The Folklore of China.* Trübner, London and China Mail, Hongkong, 1876. Orig. pub. *CR*, 1875, **3**, 269, 342; **4**, 1, 67, etc.

DIDEROT, D. (1). *Supplément au 'Voyage de Bougainville'*, ed. G. Chinard. Johns Hopkins Univ. Press, Baltimore, 1935. Eng. tr. 'Supplement to Bougainville's Voyage; or, Dialogue between A and B on the Disadvantage of attaching Moral Ideas to certain Physical Actions incompatible therewith.' In *Diderot, Interpreter of Nature*, ed. J. Stewart & J. Kemp, pp. 146 ff. Lawrence & Wishart, London, 1937.

DIELS-FREEMAN = FREEMAN, K. (1). *Ancilla to the Pre-Socratic Philosophers; a complete translation of the Fragments in Diels' 'Fragmente der Vorsokratiker'.* Blackwell, Oxford, 1948.

DIETERICI, F. (1) (tr.). *Die Philosophie der Araber im IX u. X Jahrhundert n. Chr., aus der Theologie des Aristoteles, den Abhandlungen Alfarabis und den Schriften der Lautern Brüder [Rasā'il Ikhwān al-Ṣafā'].* Hinrichs, Leipzig, 1858–95. 1, Einleitung und Makrokosmos, 1876; 2, Mikrokosmos, 1879; 3, Propaedeutik, 1865; 4, Logik und Psychologie, 1868; 5, Naturanschauung und Naturphilosophie, 1861, 1876; 6, Der Streit zwischen Thier und Mensch, 1858; 7, Anthropologie, 1871; 7, Lehre von der Weltseele, 1872.

DODDS, E. R. (1). *The Greeks and the Irrational.* Univ. of California Press, 1951. (Sather Classical Lectures, no. 25.)

DOOLITTLE, J. (1). *A Vocabulary and Handbook of the Chinese Language.* Fuchow, 1872.

DORÉ, H. (1). *Recherches sur les Superstitions en Chine.* 15 vols. T'u-Se-Wei Press, Shanghai, 1914–29.
Pt. I, vol. 1, pp. 1–146: 'Superstitious' practices, birth, marriage and death customs (*VS*, no. 32).
Pt. I, vol. 2, pp. 147–216: talismans, exorcisms and charms (*VS*, no. 33).
Pt. I, vol. 3, pp. 217–322: divination methods (*VS*, no. 34).
Pt. I, vol. 4, pp. 323–488: seasonal festivals and miscellaneous magic (*VS*, no. 35).
Pt. I, vol. 5, sep. pagination: analysis of Taoist talismans (*VS*, no. 36).
Pt. II, vol. 6, pp. 1–196: Pantheon (*VS*, no. 39).
Pt. II, vol. 7, pp. 197–298: Pantheon (*VS*, no. 41).
Pt. II, vol. 8, pp. 299–462: Pantheon (*VS*, no. 42).
Pt. II, vol. 9, pp. 463–680: Pantheon, Taoist (*VS*, no. 44).
Pt. II, vol. 10, pp. 681–859: Taoist celestial bureaucracy (*VS*, no. 45).
Pt. II, vol. 11, pp. 860–1052: city-gods, field-gods, trade-gods (*VS*, no. 46).
Pt. II, vol. 12, pp. 1053–1286: miscellaneous spirits, stellar deities (*VS*, no. 48).
Pt. III, vol. 13, pp. 1–263: popular Confucianism, sages of the Wên Miao (*VS*, no. 49).
Pt. III, vo'. 14, pp. 264–606: popular Confucianism historical figures (*VS*, no. 51).
Pt. III, vol. 15, sep. pagination: popular Buddhism, life of Gautama (*VS*, no. 57).

DORSEY, G. L. (1). 'Two objective Bases for a Worldwide Legal Order.' In *Ideological Differences and World Order*, ed. F. S. C. Northrop. Yale Univ. Press, 1949.

DOUGLAS, R. K. (1) (tr.). 'Early Chinese Texts, 1, The Calendar of the Hsia Dynasty.' *OAA*, 1882, **1**, 1.

DRYDEN, JOHN, GARTH, S. *et al.* (1) (tr.). *Ovid's 'Metamorphoses' in 15 Books, translated by the most eminent Hands.* Tonson, London, 1717.

Dubos, R. (1). 'Studies on a Bactericidal Agent extracted from a Soil Bacillus.'
I. 'Preparation of the Agent and its Activity *in vitro*.' *JEM*, 1939, **70**, 1; *SRIMR*, 1939, **113**, 337.
II. 'Protective Effect of the Agent against experimental *Pneumococcus* infections in Mice.' *JEM*, 1939, **70**, 11; *SRIMR*, 1939, **113**, 347.
III. (With C. Cattaneo) 'Preparation and Activity of a Protein-free Fraction.' *JEM*, 1939, **70**, 249; *SRIMR*, 1940, **114**, 377.

Dubos, R. & Avery, O. T. (1). 'Decomposition of the Capsular Polysaccharide of *Pneumococcus* Type III by a Bacterial Enzyme.' *JEM*, 1931, **54**, 51.

Dubs, H. H. (2) (tr., with assistance of Phan Lo-Chi and Jen Thai). '*History of the Former Han Dynasty*', *by Pan Ku, a Critical Translation with Annotations*. 2 vols. Waverly, Baltimore, 1938.

Dubs, H. H. (3). 'The Victory of Han Confucianism.' *JAOS*, 1938, **58**, 435. (Reprinted in Dubs (2), pp. 341 ff.)

Dubs, H. H. (5). 'The Beginnings of Alchemy.' *ISIS*, 1947, **38**, 62.

Dubs, H. H. (7). *Hsün Tzu; the Moulder of Ancient Confucianism*. Probsthain, London, 1927.

Dubs, H. H. (8) (tr.). *The Works of Hsün Tzu*. Probsthain, London, 1928.

Dubs, H. H. (9). 'The Political Career of Confucius.' *JAOS*, 1946, **66**, 273.

Dubs, H. H. (10). 'The Attitude of Han Kao-Tsu to Confucianism.' *JAOS*, 1937, **57**, 172.

Dubs, H. H. (11). 'The Date and Circumstances of Lao Tzu.' *JAOS*, 1941, **61**, 215.

Dubs, H. H. (12). 'The Date and Circumstances of Lao Tzu.' *JAOS*, 1942, **62**, 300.

Dubs, H. H. (13). 'An Ancient Chinese Mystery Cult.' *HTR*, 1942, **35**, 221.

Dubs, H. H. (14). 'The Development of Altruism in Confucianism.' *Proc. Tenth Internat. Congress of Philosophy*, p. 156. Amsterdam, 1948.

Dubs, H. H. (15). '[Human] "Nature" in the Teachings of Confucius.' *JAOS*, 1930, **50**, 233.

Dubs, H. H. (16). 'Han Yü and the Buddha's Relic; an episode in medieval Chinese religion.' *RR*, 1946, 5.

Dubs, H. H. (17). 'Did Confucius study the Book of Changes?' *TP*, 1927, **25**, 82.

Dubs, H. H. (18). 'The Date of Confucius' Birth.' *AM*, 1949 (n.s.), **1**, 139.

Dubs, H. H. (19). 'Taoism.' In *China*, ed. H. F. McNair, p. 266. Univ. of California Press, 1946.

Dudgeon, J. (1). 'Kung-Fu, or Medical Gymnastics.' *JPOS*, 1895, **3**, 341.

Dukes, E. J. (1). 'Fêng-Shui.' *ERE*, vol. 5, p. 833.

Dumont, P. E. (1). 'Primitivism in Indian Literature.' Supplementary essay no. 2 in Lovejoy & Boas (1), *q.v.*

Durkheim, A. & Mauss, M. (1). 'De Quelques Formes Primitives de Classifications.' *AS*, 1901, **6**, 1. Eng. tr. by R. Needham, Cohen & West, London 1963.

Duyvendak, J. J. L. (3) (tr.). '*The Book of the Lord Shang*'; *a Classic of the Chinese School of Law*. Probsthain, London, 1928.

Duyvendak, J. J. L. (4). 'Hsün Tzu on the Rectification of Names' (tr. of *Hsün-Tzu*, ch. 22). *TP*, 1924, **23**, 221.

Duyvendak, J. J. L. (5). Comments on Wulff's translations of certain chapters of the *Tao Tê Ching*. *TP*, 1948, **38**, 332.

Duyvendak, J. J. L. (6). Comments on Pasquale d'Elia's *Galileo in Cina*. *TP*, 1948, **38**, 321.

Duyvendak, J. J. L. (7). (*a*) 'The Philosophy of *Wu Wei*.' *ASEA*, 1947, **3/4**, 81. (*b*) 'La Philosophie du Non-Agir.' *Conferenze d. Istituti Ital. per il Medio ed Estremo Oriente*. Rome, 1951, **1**, 1.

Duyvendak, J. J. L. (18) (tr.). '*Tao Tê Ching*', the Book of the Way and its Virtue. Murray, London, 1954 (Wisdom of the East series), Crit. P. Demiéville, *TP*, 1954, **43**, 95; D. Bodde, *JAOS*, 1954, **74**, 211.

Duyvendak, J. J. L. (20). 'A Chinese *Divina Commedia*.' *TP*, 1952, **41**, 255. (Also sep. pub. Brill, Leiden, 1952.)

Eberhard, W. (4). 'Typen chinesischen Volksmärchen.' *FFC*, no. 120, p. 98. Helsinki, 1937.

Eberhard, W. (5) (coll. and tr.). *Chinese Fairy Tales and Folk Tales*. Kegan Paul, London, 1937.

Eberhard, W. (6). 'Beiträge zur kosmologischen Spekulation Chinas in der Han Zeit.' *BA*, 1933, **16**, 1.

Eberhard, W. (7). *Mazdaizm ve Maniheizm hakkinda notlar*. *Ülkü* (Ankara), June 1941, p. 295.

Eberhard, W. (8). Criticism of Dubs' theory of the influence of Zoroastrianism (Mazdaism) on religious Taoism. *OR*, 1949, **2**, 191.

Eberhard, W. (9). *A History of China from the Earliest Times to the Present Day*. Routledge & Kegan Paul, London, 1950. Tr. from the German ed. (Swiss pub.) of 1948 by E. W. Dickes. Turkish ed. *Čin Tarihi*, Istanbul, 1946. Crit. K. Wittfogel, *AA*, 1950, **13**, 103; J. J. L. Duyvendak, *TP*, 1949, **39**, 369; A. F. Wright, *FEQ*, 1951, **10**, 380.

Eberhard, W., Müller, R. & Henseling, R. (1). 'Beiträge z. Astronomie d. Han-Zeit. II.' *SPAW/PH*, 1933, **23**, 937.

EDGERTON, W. F. (1). (a) 'The Upanishads, what do they seek and why?' *JAOS*, 1929, **49**, 97. (b) 'Sources of the Philosophy of the Upanishads.' *JAOS*, 1917, **36**, 197. (c) 'Dominant Ideas in the Formation of Indian Culture.' *JAOS*, 1942, **62**, 151.

EDKINS, J. (3). 'Astrology in Ancient China.' *CR*, 1885, **14**, 345.

EDKINS, J. (4). *Chinese Buddhism*. Kegan Paul, London, 1879.

EDKINS, J. (10). 'On the Poets of China during the Period of the Contending States and of the Han Dynasty' (Chhü Yuan, etc.). *JPOS*, 1889, **2**, 201.

EDKINS, J. (14). '[Glossary of] Terms used in [Chinese] Geomancy.' In J. Doolittle (1), vol. 2, p. 515.

EDKINS, J. (16). 'A Sketch of the Taoist Mythology in its Modern Form.' *JRAS/NCB*, 1859, 309.

EDMUNDS, A. J. & ANESAKI, MASAHARU. *Buddhist and Christian Gospels, now first compared from the Originals; being 'Gospel Parallels from Pāli texts' reprinted with additions*. Innes, Philadelphia; Luzac, London; Harrassowitz, Leipzig, 1914.

EGERTON, C. (1) (tr.). *The Golden Lotus [Chin Phing Mei]*. 4 vols. (complete Eng. tr. but some passages in Latin). Routledge, London, 1939, repr. 1954.

EGGLESTON, SIR F. (1). *Search for a Social Philosophy*. University Press, Melbourne, 1941.

EICHHORN, W. (1) (tr.). 'Ein Beitrag zur Kenntnis der chinesischen Philosophie, der *T'ŭng-Šū* des Čeŭ-Tsï....' (Chou Tun-I's *I Thung Shu*, chs. 21–40.) *AM*, 1932, **8**, 442, 501, 541, 589. Issued with the earlier chapters (Grube, 5) in *Asia Major*, China Bibliothek series, 1932, no. 3.

EICHHORN, W. (2). 'Chou Tun-I, ein chinesische Gelehrtenleben a. d. 11. Jahrhundert' (*Abhdl. f. d. Kunde d. Morgenlandes*, **21**, no. 5). Brockhaus, Leipzig, 1936. Crit. J. J. L. Duyvendak, *TP*, 1937, **33**, 100; W. Franke, *OLZ*, 1938, 126.

EICHHORN, W. (3) (tr.). 'Die *Westinschrift* des Chang Tsai, ein Beitrag z. Geistesgeschichte d. Nordl. Sung' (*Abhdl. f. d. Kunde d. Morgenlandes*, **22**, no. 7). Brockhaus, Leipzig, 1937. Crit. H. Maspero, *OLZ*, 1942, 378.

EISLER, ROBERT (1). *The Royal Art of Astrology*. Joseph, London, 1946. Crit. H. Chatley, *O*, 1947, **67**, 187.

EISLER, RUDOLF (1). *Wörterbuch der philosophischen Begriffe*. 3 vols. Mittler, Berlin, 1929.

EITEL, E. J. (1) (tr.). 'Travels of the Emperor Mu.' *CR*, 1888, **17**, 233, 247.

EITEL, E. J. (2). *Fêng-Shui; Principles of the Natural Science of the Chinese*. Hongkong; Trübner, London, 1873. French tr. by L. de Milloué, *AMG*, 1880, **1**, 203.

EITEL, E. J. (3). 'Chinese Philosophy before Confucius.' *CR*, 1878, **7**, 388.

EITEL, E. J. (4). 'Fragmentary Studies in Ancient Chinese Philosophy.' *CR*, 1887, **15**, 338; 1888, **17**, 26.

EITEL, E. J. (5). 'Spirit Rapping in China.' *NQCJ*, 1867, **1**, 164.

D'ELIA, PASQUALE (1). 'Echi delle Scoperte Galileiane in Cina vivente ancora Galileo (1612–1640).' *AAL/RSM*, 1946 (8e sér.), **1**, 125. Republished in enlarged form as 'Galileo in Cina. Relazioni attraverso il Collegio Romano tra Galileo e i gesuiti scienzati missionari in Cina (1610–1640).' *Analecta Gregoriana*, **37** (Series Facultatis Missiologicae A (N/1)), Rome, 1947. Revs.: G. Loria, *A/AIHS*, 1949, **2**, 513; J. J. L. Duyvendak, *TP*, 1948, **38**, 321; G. Sarton, *ISIS*, 1950, **41**, 220.

ELIADE, MIRCEA (1). *Le Mythe de l'Eternel Retour; Archétypes et Répétition*. Gallimard, Paris, 1949.

ELIADE, MIRCEA (2). *Traité d'Histoire des Religions*. Payot, Paris, 1949.

ELIADE, MIRCEA (3). *Le Chamanisme et les Techniques Archaïques de l'Extase*. Payot, Paris, 1951.

ELLIS, HAVELOCK (1). *Affirmations*. London, 1898.

ELTON, C. (1). *Animal Ecology*. Sidgwick & Jackson, London, 1927.

ENGELS, F. (1). *The Origin of the Family, Private Property, and the State*. Kerr, Chicago, 1902.

ENGELS, F. (2). *Socialism, Utopian and Scientific*. Allen & Unwin, London, 1892 (1936).

ENGELS, F. (3). *Dialectics of Nature*. Ed. and tr. C. Dutt, with preface and notes by J. B. S. Haldane. 2nd ed. Lawrence & Wishart, London, 1946.

ENGLISH HYMNAL. See Birkbeck, Dearmer *et al.*

ERKES, E. (1) (tr.). 'Das Weltbild d. *Huai-nan-tzu*' (tr. of ch. 4). *OAZ*, 1918, **5**, 27.

ERKES, E. (3). 'The cosmogonic myth in *Tao Tê Ching* ch. 42.' *AA*, 1940, **8**, 16.

ERKES, E. (4). 'Ho Shang Kung's Commentary on Lao Tzu.' *AA*, 1945, **8**, 119; 1946, **9**, 197.

ERKES, E. (5). 'Mystik und Schamanismus.' *AA*, 1945, **8**, 197.

ERKES, E. (6). Comments on Waley's translation of the *Tao Tê Ching*. *AA*, 1935, **5**, 288.

ERKES, E. (7). 'Lü Dsus *Lied vom Talgeist*.' *SA*, 1933, **8**, 94.

ERKES, E. (8). 'Chhü Yüan's *Thien Wên*.' *MS*, 1941, **6**, 273.

ERKES, E. (9). 'Zur Textkritik d. *Chung Yung*.' *MSOS*, 1917, **20**, 142.

ERKES, E. (10) (tr.). 'Das Mädchen vom Hua-Shan, von Han Yü.' *AM*, 1933, **9**, 591.

ERKES, E. (11). Observations on Karlgren's 'Fecundity Symbols in Ancient China' (9). *BMFEA*, 1931, **3**, 63.

ERKES, E. (13). 'Der Druck der Taoistischer Kanon.' *GUJ*, 1935, 326.

ERKES, E. (14). 'Die Anfänge des dauistischen Mönchstums.' *SA*, 1936, **11**, 36.

ERKES, E. (15). 'Das Primat des Weibes im alten China.' *SA*, 1935, **10**, 166.

ESCARRA, J. (1). (*a*) *Le Droit Chinois*. Vetch, Peiping, 1936; Sirey, Paris, 1936. (*b*) *La Conception Chinoise du Droit*. *APDSJ*, 1935, **5**, 7. (Identical with the earlier part of the book.) (*c*) 'Chinese Law.' *ESS*, vol. 9, p. 249.

ESCARRA, J. (2). *Loi et Coutume en Chine*. Etudes de Sociol. et d'Ethnol. Juristique. 1931.

ESCARRA, J. (3). 'Western Methods of Research into Chinese Law.' *CSPSR*, 1924, **8** (no. 1), 227.

ESCARRA, J. (4). 'La Chine et l'Esprit Juridique.' *SCI*, 1938, **63**, 99.

ESCARRA, J. & GERMAIN, R. (1) (tr.). *La Conception de la Loi et les Théories des Légistes à la Veille des Ts'in [Chhin]*. (Tr. of chs. 7, 13, 14, 15 and 16 of Liang Chhi-Chhao, *5*). Preface by G. Padoux. China Booksellers, Peking, 1926.

ESPINAS, A. (1). *Les Origines de la Technologie*. Alcan, Paris, 1897.

EVANS, E. P. (1). *The Criminal Prosecution and Capital Punishment of Animals*. Heinemann, London, 1906.

FARRINGTON, B. (1). *Science in Antiquity*. Butterworth, London, 1936.

FARRINGTON, B. (2). *The Civilisation of Greece and Rome*. Gollancz, London, 1938.

FARRINGTON, B. (3). *Science and Politics in the Ancient World*. Allen & Unwin, London, 1939. Crit. F. M. Cornford (5); reply and extension, B. Farrington (9).

FARRINGTON, B. (4). *Greek Science (Thales to Aristotle); its meaning for us*. Penguin Books, London, 1944.

FARRINGTON, B. (5). *Head and Hand in Ancient Greece; Four Studies in the Social Relations of Thought*. Watts, London, 1947.

FARRINGTON, B. (6). *Francis Bacon; Philosopher of Industrial Science*. Schuman, New York, 1949.

FARRINGTON, B. (7). '*Temporis Partus Masculus*; an untranslated Writing of Francis Bacon.' *CN*, 1951, **1**, 193.

FARRINGTON, B. (9). 'Second Thoughts on Epicurus' (Cornford Lecture at Cambridge, 1953). *SS*, 1953, **17**, 289.

FARRINGTON, B. (10). 'Vita Prior in Lucretius.' *HMA*, 1953, **81**, 59.

FARRINGTON, B. (11). 'The Meanings of Voluptas in Lucretius.' *HMA*, 1952, **80**, 26.

FARRINGTON, B. (12). 'Lucretius and Manilius on Friendship.' *HMA*, 1954, **83**, 10. 'La Amistad Epicurea.' *NEF*, 1952, **3**, 105.

FARRINGTON, B. (13). 'Epicureanism and Science.' *SCI*, 1954, **48**.

FARRINGTON, B. (14). 'On Misunderstanding the Philosophy of Francis Bacon.' In *Science, Medicine and History* (Singer Presentation Volume), ed. E. A. Underwood, Oxford, 1954, vol. 1, p. 439.

FAVRE, B. (1). *Les Sociétés Sécrètes en Chine; origine, rôle historique, situation actuelle*. Maisonneuve, Paris, 1933.

FECHNER, G. (1). *Zend-Avesta*. Leipzig, 1851–4. Ed. K. Lasswitz, Hamburg and Leipzig, 1906.

FEI HSIAO-TUNG (1). 'The Problem of the Chinese Relationship System.' *MS*, 1936, **2**, 125.

FEIFEL, E. (1) (tr.). *Pao Phu Tzu, Nei Phien*, chs. 1 to 3. *MS*, 1941, **6**, 113.

FEIFEL, E. (2) (tr.). *Pao Phu Tzu, Nei Phien*, ch. 4. *MS*, 1944, **9**, 1.

FELDHAUS, F. M. (1). *Die Technik der Vorzeit, der geschichtlichen Zeit, und der Naturvölker* (technological encyclopaedia). Engelmann, Leipzig and Berlin, 1914.

FÊNG HAN-CHI (H. Y. FÊNG) (1). 'The Origin of Yü Huang.' *HJAS*, 1936, **1**, 242.

FÊNG HAN-CHI (H. Y. FÊNG) (2). 'The Discovery and Excavation of the Royal Tomb of Wang Chien.' *QBCB/E*, 1944 (n.s.), **4**, 1. Reissued as *Occasional Papers of the Szechuan Museum*, no. 1. Chhêngtu, 1944.

FÊNG HAN-CHI (H. Y. FÊNG) & SHRYOCK, J. K. (1). 'Chinese Mythology and Dr Ferguson.' *JAOS*, 1933, **53**, 53.

FÊNG HAN-CHI (H. Y. FÊNG) & SHRYOCK, J. K. (2). 'The Black Magic in China Known as Ku.' *JAOS*, 1935, **65**, 1.

FÊNG YU-LAN (1). *A History of Chinese Philosophy*. Vol. 1, *The Period of the Philosophers (from the Beginnings to c. −100)*, Vetch, Peiping, 1937; Allen & Unwin, London, 1937. Vol. 2, *The Period of Classical Learning (from the −2nd Century to the +20th Century)*, Princeton Univ. Press, Princeton, N.J., 1953. Tr. D. Bodde; crit. Chhen Jung-Chieh (Chan Wing-Tsit), *PEW*, 1954, **4**, 73; J. Needham, *SS*, 1955, **19**, 268. Translations of parts of vol. 2 also appeared in *HJAS*; see under Bodde. See also Fêng Yu-Lan (*1*).

FÊNG YU-LAN (2). *The Spirit of Chinese Philosophy*, tr. E. R. Hughes. Kegan Paul, London, 1947.

FÊNG YU-LAN (3). 'The Origin of Ju and Mo.' *CSPSR*, 1935, **19**, 151.

FÊNG YU-LAN (4). 'The Place of Confucius in Chinese History.' *CSPSR*, 1932, **16**, 1.

FÊNG YU-LAN (5) (tr.). *Chuang Tzu; a new selected translation with an exposition of the philosophy of Kuo Hsiang*. Com. Press, Shanghai, 1933.

FÊNG YU-LAN (6). 'Mao Tsê-Tung's "On Practice", and Chinese Philosophy.' *PC*, 1951, **4** (no. 10), 5.

FÊNG YU-LAN & PORTER, L. C. (1). Various translations in Porter's *Aids to the Study of Chinese Philosophy* (1), q.v.

FERGUSON, J. C. (2). *Survey of Chinese Art*. Com. Press, Shanghai, 1940.

FERRAND, G. (1). *Relations de Voyages et Textes Géographiques Arabes, Persans et Turcs relatifs à l'Extrême Orient, du 8ᵉ au 18ᵉ siècles, traduits, revus et annotés etc.* 2 vols. Leroux, Paris, 1913.

FESTUGIÈRE, A. G. (1). *La Révélation d'Hermès Trismégiste, I. L'Astrologie et les Sciences Occultes*. Gabalda, Paris, 1944. (See Filliozat, 5.)

FIGGIS, J. N. (1). *Studies in Political Thought from Gerson to Grotius*. Cambridge, 1916.

FILLIOZAT, J. (1). *La Doctrine Classique de la Médecine Indienne*. Imp. Nat., CNRS and Geuthner, Paris, 1949.

FILLIOZAT, J. (2). 'Les Origines d'une Technique Mystique Indienne.' *RP*, 1946, **136**, 208.

FILLIOZAT, J. (3). 'Taoisme et Yoga.' *DVN*, 1949, **3**, 1.

FILLIOZAT, J. (5). Review of Festugière (1), q.v. *JA*, 1944, **234**, 349.

FLUDD, ROBERT (1). *Medicina Catholica*. Frankfurt, 1629.

FLÜGEL, G. (1). 'Über Inhalt und Verfasser d. arabischen Encyclopädie' (*Rasā'il Ilkhwān al-Ṣafā'*). *ZDMG*, 1859, **3**, 11.

FORDE, C. DARYLL (1). *Habitat, Economy and Society; a geographical introduction to Ethnology*. Methuen, London, 1934.

FORKE, A. (2) (tr.). 'Yang Chu the Epicurean in his relation to Lieh Tzu the Pantheist.' *JPOS*, 1893, **3**, 203. Repr. as *Yang Chu's Garden of Pleasure*, with introduction by H. Cranmer-Byng. Murray, London, 1912. (Wisdom of the East series.)

FORKE, A. (3) (tr.). *Mo Ti des Sozialethikers und seiner Schüler philosophische Werke*. Berlin, 1922. (*MSOS*, Beibände, **23** to **25**.)

FORKE, A. (4) (tr.). '*Lun Hêng*', *Philosophical Essays of Wang Chhung*. Pt. I, 1907, Kelly & Walsh, Shanghai; Luzac, London; Harrassowitz, Leipzig. Pt. II, 1911 (with the addition of Reimer, Berlin). (*MSOS*, Beibände, **10** and **14**. Orig. pub. 1906, **9**, 181; 1907, **10**, 1; 1908, **11**, 1; 1911, **14**, 1.)

FORKE, A. (5). 'The Chinese Sophists' (includes complete tr. of *Têng Hsi Tzu, Hui Tzu* and other paradoxes, *Kungsun Lung Tzu*). *JRAS/NCB*, 1902, **34**, 1.

FORKE, A. (6). *The World-Conception of the Chinese; their astronomical, cosmological and physico-philosophical Speculations* (Pt. 4 of this, on the Five Elements, is reprinted from Forke (4) vol. 2, App. I). Probsthain, London, 1925. German tr. *Gedankenwelt des chinesischen Kulturkreis*. München, 1927. Chinese tr. *Chhi-Na Tzu-Jan Kho-Hsüeh Ssu-Hsiang Shih*. Crit. B. Schindler, *AM*, 1925, **2**, 368.

FORKE, A. (7). 'The Philosopher Yang Hsiung.' *JRAS/NCB*, 1930, **61**, 108.

FORKE, A. (8). 'Wang Chhung and Plato on Death and Immortality.' *JRAS/NCB*, 1896, **31**, 40.

FORKE, A. (9). *Geschichte d. neueren chinesischen Philosophie* (i.e. from beg. of Sung to modern times). De Gruyter, Hamburg, 1938. (Hansische Univ. Abhdl. a.d. Geb. d. Auslandskunde, no. 46 (Ser. B, no. 25).)

FORKE, A. (10). 'Die chinesischen Skeptiker.' *SA*, 1939, **14**, 98.

FORKE, A. (11). Critique of Bruce (1) and (2), stating the case against translating Neo-Confucian *Li* as 'law'. *AM*, 1924, **1**, 186.

FORKE, A. (12). *Geschichte d. mittelalterlichen chinesischen Philosophie* (i.e. from beg. of Former Han to end of Wu Tai). De Gruyter, Hamburg, 1934. (Hamburg Univ. Abhdl. a.d. Geb. d. Auslandskunde, no. 41 (Ser. B, no. 21).)

FORKE, A. (13). *Geschichte d. alten chinesischen Philosophie* (i.e. from high antiquity to beg. of Former Han). De Gruyter, Hamburg, 1927. (Hamburg. Univ. Abhdl. a.d. Geb. d. Auslandskunde, no. 25 (Ser. B, no. 14).)

FORKE, A. (14). 'Die Anfänge des Idealismus in der chinesischen Philosophie.' *AM*, 1933, **9**, 141.

FÖRSTER, E. (1). *Roger Bacon's 'De Retardandis Senectutis Accidentibus et de Sensibus Conservandis' und Arnold von Villanova's 'De Conservanda Juventutis et Retardanda Senectute'*. Inaug. Diss., Leipzig, 1924.

FRANCK, A. (1). *La Kabbale; la Philosophie Réligieuse des Hébreux*. Hachette, Paris, 1843.

FRÄNGER, W. (1). *The Millennium of Hieronymus Bosch*. Faber, London, 1952.

FRANK, J. N. (1). *Courts on Trial; Myth and Reality in American Justice*. Princeton Univ. Press, N.J., 1949.

FRANK, J. N. (2). *Fate and Freedom*. Simon & Schuster, New York, 1945.

FRANK, J. N. (3). *Law and the Modern Mind*. Stevens, London, 1930 (6th ed. 1949).

FRANK, J. N. (4). 'Modern and Ancient Legal Pragmatism.' *NDL*, 1949, **25**, 207 and 460.

FRANKE, H. (3). 'Volksaufstände in d. Geschichte Chinas.' *GWI*, 1951, **1**, 31.

FRANKE, H. (5). *Sinologie* (review of literature from about 1935 onwards). 1st part of 'Orientalistik' Section forming vol. 19 (pp. 1–219) of *Wissenschaftliche Forschungsberichte (Geisteswissenschaftliche Reihe)*, ed. K. Hönn. A. Francke, Bern, 1953.

FRANKE, O. (5). 'Zur Frage der Einführung des Buddhismus in China.' *MSOS*, 1910, **13**, 295.

FRANKE, O. (6). 'Der kosmische Gedanke in Philosophie und Staat d. Chinesen.' *VBW* (1925/1926), 1928, p. 1. Reprinted in Franke (8), p. 271.

FRANKE, O. (7). 'Leibniz und China.' *ZDMG*, 1928, **82** (NF **7**), 155. Reprinted in Franke (8), p. 313.

FRANKE, O. (8). *Aus Kultur und Geschichte Chinas, Vorträge und Abhandlungen aus den Jahren 1902–1942*. Deutschland Institut, Peking, 1945.

FRANKFORT, H. (1). *Before Philosophy*. London, 1949.

FRAZER, SIR J. G. (1). *The Golden Bough*. 3 vol. ed. Macmillan, London, 1900; superseded by 12 vol. ed. (here used), Macmillan, London, 1913–20. Abridged 1 vol. ed. Macmillan, London, 1923.

FREEMAN, K. (2). *The Pre-Socratic Philosophers, a companion to Diels' 'Fragmente der Vorsokratiker'*. Blackwell, Oxford, 1946. Cf. Diels-Freeman.

FREEMAN, M. (1). 'The Philosophy of Tai Tung-Yuan.' *JRAS/NCB*, 1933, **64**, 50.

FRIEDLÄNDER, P. (1). 'Pattern of Sound, and Atomistic Theory, in Lucretius.' *AJP*, 1941, **62**, 16.

FRIEDMANN, H. & WEBER, W. A. (1). 'The Honey-Guide; a Bird that eats Wax.' *NGM*, 1954, **105**, 551.

FRUTON, J. S. & SIMMONDS, S. (1). *General Biochemistry*. Wiley, New York, 1953; Chapman & Hall, London, 1953.

V.D. GABELENTZ, G. (2) (tr.). '*Thai-Kih-Thu*' ['*Thai Chi Thu*'], des *Tscheu-Tsi* [*Chou Tzu*]; *Tafel des Urprinzipes mit Tschu-Hi's* [*Chu Hsi's*] *Commentare nach dem* '*Hoh-Pih-Sing-Li*'. Chinesisch mit mandschurischer und deutscher Übersetzung, Einleitung und Anmerkungen. Zahn, Dresden, 1876.

GABOR, D. (1). 'Communication Theory, Past, Present and Prospective.' In *Symposium on Information Theory*, p. 2. Min. of Supply, London, 1950 (mimeographed).

GALE, E. M. (1) (tr.). *Discourses on Salt and Iron* ('*Yen Thieh Lun*'), *a Debate on State Control of Commerce and Industry in Ancient China, chapters 1–19*. Brill, Leiden, 1931. (Sinica Leidensia, no. 2.)

GALE, E. M., BOODBERG, P. A. & LIN, T. C. (1) (tr.). 'Discourses on Salt and Iron (*Yen Thieh Lun*), Chapters 20–28.' *JRAS/NCB*, 1934, **65**, 73.

LE GALL, S. (1). *Le Philosophe Tchou Hi, Sa Doctrine, son Influence*. T'u-Se-Wei, Shanghai, 1894 (*VS*, no. 6). (Incl. tr. of part of ch. 49 of *Chu Tzu Chhüan Shu*.)

GALLAGHER, L. J. (1) (tr.). *China in the 16th Century; the Journals of Matthew Ricci, 1583–1610*. Random House, New York, 1953. [A complete translation, preceded by inadequate bibliographical details, of Nicholas Trigault's *De Christiana Expeditione apud Sinas* (1615).] Based on an earlier publication: *The China that Was; China as discovered by the Jesuits at the close of the 16th Century: from the Latin of Nicholas Trigault*. Milwaukee, 1942. [Identifications of Chinese names in Yang Lien-Shêng (4).] Crit. J. R. Ware, *ISIS*, 1954, **45**, 395.

VON GARBE, R. K. (1). 'Yoga.' *ERE*, vol. 12, p. 831.

VON GARBE, R. K. (2). *Die Saṃkhya Philosophie, eine Darstellung des Indischen Rationalismus nach den Quellen bearbeitet*. Haessel, Leipzig, 1894.

GARVIE, A. E. (1). 'Pantheism' (Introductory Section). *ERE*, vol. 9, p. 611.

GAUCHET, L. (1). 'Un Livre Taoïque, le *Chêng Chen King* [*Sêng Chen Ching*], sur la Génération des Esprits dans l'Homme.' *BUA*, 1949 (3ᵉ sér.), **10**, 63.

GAUCHET, L. (2). 'Contribution à l'Étude du Taoisme' (*Tao Tsang*). *BUA*, 1948 (3ᵉ sér.), **9**, 1.

GAUCHET, L. (3). 'A Travers le Canon Taoïque, quelques Synonymes du Tao.' *BUA*, 1942 (3ᵉ sér.), **3**, 303.

GAUCHET, L. (4). 'Le *Tou-jen King* [*Tu Jen Ching*] des Taoistes; son Texte Primitif et sa Date Probable.' *BUA*, 1941 (3ᵉ sér.), **2**, 511.

GEDEN, A. S. (1). 'Hindu Astrology.' *ERE*, vol. 12, p. 83.

GÉNY, V. (1). *Méthode d'Interprétation et Sources en Droit Positif*. Paris, 1919. *Science et Technique en Droit Privé Positif*. Paris, 1924.

GERARD, R. W. (1). 'Organism, Society and Science.' *SM*, 1940, **50**, 340, 403 and 530.

GERNET, L. (1). *Recherches sur le Développement de la Pensée Juridique et Morale en Grèce*. Paris, 1917.

GIBB, H. A. R. (2). 'An Interpretation of Islamic History.' *JWH*, 1953, **1**, 39.

GIERKE, OTTO (1). *Natural Law and the Theory of Society, 1500–1800*. Tr. E. Barker. Cambridge, 1934.

GIERKE, OTTO (2). *Political Theories of the Middle Ages*. Tr. F. W. Maitland. Cambridge, 1900.

GILES, H. A. (1). *A Chinese Biographical Dictionary*. 2 vols. Kelly & Walsh, Shanghai, 1898; Quaritch, London, 1898. Supplementary Index by J. V. Gillis & Yü Ping-Yüeh, Peiping, 1936. Account must be taken of the numerous emendations published by von Zach (4) and Pelliot (34), but many mistakes remain. Cf. Pelliot (35).

GILES, H. A. (4) (tr.). '*San Tzu Ching*', *translated and annotated*. Kelly & Walsh, Shanghai, 1900.

GILES, H. A. (5). *Adversaria Sinica*: 1st series, no. 1, pp. 1–25 (1905); no. 2, pp. 27–54 (1906); no. 3, pp. 55–86 (1906); no. 4, pp. 87–118 (1906); no. 5, pp. 119–44 (1906); no. 6, pp. 145–88 (1908); no. 7, pp. 189–228 (1909); no. 8, pp. 229–76 (1910); no. 9, pp. 277–324 (1911); no. 10, pp. 326–96 (1913); no. 11, pp. 397–438 (with index) (1914). Kelly & Walsh, Shanghai.
2nd series, no. 1, pp. 1–60. Kelly & Walsh, Shanghai, 1915.

GILES, H. A. (7) (tr.). 'The *Hsi Yüan Lu* or "Instructions to Coroners" translated from the Chinese.' *PRSM*, 1924, **17**, 59.

GILES, L. (2). *An Alphabetical Index to the Chinese Encyclopaedia* (*Chhin Ting Ku Chin Thu Shu Chi Chhêng*). British Museum, London, 1911.

GILES, L. (4) (tr.). *Taoist Teachings from the Book of 'Lieh Tzu'*. Murray, London, 1912; 2nd ed. 1947. (Wisdom of the East series.)

GILES, L. (6). *A Gallery of Chinese Immortals* ('*hsien*'); *selected biographies translated from Chinese sources* (*Lieh Hsien Chuan, Shen Hsien Chuan*, etc.). Murray, London, 1948. (Wisdom of the East series.)

GILES, L. (7). 'Wizardry in Ancient China.' *AP*, 1942, **13**, 484.

GILES, L. (11) (tr.). *Sun Tzu on the Art of War* [*Sun Tzu Ping Fa*]; *the oldest military Treatise in the World*. Luzac, London, 1910 (with original Chinese text). Repr. without notes Nan-fang, Chungking, 1954; Eng. text only repr. in *Roots of Strategy*, ed. Phillips, T. R. (*q.v.*).

GILL, W. (1). *The River of Golden Sand, being the narrative of a Journey through China and Eastern Tibet to Burmah*, ed E. C. Baber and H. Yule. Murray, London, 1883.

GINSBERG, M. (1). 'The Concepts of Juridical and Scientific Law.' *P*, 1939, **4**, 1.

GINZBERG, L. (1). 'Jewish Folklore, East and West.' In *Independence, Convergence and Borrowing, in Institutions, Thought and Art*, p. 89. Harvard Tercentenary Publication, Harvard Univ. Press, 1937.

GLANVILL, JOSEPH (1). *Scepsis Scientifica; or Confest Ignorance the Way to Science, in an Essay on the Vanity of Dogmatising and Confident Opinion*. London, 1661; 2nd ed. 1665. Repr. and ed. J. Owen, Kegan Paul, London, 1885.

GLICK, C. & HUNG SHÊNG-HUA (1). *Swords of Silence; Chinese Secret Societies, past and present*. Brill, Leiden, 1947.

GOLDAMMER, K. (1). *Paracelsus; Sozial-Ethische und Sozial-Politische Schriften; aus dem theologisch-religionsphilosophischen Werk ausgewählt, eingeleitet, und mit erklärenden Anmerkungen herausgegeben.*... Mohr, Tübingen, 1952. (Civitas Gentium series, ed. Max Graf zu Solms.)

V.D. GOLTZ, F. (1). 'Zauberei u. Hexenkunste, Spiritismus u. Schamanismus in China.' *MDGNVO*, 1893, **6**, 1.

GONNARD, R. (1). *La Légende du Bon Sauvage*. De Medicis, Paris, 1946.

GOOCH, G. P. (1). *English Democratic Ideas in the Seventeenth Century*, ed. H. J. Laski. 2nd ed. Cambridge, 1927.

GRAF, O. (1). 'Chu Hsi and Spinoza.' *Proc. Xth Internat. Congress of Philosophy*, vol. 1, p. 238. Amsterdam, 1949.

GRAF, O. (2) (tr.). '*Djin-Si Lu*' [*Chin Ssu Lu*]; *die Sungkonfuzianische Summa mit dem Kommentar des Yeh Tsai*. 3 vols. Sophia University Press, Tokyo, 1953–4. (Mimeographed.) Vol. 1, 'Einleitung'; vol. 2 (pts. 1 and 2), 'Text'; vol. 3, 'Anmerkungen'. (*MN* Monographs, no. 12.)

GRAHAM, A. C. (1). 'The Philosophy of Chhêng I-Chhuan (+1033/+1107) and Chhêng Ming-Tao (+1032/+1085).' Inaug. Diss., London, 1953.

GRANET, M. (1). *Danses et Légendes de la Chine Ancienne*. 2 vols. Alcan, Paris, 1926.

GRANET, M. (2). *Fêtes et Chansons Anciennes de la Chine*. Alcan, Paris, 1926; 2nd ed. Leroux, Paris, 1929.

GRANET, M. (3). *La Civilisation Chinoise*. Renaissance du Livre, Paris, 1929; 2nd ed. Albin Michel, Paris, 1948. (Evol. de l'Hum. series, no. 25.) Crit. Ting Wên-Chiang, *MSOS*, 1931, **34**, 161.

GRANET, M. (4). *La Religion des Chinois*. Gauthier Villars, Paris, 1922.

GRANET, M. (5). *La Pensée Chinoise*. Albin Michel, Paris, 1934. (Evol. de l'Hum. series, no. 25 bis.)

GRANET, M. (6). *Études Sociologiques sur la Chine*. Presses Univ. de France, Paris, 1953.

GRIFFITH, E. F. (1). *Modern Marriage*. Methuen, London, 1946.

DE GROOT, J. J. M. (1). *Chinesische Urkunde z. Geschichte Asiens*, (*a*) *Die Hunnen d. vorchristlichen Zeit*; (*b*) *Die Westlände Chinas in d. vorchristl. Zeit*, ed. O. Franke. De Gruyter, Berlin, 1921. Crit. E. von Zach, *AM*, 1924, **1**, 125.

DE GROOT, J. J. M. (2). *The Religious System of China*. Brill, Leiden, 1892.
Vol. 1, Funeral rites and ideas of resurrection.
Vols. 2, 3, Graves, tombs, and *fêng-shui*.
Vol. 4, The soul, and nature-spirits.
Vol. 5, Demonology and sorcery.
Vol. 6, The animistic priesthood (*wu*).

DE GROOT, J. J. M. (3). 'On Chinese Divination by Dissecting Written Characters.' *TP*, 1890, **1**, 239.

GRUBE, W. (3). *Die Religion der alten Chinesen*. Mohr-Siebeck, Tübingen, 1911. (Part of the *Religionsgeschichtliches Lesebuch*.)

GRUBE, W. (4). 'Beiträge z. chinesische Grammatik: die Sprache des Liet-tsï' (Lieh Tzu). *BVSAW/PH*, 1889, **41**, 155.

GRUBE, W. (5) (tr.). *Ein Beitrag zur Kenntnis der chinesischen Philosophie, 'T'ung-Šu' ['I Thung Shu'] des Ceu-Tsi [Chou Tun-I], mit Cu-Hi's [Chu Hsi's] Kommentar*. Chs. 1–8, Halzhausen, Vienna, 1880; chs. 9–20, Leipzig, 1881. Completed by Eichhorn (1).

GUDGER, E. W. (1). On so-called legends of wild animal behaviour. *SM*, 1935, **40**, 415.

GUDGER, E. W. (2). 'On certain small Terrestrial Mammals that are alleged to fish with the Tail.' *AMLN*, 1953, **50**, 189.

GUDGER, E. W. (3). 'Does the Jaguar use its Tail as a Lure in Fishing?' *JMLOL*, 1946, **27**, 37.

GUDGER, E. W. (4). 'How the Cassowary goes Fishing.' *NH*, 1927, **27**, 485.

GUÉRIN, P. (1). *L'Idée de Justice dans la Conception de l'Univers chez les premiers Philosophes Grecs; de Thalès à Heraclite*. Alcan, Paris, 1934.

VAN GULIK, R. H. (3). *Erotic Colour Prints of the Ming Period with an Essay on Chinese Sex Life from the Han to the Chhing Dynasty (−206 to +1644)*. 3 vols. in case. Privately printed, Tokyo, 1951 (50 copies only, distributed to fifty of the most important libraries of the world). Crit. W. L. Hsü, *MN*, 1952, **8**, 455; E. B. Ceadel, *AM*, 1954.

VAN GULIK, R. H. (4). 'The Mango "Trick" in China; an essay on Taoist Magic.' *TAS/J*, 1952 (3rd ser.), **3**.

HAAS, H. (1). 'Tsungmi's *Yuen-Zen-Lun* [*Yuan Jen Lun*], eine Abhandlung ü.d. Ursprung d. Menschen a.d. Kanon d. chinesischen Buddhismus.' *ARW*, 1909, **12**, 491.

HACKMANN, H. F. (1). 'Die Mönchsregeln des Klostertaoismus.' *OAZ*, 1919, **8**, 142.

HACKMANN, H. F. (2). *Chinesischen Philosophie*. Reinhardt, München, 1927.

HACKMANN, H. F. (3). 'Alphabetisches Verzeichnis zum *Kao Sêng Chuan* [biographies of Buddhist monks].' *AO*, 1923, **2**, 81.

HAENISCH, E. (1). 'Der Aufstand von Chhen Shê im jahre 209 v. Chr.' *AM*, 1951 (n.s.), **2**, 71.

HAGERTY, M. J. (1) (tr.). 'Han Yen-Chih's *Chü Lu* (Monograph on the Oranges of Wên-Chou, Chekiang),' with introduction by P. Pelliot. *TP*, 1923, **22**, 63.

HALDANE, J. B. S. (1). 'Animal Ritual and Human Language.' *DIO*, 1953, **4**, 61. 'La Signalisation Animale.' *ANNB*, 1954, **30**, 89.

HALOUN, G. (2). Translations of *Kuan Tzu* and other ancient texts made with the present writer. MS. 1938–1941.

HALOUN, G. (3). 'Contribution to the Theory of Clan Settlement.' *AM*, 1924, **1**, 76 and 587.

HALOUN, G. (5). 'Legalist Fragments, I; *Kuan Tzu*, ch. 55, and related texts.' *AM*, 1951 (n.s.), **2**, 85.

HAMADA, KOSAKU & UMEHARA, SUEJI (1). *A Royal Tomb, 'Kuikan-Tsuka' or 'Gold-Crown' Tomb, at Keishu (Korea) and its Treasures*. 2 vols. text, 1 vol. plates. Sp. Rep. Serv. Antiq. Govt. Gen. Chosen, 1924, no. 3.

HAMILTON, C. H. (1) (tr.). *'Wei Shih Erh-shih Lun'; or, the Treatise in Twenty Stanzas on Representation-Only [Mere Ideation]; translated by Hsüan-Chuang (+596/+664) into Chinese, from Vasubandhu's (+420/+500) 'Vijñapti-mātratā-siddhi Viṃsatikā'*. Amer. Orient. Soc. New Haven, 1938. (Amer. Orient. Ser. no. 13.)

HAMILTON, C. H. (2). 'Buddhism' (in China). In *China*, ed. H. F. McNair, p. 290. Univ. of California Press, 1946.

HANSFORD, S. H. (1). *Chinese Jade Carving*. Lund Humphries, London, 1950.

HANSON, N. R. (1). 'Causal Chains.' *M*, 1955, **64**, 289.

HARLAND, W. A. (2). 'Forensic Medicine in China [the *Hsi Yuan Lu*].' *JRAS (Trans.)/NCB*, 1854, **1** (no. 4), 87.

DE HARLEZ, C. (1). *Le Yih-King [I Ching], Texte Primitif Rétabli, Traduit et Commenté*. Hayez, Bruxelles, 1889.

DE HARLEZ, C. (2) (tr.). *'Kong-Tze-Kia-Yu' ['Khung Tzu Chia Yü']; Les Entretiens Familiers de Confucius*. Leroux, Paris, 1899; and *BOR*, 1893, **6**; 1894, **7**.

DE HARLEZ, C. (3) (tr.). 'Textes Taoistes' (*Tao Tê Ching, Pao Phu Tzu, Wên Tzu, Han Fei Tzu, Huai Nan Tzu, Chuang Tzu, Lieh Tzu, Huang Ti Nei Ching, Chang Tzu*; all fragmentary translations). *AMG*, 1891, **20**, 1.

DE HARLEZ, C. (4) (tr.). *Livres des Esprits et des Immortels* (transl. of '*Shen Hsien Chuan*'). Hayez, Brussels, 1893.

DE HARLEZ, C. (5) (tr.). '*Kuo Yü*' (partially). *JA*, 1893 (9ᵉ sér.), **2**, 37, 373; 1894 (9ᵉ sér.), **3**, 5. Later parts published separately, Louvain, 1895.

HARNACK, A. (1). (*a*) *Gesch. d. kgl. preuss. Akademie d. Wissenschaften zu Berlin*. Berlin. (*b*) *Sämtliche Schriften und Briefe d. G. W. Leibniz*. Pr. Akad. Wiss. Darmstadt, 1931.

HARTREE, D. R. (1). (*a*) 'Electronic Numerical Integrator and Computer.' *N*, 1946, **158**, 500. (*b*) *Calculating Instruments and Machines*. Univ. of Illinois Press, Urbana, 1949; Cambridge, 1949.

HARVEY, E. D. (1). *The Mind of China*. Yale Univ. Press, 1933.

HAUPT, J. T. (1). *Neue u. vollständige Auslegung des von dem Stifter u. ersten Kaiser des chinesischen Reiches Fohi hinterlassenen Buches, 'Ye-Kim' genannt*. 1753.

HAYES, L. N. (1). 'The Gods of the Chinese.' *JRAS/NCB*, 1924, **55**, 84.

HEARN, LAFCADIO (1). 'The Idea of Pre-Existence.' In *Kokoro; Hints and Echoes of Japanese Inner Life*. Gay & Bird, London, n.d. (1896).

HEGEL, G. W. F. (1). *Logic*. Translated from the *Encyclopaedia of the Philosophical Sciences* by W. Wallace, Oxford (2nd ed.), 1892.

HEGEL, G. W. F. (2). *The Philosophy of History*, tr. J. Sibree. Willey, New York, 1944.

HEGEL, G. W. F. (3). *Sämtliche Werke*, ed. G. Lasson *et al.* 29 vols. Meiner, Leipzig and Hamburg, 1928–54.

HEGEL, G. W. F. (4). *Vorlesungen ü.d. Geschichte d. Philosophie*. 3 vols., ed. C. L. Michelet. Duncker & Humblet, Berlin, 1840 (vols. 13–15 of the complete works). Eng. tr. E. S. Haldane, Kegan Paul, London, 1892.

HEIDEL, W. A. (1). 'Peri Physeos.' *PAAAS*, 1910, **45**, 77.

HEINIMANN, F. (1). *Nomos und Physis; Herkunft und Bedeutung einer Antithese im griechischen Denken des 5 Jahrhunderts*. Basel, 1945. (Schweizerische Beiträge z. Altertumswiss. no. 1.)

VAN HELMONT, F. M. (1). *The Paradoxal Discourses of F. M. van Helmont concerning the Macrocosm and Microcosm, or the Greater and Lesser World, and their Union, set down in writing by J. B. and now published*. Kettlewel, London, 1685.

HENKE, F. G. (1). 'A Study of the Life and Philosophy of Wang Yang-Ming.' *JRAS/NCB*, 1913, **44**, 46.

HENKE, F. G. (2). *The Philosophy of Wang Yang-Ming*. Open Court, Chicago, 1916.

HENTZE, C. (1). *Mythes et Symboles Lunaires (Chine Ancienne, Civilisations anciennes de l'Asie, Peuples limitrophes du Pacifique)*, with appendix by H. Kühn. De Sikkel, Antwerp, 1932. Crit. *OAZ*, 1933, **9 (19)**, 33.

HENTZE, C. (2). 'Schamanenkronen z. Han-Zeit in Korea' [from the Hamada & Umehara excavations]. *OAZ*, 1933, **9 (19)**, 156.

HENTZE, C. (3). 'Le Culte de l'Ours ou du Tigre et le Thao-Thieh.' *Z*, 1938, **1**, 50.

VON HERDER, J. G. (1). *Ideen zur Philosophie der Geschichte der Menschheit* (1784). Hempel, Berlin, 1879.

DE HETRELON, R. (ps.) (1). 'Essai sur l'Origine des Différences de Mentalité entre Occident et Extrême-Orient.' *FASIE*, 1954, **10**, 815 (article written 1951).

HETT, G. V. (1). 'Some [Confucian] Ceremonies at Seoul.' *GGM*, 1936, **3**, 179.

HIGHTOWER, J. R. (1). *Topics in Chinese Literature; Outlines and Bibliographies*. Harvard Univ. Press, 1950.

HIRTH, F. (2) (tr.). 'The Story of Chang Chhien, China's Pioneer in West Asia.' *JAOS*, 1917, **37**, 89. (Translation of ch. 123 of the *Shih Chi*, containing Chang Chhien's Report; from §18–52 inclusive and 101 to 103. §98 runs on to §104, 99 and 100 being a separate interpolation. Also tr. of ch. 111 containing the biogr. of Chang Chhien.)

HITCHCOCK, E. A. (1). *Remarks upon Alchemy and the Alchemists*. Boston, 1857.

HITTI, P. K. (1). *History of the Arabs*. 4th ed. Macmillan, London, 1949.

HOBBES, THOMAS (1). *Leviathan, or, the Matter, Forme and Power of a Commonwealth, Ecclesiasticall and Civil*. 1651. Ed. M. Oakeshott, Blackwell, Oxford, n.d. (but after 1934).

HOBSBAWM, E. J. (1). 'The Machine Breakers.' *PP*, 1952, **1**, 57.

HOCKING, W. E. (1). 'Chu Hsi's Theory of Knowledge.' *HJAS*, 1935.

HODOUS, L. (1). *Folkways in China*. Probsthain, London, 1929.

HOEPPLI, R. & CHHIANG I-HUNG (1). 'The Doctrine of Spontaneous Generation of Parasites in Old-Style Chinese and Western Medicine; a Comparative Study.' *PNHB*, 1950, **19**, 375.

HOEPPLI, R. & CHHIANG I-HUNG (2). 'Similar Superstitions concerning Parasites in Old-Style Chinese and Early Western Medicine.' *PNHB*, 1951, **20**, 209.

HOLLAND, T. E. (1). *Elements of Jurisprudence*. Oxford, 1880; new ed. 1928.

HONIGSHEIM, P. (1). 'The American Indian in the Philosophy of the English and French Enlightenment.' *OSIS*, 1952, **10**, 91.

HOOVER, H. C. & HOOVER, L. H. (1) (tr.). *Georgius Agricola 'De Re Metallica', translated from the first Latin edition of 1556, with biographical introduction, annotations and appendices upon the development of mining methods, metallurgical processes, geology, mineralogy and mining law from the earliest times to the 16th century.* 1st ed. Mining Magazine, London, 1912; 2nd ed. Dover, New York, 1950.

HOPKINS, E. W. (1). 'The Social and Military Position of the Ruling Class in India, as represented by the Sanskrit Epic.' *JAOS*, 1889, **13**, 57–372 (with index) (military techniques, pp. 181–329).

HOPKINS, L. C. (2). (*a*) 'The Shaman or Wu.' *NCR*, 1920, **2**, 423. (*b*) 'The Shaman or Chinese Wu; his inspired dancing and versatile character.' *JRAS*, 1945, 3.

HOPKINS, L. C. (3). *The Development of Chinese Writing.* China Society, London, n.d.

HOPKINS, L. C. (5). 'Pictographic Reconnaissances, I.' *JRAS*, 1917, 773.

HOPKINS, L. C. (6). 'Pictographic Reconnaissances, II.' *JRAS*, 1918, 387.

HOPKINS, L. C. (7). 'Pictographic Reconnaissances, III.' *JRAS*, 1919, 369.

HOPKINS, L. C. (8). 'Pictographic Reconnaissances, IV.' *JRAS*, 1922, 49.

HOPKINS, L. C. (10). 'Pictographic Reconnaissances, VI.' *JRAS*, 1924, 407.

HOPKINS, L. C. (11). 'Pictographic Reconnaissances, VII.' *JRAS*, 1926, 461.

HOPKINS, L. C. (12). 'Pictographic Reconnaissances, VIII.' *JRAS*, 1927, 769.

HOPKINS, L. C. (13). 'Pictographic Reconnaissances, IX' (and Index to Chinese characters examined in this series and other papers in the same Journal 1916–28). *JRAS*, 1928, 327.

HOPKINS, L. C. (14). 'Archaic Chinese Characters, I.' *JRAS*, 1937, 27.

HOPKINS, L. C. (15). 'Archaic Chinese Characters, II.' *JRAS*, 1937, 209.

HOPKINS, L. C. (17). 'The Dragon Terrestrial and the Dragon Celestial; I, A Study of the *Lung* (terrestrial).' *JRAS*, 1931, 791.

HOPKINS, L. C. (18). 'The Dragon Terrestrial and the Dragon Celestial; II, A Study of the *Chhen* (celestial).' *JRAS*, 1932, 91.

HOPKINS, L. C. (19). 'The Human Figure in Archaic Chinese Writing; a Study in Attitudes.' *JRAS*, 1929, 557.

HOPKINS, L. C. (20). 'The Human Figure in Archaic Chinese Writing.' *JRAS*, 1930, 95.

HOPKINS, L. C. (22). 'The Wind, the Phoenix, and a String of Shells.' *JRAS*, 1917, 377.

HOPKINS, L. C. (26). 'Where the Rainbow Ends.' *JRAS*, 1931, 603.

HOPKINS, L. C. (27). 'Archaic Sons and Grandsons; a Study of a Chinese Complication Complex.' *JRAS*, 1934, 57.

HOPKINS, L. C. (28). 'Symbols of Parentage in Archaic Chinese, I.' *JRAS*, 1940, 351.

HOPKINS, L. C. (29). 'Symbols of Parentage in Archaic Chinese, II.' *JRAS*, 1941, 204.

HOPKINS, L. C. (33). 'The Bearskin, another Pictographic Reconnaissance from Primitive Prophylactic to Present-Day Panache; a Chinese Epigraphic Puzzle.' *JRAS*, 1943, 110.

HOPKINS, L. C. (37). 'Eclectic Preferences; a fragmentary Study in Chinese Palaeography.' *JRAS*, 1949, 188.

HOPKINS, L. C. (39). 'The Archives of an Oracle.' *JRAS*, 1915, 49.

HOPKINS, L. C. (40). 'Working the Oracle.' *NCR*, 1929, **1**, 111, 249.

HOPKINS, L. C. (41). 'Dragon and Alligator; being Notes on some ancient Inscribed Bone Carvings.' *JRAS*, 1913, 545.

HORA, S. L. (1). 'The History of Science and Technology in India and South-East Asia.' *N*, 1951, **168**, 1047.

D'HORMON, A. (1) (ed.). *Lectures Chinoises.* École Franco-Chinoise, Peiping, 1945–.

HOWELL, E. B. (1) (tr.). '*Chin Ku Chhi Kuan*; story no. XIII; the Persecution of Shen Lien.' *CJ*, 1925, **3**, 10.

HSIAO CHING-FANG (SIAO KING-FANG) (1). *Les Conceptions Fondamentaɩɛs ·'ɩ Droit Public dans la Chine Ancienne.* Sirey, Paris, 1940. (Biblioth. d'Hist. Polit. et Constitutionelle, no. 4.)

HSIAO KUNG-CHHÜAN (1). 'Anarchism in Chinese Political Thought.' *TH*, 1936, **3**, 249.

HSÜ, FRANCIS. See Hsü Lang-Kuang.

HSÜ LANG-KUANG (1). *Religion, Science and Human Crises; a Study of China in Transition and its Implications for the West.* Routledge & Kegan Paul, London, 1952.

HSÜ SHIH-LIEN (1). *The Political Philosophy of Confucianism.* Routledge, London, 1932. Crit. O. Franke, *OAZ*, 1933, **9** (**19**), 38.

HSÜ SHIH-LIEN (2). 'Ta Thung; the Confucian [*sic*] Concept of Progress.' *CSPSR*, 1926, **10**, 582.

HU SHIH (1). *The Chinese Renaissance.* Univ. of Chicago Press, 1934.

HU SHIH (2). *The Development of the Logical Method in Ancient China.* Oriental Book Co., Shanghai, 1922.

HU SHIH (3). 'Religion and Philosophy in Chinese History,' art. in *Symposium on Chinese Culture*, ed. Sophia H. Chen Zen, p. 31. IPR, Shanghai, 1931.

HU SHIH (4). 'Buddhist Influence on Chinese Religious Life.' *CSPSR*, 1925, **9**, 142.

HU SHIH (5). 'A Note on Chhüan Tsu-Wang, Chao I-Chhing and Tai Chen; a Study of Independent Convergence in Research as illustrated in their works on the *Shui Ching Chu*.' In Hummel (2), p. 970.

HU SHIH (6). 'Chinese Thought.' In *China*, ed. H. F. McNair, p. 221. Univ. of California Press, 1946.

HU SHIH (7). 'The Indianization of China; a Case Study in Cultural Borrowing.' In *Independence, Convergence and Borrowing, in Institutions, Thought and Art*, p. 219. Harvard Tercentenary Publication, Harvard Univ. Press, 1937.

HU SHIH (8). 'Der Ursprung der *Ju* und ihre Beziehung zu Konfuzius und Lau-dsï [Lao Tzu]', tr. from Hu Shih (8) by W. Franke. *SA* (Sonderausgabe), 1935, 141; 1936, 1.

HU SHIH (9). 'The Natural Law in the Chinese Tradition.' *NLIP*, 1953, **5**, 119.

HU YEN-MÊNG (HU YAN-MUNG) (1). *Etude Philosophique et Juridique de la Conception de 'Ming' et de 'Fên' dans le Droit Chinois*. Paris, 1932. *ESEJ*, **17**, 1–141. (Etudes de Sociologie et d'Ethnologie Juridique, Institut de Droit Comparé, no. 17.)

HUANG FANG-KANG (1). 'Szechuan Taoism.' Paper read before the West China Border Research Society (Chhêngtu), 1936, and given to the author in MS. form.

HUANG HSIU-CHI (1). *Lu Hsiang-Shan, a twelfth-century Chinese idealist Philosopher*. Amer. Orient. Soc. New Haven, 1944. (Amer. Oriental Series, no. 27.)

HUARD, P. (1). 'La Science et l'Extrême-Orient' [mimeographed]. Ecole Française d'Extr. Orient, Hanoi, n.d. [1950]. (Cours et Conférences de l'Ec. Fr. d'Extr. Or., 1948–9.) This paper, though admirable in choice of subjects and intention, is full of serious mistakes and should only be used with circumspection. Crit. L. Gauchet, *A/AIHS*, 1951, **4**, 487.

HUARD, P. (2). 'Sciences et Techniques de l'Eurasie.' *BSEIC*, 1950, **25** (no. 2), 1. This paper, though correcting a number of errors in Huard (1), still contains many mistakes and should be used only with care; nevertheless it is again a valuable contribution on account of several original points.

HUARD, P. & DURAND, M. (1). *Connaissance du Viêt-Nam*. Ecole Française d'Ext. Orient, Hanoi, 1954; Imprimerie Nationale, Paris, 1954.

HUBERT, H. & MAUSS, M. (1). 'Esquisse d'une Théorie Générale de la Magie.' *AS*, 1904, **7**, 56.

HUBERT, H. & MAUSS, M. (2). *Mélanges d'Histoire des Religions*. Alcan, Paris, 1929. Contains: (a) 'Essai sur la Nature et la Fonction du Sacrifice'; (b) 'L'Origine des Pouvoirs Magiques'; (c) 'Etude Sommaire de la Répresentation du Temps dans la Magie et la Religion.'

HÜBOTTER, F. (2) (tr.). 'Die Sūtras über Empfängnis und Embryologie [the *Mahāratnakūṭa Sūtra (Fo Shuo Pao Thai Ching)*].' *MDGNVO*, 1932, **26**c.

HÜBRIG, H. (1). 'Fung Schui [*Fêng Shui*], oder chinesische Geomantie.' *SBGAEU*, 1879 (no. 2), 5.

HUDSON, G. F. (1). *Europe and China; A Survey of their Relations from the Earliest Times to 1800*. Arnold, London, 1931.

HUGHES, E. R. (1). *Chinese Philosophy in Classical Times*. Dent, London, 1942. (Everyman Library, no. 973.)

HUGHES, E. R. (2) (tr.). *The Great Learning and the Mean-in-Action*. Dent, London, 1942.

HUGHES, E. R. (5). 'A Historical Approach to Chinese Epistemology' (mimeographed). Paper read at the East-West Philosophers' Conference, University of Hawaii, June 1949.

HUGHES, E. R. (6). 'Law and Government in China.' In *Law and Government in Principle and Practice*, p. 307. Ed. J. L. Brierly. Odhams, London, n.d. (1949).

HUGHES, E. R. (7) (tr.). *The Art of Letters, Lu Chi's 'Wên Fu', A.D. 302; a Translation and Comparative Study*. Pantheon, New York, 1951. (Bollingen Series, no. 29.)

HULSEWÉ, A. F. P. (1). *Remnants of Han Law*. Brill, Leiden, 1955 (Sinica Leidensia, no. 9). Incl. tr. chs. 22, 23 of the *Chhien Han Shu*.

HUMMEL, A.W. (2) (ed.). *Eminent Chinese of the Chhing Period*. 2 vols. Library of Congress, Washington, 1944.

HUMMEL, A. W. (3). 'The Case against Force in Chinese Philosophy.' *CSPSR*, 1925, **9**, 334.

HUMMEL, A. W. (5). 'A Late Ming Miscellany' (notes on Chang Sui (Ming), sceptical historian). *ARLC/DO*, 1938, 233.

HUTCHINSON, A. B. (1) (tr.). 'The Family Sayings of Confucius.' *CRR*, 1878, **9**, 445; 1879, **10**, 17, 96 175, 253, 329, 428.

HUTCHINSON, G. EVELYN (1). Review of M. Mead's *Male and Female*. *AMSC*, 1950, **38** (no. 2), in, *Marginalia*.

HUXLEY, L. (1). *Life and Letters of Thomas Henry Huxley*. 3 vols. Macmillan, London, 1903.

HUXLEY, T. H. (1). *Evolution and Ethics*. Macmillan, London, 1895.

HYDE, W. W. (1). 'The Prosecution and Punishment of Animals and Lifeless Things in the Middle Ages and Modern Times.' *UPLR/ALR*, 1916, **64**, 696.

INAMI, HAKUSUI (1). *Nippon-Tō, the Japanese Sword*. Cosmo Pub. Co., Tokyo, 1948.

INGALLS, D. H. H. (1). *Materials for the Study of Navya-Nyāya [New Nyāya] Logic.* Harvard Univ. Press, Cambridge, Mass. 1951 (Harvard Oriental Series, no. 40). Crit. J. Brough, *JRAS*, 1954, 87.

INGE, W. R. (1). *Christian Mysticism.* Methuen, London, 1921 (5th ed.). (Bampton Lectures, Oxford, 1899.)

INGE, W. R. (2). 'Logos.' *ERE*, vol. 8, p. 133.

INOUYE, T. (1). 'Die Streitfrage d. chinesischen Philosophie ü.d. menschlichen Natur.' *Proc. VIIIth Orientalist Congress*, Stockholm, 1889, Sect. Asie Centr. et Extr. Or., p. 3.

INTORCETTA, P., HERDTRICH, C., ROUGEMONT, F. & COUPLET, P. (1) (tr.). *Confucius Sinarum Philosophus, sive Scientia Sinensis, latine exposita...; adjecta est: Tabula Chronologica Monarchiae Sinicae juxta cyclos annorum LX, ab anno post Christum primo, usque ad annum praesentis Saeculi 1683* (the latter by P. Couplet, pr. 1686). Horthemels, Paris, 1687.

ISAACS, N. (1). 'The Influence of Judaism on Western Law.' In *Legacy of Israel*, ed. E. R. Bevan & C. Singer. Oxford, 1928.

JABŁOŃSKI, W. (1). 'Marcel Granet and his Work.' *YJSS*, 1939, **1**, 242.

JABŁOŃSKI, W., CHMIELEWSKI, JANUSZ, WOJTASIEWICZ, O. & ŻBIKOWSKI, T. (1) (tr.). *Czuang-Tsy, 'Nan-Hua-Czên-King' [Chuang Tzu; 'Nan Hua Chen Ching'], Prawdziwa Księga Południowego Kwiatu.* Państwowe Wydawnictwo Naukowe, Warsaw, 1953.

JACOB, E. F. (1). 'Political Thought.' In *Legacy of the Middle Ages*, p. 527, ed. C. G. Crump & E. F. Jacob. Oxford, 1926.

JACOBI, H. (1). 'Der Ursprung d. Buddhismus aus d. Saṃkhya-Yoga.' *NGWG/PH*, 1896, 43.

ABD AL-JALIL, J. M. (1). *Brève Histoire de la Littérature Arabe.* Maisonneuve, Paris, 1943; 2nd ed. 1947.

JAMES, WILLIAM (1). *Varieties of Religious Experience; a study in Human Nature.* Longmans Green, London, 1904. (Gifford Lectures, 1901–2.)

JASTROW, M. (1). *The Civilisation of Babylonia and Assyria.* Lippincott, Philadelphia, 1915.

JASTROW, M. (2). *Religion of Babylonia and Assyria.* Boston, 1898. *Die Religion Babyloniens und Assyriens.* Giessen, 1905.

JELLINEK, A. (1) (tr.). *Der Mikrokosmos, ein Beitrag z. Religionsphilosophie und Ethik* [the *Sefer Olam Katan* of Joseph ben Zaddiq (+1149)]. Leipzig, 1854.

JOHNSTON, R. F. (1). *Confucianism and Modern China.* Gollancz, London, 1934.

JOHNSTON, R. F. (2). *Buddhist China.* Murray, London, 1913.

JOLOWICZ, H. F. (1). *Historical Introduction to the Study of Roman Law.* Cambridge, 1932.

JOYCE, T. A. (1). *Mexican Archaeology.* London, 1914.

KALTENMARK, M. (1). 'Les *Tch'an Wei* [*Chhan Wei*].' *HH*, 1949, **2**, 363.

KALTENMARK, M. (2) (tr.). *Le 'Lie Sien Tchouan' ['Lieh Hsien Chuan']; Biographies Légendaires des Immortels Taoistes de l'Antiquité.* Centre d'Etudes Sinologiques Franco-Chinois (Univ. Paris), Peking, 1953. Crit. P. Demiéville, *TP*, 1954, **43**, 104.

KARLGREN, B. (1). 'Grammata Serica; Script and Phonetics in Chinese and Sino-Japanese.' *BMFEA*, 1940, **12**, 1. (Photographically reproduced as separate volume, Peiping, 1941.)

KARLGREN, B. (2). 'Legends and Cults in Ancient China.' *BMFEA*, 1948, **18**, 199.

KARLGREN, B. (8). 'On the Authenticity and Nature of the *Tso Chuan*.' *GHA*, 1926, **32**, no. 3. Crit. H. Maspero, *JA*, 1928, **212**, 159.

KARLGREN, B. (9). 'Some Fecundity Symbols in Ancient China.' *BMFEA*, 1930, **2**, 1.

KARLGREN, B. (12) (tr.). 'The Book of Documents' (*Shu Ching*). *BMFEA*, 1950, **22**, 1.

KARLGREN, B. (14) (tr.). *The Book of Odes; Chinese Text, Transcription and Translation.* Museum of Far Eastern Antiquities, Stockholm, 1950. (A reprint of the translation only from his papers in *BMFEA*, **16** and **17**.)

KARPPE, S. (1). *Études sur l'Origine et la Nature de 'Zohar'.* Paris, 1901.

KEITH, A. BERRIEDALE (1). *Buddhist Philosophy in India and Ceylon.* Oxford, 1923.

KEITH, A. BERRIEDALE (2). *Indian Logic and Atomism.* Oxford, 1921.

KEITH, A. BERRIEDALE (3). *The Saṃkhya System; a History of the Saṃkhya Philosophy.* Ass. Press, Calcutta, 1918.

KELLER, P. J. & CORI, G. T. 'Purification and Properties of the Phosphorylase-Rupturing Enzyme.' *JBC*, 1955, **214**, 127.

KENNEDY, D. (1). *England's Dances.* Bell, London, 1949.

KENT, C. F. & BURROWS, M. (1). *Proverbs and Didactic Poems.* Hodder & Stoughton, London, 1927.

KEYES, C. W. (1) (tr.). *Cicero's 'De Legibus'.* (Loeb Classics series), Heinemann, London, 1928.

KEYNES, J. M. (Lord Keynes) (1) (posthumous). 'Newton the Man.' Essay in *Newton Tercentenary Celebrations* (July 1946). Royal Society, London, 1947. Reprinted in *Essays in Biography*.

KIMM CHUNG-SE (1) (tr.). 'Kuei Ku Tzu.' AM, 1927, 4, 108.

KING CHIEN-KÜN. See Lyall & Ching Chien-Chün.

KING, H. (1) (tr.). The 'Metamorphoses' of P. Ovidius Naso. Blackwood, Edinburgh, 1871.

KOCH, H. L. (1). Materie und Organismus bei Leibniz. Niemeyer, Halle, 1908. (Abhdl. z. Philos. u. ihrer Gesch. no. 30.)

KOHLER, J. (1). Das chinesische Strafrecht; Beitrag z. universal-Geschichte d. Strafrechts. Würzburg, 1886.

KOHLER, J. (2). Rechtsvergleichende Studien ü. Islamitisches Recht, Recht d. Berbern, Chinesische Recht, u. Recht auf Ceylon. Heymann, Berlin, 1889.

KÖHLER, W. (1). Gestalt Psychology. Bell, London, 1930.

KOMENSKY. See Comenius.

KÖNIG, H., GUSINDE, M., SCHEBESTA, P. & DIETSCHY, H. (1). 'Le Chamanisme.' CIBA/M, 1947, no. 60, 2145ff.

DE KORNE, J. C. (1). The Fellowship of Goodness. Pr. pub. Grand Rapids, 1941.

KÖRNER, B. (1). 'Nan Lao Chhüan; eine Flutsage aus West China, und ihre Auswirkung auf örtliches Brauchtum.' ETH, 1950, 15, 46.

KORTHOLT, C. (1) (ed.). Viri Illustris Godefridi Guil. Leibnitii Epistolae ad Diversos, Theologici, Juridici, Medici, Philosophici, Mathematici, Historici, et Philologici Argumenti, e Msc. Auctoris cum Annotationibus suis primum divulgavit.... 2 vols. Breitkopf, Leipzig, 1735.

KOU, IGNACE. See Ku Pao-Ku.

KRAMERS, R. P. (1) (tr.). 'Khung Tzu Chia Yü'; the School Sayings of Confucius [chs. 1 to 10]. Brill, Leiden, 1950. (Sinica Leidensia no. 7.)

KRAUSS, F. S., SATO, TOMIO & IHM, H. Das Geschlechtsleben im Glauben, Sitte, Brauch und Gewohnheitrecht der Japaner (2nd ed.). Ethnol. Verlag, Leipzig, 1911 (Beiwerke zum Studium der Anthropophyteia; Jahrbücher f. folkloristische Erhebungen und Forschungen zur Entwicklungsgesch. d. geschlechtlichen Moral, no. 2). Vol. 2 (by T. Sato & H. Ihm) Abhandlungen und Erhebungen über das Geschlechtsleben des Japanischen Volkes; Folkloristische Studien. Ethnol. Verlag, Leipzig, 1931 (Beiwerke zum Studium der Anthropophyteia, no. 1).

KROEBER, A. L. (1). Anthropology. Harcourt Brace, New York, 1948.

DE KRUIF, P. (1). (a) Microbe Hunters. Harcourt Brace, New York, 1926. (b) Hunger Fighters. Harcourt Brace, New York, 1928.

KU HUNG-MING (1) (tr.). The Discourses and Sayings of Confucius. Kelly & Walsh, Shanghai, 1898.

KU PAO-KU (1) (tr.). Deux Sophistes Chinois; Houei Che [Hui Shih] et Kong-souen Long [Kungsun Lung]. Presses Univ. de France (Imp. Nat.), Paris, 1953. (Biblioth. de l'Instit. des Hautes Etudes Chinoises, no. 8.) Crit. P. Demiéville, TP, 1954, 43, 108.

KÜHN, A. (1). Berichte ü.d. Weltanfang bei d. Indochinesen und ihren Nachbar-Völkern; ein Beitrag z. Mythologie des Fernen Ostens. Harrassowitz, Leipzig, 1935.

KUHN, F. W. (1) (tr.). Mondfrau und Silbervase. Berlin, 1939. (A translation of Hsü Chin Phing Mei.)

KUHN, F. W. (2) (tr.). Goldene Lotus ('Chin Phing Mei'). Leipzig, 1939. Eng. tr. by B. Miall entitled The Adventurous History of Hsi Mên and his Six Wives. Introduction by A. Waley. London, n.d. (1940).

KUWABARA, JITSUZO (1). 'On Phu Shou-Kêng, a man of the Western Regions, who was the Superintendent of the Trading Ships' Office in Chhüan-Chou towards the end of the Sung Dynasty, together with a general sketch of the Trade of the Arabs in China during the Thang and Sung eras.' MRDTB, 1928, 2, 1; 1935, 7, 1. Revs. P. Pelliot, TP, 1929, 26, 364; S.E[lisséev], HJAS, 1936, 1, 265.

LACH, D. F. (1). 'Leibniz and China.' JHI, 1945, 6, 436.

LACH, D. F. (3). 'China and the Era of the Enlightenment.' JMH, 1942, 14, 209.

DE LACOUPERIE, TERRIEN (1). The Western Origin of Chinese Civilisation. London, 1894.

LAIGNEL-LAVASTINE, M. (1) (ed.). Histoire Générale de la Médecine, de la Pharmacie, de l'Art Dentaire et de l'Art Vétérinaire. 2 vols. Albin Michel, Paris, 1938.

LAMOTTE, E. (1) (tr.). 'Mahaprajñaparamita-sāstra'; Le Traité (Madhyamika) de la Grande Vertu de Sagesse, de Nāgārjuna. 3 vols. Louvain, 1944. (rev. P. Demiéville, JA, 1950, 238, 375.)

DE LANDA, DIEGO (1). Relación de las Cosas de Yucatán. French tr. J. Genet, Genet, Paris, 1928; Eng. tr. W. Gates, Baltimore, 1937.

LAPEYRONIE, M. (1). 'Observation sur les petits œufs de Poule sans jaune, que l'on appelle vulgairement œuf de Coq.' MRASP, 1710, 553.

LAROCHE, E. (1). 'Les Noms Grecs de l'Astronomie.' RPLHA, 1946 (3e sér.), 20, 118.

LATTA, R. (1). *Leibniz; the Monadology and other Philosophical Writings, with Introduction and Notes.* Oxford, 1898. (2nd ed.) 1925.

LATTIMORE, O. (6). *Manchuria, Cradle of Conflict.* New York, 1932.

LAU, D. C. (1). 'Some Logical Problems in Ancient China.' *PAS*, 1953, 189.

LAU, D. C. (2). 'Theories of Human Nature in Mêng Tzu and Hsün Tzu.' *BLSOAS*, 1953, **15**, 541.

LAUBRY, C. & BROSSE, T. (1). 'Documents recueillis aux Indes sur les Yoguis par l'enregistrement simultané du pouls, de la respiration et de l'electrocardiogramme.' *PM*, 1936, no. 83 (14 Oct.).

LAUFER, B. (4). 'The Prehistory of Aviation.' *FMNHP/AS*, 1928, **18**, no. 1 (pub. no. 253). Cf. 'Mitt. ü.d. angeblicher Kenntnis d. Luftschiffahrt bei d. alten Chinesen.' *OLL*, 1904, **17**; *OB*, 1904, no. 1489, p. 78; also *OC*, 1931, **45**, 493.

LAUFER, B. (5). 'Origin of the Word Shaman.' *AAN*, 1917, **19**.

LAUFER, B. (6). 'The Story of the *Pinna* and the Scythian Lamb.' *JAFL*, 1915, **28**, 103.

LAUFER, B. (8). 'Jade; a Study in Chinese Archaeology and Religion.' *FMNHP/AS*, 1912. Repub. in book form, Perkins, Westwood & Hawley, South Pasadena, 1946. (rev. P. Pelliot, *TP*, 1912, **13**, 434.)

LAUFER, B. (17). 'Historical Jottings on Amber in Asia.' *MAAA*, 1906, **1**, 211.

LEDLIE, J. C. (1). 'Ulpian.' *JSCL*, 1905 (n.s.), **5**, 14.

LEFEBVRE, H. (1). *Descartes.* Ed. Hier et Aujourd'hui, Paris, 1947.

LEGGE, J. (1). (*a*) *The Texts of Confucianism, translated.* Pt. I, The '*Shu Ching*', the Religious portions of the '*Shih Ching*', the '*Hsiao Ching*'. Oxford, 1879. (*SBE*, no. 3; reprinted in various eds. Com. Press, Shanghai.) (*b*) Full version of the *Shu Ching*, with Chinese text and notes, in *The Chinese Classics*, Vol. 3, Pts. 1 and 2, Legge, Hongkong, 1865; Trübner, London, 1865.

LEGGE, J. (2). *The Chinese Classics, etc.*: Vol. 1. *Confucian Analects, The Great Learning, and the Doctrine of the Mean.* Legge, Hongkong, 1861; Trübner, London, 1861.

LEGGE, J. (3). *The Chinese Classics, etc.*: Vol. 2. *The Works of Mencius.* Legge, Hongkong, 1861; Trübner, London, 1861.

LEGGE, J. (4) (tr.). *A Record of Buddhistic Kingdoms; an account by the Chinese monk Fa-Hsien of his travels in India and Ceylon* (+399 to +414) *in search of the Buddhist books of discipline.* Oxford, 1886.

LEGGE, J. (5) (tr.). *The Texts of Taoism.* (Contains (*a*) *Tao Tê Ching*, (*b*) *Chuang Tzu*, (*c*) *Thai Shang Kan Ying Phien*, (*d*) *Chhing Ching Ching*, (*e*) *Yin Fu Ching*, (*f*) *Jih Yung Ching*.) 2 vols. Oxford, 1891; photolitho reprint, 1927. (*SBE*, nos. 39 and 40.)

LEGGE, J. (6). 'Imperial Confucianism.' *CR*, 1877, **6**, 148, 223, 299, 363.

LEGGE, J. (7) (tr.). *The Texts of Confucianism, Pt. III. The '*Li Chi*'.* 2 vols. Oxford, 1885; repr. 1926. (*SBE*, nos. 27 and 28.)

LEGGE, J. (8) (tr.). *The Chinese Classics, etc.*: Vol. 4, Pts. 1 and 2. *The Book of Poetry.* Lane Crawford, Hongkong, 1871; Trübner, London, 1871. Repr. Com. Press, Shanghai, n.d.

LEGGE, J. (9) (tr.). *The Texts of Confucianism, Pt. II. The '*Yi King*'* ('*I Ching*'). Oxford, 1899. (*SBE*, no. 16.)

LEGGE, J. (11). *The Chinese Classics, etc.*: Vol. 5, Pts. 1 and 2. *The '*Ch'un Ts'eu*' with the '*Tso Chuen*'* ('*Chhun Chhiu*' and '*Tso Chuan*'). Lane Crawford, Hongkong, 1872; Trübner, London, 1872.

LEIBNIZ, G. W. (1). *Novissima Sinica; Historiam Nostri Temporis Illustratura.* Hanover, 1697.

LEIBNIZ, G. W. (2). *Ars Combinatoria* (1666). Frankfurt, 1690.

LEIBNIZ, G. W. (3). *Monadology.* Ed. E. Boutroux, Paris, 1930. See Carr (1) and Latta (1).

LEIBNIZ, G. W. (4). 'Explication de l'Arithmétique Binaire, qui se sert des seuls caractères o et 1, avec des Remarques sur son Utilité, et sur ce qu'elle donne les sens des anciennes Figures Chinoises de Fohy.' *MRASP*, 1703, **3**, 85.

LEIBNIZ, G. W. (5). *Die Philosophische Schriften*, ed. C. I. Gerhardt. 7 vols. Berlin, 1875–90.

LEIBNIZ, G. W. (6). *Discourse on Metaphysics.* See Lucas & Grint (1).

LEMAITRE, S. (1). *Les Agrafes Chinoises jusqu'à la fin de L'Epoque Han.* Art et Hist., Paris, 1939.

LENORMANT, F. (1). *La Divination et la Science des Présages chez les Chaldéens.* Maisonneuve, Paris, 1875.

LENORMANT, F. (2). *La Magie chez les Chaldéens et ses Origines Accadiennes.* Maisonneuve, Paris, 1874. Eng. tr. *Chaldean Magic* (enlarged), Bagster, London, 1877.

LEONARD, W. E. (1) (tr.). *T. Lucretius Carus, '*Of the Nature of Things*'; a new metrical translation.* Dent, London, 1916; Everyman ed. 1921.

DE LÉRY, JEAN (1). *Histoire d'un Voyage fait en la Terre du Brésil.* Vignon, Geneva, 1586. Ed. C. Clerc. Payot, Paris, 1927.

LESLIE, D. (1). *Man and Nature; Sources on Early Chinese Biological Ideas* (especially the *Lun Hêng*). Inaug. Diss., Cambridge, 1954.

LESLIE, D. (2). 'The Problem of Action at a Distance in Early Chinese Thought' (discussion on lecture by J. Needham). *Actes du VIIe Congrès International d'Histoire des Sciences, Jerusalem 1953* (1954), p. 186.

LÉVI, S. (3). Obituary notice of T. Ganaspati Sastri of Trivandrum referring to a technological book, the *Samarāṅgaṇa Sūtradhāra* of +11th century, attrib. to King Bhoja. *JA*, 1926, **208**, 379.

LÉVI, S. (4). 'On a Tantric Fragment from Kucha.' *IHQ*, 1936, **12**, 207.

LEVIS, J. H. (1). *Foundations of Chinese Musical Art*. Vetch, Peiping, 1936. Crit. Ying Shang-Nêng, *TH*, 1937, **4**, 317.

LÉVY-BRUHL, L. (1). *Les Fonctions Mentales dans les Sociétés Inferieures*. Alcan, Paris, 1928 (2nd ed.). Eng. tr. of 1st ed. by L. A. Clare, *How Natives Think*, Allen & Unwin, London, 1926.

LEWIS, C. I. (1). *A Survey of Symbolic Logic*. Univ. of California Press, 1918.

LEWIS, J., POLANYI, K. & KITCHIN, D. (1) (ed.). *Christianity and the Social Revolution*. Gollancz, London, 1935.

LEWIS, SINCLAIR (1). *Martin Arrowsmith*. Cape, London, 1925.

LI CHHIAO-PHING (1). *The Chemical Arts of Old China*. Journ. Chem. Educ., Easton, Pa. 1948. Crit. J. R. Partington, *ISIS*, 1949, **40**, 280.

LI CHI (1). 'Archaeology [in China, and its History],' art. in *Symposium on Chinese Culture*', ed. Sophia H. Chen Zen, p. 184. IPR, Shanghai, 1931.

LI SHIH-I (1). 'Wang Chhung.' *TH*, 1937, **5**, 162 and 290.

LI THAO (1). 'Achievements of Chinese Medicine in the Northern Sung Dynasty (A.D. 960–1127).' *CMJ*, 1954, **72**, 65.

LI THAO (2). 'Achievements of Chinese Medicine in the Southern Sung Dynasty (A.D. 1127–1279).' *CMJ*, 1954, **72**, 225.

LIANG CHHI-CHHAO (2). *La Conception de la Loi et les Théories des Légistes à la Veille des Ts'in [Chhin]*. Tr. J. Escarra & R. Germain from the relevant chapters of Chinese edition of Liang (1). With an introduction by G. Padoux. China Booksellers, Peking, 1926.

LIAO WÊN-KUEI (1) (tr.). *The complete Works of Han Fei Tzu; a Classic of Chinese Legalism*. 2 vols. Probsthain, London, 1939. (Only the first volume published.)

LIEBENTHAL, W. (1) (tr.). *Sêng-Chao's 'Chao Lun', or the Book of Chao*. Fu-Jen Univ., Peiping, 1948. (Monumenta Serica Monograph Ser. no. 13.) Crit. A. F. Wright, *JAOS*, 1950, **50**, 324; A. Waley, *JRAS*, 1950, 80.

LILLEY, S. (1). 'Mathematical Machines.' *N*, 1942, **149**, 462; *D*, 1945, **6**, 150, 182; 1947, **8**, 24.

LIN THUNG-CHI (1). 'The Taoist Substratum of the Chinese Mind.' *JHI*, 1947, **8**, 259.

LIN YÜ-THANG (1) (tr.). *The Wisdom of Lao Tzu [and Chuang Tzu] translated, edited and with an introduction and notes*. Random House, New York, 1948.

LIN YÜ-THANG (2). *Moment in Peking*. Day, New York, 1939.

LIN YÜ-THANG (3). *My Country and My People*. Heinemann, London, 1936. Crit. Wu Ching-Hsiung, *TH*, 1935, **1**, 468.

LIN YÜ-THANG (4). *The Importance of Living*. Heinemann, London, 1938.

LIN YÜ-THANG (5). *The Gay Genius; Life and Times of Su Tung-Pho*. Heinemann, London, 1948.

LIN YÜ-THANG (6). 'Feminist Thought in Ancient China.' *TH*, 1935, **1**, 127.

VON LIPPMANN, E. O. (1). *Entstehung und Ausbreitung der Alchemie...Ein Beitrag zur Kulturgeschichte*. Springer, Berlin, 1919.

LLOYD-MORGAN, C. (1). *Emergent Evolution*. London, 1923 (Gifford Lectures). *Life, Mind, and Spirit*. London, 1926 (Gifford Lectures).

LOCKE, L. L. (1). *The Quipu*. Amer. Mus. Nat. Hist., New York, 1923.

LONGOBARDI, N. (1). *Traité sur Quelques Points de la Religion des Chinois* (1701). Reprinted in Kortholt (1).

LOTKA, A. J. (1). *Elements of Physical Biology*. Williams & Wilkins, Baltimore, 1925.

LOVEJOY, A. O. (1). 'The supposed Primitivism of Rousseau's *Discourse on Inequality*.' In *Essays in the History of Ideas*, p. 14. Johns Hopkins Univ. Press, Baltimore, 1948.

LOVEJOY, A. O. (2). 'Nature as Aesthetic Norm.' In *Essays in the History of Ideas*, p. 69. Johns Hopkins Univ. Press, Baltimore, 1948. Cf. Lovejoy & Boas (1), p. 447.

LOVEJOY, A. O. (3). 'The Chinese Origin of a Romanticism.' In *Essays in the History of Ideas*, p. 99. Johns Hopkins Univ. Press, Baltimore, 1948. Also *JEGP*, 1933, **32**, 1.

LOVEJOY, A. O. (4). 'The Communism of St Ambrose.' *JHI*, 1942, **3**, 458.

LOVEJOY, A. O. & BOAS, G. (1). *A Documentary History of Primitivism and Related Ideas*. Vol. 1. *Primitivism and Related Ideas in Antiquity*. Johns Hopkins Univ. Press, Baltimore, 1935.

LOEWE, H. (1). 'Kabbalah.' *ERE*, vol. 7, p. 622.

LOEWENSTEIN, P. J. (1). 'Swastika and Yin-Yang.' *China Society Occasional Papers* (n.s.), no. 1. China Society, London, 1942.

LU, GWEI-DJEN & NEEDHAM, JOSEPH (1). 'A Contribution to the History of Chinese Dietetics.' *ISIS*, 1951, **42**, 13 (submitted 1939, lost by enemy action; again submitted 1942 and 1948). Mem. Sarton (1), vol. 3, p. 905.

DE LUBAC, H. (2). *La Rencontre du Bouddhisme et de l'Occident*. Paris, 1952.

LUCAS, P. G. & GRINT, L. (1) (tr. and ed.). *Leibniz' 'Discourse on Metaphysics'*. Manchester Univ. Press, 1953.

LUCRETIUS. See Leonard.

LYALL, L. A. (1) (tr.). *Mencius*. Longmans Green, London, 1932.

LYALL, L. A. (2) (tr.). *The Sayings of Confucius* ['*Lun Yü*']. Longmans Green, London, 1935 (this edition superseded earlier editions).

LYALL, L. A. & CHING CHIEN-CHÜN [KING CHIEN-KÜN] (1) (tr.). *The 'Chung Yung'*. Longmans Green, London, 1927.

LYELL, Sir CHARLES (1). *Principles of Geology*. 2 vols. Murray, London, 1872 (11th ed.).

McCULLOGH, J. A. (1). 'Shamanism.' *ERE*, vol. 11, p. 441.

McGOVERN, W. M. (2). *Manual of Buddhist Philosophy; I, Cosmology* (no more published). Kegan Paul, London, 1923.

McGOWAN, D. J. (2). 'The Movement Cure in China' (Taoist medical gymnastics). *CIMC/MR*, 1885, no. 29, 42.

McMURRAY, J. (1). *The Philosophy of Communism*. Faber, London, 1933.

McNAIR, H. F. (1) (ed.). *China* (collective essays). Univ. of California Press, 1946.

MAIMONIDES, MOSES [MOSHE BEN MAIMON] (1). *Guide for the Perplexed*, tr. M. Friedländer. Routledge, London, 1904.

MAINE, SIR HENRY (1). *Ancient Law*. Murray, London, 1861. Repr. 1916, with editorial notes by F. Pollock.

MAINE, SIR HENRY (2). *Lectures on the Early History of Institutions*. Murray, London, 1914.

MAINE, SIR HENRY (3). *Village Communities in the East [largely in India] and West*. Murray, London, 1887.

MALYNES, G. (1). *Consuetudo, vel Lex Mercatoria; or, The Antient Law-Merchant*. London, 1622.

MAO, C. (1). 'Deux Principes Fondamentaux de Confucianisme.' *Proc. Xth Internat. Congr. Philos.* vol. 1, p. 231. Amsterdam, 1949.

MARIADASSOU, PARAMANANDA (1). *Médecine Traditionelle de l'Inde; Histoire de la Médecine Hindoue.* 2 vols. Pondicherry, 1943.

MARSHALL, A. J. (1). 'Bower Birds.' *BR*, 1954, **29**, 1.

MARTIN, W. A. P. (3). *Hanlin Papers*. 2 vols. Vol. 1, Trübner, London, 1880; Harper, New York, 1880. Vol. 2, Kelly & Walsh, Shanghai, 1894.

MARTIN, W. A. P. (5). 'Isis and Osiris; or, Oriental Dualism.' *CRR*, 1867. Repr. in Martin (3), vol. 1, p. 203.

MARTIN, W. A. P. (6). 'The Cartesian Philosophy before Descartes' (centrifugal cosmogony in Neo-Confucianism). *JPOS*, 1888, **2**, 121. Repr. in Martin (3), vol. 2, p. 207.

MARTIN, W. A. P. (7). 'Remarks on the Ethical Philosophy of the Chinese' (based on a set of diagrams resembling the *Sheng Mên Shih Yeh Thu* of Li Kuo-Chi). *PR*, 1862. Repr. in Martin (3), vol. 1, p. 163.

MARX, K. (1). *Das Kapital*. Meissner, Hamburg, 1922.

MASON, S. F. (1). 'The Idea of Progress and Theories of Evolution in Science.' *CN*, 1953, **3**, 90.

MASPERO, H. (1). 'La Composition et la Date du *Tso Chuan*.' *MCB*, 1931, **1**, 137.

MASPERO, H. (2). *La Chine Antique*. Boccard, Paris, 1927. (Histoire du Monde, ed. E. Cavaignac, vol. 4); rev. B. Laufer, *AHR*, 1928, **33**, 903.

MASPERO, H. (5). (*a*) 'Le Songe et l'Ambassade de l'Empereur Ming.' *BEFEO*, 1910, **10**, 95, 629. (*b*) 'Communautés et Moines Bouddhistes Chinois au 2e et 3e siècles.' *BEFEO*, 1910, **10**, 222.

MASPERO, H. (7). 'Procédés de "nourrir le principe vital" dans la Religion Taoïste Ancienne.' *JA*, 1937, **229**, 177 and 353.

MASPERO, H. (8). 'Légendes Mythologiques dans le *Chou King*' (*Shu Ching*). *JA*, 1924, **204**, 1.

MASPERO, H. (9). 'Notes sur la Logique de Mo-Tseu et de Son École.' *TP*, 1928, **25**, 1.

MASPERO, H. (10). 'Le Serment dans la Procèdure Judiciaire de la Chine Antique.' *MCB*, 1935, **3**, 257.

MASPERO, H. (11). 'The Mythology of Modern China; The Popular Religion and the Three Religions.' In *Asiatic Mythology; a Detailed Description and Explanation of the Mythologies of all the Great Nations of Asia*, ed. P. L. Couchoud. Harrap, London, 1932.

MASPERO, H. (12). 'Les Religions Chinoises.' In *Mélanges Posthumes sur les Religions et l'Histoire de la Chine*, vol. 1, ed. P. Demiéville. Civilisations du Sud, Paris, 1950. (Publ. du Mus. Guimet, Biblioth. de Diffusion, no. 57.)

MASPERO, H. (13). 'Le Taoisme.' In *Mélanges Posthumes sur les Religions et l'Histoire de la Chine*, vol. 2, ed. P. Demiéville. Civilisations du Sud, Paris, 1950. (Publ. du Mus. Guimet, Biblioth. de Diffusion. no. 58); rev. J. J. L. Duyvendak. *TP*, 1951, **40**, 366.

MASPERO, H. (19). 'Communautés et Moines Bouddhistes Chinois aux 2e et 3e siecles' *BEFEO*, 1910, **10**, 222.

MASPERO, H. (20). 'Les Origines de la Communanté Bouddhiste de Loyang'. *JA*, 1934, **225**, 87.

MASPERO, H. (26). 'Le Saint et la Vie Mystique chez Lao-tseu et Tchouang-Tseu [Lao Tzu and Chuang Tzu].' *BAFAO*, 1922, no. 3, 69 (73). Repr. in (13), p. 227 (230).

MASPERO, H. (27). 'Les Dieux Taoistes; comment on communique avec eux.' *CRAIBL*, 1937, 362.

MASSIGNON, L. (1). 'Sufism.' *EI*, vol. 4, p. 681.

MASSIGNON, L. (2). *Essai sur les Origines du Lexique Technique de la Mystique Mussulmane*. Paris, 1922.

MASSIGNON, L. (3). 'The Qarmatians.' *EI*, vol. 2, p. 767.

MASSON-OURSEL, P. (1). 'Etudes de Logique Comparée.' *RMM*, 1912, **20**, 811; 1916, **23**, 343. *RHR/AMG*, 1913, **67**, 49. *RP*, 1917, **83**, 453; 1918, **84**, 59; 1918, **85**, 148; 1920, **90**, 123.

MASSON-OURSEL, P. & CHU CHIA-CHIEN (1) (tr.). 'Yin Wên Tzu'. *TP*, 1914, **15**, 557.

MATHEWS, R. H. (1). *Chinese-English Dictionary*. China Inland Mission, Shanghai, 1931; Harvard-Yenching Inst., Harvard, 1947.

MAUBLANC, R. (1). *La Philosophie du Marxisme*. Paris, 1935.

MAUNIER, R. (1). 'Le Folklore Juridique.' In *Travaux du Ier Congrès International de Folklore*, p. 185. Paris, 1937; Arrault, Tours, 1938. (Pub. du Dép. et du Mus. Nat. des Arts et Trad. Populaires.)

MAUPOIL, B. (1). 'La Géomancie à l'ancienne Côte des Esclaves.' *TMIE*, 1943, **42**, 1.

MAVERICK, L. A. (1). *China a Model for Europe* (photolitho typescript). Anderson, San Antonio, Texas, 1946. Vol. 1: 'China's Economy and Government admired by seventeenth and eighteenth century Europeans.' Vol. 2: '*Despotism in China*, a translation of François Quesnay's *Le Despotisme de la Chine* (Paris, 1767).' Issued bound together in one.

MAYERS, W. F. (1). *Chinese Reader's Manual*. Presbyterian Press, Shanghai, 1874; repr. 1924.

MAYERS, W. F. (2). 'Bibliography of the Chinese Imperial Collections of Literature' (i.e. *Yung-Lo Ta Tien*; *Thu Shu Chi Chhêng*; *Yuan Chien Lei Han*; *Phei Wên Yuan Fu*; *Phien Tzu Lei Pien*; *Ssu Ku Chhüan Shu*). *CR*, 1878, **6**, 213, 285.

MAYERS, W. F. (4). 'Comparative Table Illustrating the Chinese Scheme of Physics.' *NQCJ*, 1867, **1**, 146.

MAYOR, R. J. G. (1). *Virgil's Messianic Eclogue*. London, 1907.

MAZAHERI, A. (1). Review of de Menasce (1). *A/AIHS*, 1950, **3**, 170.

MEAD, M. (1). *Sex and Temperament in Three Primitive Societies*. Routledge, London, 1935.

MEDHURST, W. H. (1) (tr.). The 'Shoo King' ['Shu Ching'], or Historical Classic (Ch. text and Eng.) Mission Press, Shanghai, 1846.

MEIJER, M. J. (1). *Modern Chinese Constitutional Law*. Sinologisch Instituut, Batavia, 1950.

MEI YI-PAO (1) (tr.). *The Ethical and Political Works of Mo Tzu*. Probsthain, London, 1929.

MEI YI-PAO (2). *Mo Tzu, the Neglected Rival of Confucius*. Probsthain, London, 1934.

MEI YI-PAO (3) (tr.). 'The *Kungsun Lung Tzu*, with a translation into English.' *HJAS*, 1953, **16**, 404.

DE MENASCE, J. (1). 'Une Apologetique Mazdéene du 9ᵉ siècle; Skand-Gumanik Vicâr (La Décision Décisive des Doutes).' *Collectanea Fribourgensia* (Pub. de l'Univ. de Fribourg, Suisse), 1945, no. 30.

DE MENDOZA, JUAN GONZALES (1). *Historia de las Cosas mas notables, Ritos y Costumbres del Gran Reyno de la China, sabidas assi por los libros de los mesmos Chinas, como por relación de religiosos y oltras personas que han estado en el dicho Reyno*. Rome, 1585 (in Spanish). Eng. tr. Robert Parke, 1588 (1589), *The Historie of the Great & Mightie Kingdome of China and the Situation thereof; Togither with the Great Riches, Huge Citties, Politike Gouvernement and Rare Inventions in the same* [undertaken 'at the earnest request and encouragement of my worshipfull friend Master Richard Hakluyt, late of Oxforde']. Reprinted in Spanish, Medina del Campo, 1595; Antwerp, 1596 and 1655; Ital. tr. Venice (3 editions), 1586; Fr. tr. Paris, 1588 and 1589; Germ. and Latin tr. Frankfurt, 1589. Ed. G. T. Staunton, Hakluyt Soc. Pub. London, 1853.

MÉNEBRÉA, L. (1). *De l'Origine de la Forme et de l'Esprit des Jugements rendus au moyen-âge contre les animaux*. Chambéry, 1846.

MERKEL, R. F. W. (1). *Die Anfänge der protestantischen Missionsbewegung: G. W. von Leibniz und die China Mission*. Leipzig, 1920.

MERKEL, R. F. W. (2). 'Leibniz und China.' In *Leibniz zu seinem 300 Geburtstag, 1646–1946*, ed. E. Hochstetter, Lieferung 8. De Gruyter, Berlin, 1952.

MEYER, ADOLF (1). 'Das Organische und seine Ideologien.' *AGMN*, 1934, **27**, 3.

MEYER, ADOLF (2). 'Ideen und Ideale der biologischen Erkenntnis.' *BIOS*, no. 1. Barth, Leipzig, 1934.

MEYER, ADOLF (3). *Wesen und Geschichte d. Theorie vom Mikro- und Makrokosmos*. Bern, 1900. (Berner Studien z. Philos. u. ihrer Gesch. no. 25.)

MEYER, ARTHUR W. (1). *The Rise of Embryology*. Stanford Univ. Press, Palo Alto, California, 1939.

MIALL, L. C. (1). *The Early Naturalists, their Lives and Work* (1530 to 1789). London, 1912.

MICHEL, H. (5). 'Le Calcul Mécanique; à propos d'une Exposition Récente.' *JSHB*, 1947 (no. 7), 307.

MIKHAILOVSKY, V. M. (1). 'Shamanism in Siberia and European Russia.' *JRAI*, 1894, **24**, 62 and 126.

MIRONOV, N. D. & SHIROKOGOROV, S. M. (1). 'Śramaṇa and Shaman.' *JRAS/NCB*, 1924, **55**, 105.

MISCH, G. (1). *The Dawn of Philosophy*, tr. and ed. R. F. C. Hull. Routledge & Kegan Paul, London, 1950.

MONBODDO, LORD. See Burnett, James.

MONRO, C. H. (1) (tr.). *The Digest of Justinian*. Cambridge, 1904.

MORGAN, E. (1) (tr.). *Tao the Great Luminant; Essays from 'Huai Nan Tzu', with introductory articles, notes and analyses*. Kelly & Walsh, Shanghai, n.d. (1933?).

MORGAN, L. H. (1). *Ancient Society, or Researches in the Lines of Human Progress from Savagery through Barbarism to Civilisation*. Holt, New York, 1877.

MORLEY, S. G. (1). *The Ancient Maya*. Stanford Univ. Press, Palo Alto, California, 1946.

MORRISON, R. (1). *A Dictionary of the Chinese Language in Three Parts*. Macao, 1815; 2nd ed. 1819.

MORTIER, F. (1). 'Du Sens Primitif de l'antique et célèbre Figure Divinatoire des Taoistes Chinois et Japonais (Hsien Thien).' *BSRBAP*, 1948, **59**, 150.

MOULE, G. E. (2). 'Notes on the Ting-Chi, or Half-Yearly Sacrifice to Confucius.' *JRAS/NCB*, 1900, **33**, 37.

MOURAD, Y. (1). *La Physiognomie Arabe et le 'Kitāb al-Firasa' de Faqir al-Dīn al-Razī*. Geuthner, Paris, 1939.

MUELLER, HERBERT (1). 'Über das Taoistische Pantheon d. Chinesen, seine Grundlage und seine historische Entwicklung.' *ZFE*, 1911, **43**, 393. (With appendix by E. Boerschmann: 'Einige Beispiele für die gegenseitige Durchdringung der drei chinesischen Religionen', p. 429.)

MUIR, M. M. PATTISON (1). *The Story of Alchemy and the Beginnings of Chemistry*. Hodder & Stoughton, London, 1913.

MÜLLER, F. W. K. (2). 'Über d. Ausdruck "Kālasūtra".' *ENB*, 1896, **1** (no. 3), 23.

MUS, P. (1). 'La Notion de Temps Réversible dans la Mythologie Bouddhique.' *AEPHE/SSR*, 1939, 1.

MYERS, C. S. 'The Taste-Names of Primitive Peoples.' *JPS*, 1904, **1**, 117.

NAGASAWA, K. (1). *Geschichte der Chinesischen Literatur, und ihrer gedanklichen Grundlage*, transl. from the Japanese by E. Feifel. Fu-jen Univ. Press, Peiping, 1945.

NAGEL, P. (1). 'Umrechnung d. zyklischen Daten des chinesischen Kalenders in europäische Daten.' *MSOS*, 1931, **34**, 153.

NALLINO, C. A. (1). 'Muslim Astrology.' *EI*, vol. 1, p. 494; *ERE*, vol. 12, p. 88.

NANJIO, B. (1). *A Catalogue of the Chinese Translations of the Buddhist Tripiṭaka*. Oxford, 1883. (See Ross, E. D.)

NEEDHAM, JOSEPH (1). *Chemical Embryology*. 3 vols. Cambridge, 1931.

NEEDHAM, JOSEPH (2). *A History of Embryology*. Cambridge, 1934.

NEEDHAM, JOSEPH (3). *Time, the Refreshing River*. Allen & Unwin, London, 1942.

NEEDHAM, JOSEPH (4). *Chinese Science*. Pilot Press, London, 1945.

NEEDHAM, JOSEPH (5). *The Great Amphibium; four lectures on the position of Religion in a world dominated by Science*. SCM, London, 1931.

NEEDHAM, JOSEPH (6). *History is on our side: a contribution to political religion and scientific faith*. Allen & Unwin, London, 1946.

NEEDHAM, JOSEPH (7). 'Science and Social Change.' *SS*, 1946, **10**, 225.

NEEDHAM, JOSEPH (8). 'Geographical Distribution of English Ceremonial Folk-Dances.' *JEFDS*, 1936, **3**, 1.

NEEDHAM, JOSEPH (9). 'A Biologist's View of Whitehead's Philosophy.' In *The Philosophy of Alfred North Whitehead*, ed. P. A. Schilpp. Northwestern Univ. Press, Chicago, 1941. Repr. in Needham (3), p. 178.

NEEDHAM, JOSEPH (10). *Integrative Levels; A Revaluation of the Idea of Progress*. Herbert Spencer Lecture, Oxford University. Oxford, 1937; repr. 1941. Repr. in Needham (3), p. 233.

NEEDHAM, JOSEPH (11). 'The Liquidation of Form and Matter.' *WR*, 1941. Repr. as art. in *This Changing World*. London, 1942. Repr. in Needham (6), p. 199.

NEEDHAM, JOSEPH (12). *Biochemistry and Morphogenesis*. Cambridge, 1942; repr. 1950.

NEEDHAM, JOSEPH (13). *The Sceptical Biologist*. Chatto & Windus, London, 1929.

NEEDHAM, JOSEPH (14). *Man a Machine*. Kegan Paul, London, 1927; Norton, New York, 1928.

NEEDHAM, JOSEPH (15). 'Pure Science and the Idea of the Holy.' Address, 1941; repr. in Needham (3), p. 92.

NEEDHAM, JOSEPH (28). 'Biochemical Aspects of Form and Growth.' Art. in *Aspects of Form*, ed. L. L. Whyte, p. 77. Lund Humphries, London, 1951.

NEEDHAM, JOSEPH & PAGEL, WALTER (1) (ed.). *Background to Modern Science*. Cambridge, 1938; repr. 1940; Macmillan, N.Y., 1938.

NEEF, H. (1). *Die im 'Tao Tsang' enthaltenen Kommentare zu 'Tao-Tê-Ching' Kap. VI.* Inaug. Diss., Bonn, 1938.

v. NEGELEIN, J. (1). 'Die ältesten Meister d. indischen Astrologie u.d. Grundidee ihrer Lehrbücher.' *ZDMG*, 1928, **82** (n.f., **7**), 1.

NEMETH, J. (1). 'On the word "Shaman" in Turkic and Uigur.' *KS*, 1913, **14**, 240.

NETTLESHIP, H. (1). 'Jus Gentium.' *JP*, 1885, **13**, 169.

NICHOLSON, R. A. (1). 'Sufism.' *ERE*, vol. 12, p. 10.

NILSSON, N. M. P. (1). *The Rise of Astrology in the Hellenistic Age*. Historical Notes and Papers of the Observatory of Lund, no. 18, 1943.

NIORADZE, G. (1). *Der Schamanismus bei den Sibirischen Völkern*. Stuttgart, 1925.

NORMAN, E. H. (1). *Andō Shōeki and the Anatomy of Japanese Feudalism*. 2 vols. Tokyo, 1949. (*TAS/J*, 3rd ser., no. 2.)

NORTHROP, F. S. C. (1). *The Meeting of East and West; an Inquiry concerning Human Understanding*. Macmillan, New York, 1946. Crit. D. Bodde, *PA*, 1947, **20**, 199.

NOWOTNY, K. A. (1). 'The Construction of Certain Seals and Characters in the Work of Agrippa of Nettesheim.' *JWCI*, 1949, **12**, 46.

NURUL HASAN, SAYYAD (1). *The Chisti and Suhrawardi Movements in India to the Middle of the Sixteenth Century*. Inaug. Diss., Oxford, 1948.

OHLMARKS, A. (1). *Studien zum Problem des Schamanismus*. Gleerup, Lund, 1939.

OLDENBERG, H. (1). *Buddha, sein Leben, seine Lehre, seine Gemeinde*. Cotta, Stuttgart, 1921. Eng. tr. W. Hoey (from earlier ed.), Williams & Norgate, London, 1882.

OLSVANGER, I. (1). *Fu-Hsi, the Sage of Ancient China* (Binary arithmetic, magic squares, and the *Book of Changes*). Massadah, Jerusalem, 1948.

OLTRAMARE, P. (1). *La Formule Bouddhique des Douze Causes, son sens originel et son interprétation théologique*. Georg, Geneva, 1909. (Mémoire publiée à l'occasion du Jubilé de l'Université de Génève.)

OSBORN, H. F. (1). *From the Greeks to Darwin; an Outline of the development of the Idea of Evolution*. Columbia Univ. Press, New York, 1894.

OTTO, RUDOLF (1). *The Idea of the Holy*. Oxford, 1923.

PADOUX, G. (1). Preface to Escarra & Germain (1).

PAGEL, WALTER (1). 'Religious Motives in the Medical Biology of the Seventeenth Century.' *BIHM*, 1935, **3**, 97.

PAGEL, WALTER (2). 'The Religious and Philosophical Aspects of van Helmont's Science and Medicine.' *BIHM*, Suppl. no. 2, 1944.

PAGEL, WALTER (3). 'The Debt of Science and Medicine to a devout Belief in God; illustrated by the work of J. B. van Helmont.' *JTVI*, 1942, **74**, 99.

PAGEL, WALTER (4). 'William Harvey; Some Neglected Aspects of Medical History.' *JWCI*, 1944, **7**, 144.

PAGEL, WALTER (5). 'The Vindication of "Rubbish".' *MHJ*, 1945.

PAGEL, WALTER (6). 'A Background Study to Harvey.' *MBH*, 1948, **2**, 407.

PAGEL, WALTER (7). 'Prognosis and Diagnosis; a Comparison of Ancient and Modern Medicine.' *JWCI*, 1939, **2**, 382.

PAGEL, WALTER (8). 'J. B. van Helmont's *De Tempore*, and Biological Time.' *OSIS*, 1949, **8**, 346.

PARACELSUS [VON HOHENHEIM, THEOPHRASTUS] (1). *Sämtliche Werke*. Zollikofer, St Gallen, 1944.

PASCAL, BLAISE (2). *Pensées* (1670). 2 vols. Larousse, Paris, n.d. (1926).

PATAI, RAPHAEL (1). *Man and Temple in Ancient Jewish Myth and Ritual*. Nelson, London, 1947.

PATRITIUS, FRANCISCUS (1). *Nova de Universalis Philosophia*. Venice, 1593.

PAULSON, J. (1). *Index Lucretianus*. Gothenburg, 1911; Leipzig, 1926.

PEARSON, KARL (1). *The Grammar of Science*. Black, London, 1900.

PECK, A. L. (1) (tr.). *Aristotle; 'The Generation of Animals'*. Heinemann, London, 1943. (Loeb Classics series.)

PECK, A. L. (2) (tr.). *Aristotle; 'The Parts of Animals'*. Heinemann, London, 1937. (Loeb Classics series.)

PEET, E. (1). *The Wisdom Literature*. London, 1930.

PEILLON, M. (1). 'Gymnastique et Massage.' In Laignel-Lavastine (1), *Histoire Générale de la Médecine, q.v.* (vol. 3, p. 627).

PELLIOT, P. (8). 'Autour d'une Traduction sanskrite du *Tao-tö-king*' (*Tao Tê Ching*). *TP*, 1912, **13**, 350.

PELLIOT, P. (11). 'Sur quelques mots d'Asie Centrale attestés dans les textes Chinois.' *JA*, 1913 (11e sér.), **1**, 466.

PELLIOT, P. (12). On the *Hua Hu Ching*. *BEFEO*, 1903, **3**, 322; 1906, **6**, 379; 1908, **8**, 515.

PELLIOT, P. (13). 'Notes de Bibliographie Chinoise; II, Le Droit Chinois.' *BEFEO*, 1909, **9**, 123.

PELLIOT, P. (14). '*Meou-Tseu [Mou Tzu]*, ou les Doutes Levés.' *TP*, 1920, **19**, 255.

642   BIBLIOGRAPHY C

PELLIOT, P. (15). Criticism of Waley (9). *TP*, 1922, **21**, 90.
PELLIOT, P. (22). 'Note on the *Ku Yü Thu Phu*.' *TP*, 1932, **29**, 199.
PELLIOT, P. (34). 'À propos du *Chinese Biographical Dictionary* de Mons. H. Giles.' *AM*, 1927, **4**, 377.
PELLIOT, P. (35). 'Les *Yi Nien Lou*' [*I Nien Lu*; Discussions of Doubtful Dates]. *TP*, 1927, **25**, 65.
PELLIOT, P. (37) '*Le Prétendu Album de Porcelaines de Hsiang Yuan-Pin* TP 10, 36 **32** 15, 1937 **33** 91.
PELSENEER, J. (1). *L'Evolution de la Notion de Phenomène Physique, des Primitifs à Bohr et Louis de Broglie*. Office Internat. de Librairie, Brussels, n.d. (1947?).
PEREIRA, GALEOTE. See next entry.
PEREYRA, GALEOTTI (1). 'Certayne Reportes of the Province China, learned through the Portugalles there imprisoned, and chiefly by the Relation of Galeotto Perera, a gentleman of good credit, that lay prisoner in that countrey many Yeeres.' In R. Eden, *The History of Travayle in the West and East Indies and other Countreys lying either Way.... London, 1577. See W[illis] (1). Also in Purchas, vol. 2, pt. 1, ch. 11, p. 199.
PERLEBERG, M. (1) (tr.). *The Works of Kungsun Lung Tzu, with a Translation from the parallel Chinese original text, critical and exegetical notes, punctuation and literal translation, the Chinese commentary, prolegomena and Index*. Pr. pub. Hongkong, 1952. Crit. J. J. L. Duyvendak, *TP*, 1954, **42**, 383.
PETROV, A. A. (1). *Wang Pi (+226 to +249); his place in the History of Chinese Philosophy* (in Russian). Inst. Orient. Stud., Moscow, Monogr. no. 13, 1936. Eng. résumé by A. F. Wright, *HJAS*, 1947, **10**, 75.
PETRUCCI, R. (2). *La Philosophie de la Nature dans l'Art d'Extreme-Orient*. Laurens, Paris (1910).
PETTAZZONI, R. (1). 'Le Corps Parsemé d'Yeux.' *Z*, 1938, **1**, 3.
PFISTER, L. (1). *Notices Biographiques et Bibliographiques sur les Jésuites de l'Ancienne Mission de Chine (+1552 to +1773)*. 2 vols. Mission Press, Shanghai, 1932 (*VS*, no. 59).
PFIZMAIER, A. (19) (tr.). 'Keu-Tsien, Konig von Yue, und dessen Haus' (Kou Chien of Yüeh, and Fan Li). *SWAW/PH*, 1863, **44**, 197. Tr. *Shih Chi*, ch. 41; cf. Chavannes (1), vol. 4.
PFIZMAIER, A. (22) (tr.). 'Der Landesherr von Schang' (Shang Yang). *SWAW/PH*, 1858, **29**, 98. Tr. *Shih Chi*, ch. 68; not in Chavannes (1). Cf. Duyvendak (3); Liao (1).
PFIZMAIER, A. (23) (tr.). 'Das Rednergeschlecht Su' (Su Chhin). *SWAW/PH*, 1860, **32**, 642. Tr. *Shih Chi*, ch. 69; not in Chavannes (1).
PFIZMAIER, A. (24) (tr.). 'Der Redner Tschang I und einige seiner Zeitgenossen' (Chang I and Chhu Li Tzu). *SWAW/PH*, 1860, **33**, 525, 566. Tr. *Shih Chi*, chs. 70, 71; not in Chavannes (1).
PFIZMAIER, A. (26) (tr.). 'Zur Geschichte von Entsatzes von Han Tan.' *SWAW/PH*, 1859, **31**, 65, 87, 104, 120. Tr. *Shih Chi*, chs. 75, 76, 78, 83; includes life of the Prince of Phing-Yuan; not in Chavannes (1).
PFIZMAIER, A. (30) (tr.). 'Li Sse, der Minister des ersten Kaisers' (Li Ssu). *SWAW/PH*, 1859, **31**, 120, 311. Tr. *Shih Chi*, chs. 83, 87; not in Chavannes (1). Cf. Bodde (1).
PFIZMAIER, A. (36) (tr.). 'Sse-ma Ki-Tschü, der Wahrsager von Tschang-ngan' (Ssuma Chi-Chu, in the chapter on diviners, Jih Chê Lieh Chuan). *SWAW/PH*, 1861, **37**, 408. Tr. *Shih Chi*, ch. 127; not in Chavannes (1).
PFIZMAIER, A. (39) (tr.). 'Die Könige von Hoai Nan aus dem Hause Han' (Huai Nan Tzu). *SWAW/PH*, 1862, **39**, 575. Tr. *Chhien Han Shu*, ch. 44.
PFIZMAIER, A. (40) (tr.). 'Das Erreigniss des Wurmfrasses der Beschwörer.' *SWAW/PH*, 1862, **39**, 50, 55, 58, 65, 76, 89. Tr. *Chhien Han Shu*, chs. 45, 63, 66, 74.
PFIZMAIER, A. (45) (tr.). 'Die Antworten Tung Tschung-Schü's [Tung Chung-Shu] auf die Umfragen des Himmelssohnes.' *SWAW/PH*, 1862, **39**, 345. Tr. *Chhien Han Shu*, ch. 56.
PFIZMAIER, A. (56) (tr.). 'Über einige Wundermänner Chinas' (magicians and technicians such as Chhen Hsün, Tai Yang, Wang Chia, Shunyu Chih, etc.). *SWAW/PH*, 1877, **85**, 37. Tr. *Chin Shu*, ch. 95.
PFIZMAIER, A. (64) (tr.). 'Die fremdländischen Reiche zu den Zeiten d. Sui.' *SWAW/PH*, 1881, **97**, 411, 418, 422, 429, 444, 477, 483. Tr. *Sui Shu*, chs. 64, 81, 82, 83, 84.
PFIZMAIER, A. (65) (tr.). 'Die Classe der Wahrhaftigen in China.' *SWAW/PH*, 1881, **98**, 983, 1001, 1036. Tr. *Sui Shu*, chs. 71, 73, 77.
PFIZMAIER, A. (67) (tr.). 'Seltsamkeiten aus den Zeiten d. Thang' I and II. I, *SWAW/PH*, 1879, **94**, 7, 11, 19. II, *SWAW/PH*, 1881, **96**, 293. Tr. *Hsin Thang Shu*, chs. 34–6 (Wu Hsing Chih), 88, 89.
PFIZMAIER, A. (68) (tr.). 'Darlegung der chinesischen Ämter.' *DWAW/PH*, 1879, **29**, 141, 170, 213; 1880, **30**, 305, 341. Tr. *Hsin Thang Shu*, chs. 46, 47, 48, 49A; cf. des Rotours (1).
PFIZMAIER, A. (69) (tr.). 'Die Sammelhäuser der Lehenkönige Chinas.' *SWAW/PH*, 1880, **95**, 919. Tr. *Hsin Thang Shu*, ch. 49B; cf. des Rotours (1).
PFIZMAIER, A. (70) (tr.). 'Über einige chinesische Schriftwerke des siebenten und achten Jahrhunderts n. Chr.' *SWAW/PH*, 1879, **93**, 127, 159. Tr. *Hsin Thang Shu*, chs. 57, 59 (in part: I Wên Chih including agriculture, astronomy, mathematics, war, five-element theory).

PFIZMAIER, A. (71) (tr.). 'Die philosophischen Werke Chinas in dem Zeitalter der Thang.' *SWAW/ PH*, 1878, **89**, 237. Tr. *Hsin Thang Shu*, ch. 59 (in part: I Wên Chih, philosophical section, including Buddhism).

PFIZMAIER, A. (72) (tr.). 'Der Stand der chinesische Geschichtsschreibung in dem Zeitalter der Thang' (original has Sung as misprint). *DWAW/PH*, 1877, **27**, 309, 383. Tr. *Hsin Thang Shu*, chs. 57 (in part), 58 (I Wên Chih, history and classics section).

PFIZMAIER, A. (74) (tr.). 'Nachrichten von Gelehrten Chinas.' (Scholars such as Khung Ying-Ta, Ouyang Hsün, etc.) *SWAW/PH*, 1878, **91**, 694, 734, 758. Tr. *Hsin Thang Shu*, chs. 198, 199, 200.

PFIZMAIER, A. (81) (tr.). 'Chinesische Begründungen der Taolehre.' *SWAW/PH*, 1886, **111**, 801. Tr. of the *Chhüan Tao Chi* in *Tao Yen Nei Wai Pi Chüeh Chhüan Shu*.

PFIZMAIER, A. (82) (tr.). 'Über d. Schriften des Kaisers des Wên Tschang.' *SWAW/PH*, 1873, **73**, 329. Tr. pt. of *Wên Chhang Ti Chün Shu*.

PFIZMAIER, A. (84) (tr.). 'Aus dem Traumleben d. Chinesen.' *SWAW/PH*, 1870, **64**, 697, 711, 722, 733. Tr. *Thai-Phing Yü Lan*, chs. 397, 398, 399, 400.

PFIZMAIER, A. (85) (tr.). 'Geschichtliches ü. einige Seelenzustände u. Leidenschaften.' *SWAW/PH*, 1868, **59**, 248, 258, 271, 274, 289, 302, 315. Tr. *Thai-Phing Yü Lan*, chs. 469 (Furcht), 483 (Zorn), 490 (Vergesslichkeit u. Irrtum), 491 (Beschämung), 493 (Verschwendung), 498 (Hochmut), 499 (Dummheit).

PFIZMAIER, A. (87) (tr.). 'Die Taolehre v. den wahren Menschen u.d. Unsterblichen.' *SWAW/PH*, 1869, **63**, 217, 235, 252, 268. Tr. *Thai-Phing Yü Lan*, chs. 660, 661, 662, 663.

PFIZMAIER, A. (88) (tr.). 'Die Lösung d. Leichnam und Schwerter, ein Beitrag zur Kenntnis d. Taoglaubens.' *SWAW/PH*, 1870, **64**, 26, 45, 60, 79. Tr. *Thai-Phing Yü Lan*, chs. 664, 665, 666, 667.

PFIZMAIER, A. (89) (tr.). 'Die Lebensverlängerungen d. Männer des Weges' (*Tao Shih*). *SWAW/PH*, 1870, **65**, 311, 334, 346, 359. Tr. *Thai-Phing Yü Lan*, chs. 668, 669, 670, 671.

PFIZMAIER, A. (99) (tr.). 'Der Geisterglaube in dem alten China.' *SWAW/PH*, 1871, **68**, 641, 652, 665, 679, 695. Tr. *Thai-Phing Yü Lan*, chs. 881, 882, 883, 884, 887 (in part).

PFIZMAIER, A. (102) (tr.). 'Über einige Gegenstände des Taoglaubens.' *SWAW/PH*, 1875, **79**, 5, 16, 29, 42, 50, 59, 61, 68, 73, 78. Tr. *Thai-Phing Yü Lan*, chs. 929, 930 (dragons), 931, 932 (tortoises), 933, 934 (snakes), 984, 985, 986, 989, 990 (miscellaneous stones).

PIGANIOL, A. (1). 'Les Etrusques, Peuple d'Orient.' *JWH*, 1953, **1**, 328.

PINOT, V. (1). *La Chine et la Formation de l'Esprit Philosophique en France (1640–1740)*. Geuthner, Paris, 1932.

PINTO, FERNAÕ MENDES (1). *Peregrinacam de Fernam Mendez Pinto em que da conta de muytas e muyto estranhas cousas que vio e ouvio no reyno da China, no da Tartaria....* Crasbeec, Lisbon, 1614. Abridged Eng. tr. by H. Cogan: *The Voyages and Adventures of Ferdinand Mendez Pinto, a Portugal, During his Travels for the space of one and twenty years in the kingdoms of Ethiopia, China, Tartaria, etc.* Gent, London, 1653. Full French tr. by B. Figuier: *Les Voyages Advantureux de Fernand Mendez Pinto....* Cotinet & Roger, Paris, 1645. Cf. M. Collis: *The Grand Peregrination* (paraphrase and interpretation), Faber & Faber, London, 1949.

PIRES, TOMÉ (1). *The Suma Oriental of T. Pires, an account of the East from the Red Sea to Japan... written in...1512–1515*, ed. A. Cortesaõ. Hakluyt Society, London, 1944. (Hakluyt Soc. Pubs. 2nd series, nos. 89, 90.)

PIRIE, N. W. (1). 'Ideas and Assumptions about the Origin of Life.' *D*, 1953, **14**, 1. *LG*, 1954, **69** (no. 1), 10; (no. 2), 30.

PITON, C. (1). 'The Six Chancellors of Chhin.' *CR*, 1884, **13**, 102, 127, 255, 305, 365.

PLATH, J. H. (1). 'Gesetz und Recht im alten China.' *ABAW* (Phil.-philol. Klasse), 1865, **10**, 675.

PLEDGE, H. T. (1). *Science since 1500*. HMSO, London, 1939.

PLEKHANOV, G. (1). *Fundamental Problems of Marxism*. Lawrence, London, 1928.

POKORNY, J. (1). *Vergleichende Wörterbuch d. indo-germanischen Sprachen*. De Gruyter, Berlin, 1930.

POLLARD, B. W. (1). 'Circuit Components of Digital Computers,' ch. 2 in *Faster than Thought; a Symposium on Digital Calculating Machines*, ed. B. V. Bowden, Pitman, London, 1953.

POLLOCK, F. (1). Editorial notes in Sir Henry Maine's *Ancient Law*, 1916 ed.

POLLOCK, F. (2). 'History of the Law of Nature.' *JSCL*, 1900 (n.s.), **2**, 418.

POLLOCK, F. & MAITLAND, F. W. (1). *History of English Law*. 2 vols. Cambridge, 1898.

DELLA PORTA, GIAMBATTISTA (1). *Magia Naturalis*. 1st ed. 1558; enlarged 1589 and 1601. Account in Beck (1), ch. 13.

PORTER, L. C. (1). *Aids to the Study of Chinese Philosophy*. Yenching University, Peking, 1934.

PORTER, L. C. (2). 'Fêng-Shui.' *CRR*, 1920.

POTT, W. S. A. (1). 'The "Natural" Basis of Confucian Ethics.' *NCR*, 1921, **3**, 192.

POWICKE, F. J. (1). *The Cambridge Platonists*. Dent, London, 1926.

PRATT, J. BISSETT (1). Art. in *Modern Trends in World Religions*, ed. A. E. Haydon, p. 35. Univ. of Chicago Press, 1934.

PROCHASKA, GEORGE. See next entry.

PROCHÁZKA, JIŘÍ (1). *De Functionibus Systematis Nervosi Commentatio*. Prague, 1784. Reissued in facsimile, with Czech translation (*Úvaha o Funkcích Nervové Soustavy*), and abundant notes, by M. Petráň, with introduction by E. Gutmann. Acad. Sci. Bohemo-Slovenica, Prague, 1954. Eng. tr. *A Dissertation on the Functions of the Nervous System*, by T. Laycock, Sydenham Soc. London, 1851.

PRZYŁUSKI, J. (2). (*a*) 'Une Cosmogonie Commune à l'Iran et à l'Inde.' *JA*, 1937, **229**, 481. (*b*) 'La Théorie des El ments.' *SCI*, 1933.

QUISTORP, M. (1). 'Männergesellschaft und Altersklassen im alten China.' *MSOS*, 1915, **18** (no. 1), 1.

RACKHAM, H. (1) (tr.). *The Nicomachean Ethics of Aristotle*. Heinemann, London, 1926. Loeb Classics series.

RÁDL, EMANUEL (1). *Západ a Východ; Filosofické Úvahy z cest*. Laichter, Prague, 1925.

RATCHNEVSKY, P. (1). *Un Code des Yuan*. Leroux, Paris, 1937. (Biblioth. de l'Inst. des Hautes Etudes Chinoises, no. 4.)

RAVEN, C. E. (1). *Natural Religion and Christian Theology*. Gifford Lectures, 1951; 1st series: 'Science and Religion.' Cambridge, 1953.

RAY, T. (1) (tr.). *The 'Anaṅga Raṅga'* [written by Kalyāṇa Malla, for Lad Khan, a son of Ahmad Khan Lodi; *c*. +1500], pref. by G. Bose. Med. Book Co., Calcutta, 1951 (3rd ed.).

READ, BERNARD E. (1) [with LIU JU-CHHIANG]. *Chinese Medicinal Plants from the 'Pên Tshao Kang Mu' A.D. 1596...a Botanical Chemical and Pharmacological Reference List* (Publication of the Peking Nat. Hist. Bull.). French Bookstore, Peiping, 1936. (Chs. 12 to 37 of *Pên Tshao Kang Mu*); rev. W. T. Swingle, *ARLC/DO*, 1937, 191.

READ, BERNARD E. (2) [with LI YÜ-THIEN]. *Chinese Materia Medica; Animal Drugs*.

		Serial nos.	Corresp. with chaps. of *Pên Tshao Kang Mu*
Pt. I	Domestic Animals	322–349	50
II	Wild Animals	350–387	51 *A* and *B*
III	Rodentia	388–399	51 *B*
IV	Monkeys and Supernatural Beings	400–407	51 *B*
V	Man as a Medicine	408–444	52

PNHB, 1931, **5** (no. 4), 37–80; **6** (no. 1), 1–102. (Sep. issued, French Bookstore, Peiping, 1931.)

READ, BERNARD E. (3) [with LI YÜ-THIEN]. *Chinese Materia Medica; Avian Drugs*.

Pt. VI	Birds	245–321	47, 48, 49

PNHB, 1932, **6** (no. 4), 1–101. (Sep. issued, French Bookstore, Peiping, 1932.)

READ, BERNARD E. (4) [with LI YÜ-THIEN]. *Chinese Materia Medica; Dragon and Snake Drugs*.

Pt. VII	Reptiles	102–127	43

PNHB, 1934, **8** (no. 4), **297**–357. (Sep. issued, French Bookstore, Peiping, 1934.)

READ, BERNARD E. (5) [with YU CHING-MEI]. *Chinese Materia Medica; Turtle and Shellfish Drugs*.

Pt. VIII	Reptiles and Invertebrates	199–244	45, 46

PNHB (Suppl.), 1939, 1–136. (Sep. issued, French Bookstore, Peiping, 1937.)

READ, BERNARD E. (6) [with YU CHING-MEI]. *Chinese Materia Medica; Fish Drugs*.

Pt. IX	Fishes (incl. some amphibia, octopoda and crustacea)	128–198	44

PNHB (Suppl.), 1939. (Sep. issued, French Bookstore, Peiping, n.d. prob. 1939.)

READ, BERNARD E. (7) [with YU CHING-MEI]. *Chinese Materia Medica; Insect Drugs*.

Pt. X	Insects (incl. arachnida etc.)	1–101	39, 40, 41, 42

PNHB (Suppl.), 1941. (Sep. issued, Lynn, Peiping, 1941.)

READ, BERNARD E. & PAK, C. (PHU CHU-PING) (1). *A Compendium of Minerals and Stones used in Chinese Medicine, from the 'Pên Tshao Kang Mu'.* PNHB, 1928, **3** (no. 2), i–vii, 1–120. (Revised and enlarged, issued separately, French Bookstore, Peiping, 1936 (2nd ed.).) Serial nos. 1–135, corresp. with chs. of *Pên Tshao Kang Mu*, 8, 9, 10, 11.

READ, J. (1). *Prelude to Chemistry; an Outline of Alchemy, its Literature and Relationships.* Bell, London, 1936.

READ, T. T. (4). *The Early Casting of Iron; A Stage in Iron Age Civilisation.* GR, 1934, **24**, 544.

READ, T. T. (8). 'China's Civilisation Simultaneous, not Osmotic' (letter). AMS, 1937, **6**, 249.

RECINOS, A., GOETZ, D. & MORLEY, S. G. (1). *Popol Vuh; the Sacred Book of the ancient Quiché Maya.* Univ. of Oklahoma Press, Norman, Okla., 1950.

REICHELT, K.V. (1). *Truth and Tradition in Chinese Buddhism; a Study of Chinese Mahāyāna.* Com. Press, Shanghai, 1934.

REICHWEIN, A. (1). *China and Europe; Intellectual and Artistic Contacts in the Eighteenth Century.* Kegan Paul, London, 1925.

RELE, VASANT GANGARAM (1). *The Mysterious Kuṇḍalinī; the physical basis of the 'Kuṇḍali (Hatha) Yoga' in terms of Western Anatomy and Physiology.* Taraporevala, Bombay, n.d. (London, 1932?).

RÉMUSAT, J. P. A. (8). *Mémoire sur la Vie et les Opinions de Lao-Tseu [Lao Tzu].* Paris, 1823.

RENOU, L. (1). *Anthologie Sanscrite.* Payot, Paris, 1947.

REY, ABEL (1). *La Science dans l'Antiquité.* Vol. 1: *La Science Orientale avant les Grecs*, 1930, 2nd ed. 1942; Vol. 2: *La Jeunesse de la Science Grecque*, 1933; Vol. 3: *La Maturité de la Pensée Scientifique en Grèce*, 1939; Vol. 4: *L'Apogée de la Science Technique Grecque (Les Sciences de la Nature et de l'Homme, les Mathématiques, d'Hippocrate à Platon)*, 1946. Albin Michel, Paris. (Evol. de l'Hum. Sér. complémentaire.)

RHYS DAVIDS, T. W. (1). *Buddhism.* SPCK, London, 1910.

RHYS DAVIDS, T. W. (2). *Buddhism; its History and Literature.* Putnam, New York, 1907.

RHYS DAVIDS, T. W. (3). 'Cosmic Law in Ancient Thought.' PBA, 1918, **8**, 279.

RIAZANOVSKY, V. A. (1). 'Mongol Law and Chinese Law in the Yuan Dynasty.' CSPSR, 1936, **20**, 266.

RICHARDS, I. A. (1). *Mencius on the Mind.* Kegan Paul, London, 1932.

DE RIVAROL, A. (1). In Sainte-Beuve, *Lundis*, vol. 5, p. 82; no. 324 in *The Spirit of Man*, ed. Robert Bridges.

ROBERTSON, ARCHIBALD (sen.) (1). *Regnum Dei, Eight Lectures on the Kingdom of God in the History of Christian Thought* (Bampton Lectures). Methuen, London, 1901.

ROBIN, P. A. (1). *Animal Lore in English Literature.* Murray, London, 1932.

ROBINSON, K. (1). *A Critical Study of Ju Dzai-Yü's [Chu Tsai-Yü's] Account of the System of the Lü-Lü or Twelve Musical Tubes in Ancient China.* Inaug. Diss., Oxford, 1948.

ROBINSON, K. (2). 'A Possible Use of Music for Divination.' MS.

ROBSON, W. A. (1). 'A Criticism of Maine.' In *Modern Theories of Law*, ed. W. I. Jennings, p. 160. Oxford, 1933.

ROBSON, W. A. (2). *Civilisation and the Growth of Law.* Macmillan, London, 1935.

ROLL, E. (1). *History of Economic Thought.* Faber & Faber, London, 1938.

ROSCOE, H. E. & SCHORLEMMER, C. (1). *A Treatise on Chemistry.* Macmillan, London, 1923.

ROSENBERG, O. (1). *Die Probleme der buddhistischen Philosophie* (tr. from Russian by E. Rosenberg), Heidelberg, 1924.

[VON ROSENROTH, K. & VAN HELMONT, F. M.] *Kabbala Denudata, seu Doctrina Hebraeorum Transcendentalis et Metaphysica, etc.* Lichtenthaler, Sulzbach, 1677.

DE ROSNY, L. (1) (tr.). *Chan-Hai-King [Shan Hai Ching]: Antique Géographie Chinoise.* Maisonneuve, Paris, 1891.

DE ROSNY, L. (2) (tr.). *Hiao-King [Hsiao Ching]; Livre Sacré de la Piété Filiale.* Maisonneuve, Paris, 1889.

ROSS, E. D. (3). *Alphabetical List of the Titles of Works in the Chinese Buddhist Tripiṭaka.* Indian Govt., Calcutta, 1910. See Nanjio, B.

ROSS, W. D. (1). *Aristotle.* Methuen, London, 1930.

ROSTOVTZEV, M. I. (3). *Inlaid Bronzes of the Han Dynasty in the Collection of C. T. Loo [Lu].* Vanoest, Paris and Brussels, 1927.

ROTH, L. (1). *The 'Guide for the Perplexed' of Moses Maimonides.* Hutchinson, London, 1948.

DES ROTOURS, R. (1). *Traité des Fonctionnaires et Traité de l'Armée, traduits de la Nouvelle Histoire des Thang* (chs. 46–50). 2 vols. Brill, Leiden, 1948 (Biblioth. de l'Inst. des Hautes Etudes Chinoises, no. 6); rev. P. Demiéville, JA, 1950, **238**, 395.

DES ROTOURS, R. (2). *Traité des Examens traduit de la Nouvelle Histoire des Thang* (Ch. 44, 45) (Biblioth. de l'Inst. des Hautes Etudes Chinoises, no. 2). Paris, 1932.

ROUSSELLE, E. (1). 'Der lebendige Taoismus im heutigen China.' SA, 1938, **8**, 122.

ROUSSELLE, E. (2). 'Yin und Yang vor ihrem Auftreten in der Philosophie.' *SA*, 1933, **8**, 41.

ROUSSELLE, E. (3). 'Das Primat des Weibes im alten China.' *SA*, 1941, **16**, 130.

ROWLEY, H. H. (1). 'The Chinese Philosopher Mo Ti.' *BJRL*, 1948, **31**, 241.

RUBEN, W. (1). *Die Philosophie der Upanishaden.* Francke, Bern, 1947.

RUBEN, W. (2). *Eski Hind Tarihi* (sketch of Indian History). Ankara Univ. Dil ve Tarihi-Cografya Fakultesi Hindoloji Enstitusu, Monograph no. 2. Ankara, 1944. (Turkish with German summary.)

RUBEN, W. (3). 'Schamanismus im alten Indien.' *AO*, 1940, **18**, 164. *'Einführung in die Indienkunde; ein Überblick über die historische Entwicklung Indiens*, Deutsches Verlag d. Wissenschaften, Berlin 1954.

> Leiden, 1949. Vol. 2, 1952. (Sinica Leidensia, no. 6.)
> Eng. tr. C. F. Baynes, Routledge & Kegan Paul, London, 1961.
> Leiden, 1949, Vol. 2, 1952. (Sinica Leidensia, no. 6.)

RUHLAND, MARTIN (1). *Lexicon Alchemiae sive Dictionarium Alchemisticum cum obscuriorum verborum et rerum Hermeticarum tum Theophrast-Paracelsicarum Phrasium, planam explicationem continens.* 1612; 2nd ed. Frankfurt, 1661.

RUSSELL, BERTRAND (1). *The Problem of China.* Allen & Unwin, London, 1922.

RUSSELL, BERTRAND (2). *History of Western Philosophy.* Allen & Unwin, London, 1946.

RUSSELL, E. S. (1). *Form and Function; a Contribution to the History of Animal Morphology.* Murray, London, 1916.

DE SAHAGUN, BERNADINO (1). *Historia General de las Cosas de Nueva España.* Span. ed. C. M. de Bustamente; Eng. tr. F. R. Bandelier. Nashville, Tenn., 1932.

DE SAINTE-BEUVE, C. A. (1). *Causeries du Lundi.* 16 vols. Garnier, Paris, n.d. (1st ed. 1850).

DE SAINTE-MARIE, A. (1). *Traité sur Quelques Points Importans de la Mission de la Chine.* 1710. Repr. in Kortholt (1).

SALMOND, J. (1). 'The Law of Nature.' *LQR*, 1895, **11**, 121.

SALMONY, A. (2). 'The Human Pair in China and South Russia.' *GBA*, 1943 (6th ser.), **24**, 321.

SANDYS, J. E. (1). *A History of Classical Scholarship.* 3 vols. Cambridge, 1908.

SARTON, GEORGE (1). *Introduction to the History of Science*, Vol. 1, 1927; Vol. 2, 1931 (2 parts); Vol. 3, 1947 (2 parts). Williams & Wilkins, Baltimore (Carnegie Institution Publ. no. 376).

ŚĀSTRĪ, V. V. RAMAN (1). 'The Doctrinal Culture and Tradition of the Siddhas.' In *Cultural Heritage of India* (Sri Ramakrishna Centenary Memorial Volume), vol. 2, p. 303.

DE SAUSSURE, L. (1). *Les Origines de l'Astronomie Chinoise.* Maisonneuve, Paris, 1930. (Commentaries by E. Zinner, *VAG*, 1931, **66**, 21; A. Pogo, *ISIS*, 1932, **17**, 267. This book (posthumously issued) contains eleven of the most important original papers of de Saussure on Chinese astronomy (3, 6, 7, 8, 9, 10, 11, 12, 13, 14). It omits, however, the important addendum to (3), 3 *a*, as well as the valuable series (16). Unfortunately the editing was slovenly. Although the reprinted papers were re-paged, the cross-references in the footnotes were unaltered; Pogo, however (*loc. cit.*), has provided a table of corrections by the use of which de Saussure's cross-references can readily be located.)

DE SAUSSURE, L. (8). 'Les Origines de l'Astronomie Chinoise; La Série Quinaire et ses Dérivés.' *TP*, 1910, **11**, 221. Reprinted as [C] in de Saussure (1).

DE SAUSSURE, L. (10). 'Les Origines de l'Astronomie Chinoise; Le Cycle des Douze Animaux.' *TP*, 1910, **11**, 583. Reprinted as [E] in de Saussure (1).

DE SAUSSURE, L. (18). 'Origine Chinoise du Dualisme Iranien.' *JA*, 1922 (11ᵉ sér., **20**), **201**, 302. Abstract only, refers to de Saussure (19).

DE SAUSSURE, L. (19). 'Le Système Cosmologique Sino-Iranien.' *JA*, 1923 (12ᵉ sér., **1**), **202**, 235.

SAYCE, R. U. (1). *Primitive Arts and Crafts.* Cambridge, 1933.

SCHÄFER, E. H. (1). 'Ritual Exposure [Nudity, etc.] in Ancient China.' *HJAS*, 1951, **14**, 130.

VON SCHELLING, F. W. J. (1). *Ideen zu einer Philosophie der Natur* (1797), Landshut, 1803. *Von der Weltseele; eine Hypothese der höheren Physik z. Erklärung des allgemeinen Organismus* (1798), Hamburg, 1809.

VON SCHIEFNER, F. A. (1). *Tibetan Tales.* Routledge, London, n.d., p. 361; and *BAISP*, 1876, **21**, 195.

SCHINDLER, B. (1). *Das Priestertum im alten China.* Teil 1: 'Königtum und Priestertum im alten China, Einleitung und Quellen.' Staatl. Forschungsinstitut f. Völkerkunde. Leipzig, 1919.

SCHINDLER, B. (2). 'Development of the Chinese Conception of Supreme Beings.' *AM*, 1923, **1** (Hirth Presentation Volume), 298.

SCHINDLER, B. (3). 'Tetragrams like Yang Hsiung's on ancient bronzes.' *OAZ*, 1915, **3**, 456.

SCHLEGEL, G. (5). *Uranographie Chinoise, etc.* 2 vols. with star-maps in separate folder. Brill, Leiden, 1875. Crit. J. Bertrand, *JS*, 1875, 557; S. Günther, *VAG*, 1877, **12**, 28. Reply by G. Schlegel, *BNI*, 1880 (4ᵉ volg.), **4**, 350.

SCHLEGELBERGER, F. (1) (ed.). *Rechtsvergleichendes Handwörterbuch f.d. Zivil- u. Handelsrecht des In- und Auslandes.* Vahlen, Berlin, 1929.

SCHMIDT, P. (1). 'Persian Dualism in the Far East.' Contrib. to *Oriental Studies in honour of Cursetji Erachji Pavry*, ed. J. D. C. Pavry, p. 405. Oxford, 1933.

SCHMIDT, R. (1) (tr.). *Das 'Kāmasūtram' des Vātsyāyana...aus dem Sanskrit übersetzt.* Barsdorf, Berlin, 1922 (7th ed.).

SCHMIDT, R. (2). *Beiträge z. Indischen Erotik.* Barsdorf, Berlin, 1911 (2nd ed.).

SCHMIDT, R. (3) (tr.). *The 'Rati Rahasyam' of Kokkoka* (said to be +9th cent.). Med. Book Co., Calcutta, 1949. Issued with Tatojaya (1), *q.v.*

SCHNABEL, P. (1). *Berossos und die babylonisch-hellenistische Literatur.* Teubner, Leipzig, 1923.

SCHOLEM, G. G. (1). *Bibliographia Kabbalistica.* Schocken, Berlin, 1933.

SCHOLEM, G. G. (2). *Major Trends in Jewish Mysticism.* Schocken, New York and Jerusalem, 1941, reissued 1946.

SCHOTT, W. (1). 'Über den Doppelsinn d. Wortes *Schamane* u.ü.d. tungusischen Schamanen-Cultus am Hofe der Mandju-Kaiser.' *APAW*, 1842 (1844), 461.

SCHUHL, P. M. (1). *Machinisme et Philosophie.* Presses Univ. de France, Paris, 1947.

SELLARS, R. W. (1). *Evolutionary Naturalism.* Chicago, 1922.

SEVERINI, A. (1) (tr.). *Astrologia Giapponese* [translations from the *Atsume Gusa*]. Georg, Geneva, 1874.

SHADIDULLAH, M. (1). *Les Chants Mystiques de Kanha et de Saraha, les Doha-Kosa et les Canja* (*Argot Tantrique*). Paris, 1928.

SHEFTELOWITZ, J. (1). 'Is Manichaeism an Iranic Religion?' (includes material on the five Elements). *AM*, 1924, **1**, 460.

SHELFORD, V. E. (1). *Laboratory and Field Ecology.* Baillière, Tindal & Cox, London, 1929.

SHERRINGTON, C. (1). *The Integrative Action of the Nervous System.* Cambridge (reset ed.), 1947.

SHIROKOGOROV, S. M. (1). (*a*) 'What is Shamanism?' *CJ*, 1924, **2**, 275 and 368. (*b*) 'General Principles of Shamanism among the Tungus.' *JRAS/NCB*, 1923, **54**, 246.

SHRYOCK, J. K. (1). *Origin and Development of the State Cult of Confucius.* Appleton-Century, New York, 1932.

SHRYOCK, J. K. (2) (tr.). *The Study of Human Abilities; the 'Jen Wu Chih' of Liu Shao.* Amer. Orient. Soc., New Haven, 1937. (Amer. Orient. Ser. no. 11.) Crit. J. J. L. Duyvendak, *JAOS*, 1939, **59**. 280.

SIAO KING-FANG. See Hsiao Ching-Fang.

SIEG, E. (1). 'Das Märchen von dem Mechaniker und dem Maler in Tocharischer Fassung.' *OAZ*, 1919, **8**, 362.

SIMON, E. (1)., Über Knotenschriften und ähnliche Knotenschnüre d. Riukiuinseln.' *AM*, 1924, **1**, 657.

SINGER, C. (1). *A Short History of Biology.* Oxford, 1931.

SINGER, C. (3). 'The Scientific Views and Visions of St Hildegard.' In *Studies in the History and Method of Science*, ed. C. Singer, vol. 1, p. 1. Oxford, 1917.

SINGER, C. (4). *From Magic to Science; Essays on the Scientific Twilight.* Benn, London, 1928.

SINGER, C. (5). 'Historical Relations of Religion and Science.' In *Science, Religion and Reality*, ed. J. Needham. Sheldon Press, London, 1925.

SINGER, C. (6). 'Galen as a Modern.' *PRSM*, 1949, **42**, 563.

SINGER, C. & SINGER, D. W. (1). 'The Jewish Factor in Mediaeval Thought.' In *Legacy of Israel*, ed. E. R. Bevan & C. Singer. Oxford, 1928.

SINGER, D. W. (1). *Giordano Bruno; his Life and Thought, with an annotated Translation of his Work 'On the Infinite Universe and Worlds'.* Schuman, New York, 1950.

SIREN, O. (6). *History of Early Chinese Painting.* 2 vols. Medici Society, London, 1933.

SMITH, V. A. (1). *Oxford History of India, from the earliest times to 1911*, ed. S. M. Edwardes. 2nd ed. Oxford, 1923

SMITH, V. A. (2). *Aśoka.* Oxford, 1920 (3rd ed.). (Rulers of India series.)

SMUTS, J. C. (1). *Holism and Evolution.* Macmillan, London, 1926. (3rd ed.) 1936.

SOOTHILL, W. E. (3) (tr.). *'Saddharma-puṇḍarīka Sūtra'; The Lotus of the Wonderful Law.* Oxford, 1930.

SOOTHILL, W. E. (5) (posthumous). *The Hall of Light; a Study of Early Chinese Kingship.* Lutterworth, London, 1951. (On the Ming Thang, and contains discussion of the *Pu Thien Ko*.)

SOOTHILL, W. E. & HODOUS, L. (1). *A Dictionary of Chinese Buddhist Terms.* Kegan Paul, London, 1937.

SOULIÉ DE MORANT, G. (1). *La Musique en Chine.* Leroux, Paris, 1911.

SOUSTELLE, J. (1). *La Pensée Cosmologique des anciens Mexicains; Représentation du Monde et de l'Espace.* Hermann, Paris, 1940.

SOWERBY, A. DE C. (1). *Nature in Chinese Art* (with two appendices on the Shang pictographs by H. E. Gibson). Day, New York, 1940.

SPALDING, K. J. (1). *Three Chinese Thinkers* [Chuang Tzu, Mo Tzu, Hsün Tzu]. Nat. Centr. Library, Nanking, 1947.

SPEERT, H. (1). 'Supernumerary Mammae, with special reference to the *Rhesus* Monkey.' *QRB*, 1942, **17**, 59.

SPINDEN, H. J. (1). *Ancient Civilisations of Mexico and Central America.* Amer. Mus. Nat. Hist., New York, 1946.

SPIZEL, G. (1). *De Re Litteraria Sinensium Commentarius.* Leiden, 1660.

STADELMANN, H. (1). *Lao Tzu und die Biologie.* Geneva, 1935. (Schriftenreihe d. Bibliothek Sino-International Genf, no. 2.)

STANTON, W. (1). *The Triad Society or Heaven-and-Earth Association.* Kelly & Walsh, Shanghai, 1900.

STAUNTON, SIR GEORGE T. (1) (tr.). '*Ta Tsing Leu Lee*' [*Ta Chhing Lü Li*]; *being the fundamental Laws, and a selection from the supplementary Statutes, of the Penal Code of China.* Davies, London, 1810.

STCHERBATSKY, T. (SHCHERBATSKOY, F. I.) (1). *Buddhist Logic.* 2 vols. (Vol. 2 contains a translation of the short treatise on Logic by Dharmakīrti and of its commentary by Dharmottara, with notes, appendices and indexes.) Acad. Sci. U.S.S.R., Leningrad, 1930–2. (Biblioth. Buddhica, no. 26.)

STCHERBATSKY, T. (2). *The Conception of Buddhist Nirvāṇa.* (With reference to the Mādhyamika logic.) Acad. Sci. U.S.S.R., Leningrad, 1928.

STCHERBATSKY, T. (3). (*a*) *La Théorie de la Connaissance et la Logique chez les Ɔuddhistes Tardifs* (partly rewritten for the French ed.), tr. from Russian by I. de Manziarly & P. Masson-Oursel. Geuthner, Paris, 1926. (*BE/AMG*, no. 36.) (*b*) *Erkenntnistheorie und Logik nach der Lehre der späteren Buddhisten* (not rewritten), tr. O. Strauss. München, 1924.

STCHERBATSKY, T. (4). *The Central Conception of Buddhism, and the Meaning of the Word 'Dharma'.* Royal Asiat. Soc. London, 1923. (Prize Publ. Fund Ser. no. 7.)

STCHERBATSKY, T. (5). *Madhyānta-Vibhanga; Discourse on Discrimination between Middle and Extremes, ascribed to the Bodhisattva Maitreya and commented by Vasubandhu and Sthiramati. . . .* Acad. Sci. U.S.S.R., Leningrad, 1936. (Biblioth. Buddhica, no. 30.)

STEELE, R. & SINGER, D. W. (1). 'The Emerald Table' [*Tabula Smaragdina*]. *PRSM*, 1928, **21**, 41.

STEIN, L. (1). *Leibniz und Spinoza.* Berlin, 1890.

STEIN, R. A. (2). 'Jardins en Miniature d'Extrême-Orient; le Monde en Petit.' *BEFEO*, 1943, **42**, 1–104.

STEINSCHNEIDER, M. (1). 'Die europäischen Übersetzungen aus dem Arabischen bis mitte d. 17 Jahrhunderts.' *SWAW/PH*, 1904, **149**, 1; 1905, **151**, 1; *ZDMG*, 1871, **25**, 378, 384.

STERN, B. J. (1). 'Engels on the Family.' *SS*, 1948, **12**, 42.

STEWART, J. & KEMP, J. (1) (ed. and tr.). *Diderot, Interpreter of Nature; Selected Writings.* Lawrence & Wishart, London, 1937.

STONE, JULIUS (1). *The Province and Function of Law.* Assoc. Gen. Pub., Sydney, 1946; Stevens, London, 1947.

VON STRAUSS, V. (1) (tr.). *Lao-Tzu's 'Tao Tê Ching'.* Fleischer, Leipzig, 1870.

STREETER, B. H. (1). *The Buddha and the Christ.* Macmillan, London, 1932.

STUART, G. A. (1). *Chinese Materia Medica; Vegetable Kingdom* (extensively revised from Dr F. Porter Smith's work). Presbyterian Mission Press, Shanghai, 1911.

STUBBE, HENRY (1). *Legends No Histories; or, A Specimen of some Animadversions upon the History of the Royal Society...together with the 'Plus Ultra' of Mr Ioseph Glanvill reduced to a Non Plus.* London, 1670.

SUGIURA, S. (1). *Hindu Logic as preserved in China and Japan.* Univ. of Pennsylvania, Philadelphia, 1900. (Pub. Univ. Penn. Philos. Series, no. 4.)

SUMNER, W. G. (1). *Folkways; a study of the sociological importance of Usages, Manners, Customs, Mores and Morals.* Ginn, Boston, 1907.

SUZUKI, B. L. (1). *Mahāyāna Buddhism.* Buddhist Lodge, London, 1938; 2nd ed. 1948.

SUZUKI, D. T. (1). *Outlines of Mahāyāna Buddhism.* Luzac, London, 1907.

SUZUKI, D. T. (2) (tr.). *The 'Laṅkāvatāra Sūtra'.* Routledge, London, 1932.

SUZUKI, D. T. (3). *Studies in the 'Laṅkāvatāra Sūtra'.* Routledge, London, 1930.

SUZUKI, D. T. (4). *Manual of Zen Buddhism.* Kyoto, 1935.

SWANN, N. L. (1) (tr.). *Food and Money in Ancient China; the Earliest Economic History of China to A.D. 25*—'[Chhien] Han Shu' ch. 24, with related texts '[Chhien] Han Shu' ch. 91 and 'Shih Chi' ch. 129—*translated and annotated....* Princeton Univ. Press, 1950. Revs. J. J. L. Duyvendak, *TP*, 1951, **40**, 210; C. M. Wilbur, *FEQ*, 1951, **10**, 320; Yanglien-Shêng, *HJAS*, 1950, **13**, 524.

SZCZESNIAK, B. (1). 'The Penetration of the Copernican Theory into Feudal Japan.' *JRAS*, 1944, 52.

SZCZESNIAK, B. (2). 'Notes on the Penetration of the Copernican Theory into China from the 17th to the 19th Centuries.' *JRAS*, 1945, 30.

TAI KUAN-I (TAI KWEN-IH) (1). *An Enquiry into the Origin and Early Development of Thien and Shang-Ti.* Inaug. Diss., Chicago.

TAKAKUSU, J. (1) (tr.). *A Record of the Buddhist Religion as practised in India and the Malay Archipelago* (+671 to +695), *by I-Tsing* (I-Ching). Oxford, 1896.

TAKAKUSU, J. (2). *The Essentials of Buddhist Philosophy* (particularly with reference to China and Japan). Ed. Chhen Jung-Chieh & C. A. Moore. Univ. of Hawaii, Honolulu, 1947.

TAKAKUSU, J. & WATANABE, K. (1). *Tables du 'Taishō Issaikyō' (nouvelle édition (Japonaise) du Canon Bouddhique Chinoise)*. [Index to the *Tripiṭaka*; cf. Nanjio.] Fascicule Annexe de *Hobogirin* (*Dictionnaire Encyclopédique du Bouddhisme d'après les sources Chinoises et Japonaises*), ed. S. Lévi, J. Takakusu & P. Demiéville. Maison Franco-Japonaise, Tokyo, 1931.

TANNERY, P. (1). *L'Histoire de la Science Hellène*. Paris, 1887.

TATOJAYA, YATODHARMA (1) (tr.). *The 'Kokkokam' of Ativira Rama Pandian* [a Tamil prince at Madura, late +16th cent.]. Med. Book Co., Calcutta, 1949. Issued with R. Schmidt (3), *q.v.*

TAWNEY, C. H. (tr.) & PENZER, N. M. (ed.) (1). *The Ocean of Story* [Somadeva's '*Kathā Sarit Sagara*']. 10 vols., Sawyer, London, 1925.

TEICHER, J. L. (1). 'Laws of Reason and Laws of Religion; a Conflict in Toledo Jewry in the +14th Century.' In *Essays and Studies presented to Stanley A. Cook*, ed. D. W. Thomas, p. 83. Taylor, London, 1950.

TEMKIN, O. (1). 'Metaphors of Human Biology.' In *Science and Civilisation*, ed. R. C. Stauffer, p. 167. Centennial Celebration Volume of the Univ. of Wisconsin. Univ. of Wisconsin Press, Madison, 1949.

TEMPLE, SIR WILLIAM (1). 'Upon the Gardens of Epicurus, or Of Gardening' (1685). In *Essays*, vol. 2, pt. 2, p. 58. London, 1690.

TEMPLE, SIR WILLIAM (2). *Miscellanea*. Simpson, London, 1705. (Contains the essays 'On Ancient and Modern Learning' (1690) and 'Of Heroick Virtue', both of which deal with Chinese questions.)

THAN PO-FU, WÊN KUNG-WÊN, HSIAO KUNG-CHÜAN & MAVERICK, L. A. (tr.). *Economic Dialogues in Ancient China; selections from the 'Kuan Tzu (Book)'...* Pr. pub. Carbondale, Illinois, and Yale Univ. Hall of Graduate Studies, New Haven, Conn. 1954.

THANG YUNG-THUNG (1). 'Wang Pi's New Interpretation of the *I Ching* and *Lun Yü*' (tr. W. Liebenthal). *HJAS*, 1947, **10**, 124.

THANG YUNG-THUNG (2). 'On "Ko-Yi", the earliest Method by which Indian Buddhism and Chinese Thought were synthesised.' Art. in *Radhakrishnan; Comparative Studies in Philosophy presented in honour of his 60th Birthday*, ed. W. R. Inge *et al.* Allen & Unwin, London, 1951.

THIEN CHHIH-KANG (1). (TIEN TCH'EU-KANG.) *L'Idée de Dieu dans les huits Premiers Classiques Chinois; ses Noms, son Existence et sa Nature, étudiée à la Lumière des Découvertes Archéologiques*. Œuvre St Justin, Fribourg, Switzerland, 1942.

THIERENS, A. E. (1). *Astrology in Mesopotamian Culture*. Brill, Leiden, 1935.

THOMAS, E. J. (1). *The History of Buddhist Thought*. Kegan Paul, London, 1933.

THOMPSON, D'ARCY W. (1). 'Excess and Defect; or the Little More and the Little Less.' *M*, 1929, **38**, 43.

THOMPSON, R. C. (1). *Reports of the Magicians and Astrologers of Nineveh and Babylon* (in the British Museum on cuneiform tablets). 2 vols. Luzac, London, 1900.

THOMSON, GEORGE (1). *Aeschylus and Athens; a Study in the Social Origins of Drama*. Lawrence & Wishart, London, 1941.

THORNDIKE, L. (1). *A History of Magic and Experimental Science*. 6 vols. Columbia Univ. Press, New York: Vols. 1 and 2, 1923; 3 and 4, 1934; 5 and 6, 1941.

TIEN TCH'EU-KANG. *See* Thien Chhih-Kang.

TJAN TJOE-SOM. See Tsêng Chu-Sên.

TOGAN, ZAKI VALIDI (1). 'Kritische Geschichtsauffassung in d. Islamischen Welt d. Mittelalters.' *Proc. XXIInd Internat. Congress of Orientalists*. Istanbul, 1951 (1953).

TOMKINSON, L. (1). *Studies in the Theory and Practice of Peace and War in Chinese History and Literature*. Friends' Centre, Shanghai, 1940.

TRACEY, M. V. (1). *Principles of Biochemistry; a Biological Approach*. Pitman, London, 1954.

TRACEY, M. V. (2). *Proteins and Life*. Pilot Press, London, 1948.

TRIGAULT, NICHOLAS (1). *De Christiana Expeditione apud Sinas*. Vienna, 1615; Augsburg, 1615. Fr. tr.: *Histoire de l'Expédition Chrétienne au Royaume de la Chine entrepris par les PP. de la Compagnie de Jésus, comprise en cinq livres...tirëé des Commentaires du P. Matthieu Riccius, etc.* Lyon, 1616; Lille, 1617; Paris, 1618. Eng. tr. (partial): 'A Discourse of the Kingdome of China, taken out of Ricius and Trigautius.' In *Purchas his Pilgrimes*, vol. 3, p. 380. London, 1625. Eng. tr. (full): see Gallagher (1).

TSÊNG CHU-SÊN (TJAN TJOE-SOM) (1). '*Po Hu Thung*'; *The Comprehensive Discussions in the White Tiger Hall; a Contribution to the History of Classical Studies in the Han Period*. Vol. 1, Brill Leiden, 1949. (Sinica Leidensia, no. 6.)

TSÊNG CHU-SÊN (2). 'The Date of Kao Tsu's first Court Ceremonial.' In *India Antiqua, a Volume of Studies presented by his friends and pupils to J. P. Vogel...*, p. 304. 1947.

Tucci, G. (1) (ed. & tr.). *Pre-Diṅnāga Buddhist Texts on Logic from Chinese Sources*. Orient. Instit. Baroda, 1929. (Gaekwad Orient. Ser. no. 49.)

Tucci, G. (2). 'Buddhist Logic before Diṅnāga.' *JRAS*, 1929, 451.

Tucci, G. (3). *Tibetan Painted Scrolls*. 2 vols. and 1 vol. plates. Libreria dello Stato, Rome, 1949.

Tulloch, J. (1). *Rational Theology and Christian Philosophy in England in the 17th Century*. 2 vols. Blackwood, Edinburgh and London, 1872.

Tustin, A. (1). (a) 'Automatic Control Systems.' *N*, 1950, **166**, 845. (b) 'Feedback.' *SAM*, 1952, **187** (no. 3), 48.

Tyrrell, G. N. M. (1). *Apparitions*. Duckworth, London, 1953; rev. E. J. Dingwall, *N*, 1954, **173**, 912.

Vacca, G. (8). 'Sulla Storia della Numerazione Binaria.' *Atti del Congresso Internazionale di Scienze Storiche*. Rome, 1903 (1904). Vol. 12 (sect. 8), p. 63.

Vacca, G. (10). 'Alcune Idee di un filosofo Cinese del 4° Secolo avanti Cristo, Chuang-Tse.' [Tr. chs. 8, 9, 10.] *L*, 1907, **5**, 68.

de la Vallée Poussin, L. (3) (tr.). *La Siddhi de Hiuen Tsang* (Hsüan-Chuang). Paris, 1928.

de la Vallée Poussin, L. (4). *Bouddhisme, Etudes et Matériaux*. Luzac, London, 1898.

de la Vallée Poussin, L. (5). 'Buddhist Tantrism.' *ERE*, vol. 12, p. 193.

de la Vallée Poussin, L. (6) (tr.). *Śāntideva's 'Bodhicaryāvatāra'*. Paris, 1907.

de la Vallée Poussin, L. (7) (tr.). *Troisième Chapitre de 'l'Abhidharmakoṣa', Kārikā, bhāṣya et vyā-khyā...Versions et textes établis (Bouddhisme; Etudes et Matériaux; Cosmologie; Le Monde des Êtres et le Monde-Réceptacle)*. Kegan Paul, London, 1918.

de la Vallée Poussin, L. (8). *Le Dogme et la Philosophie du Bouddhisme*. Paris, 1930.

Veith, I. (1) (tr.). *'Huang Ti Nei Ching Su Wên'; the Yellow Emperor's Classic of Internal Medicine, chs. 1–34 translated from the Chinese, with an Introductory Study*. Williams & Wilkins, Baltimore, 1949. Crit. J. R. H[ightower], *HJAS*, 1951, **14**, 306. W. Hartner, *ISIS*, 1951, **42**, 265; J. R. Ware, *BIHM*, 1951, **24**, 487; reply: *BIHM*, 1951, **25**, 86.

Vidyabhusana, S. C. A. (1). 'History of the Mādhyamika Philosophy of Nāgārjuna.' *JBTS*, 1897, **5**, 3.

Vincent, I.V. (1). *The Sacred Oasis; the Caves of the Thousand Buddhas at Tunhuang*. Univ. of Chicago Press, 1953.

Vinogradov, Paul (1). *Outlines of Historical Jurisprudence*. 2 vols. Oxford, 1922.

van Vloten, J. & Land, J. P. N. (1) (ed.). *Benedict de Spinoza; Opera quotquot reperta sunt*. 3 vols. The Hague, 1882–95.

Vogel, W. (1). 'Die historischen Grundlagen des chinesischen Strafrechts.' *ZVRW*, 1923, **40**, 37–134.

Volpert, P. A. (1). 'Tsch'öng Huang [Chhêng Huang], der Schutzgott d. Städte in China.' *AN*, 1910, **5**, 991.

Waddington, C. H. *et al.* (1). *Science and Ethics*. Allen & Unwin, London, 1942.

Waley, A. (1) (tr.). *The Book of Songs*. Allen & Unwin, London, 1937.

Waley, A. (4). *The Way and its Power; a study of the 'Tao Tê Ching' and its Place in Chinese Thought* (tr. of the *Tao Tê Ching* with introduction and notes). Allen & Unwin, London, 1934. Crit. Wu Ching-Hsiung, *TH*, 1935, **1**, 225

Waley, A. (5) (tr.). *The Analects of Confucius*. Allen & Unwin, London, 1938.

Waley, A. (6). *Three Ways of Thought in Ancient China*. Allen & Unwin, London, 1939.

Waley, A. (7). 'Observations on Karlgren's "Fecundity Symbols in Ancient China".' *BMFEA*, 1931, **3**, 61. Cf. Karlgren (9).

Waley, A. (8). 'The Book of Changes.' *BMFEA*, 1934, **5**, 121.

Waley, A. (9). 'Leibniz and Fu Hsi.' *BLSOAS*, 1921, **2**, 165.

Waley, A. (10) (tr.). *The Travels of an Alchemist; the Journey of the Taoist Chhang-Chhun from China to the Hindu-Kush at the summons of Chingiz Khan, recorded by his disciple Li Chih-Chhang*. Routledge, London, 1931.

Waley, A. (12). *The Life and Times of Po Chü-I (+772 to +846)*. Allen & Unwin, London, 1949.

Waley, A. (16). *The Real Tripiṭaka* (life of Hsüan-Chuang, and other essays). Allen & Unwin, London, 1952.

Waley, A. (19). *An Introduction to the Study of Chinese Painting*. Benn, London, 1923.

Waley, A. (23). *The Nine Songs; a study of Shamanism in Ancient China* [the *Chiu Ko* attributed traditionally to Chhü Yuan]. Allen & Unwin, London, 1955.

Walleser, M. (1). *Die buddhistischen Philosophie in ihrer geschichtlichen Entwicklung*. Vol. 1: 'Die philosophische Grundlage des älteren Buddhismus.' Winter, Heidelberg, 1904; 2nd ed. 1925.

Walleser, M. (2). *Die buddhistischen Philosophie in ihrer geschichtlichen Entwicklung*. Vol. 2: 'Die mittlere Lehre (*Mādhayamika-Śāstra*) des Nāgārjuna, nach der Tibetischen Version übertragen.' Winter, Heidelberg, 1911. Vol. 3: 'Die mittlere Lehre des Nāgārjuna, nach der Chinesischen Version übertragen' (*Chung Lun*). Winter, Heidelberg, 1912.

WALTER, E. J. (1). 'Warum gab es im Altertum keine Dynamik?' *Actes du Ve Congr. Internat. d'Hist. des Sci.* Lausanne, 1947. p. 53.

WALTER, H. (1) (tr.). '*Hathayoga Pradīpikā*' *of Swatmeram Swami.* Inaug. Diss. Munich, 1893.

WANG CHHANG-CHIH (1). *La Philosophie Morale de Wang Yang-Ming.* T'ou-Se-Wei, Shanghai, 1936 (*VS*, no. 63).

WANG KUO-WEI (1). 'Archaeology in the Sung Dynasty.' *CJ*, 1927, **6**, 222.

WARD, J. S. M. & STIRLING, W. G. (1). *The Hung Society, or the Society of Heaven and Earth.* 3 vols. London, 1925–6.

WARE, J. R. (1). 'The *Wei Shu* and the *Sui Shu* on Taoism.' *JAOS*, 1933, **53**, 215. Corrections and emendations in *JAOS*, 1934, **54**, 290. Emendations by H. Maspero, *JA*, 1935, **226**, 313.

WARE, J. R. (3). 'Notes on the History of the *Wei Shu*.' *JAOS*, 1932, **52**, 35.

WARE, J. R. (4) (tr.). 'Wei Shou on Buddhism [tr. of ch. 114 of the *Wei Shu*].' *TP*, 1933, **30**, 100.

WARREN, G. G. (1). 'Was Chu Hsi a Materialist?' *JRAS/NCB*, 1924, **55**, 28.

WATTERS, T. (2). *A Guide to the Tablets in the Temple of Confucius.* Presbyt. Miss. Press, Shanghai, 1879.

WATTERS, T. (3). *Lao Tzu, a Study in Chinese Philosophy.* China Mail, Hong Kong, 1870.

WENSINCK, A. J. (1). 'The Refused Dignity.' In E. G. Browne Commemoration Volume, *A Volume of Oriental Studies*, p. 491. Ed. T. W. Arnold & R. A. Nicholson. Cambridge, 1922.

WERNER, E. T. C. (2). *Descriptive Sociology* [Herbert Spencer's]: *Chinese.* Williams & Norgate, London, 1910. See also art. 'Law' in Couling's *Encyclopaedia Sinica.*

WHEELER, W. M. (1). 'Termitodoxa' and 'Animal Societies'. In *Essays in Philosophical Biology*, ed. G. H. Parker, pp. 71 and 233. Harvard Univ. Press, Cambridge, Mass, 1939.

WHEWELL, W. (2). *Philosophy of the Inductive Sciences.* 2 vols. Parker, London, 1847.

WHITEHEAD, A. N. (1). *Science and the Modern World.* Cambridge, 1926.

WHITEHEAD, A. N. (2). *Adventures of Ideas.* Cambridge, 1933. Repr. 1938.

WHITEHEAD, A. N. (3). *Nature and Life.* Cambridge, 1934.

WHITEHEAD, A. N. (4). *Process and Reality; an Essay in Cosmology.* Gifford Lectures, 1927–8. Cambridge, 1928.

WHITEHEAD, A. N. (5). Autobiographical essay in *The Philosophy of Alfred North Whitehead*, ed. P. A. Schilpp. Northwestern Univ. Press, Chicago, 1941.

WHITEHEAD, A. N. (6). 'Mathematics and the Good.' In *The Philosophy of Alfred North Whitehead*, ed. P. A. Schilpp, p. 676. Northwestern Univ. Press, Chicago, 1941.

WHITNEY, L. (1). *Primitivism and the Idea of Progress in English Popular Literature of the Eighteenth Century.* Johns Hopkins Univ. Press, Baltimore, 1934.

WIEGER, L. (1). *Textes Historiques.* 2 vols. (Ch. and Fr.) Mission Press, Hsienhsien, 1929. Tr. many passages from the *Thung Chien Kang Mu.*

WIEGER, L. (2). *Textes Philosophiques.* Mission Press, Hsienhsien, 1930. Repr. Belles Lettres, Paris, 1953.

WIEGER, L. (4). *Histoire des Croyances Religieuses et des Opinions Philosophiques en Chine depuis l'origine jusqu'à nos jours.* Mission Press, Hsienhsien, 1917. Repr. Belles Lettres, Paris, 1953. Eng. tr. E. T. C. Werner, Hsienhsien, 1927.

WIEGER, L. (6). *Taoisme.* Vol. 1. *Bibliographie Générale*: (1) *Le Canon* (*Patrologie*); (2) *Les Index Officiels et Privés.* Mission Press, Hsienhsien, 1911. Repr. Belles Lettres, Paris, 1953. (Crit. P. Pelliot, *JA*, 1912 (10e sér.), **20**, 141.)

WIEGER, L. (7). *Taoisme.* Vol. 2. *Les Pères du Système Taoiste* (tr. selections of *Lao Tzu, Chuang Tzu, Lieh Tzu*). Mission Press, Hsienhsien, 1913. Repr. Belles Lettres, Paris, 1953.

WIEGER, L. (8). *Folklore Chinois Moderne.* Mission Press, Hsienhsien, 1909.

WIEGER, L. (9). (*a*) *Bouddhisme.* Vol. 1, *Monachisme*; Vol. 2, *Les Vies Chinoises du Bouddha.* Mission Press, Hsienhsien, 1913. (*b*) *Amidisme Chinois et Japonais.* Mission Press, Hsienhsien, 1928. Repr. Belles Lettres, Paris, 1953.

WIENER, N. (1). (*a*) *Cybernetics; or Control and Communication in the Animal and the Machine.* Wiley, New York, 1948. (*b*) 'Cybernetics...Processes common to Nervous Systems and Mathematical Machines.' *SAM*, 1948, **179** (no. 5), 14.

WIENER, P. P. (1). 'Notes on Leibniz' Conception of Logic and its Historical Context.' *PHR*, 1939, **48**, 567.

WIGMORE, J. H. (1). *Panorama of World Legal History*, vol. 1, pp. 141ff. West, St Paul, Minn., 1928.

WILHELM, HELLMUT (1). *Chinas Geschichte; zehn einführende Vorträge.* Vetch, Peiping, 1942.

WILHELM, HELLMUT (2) (tr.). 'Schriften und Fragmente zur Entwicklung der Staatsrechtlichen Theorie in der Chou-zeit' (incl. transl. of *Têng Hsi Tzu*). *MS*, 1947, **12**, 41.

WILHELM, HELLMUT (3). *Gesellschaft und Staat in China.* Vetch, Peiping, 1944. Crit. E. Balazs, *ER*, 1946, 119.

WILHELM, HELLMUT (4). *Die Wandlung; acht Vorträge zum 'I-Ging'* (*I Ching*). Vetch, Peiping, 1944.

WILHELM, HELLMUT (5). 'Leibniz and the *I Ching*.' *CCS*, 1948, 16, 205.

WILHELM, HELLMUT (6). 'Eine Chou-Inschrift über Atemtechnik.' *MS*, 1948, 13, 385.

WILHELM, RICHARD (1). *Short History of Chinese Civilisation*, tr. J. Joshua. Harrap, London, 1929.

WILHELM, RICHARD (2) (tr.). '*I Ging*' [*I Ching*]; *Das Buch der Wandlungen*. 2 vols. (3 books, pagination of 1 and 2 continuous in first volume). Diederichs, Jena, 1924. Eng. tr. C. F. Baynes (2 vols.). Pantheon, New York, 1950. (See note *a* on p. 308 above.)

WILHELM, RICHARD (3) (tr.). *Frühling u. Herbst d. Lü Bu-We* (the *Lü Shih Chhun Chhiu*). Diederichs, Jena, 1928.

WILHELM, RICHARD (4) (tr.). '*Liä Dsi*' [*Lieh Tzu*]; *Das Wahre Buch vom Quellenden Urgrund; 'Tschung Hü Dschen Ging*'; *Die Lehren der Philosophen Liä Yü-Kou und Yang Sschu*. Diederichs, Jena, 1921.

WILHELM, RICHARD (5). *Confucius and Confucianism*, tr. G. H. & A. P. Danton from the Stuttgart ed. of 1925. London, 1931.

WILHELM, RICHARD (6) (tr.). '*Li Gi*', *das Buch der Sitte des älteren und jungeren Dai* [i.e. both *Li Chi* and *Ta Tai Li Chi*]. Diederichs, Jena, 1930.

WILHELM, RICHARD & JUNG, C. G. (1). *The Secret of the Golden Flower; a Chinese Book of Life*. [Incl. tr. of the *Thai I Chin Hua Tsung Chih*.] Eng. ed. tr. C. F. Baynes. Kegan Paul, London, 1931.

W[ILLIS], R[ICHARD] (1). *Certain reports of China, learned through the Portugals there imprisoned, and chiefly by the relation of Galeote Pereira, a gentleman of good credit, that lay prisoner in that country many years. Done out of Italian into English by R[ichard] W[illis]*. From *Nuovi Avisi delle Indie di Portogallo, Venuti novamente delli R. padri della compagnia di Giesu & tradotti dalla lingua Spagnola nella Italiana*. Venice, 1565. Printed in *History of Travayle in the West and East Indies*. London, 1577. Repr. in Hakluyt and Purchas. See Boxer (1).

WILSON, H. H. (1). *Sketch of the Religious Sects of the Hindus*. Calcutta, 1846.

WILSON, J. A., SPEISER, E. A., GÜTERBOCK, H. G., MENDELSOHN, I., INGALLS, D. H. H. & BODDE, D. *Authority and Law in the Ancient Orient* (presence or absence of god-kings, and relative roles of good custom or codified law in Egypt, Mesopotamia, Hatti, Israel, India and China). *JAOS*, 1954, 74, Supplement no. 17, 1–55.

WINSTANLEY, GERRARD (1). (*a*) *The Works of Gerrard Winstanley*, ed. G. H. Sabine. Cornell Univ. Press, Ithaca, 1941; (*b*) *Selections from the Works of Gerrard Winstanley*, ed. L. Hamilton. Cresset, London, 1944.

WINTERNITZ, M. (1). 'Notes on the *Guhyasamāja-tantra* and the Age of the Tantras.' *IHQ*, 1933, 9, 1.

WITHINGTON, E. (1). 'Dr John Weyer and the Witch Mania.' In *Studies in the History and Method of Science*, ed. C. Singer, vol. 1, p. 189. Oxford, 1917

WITTFOGEL, K. A., FÊNG CHIA-SHÊNG et al. (1). *History of Chinese Society (Liao), +907 to +1125.* *TAPS*, 1948, 36, 1–650. Crit. P. Demiéville, *TP*, 1950, 39, 347; E. Balazs, *PA*, 1950, 23, 318.

WITTGENSTEIN, L. (1). *Tractatus Logico-Philosophicus*. Kegan Paul, London, 1922.

WOODGER, J. H. (1). *Biological Principles*. Kegan Paul, London, 1929.

WOODROFFE, SIR J. (pseudonym: A. Avalon) (1). *Śakti and Śakta*. Ganesh, Madras, 1929; Luzac, London, 1929.

WOODROFFE, SIR J. (pseudonym: A. Avalon) (2). *The Serpent Power* (Kuṇḍalinī Yoga). Ganesh, Madras, 1931; Luzac, London, 1931.

WOODS, J. H. (1). 'The Yoga System of Patañjali.' Cambridge, Mass., 1914. (Harvard Orient. Ser. no. 17.)

WORTLEY, B. A. (1). 'François Gény.' In *Modern Theories of Law*, ed. W. I. Jennings, p. 139. Oxford, 1933.

WOTTON, WILLIAM (1). *Reflections upon Ancient and Modern Learning*. Leake & Buck, London, 1697.

WRIGHT, A. F. (1). Eng. résumé of A. A. Petrov's monograph on Wang Pi. See Petrov (1).

WRIGHT, A. F. (2). 'Fo-Thu-Têng; a Biography.' *HJAS*, 1948, 11, 321.

WRIGHT, A. F. (3). 'Fu I and the Rejection of Buddhism.' *JHI*, 1951, 12, 33.

WU CHING-HSIUNG (JOHN) (1). Translation of the *Tao Tê Ching*. *TH*, 1939, 9, 401, 498; 1940, 10, 66.

WU CHING-HSIUNG (JOHN) (2). 'The Struggle between Government of Laws and Government of Men in the History of China.' *CHLR*, 1932, 5, 53.

WU KHANG (1). *Les Trois Politiques du Tchouen Tsieou interprétée par Tong Tchong-Chou d'après les principes de l'école de Kong-Yang*. Leroux, Paris, 1932.

WU LU-CHHIANG & DAVIS, T. L. (1) (tr.). 'An Ancient Chinese Treatise on Alchemy entitled *Tshan Thung Chhi*.' *ISIS*, 1932, 18, 210.

WU LU-CHHIANG & DAVIS, T. L. (2) (tr.). 'An Ancient Chinese Alchemical Classic; Ko Hung on the Gold Medicine, and on the Yellow and the White; being the fourth and sixteenth chapters of *Pao Phu Tzu*,' etc. *PAAAS*, 1935, 70, 221.

WU SHIH-CHHANG (1). 'A Short History of Chinese Prose Literature.' In the press.

Wu Tsê-Ling (Charles) (1). 'The Social Thought of Confucius.' *CSPSR*, 1927, **11**, 432, 594; 1928, **12**, 100, 294, 381.

Wulff, K. (1). 'Acht Kapitel des Tao-Te-King' (*Tao Tê Ching*). *KDVS/HFM*, 1942, **28**, no. 4.

Wylie, A. (1). *Notes on Chinese Literature.* 1st ed. Shanghai, 1867. Ed. here used, Vetch, Peiping, 1939 (photographed from the Shanghai 1922 ed.).

Wylie, A. (2). 'History of the Hsiung-Nu' (tr. of the chapter on the Huns in the *Chhien Han Shu*, ch. 94). *JRAI*, 1874, **3**, 401; 1875, **5**, 41.

Wylie, A. (3). 'The History of the South-western Barbarians and Chao Sëen' (Chao-Hsien, Korea) (tr. of ch. 95 of the *Shih Chi*). *JRAI*, 1880, **9**, 53.

Wylie, A. (10) (tr.). 'Notes on the Western Regions, translated from the "Ts'een Han Shoo" [*Chhien Han Shu*] Bk. 96.' *JRAI*, 1881, **10**, 20; 1882, **11**, 83. (Chs. 96A and B, as also ch. 61, pp. 1-6, being the biography of Chang Chhien, and ch. 70, the biography of Chhen Thang.)

Yang Hsing-Shun (1). *Drevenekitaiskie Philosoph Lao-Tzu i ego Uchenye* [The Ancient Chinese Philosopher Lao Tzu and his Ideas] (in Russian). Academy of Sciences, Moscow, 1950. German edn. 1955, *Die chinesische Philosoph Laudse und seine Lehre.* Deutscher Verlag d. Wiss. Berlin 1955.

Yao Shan-Yu (4). 'The Cosmological and Anthropological Philosophy of Tung Chung-Shu.' *JRAS/NCB*, 1948, **73**, 40.

Yetts, W. P. (4). 'Taoist Tales; III, Chhin Shih Huang Ti's Expeditions to Japan.' *NCR*, 1920, **2**, 290.

Yuan Cho-Ying (1). *La Philosophie Morale et Politique de Mencius.* Geuthner, Paris, 1927. (Etudes et Doc. pub. Inst. Franco-Chinois, Lyon, no. 2.)

Yule, Sir Henry (2). *Cathay and the Way Thither; being a Collection of Medieval Notices of China.* Hakluyt Society Pubs. (2nd ser.) London, 1913-15. (1st ed. 1866.) Revised by H. Cordier. 4 vols. Vol. 1 (no. 38), *Introduction; Preliminary Essay on the Intercourse between China and the Western Nations previous to the Discovery of the Cape Route.* Vol. 2, (no. 33), *Odoric of Pordenone.* Vol. 3 (no. 37), *John of Monte Corvino and others.* Vol. 4 (no. 41), *Ibn Baṭṭuṭah and Benedict of Goes.* (Photographically reproduced, Peiping, 1942.)

von Zach, E. (1) (tr.). 'Aus dem *Wên Hsüan*; Lu Chi's "Erweitete Perlenkette" (*Yen Lien Chu*) in 50 Abschnitten.' *Jubiläumsband herausgeg. v.d. deutsch. Gesellsch. f. Natur- und Volkerkunde Ostasiens*, 1933, vol. 1, p. 1.

von Zach, E. (4). 'Einige Verbesserungen zu Giles' *Chinese Biographical Dictionary*.' *AM*, 1926, **3**, 545.

von Zach, E. (5) (tr.). 'Yang Hsiung's *Fa Yen* (Sinologische Beiträge, IV).' Drukkerij Lux, Batavia, 1939.

Zeller, E. (1). *Stoics, Epicureans and Sceptics.* Longmans Green, London, 1870.

Zen, P. J. See Chhen Tai-O.

Zenker, E. V. (1). *Geschichte d. Chinesischen Philosophie.* 2 vols. Stiepel, Reichenberg, 1927.

Zilsel, E. (1). 'The Genesis of the Concept of Physical Law.' *PHR*, 1942, **51**, 245. Comment by M. Taube, *PHR*, 1943, **52**, 304.

Zimmer, H. (1). *Myths and Symbols in Indian Art and Civilisation*, ed. J. Campbell. Pantheon, New York, 1947.

ADDENDA TO BIBLIOGRAPHY C

Bielenstein, H. (1). 'An Interpretation of the Portents in the *Ts'ien Han Shu* [*Chhien Han Shu*].' *BMFEA*, 1950, **22**, 127.

Jung, C. G. (1). *Psychologie und Alchemie.* Rascher, Zürich, 1944.

Li An-Chê (1). 'Bon; the Magico-Religious Belief of the Tibetan-speaking Peoples.' *SWJA*, 1948, **4**, 31.

Pohlenz, M. (1). 'Nomos.' *PL*, 1948, **97**, 135.

Soymié, M. (1). 'L'Entrevue de Confucius et de Hsiang Tho.' *JA*, 1954, **242**, 311.

GENERAL INDEX

by Muriel Moyle

Notes

(1) Articles (such as 'the', 'al-', etc.) occurring at the beginning of an entry, and prefixes (such as 'de', 'van', etc.) are ignored in the alphabetical sequence. Saints appear among all letters of the alphabet according to their proper names. Styles such as Mr, Dr, if occurring in book titles or phrases, are ignored; if with proper names, printed following them.

(2) The various parts of hyphenated words are treated as separate words in the alphabetical sequence. It should be remembered that, in accordance with the conventions adopted, some Chinese proper names are written as separate syllables while others are written as one word.

(3) In the arrangement of Chinese words, Chh- and Hs- follow normal alphabetical sequence, and *ü* is treated as equivalent to *u*.

(4) References to footnotes are not given except for certain special subjects with which the text does not deal. They are indicated by brackets containing the superscript letter of the footnote.

(5) Explanatory words in brackets indicating fields of work are added for Chinese scientific and technological persons (and occasionally for some of other cultures), but not for political or military figures (except kings and princes).

(6) When there are many page-references under a single entry, the most important places are indicated by bold type.

Pre-established Harmony, 292, 381, 499, 502
Pre-Socratic philosophers, 27, 40, 42, 54, 93, 130,
132, 161, 163, 203, 239, 245, 255, 278, 285,
294, 368, 444, 472, 488, 533, 579
  supposed transmission of theories to China,
  374 (b)
Precepts Presented (to the Emperor). *See Shen
  Chien*
Preconceptions, avoidance of, 48, 456 (d)
Preformation, 500
Pregnancy, 422
'Preparatory raw material', 42, 480
Prester John, 129
Primary and secondary qualities, 188, 451
Primitive collectivism and communism, 104, 105
Primitive Homogeneity School. *See Hun-Tun
  Shih*
Primitive societies, 104 ff., 211, 448
  case-law in, 519
Primitivism, 113, 127
'Princes and grooms', 87
Principia Philosophiae (Principles of Philosophy),
  542
Principle of Control (in five-element theory),
  257-8
Principle of Masking (in five-element theory),
  258-9
Printing, 495
Private property, 101, 104, 105, 110 ff., 142
Prochaska, George, 54 (a)
Proclus, 297
Production for profit, 112, 167-8
'Production without possession . . .', 164
Prognostication, 249, 256, 311, 328, 346 ff., 367
Property. *See* Private property
Prophets, Hebrew, 376
Protein chemistry. *See* Chemistry
Protestantism, 538
Proto-feudalism, 3, 60, 104, 105, 119, 130, 217
Proto-science, 2, 33, 34, 63, 68-9, 85, 86, 87, 126,
164, 218, 235, 239, 247-8, 279, 367-8, 432,
443, 448, 579
Proto-Taoists, 3, 16, 117
Proverb about circumventing laws, 522
Przyłuski, J. (2), 277
Pseudo-Callisthenes, 129
Psychology, 18, 19
  of character, 386
  of dominance, 576-7
  Freudian, 364
  Gestalt, 291
Ptolemy (astronomer), 297
Ptolemy II (king of Egypt), 535
Pu-Khung. *See Amoghavajra*
Pu Shih Chêng Tsung Chhüan Shu (Encyclopaedia
  of Divination by the Tortoise-shell and the
  Milfoil), 349
Punishments, 71, 206-7, 212, 527, 528, 531, 544,
546, 549, 561
  gradations of, according to rank, age, etc., 531,
  532 (c)
Punjab, 530

Pure and applied science, 99, 112
'Pure Conversation' group. *See* Chhing Than
Pure Land sect, 407
Pure Questions of the Yellow Emperor; the
  Canon of Internal Medicine. *See Huang Ti
  Su Wên Nei Ching*
the Purple Yang Taoist, 429
Purposiveness, 454 (a)
Pus, 370
Pyrotechnics, 515
Pythagoras, 535
Pythagoreans, 162, 195, 270, 271, 278, 287, 296,
455, 484, 534

Qarmatianism, 96, 97
Quaestiones Naturales (Questions about Nature),
  295
Quakers. *See* Society of Friends
Quantitative experiments, 91
Quantitative rules, 542
Quantitative standards, 209, 214, 549
Quantitative thinking, 209 ff., 259, 260
Quesiti ed Inventioni (Problems and Inventions),
  542
Questions of a Head-Scratcher. *See Sao Shou
  Wên*
Questions about Heaven. *See Thien Wên*
Questions about Nature. *See Quaestiones Naturales*
Quipu (knotted cords for recording), 100, 327,
  556

Rack, 125
Rádl, Emanuel, 63 (c)
Ragley, 162
Rain, 190, 269, 272, 276, 282, 491
Rain-magic, and praying for rain, 119, 135, 145,
365, 427, 491
Raleigh, Sir Walter, 530 (e)
Rāmāyana, 429
Ranjit Singh, 530
Rarefaction. *See* Condensation and Dispersion
Rasā'il Ikhwān al-Ṣafā' (Epistles of the Brethren
  of Sincerity), 95, 96, 296 297
Ratchnevsky, P. (1), 524
Rates of reaction, 259
Rationalism, 2, 12, 26, 27, 30, 34, 73, 86, 89 ff.,
127, 139, 248, 365 ff.
Rats, 447
Rats and Jade. *See Shu Pho*
Rawley, William, 200
Ray, John (naturalist), 503
Read, J. (1), 278
Read, T. T. (4, 8), 267
Reality, in Chinese and European philosophy, 478
Receptacle, of Plato, 37 (b)
Record of the Class of Helical Machines. *See Lo
  Tsu Chhê Chi*
Record of Rites. *See Li Chi*
Record of the Rites of the Chou Dynasty. *See
  Chou Li*
Record of Rites of the Elder Tai. *See Ta Tai Li
  Chi*

夏	HSIA kingdom (legendary?)	c. −2000 to c. −1520
商	SHANG (YIN) kingdom	c. −1520 to c. −1030
周	CHOU dynasty (Feudal Age) { Early Chou period	c. −1030 to −722
	Chhun Chhiu period 春秋	−722 to −480
	Warring States (Chan Kuo) period 戰國	−480 to −221

First Unification 秦	CHHIN dynasty	−221 to −207
漢 HAN dynasty {	Chhien Han (Earlier or Western)	−202 to +9
	Hsin interregnum	+9 to +23
	Hou Han (Later or Eastern)	+25 to +220
三國	SAN KUO (Three Kingdoms period)	+221 to +265

First Partition	蜀 SHU (HAN)	+221 to +264
	魏 WEI	+220 to +264
	吳 WU	+222 to +280

Second Unification	晉 CHIN dynasty: Western	+265 to +317
	Eastern	+317 to +420
劉宋	(Liu) SUNG dynasty	+420 to +479

Second Partition	Northern and Southern Dynasties (Nan Pei chhao)	
	齊 CHHI dynasty	+479 to +502
	梁 LIANG dynasty	+502 to +557
	陳 CHHEN dynasty	+557 to +589
魏 {	Northern (Thopa) WEI dynasty	+386 to +535
	Western (Thopa) WEI dynasty	+535 to +556
	Eastern (Thopa) WEI dynasty	+534 to +550
北齊	Northern CHHI dynasty	+550 to +577
北周	Northern CHOU (Hsienpi) dynasty	+557 to +581

Third Unification	隋 SUI dynasty	+581 to +618
	唐 THANG dynasty	+618 to +906
Third Partition	五代 WU TAI (Five Dynasty period) (Later Liang, Later Thang (Turkic), Later Chin (Turkic), Later Han (Turkic) and Later Chou)	+907 to +960
	遼 LIAO (Chhitan Tartar) dynasty	+907 to +1124
	West LIAO dynasty (Qarā-Khiṭāi)	+1124 to +1211
	西夏 Hsi Hsia (Tangut Tibetan) state	+986 to +1227

Fourth Unification	宋 Northern SUNG dynasty	+960 to +1126
	宋 Southern SUNG dynasty	+1127 to +1279
	金 CHIN (Jurchen Tartar) dynasty	+1115 to +1234
	元 YUAN (Mongol) dynasty	+1260 to +1368
	明 MING dynasty	+1368 to +1644
	清 CHHING (Manchu) dynasty	+1644 to +1911
	民國 Republic	+1912

N.B. When no modifying term in brackets is given, the dynasty was purely Chinese. Where the overlapping of dynasties and independent states becomes particularly confused, the tables of Wieger (1) will be found useful. For such periods, especially the Second and Third Partitions, the best guide is Eberhard (9). During the Eastern Chin period there were no less than eighteen independent States (Hunnish, Tibetan, Hsienpi, Turkic, etc.) in the north. The term 'Liu chhao' (Six Dynasties) is often used by historians of literature. It refers to the south and covers the period from the beginning of the +3rd to the end of the +6th centuries, including (San Kuo) Wu, Chin, (Liu) Sung, Chhi, Liang and Chhen.